CONCRETE
PETROGRAPHY

A HANDBOOK OF
INVESTIGATIVE TECHNIQUES

SECOND EDITION

CONCRETE
PETROGRAPHY

A HANDBOOK OF INVESTIGATIVE TECHNIQUES

SECOND EDITION

ALAN B. POOLE AND IAN SIMS

CRC Press
Taylor & Francis Group
Boca Raton London New York

CRC Press is an imprint of the
Taylor & Francis Group, an **informa** business

CRC Press
Taylor & Francis Group
6000 Broken Sound Parkway NW, Suite 300
Boca Raton, FL 33487-2742

First issued in paperback 2019

© 2016 by Taylor & Francis Group, LLC
CRC Press is an imprint of Taylor & Francis Group, an Informa business

No claim to original U.S. Government works

ISBN-13: 978-1-85617-690-3 (hbk)
ISBN-13: 978-0-367-86688-4 (pbk)

Visit the Taylor & Francis Web site at
http://www.taylorandfrancis.com

and the CRC Press Web site at
http://www.crcpress.com

Contents

Foreword

In the past, many a book opened with a foreword written by a grand old man in the author's subject who took it upon himself to promote the work and to laud the author, not infrequently his protégé. Nothing could be further from the truth in the present case.

First of all, by no stretch of the imagination am I grand, although I am undoubtedly old. Furthermore, I am not a concrete petrographer, although in my study of concrete and of concrete structures I often rely on petrographic information and I have used the first edition of *Concrete Petrography*. As Poole and Sims rightly say, petrographic techniques provide insights into mineralogical, chemical and microstructural features of concrete. The very wide range of these techniques makes them very valuable but, at the same time, creates difficulties for a non-petrographer.

The authors have done an admirable job of presenting in a clear and orderly manner the petrographic methods and procedures, starting with sampling and including statistical treatment.

The concretes considered are wide-ranging, including fibre-reinforced concrete, polymer concrete, terrazzo, as well as oil-well cements. In addition, lime-based materials, of interest in the repair of historic buildings, are discussed by A. Leslie.

What impressed me is the cohesion of the book: unlike books written by several authors, this one seems seamless, and this makes it particularly valuable. It is not my task to be an advertiser or a salesman, but I have to praise the quality of the English and the rarity of spelling mistakes – a situation unusual in this day and age.

So, the book is encyclopaedic in character but very lucid and readable and I can truly recommend it to those who seriously investigate concrete, and especially problems in concrete in service. I know no better book on concrete petrography.

Adam Neville, CBE, TD, DSc, FREng, FRSE
London, England

Background note: Dr Adam. M. Neville was, before retirement, professor of civil engineering at Leeds and then Dundee Universities, also principal and vice principal at the University of Dundee, United Kingdom, and now a civil engineering consultant. He is a CBE, DSc (Eng.), DSc, PhD, FIStructE, FICE, FEng and FRSE.

Portrait of Adam Neville (2009)

Preface

Since the three authors completed the first edition of *Concrete Petrography* in 1998, there have been major advances in the understanding of concrete and its problems. The intervening years have also brought remarkable improvements in equipment and the methods of petrographic investigation relevant to concrete. Additionally, the variety of cementitious materials available and the range of applications they now cover have similarly increased dramatically.

Consequently, the first edition, though still providing a sound basic review of concrete petrography, has increasingly been in need of revision, updating and indeed improvement. With the passage of years, Don St John did not feel able to embark on a second edition. The remaining two authors, who readily recognised the important contribution Don St John made to the first edition, considered that the new petrographic information, methods and techniques, including a number of critical discoveries and developments in concrete technology since the first edition was published, merited a complete reappraisal of the original book. They discovered that the magnitude of the task they had set themselves was just as formidable as it was with the first edition. Nevertheless, thanks to the encouragement, unstinting help and advice from numerous experts from around the world, the task has been completed and provides a new and up-to-date balanced review of concrete and related materials, together with the modern petrographic approaches used in their investigation.

As with the first edition, the authors have emphasised the value and importance of the polarising microscope and the related petrographic techniques as fundamental to the understanding of cementitious materials and as basic tools for the scientific investigation of these materials and their properties. However, they have also included increased reference to the allied use of scanning electron microscopy and integral microanalysis as this powerful technique has become much more accessible to practising petrographers since the preparation of the first edition.

The authors believe that this book will be of practical value to the professional petrographer, providing both background information and details of some of the specialist techniques necessary for the petrographic investigation of cementitious materials. They hope that, as with the first edition, it will also be of value to the technologist, professional engineer and technical legal expert who may have to address materials problems with concrete and similar materials, interpret petrographic findings or decide on the most effective means of investigation of such materials.

Although in general terms cementitious materials are composed of crystalline and amorphous mineral constituents with a range of optical properties similar to those of rock-forming minerals that are familiar to all geological petrologists, differences arising from the methods of manufacture, chemical composition, time-dependent factors and environmental effects make the petrographic study of these materials different both in technical detail and in philosophy of approach.

Additional to reviewing the specialist laboratory techniques appropriate to the optical and electron-microscopical examination of cement-based materials, this book addresses the practical problems associated with the inspection and sampling of concrete structures and the special techniques appropriate to the preparation of the material for laboratory examination.

The principal area covered in this book is concerned with the use of the optical polarising microscope in the detailed study of concretes and related materials. It is anticipated that the trained petrographer will already be familiar with the use of the microscope for the examination of rocks in thin section, and so the emphasis is given to the cementitious phases and artificial materials, rather than the natural rock aggregates and fillers that are also used and encountered.

Where appropriate, case study examples taken from around the world are included in the text to illustrate particular points or to show the interrelationships between observed petrographic features, construction history and local environmental conditions.

Two undeniable disappointments of the first edition were the severe limitation on the number of colour photographs and the poor reproduction of a number of the other images in black and white. Thus, the authors are delighted that with the publisher's support, this second edition is illustrated in colour by a very much larger number of photographs excepting some older photographs and figures which remain in monochrome.

It was found that many petrographers used the first edition as a reference source in relation to a particular problem, in order to identify a particular mineral or form of deterioration. Consequently, the glossary introduced in the first edition has been retained, updated and extended with reference to photomicrographic illustrations where appropriate.

This second edition has been greatly enhanced by the contributions, advice and support from numerous internationally acknowledged experts from around the world, several of whom played a significant role in helping to update and improve particular sections of the book. The authors offer them all their sincere and very grateful thanks. These contributors' names and affiliations are recorded in the acknowledgements, and some individuals are also recognised in association with particular areas of expertise.

Notes

1. Note the use of the internationally accepted cement chemists' shorthand notation: $C = CaO$, $S = SiO_2$, $A = Al_2O_3$, and $F = Fe_2O_3$, so that, for example, $C_3S = 3CaO \cdot SiO_2$. This notation is included for individual phases in the glossary.
2. Some petrographers prefer to use the terms 'alite' and 'belite' instead of the chemical terms C_3S and C_2S respectively.

Alan B. Poole
Ian Sims

Acknowledgements

Although Don St John felt unable to embark on drafting this new revised edition, his excellent and major contributions to the first edition formed a foundation on which large parts of this edition are built. They have proved invaluable as an aid to the present authors as an initial basis for their extensive revisions, updating and enhancement of this new edition. A number of Don St John's classic case study examples are still very relevant today, and some of these have also been retained to illustrate particular points. Consequently, the present authors are indebted to Don for these examples and for his work on the petrographic examination of concrete, which has formed a valuable background that underlies this current edition.

The authors have also been most fortunate in receiving unstinting help and assistance from internationally acknowledged experts from around the world. This help has taken the form of specialist contributions to the text, advice and help with additional up-to-date information on a host of topics, innumerable new photomicrographs and many case study examples. These have been built into the text and have greatly enhanced the value of this book. All this assistance has of course been acknowledged below, and the authors owe a debt of gratitude to all concerned.

The current authors particularly thank the experts who have contributed significant sections of text or equivalent assistance in the various chapters. These contributors, many of whom helpfully also provided images for the book, include the following (in alphabetical order):

Dr Tony Asbridge (IMERYS Minerals LLP): mineral additives, including metakaolin
Paul Bennett-Hughes (formerly RSK Environment Ltd): concrete fire damage
Dr Alan Bromley (consultant, ex-University of Exeter): computer-aided petrography and mundic sulphate attack
Dr David B Crofts (RSK Environment Ltd): chemical and infrared analyses of concrete materials
Dr Alison Crossley (OMCS, Oxford University, Department of Materials): scanning electron microscopy and microanalysis
Mike A Eden (Sandberg LLP, formerly Geomaterials Research Services Ltd): sample preparation, delayed ettringite formation and micrometric analysis
Dr Paddy E Grattan-Bellew (consultant, ex-NRC Canada): general advice and damage rating index
Dr Viggo Jensen (NBTL, Norway): alkali-silica reaction
Dr Tetsuya Katayama (Taiheiyo Cement Corporation, Japan): carbonate aggregate reactivity
Dr Darrell Leek (Mott Macdonald Ltd): reinforcement corrosion
Dr Alec Leslie (ex-BGS, Edinburgh, now Historic Scotland): historic mortars

Anthony Meyers (Microscopes Plus Ltd): petrographic laboratory

Dr Mario R de Rooij (TNO Built Environment, Delft, Netherlands): fluorescence and other microscopy

Dr R G 'Ted' Sibbick (W R Grace & Co, USA, also ex-BRE, UK): general advice, plus especially thaumasite sulphate attack and aggressive carbonation

Dr Graham F True (GFT Materials Consultancy): scanning for aggregate shape and concrete air content

Nick B Winter (WHD Microanalysis Consultants Ltd): scanning electron microscopy and lime–cement mixtures.

Numerous other experts from around the world have also kindly offered advice, specific information, images and editorial comments and made valuable additions to improve sections of the text after reading early drafts. These contributions have proved vital to the authors in ensuring that this book is as accurate and as fully up to date as possible. Reading drafts is an arduous and time-consuming procedure, so the authors owe them a debt of gratitude for this invaluable assistance. In addition to the experts listed earlier, the numerous additional individual experts and companies who have helped the authors in this way include the following (in alphabetical order):

Dr Ian G Blanchard (RSK Environment Ltd)

Dr Maarten A T M Broekmans (Geological Survey of Norway)

Prof Nick Buenfeld (Imperial College, University of London)

Dr Sarah Denton, formerly Huntley, née Hartshorn (Vinci Construction UK Ltd)

Prof Isabel Fernandes (Lisbon University, ex-Porto University, Portugal)

James Ferrari (RSK Environment Ltd)

Prof Benoit Fournier (Université Laval, Québec, Canada)

Martin Grove (retired, formerly of the Cement & Concrete Association, now BCA)

Peter Laugesen (Pelcon Materials & Testing ApS, Denmark)

David F Lawrence (Technical Consultant, Sandberg LLP)

James McAneny (Logitech Ltd)

Chris A Rogers (Consultant, ex-Ministry of Transportation, Ontario, Canada)

Prof Peter Scott (University of Exeter)

Dr Ahmad Shayan (ARRB Group Ltd, Melbourne, Australia)

Prof Mike D A Thomas (University of New Brunswick, Canada)

Richard Wagner (Buehler Ltd)

Dr Barrie Wells (Conwy Valley Systems Ltd)

Readers of this book will soon be aware of the value of the illustrations and of the case history examples. Some of the colour illustrations and photomicrographs that were used in the first edition could not be bettered, so these have been reused. However, a large number of new colour illustrations were required for this edition and the authors have been most fortunate in being allowed to use photographs that other petrographers have provided, sometimes being able to select from lifetime collections. In other instances, special photographs have been taken for the authors in order to show particular features. Although acknowledgements are given where the material appears in the text, the authors thank all the organisations that have provided willing help with the photographic material.

This book could not have been produced without the technical and practical assistance of the publishing team at Taylor & Francis Group and the technical expertise of colleagues at RSK Environment Ltd. Individuals who were of great assistance to the authors during the preparation of this book include A. Hollingsworth, R. Griffiths, M. Cash, A. Moore, and

A. Blalock of Taylor & Francis Group and the tireless organisational, collating and computing skills of especially Claire Bennett, but also Clare Skakle and Siobhan Anderson of RSK Environment Ltd.

The authors have endeavoured to recognise everyone who has helped them during the several years of this revision; should we have inadvertently overlooked anyone, we apologise most sincerely to them and ask for their understanding and forgiveness.

Alan B. Poole
Ian Sims

Chapter 1

Introduction

1.1 CONCRETE PETROGRAPHY

Concrete is an essential and irreplaceable constructional material in the modern world. In recent times, technological advances in the manufacture of cements and the expansion of the use of concretes and related materials in civil engineering have led to a need for a practical, science-based understanding of these materials and the relationships between their compositions and properties.

The very nature of concrete and its innate variability on very small scales requires investigative techniques which allow examination and analysis at microscopic and submicroscopic levels. A wide range of petrographic techniques are currently available, with the polarising and scanning electron microscopes in particular being well-established geological techniques. These same methods have been extended and developed to investigate concrete and related materials. They are unique in providing insights into the mineralogical, chemical and microstructural features of these materials and thus provide a scientific basis of understanding that allows the physical and mechanical behaviour of concrete and related materials to be evaluated and their future behaviour predicted.

1.2 HISTORICAL BACKGROUND

The cementitious properties of materials such as slaked lime have been known and used since antiquity. The first record of a concrete floor comes from a hut in Yugoslavia dated 5600 BC (Stanley 1979). Slaked lime slowly carbonates and hardens in the atmosphere. It has been used, for example, as a mortar between the masonry blocks in the pyramids. The Greeks and Romans further developed the use of lime by mixing it with pozzolana (volcanic ash) and aggregate to produce the earliest durable concrete, examples of which still stand today (Figure 1.1). Although Smeaton undertook systematic studies of hydraulic lime for construction of the Eddystone Lighthouse commissioned in 1756, the beginning of the modern development of cement and concrete is usually identified with the patent taken out by Joseph Aspdin in 1824 (Gooding and Halstead 1952). Contemporary technological developments have greatly extended the variety and types of cement and concrete available for construction, while the use of cement additives and admixtures have further increased the range of applications of concrete in civil engineering. The investigation of the mineralogy and properties of all these materials form the subject of concrete petrography.

(a)   (b)

Figure 1.1 (a) The Roman Pantheon, the exterior view and (b) part of the interior of the dome, possibly the first lightweight concrete structure. The 50 m diameter dome was built with pumice aggregate in 27 BC. (Courtesy of A. Ellam.)

The optical compound microscope was first demonstrated in London by Cornelis Drebbel in 1621. It was developed and used by Robert Hooke who published his *Micrographia* in 1665. Improvements to lens design in general and the invention of the achromatic lens in particular provided a basis for the development of the petrographic microscope which first appeared in 1856. Its ability to polarise and analyse light led to its extensive use, first as a 'chemical microscope' and soon as an essential tool for geologists in their examination of rock-forming minerals. As early as 1849, Sir Henry Clifton Sorby had prepared the first rock optical thin section by hand (Figure 1.2, Humphries 1992).

The first application of the polarising microscope to cement clinkers is usually ascribed to Le Chatelier (1882). In the early twentieth century, the microscope was being used to investigate hardened concrete by scientists such as Johnson (1915), but difficulties of the

Figure 1.2 A photograph of the first rock thin section by Sir Henry Clifton Sorby, made in 1849 by manual grinding to transparency with emery and mounted in Canada Balsam complete with a cover slip. (Courtesy of the Royal Microscopical Society, Oxford Science Publications, *Microscopy Handbook*, 24, Humphries 1992.)

thin-section specimen preparation necessary for a detailed study of the matrix of such heterogeneous materials severely limited its use. Thin sections of hardened concretes and mortars first became practicable when commercially produced epoxy resins became available after 1946, but it was not until the 1960s that satisfactory large-area thin sections of concrete 20–30 μm thick were being routinely made for examination using the petrographic polarising microscope.

1.3 OVERVIEW OF PETROGRAPHIC METHODS

As has already been noted, petrographic techniques, particularly those using the polarising and scanning electron microscopes, provide a unique and essential means of investigating concretes and related materials to provide definitive information relating to mineralogy, compliance with specification, compositional features, workmanship and quality. These techniques are also used to provide forensic information relating to mechanisms and causes of degradation, decay and damage of many kinds.

The complexity and depth required in a petrographic study will of course depend on the information required from a particular investigation. This is illustrated schematically in Figure 1.3, with complexity increasing sequentially, as the interrelated stages 1–5 are outlined as follows.

Stage 1. Whatever level of information about a concrete structure or manufactured product is required, the scientific investigation will begin with an inspection of the structure and/or manufacturing process itself. In some simple situations, this investigation may provide sufficient information without recourse to more detailed study. However, the great majority of investigations require samples to be taken so that more detailed investigation can be carried out at a laboratory.

Stage 2. If the additional information required is of a simple nature, for example, the rock types and their proportions present in an aggregate sample, in a concrete slab or a mortar, visual inspection by a competent petrographer, perhaps aided by a hand lens, binocular microscope and some simple physical tests, may be all that is required to provide the information. Similarly many loose aggregate samples may be adequately categorised by visual inspection, though issues of a 'representative sample' may need to be addressed (see Chapter 3).

Stage 3. The definitive identification of some minerals and fine-grained materials, both particulate admixtures and fine-grained aggregates, require more detailed investigation and appropriate sampling and specimen preparation followed by the skilled use of more sophisticated equipment (see Chapter 3). In particular, the petrographic polarising microscope, which combined with the possible use of a scanning electron microscope, x-ray diffractometer, or other specialised equipment, becomes an essential tool for this type of investigation.

Stage 4. If quantitative information is required, then representative sampling issues will need resolution and the use of specialist equipment of types described in Chapter 2 will be required. A quantitative analysis necessitates a 'representative sample' of sufficient size to be statistically viable (see Chapter 3). Numerous methods are now available for the quantitative or modal analysis and can range from simple sieve grading of an aggregate sample, through various mechanical and automated modal analysis methods to elaborate camera-scanned image analysis of the sample controlled by sophisticated software (see Section 2.6).

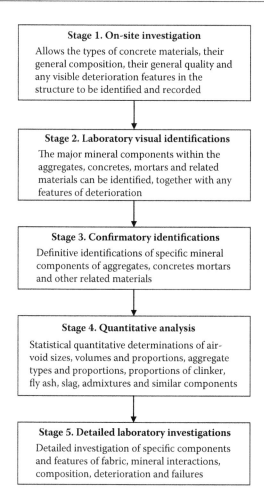

Stage 1. On-site investigation
Allows the types of concrete materials, their general composition, their general quality and any visible deterioration features in the structure to be identified and recorded

Stage 2. Laboratory visual identifications
The major mineral components within the aggregates, concretes, mortars and related materials can be identified, together with any features of deterioration

Stage 3. Confirmatory identifications
Definitive identifications of specific mineral components of aggregates, concretes mortars and other related materials

Stage 4. Quantitative analysis
Statistical quantitative determinations of air-void sizes, volumes and proportions, aggregate types and proportions, proportions of clinker, fly ash, slag, admixtures and similar components

Stage 5. Detailed laboratory investigations
Detailed investigation of specific components and features of fabric, mineral interactions, composition, deterioration and failures

Figure 1.3 **A schematic diagram illustrating the stages of complexity of petrographic investigation depending on the information and detail that is required.**

Stage 5. Where more complex investigation becomes necessary, for example, the forensic investigation of concrete failure or confirmation of the presence of specific chemical admixtures used in a particular concrete, consideration as to the approach and type of equipment used is important. Although the initial studies will use the visual and microscopic techniques described earlier, methods involving specialist equipment such as described in Chapter 2 may need to be selected in order to refine the investigation and confirm findings.

One of the most important methods of investigation involves the use of the polarising petrological microscope. This in turn requires the preparation of petrographic thin sections some 20–30 µm thick, since these are essential for any detailed study of aggregates, concretes and similar materials (see Chapter 3). Alternatively, in some circumstances, polished surfaces for use with a reflected light microscope may be a more appropriate technique and is particularly useful for identifying fine particulate materials, such as anhydrous cements, fillers and cement admixtures. Depending on the type, such sample fragments may be individually polished, but fine-grained particulate materials are sometimes cast into a resin block before polishing (see Chapter 3).

As already noted, more specialised techniques are required in some cases to confirm identifications or to explore interactions and microfabric modifications. A most important and powerful method for the identification and study of very small grains involves SEM, which is usually furnished with an energy-dispersive x-ray micro-analysis system (EDS) allowing chemical elements to be identified and, if calibrated, quantified. X-ray diffractometry (XRD) is another powerful technique for identifying crystalline materials, while infrared (IR) spectrometry or Raman spectroscopy may be used to identify molecular groupings in very small samples. These last methods are not limited to crystalline material.

There is a very wide range of modern techniques available which have application for particular specialist aspects of investigation of concrete and related materials. The practical and scientific backgrounds to these techniques are outlined in Chapter 2, and their applications and usefulness in the study of particular materials are considered in the later chapters and summarised in Tables 2.1 and 2.2.

1.4 STANDARD PROCEDURES

A number of national and international standard procedures and specifications have been developed for the petrographic evaluation of aggregates for concrete. In addition there are also numerous standard and guidance documents concerned with specific types of petrographic investigation of aggregates, for example, ASTM C 33/C33M-08, (2008), C295-09, (2009), BS 7943, (1999) and RILEM AAR-1 (2003) are concerned with specification, petrographic identification of aggregate, or potential alkali reactivity.

Much of the more routine work undertaken by petrographers evaluating aggregate for concrete complies with one or more of the national standards. However, very few standards other than ASTM C 856 (2004) cover petrographic methods relating to the detailed examination of hardened concrete, and reliance is commonly placed on available guideline and code of practice documents or on standardised procedures developed by individual specialised petrographic laboratories. The more general codes of practice for concrete and for mortars, plasters and related materials produced by the Applied Petrography Group, United Kingdom, in 2008 are typical of a number of published documents aimed at providing a unified approach to the petrographic examination of these materials. Additionally, there is a number of available and widely used documents providing standard procedures or guidelines for the petrographic examination of particular problems which may occur in hardened concrete, such as alkali–aggregate reaction. Details of these and examples of their application are given in Chapter 6.

Standards for aggregates include ASTM C 295-09 (2009) used to assess natural gravels and sands, solid rock and crushed coarse and fine aggregates. ASTM C 294-05 (2005) provides the nomenclature for the rocks and minerals identified. The European standard BS EN 932-3 (1997) is also concerned with natural and crushed rock aggregates and is generally a qualitative method. British standard BS 812-104 (1994) also deals with aggregates but is a quantitative and statistically rigorous procedure. The standard BS 7943 (1999) is more specific in providing guidance for the interpretation of petrological examinations of potentially alkali-reactive aggregates. Without exception, all these documents state or imply that the petrographic examination must be carried out by a competent and qualified petrographer.

1.5 OBJECTIVES AND COVERAGE

Since the publication of the first edition of *Concrete Petrography* (1998), major technological advances both in equipment and in computer control and analysis software have enabled the petrographer to undertake investigations and reach definitive conclusions hitherto unobtainable. Previously, problems of resolution, quantification, limits of detection and composition determination had all placed severe constraints on the results obtainable.

The use of the polarising microscope remains the basic and most important fundamental practical method for the detailed investigation of the mineralogy and microstructure of concrete and the wide range of related materials. Consequently, its correct use is described in Chapter 2. Other special techniques also described in the chapter have been selected because they are most effective in allowing the petrographer to investigate the specific chemical, compositional and structural details of these materials so that clear conclusions can be reached. Additionally, Chapter 2 describes the image analysis systems, currently available, detailing their individual practical applications to the investigation of concrete and similar materials.

Of primary importance in any investigation is the selection of representative or special samples of the appropriate material and also its preparation as a specimen for examination in the equipment to be used. The practicalities of sample selection and details of specimen preparation for the various investigative techniques are covered in Chapter 3.

Chapters 4 and 5 describe and illustrate the normal compositional, mineralogical, textural and visual features of cements, concretes, mortars and similar related materials as observed with the optical and electron microscope. Where appropriate, observational information obtained from more specialised equipment is included with these descriptions.

Chapters 6 through 9 explore, describe and illustrate the range of concretes, mortars and the more specialised but related products. They cover the range, types, mechanisms and appearance of damage, deterioration and failure that can occur in these materials and suggest the most effective methods of diagnosis in each case.

A glossary dealing with the optical and chemical properties of cementitious and related minerals which has proved valuable in the first edition has been retained and is extended, updated and is illustrated with reference to photomicrogaphs where this is appropriate in this edition.

REFERENCES

Applied Petrography Group. 2008: A code of practice mortars, plasters and related materials. APG Special Report SR1 (Available as a download from the APG web-site).

Applied Petrography Group. 2008: A code of practice for the petrographic examination of concrete. APG Special Report SR2 (Available as a download from the APG web-site).

ASTM C 33/C33M-08, 2008: *Standard Specification for Concrete Aggregates*. ASTM, Philadelphia, PA.

ASTM C294-05, 2005: *Standard Descriptive Nomenclature for Constituents of Concrete Aggregates*. ASTM, Philadelphia, PA.

ASTM C295-09, 2009: *Standard Guide for Petrographic Examination of Aggregates for Concrete*. ASTM, Philadelphia, PA.

ASTM C 856-04, 2004: *Standard Practice for Petrographic Examination of Hardened Concrete*. ASTM, Philadelphia, PA.

BS 812-104, 1994: *Testing Aggregates. Procedure for Qualitative and Quantitative Petrographic Examination of Aggregates*. BSI, London, U.K.

BS 7943, 1999: *Guide to the Interpretation of Petrological Examinations for Alkali-Silica Reactivity*. BSI, London, U.K.

BS EN 932-3, 1997: *Tests for the General Properties of Aggregates. Procedure and Terminology for Simplified Petrographic Description*. BSI, London, U.K.

Gooding, P. and Halstead, P.E. 1952: The early history of cement in England. In *Proceedings of the 3rd Symposium on the International Chemistry of Cement*, London, U.K., Cement and Concrete Association, pp. 1–29.

Humphries, D.W. 1992: The preparation of thin sections of rocks minerals and ceramics. *Royal Microscopical Society Handbook 24*. Oxford University Press.

Johnson, N.C. 1915: The microstructures of concretes. *Proceedings of the American Society of Testing and Materials* 15, Part II, 171–213.

Le Chatelier, H. 1882: *Comptes Rendus hebdomadaires des séances de l'académie des Sciences* 94, 13.

RILEM AAR-1. 2003: RILEM recommended test method AAR-1. Detection of potential alkali-reactivity aggregates' Petrographic methods. (Available as a download from the RILEM web-site: www.rilem.net).

Stanley, C.C. 1979: *Highlights in the History of Concrete* (reprinted 1982). Cement and Concrete Association, Slough, U.K.

Chapter 2

Petrographic equipment and methods

2.1 PETROGRAPHIC EXAMINATION OF CONCRETE

As outlined in Chapter 1, the petrographic study of concrete, mortar and related materials can be undertaken in a series of interrelated stages of increasing complexity. The extent of any such investigation will of course be dependent on the nature and detail of the information required.

Although valuable information is obtained from on-site investigation and monitoring, the majority of investigations involve samples being dispatched to a petrographic laboratory where a more detailed examination of the material can be undertaken. Such laboratory studies will begin with a visual inspection and recording of the received samples, but examination using a good-quality binocular microscope is an almost essential aid to this initial inspection. A more detailed study will then require the samples to be prepared appropriately for the particular examination (see Chapter 3).

Perhaps the most valuable and useful piece of equipment in a petrographic laboratory is the petrographic polarising microscope. However, in many complex investigations, the examination of material using the polarising microscope is only a first step which is then extended to investigate specific aspects of the material involving the use of specialist equipment or methods such as scanning electron microscopy, chemical analysis or x-ray diffractometry.

This chapter will deal in detail with the polarising microscope's use and adjustment. It will also provide summaries of the use, application and scope of the complimentary techniques commonly used by petrographers, but the reader is referred to the relevant technical publications for more detailed information concerning the technical and theoretical basis of this specialist equipment.

Tabular summaries of the most appropriate equipment or techniques that may be used for the investigation of the various particulate or solid samples commonly encountered by petrographic laboratories are given in Figures 2.1 and 2.2.

2.2 INITIAL LABORATORY EXAMINATION

Samples which arrive at the laboratory for petrographic analysis range from kilogramme quantities of coarse aggregate to small quantities of fine powder or to large samples taken from concrete structures. The most satisfactory type of concrete samples for most general petrographic investigations is the 100 mm diameter drilled cores. In some circumstances, larger-diameter cores of 150 mm are available; these will provide larger areas to be examined and be more representative of the concrete. However, smaller-diameter cores of 75 mm, or even 50 mm, are sometimes provided because of constraints, such as reinforcement

		On-site visual inspection	Laboratory visual inspection	Ground/polished slices/specimens	Thin-section microscopy	Reflected light microscopy	Fluorescence microscopy	Stain or etch methods	SEM/EDS methods	X-ray diffraction methods	FTIR methods	Thermal analysis methods	Chemical analysis methods	Physical testing methods	Quantitative point-count analysis
Concrete aggregate materials	Coarse	I3	I2	A2M	P1	A1		A1						P2M	M
	Fine	I3	I3	A2	P1M	A1		A1						P2M	M
	Recycled/secondary coarse aggregate	I2	I2	A2M	P1		A1			A2				P1M	M
	Artificial coarse aggregate	I2	I1	A2M	P1		A1			A2				P1M	M
	Expanded coarse aggregate	I2	I1	A2M	P1		A1							P1M	M
	Fillers	I4	I1	A4	P1	A1		A1	A1M	A1M	A1		A1	P1M	M
Unhydrated cementitious materials	Unground cement clinkers	I3	I3		A2	P1		A1	P1						
	Portland cements	I4	I3		A2	P1		A1	P1	A2		A2	A1		
	Special cements	I4	I3		A2	A2			P1	A1	A1	A1	A1		
	High-alumina cement/calcium aluminate cement	I4	I3		P1	P1			P1	A2		A1	A1		
	Non-Portland-type cements	I3	I3		P2			A1	P1	A1	A1	A1	A1		
Unhydrated cement–admixtures and replacements	ggbs	I4	I3		P1	A1			A1	A1			A1		
	PFA	I4	I3		P1	A1			A1	A1					
	Natural pozzolanas	I4	I3		P1	A1			A1						
	Silica fume		I4		P2	A2			P1						
	Chemical admixtures				A4				A2		P1				

Relevance of the method to the Identification of components

I = Initial/tentative/partial identifications

P = Principal method of identification

A = Alternative/complementary method

M = Possible quantitative method/modal analyses possible

The likely quality of the identifications obtainable

1 = Excellent definitive identifications

2 = Good comprehensive identifications

3 = Fair or partial component identification

4 = Poor identification requiring confirmation

Figure 2.1 Guide to the petrographic methods appropriate for the identification and evaluation of particulate materials.

		On-site visual inspection	Laboratory visual inspection	Ground/polished slices	Thin-section microscopy	Reflected light microscopy	Fluorescence microscopy	Stain or etch methods	SEM/EDS methods	X-ray diffraction methods	FTIR methods	Thermal analysis methods	Chemical analysis	Physical testing methods	Quantitative point-count analysis
Hardened concretes and mortars	Cement types (hydrated)	I4	I3		P2M			A1	P1M	A2	A2	A2	A2M		M
	Aggregate types	I3	I2	P1M	P1M			A2	A1		A1				M
	Air voids	I3	I2	P1M	P1M		A1		P1						M
	Textures/fabric	I4	I3	P2	P1		A2								
	PFA admixtures	I4	I3	A2	P1		A1M		P1M						
	ggbs admixtures	I4	I3	A2	P1		A1M		P1M						
	Water/cement ratio	I4	I3	I3	P1		P1		A1				P3		
	Natural pozzolanas	I4	I4		P1				P1M						
	Silica fume admixtures	I4	I4		P1				P1						
	Chemical admixtures	I4	I4		A3						A2		A2		
	Mix proportions	I4	I3	P2M	P1M		A1						A3		M
Possible causes of deterioration	Chemical attack	I2	I2	A3	P1				P1	A3					
	Delayed ettringite formation (DEF)	I4	I4	I3	P1				P1	A2					
	Aggregate shrinkage	I4	I4	I3	P1				A2						
	Salt attack	I2	I2	A3	P1				P1	A2			A1		
	Carbonation	I2	I1	A1	P1			P1	A1			A2			
	Reinforcement corrosion	I1	I1	A1	P2	P1									
	Sulphate attack	I2	I2	A2	P1		A1		A1	A2			A3		
	ASR/ACR	I2	I1	A1	A1		A1	A2	A1				A2		
	Frost damage	I2	I2	P1	A1	P1	A1								
	Fire damage	I2	I2	I2	P1		A1								
	Workmanship	I2	I2	P1	P1		A1						A2		

Relevance of the method to the Identification of components

I = Initial/tentative/partial identifications

P = Principal method of identification

A = Alternative/complementary method

M = Possible quantitative method/modal analyses possible

The likely quality of the identifications obtainable

1 = Excellent definitive identifications

2 = Good comprehensive identifications

3 = Fair or partial component identification

4 = Poor identification requiring confirmation

Figure 2.2 Guide to the petrographic methods appropriate to the investigation of concrete, mortar and related materials.

spacing. Drilling damage, surface alteration, core recovery and volume of sample make these small-diameter cores much less suitable for petrographic study.

Cores can vary from a few millimetres to many metres in length. Thus, one of the problems in the initial laboratory examination of such material, once it has been unwrapped, can be handling of the sheer size and weight of the larger samples. Irrespective of the examination procedures to be employed, all specimens must be first photographed as received, for record purposes. The photographic methods required are outlined in Section 2.3.4. Unless special facilities are devised, it is difficult to examine a long concrete core without sawing it into more manageable lengths. For most purposes, this is acceptable, providing that all the pieces are clearly marked for identification and orientation, and care is taken to retain evidence.

The purpose of the initial examination is to observe any unusual features and provide a description of the general material characteristics of the sample. With concrete specimens, the examination method may have to cover large surface areas but may also include minute crystals present in pores and cracks. This requires inspection with the unaided eye, perhaps assisted by a hand lens and with a stereomicroscope. The careful and combined use of both the unaided eye and the stereomicroscope is vital, as the unaided eye will often fail to see fine cracks and details until they are observed through the stereomicroscope. On the other hand, some features such as surface discoloration clearly visible with the unaided eye tend to 'disappear' under the stereomicroscope. An important consideration for such observations is to have sufficient illumination and at a suitable orientation. Strong daylight for the unaided eye is superior to artificial light, while a well-adjusted bench or nose lamp will be required for the stereomicroscope. Generally, oblique lighting at a 45° angle to the surface is appropriate, but overhead illumination from a ring light may reveal additional detail.

Large samples required for thin sectioning, determination of cement content and chemical analyses are easily removed after this initial examination using a diamond saw. For sampling local areas of hardened cement paste and other materials, a modified dental drill, or a suitable workshop motor with flexible shaft and handpiece, is a useful tool. Modern turbine-powered dental drills can achieve speeds of 300,000–400,000 rpm, but for concrete sampling, a lower shaft speed of 12,000 rpm or above will allow diamond-tipped grinders and cutters to be used. It should be remembered that high-speed drilling will generate considerable localised heating which may alter the mineralogy adjacent to the cut.

Sampling efflorescence, alkali–silica gel and small surface deposits from pores and cracks require manipulative skill and the use of pointed scalpels and probes. Particles are often less than 100 μm in size and in many cases, the material is very fragile and deposits may be present only in limited quantities. Fine probes may be made from tungsten wires sharpened with molten sodium nitrate as described by McCrone and Delly (1973), or alternatively some types of needles and dental probes may be used. As the quantity of material available for collection becomes less than a few milligrams, careful collection methods and special sampling techniques are required. These need to be devised specifically for the particular sampling problem. Again, some of these special techniques have been described by McCrone and Delly (1973).

2.3 PETROGRAPHIC LABORATORY

The petrographic laboratory should be equipped with a sufficient bench and storage space to allow samples to be laid out and examined. In addition, there should also be a space to store equipment and any current samples/specimens in a clean dry environment. The laboratory will contain suitable photographic/electronic imaging equipment to record and document samples and specimens as appropriate (see Chapter 3) and should also house essential reference books and charts.

Highest-quality optical microscopes are central to the laboratories' operation. Such a laboratory will contain at least one stereo-zoom microscope that will be used in the initial stages of sample examination (see Sections 2.2 and 2.3.1) together with a high-quality polarising petrographic microscope, which should be housed in a separate dust-free clean room and ideally within the laboratory.

Additionally, there will be photographic equipment and electro/mechanical point-counting devices for use with the microscopes and appropriate computer facilities for recording and manipulating data and for image analysis together with the relevant software (see Section 2.6).

A flat-bed scanner and associated computer imaging software are also in common laboratory use for recording details visible on cut, ground or polished concrete surfaces and can provide excellent resolution of detail (see Section 2.6.1). With a little ingenuity, a system for rotating a core and synchronising its rotation with the scanning process will allow the whole outer surface of the core cylinder to be imaged and recorded.

2.3.1 Low-power stereomicroscope

In practice, the stereo-zoom microscope is simple to use, and this is why it finds such a wide use in the laboratory. There is no doubt that for manipulative procedures, the stereomicroscope with its crisp, erect image is an excellent tool. Thus, it is an essential item of equipment for the concrete petrographer and will also be required if the air content of concrete is to be measured following the widely used ASTM C457- 10a which is a standard practice for microscopical determination of air void content and parameters of the air void system in hardened concrete.

The modern, low-power, stereo-zoom microscope employs 1:4 to 1:8 zoom ranges dependent on the make. With the use of 10× eyepieces, the range of magnification is about 7×–50× but can be extended by accessories. Manufacturers offer a range of eyepieces from 5× to 20×, but for most purposes, the 10× eyepiece is ideal as the image on a stereo-zoom microscope is clear and crisp up to 50× magnification. Above this, the quality of the image is dependent on the optics used, the lighting and the type of surface being examined. For many types of concrete samples, the image is degraded at 100×. One eyepiece should be fitted with a micrometre scale for length measurement. A detachable stereo-zoom microscope body is essential so that it can be used with a variety of stands and with air void equipment. The basic stand commonly supplied cannot accommodate concrete specimens of large size, so for this purpose, a long swinging-arm type of stand is appropriate as it provides the horizontal and vertical adjustments necessary for examining large core specimens. The stereo-zoom microscope is usually provided with a 20 W halogen lamp illuminator, or LEDs of the same brightness, mounted either on the microscope objective or on a separate stand; fibre optic and LED ring illuminators are also available if a uniform field of light is required.

If the eyepiece head is detachable from the body of the stereo-zoom microscope, it is possible to mount a camera on a slide track to produce stereophotographic pairs.

There are many models of the stereo-zoom microscope available. When purchasing this type of microscope, it is necessary to keep clearly in mind that it is a low-power microscope intended for general viewing and manipulation. However, the highest-quality optical system should be sought, because if it is to be used with polished or ground slices for quantitative analysis, such as air void determination, the quality of the image is most important for the accurate evaluation of fine details.

2.3.2 Petrographic polarising microscope

The polarising microscope is designed to analyse light transmitted through or reflected from substances, and this requires strain-free optics aligned with the optic axis of the microscope. Some manufacturers publish short texts on microscopy for the beginner, and there are a number of more detailed texts that are helpful in explaining the practical use of the various components of the microscope including a range of technical handbooks: one is an introductory text on microscopy (Bradbury and Bracegirdle 1998), a second on polarised light microscopy (Robinson and Bradbury 1992) and another on fluorescence microscopy (Ploem and Tanke 1987), which are all produced by the Royal Microscopical Society. One of the more detailed texts is Loveland's (1970) book on photomicrography which is still relevant today. The following brief description lists the parts of a typical universal petrographic microscope for transmitted light observations which is suitable for concrete petrography.

Before using a new microscope, it is essential to read the manufacturer's instructions on its assembly and operation. It is necessary to know how to assemble the microscope and to identify how individual parts relate to each other. Many microscope accessories are only marked with the manufacturers name and code number so that a descriptive list must be drawn up for future users of the instrument.

This is especially the case with older microscopes where the manufacturer no longer exists and the information on code numbers is no longer available.

The quality modern microscope is a precision instrument and will only operate properly if it is looked after carefully. The microscope room should be kept at a constant temperature as this reduces dust being pumped into the interior optics. A dust cover should always be kept on the microscope when not in use. Objective(s) should always be left in place and slot covers closed. Periodically, the instrument should be dusted with a soft cloth and exterior surfaces of glass elements cleaned with a soft airbrush. If marks or more persistent dirt are present on these surfaces, they can be gently cleaned with absolute alcohol and soft lens tissue. Where marks or dirt on lens surfaces do not respond to this treatment, the item should be returned to the manufacturer for inspection and repair.

Objectives, oculars, condensers, stage mechanisms and other precision-adjusted items should only be disassembled by a qualified microscope technician. With time, it may be found that movement of the stage or focusing controls becomes uneven or stiff, and this is a sign that the microscope requires a routine overhaul. Oil immersion lenses and condenser caps should be carefully wiped clean with a lens tissue *immediately* after use. Periodically, they should also be cleaned with a lens tissue moistened with absolute alcohol. Once immersion oil has dried on lens surfaces, its removal is a job for a specialist.

2.3.2.1 Adjustment of the polarising microscope for transmitted light

1. Adjust the seat height so that the head is comfortable and in a relaxed position when looking down the microscope. The forearms should be able to rest comfortably on the microscope bench, and the seat should be armless and with an adjustable back. To avoid discomfort, fully adjustable seating, such as a 'typist's chair', is essential for extended sessions at the microscope.
2. Switch on the light source, adjusting it to ensure that it is not too bright for comfort; remove any filters from the light path.
3. Set the eyepiece tubes to the 'zero' or middle position for each eyepiece.
4. Remove the eyepiece with the micrometre graticule. Look through it against bright daylight and focus the graticule by adjusting the front lens and then replace the eyepiece in the right-hand eyepiece tube.

5. Place a thin section on the microscope stage, open the field and condenser aperture diaphragms and, using a low-power objective (10×), focus the microscope using the micrometre eyepiece only. Check the focus in the left eyepiece and correct if necessary by means of the adjustment on the left eyepiece tube.

6. Set the distance between the oculars to your interpupillary distance so that no double image is observed; note this setting for future use (both eyepieces are now adjusted to compensate for the individual variation in your eyes and are essential to relaxed viewing and should *always be checked prior to use*).

7. Fully close the field diaphragm and adjust the height of the condenser until the sharpest image of the iris diaphragm aperture is seen. Centre the iris in the field of view by using the centring screws on the condenser system. Open the field diaphragm until the edges of the iris are just outside the field of view. The lighting is now adjusted to give 'Kohler illumination'.

As a 'rule of thumb', the condenser diaphragm should only be shut down by about two-thirds (best accomplished by closing it as far as possible and then opening it one-third). The setting of the aperture diaphragm is critical to viewing and should be varied according to the object being viewed to give optimum contrast.

2.3.2.2 *Centring the objective lenses*

Centre the objective by focusing on a small spot on the slide and rotating the stage. Most polarising microscopes have objective nosepiece mechanisms which allow each objective to be individually and more accurately centred on the thin section. However, the manufacturer's instructions should be followed to ensure correct set-up and usage. This centring needs to be carried out whenever an objective is changed, but existing objectives should also be checked from time to time.

2.3.2.3 *Calibrating the eyepiece micrometre graticule*

Place a stage micrometre in the field of view. Take each objective in turn, adjust as described earlier and calibrate the micrometre in the eyepiece with the stage micrometre. Measure both the small divisions of the eyepiece micrometre and also the diameter of the field of view. Calculate the effective magnification of each objective, and assemble a table of effective magnifications, micrometre calibrations and field sizes for each of the objectives in use. This is an essential data table for detailed examinations and quantitative microscopy and should be kept close to the microscope for easy reference (see Section 2.4).

2.3.2.4 *Reflected or incident-light illumination*

Most modern petrographic microscopes will accommodate a reflected light module comprising of a light source, collimator system and optics to allow the microscope to be switched between transmitted and incident-light operation without the need for readjustment. The light source is usually adjustable for brightness and for centring. Removable filters and an adjustable field diaphragm in the optic path allow a specimen on the microscope stage to be viewed in optimum conditions. Incident illumination using high-intensity ultraviolet (UV) light is used in a number of applications. This causes yellow fluorescent dyes in resin-impregnated thin sections and polished surfaces to fluoresce (see Section 3.5.4). The UV light is produced by a high-pressure mercury vapour lamp which usually requires its own separate power source and lamp housing. A filter in the optic path is used to limit the incident light to an appropriate wavelength range to produce maximum fluorescence and must be matched to the dye type (see also Section 2.5.1). Note that national guidelines as to the safe use of UV light should be followed.

Reflected light illumination is usually accomplished by directing the light beam through a Berek prism or an inclined plane glass plate directly down onto the specimen from around the objective lens. The Berek prism is better at low magnifications giving a brighter image and greater depth of focus (but is often not available in modern microscopes), while the glass plate is better at examining polished surfaces at high magnifications. With an inclined glass plate, the field aperture diaphragm should be as wide as possible commensurate with avoiding 'flare'. The quality of the image requires careful adjustment of the incident-light optics and is critically dependent on the surface finish of the specimen.

Reflected light high-intensity UV illumination is commonly used in conjunction with image processing for quantitative studies (see Section 2.6). Some good examples of faults that can occur in polished cement clinker specimens are illustrated by Campbell (1986).

2.3.3 Construction features of the polarising microscope

The details of construction of petrographic polarising microscopes vary according to the make and model, but most are built up from a series of basic modules. In general terms, the essential details of a research quality polarising petrographic microscope can be itemised as follows:

1. *Universal stand* (Figure 2.3). The stand must be robust enough to provide sufficient stability to support heavy accessories. It will contain graduated focusing controls and provision for the attachment of transmitted and incident lighting systems.
2. *Lighting systems* (Figure 2.3). Work in concrete petrography requires a transmitted light source with centring controls and built-in condenser system. Typically, it will use a 50 or 100 W tungsten halogen lamp which has an adjustable power supply. Besides the daylight blue, green and diffusion filters usually supplied, a range of neutral density filters are also necessary.
3. *Object stage* (Figure 2.4). The petrographic stage is a circular rotating platform, graduated in degrees around its perimeter, which can be clamped and has provision for 45° interval click stops. A detachable ring plate which allows for the fitting of a universal stage is replaced by a solid ring plate in the incident-light mode. There are holes in the top of the stage for stage clips and also threaded holes for attaching micrometre and electromechanical point-counting stages. An attachable mechanical micrometre stage for use with slides up to 75 × 50 mm can be purchased but is not really necessary for general work. A rack and pinion with yoke is located underneath the stage to which the condenser substage assembly is attached.
4. *Condenser system* (Figure 2.3). The condenser system for a petrographic microscope is a strain-free, achromatic two-lens Abbe system with a numerical aperture of approximately 0.25. Mounted on the condenser is a swing-in holder into which screws the strain-free top lens. The purpose of the screw-in top lens is to match the numerical aperture of the condenser to that of the higher-power objectives. Condenser top lenses commonly supplied provide a numerical aperture between 0.9 and 1.4 and can be used with oil immersion lenses. The alignment of the condenser system must be adjustable. An iris diaphragm for aperture control is mounted between the polariser and the condenser lenses, and at the bottom of the main condenser lens, there is a graduated, rotatable polarising plate which is often pivoted so that it can be removed from the light path if required. Often, there is a special filter slot provided just above the

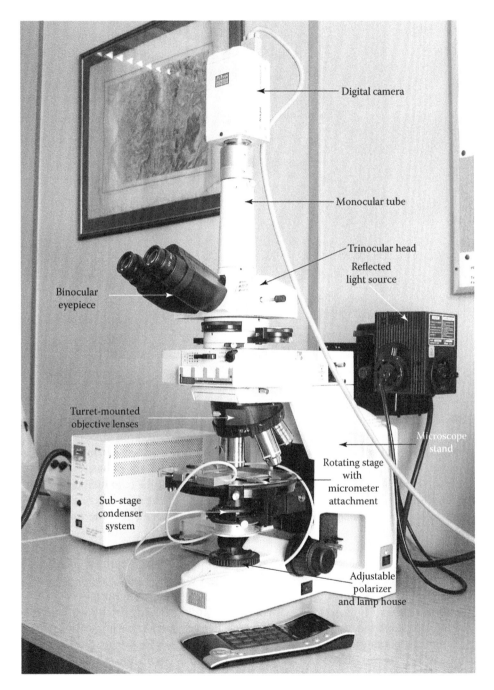

Figure 2.3 A modern petrographic polarising microscope, Nikon Eclipse E600POL. (Photograph by A. Bromley.)

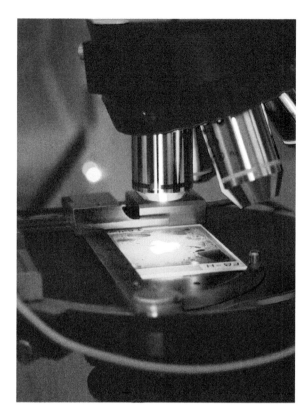

Figure 2.4 **A thin section of concrete held in a micrometre stage attachment in UV incident illumination.** (Courtesy of A. Bromley.)

polariser for the insertion of a 1/4 wave circular polarising filter. The type of polarising plate used is especially important for concrete petrography and must be of high quality and provide intense light sufficient to flood the full aperture of the objective. Nicol prism polarisers and stacked glass plate polarisers are designed to protect the optics against heat from strong light sources, but should be avoided for general work as they limit the numerical aperture. It is necessary for the entire condenser system to be able to rack independently of the stage to allow critical adjustment for Kohler illumination.

5. *Objective lenses* (Figure 2.4). Most current models accept five or more objectives in a rotating nose turret. Each objective will have collars or rings as a means of individual centring. Some older microscopes may be fitted with a clutch nosepiece system that accepts a range of single interchangeable objective lenses.

6. *A trinocular head* (Figure 2.3). Is a common feature of many modern research microscopes and is designed to provide binocular viewing together with provision for the use of digital cameras for photography and for image analysis for processing or display. These are all placed on the vertical monocular tube. The inclined binocular tubes are adjustable to match the interpupillary distance of the eyes. When the binocular tubes are adjusted, some correction of the phototube length may be necessary to correct for the 'parfocality' of the camera if fitted so that the camera on the monocular tube remains in focus. Advanced polarising microscopes typically have individual focusing on both eyepiece tubes and specially aligned slots at 45° for the use of an ocular with accurately fixed cross hairs. A focusing and centring Bertrand lens, which is used for

conoscopic observation of interference figures, and a polarising plate called the analyser are housed in an 'analyser unit' mounted between the stand and the binocular/trinocular head. Most analyser systems have full rotation capability 0°–360° and are calibrated in degrees.

7. *Eyepieces* (Figure 2.3). The choice of eyepieces is limited to purchasing those matched to the objectives in use because the eyepieces may contain part of the correction designed into the optical system. Since the wearing of spectacles is common, the purchase of some high-point eyepieces which allow the user to wear spectacles while operating the microscope is almost obligatory. These are now available in a wide field option and can be used with or without spectacles by unrolling the soft rubber caps. In modern microscopes, both eyepieces are focusable and able to take graticules unless they contain fixed cross hairs. It is convenient if a number of eyepieces for interchangeable graticules are purchased. For precise work, an eyepiece containing accurately aligned cross hairs with a registration detent to fit the slots in the ocular tubes is also necessary. A wide range of graticules are available. The two that are most essential are cross hairs with a micrometre scale along one axis and another for focusing. The amount of useful magnification provided by the eyepieces is a subject often discussed in texts where it states that the overall magnification should not exceed 100× the numerical aperture of the objective (i.e. 40× objective with 65NA = 650× maximum magnification). For practical purposes, the 10× or 12.5× eyepieces give optimum magnification, and the use of higher-power eyepieces is rarely necessary.

2.3.4 Photomicrographic equipment

A wide range of electronic digital cameras are available and their resolution in megapixels continues to improve. In addition to a general-purpose camera capable of taking high-resolution photographs of samples as received and detailed features on them, a camera mounted on the polarising microscope is essential. A typical digital camera attached to the monocular tube is illustrated in Figure 2.5. High-definition (HD) video cameras are also available, and these can produce high-resolution images in real time (see Section 2.6.1). An introductory guide to the practical applications of digital light microscopy imaging is given by Entwistle (2003, 2004).

The two most common requirements of photomicrography are to provide high-resolution colour photomicrographs for illustrating research or petrographic reports and to provide an input allowing an image of the microscope's field of view to be enlarged and displayed on a VDU so that additional observers or students can view the image as seen by the petrographer.

A computer software that is compatible with the camera may also be available to allow image enhancement, component identification and modal analyses to be undertaken. Further details of computer-aided image analysis are given in Section 2.6.1.

2.4 QUANTITATIVE METHODS OF COMPONENT ANALYSIS

The quantitative determinations of dimensions and/or proportions of components in a sample are an essential part of concrete petrography. Bradbury (1991) has discussed the basic measurement techniques for light microscopy. Most microscopes have a fine focus control knob engraved with a scale graduated in microns. This also allows the thickness of a section or particle to be estimated provided that it is possible to focus accurately on the top and the bottom of the object. To obtain the correct measurement, the distance traversed by the focus must be multiplied by the refractive index of the object. This method works well for particles

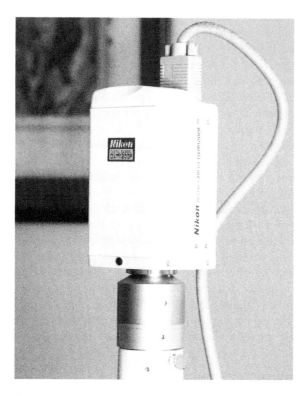

Figure 2.5 Typical of modern cameras, a Nikon DXM 1200F 12 megapixel digital camera is mounted on the phototube of the trinocular head of a polarising microscope. (Photograph courtesy of A. Bromley.)

in liquid immersion, but is less applicable to sections where thickness is traditionally and more easily estimated by observing the maximum birefringence of minerals such as quartz.

The essential item required for the measurement of linear dimension is the eyepiece micrometre. This is a disc of glass engraved with a scale, for example, perhaps 100 scale divisions might equal 10 mm. The eyepiece should also include a cross hairs disc for centring and the disc fits into an eyepiece designed for its use. Using a stage micrometre, the eyepiece micrometre must be calibrated for each objective and results recorded in a table (see Section 2.3.2). If more accurate measurements are required, a screw micrometre eyepiece can be used.

2.4.1 Standard modal analysis methods

To determine the volumes of components in a thin section, a ground slab or numbers of particles in grain mounts require the use of counting techniques using a flat-bed scanner, or optical microscope. This is known as modal analysis. A full account of the range of modal analysis methods that are used is given in Section 4.5.2. The two basic methods for concrete are the Chayes (1956) point count and the Rosiwal (1898) lineal traverse techniques. ASTM C457/C457M-10a (2010) and ASTM C295-08 (2008) briefly describe these methods. Although these methods remain in current use, they are being replaced by computer processing and image analysis software which is readily available and allows component sizes and proportions to be determined quickly and accurately from optical and electron microscope images (Section 2.6) and from computer-scanned ground concrete surfaces (see Sections 3.3.6 and 4.5.2).

Systematic identification and counting of particles, voids or minerals in a finely ground surface or thin section to estimate volume proportions rely on the assumptions that the surface or section is cut through randomly orientated particles and also that the area is large enough to be representative of the material. A further assumption is made that the areas of the components measured per unit of the test area surface also estimate the volume of the components per unit volume, that is, that the relationship proposed by Delesse (1848) holds. Modal analysis, which forms part of the general subject known as stereology, has long been accepted as a standard method in petrography and materials science. It is a complex subject and for further information, DeHoff and Rhines (1968) should be consulted.

The two requirements for point counting are the ability both to lay down an orthogonal grid on the surface under examination and to identify the component of the sample that falls under each intersection point of this grid. The devices described here are all concerned with generating the orthogonal grid.

Hand-operated point counters which attach to the microscope stage are readily available but these tend to be limited to working areas of 20 × 20 mm. That is, they are solely intended for the standard-sized, geological thin section. Similarly, the now obsolete Swift automatic point counter operated the specimen stage when any one of the thirteen buttons on the tally counting unit was depressed, but this instrument had a total movement of only 30 × 30 mm. Dials on the counting unit record the number of times each button is operated. This point counter could be adjusted to traverse in intervals in the horizontal direction, but required manually resetting orthogonally at the end of each traverse. In this method, the point-counting stage is moved incrementally along a lineal traverse and the component observed under the cross hairs in the ocular identified and tallied at each increment. A series of such lineal traverses could be measured in this way so that the points examined comprise a series of regular grid intersections.

This electromagnetic point counter was a major improvement on the hand-operated stage counters, but has now been superseded by electronic versions which tally the components identified in a similar way and automatically calculate the volume percentages. These devices are designed to accommodate larger thin sections; a 60 mm × 40 mm thin section is typical. Stepping motors are used to move the stage incrementally in one or both directions, and the associated software can be used to define the size of the area to be examined and the number of points to be accumulated from that area. A typical example of a point-counter stage is illustrated in Figure 2.6. This device is itself being superseded in many applications by computer processing methods where the complete field or series of fields of view are evaluated and mineral proportions, sizes and shapes calculated automatically.

Concrete, because of its inherent variability, and the use of aggregate which can range from micron-sized dust to aggregates whose maximum size commonly is in the range of 20–40 mm, requires specimens with large surface areas, or a series of thin sections in order to be representatives if the coarse aggregate proportions are to be determined. ASTM C457/C457M-10a (2010) specifies and illustrates a mechanical counting table, which is suitable for point counting or making lineal traverses of large, ground surfaces of concrete. Using a stereo-zoom microscope in the range of magnifications of 60×–80×, the volume of air voids and hardened cement paste can be determined. The problem with this method is that at this magnification, the stereo-zoom microscope does not image the ground surface of the concrete (even when finished with 5 µm grit) with sufficient clarity to be able reliably to distinguish remnant cement grains from small grains of aggregate in the cement paste. Provided the microscope is of high quality, this is not a problem with air voids which are usually clearly distinguishable and may require magnifications of as much as 120× to be able to count the smallest bubbles. In this case, accuracy of identification is a function of the sharpness of the void margins and quality of the microscope. Increasingly, image analysis techniques are used to determine percentages of air voids and other components in ground concrete surfaces (see Section 2.6).

Another method is to use a counting grid inserted in the microscope eyepiece. The component at each intersection is counted and then a mechanical stage is used to move a set

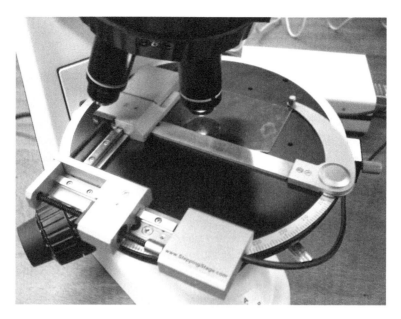

Figure 2.6 A modern point-counter stage mounted on a petrological microscope. Two stepping motors control the *x* and *y* movement directions. (Courtesy of B. Wells, Conwy Valley Systems, Conwy, United Kingdom.)

distance to a new field until the specimen surface has been traversed. This method as applied to cement clinkers is discussed by Hoffmanner (1973).

As an alternative to using a counting grid in the microscope eyepiece, a regular grid printed onto a transparent film can be taped over the specimen to allow the identification of components at the grid intersects to be tallied in a similar way to other point-counting devices. A photographically or computer-reduced grid of an appropriate size can be fixed and printed side down on top of the cover glass of a thin section, and using a magnification in the range 30×–70×, the whole section can be moved with the fingers from intersection to intersection quickly traversing the section backwards and forwards until all grid points have been counted. For counting a very large section, or ground concrete surface, the grid can be re-taped over adjacent areas of the surface. Some blurring of the section and the grid occurs but this is not usually sufficiently serious to interfere with the counting. However, for the petrographer who only needs to count a few samples each year, this simple and inexpensive method should suffice.

The question of how many sample specimens are required and how many points to count depends on the adequacy of sampling and the accuracy required (see also Section 3.3.2). A working rule for concrete is that it is best to count over as many samples and the largest surface area possible because of the natural variability of concrete.

Several sources of error are inherent in these point-counting procedures in addition to the size of the surfaces sampled. The Delesse relationship assumes that the ratio of the area occupied by a single component in a randomly cut surface to the total measurement area is a consistent estimate of the volume percentage of the component in the whole sample. Since an analysis made on a randomly cut surface is essentially the analysis of an area, the relation holds, but the volume concerned will be dependent on the homogeneity of the material. Preferential orientation, segregation and banding of components, grain size variation and other forms of heterogeneity will introduce errors to point counting. When coloured and transparent minerals or components occur adjacent to each other in a thin section, the coloured materials tend to be overestimated and the transparent ones underestimated as illustrated diagrammatically in Figure 2.7.

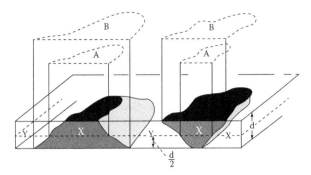

Figure 2.7 The dark grains in the section thickness d should be estimated to cover areas A, whereas an observer will see them as covering areas B.

The precision of point-counting methods depends on the number of points counted, the grain size or sizes of the component particles in addition to the size of the area traversed which must be representative of the volume to be sampled. Van der Plas and Tobi (1965) devised a chart, Figure 2.8, for estimating the error to be expected in the volume percentage of a particular component given the total number of points counted.

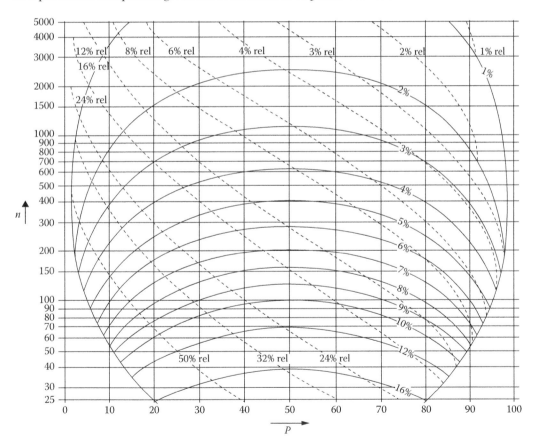

Figure 2.8 Chart for estimation of error in point counting. The total number of points n counted on the specimen is shown on the y-axis and the percentage P of a component on the x-axis. At the intersection point of n and P, the relative error is read from the dotted lines and standard deviation at 95% probability from solid curved lines.

The Delesse relationship applies only if the step size of the measuring grid is larger than the individual particle sizes of the components being measured. When carrying out measurements on concrete, it is not always possible to comply with this requirement because of the large size of the coarse aggregate in relation to the smaller components being counted, but the counting errors involved are usually not serious.

As an example in the use of this chart, suppose that some 2500 (n) points had been tallied and that the component of interest is represented by half the points. That is 50% (P) of the volume. From Figure 2.8 for a 95% probability, the relative error (2σ expressed as a fraction of P) shown by the dashed lines is 4%. The solid lines give the error at 95% probability (2σ) as 50% ± 2%.

2.5 COMPLEMENTARY AND SPECIALISED TECHNIQUES

Optical petrography is an essential first step in the detailed examination of concrete and related materials and is referred to in Chapter 1 as a 'Stage 3' level of an investigation. However, in many cases, more specialised methods, complimentary to the initial optical microscopy, may be essential in providing the conclusive evidence required to complete an investigation. Perhaps the most powerful and widely used of these techniques is scanning electron microscopy. These instruments normally include some forms of x-ray fluorescence microanalytical system, most usually based on an x-ray energy analysis, an energy dispersive system (EDS). The scanning electron microscope and its related equipment will be described in Section 2.5.2. Other specialised techniques and instruments, such as x-ray diffraction, are also available and are in common use for particular applications in the investigation of concrete and related materials. Such equipment used in these applications are also outlined in this chapter, but the relevant references should be consulted for more detailed accounts of their theoretical basis and application.

2.5.1 Petrographic examination under ultraviolet illumination

In order to enhance features such as microcracks, voids and porosity of the cement paste of a concrete or mortar, it has become almost universal practice to include a yellow fluorescent dye in the resin impregnating the specimen during the manufacture of the thin section (see Sections 3.4.4 and 3.5.4). By illuminating the thin section either in transmitted or reflected high-intensity UV light (200 W), the yellow dye fluoresces allowing these features to stand out clearly (see also Section 2.3.2). This effect provides the high contrast which simplifies modal analysis and is particularly advantageous for image analysis processing (see Section 2.6).

One important application makes the comparison of the fluorescent intensity obtained through an area of cement paste in a hardened concrete in thin sections against intensities from a series of standard concrete thin sections made to known water/cement ratios and of similar composition to the unknown. This allows the water/cement ratio of the unknown to be estimated (see Section 4.4). However, many variables such as temperature and conditions of curing, aggregate type and grading or the presence of admixtures can all cause modifications to the microporosity of the cement paste and consequently lead to reduction in precision of the result.

2.5.2 Scanning electron microscopy and microanalysis

The high resolution and magnification of electron microscopy make it invaluable as an aid to detailed petrographic investigations of concrete and related materials and forms an invaluable compliment to optical polarised light microscopy. In the simplest terms, there are two broad groups of electron microscopes, the first is the transmission electron microscope (TEM). This instrument images a very thin specimen by passing a focused beam of electrons through it and displaying the transmitted image. It is capable of very high resolutions and magnifications but is little used for the examination of concrete specimens, so will not be considered in detail here, though introductory reviews of both types are given by Flegler et al. (1997) and Reed (2005).

The second type is referred to as a scanning electron microscope (SEM) and is widely used by petrographers to augment the data obtained by optical microscopy. It images a solid specimen by rastering the incident electron beam across it and detecting the electrons (and other radiations) reflected, backscattered or emitted from the surface. The signals detected are displayed on a VDU system which is synchronised with the beam raster. Secondary electron (SE) images show the topographic features of the specimen, while backscattered electron (BSE) images can give information about the atomic composition of components in the specimen based on atomic mass (Reed 2005).

Since the majority of concrete and related specimens are not electrically conducting, they usually require a thin conducting coating to be applied to their surfaces prior to insertion into the microscope. Carbon is commonly used and is applied from a carbon arc under a vacuum of better than 10^{-4} torr. The carbon particles move in straight lines so will coat a polished surface uniformly, but irregular surfaces need to be rotated to be uniformly coated. Gold and various other metal coatings are also used (see Section 3.5.1), but carbon is preferred if x-ray analysis is required because of its low atomic number and x-ray absorption characteristics, but thickness of the coating (usually about 20 nm) is also an important factor.

Both types of electron microscope can be fitted with an x-ray detecting system in order to analyse the x-ray radiations emitted from the specimen. These x-rays will be characteristic of the elements in the specimen at the point where the electron beam strikes the specimen. Scanning electron microscopes may be fitted with either one or more wavelength-dispersive x-ray spectrometers and are usually referred to as electron probe microanalysers (EPMAs). The alternative system uses a multichannel analyser detector to measure the emitted x-radiations in terms of their energies and is usually referred to as an EDS and is more commonly used with SEMs. The EPMA spectrometers analyse the x-radiation in terms of its wavelength and are more sensitive and have better line resolution and lower limits of detection than the EDS, though the EDS has the advantage of building up the whole x-ray energy spectrum at once, and this can be accumulated over a number of seconds and displayed directly on a VDU.

CARBON SPECIMEN COATING THICKNESSES

The thicker the specimen surface coating, the greater the absorption of x-ray emissions (particularly of low-energy, long-wavelength x-rays characteristic of light elements). Coating thicknesses can be determined by their electrical resistivity, but visual colour estimation of thickness is more commonly used. The thickness related to colour on a polished brass sample is a useful guide in this context and is as follows (Kerrick et al. 1973):

Orange	~15 nm
Indigo red	~20 nm
Blue	~25 nm
Bluish green	~30 nm

2.5.2.1 Components of a scanning electron microscope

The general configuration of an SEM is shown diagrammatically in Figure 2.9. A typical scanning microscope and its sample chamber are shown in Figure 2.10a and b.

It will be noted the sample chamber of the scanning microscope is usually relatively large, and this is because a major application of these instruments concerns the examination of electronic components and circuit boards in the computer industry:

1. *The electron gun* comprises of a 'hairpin'-shaped electrically heated filament usually of tungsten. This type is almost invariably used, though much more expensive indirectly heated filaments and field emission filaments are available. They provide greater brightness, but require higher vacuums and require great care in handling. The tungsten filament is heated to temperatures around 2700 K and, if carefully adjusted to the aperture in the Wehnelt cylinder cathode, should last for several weeks, though filament life is critically temperature dependent. In the evacuated column just below the electron gun assembly is an earthed anode plate with an aperture allowing the beam to pass through. An accelerating voltage (3–40 kV) is applied between the cathode (the specimen) and anode. Beam direction is critically dependent on the position of the filament which wanders with time in the Wehnelt aperture. Below the electron gun, there are beam alignment coils which can adjust the beam direction by adjusting the current input to these coils (see Figure 2.9).

2. *The microscope column* houses the electron gun, the alignment coils, electromagnetic electron condenser lens system and scanning coils. The focused electron beam

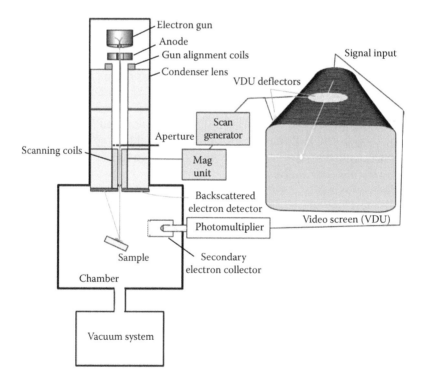

Figure 2.9 A diagrammatic representation of a typical scanning electron microscope. Details of various emissions from the sample are shown in Figure 2.11. (Courtesy of A. Crossley, Oxford Materials, Oxford, United Kingdom.)

(a)   (b)

Figure 2.10 (a) General view of a conventional scanning electron microscope. (b) A close-up view of concrete specimens mounted on the sample holder ready to be inserted into the sample chamber of the microscope. (Photographs courtesy of M. Eden, Sandberg Ltd., London, United Kingdom.)

will exhibit some spherical aberration, usually corrected by a beam limiting aperture. Another aberration is astigmatism caused by lens imperfections or contamination on apertures. This can be corrected by 'stigmator' coils which can be adjusted usually by observation of the scanned image to cancel the effect.

3. *Scanning and display system* makes use of scanning coils, often set inside the final lens coils, to sweep the beam across the specimen in a linear or raster pattern. Analogue scanning systems are replaced in modern instruments with an analogue to digital converter and displayed on a computer monitor (Figure 2.9).

4. *The specimen stage* is housed in the sample chamber which is pumped down to a high vacuum of 10^{-5} mbar or better. Consequently, small specimens with minimum porosity will allow the chamber to pump down more rapidly than large porous ones. The specimen and holder are usually mounted on a carriage mechanism which can be adjusted in the x and y directions perpendicular to the column axis to allow different areas of the specimen to be imaged. Movement in the z direction allows the specimen surface to be adjusted to the correct beam focus height, while tilt and rotation movement allows adjustment relative to the beam and detectors. Typically, for quantitative x-ray analysis studies, the detachable specimen holder will have locations at one edge for several reference standards to be inserted coplanar with the specimen surface and so can receive the same thickness of conductive coating as the specimen during the coating process (see Section 3.5.1).

5. *The vacuum system* must be capable of evacuating the column and sample chamber to better than 10^{-5} mbar in order to avoid damage to the filament, avoid high-voltage breakdown in the gun and allow the electrons to reach the specimen without being scattered. Usually, this evacuation is accomplished in two stages; once the specimen has been inserted and the sample chamber closed, a rotary pump reduces the pressure to about 1 millibar. The rotary pump then switches automatically from the chamber but continues to back the diffusion pump or pumps which complete the evacuation. In some instruments, isolation valves and airlocks are used to allow sample or filament changes to be made without venting the whole instrument.

UNITS OF PRESSURE

The SI unit of pressure is the pascal:

1 pascal = 1 Nm^{-2}

130 pascals = 1 torr = 1 mmHg = 1.3 millibar

1 atmosphere = 1013.25 millibar = 760 mmHg

6. *Detectors* are devised to measure the signals generated from the electron beam striking the surface of a specimen. These signals and their typically sampling volumes are
illustrated diagrammatically in Figure 2.11.

The three most important emissions for investigation of concrete and related materials
are the secondary electrons (SEs) with low energies (3–25 eV) which can only escape from
a shallow sample depth of 5–10 nm. Backscattered electrons with energies of between
60% and 80% of the incident beam energy escape depths inversely proportional to their
atomic number. The third type of emissions are the characteristic x-radiations with energies which are proportional to the atomic number of the elements under the beam.

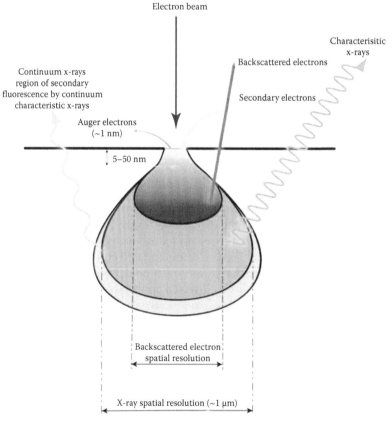

Figure 2.11 Schematic diagram showing the signals emitted and their sampling volumes resulting from
an electron beam striking a specimen (see Figure 2.9 for an overall diagram of a typical SEM
apparatus) surface in scanning electron microscopy and microanalysis. (Diagram courtesy of
A. Crossley, Oxford Materials, Oxford, United Kingdom.)

ENVIRONMENTAL SCANNING ELECTRON MICROSCOPES

Recent developments of 'environmental scanning electron microscopes' which work by means of differential pumping between the main microscope column and a separate specimen chamber. This allows a direct examination of hydrous and 'moist' materials. They do not need an electrically conducting surface coating as the gas phase surrounding the specimen provides the electrical conductivity. This type of microscope has been used extensively for studying plant and animal materials, but there is a potential for application to cementitious and other gel-like materials. The principal disadvantage is the limitation on the distance the electron beam can travel through the gaseous phase surrounding the specimen, and this may be only a fraction of a millimetre. Also, the aperture between the specimen and main chamber of the microscope is very small. Consequently, the high-magnification image can only cover a very small area of the specimen.

The detection of SEs is usually accomplished by the electrons impinging on a scintillator which may have a conducting mesh placed in front of it to control electron collection and be backed by photomultiplier to amplify the signal. This Everhart-Thornley type of detector measures SE emissions, and as the beam is rastered across the specimen, it produces topographic 3-D images of the specimen surface. BSEs are of higher energies so that if a negative voltage is applied to the grid in front of the detector, the SEs are repelled but the BSEs because of their higher energies are detected. However, detection efficiency is low because of the small solid angle subtended by this type of detector.

A more efficient method involves locating a scintillator plate directly above the specimen. This plate has a central aperture so that the incident electron beam can pass through. This 'Robinson' detector is usually mounted on a retractable arm and because of the large solid angle, it can provide a large, relatively noise-free signal. An efficient alternative is a solid-state detector mounted coaxially with the beam directly above the specimen. This is often divided into segments around the central beam aperture, enabling different types of signal to be collected simultaneously.

As can be inferred from Figure 2.11, the resolution of the SEM cannot be better than the diameter of the sampling volume; hence, the resolution is determined by the electron beam diameter. This in turn is controlled by the quality of the lens system, the operating conditions and the final aperture size. With digitised displays, the image will typically be a few megapixels in size. The individual pixel should not be larger than the minimum achievable spot size for maximum resolution. In practical terms, for an SEM operating at 20 kV, the spot size may be between 2 and 2.5 nm. To obtain the best resolution, the sequence of operations is as follows:

- Select the best final aperture size (usually the smallest).
- Obtain an image at the highest magnification.
- Reduce the spot size gradually by increasing first condenser lens excitation.
- As the image becomes noisier, compensate by scanning more slowly.
- The problem of long 'refresh' times can be compensated by using a selected scan area or an automated computer data storage facility.
- Minimise astigmatism by adjusting the 'stigmator' coils.

2.5.2.2 Elemental x-ray microanalysis

As already noted, emissions generated by the electron beam striking the specimen include BSEs and x-rays. The fraction of beam that is backscattered increases in part with increasing atomic number (Z) of the elements in the specimen under the beam. Consequently, the brightness of individual components imaged in the rastered area depends on the atomic number of the elements contained in them. The scanning electron microscope images are produced on a greyscale, but the possibility of segmenting this scale and allotting false colours to the segments can assist in identifying the various components (see Figure 2.12).

Quantitative analysis of elemental compositions relies on a comparison of the x-ray intensity of the element of interest (measured as an accumulated count over a fixed time interval) with the intensity obtained from a known standard under identical conditions. As may be seen from Figure 2.11, the sampling volume for x-rays is of the order of 1 μm and is a practical minimum if sufficient x-ray photons are to be generated and detected. Various correction factors need to be applied to the raw data from the unknown; these include corrections for the specimen matrix, atomic number (Z), absorption (A) and fluorescence (F) (ZAF corrections); in addition, corrections for line overlaps are necessary in some cases. An explanation of these correction factors is given in standard texts for x-ray analysis such as Reed (2005) or Goldstein et al. (1992).

Various computer software packages such as 'CITZAF' (Armstrong 1991) and those described by Heinrich and Newbury (1991) are included in commercial EDS systems together with databases for identifying lines in the x-ray spectrum. The software will automatically compare intensity data from elements in the specimen against stored reference standard intensities (see Reed 2005). It will then output a quantitative analysis of the material under the electron beam. It must be stressed that the stored reference data will originally be taken from polished standards, so comparison with the unknown relies on it being polished to the same level of finish. Usually, coating thickness reference is achieved by coating a reference standard (often cobalt) along with the specimen to provide the comparison with the original standards.

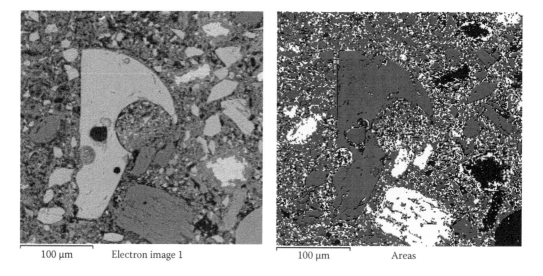

| 100 μm | Electron image 1 | 100 μm | Areas |

Figure 2.12 BSE image of a ggbs particles in a cement matrix. Left in greyscale and right in 'false colour' with ggbs in red. (Courtesy of M. Eden, Sandberg Ltd., London, United Kingdom.)

ATOMIC NUMBER AND Kα X-RAY INTENSITIES

The characteristic x-ray wavelength (Kα) emitted by an element reduces, and its energy increases with increasing atomic number (as do the Lα series). The fluorescent yield (the proportion of x-ray photons emitted for a given excitation energy) drops sharply below $Z = 20$. Consequently, it becomes increasingly difficult to obtain quantitative data from x-ray emissions as atomic number decreases, partly because the low-energy (long-wavelength) x-ray photons are readily absorbed before they reach the detector and partly because the output of characteristic x-rays for a given electron beam voltage decreases markedly for the lightest elements. This implies that elements lighter than Mg ($Z = 12$) will require longer counting times to accumulate sufficient statistically significant counts (peak/noise) at the detector. This typically gives rise to higher values quoted for the limits of detection and a reduction of analytical precision for the lightest elements.

When elemental compositions and particularly if quantitative data are required, the x-ray emission resulting from electron beam excitation of the specimen compared against standards is the usual method used. In the great majority of studies concerned with concretes and mortars, an EDS rather than the wavelength-dispersive spectrometers of an EPMA instrument is used to obtain this analytical information. The electron beam 'spot size' limits the minimum zone for x-ray analysis to a volume of about 1 μm³ (see Figure 2.11), but larger areas can be analysed by adjusting the rastered area appropriately, and analysis along a line traverse is also a simple procedure.

Modern windowless solid-state x-ray detectors are capable of detecting characteristic x-rays from elements atomic number above $Z = 3$ (Li), and for many elements of interest to the petrographer, quantitative analyses with precisions of the order of ±1% are obtainable with short counting times of 1–5 min. Also, detection limits are such as to allow most minor and trace elements to be determined. Since the emitted x-ray radiation, particularly the longer wavelengths (characteristic of light elements), is absorbed by the conducting coating on the specimen, this should be of a low absorbing light element (usually carbon) and as thin as possible.

An alternative, the so-called 'standardless' method of analysis, compares the x-ray intensity from the specimen with theoretically calculated or empirical values of the intensities of pure elements. This option is available for use on many commercial instruments. Some of these methods are described by Labar (1995) and by Fournier et al. (1999, 2000), but this approach is generally considered to be less reliable than those methods that are based on comparisons with data from known physical standards.

2.5.2.3 Mineral and element recognition and mapping

Under favourable circumstances, the backscattered image obtained from a scanned area of a specimen will be able to differentiate the mineral components in that area on the basis of the atomic numbers of the elements they contain (see Figure 2.12). However, the characteristic x-ray emissions from the elements in the scanned area can also show the spatial distribution of an element, or elements within the scanned area with a limit of resolution of about 1 μm, displayed as a spectra, Figure 2.13, or as a map, Figure 2.14. Similarly, if the electron beam is limited to scanning along a line, the variation of element composition along that line can be displayed.

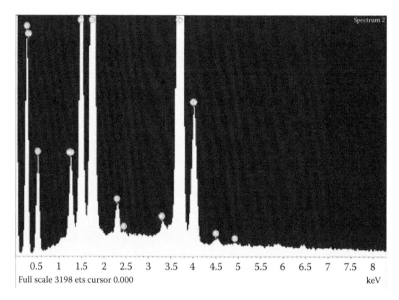

Spectrum 2

Full scale 3198 ets cursor 0.000   keV

Figure 2.13 A typical spectral analysis of a ggbs particle, peaks representing the characteristic x-ray lines of the elements present. (Courtesy of M. Eden, Sandberg Ltd., London, United Kingdom.)

DETECTORS AND 'DEAD TIME'

Detectors and the electronics take a small but finite time to process and record the energy of an x-ray photon striking the detector. Consequently, while one photon is being recorded, the detector will be 'dead' and unable to record the arrival of another while the first is being processed. Further, the second one will extend the dead time of the detector by an additional small amount. As the rate of arrival of incident photons increases, the dead time will also increase. This dead time will be insignificant for the typical solid-state detector used with EDS up to about 1,000 counts/sec but increases rapidly beyond that so that the dead time can be extended by as much as 40%–50% at 10,000 counts/s. This can be compensated automatically by adding an additional 'live time' to allow for the detector dead time, but at very high levels, the detector becomes saturated and will cease to function.

Computer control of scanning electron microscopes with an EDS attachment allows x-ray intensity to be accumulated over fixed times as the beam rasters across the specimen. Since the EDS system accumulates the x-ray energy spectrum, the background noise and counter dead time at high count rates need to be considered in assessing the quality of the intensity data, particularly if quantitative data are required. The pixel size and 'dwell time' at each point are also important considerations for statistical validity and image quality. Possible line overlaps and ZAF corrections will also need to be addressed if quantitative analyses are required. The data can then be displayed as a 'map' in terms of relative brightness for particular elemental concentrations on greyscales, or, alternatively, if imaged as a series of elemental intensities or element groups and then displayed in false colours on the computer screen, the data will be easier to interpret (see Figure 2.14). It is also possible to carry out automated modal analyses using the image processing techniques as described in Section 2.6 with x-ray intensity maps of this kind.

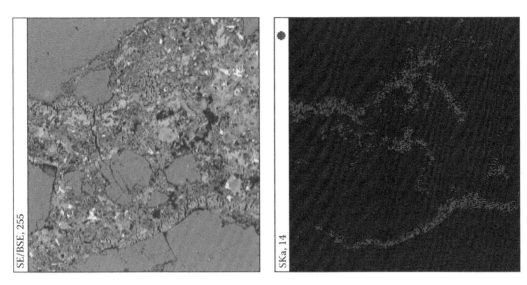

Figure 2.14 X-ray mapping of a sulphur distribution map in an area of cement paste. The image on the left is a BSE image of the paste. On the right, the sulphur distribution is highlighted in red. (Courtesy of M. Eden, Sandberg Ltd., London, United Kingdom.)

2.5.3 X-ray powder diffraction techniques

X-ray diffraction spectrometry (XRD) is one of the few methods available for the direct identification of crystalline phases in substances and is widely used in the material sciences (Klug and Alexander 1974). It is to be noted that phases must be crystalline as a unique diffraction pattern is not obtained from amorphous substances, although background intensity maxima can provide an indication of the presence of amorphous phases and the average atomic spacings. The usual method of analysis is to grind a sample (typically approximately 0.5 g is required) to less than a 50 μm particle size; further details of sample preparation are given in Section 3.5.2 and in Buhrke et al. (1998). The sample is then mounted and inserted into a diffractometer where the x-ray diffraction data are collected electronically and stored in a computer and displayed on a VDU or recorded on a strip chart. The identification of the phases present is made by comparison with a library of known patterns. The major source of diffraction data relating to materials is the Powder Diffraction File (PDF), which was initiated by ASTM, continues to be updated and is now available from the independent International Centre for Diffraction Data (ICDD.com). Various subsets of these data are available and are often built into the x-ray diffractometer data processing software packages so that crystalline components can be identified immediately.

The basic diffractometer design has a flat powder specimen placed at the centre of a goniometer system with an x-ray detector mounted concentrically with the specimen. Both rotate, with the detector geared to rotate at twice the speed of rotation of the specimen. Although fundamental design has changed little with time, improvements in control and precision of the goniometer, improvements to the detector and the software control of the instrument and the processing and the display of the intensity data have greatly enhanced the sensitivity and flexibility of the instrument.

The diffractometer line spectra of a mixture of crystalline materials can provide four types of information:

- Diffraction line positions – crystal species present
- The line intensity (height – background) – crystal composition and amount

- Preferred orientations of particles in the specimen
- The line width – degree of crystallinity and specimen particle size distribution

This information is most readily interpreted by comparison with appropriate standards that may require careful preparation to be directly comparable with the material under investigation. Care must be exercised particularly in relation to particle size distribution, preferred particle orientation and compaction of the specimen in preparing these standards.

Quantitative analysis of crystalline minerals in a mixture relies either on comparison of particular line intensities from the unknown and a series of comparable standards or a technique involving inclusion of an internal standard in the sample under investigation, usually referred to as 'standard addition' methods. The increased line intensities obtained with the addition present compared to those obtained from the unmodified specimen allow the amount present in the unknown to be calculated (Brindley and Brown 1980).

If only small amounts of sample are available, the diffractometer can still be used. Very thin layers of powder can be placed on double-sided adhesive tape or special backing plates, though a preferred orientation of the particles is likely in these circumstances. If only a few milligrams of material is available, a photographic technique such as the Gandolfi camera may be used. In this way, the identification of a few grains of material, perhaps selected from a larger sample, may be achieved.

In the case of cement clinkers and anhydrous cement powders, the principal mineral phases can usually be readily identified from a diffraction trace. Quantitative estimates of the cement phases are possible by using selective extraction of the silicates by salicylic acid in methanol (SAM) and extraction of the alumina bearing phases by potassium hydroxide and sugar solutions (KOSH) (Takashima 1958, 1972; Gutteridge 1979). The identification of the cement phases present in concrete and mortars is more complicated because the calcium silicate hydrates are poorly crystalline and give weak broad diffraction peaks which tend to be swamped by the diffraction peaks from calcium hydroxide. In addition, peaks from minerals present in the aggregates may also interfere.

The practical limits for detecting phases depend principally on the crystallinity and ordering of the phase. For quartz, as little as 0.1% may be detected, but for many cement hydrates, as much as a 5% concentration may be required for detection. As a rough guide, if less than 2% of a phase is present, x-ray diffraction analysis may have difficulty in detecting that phase. In general, results obtained from x-ray diffraction techniques are at best only semi-quantitative, unless careful standardisation procedures are used.

The value of x-ray diffraction methods for research studies of cement hydration has been extended by the use of synchrotron energy dispersive x-ray diffraction techniques (Barnes et al. 1992). This powerful technique enables real-time investigations of early hydration reactions to be undertaken such as the synthesis of ettringite (Muhamad et al. 1993) and the rapid conversion of calcium aluminate cement hydrates (Rashid et al. 1992, 1994).

2.5.4 Fourier transform infrared spectroscopy

Fourier transform infrared spectroscopy (FTIR) is an analytical technique used to identify characteristic chemical groups (usually organic) in a material. FTIR measures the absorption of infrared (IR) radiation by the sample material versus the wavelength of the radiation. The wavelengths that are absorbed by the sample are characteristic of its molecular structure.

The frequency of IR radiation used for spectroscopy is in the range 4,000–400 cm^{-1} which equates to a wavelength of 2,500–25,000 nm. The reciprocal centimetre units (wave number) are used rather than Hz since they provide more manageable numbers. The photon energies in this range are not great enough to excite electrons but can resonate with the vibrational energies associated with molecular bonds.

Figure 2.15a shows a typical FTIR instrument, main specimen chamber and holder to the right with the microscope specimen stage to the left and Figure 2.15b shows a close-up view of the chamber and specimen holder.

In IR spectroscopy, a broad spectrum source of IR radiation, usually generated by electrically heating a ceramic until it glows white hot, is then passed through a sample and detected by an IR detector such as a bolometer or solid-state semiconductor detector. Dispersive instruments would use an IR transparent prism (typically sodium chloride or potassium bromide) to disperse the IR source radiation and scan through the wavelength measuring sample absorption at each frequency, whereas modern Fourier transform (FT) instruments 'collect' the entire spectrum at once which accumulates over a period of time.

The FTIR spectrometer uses an interferometer to modulate the wavelength from the broadband IR source. A detector measures the intensity of transmitted or reflected radiation as a function of its wavelength. The signal obtained from the detector is an interferogram, which is converted mathematically using Fourier transformation to obtain a single-beam IR spectrum. The FTIR spectrum is usually presented ratioed against a background as a plot of intensity versus wave number (in cm^{-1}). The intensity can be plotted as the percentage of transmittance or absorbance at each wave number. The intensity of any particular peak is related to the concentration of the species responsible for the absorption. A detailed review of IR spectroscopy is given by Cross (1960), and general details of IR spectra may be obtained from http://www.chem.ucla.edu/~webspectra/.

2.5.4.1 Sample preparation

Solid samples can be prepared for IR transmission measurement using a variety of techniques. One commonly used is the 'in disc method' with potassium bromide as the binding agent. A very small amount of the solid sample is ground together with KBr using an agate pestle and mortar. The resulting fine powder is pressed into a transparent disk using a suitable die and press.

A second method is a thin film technique. A small amount of the solid sample is dissolved in a suitable solvent. The resulting solution is deposited on an IR window and the solvent allowed to evaporate leaving behind a thin film of the original sample. A third method is referred to as the Nujol mull technique. Here, a small amount of the solid sample is ground into a fine powder using an agate pestle and mortar. This fine powder is then made into a suspension or mull using a liquid paraffin such as nujol. The resultant paste is spread on an IR window and the spectrum acquired using nujol only as the background. However, some materials, for example, polymeric samples, can be examined directly by preparing the sample as a thin film by hot-pressing or extrusion.

Solid samples may also be examined using reflection IR techniques, such as diffuse reflectance (DRIFTS). The sample is ground to a fine powder and can be measured directly in the DRIFTS cell, or the sample may be diluted with KBr to facilitate the measurement. In either case, the data collected are usually presented as wave number versus reflectivity (as Kubelka–Munk units). This technique can also be used on solid samples with rough surfaces.

The example involving the identification of asbestos minerals given in Section 7.7.1 made use of attenuated total internal reflection (ATR). The FTIR spectra (Figure 7.12) were obtained using a permanently aligned diamond window ATR sampling attachment (Specac Golden Gate). In ATR, the sample is pressed against a diamond window which forms the flat face of a prism. The probe IR radiation couples into the prism at the Brewster angle and

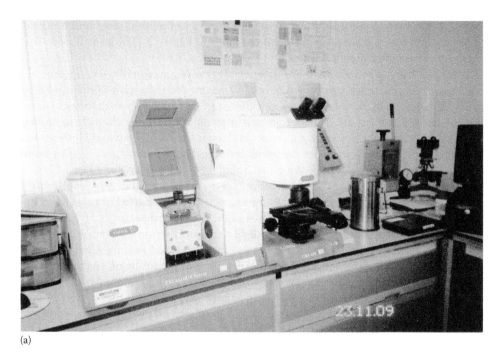

(a)

(b)

Figure 2.15 A typical modern FTIR instrument, a Varian Excalibur Series UMA 600. (a) Shows two specimen
locations, one on the microscope stage to the right and the main specimen chamber and specimen holder to the left. (b) A close-up view of the main chamber and specimen holder.

is totally internally reflected. The contact face propagates an evanescent wave which interacts with the material to a depth of ca. 1–10 μm. The data are usually presented as wave number versus transmission (or absorbance).

2.5.4.2 Spectral interpretation

IR spectroscopy is used mainly to interpret organic molecules; however, certain inorganic species also exhibit relatively strong IR absorptions. Each chemical group has its own position within the IR spectrum, and numerous correlation charts have been published that show the range and strength of absorption bands across the IR spectrum for a large number of molecular species. In the study of asbestos samples illustrated in Chapter 7, the series of bands between ca. 3600 and 3800 cm^{-1} are all characteristic of hydroxyl (–OH) groups being the vibration of the oxygen–hydrogen bond. The fact that more than one band exists in most of the asbestos samples analysed indicates that these hydroxyl groups are in turn bonded to different metal atoms; for example, in the IR spectrum of chrysotile, there are two strong OH absorptions at 3685 and 3645 cm^{-1} reflecting that the hydroxyl groups are bonded to either different metal atoms (e.g. silicon or magnesium) or are in different structural environments. In the lower frequency end of the spectrum for asbestos, the bands are mainly due to metal-oxygen vibrations such as Si–O and Mg–O.

The use of IR spectra as a technique for identifying constructional materials on the basis of their unique spectral 'fingerprint' is becoming increasingly common. One example from the work of Mangabhai et al. (2010) is the use of FTIR attenuated total reflectance (ATR) methods to successfully differentiate certain superplasticisers, retarders, cement types, calcium aluminate cement and ggbs. A source providing IR spectra of polymer materials such as acrylate, alkyd, urethane and latex may be downloaded from http://sdbs.db.aist. go.jp/sdbs/cgi-bin/cre_index.cgi?lang=eng and from the Spectral Database for Organic Compounds (SDBS) (see Section 2.5.6.3).

2.5.5 Thermal methods of analysis

This technique consists of the programmed heating of a small sample in a special furnace and then measuring energy or weight changes which take place with time. In some instruments, the atmosphere can be controlled or pressurised if required. The methods where energy changes are measured are known as differential thermal analysis (DTA), while differential scanning calorimetry (DSC) measures heat flow rates. In thermogravimetric analysis (TGA), weight changes are measured against time and temperature, and with thermal and dynamic mechanical analyses (TMA and DMA), dimensional changes to the sample are monitored. Separate instruments are normally used for these techniques, although with improvements in instrumentation, particularly with sensors and data acquisition, combined models often built up on a modular basis to meet particular requirements are now available.

TGA can be applied as a quantitative method of analysis, while DTA is usually qualitative. If quantitative DTA data are required, the related technique, DSC, may be used. Commercial DSC instruments are typically limited to a maximum temperature of approximately 700°C. However, there are variants of DSC which can be made semi-quantitative and if required which will operate to higher temperatures.

These techniques are widely used in the analysis of plastics, polymer composites, fibres, textiles and metals. Thermal analysis techniques have been widely used in the clay, ceramic and materials sciences. Their principal use in cement chemistry is for measuring dehydration reactions and for detecting phases such as gypsum and organic materials.

The samples typically of a few milligrams are prepared in powder form, but liquids and some types of solids can also be analysed. The techniques span the temperature range of −195°C to 1600°C but for the type of materials likely to be analysed in cement and concrete technology, the range ambient to 1150°C is applicable.

There are relatively few published reports of thermal techniques used in the evaluation of concretes, mortars and related materials. Bhatty (1991) has reviewed the application of thermal analysis to cement–admixture systems. Reviews of thermal analysis applied to cement and concrete have been made by Ramachandran (1969), Barta (1972) and Ben-Dor (1983). Smillie et al. (1993) have applied DTA to the detection of syngenite [$K_2(CaSO_4)_2 \cdot H_2O$]. Ellis (2010) has published a series of DTA charts for mortars, including historic mortars from 18 buildings in the United Kingdom, but points out that great caution is needed in the identification of existing components in these materials. Also, an evaluation of the hydration of Portland cement with carbonate additions using TGA methods has been published by Gabrovsek et al. (2006). A particular application of DTA is its special use in detecting the conversion ratio of the hexagonal aluminate hydrate to the low-strength cubic phase in concrete containing high-alumina cement (Midgley and Midgley 1975).

TGA is one of the few methods for estimation of calcium oxide in cements. The difference between the amount of calcium hydroxide found by TGA and the free lime determined by chemical analysis is equivalent to calcium oxide. DSC is now the preferred thermoanalytical method because of its high sensitivity especially for the detection of organic materials in cements (Portilla 1989). It can be used to detect the small changes that cause flash and false sets in cements and easily detects the first stages of hydration, ettringite, gypsum and syngenite. However, like all highly sensitive thermoanalytical systems that use low-mass sample holders, it does not satisfactorily discriminate between gypsum and hemihydrate. The older type of large sample holder with thermocouples that protrude into the sample is required for this purpose.

2.5.6 Chemical methods of analysis

There are a number of different aims in analysing a sample of cement or concrete. In many of the methods already described, the aim of the analysis is to identify an unknown component. This may be carried out directly by x-ray diffraction, microscopy, FTIR, thermal analysis and other techniques depending on the nature of the sample. These techniques typically identify the mineral species rather than the chemical composition, so this can only be inferred from their known chemical formulae.

Another objective of analysis is to determine precisely the composition of a sample whose identity is already known, for the purposes of controlling quality and properties. As usually plenty of samples are available, the methods used are largely dictated by the degree of accuracy required. Recommended methods for the chemical analysis of cementitious materials will be found in most national standards. 'Wet chemical methods' are in many cases being superseded by x-ray fluorescence, which is capable of producing results of a similar reliability (Norrish and Chappell 1977).

2.5.6.1 Chemical analysis of unhydrated blended cements

The analysis of blended cements presents analytical problems, many of which still have no satisfactory solution. Materials such as limestone, granulated blast furnace slag, fly ash, silica fume and a range of natural pozzolana are added to cement by either intergrinding or addition during mixing. The use of these materials is common, and if the specifier wishes to determine that a claimed blending material has been used at the specified addition proportion, this can only be reliably determined if samples of the cement and the blending material are also supplied to the analyst. When the analyst is supplied a blended cement without any data on the materials which it contains, the probability of obtaining a reliable chemical analysis is more limited without recourse to other methods. Identification of the blended material presents few problems. Either examination of residues after acid extraction or examination of the cement by optical microscopy or SEM will usually identify the blended material, and an EDS analysis will allow the chemical composition to be determined. A general discussion of methods has been given by Papadoulos and Suprenant (1988).

The methods required to determine the quantity of blended material will vary according to the material added. In the case of Portland/limestone cement, it may be possible to estimate the limestone content by using a combination of TGA and XRD methods to determine calcium carbonate and then checking these results by a lithium gluconate extraction of the cement fraction. Where granulated blast furnace slag is used, a selective dissolution technique has been proposed by Demoulian et al. (1980) and has been further investigated by Luke and Glasser (1987). Investigation by Goguel (1995) found that Demoulian's technique was not applicable to all types of granulated slag and that selectivity ratios have to be increased by the use of large cations in the leachant. Similar techniques may possibly be applied to Class F fly ash but for the Class C high lime fly ashes, no reliable analytical technique is available for their determination in blended cements without the original material also being available.

The greater portion of silica fume and natural pozzolana are not soluble in either hydrochloric or nitric acid, and a reasonable estimate may be made on the basis of the amount of residue left after an acid extraction of the blended cement. Where the use of blended cements is specified, it is prudent for the specifier to obtain representative samples of both the cement and blending material if check analyses of the blended cement are to be carried out. An accurate estimation of the amount of blending material present may not be possible if only the blended cement is available.

2.5.6.2 Chemical analysis of hardened cement pastes in concrete

The identification and estimation of the materials present in hardened concrete are not a simple matter and may give results which differ from those originally used in the concrete mix. Even concrete mixed under close control is a somewhat variable material, and this variability may be further increased by incomplete compaction and areas of local inhomogeneity caused by segregation and settlement in the plastic state. It must be remembered that the setting of concrete is a series of chemical reactions that can continue far into the life of the structure. Chemical and mineral admixtures if present also take part in these reactions, and in the process, they may be partially or completely consumed as they become incorporated into the hardened matrix of the concrete. Concrete also reacts with its environment. Moisture and alkalis can move both into and out of the concrete, while carbon dioxide from the atmosphere and ions such as chloride and sulphate can penetrate the outer layers. Thus, it may be impossible to determine all of the components used in the original mix.

Methods for the analysis of concrete were detailed and discussed by Figg and Bowden (1971) and their publication remains a useful reference. A working party of the Concrete Society and the Society of the Chemical Industry (Concrete Society 1989) updated and discussed methods of analysis, but some of the useful pictorial information given in Figg and Bowden has not been included. A working party to undertake a further review has recently been formed, although the results of the review will not be available before 2014. These two publications are good sources of information for the reader who does not wish to consult the specialised literature. BS 1881-124 (1988), currently under review, also details some methods for the analysis of hardened concrete.

The component most likely to require determination is the cement content as this information is vital where there is a dispute over the quality of the concrete. The classical method involves acid dissolution of the concrete to determine the calcium and silicon oxides in the hardened paste and then uses these results to determine the cement content from an assumed composition of the cement. Acid dissolution also attacks aggregate, the amount of attack varying considerably depending on the aggregate type. The problem is that it may be difficult to satisfactorily estimate the unwanted contribution made by the aggregate to the analytical results and large errors can occur. Determination of hardened paste volumes by microscopic point counting in thin sections avoids the problem of contamination from aggregate and allows estimation of the cement content provided the water/cement ratio can be determined. It is claimed by French (1991) that microscopical point-counting methods are quicker and more accurate than chemical analysis in the determination of the cement content of concrete. Errors in the chemical determination of cement content have been estimated at the 95% confidence level as ± 25 kg/m^3 for sampling, ± 30 kg/m^3 in analysis and 40 kg/m^3 for sampling and analysis combined (Concrete Society 1989). Confidence levels for microscopic point counting are not available.

Where high-alumina cement has been used in the concrete, the cement content is usually estimated from the alumina content, which is prone to error. Chinchon et al. (1994) have proposed a method based on the x-ray fluorescence analysis of the iron content of the concrete corrected for the iron in the aggregates, which they claim reduces the errors.

The older method for the determination of water/cement ratio, BS 1881-124 (1988), is complex and does not always give satisfactory results, in particular when the binder is not a simple CEM I (Portland-type) cement. However, individual laboratories have found that repeatability can be achieved with undamaged specimens and by using tetrachloroethene as the solvent for determining capillary porosity (the method in BS 1881-124 [1988] used 1.1.1-trichloroethane, which is no longer commercially available owing to environmental concerns). Determination of the porosity of the hardened paste-based fluorescence techniques as described in Sections 4.4 and 2.5.1 is used in some European countries to determine water/cement ratios. Another method of relating porosity to water/cement ratio as measured by porosimetric data has been proposed by Tenoutasse and Moulin (1990).

A method for determining the origin of the cement used in a hardened concrete, providing limestone aggregates are not present, has been described by Goguel and St John (1993a,b). Once the identity of the works which manufactured the cement is known, it is then possible to use historical chemical analyses to estimate the composition of the cement used. Because the composition of cement from the works does not vary greatly, the original alkali content of the concrete and other compositional details can be calculated if the cement content is estimated as detailed earlier. Where construction records are no longer available, this is the only method of estimating the alkali content in the concrete as originally mixed and is an effective method of resolving disputes on this issue.

The determination of the alkali content in a concrete that has been subject to weathering is possible, but the results may not relate to the original alkali content as mixed. Not only can acid dissolution of concrete extract alkalis from some aggregates, but the alkalis themselves are mobile and may be deeply leached from the concrete by weathering and in some cases exchange into the aggregates. These are competing effects, which can be partially resolved by dissolution of the hardened paste with alkaline EDTA instead of acid to determine the alkalis and also by determining the silica and alumina contents to interpret the results (St John and Goguel 1992).

Chemical admixtures, such as plasticisers, superplasticisers, air entrainers, retarders, accelerators and integral waterproofers, may be identified in concrete (Concrete Society 1989), provided that it is not more than a few days old. In older concretes, this is not always possible because chemical admixtures based on organic materials may be broken down by the alkaline pore solution (Milestone 1984). Investigation of some of the published methods (Rixom 1978) indicates that they do not appear to work for older concretes (St John 1992).

Identification of mineral additions in old concretes is usually possible by dissolution of the hardened cement paste and examination of the residues. Sufficient of the mineral addition has to remain in the concrete to allow identification of the residue by optical microscopy, scanning electron microscopy or, if crystalline, x-ray diffraction analysis. Accurate estimation of the quantity of the mineral addition used is possible only for very young concretes and even then may be difficult. In older concretes, a considerable portion of the mineral addition is no longer present because it will have reacted to form hydrates of a similar composition to those formed from the cement and as a result estimates will be unreliable. French (1991) has proposed an electron microscope method for the determination of slag and fly ash in concrete.

When concrete is attacked by sulphate or chlorides, the amount of attack can be estimated by analysis of the concrete. The methods commonly used are detailed in BS 1881-124 (1988). It is necessary to have some idea of the sulphate content of the cement so that this can be subtracted from the results to give the amount of external sulphate that has penetrated the concrete. Where estimation of ettringite is required, an XRD method has been reported by Ludwig and Rudiger (1993). It should be noted that ettringite is only one of the sulphate compounds formed in sulphate attack on concrete.

Total soluble chloride can be determined by acid extraction of a powdered concrete sample. Work by Dhir et al. (1990) shows that to extract chlorides fully, the sample should be boiled in the acid for a longer period than specified in BS 1881-124. On chemical grounds, extraction would also be assisted by using acid concentrations of 1 M or less, and for the lower chloride levels, colorimetric determination is superior to titration. Dhir et al. (1990) also concluded that the method used to determine water-soluble chlorides is unimportant provided that the extraction time exceeds 24 h. A rapid method for measuring acid-soluble chloride in powdered concrete samples has been proposed by Weyers et al. (1993).

2.5.6.3 Identification of polymer additions in mortars and concrete

Chemical admixtures including a wide range of organic polymers are commonly used to modify the properties of concretes and similar materials. Problems and failures can arise through the incorrect use (particularly overdosing) of these admixtures in the mix. Consequently, the petrographer may be required to identify the polymer used and possibly estimate the amounts present in a particular concrete.

In some cases, careful examination of the cementitious matrix in a thin section will provide an indication of the polymer either from the morphology and composition of the matrix itself or from remnant particles of polymer present (see Figure 8.6).

Identification of the polymer can be achieved in favourable circumstances using FTIR equipment (see Section 2.5.4). Since such instruments require only milligram quantities of material as specimen, it may be possible to hand-pick a small amount of crushed material so as to exclude aggregate and concentrate the cement/polymer component.

The methods of extracting organic components from a crushed and powdered sample are outlined in ASTM STP 395 (1965). Three methods of extraction are suggested, a chloroform extraction, an acid/chloroform extraction and a sodium carbonate extraction procedure suitable for lignosulfonate extraction. Once the polymer has been extracted, the solvent can be evaporated and the polymer identified using FTIR by comparison with a control specimen. Estimation of amounts in the sample is difficult but may be attempted with young concretes if careful record is kept of the masses of original sample and of the amounts of polymer extracted and if the IR indicates adequate purity of the product.

The identification of certain superplasticisers and retarders has been carried out by Mangabhai et al. (2010) (see also Section 2.5.4). Sources of reference data providing IR spectra of typical polymers used in mortars, concretes and polymer cement products (see Section 7.8), for example, methylmethacrylate, may be obtained from the SDBS of the Japanese Institute of Advanced Science and Technology. Unfortunately, many polymers used in cementitious materials are mixtures of materials, making identification of generic spectra difficult. They may also be modified over time by reactions in the alkaline environment of the hydrated cement. A good general source of reference data for polymer analysis is in the book edited by Stuart (2007).

2.6 COMPUTER-AIDED PETROGRAPHIC METHODS

The use of computer programmes to control, process and output data from all types of scientific equipment is now commonplace, and they also have become a common means of manipulating data and applying appropriate corrections and statistical treatments. In many applications, this automation has greatly improved both the speed of acquisition of data and the accuracy of the results obtained. A useful introduction to SEM-based applications of these methods has been provided by Pirrie and Rollinson (2011). An important but more general application of these computer techniques might be the particle sizes and aggregate gradings that are discussed in Section 4.3.2.

The petrographer who is investigating concrete, aggregate and similar materials will typically be concerned with one or more of the following:

- The statistical evaluation of the sample material components
- Manipulation and processing of chemical data
- Evaluation of optical and electron microscope images
- Quantitative modal analysis of components

He or she will also require and expect the data generated by equipment such as an x-ray diffractometer or scanning electron microscope to have been corrected as necessary by the software associated with the instrument itself. However, it must be remembered that the 'corrected' data obtained, irrespective of the apparent precision offered by the output, will be entirely dependent on the quality of the specimen being examined, on the quality and validity of the reference standard data used as calibration and on the limitations of the instrument itself.

The petrographer is principally concerned with the interpretation of images obtained from observation of thin sections and cut, ground or polished surfaces. As a consequence, image analysis is perhaps the most important of all the computer-aided petrographic techniques used.

2.6.1 Quantitative image analysis

Image analysis is concerned with the extraction of data from digital images. Imaging systems include astronomical telescopes, satellite photography, conventional video and still cameras, scanners and many kinds of optical and electron microscopes. Quantitative image analysis developed during the late 1960s and early 1970s, before the advent of the digital camera and personal computer. Early applications were mainly in astronomy, metallurgy and biology. Image analyzers were hardware driven, very large and expensive. Processing and measurement techniques were extremely limited. The advent of the digital cameras, cheap scanners and personal computers of immense processing power has enabled enormous growth in the scope of image analysis with a concomitant, though even more remarkable, reduction in cost.

A wide variety of image processing and analysis software is currently available. Advanced stand-alone packages will accept many types of input and usually have add-ins to control automatic microscope stages and a variety of cameras. They provide routines for all stages of processing and analysis and are often equipped with sophisticated databases and report generators. They are also expensive. A more affordable, but still very powerful approach, is offered by companies such as Reindeer Graphics (reindeergraphics.com) who market suites of 'plug-ins' that may be used inside photo-editing software such as Adobe Photoshop. Indeed, a little searching on the 'web' will reveal an abundance of freeware and shareware, some as stand-alone programmes and some in the form of 'plug-ins'. This is probably the best way to get a feel for the possibilities of image analysis in concrete petrography and allied fields of interest.

In spite of the enormous progress made in image analysis over the past 50 years, the principal developments have remained in the fields of astronomy, in the biological sciences and metallurgy. There has been some progress in geologically-based applications, such as automated mineralogical analysis in ore processing, but image analysis has made scarcely any inroads into the investigation of building stones, aggregates or concrete and allied materials. There are many potentially fruitful areas for research and development.

One such area in petrography is concerned with image analysis. This procedure involves four basic steps:

- Acquisition
- Enhancement
- Segmentation and thresholding
- Measurement

2.6.1.1 Acquisition

In concrete petrography, the fundamental tool for image acquisition is, of course, the polarising microscope. Valuable additional tools are the flat-bed scanner and the film scanner, sometimes combined into a single instrument.

Flat-bed scanners can produce high-quality images of ground surfaces of large specimens. These are often useful for illustrative purposes alone and are more quickly and easily produced than conventional photographs. Flat-bed scanner images are ideal for operations involving measurements of coarse aggregates and voids and the analysis of fractures. At a simple level, a computer-generated grid superimposed on an image of a large polished slab provides the basis for manual estimation of coarse aggregate abundance that is as quick and is more accurate than point counting of thin sections (Figure 2.16).

Film scanners are easily adapted to produce high-resolution, low-magnification digital images of whole thin sections (Figure 2.17). Some scanners operate at scan resolutions up to 4000 dpi (dots per inch) and can create images capable of very high magnification. With a little ingenuity, polarising plates can be fitted to the section carrier, adding to the range of image types available for processing.

The polarising microscope is usually the main acquisition system. The facility to attach one or more cameras is, of course, essential. The microscope should have transmitted and

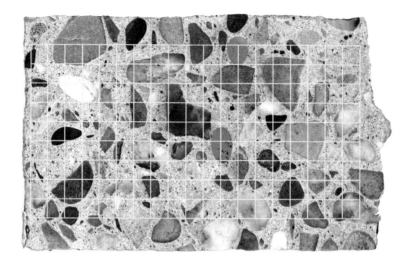

Figure 2.16 Flat-bed scanner image of polished surface of 75 mm diameter core sawn parallel with its axis. A counting grid with 200 intersections is superimposed on the image. It is suitable for manual point counting of coarse aggregate and offers an area approximately equivalent to that of three 75 mm × 50 mm thin sections.

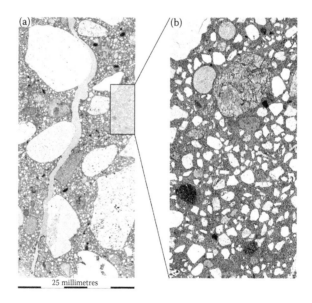

Figure 2.17 (a) High-resolution image of part of a thin section produced with a Nikon Super Coolscan 9000 film scanner. The scan resolution is 4000 dots/in. The concrete is from a floor slab in the West Midlands. The coarse aggregate is quartz-dominated gravel (mainly sandstone, quartzite and vein quartz); the sand is mainly quartz (white). The binder appears brown. Voids and fractures are filled with epoxy resin containing yellow dye. The vertical fracture is thought to be a result of long-term drying shrinkage. (b) Shaded portion of Figure 2.17a enlarged between 5 and 6 times. Because of the high scan resolution, there is enough image detail to enable substantial enlargement on the computer. Enlarged images may be used, for example, to estimate the proportions of fine aggregate and binder (either manually or by further image processing and measurement), investigating the distribution and morphology of voids and analysing microfractures.

incident-light illuminators. An incident, high-intensity UV illuminator is also very desirable. If thin sections of concrete are impregnated with fluorescent dye, UV images offer several methods of estimating water/cement ratio (see Section 4.4), but because they are monochrome and often have very high contrasts, they can also be used for modal analysis and quantitative investigations of void distribution, microfractures and many other applications.

Ideally, the microscope should have a stepping stage controlled by the image analysis software. If it has *z-axis* (focus) control, large-scale automation may be possible. Data from the whole area of a thin section may be acquired and processed without operator intervention.

Modern HD video cameras produce high-resolution images in real time. They are ideal in combination with software-controlled stepping stages for fully automated acquisition systems. Semi-automatic and manual systems may utilise still cameras. A series of images can be accumulated from one or many samples, and these can often be prepared for subsequent batch processing. A perfectly satisfactory low-cost system can be built using a simple mechanical stage and an ordinary digital still camera fitted to the microscope with a suitable adapter.

2.6.1.2 Enhancement

The measurement of images demands that features are clearly delimited, either by unique and uniform brightness or colour or by sharply defined edges. Image enhancement is usually carried out in the spatial domain (the array of pixels that make up the conventional view of the image) or in other domains such as frequency space. Nowadays, most people are familiar with image enhancement through widely used photo-editing software like Adobe Photoshop or Corel Paint Shop Pro. This type of software has very powerful facilities for manipulating contrast, brightness, saturation, hue, routines to perform smoothing, noise reduction and various image sharpening techniques. Some photo-editing software also includes routines for correcting uneven illumination and lens distortion and removing colour fringes. Similar techniques are available in most dedicated image analysis software, but they will not necessarily produce better results. It is always worth running a familiar photo-editing package alongside dedicated analytical software.

2.6.1.3 Segmentation and thresholding

Segmentation is the most widely used process for reducing digital images to information. Its purpose is to divide the image into regions that are intended to correspond with unique features or structural units. Thresholding simply involves isolating features in the image prior to making measurements. The simplest method of thresholding is by defining a range of brightness values in the image. Pixels that fall within the range belong to the foreground; the rejected pixels make up the background. The brightness histogram of the image is commonly used for setting thresholds. The histogram is a plot of the number of pixels in the image having each brightness level. An 8-bit monochrome image has 256 grey levels. Peaks in the histogram often correspond with homogeneous phases in the image; one or more thresholds can be set, either manually or automatically between the peaks. More complex procedures may be used to threshold coloured images or those with phases having different textural characteristics or orientation (see Figure 2.18d). Boolean logic can be used at pixel and feature levels to refine thresholding even further. For example, it is possible to select for measurement only those grains of a particular mineral that have inclusions of a second mineral.

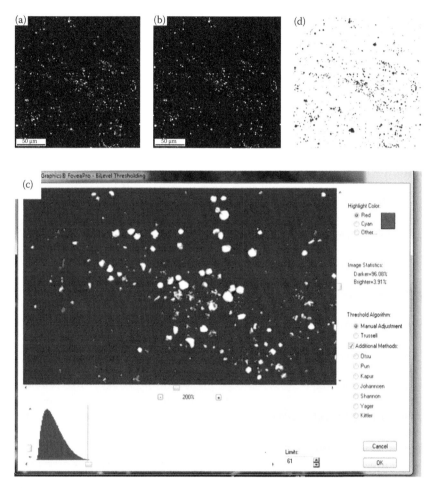

Figure 2.18 Simple modal analysis. Determination of pyrite content of mudstone aggregate. (a) Mudstone with fine-grained, disseminated pyrite (almost white). Limonite is pale grey; mudstone matrix is dark grey–black. (Polished thin section photographed in linearly polarised reflected light.) (b) Application of contrast enhancement to improve differentiation between pyrite and other phases. (c) Using brightness histogram to set threshold between pyrite and all other phases. (d) Detected pyrite (black).

2.6.1.4 Measurement

The difference between image processing and image analysis is that the latter seeks to extract information, usually numerical, from the image. The isolation of features prior to measurement may require removal of acquisition defects, increasing the visibility of particular phases or structures, isolating (thresholding) them from their background and perhaps other operations such as separation of touching objects or filling of holes.

The measurements that may be performed on features in a segmented image may be grouped into five classes:

1. *Brightness*. Thresholding on brightness, and allied properties that depend on colour, is the main technique by which the abundance of features is measured. It forms the basis of most automated modal analyses. The direct measurement of brightness or optical density provides the concrete petrographer with a rapid means of estimating approximate water/cement ratio (see Section 4.4).

2. *Size*. All image processing software offer a range of specific measurements including area, perimeter and various derived parameters such as roundness and aspect ratio.
3. *Shape*. Using two or more basic measurements to generate dimensionless form factors, it is possible to classify features on the basis of their shape, for example, roundness, aspect ratio and elongation.
4. *Count*. The ability to count features, especially in combination with other measurements, offers the means to classify objects and produce distributions of size, shape and even orientation.
5. *Position*. The ability to define the position of selected features offers scope for investigating clustering (nearest neighbour analysis), which has application in studies of void distribution and in fracture orientation studies.

2.6.2 Case study applications of image analysis

2.6.2.1 Estimation of pyrite content in mudstone aggregate

Measurement of a single phase with colour or brightness that is markedly different from that of all other components in a system is probably the simplest application of quantitative image analysis. Figure 2.18a shows a polished thin section of a mudstone aggregate viewed under plane-polarised reflected light. Disseminated pyrite (almost white) is much brighter than any other components, including limonite (light grey) and the silicate minerals that make up most of the rock. In Figure 2.18b, contrast enhancement has been used to improve differentiation between pyrite and other components even further. The brightness histogram (Figure 2.18c) is used to set a threshold to separate pyrite from all other components before measurement. Here, the brightness threshold is set manually by inspection of the histogram, but most software offers several algorithms for setting thresholds, ensuring good reproducibility between successive fields of measurement.

Figure 2.18d shows the detected pyrite (black). In this example, the measured pyrite content (averaged over many fields of view) is 3.91%, whereas the chemically determined pyrite equivalent (total sulphur × 1.86) is 4.30%, suggesting some pyrite grains are too small to be visible under the optical microscope. Subsequent investigation by scanning electron microscopy confirmed that part of the pyrite occurs as grains less than 0.5 μm in diameter.

2.6.2.2 Modal analysis of a mortar specimen

In thin sections of concrete, there is rarely enough contrast between aggregates, binder and voids to permit satisfactory thresholding. Several methods have been proposed for filling air voids in polished blocks with a strongly coloured or reflecting phase (Laurencot et al. 1992; Cahill et al. 1994) and for improving the contrast of the paste by acid etching (Efes 1988; Shi 1988). In many cases, adequate discrimination may be achieved by impregnating polished thin sections with fluorescent dye (double vacuum impregnation) and carrying out measurements under reflected high-intensity UV illumination. Figure 2.19 illustrates a simple example of modal analysis of bedding mortar. Figure 2.19a shows a polished thin section under conventional transmitted plane-polarised illumination. In Figure 2.19b, the same field of view is seen under high-intensity reflected UV illumination. Impermeable sand grains (crystalline quartz) have not absorbed any dye and appear black. Voids, filled with resin, are bright green. The binder, with slightly variable capillary porosity, shows a range of dark green colours. Manipulation of brightness, contrast and gamma (Figure 2.19c) produces an image in which the differences between the three phases are more clearly defined. The brightness histogram (Figure 2.19d) shows three clearly separated peaks which correspond to sand grains (black, left), binder (middle) and voids (right). Thresholds are set,

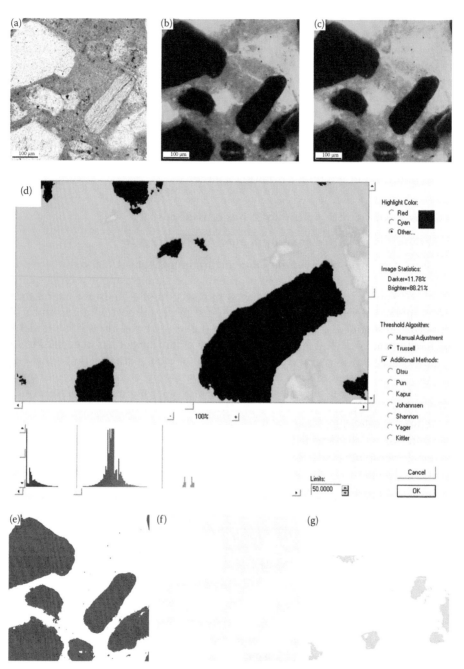

Figure 2.19 Image analysis is a useful technique for rapid modal analysis of hardened mortar and concrete using images of polished thin sections impregnated with fluorescent dye viewed under reflected UV illumination. (a) A thin section of bedding mortar seen under conventional plane-polarised transmitted illumination. (b) Under high-intensity reflected UV illumination, impermeable sand grains appear black and resin-filled voids are bright green. The binder with variable capillary porosity shows a range of darker greens. (c) Manipulation of image brightness, contrast and gamma produces an image in which the differences between the three phases are more clearly defined. (d) The brightness histogram shows three clearly defined peaks corresponding to sand (left), binder (middle) and voids (right). Thresholds are set, either manually or automatically to separate the three phases (e–g). Sand binder and voids separated and artificially coloured.

either by inspection or automatically, to separate the three phases (Figures 2.19e–g). In this example, 100 fields of view were measured in 10–15 min. The average composition compares favourably with that obtained by conventional point counting (500 points). Further discrimination may be achieved, for example, to separate coarse aggregate and sand, by applying a size threshold, or to classify voids on the basis of a dimensionless shape factor.

Method	Sand	Binder	Voids
Image analysis	45.5	49.4	5.1
Point counting	47.1	47.5	5.4

2.6.2.3 Quantitative investigation of fractures

Image analysis provides several useful techniques for quantitative investigation of fractures on all kinds of scales. Images may include macrophotographs of dam walls or abutments, polished slabs of concrete imaged on the flat-bed scanner, microscopic preparations and scanning electron microscope images.

In this example, fracture density and length in a 150-year-old dimension stone (granite) is compared at 10 and 50 mm, respectively, below the exposed surface. The technique is readily applicable to investigation of all types of fractures in concrete, for example, in studies of alkali–silica reaction (ASR) and fire and frost damage. Common measurements include total fracture area (fracture porosity), total length, average and maximum aperture and orientation.

A series of images of dye-impregnated polished slabs were acquired under reflected, high-intensity UV illumination. Fractures, filled with fluorescent dye penetrant, appear bright green. Crystalline minerals in the granite are black. Figure 2.20a shows an area of fractures at a depth of about 50 mm; Figure 2.20b shows an area about 10 mm from the exposed surface of the granite.

Segmentation, using bi-level thresholding, is used to isolate the fractures prior to measurement.

Further processing, using skeletonisation, enables direct measurement of the total fracture length per unit area. Each detected feature is reduced to a median line one pixel wide. Provided the image is suitably calibrated, the total number of detected pixels can be directly converted to fracture length.

The measurements suggest fracture porosity has increased about sevenfold, and total fracture length has increased 10-fold as a result of weathering over a period of about 150 years.

2.6.2.4 Evaluation of shape and distribution of voids

In concrete, mortar and allied materials voids are the feature most readily investigated by image analysis (Chatterji and Gudmundsson 1977; Roberts and Scali 1984; Laurencot et al. 1992; Cahill et al. 1994). By suitable impregnation, they are easily differentiated and capable of volumetric, morphological and spatial measurements.

The following example illustrates classification of void morphology and distribution in bedding mortar. Figure 2.21a is a digital image of bedding mortar obtained from a film scanner (Nikon Super Coolscan 900) with film holder modified to accommodate 75 mm × 50 mm thin sections. The mortar was doubly impregnated with epoxy resin containing yellow dye to ensure complete filling of voids. Firstly, the voids are isolated by bi-level thresholding using the brightness histogram (Figure 2.21b). Their percentage area is 14.4%. The morphology of voids is classified using a dimensionless form factor ($4\pi \cdot area/perimeter^2$) and colour coded (Figure 2.21c). Nearly spherical voids have high form factor. With increasing irregularity, the

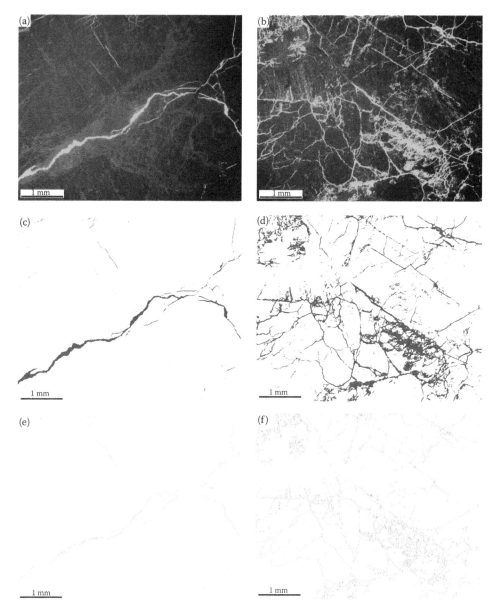

Figure 2.20 (a and b) Segmentation, using bi-level thresholding, used to isolate the fractures in a granite slab prior to measurement. (c) Area measurement of fractures at 50 mm depth. Area fraction (fracture porosity) = 2.6%. (d) Area measurement of fractures at 10 mm depth. Area fraction (fracture porosity) = 18.6%. (e) Fractures 50 mm below surface. Total fracture length = 6.25 mm. (f) Fractures 10 mm below surface. Total fracture length = 61.25 mm.

form factor value decreases. Figure 2.21d is a form factor histogram for all voids with a mean diameter > 200 μm. Form factors of the larger voids have an almost normal distribution. A form factor histogram for voids with a mean diameter < 200 μm is shown in Figure 2.21e. The distribution is strongly skewed towards high form factors. Small voids have predominantly spherical morphologies.

The same image can be used to describe the void distribution using nearest neighbour analysis (Figure 2.21f). The software computes the mean nearest neighbour distance for all

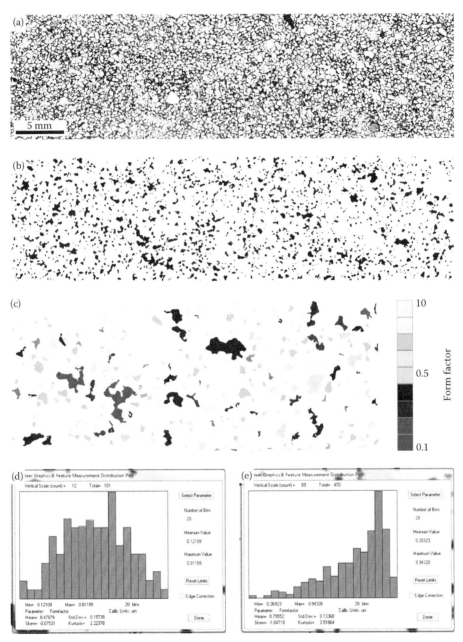

Figure 2.21 Shape and distribution of voids. This example illustrates classification of void morphology and distribution in bedding mortar. (a) Digital image of bedding mortar obtained from film scanner (Nikon Super Coolscan 900) with film holder modified to accommodate 75 mm × 50 mm thin sections. The mortar was doubly impregnated with epoxy resin containing yellow dye to ensure complete filling of voids. (b) The voids are isolated by bi-level thresholding using the brightness histogram. Their percentage area is 14.4%. (c) The morphology of voids is classified using a dimensionless form factor ($4\pi \cdot$ area/perimeter^2) and colour coded. Nearly spherical voids have high form factor. With increasing irregularity, the form factor value decreases. (d) Form factor histogram for all voids with mean diameter > 200 μm. Form factors of the larger voids have an almost normal distribution. (e) Form factor histogram for voids with mean diameter < 200 μm. The distribution is strongly skewed towards high form factors. Small voids have predominantly spherical morphologies. *(Continued)*

(f)

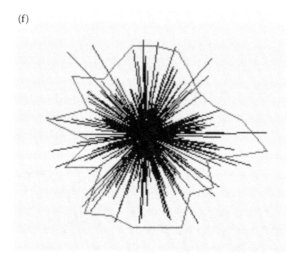

Figure 2.21 (Continued) Shape and distribution of voids. This example illustrates classification of void morphology and distribution in bedding mortar. (f) Summarises the nearest neighbour distribution for 322 voids in the bedding mortar shown in (a and b). Mean nearest neighbour distance = 303 μm. Mean distance for random distribution = 206 μm.

detected features (voids). It compares this with the mean nearest neighbour distance for a random distribution of the same voids in the same area. If the mean nearest neighbour distance is greater than that for a random distribution, the voids are evenly spaced. Values less than the mean value will indicate a tendency towards clustering. The smaller the value, the greater is the tendency towards an irregular void distribution.

Figure 2.21f summarises the nearest neighbour distribution for 322 voids in the bedding mortar shown in Figures 2.21a and b:

Mean nearest neighbour distance = 303 μm
Mean distance for random distribution = 206 μm
No preferred dimensional orientation

The data indicate that voids are evenly distributed through the mortar with no tendency to form clusters. The orientation envelope (red line) shows there is no preferred orientation.

Image analysis provides several useful techniques for quantitative investigation of fractures on all kinds of scales. Images may include macrophotographs of dam walls or abutments, polished slabs of concrete imaged on the flat-bed scanner, microscopic preparations and scanning electron microscope images.

Other examples using quantitative image processing techniques are given elsewhere in this book; see Section 4.3 for concrete aggregate, Section 4.4 for water/cement ratio and Section 4.5.2 for air void determination.

REFERENCES

Armstrong, J.T. 1991: Quantitative elemental analysis of individual microparticles with electron beam instruments. In *Electron Probe Quantitation*, Heinrich, K.J.F. and Newbury, D. (eds.). Plenum Press, New York.

ASTM C90-09. 2009: *Standard Specification for Loadbearing Concrete Masonry Units.* ASTM, Philadelphia, PA.

ASTM C295-08. 2008: *Standard Guide for Petrographic Examination of Aggregates for Concrete.* ASTM, Philadelphia, PA.

ASTM C457/C457M-10a. 2010: *Standard Test Method for Microscopical Determination of Parameters of the Air-Void System in Hardened Concrete.* ASTM, Philadelphia, PA.

ASTM STP 395. 1965: *Analytical Techniques for Hydraulic Cements and Concrete 68th Annual Meeting Symposium,* Lafayette, LA, Hansen, W. (ed.). ASTM, Philadelphia, PA.

Barnes, P., Clark, S.M., Fentiman, C.H. et al. 1992: The rapid conversion of high alumina cement hydrates as revealed by synchrotron energy-dispersive diffraction. *Advances in Cement Research* 4 (14), 61–67.

Barta, R. 1972: Cements. In *Differential Thermal Analysis,* Vol. 2, Mackenzie, R.C. (ed.). Academic Press, London, U.K., pp. 207–228.

Ben-Dor, I. 1983: Thermal methods. In *Advances in Cement Technology,* Gosh, S.N. (ed.). Pergamon Press, Oxford, U.K.

Bhatty, J.I. 1991: A review of the application of thermal analysis to cement-admixture system. *Thermochemica Acta* 189, 313–350.

Bradbury, H.S.M. and Bracegirdle, B. 1998: *Introduction to Light Microscopy.* Handbook 42, Royal Microscopical Society, Oxford, U.K.

Bradbury, S. 1991: *Basic Measurement Techniques for Light Microscopy.* Oxford University Press/ Royal Microscopical Society, Oxford, U.K.

Brindley, G.W. and Brown, G. (eds.) 1980: *Crystal Structures of Clay Minerals and Their Identification.* The Mineralogical Society, London, U.K., Chapter 7, pp. 411–438.

BS 1881-124. 1988: *Testing Concrete Part 124. Methods of Analysis of Hardened Concrete.* BSI, London, U.K.

Buhrke, V.E. Jenkins, R. and Smith, D.K. 1998: *Preparation of Specimens for X-Ray Fluorescence and X-Ray Diffraction Analysis.* Wiley, New York.

Cahill, J., Dolan, J.C. and Inward, P.W. 1994: The identification and measurement of entrained air in concrete using image analysis. *Petrography of Cementitious Materials.* ASTM STP 1215, Greensburg, PA, pp. 111–124.

Campbell, D.H. 1986: *Microscopical Examination of Interpretation of Portland Cement and Clinker.* Construction Technology Laboratories, Portland Cement Association, Skokie, IL.

Chatterji, S. and Gudmundsson, H. 1977: Characterisation of entrained air bubble systems in concrete by means of image analysing microscope. *Cement and Concrete Research* 7, 423–428.

Chayes, F. 1956: *Petrographic Modal Analysis.* John Wiley & Sons, New York.

Chinchon, S., Guirado, F., Gali, S. and Vazques, E.1994: Cement content in concrete made with alumi-nous cement. *Materials and Structures* 27 (169), 285–287.

Concrete Society. 1989: Analysis of hardened concrete – A guide to tests, procedures and interpreta-tion of results. Report of a Joint Working Party of the Concrete Society, Technical Report No. 32. Surrey, U.K.

Cross, A.D. 1960: *Introduction to Practical Infrared Spectroscopy.* Butterworths, London, U.K.

DeHoff, R.T. and Rhines, F.N. 1968: *Quantitative Microscopy.* McGraw-Hill Books, New York.

Delesse, A. 1848: Procédé mechaniquepour determiner la composition des roches. *Annal. des Mines* 1, 379–388.

Demoulian, E., Gourdin, P., Hawthorn, F. and Vernet, C. 1980: Influence of slags chemical composition and texture on their hydraulicity. In *Proceedings of the Seventh International Congress on the Chemistry of Cement,* Paris, France, Vol. III, pp. 89–94.

Dhir, R.K., Jones, M.R. and Amned, H.E.H. 1990: Determination of total and soluble chlorides in con-crete. *Cement and Concrete Research* 20, 579–590.

Efes, Y. 1988: Determination of the composition of hardened concrete by image analysis. *Betonwerk and Fertigteill-Technik* Heft 11, 86–91 and 12, 62–68.

Ellis, P.R. 2010: Analysis of mortars. Available as a download from: www.rose-of-jericho.demon.co.uk.

Entwistle, A. 2003: Basic digital light micrography. *Proceedings of the Royal Microscopical Society* 38, 2, 74–78.

Entwistle, A. 2004: Basic digital light micrography, Part 2. *Proceedings of the Royal Microscopical Society* 39, 1, 15–24.

Figg, J.W. and Bowden, S.R.1971: *The Analysis of Concretes*. Building Research Establishment, HMSO, London, U.K.

Fleger, S.L., Heckman, Jr. J.W. and Komparens, K.L.1997: *Scanning and Transmission Electron Microscopy: An Introduction*. Oxford University Press, Oxford, U.K.

Fournier, C., Merlet, C. and Dugne, O. 2000: An expert system for EPMA. *Microchimica Acta* 132, 531–539.

Fournier, C., Merlet, C., Dugne, O. and Fialin, M. 1999: Standardless semi-quantitative analysis with WDS-EPMA. *Journal of Analytical Atomic Spectrometry* 14, 381–386.

French, W.J.1991: Comments on the determination of the ratio of ggbs to Portland cement in hardened concrete. *Concrete (The Journal of the Concrete Society, UK)* 25 (6), 33–36.

Gabrovsek, R., Vut, T. and Kaucic, V. 2006: Evaluation of the hydration of Portland cement containing various carbonates by means of thermal analysis. *Acta Chimica Slovenica* 53, 159–165.

Goguel, R.L. 1995: A new consecutive dissolution method for the analysis of slag cements. *Cement, Concrete and Aggregates, CCAGP* 17 (1), 59–68.

Goguel, R.L. and St. John, D.A.1993a: Chemical identifications of Portland cements in New Zealand concretes – I. Characteristic differences among New Zealand cements in minor and trace element chemistry. *Cement and Concrete Research* 23 (1), 59–68.

Goguel, R.L. and St. John, D.A.1993b: Chemical identification of Portland cements in New Zealand concretes. II. The Ca-Sr-Mn plot in cement identification and the effect of aggregates. *Cement and Concrete Research* 23 (2), 283–293.

Goldstein, J.I., Newbury, D.E., Echlin, P. and Joy, D.C. 1992: *Scanning Electron Microscopy and X-Ray Microanalysis*, 2nd ed. Plenum Press, New York.

Gutteridge, W.A. 1979: On the dissolution of the interstitial phases in Portland cement. *Cement and Concrete Research* 9 (3), 319–324.

Heinrich, K.F.J. and Newbury, D.E.1991: *Electron Probe Quantitation*. Plenum Press, New York.

Hime, W.G., Mivelaz, W.F. and Connolly, J.D. 1965: *Use of Infrared Spectrophotometry for the Detection and Identification of Organic Additions in Cement and Admixtures in Hardened Concrete, Analytical Techniques for Hydraulic Cements and Concrete*. ASTM STP 295, American Society for Testing Materials.18. ASTM International, Philadelphia, PA.

Hoffmanner, F. 1973: *Microstructure of Portland Cement Clinker*. Holderbank Management & Consulting, Holderbank, Switzerland, pp. 33–37.

Kerrick, D.M., Eminhizer, L.B. and Villaume, J.F. 1973: The role of carbon film thickness in electron microprobe analysis. *American Mineralogist* 58, 920–925.

Klug, H.P. and Alexander, L.E. 1974: *X-Ray Diffraction Procedures for Polycrystalline and Amorphous Material*. Wiley Intersciences, New York.

Labar, J.L. 1995: Standardless electron probe x-ray analysis of non-biological samples. *Microbeam Analysis* 4, 65–83.

Laurencot, J.L., Pleau, R. and Pigeon, M. 1992: The microscopical determination of air voids' characteristics I hardened concrete: Development of an automatic system using image analysis techniques applied to micro-computers. In *Proceedings of the 14th International Conference on Cement Microscopy*, Orange Country, CA, pp. 259–273.

Loveland, R.P. 1970: *Photomicrography. A Comprehensive Treatise*. Vols. 1 and 2. Wiley, New York.

Ludwig, U. and Rudiger, I. 1993: Quantitative determination of ettringite in cement pastes, mortars and concretes. *Zement-Kalk-Gips* 46 (5), E153–E156.

Luke, K. and Glasser, F.P. 1987: Selective dissolution of hydrated blast furnace slag cements. *Cement and Concrete Research* 17, 273–282.

Mangabhai, R., Cave, S. and Wills, R. 2010: Application of FTIR for material identification in the construction industry. *Concrete* 44, 1, 37–38.

McCrone, W.C. and Delly, J.G. 1973: *The Particle Atlas*. Vols. 1 and 2. Ann Arbor Science Publishers, Ann Arbor, MI.

Midgley, H.G. and Midgley, A. 1975: The conversion of high alumina cement. *Magazine of Concrete Research* 27 (91), 77–82.

Milestone, N.B.1984: Identification of concrete admixtures by differential thermal analysis in oxygen. *Cement and Concrete Research* 14 (5), 207–214.

Muhamaed, M.N., Barnes, P., Fentiman, C.H., Hausermann, D., Pollmann, H. and Rashid, S. 1993: A time resolved synchrotron energy-dispersive diffraction study of the dynamic aspects of the synthesis of ettringite during minepacking. *Cement and Concrete Research* 23, 267–272.

Norrish, K. and Chappell, B. 1977: X-ray fluorescence spectrometry. In *Physical Methods of Determinative Mineralogy*, 2nd ed., Zussman, J. (ed.). Academic Press, London, U.K., pp. 201–272.

Papadoulos, G. and Suprenant, B. 1988: Identifying fly ash, slag and silica fume in blended cements and hardened concretes. In *Proceedings of the 10th International Conference on Cement Microscopy*, San Antonio, TX, pp. 344–356.

Pirrie, D. and Rollinson, G.K. 2011: Unlocking the applications of automated mineral analysis. *Concrete Today* 27 (6), 226–235.

Ploem, J.S. and Tanke, H.J. 1987: *Introduction to Fluorescence Microscopy.* Handbook 10, Royal Microscopical Society, Oxford, U.K.

Portilla, M. 1989: Enthalpy of dehydration of Portland and pozzolanic cements. *Cement and Concrete Research* 19, 319–326.

Ramachandran, V.S. (ed.) 1969: *Applications of Differential Thermal Analysis in Cement Chemistry.* Chemical Publishing Co., New York.

Rashid, S., Barnes, P., Benstead, J. and Turrillas, X. 1994: Conversion of calcium aluminate cement hydrates re-examined with synchrotron energy-dispersive diffraction. *Journal of Materials Science Letters* 13, 1232–1234.

Rashid, S. Barnes, P. and Turrillas, X. 1992: The rapid conversion of high alumina cement hydrates, as revealed by synchrotron energy-dispersive x-ray diffraction. *Advances in Cement Research* 4 (14), 61–67.

Reed, S.J.B. 2005: *Electron Microprobe Analysis and Scanning Electron Microscopy in Geology*, 2nd ed. Cambridge University Press, Cambridge, U.K.,192pp.

Rixom, M.R.1978: *Chemical Admixtures for Concrete.* E & F.N. Spon, London, U.K., 234pp.

Roberts, L.R. and Scali, M.J. 1984: Factors affecting image analysis for measurement of air content in hardened concrete. In *Proceedings of the Sixth International Conference on Cement Microscopy*, Albuquerque, NM, pp. 402–409.

Robinson, P.C. and Bradbury, H.S.M. 1992: *Quantitative Polarized Light Microscopy.* Handbook 9, Royal Microscopical Society, Oxford, U.K.

Rosiwal, A. 1898: On geometric rock analysis. A simple surface measurement to determine the quantitative content of mineral constituents of a stony aggregate. *Verhandl. K. K. Geol. Recib. Wein* 5–6, 143.

Russ, J.C. 2006: *The Image Processing Handbook*, 5th ed. CRC Press, Baco Raton, FL.

Russ, J.C. and Russ, J.C. 2007: *Introduction to Image Processing and Analysis.* CRC Press, Baco Raton, FL.

Shi, D.1988: An automated quantitative analysis system for cement and concrete research. *Computer Applications in Concrete Technology.* ACI AP 106. American Concrete Institute, Detroit, MI, pp. 139–157.

Smillie, S., Moulin, E., Macphee, D.E. and Glasser, F.P. 1993: Freshness of cement conditions for syngenite $CaK_2(SO_4)_2 \cdot H_2O$ formation. *Advances in Cement Research* 5 (19), 93096.

St. John, D.A.1992: AAR investigation in New Zealand – Work carried out since 1989. In *Proceedings of the Ninth International Conference on Alkali-Aggregate Reaction*, The Concrete Society, London, U.K., Vol. 2, pp. 885–893.

St. John, D.A. and Goguel, R.L.1992: Pore solution/aggregate enhancement of alkalies in hardened concrete. In *Proceedings of the Ninth International Conference on Alkali-Aggregate Reaction*, The Concrete Society, London, U.K.

Stuart, B.H. (ed.). 2007: *Identification, in Analytical Techniques in the Sciences: Polymer Analysis.* Wiley & Sons Ltd., Chichester, U.K.

Takashima, S. 1958: Systematic dissolution of calcium silicate in commercial Portland cement by organic acid solution. In *Review of the 12th General Meeting*, Cement Engineering Association, Tokyo, Japan, pp. 12–13.

Takashima, S. 1972: Studies on belite in Portland cement. In *Review of the 26th General Meeting, Technical Session*, Cement Engineering Association, Tokyo, Japan, pp. 27–28.

Tenoutasse, N. and Moulin, E. 1990: Influence of water-cement ratio on concrete microstructure. In *Proceedings of the 12th International Conference on Cement Microscopy*, Gouda, G.R., Bayles, J. and Nisperos, A. (eds.). International Cement Microscopy Association, Vancouver, British Columbia, Canada, pp. 323–338.

UK Government Web-Site. 2011: http://infrared.als.lbl.gov/content/web-links/60-ir-band-positions.

Van der Plas, L. and Tobi, A. 1965: A chart for judging the reliability of point counting results. *American Journal of Science* 236 (1), 87–90.

Weyers, R.E., Brown, M., Al-Quaida, I.L. and Henry, M. 1993: A rapid method for measuring the acid-soluble chloride content of powdered concrete samples. *Cement and Concrete Aggregates* 15 (1), 3–13.

Chapter 3

Sampling and specimen preparation

3.1 SAMPLING CONCRETE AND RELATED MATERIALS

Samples collected in order to carry out a laboratory-based petrographic examination may be taken from a concrete structure or from a source of raw material such as aggregate, cement, mineral addition or an admixture. They will either have been selected to be representative of the material under investigation as a whole or selected specifically to investigate some particular identified feature in the concrete or other material. In either case sufficient material needs to be collected for all the necessary laboratory tests to be carried out.

It is generally recognised that it is important for the petrographer to be present during the site inspection and the collection of samples. However, the petrographer is sometimes required to investigate samples provided by a client, perhaps from abroad without the advantage of being involved in the selection process. Thus, the investigation is likely to be hampered by lack of information relating to

- Location of the sample in the structure or source of material
- Method of selection chosen
- Method of collection
- The scheme of wrapping and labelling samples
- Size of the sample received
- Information on local environmental conditions
- Information about the structure or source and its history

In these difficult circumstances, the petrographer must endeavour to ensure that if at all possible, the client selects, collects and wraps the sample appropriately in accordance with the petrographer's instructions before dispatch and also provides as much information as possible about the structure or material itself and the local environment, so as to maximise the value of any conclusions to be drawn from a factual petrographic investigation of the sample provided.

3.2 INSPECTION OF STRUCTURES

As noted earlier, petrologists and engineers will appreciate the importance of visiting and carefully examining any site which is the subject of an investigation. If samples are to be taken for petrographic examination, it is important for the petrographer to be involved in the sample selection. Where precast units are to be investigated, it is often very helpful to observe the manufacturing process and to evaluate the raw materials used. The level of detail and the time spent on a particular site inspection will of course be dictated not only by the objectives but also by economic considerations and the size and complexity of the structure.

Problems with concrete structures may range from defects in curbstones and roof tiles to deterioration of major viaducts or dams. In all cases, the importance of carrying out an initial investigation on-site, obtaining details of the materials used and obtaining data on the methods and date of construction cannot be over emphasised. Study of other structures in the local area constructed of similar materials can also be helpful in providing 'case-study' evidence, particularly if those structures vary in age from the one under investigation. Account must also be taken of the local climatic and other external environmental factors. These will probably vary even for different parts of the same structure. Detailed record should be kept, perhaps by recording a scale of severity of conditions on detailed plans or drawings, because factors such as degree of exposure to weather, freeze–thaw conditions or localised moisture ingress may be of considerable importance when assessing the causes of deterioration.

Record of all observations should be made as the inspection proceeds. Drawings can be marked, coloured or shaded to indicate the local severity of each feature. Features which commonly need recording include cracking which can vary widely in nature and style depending on the causative mechanism, surface pitting and spalling, surface staining, differential movements or displacements, variation in algal or vegetative surface growths, surface voids, honeycombing, bleed marks, construction and lift joints and exudations or efflorescence. Certain parts of a structure may be considered to merit a more detailed inspection than others prior to sampling. It is at this stage that recourse may be made to detailed mapping, the measurement of cracks and photographic or video recording, while surveying techniques might be used to check any apparent misalignments.

Reference should be made to the appropriate guides and codes of practice if additional or specific details relating to the inspection of particular types of concrete structure are required. There are many excellent reference works and guidance documents now available to assist the petrographer or engineer in examining structures and identifying and quantifying the various features of deterioration or damage. These include the *Concrete Societies* series of reports, *Non-structural cracking in concrete* (1992), *Diagnosis of deterioration in concrete structures* (2000), *Assessment, design and repair of fire damaged concrete structures* (2008), *Alkali-silica reaction* (1999) and *Guide to concrete petrography* (2010). Other useful guides include *Crack mapping* (Fookes et al. 1982), *Fire damage* (Green 1976), *Concrete deterioration and repair–discussion* (Fookes et al. 1981) and technical reports relating to appraisal of structures affected by the alkali–aggregate reaction by the Institution of Structural Engineers (1992) and by RILEM AAR-6.1 (2013).

The completed site survey with its detailed observations should allow a preliminary assessment of the structure to be made and will have identified any areas requiring more detailed examination and testing. This further work will include locating the most appropriate sampling sites, since except for very minor cases it is probable that some laboratory work will be required. It may also involve on-site non-destructive testing or longer-term monitoring of parts of the structure. However, it is important to always take additional representative samples from undamaged parts of a structure to provide a comparison with samples selected to investigate damage or deterioration.

3.2.1 Surface expression of concrete deterioration

The commonest defects observed on concrete surfaces are cracking, misalignment of elements, exudations, surface discoloration and degradation. Cracking can range from hairline superficial cracks or crack networks to large open cracks several millimetres wide and some

(a)   (b)

Figure 3.1 (a) The Elgeseter Bridge, Trondheim, constructed in 1951. (b) The top of a concrete support column of the bridge which has been cut free and repositioned to take account of an expansion of over 200 mm due to ASR in the 220 m long bridge deck.

which exhibit differential movements. Some multiple expansive microcracking in a concrete element can result in a significant measurable overall expansion of the concrete (Figure 3.1).

The reasons why cracks develop vary widely, but because they are the surface expression and perhaps the only indication of some form of distress in the structure, they have become the subject of numerous articles, for example, Fookes et al.'s (1982) reviews in textbooks and special publications such as the Concrete Societies Technical Report TR22. The petrographer will need to examine both macroscopic and microscopic cracks and crack patterns during the course of the investigations. Some of these cracks exhibit characteristic features which relate to the original cause of cracking, though their interpretation should be tempered with caution. Cracking which develops subparallel to the concrete surface may progress to an extent where spalling occurs. Section 5.8 deals specifically with cracking in concrete and describes their appearances, widths and causes. The relationship between concrete deterioration and crack development is discussed more fully in Section 6.8.3, and examples of cracking due to reinforcement corrosion are illustrated in Figure 6.17.

Other features of deterioration include surface degradation, exudations or efflorescence. By far the most common exudation is calcium carbonate derived from the gradual leaching of calcium hydroxide from the cement paste, its subsequent evaporation at a surface and the carbonation of the hydroxide solution by the air. Nevertheless, exudations can be used to provide evidence of the nature of a deterioration mechanism since they may contain salts produced within the concrete or, as in the case of deterioration due to alkali–silica reactions, gels produced by the reaction.

3.3 REPRESENTATIVE SAMPLE

3.3.1 Particulate materials

Representative sampling of a solid concrete can be very difficult because of its variability, but sampling of loose particulate materials such as aggregate or cement should not present a problem since procedures are based on statistical principles, some of which are described in appropriate texts and standards, for example, Gy (1998) and BS 812 part 102 (1989), now superseded by BS EN 932-1 (1997).

Particulate building materials can vary in size from boulders to sub-micron-sized powder. Coarse-grained particulates are usually sampled on the basis of counting numbers of particles, but when large numbers of particles are to be considered, counting becomes impractical and it becomes preferable for sampling to be by mass. In some cases with coarse material, a crushing and sample splitting scheme will be necessary to allow a smaller and manageable subsample to be prepared, provided that the original particle size distribution is not required.

Unless the sample is specially selected to investigate a particular feature, a sample whether taken from an aggregate stockpile or as solid material from a concrete structure, it should be taken so as to be representative of the material as a whole.

The initial sampling strategy is fundamental to the value of the numerical data obtained from the sample. Thus, the sample must have the following characteristics:

- It must be truly a representative of the material as a whole, or the feature of interest.
- It must avoid all sampling bias.
- It must be of adequate size to meet the precision requirement of the objective.
- It must be packaged to avoid modification during transport and handling.
- If a subsampling procedure is necessary, it must be undertaken in such a manner that the subsample remains representative of the whole sample.

Whether the petrographer is required to make a quantitative determination of the proportions of a particular rock species in a sample of aggregate or in a cut surface of concrete or whether some other feature, such as air void content of a concrete or the particle size grading of an aggregate, is required, it is essential that the sample size is sufficient to meet the statistical precision requirement of the particular analysis. The size of representative bulk aggregate or other particulate sample that is recommended for dispatch to a laboratory for petrographic analyses in BS 812-104 is included in Table 3.2.

The minimum size of a representative sample mass will depend on the test precision requirement and on the particle sizes involved; the mass increases rapidly with increasing particle size as indicated by Table 3.1. However, segregation within a stockpile or variation in a conveyor feed must also be taken into account when sampling. Specific sampling procedures given in national standards have been devised to minimise the effects of such variations and to ensure that representative samples are obtained. Methods of sampling aggregate stockpiles are given by BS EN 932-1 (1997), which replaces BS 812-102 (1989), and in ASTM D75/D75M (2009).

A representative subsample may need to be prepared from large particulate samples in order to undertake particular laboratory test procedures. It may also involve further crushing of the material to provide a fine powder for analysis, but it must remain uncontaminated and representative. The particle size of the crushed powder specimen is not very critical for thermal analysis, but for the other analytical techniques, the sample may require grinding to a very fine particle size. The method for the preparation of test specimens for the analysis of

Table 3.1 Minimum size of bulk aggregate sample for dispatch to a laboratory for petrographic analysis

Nominal maximum particle size of the aggregate	Number of subsample increments to be combined	Minimum mass for dispatch (kg)	Minimum mass for test portion[a] (kg)
40 mm	80	200	50
20 mm	20	50	6
10 mm	10	25	1
5 mm and smaller	10 (half scoops)	10	0.1

Source: After BS 812-104, *Testing Aggregates. Procedure for Qualitative and Quantitative Petrographic Examination of Aggregates*, BSI, London, U.K., 1994.

[a] From Harris and Sym (1990) for a precision of ±10% relative.

concrete is given in BS 1881-124 (1988), and the modes of operation of crushing and grinding equipment are detailed in Allman and Lawrence (1972) and manufacturers' literature.

The heterogeneity of concrete and aggregate requires that several kilograms be crushed in order to obtain a representative subsample for chemical analysis. As discussed in Section 3.3.2, the initial large samples from a sample site must be representative. These are then reduced to representative subsamples of 100–200 g, passing a 150 μm sieve. As an example, BS 1881-124 (1988) requires the minimum linear dimension of a solid concrete sample for petrographic examination to be at least five times that of the nominal maximum dimension of the aggregate particles it contains and also to be not less than 1 kg if an analysis of the concrete is required. It also points out that a major source of error in the analysis of hardened concrete is inadequate sample preparation.

In those circumstances where the analysis of concrete makes it necessary to crush a large concrete sample, the initial breaking is performed using either a mechanical knife edge hydraulic splitter or hammer and bolster. In reducing these fragments to a powder passing a 150 μm sieve, it is important to take steps to limit any contamination; restrict loss of material, principally as dust; and carry out the reduction as quickly as possible to limit exposure to atmospheric carbon dioxide. Reciprocating jaw crushers are commonly used for crushing materials to pass a 5 mm sieve as they can handle large quantities of material rapidly. Iron is the most likely contaminant from a jaw crusher, but this should be minimal if the jaws are made from specialised hard manganese steel alloys. Cross-contamination from previous samples is a much more serious problem which is minimised by thorough wire brushing and cleaning with compressed air between samples and by initially running a small amount of the sample through the crusher and discarding it prior to crushing the main bulk sample.

If carried out carefully, jaw crushing produces particles of 5 mm and smaller that can be subsampled at this stage. There is a range of mechanical equipment available that can be used to avoid hand grinding and to reduce a –5 mm sample to coarse powder. The commonest of these are various types of disc mills. BS 1881-124 (1988) specifies that before this grinding is done, the sample be split down to 500–1000 g using a riffle box and that, as the material is reduced in size, further sample splitting be carried out for material passing the 2.36 mm and 600 μm sieves. The final sample required is 100–200 g of material passing a 150 μm sieve which is sufficient for the determination of cement, aggregate and original water contents.

If quantities of less than 1 g of material are required, hand grinding with a pestle and mortar is the simplest and most effective method. However, there is a range of mechanical equipment such as laboratory ball mills or disc mills that can be used to avoid hand grinding and will reduce the –5 mm sample to a powder. Where a powder finer than 150 μm is required, a small subsample, often of only a few grams, is ground in a specialised laboratory mill. There is a wide range of this type of equipment available with the type and method of grinding being dictated by the analytical technique to be used and the possibilities of contamination. There are no general references available, but several manufacturers of milling equipment produce informative catalogues.

3.3.2 Statistical considerations

A truly representative sample would be the 'whole lot', but for practical purposes, this is not a possibility, and where very large numbers of particles are to be considered, the sample size is best measured in terms of mass as indicated in Table 3.1.

A truly representative random sampling method is amenable to mathematical analysis as is discussed by Harris and Sym (1990) who demonstrate how the reliability of the data improves with increasing bulk sample size. The maximum size of the particles present will be the controlling factor in the determination of the actual size of the particular bulk sample

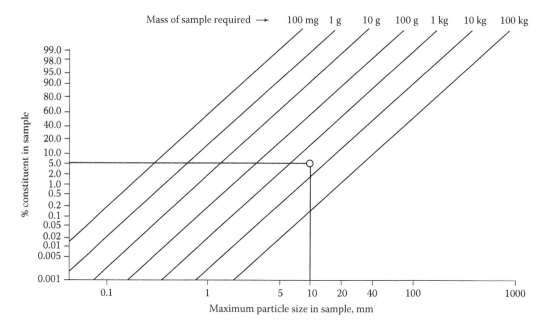

Figure 3.2 Mass of test portion necessary to achieve accuracy of ±10% relative (single-sized, fully liberated particles) modified from BS 812 Part 104.

required for a given precision. They provide tables and nomograms relating individual particle size, sample size and sampling error. The mass of a test portion necessary for a precision of ±10% relative for the estimated proportion of a given constituent is given in Figure 3.2; using as an example, a 10 mm aggregate with the constituent of interest representing 5% of the total sample would require a sample mass of about 8 kg.

Theoretical reviews concerned with the size of random particulate samples and the probabilities associated with selection of a particular constituent from them normally deal with ideal single-sized spherical particles which are all of the same density. In practical sampling procedures, allowance must be made for deviation from such ideal material. If there is a wide distribution of particle sizes, it may be necessary to devise a separate sampling procedure for each size fraction.

CONFIDENCE LIMITS

Considerations as to the size of sample required (n) make the assumption that the sample will form a normal or Gaussian distribution about a mean value ($\bar{x}$) of the particular feature of interest in the sample. This allows an estimated standard error (σ_{se}) to be calculated for a sample of n particles using the following formula:

$$\text{Estimated standard error } \sigma_{se} = \frac{\bar{x} \pm k\sigma_{se}}{\sqrt{\bar{x}(100-x)/n}}$$

The formula indicates that 99.7% of the value a will be within $3\sigma_{se}$ ($k = 3$), 95% of the value a will be within $2\sigma_{se}$ ($k = 2$) and 67% of the value a will be within $1\sigma_{se}$ ($k = 1$).

In many practical applications $2\sigma_{se}$, 95% confidence limits are chosen as satisfactory.

Table 3.2 Minimum masses of bulk aggregate samples

Nominal particle diameter range (mm)	Minimum mass of bulk sample (kg)
28 and larger	50
5–28	25
5 and smaller	10

Source: After BS 812-102, *Testing Aggregates. Methods for Sampling*, BSI, London, U.K., 1989.

Note: BS 812-02 is currently under review.

To accommodate practical factors such as particle shape, composite particles and particle size distribution, a comprehensive sampling method has been developed by Gy (1982, 1998) which is widely used in the metal mining industry for estimating the minimum mass of sample required. Gy's methods are considered in detail by Pitard (1989) and reviewed by Francois-Bongarcon (1998) and Harris and Sym (1990). In BS 812-102 (1989) an application of Gy's formula is applied to estimating aggregates and fillers as in Table 3.2. (Note: BS 812-102 is now replaced by BS EN 932-1 [1997].) These figures are similar or slightly smaller than the recommendations given in ASTM D75/D75M (2009). The simplified version of the formula is used in BS 812-102 and assumes likely values for the numerical factors dealing with shape, size distribution, composite grains (liberation factor) and density which are used in the original formula (see box). Also, the approximate proportion of the constituent of interest must be known or inferred from previous results to use the formula.

The minimum test portion M (kg) is related to the estimated proportion of a particular constituent, p (%), whose nominal maximum diameter is d (mm) in the following formula:

$$M = 0.0002(100 - p)d^3/p$$

Alternatively, for general petrographic analysis, smaller samples (significantly smaller than those calculated from this formula) may be used following values given in Table 3.1 after Harris and Sym (1990) who substitute the most likely values to the parameters in Gy's formula and assume the material to be narrowly graded. They calculate the results to have a precision of ±10% relative.

In the case of cements, BS EN196-7 (2007) recommends a minimum sample of 5 kg and ASTM C183 (2008) also gives 5 kg, as a minimum representative bulk sample from a defined batch. Taken together, these recommendations suggest that for the purposes of sampling building materials, the size bulk sample of powders (i.e. particulate materials with maximum particle diameters less than 100 μm) should be a minimum of 5 kg. Specific sampling procedures are described in detail in national standards and a good review of sampling procedures for aggregates is given by Smith and Collis (2001).

In most cases bulk representative particulate samples are supplied to the petrographer who should ensure that unambiguous identification is marked on the sample bags and in addition loose identification tags are present inside the bags. On receipt at the laboratory, each sample should be logged and given a laboratory reference number. Standard methods of sampling aggregates are given in BS 812-102 (1989), now partially replaced by BS EN 932-1 (1997). It recommends that minimum masses of bulk samples in kilograms for normal-density aggregates may be calculated from the empirical relationship of six times the square root of the maximum particle size (d) in mm times the loose bulk density (ρ), or $M = 6\sqrt{d} \cdot \rho$.

The next step involves reduction of the bulk sample to sizes suitable for laboratory testing and petrographic examination. In the case of powders, this reduction may involve the separation of a few mg of material from as much as a 5 kg sample. The most reliable method of carrying out this type of sample reduction is to use a sample splitter, such as a riffle box or preferably a spinning riffler. Harris and Sym (1990) reviewed the various methods of sample reduction for powders and concluded that the cone and quartering method is subject to considerable sampling error.

Some national and international standards recommend minimum numbers of particles in samples of loose materials such as aggregates for petrographic examination. The Canadian Standard CSA A23.2-09 recommends 300 particles for a –20 mm aggregate fraction, while ASTM C295 recommends 45 kg or 300 particles, whichever is larger, but BS 812-104 recommends a minimum test portion mass of 6.4 kg for ±10% relative error for a constituent of interest estimated to be 20% of the sample. The precision of petrographic modal analyses are dependent on the proportions of the constituents of interest in the sample and are considered more fully in Section 2.4 and Figure 2.6.

GY'S SIMPLIFIED SAMPLING FORMULA (1982, 1998)

The requirement to determine proportions of specific items of interest within a sample (originally concerned with ore samples) is given by Gy's simplified formula expressing the relative variance of a sampling error (σ_{se}) in terms of the nominal size of the particles (d_N) and the mass of the sample taken (M_s):

$$\sigma_{se}^2 = \left(cl\right)\left(fgd_N^3\right)/M_s$$

The term $\left(fgd_N^3\right)$ is referred to as the 'volume group' and (cl) the 'density variance group'. c is a mineralogical factor related to mineral grade and density difference between mineral and gangue, but simplifies to $c = \rho/t$ where t is the proportion of mineral of interest present in the sample and ρ the relative density if all particles are similar, as is common with aggregates. l is called the 'liberation factor' and relates to size distribution and arrangement of mineral grains within rock particles. l is taken as 1 for completely separated mineral (fully 'liberated') species and 0 for completely homogenous particles. f and g are empirical constants related to shape and size grading of the particles. The f factor transforms the cube of any fragment size into a true fragment volume. These convenience factors are often taken as $f = 0.5$ for closely sized aggregate particles and 0.25 for unsized material. g is taken as 0.4 if particles are similar in size.

The formula given in BS 812-102 (1989) is based on Gy's formula and is given as

Test portion mass (kg) $= M = 0.0002(100 - p)d^3/p$

p is the constituent of interest, %
d is the nominal maximum particle size, mm

This makes the assumptions that $f = 0.5$, $g = 0.4$ and $l = 1$ and relative density of aggregate $\rho = 2650 \times 10^{-9}$ kg/mm^3 for the $2\sigma_{se}$ case. The coefficient 0.0002 is estimated based on these assumptions for a value of $p = 20\%$ (see also Harris and Sym 1990).

3.3.3 Examination of particulate materials

Where the identification and estimation of constituents in particulate materials can be determined by external visual appearances, examination of the loose material may be carried out directly with the aid of a low-power stereomicroscope. The only specimen preparation required in such cases is to ensure that a sufficient number of particles are examined as discussed earlier. To assist identification, selected particle grains may be crushed and placed in immersion oil mounts for examination by the petrographic microscope. Alternatively, definitive identification of selected particles may be made using a scanning electron microscope by mounting them directly onto specimen stubs. These are routine petrographic procedures, and for aggregates, the methods required are discussed in ASTM C295 (2009), BS 812-104 (1994) and in Dolar-Mantuani (1983), Klein and Hurlburt (1993) and Larsen and Berman (1934). When the material is in the form of a fine powder, most standards are restricted to detailing methods of measuring fineness and surface area, for example, as specified in ASTM C204-07 (1996), ASTM C786-96 (2007) and BS EN 196-6 (2010). General methods for the microscopic determination of particle sizes in fine powders is given in ASTM E20-85 (1985), but this standard was withdrawn without replacement in 1994.

If particles are too small or difficult to identify by direct visual means, for example, fine sands and powders, a useful specimen preparation method is to cast them into blocks of resin which are then thin-sectioned for microscopic examination by transmitted light. The practical procedures for various types of specimen preparation are summarised in Section 3.4. If the material is too fine for examination by optical microscopy, then in addition to the standard procedures noted earlier, the material can be examined on a suitable mount or on a polished surface using scanning electron microscopy. Modern SEM instruments will readily resolve sub-micron-sized particles, and if, as is usual, they include an x-ray energy-dispersive analytical system, it can provide an indication of the elemental composition of the particle. Alternatively, x-ray diffraction techniques can provide a definitive identification of crystalline material, but unlike the SEM cannot provide much information about particle morphology. If quantitative elemental compositions are required, then the specimen should be polished to allow a direct comparison to the polished reference standards that will have been used for calibrating the EDS software. X-ray fluorescence analysis of polished mounts, or thin sections, may be required for determining bulk elemental compositions of powders. Further details of SEM/EDS equipment are given in Section 2.5.2. Campbell (1986) describes the use of polished thin sections for the examination of cement and clinker by incident, transmitted light and SEM.

3.3.4 Solid samples

Hardened concrete and mortar samples are mixtures of several constituents, principally aggregate particles, hardened cement paste, air voids and often mineral admixtures. Consequently, the same strategy for sampling that is used with particulates can also be applied to provide data that are representative of the material in the sample provided that:

- The components are uniformly distributed
- Sufficient sample can be collected to meet the statistical precision required
- Localised features such as deterioration are not present

In many cases, these requirements are often difficult to achieve in practice, unless an impractically large number of samples are taken. A careful review of all the required objectives of the investigation and the precisions required may lead the petrographer to conclude that a better and more economically effective approach is to make a careful selection of particular samples from the structure or element together with appropriate 'control' samples. The subject of modal analysis of concrete is considered further in Sections 2.4.1 and 4.5.

In some limited circumstances, for example, when the objective is to only to identify the nature of the aggregate, a broken or spalled piece of concrete or mortar will suffice provided it is large enough to be a representative. In some cases, the petrographer is only provided with broken or loose material with possible surface weathering, cracking and other artefacts introduced by the sampling procedure. Consequently the information that can be obtained is likely to be both limited and subject to restricting caveats. In such cases, the limitations relevant to the data must also be clearly recorded in the petrographers' report.

The preferred method of sampling concrete for petrographic examination that is generally accepted worldwide is to drill a number of cores from the structure. Descriptions of coring the techniques for coring structures can be found in Andersen and Petersen (1961), the U.S. Department of the Interior, Concrete Manual (1981), ASTM C42/C42M-04 and the Concrete Society Technical Report 30 (1999). Obviously the diameter of the core is vitally important in relation to the size of the constituents present in the concrete which in practice means that the core diameter should be not less than three times the maximum size of the aggregate (see Figure 3.3). In the majority of structural concretes, the maximum size of aggregate does not exceed 40 mm and most are 20 mm or less, so that 100 mm diameter cores are normally adequate. In mass concretes larger aggregates are used and 150 mm or larger-diameter cores may be necessary. However, scattered through the literature are many statements which claim that 75 mm and even 50 mm or smaller-diameter cores may be used. The justifications for these statements are that potential damage to critical reinforcement may prevent the drilling of larger-diameter cores and the high costs involved in drilling large cores.

It is also argued that since deterioration is often confined to the hardened cement paste, it is not necessary to use larger-diameter cores. This does not take account of the fact that core recovery is much poorer for smaller-diameter cores and particularly for deteriorated concrete, as is pointed out in ASTM C856-04 (2004). Also, with the smaller-diameter cores, there will be a greater possibility of damage occurring as a result of the coring process. There are strong arguments for the use of larger-diameter cores, and as a general rule, cores of less than 100 mm in diameter should not be used for petrographic examination of structural concretes. If circumstances force the petrographic examination of smaller-diameter cores, longer and/or additional cores should be taken and the report should include the reasons for the use of the smaller cores and their likely impact on the results obtained. Figure 3.3 shows the difficulty in obtaining statistically representative aggregate proportions if the core diameter is small compared to the maximum aggregate size.

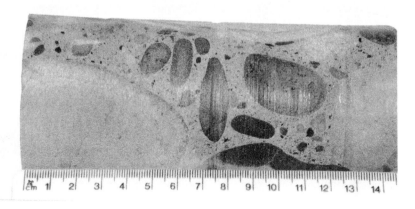

Figure 3.3 **A 67 mm diameter concrete core with rounded aggregate particles of sizes greater than 60 mm. This core would be inappropriate for determining coarse aggregate proportions. (Courtesy of M. Grove.)**

The length of the concrete cores will be controlled by the type of examination, tests required and the size of the concrete element sampled. The outer zone of a reinforced concrete, the 'covercrete', is usually less than 75 mm in depth. Even where severe AAR has occurred, the altered outer zone rarely exceeds 100 mm in depth beyond which any effects of reaction become more uniform. Thus, below a depth of 100 mm, the core usually represents the bulk concrete. For petrographic purposes, this suggests that the outer 100 mm must be examined plus a further 100–150 mm below this depth to characterise the bulk concrete. Core depths of 250 mm are usually adequate which accords with ASTM C856-04 (2004). For general sampling of concrete, ASTM C823/C823M-07 (2007) recommends lengths between 150 and 600 mm are taken dependent on the type of structural element being sampled. Where physical tests will be required, sufficient length of core should be taken to allow for both petrographic examination and the test specimens. Preferably separate cores should be taken for the petrographic examination and for the physical tests. An example of a short 100 mm concrete core taken from a structure photographed and marked up and with a diagram indicating how the core should be cut for detailed study is shown in Figure 3.4.

In summary, an adequate petrographic examination requires that deteriorated structural concrete should be sampled by drilling 100 mm or greater diameter cores to a depth of not less than 250 mm. Where these sampling conditions cannot be met, the petrographic report must take account of the effect of the sampling conditions on the results of the examination.

In the case of precast concretes, mortars, screeds and special products, coring may be an inappropriate method of sampling. Cutting sections, blocks or slabs with a mechanical saw or diamond wheel may be preferable, and recently experimental cutting using a very-high-pressure water jet has been reported, but this is not very satisfactory for concrete because of the inherent differences in hardness between aggregate and matrix. It should be remembered that all methods of coring and cutting will introduce damage to the sample.

This damage can be categorised under three general headings:

1. Introduced new fractures and development of pre-existing cracks
2. Friction-generated thermal effects on the fabric and mineralogy of the sample
3. Cutting lubricants causing solution, dissolution and/or recrystallisation of soluble constituents

In detail the severity of these effects will depend on the sampling method adopted and will be most severe close to the cut surface. Consequently, it is necessary to assess the types and severity of damage likely to be introduced by the specific coring or cutting method and to minimise these effects by careful control of the cutting equipment, selection of cutting speeds and lubricating fluids.

As noted in Section 3.2, the number and locations of samples to be taken will depend on the type of test or examination required. Where concrete is to be tested for a physical property or attribute, representative sampling as specified by an appropriate national standard such as ASTM C823/C823M-07 (2007) should be applied. This type of sampling assumes that the concrete or identifiable portions of the concrete are all of a similar condition and that a probability sampling plan can be formulated. Guidance for the formulation of probability sampling is given in ASTM E 105-04 (2004). For compliance with construction specifications, ASTM C823/C823M-07 (2007) specifies that not less than five samples shall be taken from each category of concrete for each test procedure stipulated. It will be apparent that if representative sampling is carried out, the coring, testing and examination of concrete from a large structure are costly exercises.

If the assumption that the concrete is in a 'similar condition throughout' cannot be made, the number and positioning of samples for petrographic examination becomes a matter of

Petrographic examination of hardened concrete – ASTM C856-04

Record digital photograph			
RSK STATS sample ref.		Client sample ref.	

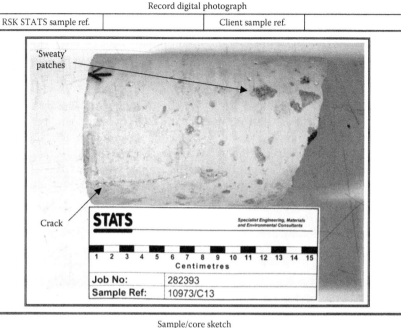

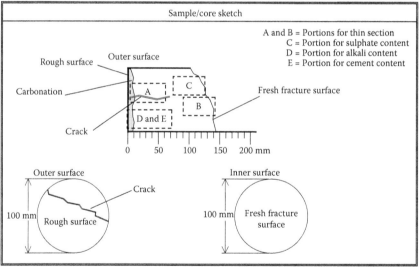

Figure 3.4 **Example of a concrete core with a sketch diagram indicating how it is to be cut for detailed laboratory investigation. (Courtesy of P. Bennett-Hughes, RSK Environment Ltd.)**

selective judgement based on skill and experience. Where deterioration is being investigated, samples must be critically sited, based on careful observation of cracking, exfoliation, efflorescence, exudation, staining and other changes visible on the concrete surfaces. ASTM C856-04 (2004) specifies that for a detailed petrographic examination, 'the minimum size sample should amount to at least one core for each mixture or condition or category of concrete'.

In practice this translates into the following type of situation. As an example, when investigating suspected alkali–aggregate reaction, it is usual to define three sample areas consisting of most severely cracked, averagely cracked and uncracked concrete from which duplicate cores should be extracted. The exact location of each core in the three designated areas is solely up to the judgement of the engineer or petrographer. If this judgement is faulty, the results of the petrographic examination will be unreliable, and this is the reason why the petrographer should be present when sample sites are selected. It is not unusual to find when investigating alkali–aggregate reaction that the severity of external cracking does not correlate well with the severity of the reaction internally.

It should be apparent that taking one core sample from each designated condition of concrete is not a satisfactory method of sampling. The possibility of error is large and the use of a single core to characterise unaffected concrete is not usually sufficient. A more desirable sampling regime would be to drill at least two critically sited cores from each designated areas of cracking and the number of cores drilled from the unaffected concrete to be controlled by the specific attributes of the concrete structure to be investigated as designated by ASTM C823/C823M-07 (2007), or an appropriate national standard.

Under some circumstances, as already noted, fragments such as spalled or broken pieces of parts of the structure may be the only samples available. It is possible to provide limited information from broken samples, but samples of this type should be avoided because mechanical breakage may mask or complicate in situ deterioration processes. In addition, samples of this type usually represent only external portions of the concrete, and if they have lain around, their structure will have been subject to additional surface weathering and leaching. Inspecting engineers need to accept that the reliability of results from petrographic examination is subject to the same sampling methodology as applies, for instance, to the measurement of compressive strength. It is regrettable that in too many cases, the petrographer is expected to produce meaningful results based on examination of a limited number of undersized samples without any consultation over the selection or location of the sample sites.

3.3.5 Practical sampling methods for solids

ASTM C42/C42M-04 (2004) gives standard procedures for cutting concrete cores and beams for physical testing. Water-cooled diamond-bit coring is the usually recommended method of sampling concrete for petrographic examination. Where smaller-diameter cores are drilled, that is, 75 mm or lesser diameter, the core bits used should be designated by internal diameter as the loss of core diameter on a nominal 50 mm diameter coring bit is considerable. It should also be remembered that damage and solution will be most severe near the cut surfaces. A recent development in coring, which is suitable for unreinforced concrete, makes use of a 5–6 atmosphere/pressure of air flush to cool the bit with a water mist spray which keeps the swarf as a paste and prevents clogging and glazing of the cutting edge.

Concrete cores can also be dry cut using either diamond or tungsten carbide bits. Dry cut diamond bits as large as 150 mm are available, but drilling depths are limited to about 350 mm. The diameter of tungsten carbide bits is restricted to about 50 mm and depths to about 100 mm. These coring bits are useful when no water is available or leaching of core surfaces by cooling water is unacceptable such as in the determination of chloride levels. However, they all suffer from the disadvantage that there is no cooling and considerable heat can be generated during drilling.

A coring bit larger than 75 mm in diameter cannot be used in a handheld drill. The amount of torque necessary to drive the 75 mm and larger coring bits requires a motor fixed in a frame that is rigidly clamped to the structure. This ensures a clean core with minimum damage due to chattering of the bit. This fixed drill procedure minimises damage and provides clean, larger-diameter cores suitable for petrographic examination to be obtained.

For concrete of restricted dimension, such as pavements and slabs or precast products, wet diamond sawing of samples may be preferable to coring as it is less likely to cause damage. The use of dry silicon carbide saw blades is not recommended except for subsampling. There are cases where wet cut core or prism samples also need to be analysed for alkalis by obtaining a dry portion from the centre of the sample where the concrete will not have been leached. This subsampling can be carried out by using a dry silicon carbide blade mounted on a masonry saw.

Where powder samples are required for analysis of chlorides, alkalis or sulphates or where leaching is suspected, dry drilling is essential. It is necessary to use a tungsten carbide–tipped masonry drill to collect dry powder from successive depths as the drill penetrates the concrete. However, since there is little control over whether the drill encounters large aggregate particles, a large number of samples from different locations are required. The technique is described by Roberts (1986).

When examining concrete where the presence of a damaged, possibly friable or heavily cracked outer surface is important to the investigation, the surface itself must be examined using a portable field microscope, sampled and photographed in situ. Unless it can be sawn directly from the structure without damage, it may be necessary to impregnate the surface with a low-viscosity epoxy resin which may then be reinforced using fibre glass matting and further coats of resin. Once the resin has cured, the concrete can be cored through the protective layers in the normal way retaining the damaged surface. Specialist concrete repair companies have the necessary expertise for applying epoxy coatings to difficult areas such as inverts.

In the special case of taking core samples from in situ concrete pipes such as sewers, coring done from the outside should be stopped 3–5 mm short of the inner surface. An appropriate handle should then be epoxy cemented to the core sample which should then be broken free from the pipe using a chisel or by drilling successive holes round the cut using a handheld drill with a small-diameter tungsten carbide bit.

Once a core is broken out of the structure, it should be labelled and its orientation marked. After inspection, it should be wrapped immediately in several layers of plastic film, sealed into a polythene bag to preserve its 'on-site' moisture condition and the details of the core recorded. A suitable record sheet is illustrated in Figure 3.5 which should accompany the core to the receiving laboratory.

3.3.6 Preparation of subsample specimens for investigation

The coring and cutting processes open up large surface areas of fresh concrete which, if they remained wrapped for any length of time, may show an increase in reaction products and exudations on surfaces and may allow local leaching of the concrete. Thus, the cores should be transported to the receiving laboratory, and the staff should then unwrap, inspect, photograph and 'log in' the cores without delay. It is important that the petrographer is present when the cores are unwrapped so that an initial petrographic assessment can be made and the appropriate positions of planes for cutting can be marked onto the sample prior to the preparation of the ground surfaces and thin sections (see Figure 3.4).

The initial petrographic examination should be made visually and by using a low-power stereomicroscope, preferably with a 'zoom' magnification facility, and all features of particular interest photographed and recorded.

A recent technique for recording the macroscopic surface features of a cylindrical core is to roll the core on an electronic scanning device and synchronising this motion to a computer output display to provide an image of the whole surface of the core. Depending on the quality of the core surface, excellent resolution of surface features can be obtained by this technique (see also Section 4.5.2 and Figures 4.74 and 4.75).

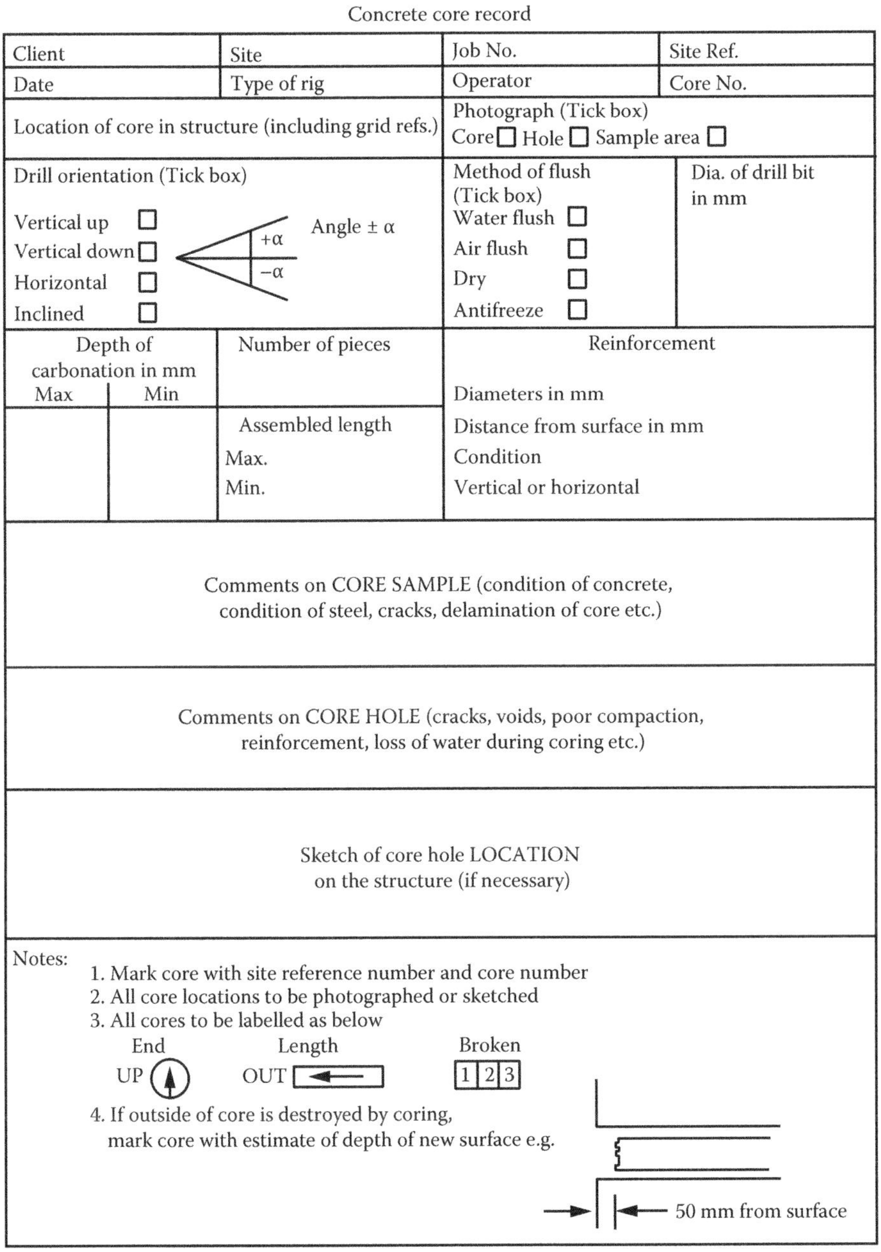

Concrete core record

Client	Site	Job No.	Site Ref.
Date	Type of rig	Operator	Core No.
Location of core in structure (including grid refs.)		Photograph (Tick box) Core ☐ Hole ☐ Sample area ☐	

Drill orientation (Tick box) Vertical up ☐ Vertical down ☐ Horizontal ☐ Inclined ☐ Angle ± α +α −α	Method of flush (Tick box) Water flush ☐ Air flush ☐ Dry ☐ Antifreeze ☐	Dia. of drill bit in mm

Depth of carbonation in mm Max Min	Number of pieces Assembled length Max. Min.	Reinforcement Diameters in mm Distance from surface in mm Condition Vertical or horizontal

Comments on CORE SAMPLE (condition of concrete,
condition of steel, cracks, delamination of core etc.)

Comments on CORE HOLE (cracks, voids, poor compaction,
reinforcement, loss of water during coring etc.)

Sketch of core hole LOCATION
on the structure (if necessary)

Notes:
1. Mark core with site reference number and core number
2. All core locations to be photographed or sketched
3. All cores to be labelled as below

     End   Length   Broken

    UP (↑)   OUT [←]   [1][2][3]

4. If outside of core is destroyed by coring,
   mark core with estimate of depth of new surface e.g.

→ |←— 50 mm from surface

Figure 3.5 **A typical concrete core sample record sheet for use on-site.**

Features such as exudations and hygroscopic patches on surfaces may be indicative of alkali–aggregate reaction. These patches are most clearly apparent as a core dries out from its initial moist condition and should be photographed and sampled at this time. Void or crack infillings and other surface deposits of interest are also best sampled during this initial examination. Particular attention should be paid both to the external cut surface and the fracture surface where the core has been broken out from the drill hole. These surfaces may provide useful information relating to aggregate–cement bonding, microcracking and other detailed textural features, so a comparison of cut and fractured surfaces should not be overlooked during the examination. Once the initial inspection and any wet cutting has been completed, the cores for petrographic analysis should be dried at 40°C. This process may require several days for larger samples, but may be assisted by enclosure in a chamber containing a drying agent such as silica gel.

The subsamples produced for petrographic examination of a drilled core or saw cut sample may consist of plane surfaces, slabs and thin plates cut for the preparation of ground or polished surfaces and thin sections. The location and size of these surfaces as subsamples are no less important than the original bulk sampling of the structure. The size of plane surfaces required for the determination of air content in hardened concrete is specified in ASTM C 457-10 (2010) and summarised in Table 3.3. This table indicates that as the aggregate size increases and the paste volume decreases, the effect is compensated by the requirement for an increased area of examination. It is to be noted that this table is based on the concrete being reasonably homogeneous and well compacted. It is recommended that if the concrete is heterogeneous in distribution of aggregate, or for large air voids, the measurements should be made on a proportionately larger area of ground surface.

These recommendations for minimum surface area are often used as a guide for petrographic examination in general, so it is important that their basis is understood.

In a typical concrete containing an air content of say 5% and assuming a uniform distribution of air voids of 50 μm in diameter, a 150 cm² sample plane will cut about 300,000 voids. This sample population is large enough to provide a realistic estimate of the percentage of air voids in the structure providing the concrete is reasonably homogeneous. Obviously the whole population is not counted, but provided the statistical requirements of modal analysis are met, reliable estimates of the air content present in the sample can be made. In this respect ASTM C 457-10 (2010) includes an excellent discussion of the types of errors involved in the measurements.

If the same sample plane is now used to examine the coarse aggregate, which might be assumed to comprise say 50% of the concrete volume, to be spherical and to average 12.5 mm in diameter, the 150 cm² sample plane will cut a population of less than 50 aggregate particles which is clearly inadequate as a representative sample of the coarse aggregate in the structure. Even for 5 mm aggregates, the population sampled is only about 300. While the argument used is simplistic in some of its assumptions, it clearly indicates a fundamental problem in the petrographic examination of concrete where examination is typically carried out on constituents ranging in size from 40 mm to 5 μm. These constituents may be aggregate, a crack system or any other quantifiable feature of the concrete. As the size of the constituent of interest increases, the extent of the sample surfaces required for adequate characterisation increases dramatically.

Table 3.3 Minimum area of finished surface for microscopical determination of air content as specified by ASTM C 457

Nominal maximum size of aggregate in the concrete (mm)	Minimum area of surface for measurement (cm²)
150	1613
75	419
37.5	155
25	77
19	71
12.5	65
9.5	58
4.8	45

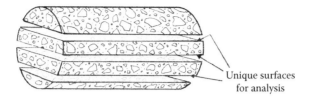

Unique surfaces
for analysis

Figure 3.6 Two diametral slices cut from a concrete core can only provide three unique surfaces for analysis and two will be between 5% and 10% narrower than the maximum core diameter.

The standard also specifies that the distance between section planes shall exceed the largest diameter of aggregate present to ensure that each plane intersects new material. Since adjacent cut surfaces will be simple 'reflections' of each other, the unique area available for analysis can only be doubled by taking a single thick diametral slice and trebled with two slices.

This requirement is illustrated in Figure 3.6. For example, a 300 mm long, 100 mm diameter core is cut diametrically, then the surface area for one cut is 30,000 mm², and for three surfaces as shown, it is perhaps 80,000 mm² while the total area of the cylinder surface would be approximately 94,260 mm², and each cross section cut perpendicular to the length of the core would be 7,855 mm².

As can be seen from the previous discussion, the sizes of the sample surfaces recommended by ASTM C 457-10 (2010) are only applicable to those cases where the constituent being measured is less than about 1 mm. Where the constituents to be quantified have average sizes much greater than 1 mm, the area of surface required for examination needs to be increased significantly if a precise, quantitative and representative modal analysis is to be attempted. In many cases the requirement for a representative modal analysis implies that as much of the sample as is practical must be prepared as plates with finely ground surfaces or as thin sections and attention must be given to the precision of the measurements made.

While a single cut subsampling provides an adequate area for detecting the presence of features associated with the fine aggregate, this is not satisfactory if interest is restricted to features associated with coarse aggregate particles, or crack patterns. A petrographic examination may detect the presence of alkali–aggregate reaction qualitatively, other observations on the reaction will be limited and may be biased. Where the concrete needs to be fully characterised, extra slices and sections should be prepared so that an adequate area can be examined.

Three approaches to providing sufficient sample area are commonly used in petrographic laboratories which are largely dictated by the facilities available (see Sections 3.4 and 3.5). Ideally perhaps the most satisfactory method is to prepare large-area thin sections with edge dimensions of 100 mm or larger. This allows full and uninterrupted study of any variations or segregation from the surface to the interior along the diametral axis of the core to at least a depth of 150 mm without edge losses.

However, there are several disadvantages associated with large-area thin sections. They are usually too large for the microscope stage and may foul the microscope stand when rotating the stage. They are too large for standard mechanical and point counting stages. They also require thicker glass mounting plates requiring the use of a longer focus condenser cap on the substage assembly of the microscope. Also, the time taken to manufacture thin sections increases very rapidly with size and the thin-section margins become increasingly difficult to retain at correct thicknesses.

Consequently, a second approach, commonly adopted and often preferred, is to produce several smaller thin sections of up to 75 × 50 mm which can be selected at the initial cutting stage from relevant areas on the core or can be taken immediately adjacent to each other if changes over a large area are to be followed. Care and consideration must be given as to the selection of plates for thin-section production in such cases, both because of edge losses due to the cutting and to the need to produce sufficient area for a viable modal analysis.

The third approach is applicable where the objective of the petrographic study is more limited or specialised, for example, when only aggregate identification is required or steel corrosion products adjacent to reinforcement are the subject of the investigation, smaller thin sections may be appropriate, although the use of the 25 × 30 mm sized thin section traditionally used by geologists is not acceptable for the general petrographic examination of concrete. However, there are specialised situations that require selected polished thin sections for examination by transmitted light supplemented by an SEM study. In such studies small thin sections will be appropriate in part because of the difficulties of producing a highly polished and flat surface larger than 30 mm diameter.

A general approach, favoured by many laboratories in North America and Europe, is to prepare finely ground surfaces on plates or slices with sides of 100 mm or greater. These are examined by incident light as described in ASTM C 856-04 (2004) and the maximum information gathered from them using a low-power stereomicroscope.

An alternative to the stereomicroscope is to make use of a flat-bed digital computer scanner to produce detailed computer-generated images of cut and ground concrete surfaces. This simple technique can be further adapted for recording the macroscopic surface features of a cylindrical core. The procedure is to roll the core on the electronic scanning device and synchronising this motion to a computer output display to provide an image of the whole surface of the core. Depending on the quality of the core surface, excellent resolution of surface features can be obtained by this technique (see also Section 4.5.3.2 and Figure 4.73).

Areas of particular interest can be located by some of these methods, and the matching opposing surfaces are then used to prepare thin sections and specimens for optical, chemical, x-ray diffraction and other examinations. No indication is given in ASTM C 856 about the dimensions of the thin sections to be used, but in practice these were usually 25 × 25 mm in size. This approach has the advantage that the ground surfaces are relatively easy and cheap to produce and the costs of thin sectioning are restricted. The difficulty is that only a limited amount of information can be gained from examination of ground surfaces and is also very dependent on the quality of both the ground surface and the stereomicroscope, or the digital scanner. Also, the small sections cut from the matching surfaces are likely to present many problems of interpretation because of their small size.

There is general agreement that where possible the orientation of the sample plane through the core should be controlled by any zonation present in the concrete. The outer zones of concrete are variable because of inhomogeneities arising during the initial mixing, placement and subsequent weathering; thus, the main axis of sectioning should always include the outside surface. The other common inhomogeneity in concrete is due to plastic settlement causing voids beneath aggregates which are best revealed by sections cut in a vertical plane. Where the core is drilled vertically, the orientation of the diametral axis is not important (unless dictated by a surface feature), as all diametral planes will be vertical. For horizontal cores whose orientation should always be marked, the vertical diametral plane should be chosen unless surface features, such as the orientation of a crack, dictate otherwise. All edges, especially the external surface, must be fully retained on the thin section so that surface features such as carbonation and any gradation from the surface are available for study. With modern methods of thin-sectioning given in Section 3.4, this is readily achievable.

A fresh fracture surface should always be examined and can be produced by simply breaking open a portion of the core. When the examination of the thin sections and other petrographic studies has been carried out, the completed diagnosis should always be confirmed by examination of new fresh fracture surfaces. Such examinations may provide valuable information, for example, crack or void infillings are readily observed on a new freshly exposed surface, but with time they dry out and become almost invisible.

There are a wide range of products such as pipe, sheet, tiles and cement renderings as well as laboratory specimens which may require sampling for petrographic examination. Some of the dimensions of these products may only be a few millimetres. However, the same sampling principles discussed earlier are applicable although they will be modified by the nature of the product. Cross sections of thin sheet products can be cut and glued together in a stack to produce an acceptable area for sectioning. Similarly mortar bars and small laboratory specimens can be cut and cemented together so that several cross sections can be included in a single thin section for examination. Surface alteration is often important in thin-walled products so that as much of these surfaces should be included in the sample as possible.

In other situations, for example, in forensic studies, the petrographer may only have fragmentary samples to work with. In such circumstances, the combination of ground surface and thin-section examination will usually provide valuable information not obtainable by other means, but this should then be augmented as appropriate by data obtained by other techniques such as those referred to in Sections 2.5.1 through 2.5.5.

3.4 PREPARATION OF THIN SECTIONS AND FINELY GROUND OR POLISHED SURFACES

There is a clear distinction between finely ground and polished surfaces which is important. Finely ground surfaces are produced when the smallest grinding medium used is limited to about 5 μm, whereas polished surfaces require that the final grinding media, usually diamond, is 1 μm or less in size. It is relatively easy to produce finely ground surfaces of up to 100 × 100 mm or larger in size, but it is difficult and extremely time-consuming to polish areas greater than 30 mm in diameter. With ceramics and metals, the size of the diamond grinding medium may go down to 0.25 μm, but because of the variable hardness of the constituents in mortars and concretes, it is found that 1 μm gives a satisfactory finish without undue 'relief' and this is not further improved by using finer grinding agents.

SILICON CARBIDE GRIT PARTICLE SIZES AFTER CARBORUNDUM CO. LTD. 1972

Designation	Obsolete Designation	Average Grit Size[a] (μm)	Minimum (μm)	Maximum (μm)
F230	240	53	34	82
F320	320	29	16.5	49
F600	600	9	3	19
F800	700	7	2	14
F1000	800	5	1	10

[a] The maxima and minima are extreme limits; 90% of particles are near average values.

When finely ground and polished surfaces are compared under the petrographic microscope at 200× magnifications, large differences in the surface roughness can be observed. The finely ground surface is dark and rough due to light scattering with blurring, smearing and poor differentiation of constituents. On the other hand, a well-polished surface is bright and crisp, almost like a mirror, with sharp edges and good differentiation between components. This indicates that observation of a finely ground surface is best restricted to a good-quality low-power stereomicroscopy using magnifications not exceeding 100×. Where surfaces are required for examination by incident light using higher magnifications, such as for etched cement clinker or for SEM and x-ray fluorescence analysis, a high-quality polished specimen surface is essential.

Finely ground surfaces are specified for the determination of air content in hardened concrete according to ASTM C457-10 (2010). It should be noted that the level of detail on ground surfaces that is observed is critically dependent on the quality of the optical system of the microscope used. Ground surfaces are also useful for determining broad textural detail of aggregates, gross inhomogeneity and crack systems and for staining alkali–silica gel. In some respects, it is easier to observe these broad details on a ground surface than in thin section, and as such, they can be an aid to the petrographic examination of thin sections. Where facilities for preparing the large-sized thin sections are not available, the petrographic examination of ground surfaces forms a valuable precursor to the selection of the locations for and manufacture of appropriate smaller-sized thin sections.

3.4.1 Laboratory methods of cutting concrete

The petrographic sectioning of concrete initially requires the cutting of large slabs and surfaces from samples (Figure 3.7). Concrete can only be cut satisfactorily with diamond impregnated saws. Silicon carbide abrasive blades produce poor-quality, uneven cuts and

Figure 3.7 The initial cutting of a large concrete sample. Note the slotted diamond-rimmed wheel. (Photograph courtesy of M. Eden.)

tend to damage surfaces. Their use should be restricted to the dry removal of portions of sample which will be later crushed and ground for chemical analysis.

A review of diamond cutting equipment and a discussion of its application is given by Allman and Lawrence (1972), and more recent information may be obtained from the technical literature available from diamond tool suppliers and firms specialising in petrographic and metallographic supplies. In general terms, diamond-rimmed wheels are available in a range of sizes depending on application. The diamonds are embedded in the steel rims of larger wheels which are usually slotted (to allow removal of slurry), and in laboratory equipment, wheel sizes range up to 1200 mm. Smaller wheels typically have the diamonds embedded in a bronze rim and range from very small bench top models with effective cutting depths (allowing for the axial bearing flange) of 50–100 mm to larger wheels with cutting depths of up to 500 mm (see Figure 3.8).

Wheel rotation speeds depend on diameter, because the critical factor is the peripheral speed of the cutting rim. This can range from 20 m/s for very hard materials such as ceramics or quartz up to 70 m/s for soft materials such as chalk (Allman and Lawrence 1972). Lubrication of the cutting edge is important and a range of lubricating fluids are available and many are water based. Water is normally used as the basis of cutting fluids for diamond sawing and will contain either a rust inhibitor or an oil/water emulsion.

The cutting wheel is either lubricated directly by a drip feed system or dips into a bath of lubricant as it rotates. Initial cutting of concrete into slabs or plates is almost always undertaken using a water-based lubricant. However, an oil, glycerol or paraffin oil (kerosene) lubricant is more appropriate for the final cutting and finishing of subsamples prior to thin section or polished surface preparation, since these avoid problems of dissolution of soluble components, but adequate vapour extraction systems must be installed. It should be noted that some oil/water–based cutting fluids are demulsified by concrete swarf unless the blade dips into a bath of cutting fluid. It is also possible to fill the bath with a straight oil cut back with a kerosene. This is preferable for the larger saws where the rate of cutting is slow as it minimises leaching from

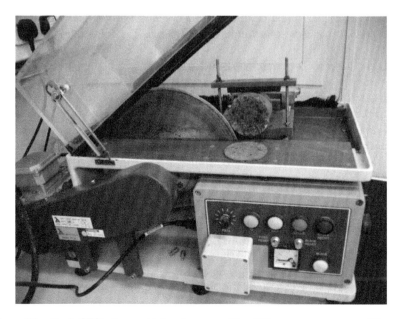

Figure 3.8 A small Logitech GTS1 diamond trimming saw with a 100 mm concrete core. (Photograph courtesy of J. McAneny, Logitech Ltd., U.K.)

the sample. With dense concretes the extent of water leaching rarely extends beyond 1 mm, but in more porous concretes this may become a problem especially for chemical analyses.

Cutting fluids may be flammable and carcinogenic, particularly in the form of vapour or aerosol. Consequently it is essential that appropriate fire precautions and extraction systems are installed which comply with national safety regulations.

It is the petrographer's responsibility to discuss with the technician how the sample is to be handled. Where the concrete is particularly friable, highly porous or broken up, it will first require consolidation by gluing, impregnation or casting with an epoxy or poly-ester resin, and there should be agreement at what stage and how this is to be carried out. It should be noted that the polymerisation of resin is an exothermic process, and if large amounts of resin are involved, cooling will be necessary. Similarly, it is necessary for the petrographer to indicate how the core should be cut especially where a range of tests are to be carried out. It is desirable to slice a core along the complete length of its diametral axis, but it may be found that the cut is limited by the diameter of the saw blade available. Where cores exceed this length, the petrographer will be required to indicate where transverse cuts are to be made and the extent to which continuity of cutting planes is to be maintained. In practice, the slabs of concrete cut for later thin sectioning should not be less than 10 mm and preferably be about 15 mm thick to ensure sufficient mechanical strength, prevent bowing and allow subsequent handling and processing. If the cut makes use of a water-based lubri-cant, the thicker slices will allow the water-leached and mechanically damaged surfaces to be ground away during the later processing.

Provided that the petrographer clearly indicates the requirements, there are few problems for the skilled technical staff in a well-equipped preparation laboratory in cutting and pre-paring a wide range of concrete samples.

3.4.2 Laboratory preparation of ground surfaces

The cut concrete slice which is to be ground must be thoroughly dried in a thermostated drying cabinet at a low temperature preferably at 40°C, because a number of minerals such as gypsum will alter or become unstable at higher temperatures. Certainly drying tempera-tures for mortars and concrete should not exceed 60°C. The minimum drying time is over-night and for larger specimens longer times should be allowed.

There are several methods of preparing and grinding cut specimen surfaces depending on the end use of the specimen. In the first method the sample is ground in retaining rings by successively decreasing sizes of loose abrasive powders on a rotating, cast-iron lap (Figures 3.9 and 3.10) until the desired quality of finish is achieved. The only consolidation of the sample carried out is surface impregnation with wax to maintain the sharpness of the margins of air voids. This method is described in ASTM C 457-09 (2009). The method recommends that the abrasive silicon carbide is used ranging in five steps from No. 100 grit (122 μm) to No. 800 grit (12 μm). However, continuing the grinding on a lap beyond the specified 12 μm grit to a 5 μm abrasive will give a smoother final finish allowing a considerable improvement in the sharpness of detail observable.

A second method is usually employed when the sample is to be subsequently polished, thin-sectioned or required for delineation of crack systems or determining porosity. In this case the sample is impregnated with a low-viscosity epoxy resin to consolidate it prior to grinding (see Section 3.4.4). Some resin types that are in common use and the solvents used for modifying their solubility are given in Table 3.4. Soluble dye can be incorporated in the epoxy resin which makes original cracks, voids and porosity clearly visible and differenti-ates them from any damage which may occur during subsequent preparation. The vacuum impregnation procedures required are described in Section 3.4.3.

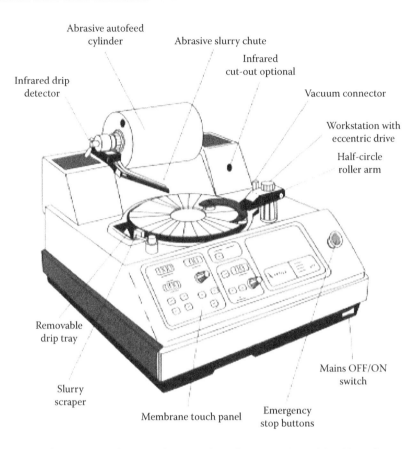

Abrasive autofeed cylinder

Abrasive slurry chute

Infrared cut-out optional

Infrared drip detector

Vacuum connector

Workstation with eccentric drive

Half-circle roller arm

Removable drip tray

Slurry scraper

Membrane touch panel

Emergency stop buttons

Mains OFF/ON switch

Figure 3.9 Diagram of the essential features of modern large-format mechanical lap. Note slow rotation of the lap prevents the specimen 'aquaplaning' on the grinding medium. (Courtesy of Logitech Ltd., U.K.)

Figure 3.10 A concrete core being lapped to provide a ground surface on a typical lapping machine. (Photograph courtesy of M. Eden.)

Table 3.4 Some characteristics of proprietary resins in use in petrography laboratories for basic resin family, epoxy co-polymers, polyester co-polymers and urethane/acrylic

Basic resin family	Epoxy co-polymers	Polyester co-polymers	Urethane/acrylic
Basic hardener material/method	Bisphenol-A, or polyamide monomer	Organic peroxide, e.g. benzoyl peroxide	UV radiation ~365 nm
Solvents for resin	Alcohol/acetone/amyl acetate	Alcohol/acetone/hexane/amyl acetate/toluene/acetone/ dichloromethane	Alcohol/methylene chloride
Curing time (25°C)	12–24 h	30 min to 2 h	1–2 min
Cured colour	Clear transparent/yellow	Clear transparent/yellow	Transparent/amber
Viscosity range (25°C)	Low to high[a]	Very low to medium	Low to medium
Refractive index	1.54–1.66	~1.5–1.6	1.5–1.66
Application	Impregnation/adhesive	Impregnation/encapsulation	Mounting adhesive

[a] Depends on the formulation; low viscosity for impregnation, higher for adhesives.

A variety of grinding machines are manufactured. Most rely on the basic principle of a slowly rotating lap as shown in Figures 3.9 and 3.10, although there are also a range of equipment which uses a diamond impregnated cup wheel to traverse the specimen fixed on a flat bed.

A recent improvement makes use of diamond impregnated plastic mats which will allow a cut slab to be ground to a 6 μm finish in a sequence of steps, and if further polishing is required, a plastic mat fed with a water/diamond slurry will then allow the slab to be polished to a 1 μm finish or better. One such lapping system is manufactured by Struers Ltd., and although costly, it has the advantage of avoiding contamination of the slab surface with silicon carbide or carborundum.

The main features of a typical mechanical polishing lapping machine are illustrated in Figure 3.11 and the jigs for lapping slabs and modern large-format thin sections in Figures 3.12 and 3.13. The details of grinding and lapping methods have been well described in the literature

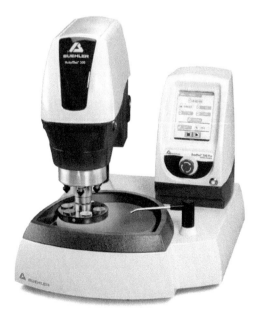

Figure 3.11 A modern polishing lap with a six-position cylindrical 30 mm diameter specimen holder. (Photograph courtesy of R. Wagner, Buehler Ltd., Coventry, U.K.)

Figure 3.12 The Buehler 6 sample polishing mount modified to hold up to a 40 × 60 mm thin section. (Courtesy of R. Wagner, Buehler Ltd., Coventry, U.K.)

Figure 3.13 Large format jigs for grinding slices or thin sections. Note the large thin section resting against the jig. (Courtesy of Logitech Ltd., U.K.)

of Allman and Lawrence (1972) and Hutchinson (1974). For cementitious materials all lapping and grinding must be carried out with non-aqueous fluids. An important requirement is that the grinding surfaces of the laps remain flat. As the laps wear, they will need to be resurfaced and checked for flatness periodically. In most modern equipment, the surfaces are checked for flatness electronically and automatically corrected as necessary. Manual checking relies on the skill of an experienced technician.

Whatever grinding method is used, it is essential that no surface relief is induced in the specimen due to differences in hardness between aggregate particles and the cement paste. The flatness of the surface can be checked by viewing it at a low angle of incidence in the light from a fluorescent tube, or alternatively by viewing with a low-power stereomicroscope. The clarity of the petrographic observations will depend on the quality of this finished surface and of course the optical system of the microscope used to examine it.

3.4.3 Laboratory preparation of polished surfaces

Polishing laps are similar in operation to laps used for producing ground surfaces. A series of diamond impregnated mats of increasing fineness are used, but very careful cleaning of both the equipment and the specimen being polished is essential at each stage and the equipment should be kept in a clean environment. Keeping the polished surface flat is difficult if a high-grade polished surface is required. A typical modern polishing lap is shown in Figure 3.10. The area of surface which can be successfully and economically polished is limited to a maximum diameter of about 30 mm. Larger polished surfaces can only be justified where detailed studies of concrete are required using analysis by transmitted light to obtain textural relationships backed up by electron microanalysis. As the surface area to be polished increases, the time required for polishing increases exponentially. It becomes increasingly difficult to maintain a flat, scratch-free surface without relief or smearing occurring. In the polishing of architectural stone, the final polish achieved depends on smearing of minerals induced by localised heat during polishing. Large areas can be polished in this way, but this type of polished surface is useless for petrographic purposes, and this distinction needs to be clearly understood.

Polished surfaces are required where the surface will be examined with a high-power reflected light microscope or where microanalytical techniques using energy or wave-dispersive x-ray methods are to be used. Specimens whether solid or particulates consist of polished mounts or polished thin sections. In a polished mount the sample is embedded in a cylindrical block of resin of which one face is polished. Optical examination of a polished block of this kind using reflected light can provide a wealth of information (see Figure 3.14) which may be further augmented by etch procedures (Section 3.5.6) and from a SEM study of the same specimen.

A polished thin section is merely a thin section of which the top face has been polished and the mounting of the protective cover glass omitted. When examined by transmitted light, a removable acrylic-based 'spray-on' cover slip, or a temporary cover glass, is attached with an immersion oil. (*Note:* A flat uniform 'spray-on' cover is difficult to achieve.) These can later be removed with a solvent before coating for SEM examination.

3.4.4 Resins for impregnation and mounting concrete specimens

Three generic types of resin are used for mounting and impregnation of specimens for the preparation of cut and ground slices, thin-section manufacture and preparation of polished specimens for reflected light or SEM microanalysis. They are based on acrylic, epoxy or polyester resins and are two-part systems comprising of a resin formulation which is often a complex mixture and a hardener system which causes the polymerisation solidification. The polymerisation process is an exothermic reaction producing sufficient heat to damage the specimen unless appropriate care is taken.

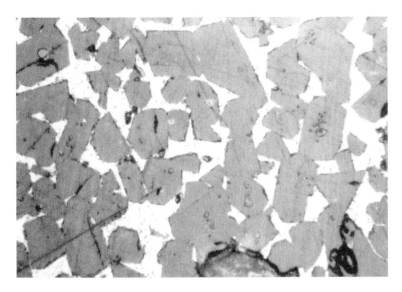

Figure 3.14 An example of a polished and HF-etched surface of anhydrous OPC clinker in reflected light. Brown pseudohexagonal C_3S crystals with interstitial C_3A and dendritic ferrite. Width of field 180 μm. Compare with etched surface of OPC clinker, Figures 3.19 and 3.20. (Courtesy of M. Grove.)

Preparing concrete as a specimen places several special constraints on the methods and materials used for consolidation/impregnation and mounting specimens as indicated in the following:

- The sample must be dry and unleached by water as some components are soluble.
- The sample should not be heated above 40°C in order to avoid the possible decomposition of components.
- Grinding/polishing needs to be carried out without smearing of the components and uneven relief due to the differential hardnesses of the components or of the impregnating material itself.
- The specimen on completion needs to be mechanically robust and chemically resistant and have long-term stability.
- The impregnating resin should be capable of accepting a coloured dye.
- The impregnating material should cure and harden within an acceptable time period.
- There should be no undue shrinkage/expansion during curing and hardening.
- The resin should have low viscosity to allow adequate impregnation of porous materials.

Epoxy resins in various formulations (sometimes containing acrylic) are by far the most widely used of the resins for concrete sample preparation both for impregnation and as a mounting adhesive. The viscosity of these resins at ambient temperatures can be modified by the detailed formulation, but organic solvents such as amyl acetate, alcohol or acetone are sometimes used, while the curing rates can be adjusted by the type and amount of hardener used and temperature. Like other resin types, polymerisation is an exothermic process and the adjustment of proportion of hardener to resin used is critical. In many cases there is also a need to cool the sample in water during curing. If too much hardener is used, exothermic heating can be sufficient to cause the resin to boil, or even catch fire. Provided exothermic heating is controlled and the viscosity is appropriate, epoxy resin may be used both for impregnation of specimens at low viscosity and as a mounting adhesive at higher viscosity.

Curing at low temperature takes at least 24 h, but a further period in an oven at 30°C is necessary to harden the resin completely so that the specimen may be cut, ground or polished.

Polyester resins are commonly used for impregnating and stabilising friable or porous materials and for embedding or encapsulating specimens. They have the advantage that they take minutes rather than hours to cure, they can be modified by organic solvents, and the mix proportion of resin to hardener is not as critical as with epoxy. A major disadvantage is that they remain sufficiently soft and plastic after curing for particles of grinding medium to become embedded in the ground surface.

A complex urethane polymethylmethacrylate (PMMA) formulation that is cured in seconds at room temperature on exposure to UV radiation with a spectral energy distribution around 365 nm is an alternative to epoxy adhesive and is very widely used for mounting thin sections. The use of this one-part resin bonding system has proved invaluable in speeding up the manufacture of optical thin sections and mounting specimens for SEM and reflected light studies. The resin must of course be accessible to the UV radiation so that glass or other transparent mounts are essential and 'over-curing' may lead to local crystallisation.

Some of the characteristics and applications of the resins relevant to the impregnation of porous or cracked samples and to the preparation of thin sections are given in Table 3.4. Resins for thin-section preparation should ideally have refractive indices close to the soda-glass slides (~1.54) in common use for thin sections. Proprietary formulations of both resin and hardener are often complex mixtures of organic components designed to provide appropriate refractive index, viscosity and curing properties for the different applications.

The majority of solid specimens will require resin impregnation prior to polishing, while particulate samples will first require casting into a solid resin block before grinding and polishing. A well-equipped laboratory will have several types of resins available for use depending on the particular application and specimen.

Reusable polyethylene (polythene) moulds such as those used for casting metallurgical specimens are suitable for casting resin blocks. If these are not available, polyethylene beakers may be substituted. For the larger specimens, where finishing will be restricted to fine grinding on a metal lap, reusable polyethylene moulds are unlikely to be available and moulds can be constructed from cardboard, paper cups, polyethylene food containers, sheet metal or any other suitable disposable material.

The way in which a particulate subsample is embedded in resin is dictated by the amount of material available, its porosity and whether it contains fine material that can separate out. Firstly, all materials must be thoroughly dried, generally at 105°C or lower depending on the material. Hydrated cementitious materials should be restricted to 60°C or less. Relatively coarse powders, sands and aggregates may be mixed with either polyester or epoxy resin (see Table 3.4) including as much material as possible so that the maximum number of particles will be intersected by the plane of grinding and polishing. The resin used should then be mixed with its hardener according to the manufacturer's instructions and poured into the mould leaving at least 5 mm free space for frothing. Note that if large amounts of resin are used, the specimen may have to be cooled during initial hardening to prevent overheating and flash hardening from the heat of polymerisation. The mould containing the mixture is placed in a vacuum desiccator and a vacuum of at least 10 mm Hg is applied. The surface of the resin should be allowed to froth and de-gas for a number of minutes and the vacuum released to force the resin into the pores of the particles. This procedure is repeated several times until the air has been removed from the sample. Care must be exercised to avoid too high a vacuum which may cause the resin to boil.

The mould can then be removed from the vacuum desiccator and placed in a warm environment to allow the resin to harden. In the case of polyester resin, it can, if required, be cured in about 30 min by placing the sample in an oven at 60°C. Epoxy resins should be left overnight to harden in a warm room and then fully cured by heating to 30°C–60°C for 1–2 h. This additional curing is important to harden the resin so that it takes a good polish. If vacuum moulding has been performed correctly, no air bubbles will be visible in the cast and the more accessible pores in the particles will be filled. In general, polyester resin is more suitable for the smaller specimens as it is easier to handle and impregnates better because of its lower viscosity. Epoxy resin should always be used for larger specimens as it is stronger and harder.

An alternative method in common use and essential where porous particles are being cast involves the following modification. The dry particles are packed into the mould which is then transferred to the vacuum desiccator. A vacuum is applied and the sample left to evacuate for a few minutes. In this case the desiccator is fitted with a device to allow the liquid resin to be injected into the mould under vacuum so that the resin is not displacing air and is better able to fill the voids. After a few minutes the vacuum is released and the previously described procedure of applying and releasing the vacuum is followed until the resin is degassed.

If very fine material is present in the sample that might possibly separate out, it may be preferable to mix it with the resin to form a slurry. After casting, the block is then cut transversely prior to polishing so that any settling can be examined directly. Alternatively, if the amount of material available is limited, a modification of one of the previously mentioned procedures can be used. In such cases only a thin layer of resin and sample is placed on the bottom of the mould. This can then be sectioned transversely when set and cemented 'back to back' to provide a larger area for examination, or at the stage when the mould is taken from the vacuum desiccator, it is allowed to stand until the resin is just starting to gel. Further resin is then added over the top of the gelled layer, they will not intermix, and the sample can be degassed as described earlier. This produces a specimen with a thin layer of embedded sample backed by clear resin.

Both epoxy and polyester resins can be coloured by dyes added to them which will delineate cracks and pores in the particles. As summarised in Section 3.5.4, a range of dyes are available from resin manufacturers, but some of these may interfere with staining techniques or be difficult to detect. Yellow fluorescent dyes are widely used for general petrographic work because of their high visibility and difference in colour from colours produced by the majority of staining techniques used for petrographic identification of minerals (see Section 3.5.3). These dyes also have the advantage that they can be made to fluoresce as described in Section 2.5.1 and if mixed to a standard specification may be used for the estimation of water/cement ratio and porosity in concrete (see Section 4.4). All dyes used in thin sections should be used with discretion as they tend to obscure fine textural detail. It is very important that the dye forms a true solution in the resin as particulate dyes often produce an uneven patchy colouration.

The various methods for polishing cement clinkers, concrete and building materials have been described by Ahmed (1991), and Walker (1979, 1980) describes methods for the production of polished ultra-thin sections. Polishing is usually not a difficult technique but it is essential that the operator has the necessary skill and experience. Concrete is one of the more difficult materials to polish because of its heterogeneous nature and the range of hardness of the components. Another problem during sample preparation is the introduction of cracks into the hardened cement paste. A method for overcoming this problem has been proposed by Hornain et al. (1995). It is strongly recommended that polishing be carried out by a skilled and experienced technician, because the quality of the polished surface is important for both reflected light and SEM studies.

In electron probe microanalysis, the basis of the method relies on the comparison between x-ray intensity obtained from the selected area of the specimen surface and the intensities obtained from a polished standard. To minimise sources of error it is important that both specimen and standard are polished to an equally high quality of finish.

3.4.5 Preparation of petrographic thin sections

The preparation of thin sections of rock for study with the petrological microscope has been an important part of geological laboratory work for more than a century (Reed and Mergner 1953). Essentially a thin section is produced by cutting about a 10 mm thick slice from a rock with a diamond saw. The slice must be thick enough to resist bowing during handling, mounting and grinding. After suitable impregnation to stabilise the slice, one face is ground flat and smooth, finishing with 16 or 12 μm silicon carbide grit. This face is then glued to a slide and the rock slice is then reduced in thickness until it is approximately 30 μm thick (Figure 3.15). The upper surface is finished by fine lapping before gluing a 0.17 mm thick cover glass over the top surface for protection. In special cases the upper surface is polished and the cover glass omitted (Moreland 1968). It is possible to produce a rock thin section entirely by manual methods and typically the 'standard' rock thin section will measure about 20 × 30 mm. An important problem with hand finishing is the difficulty of retaining the original edges of the rock slice in the section. Traditionally the ragged thinned edges produced by the hand finishing were cut away to produce a clean edge. Also it is difficult to produce a section by hand much greater in size than 75 × 50 mm. This loss of edges and the small size has generally been acceptable to the petrographer when examining rocks but is completely unacceptable for concrete samples.

However, the manual procedures have now largely been supplanted by mechanical methods as a result of advances in the technology of adhesives and also of precision cutting and grinding equipment. These improvements have simplified and speeded up thin-section manufacture, although many laboratories still finish concrete sections by hand (Figure 3.16) and check them optically with a polarising microscope during the final stages. Increasingly, preparation facilities are using automated or semi-automated specialized apparatus; an example is shown in Figure 3.17, including details of the equipment and a summary checklist of the steps involved in preparing thin sections of samples impregnated with flourescent resin.

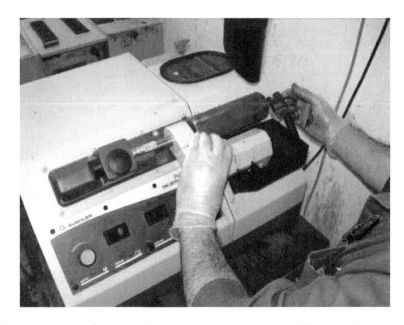

Figure 3.15 Precision grinding of a mounted thin slice on a Logitech machine. (Photograph courtesy of M. Eden.)

Figure 3.16 The hand finishing of a concrete thin section using a glass plate. (Photograph courtesy of M. Eden.)

Most manuals on petrology and optical mineralogy have included some type of description of thin-section preparation, and there are numerous other descriptions in the literature. Two references that include good descriptions of the techniques as well as a range of ancillary techniques are Allman and Lawrence (1972) and Hutchinson (1974). Campbell (1986) has described techniques for sectioning cement clinkers and Wilson (1973) for producing sections on semi-automatic lapping equipment. In Europe, some petrographers now follow methods outlined within the CUR-Recommendation 89, Appendices A, B and C (2008). These specify procedures for preparation of thin sections and thin-section petrography, fluorescent epoxy and epoxy-casting of particulate materials, respectively.

Concrete, mortar and similar materials present special problems in their preparation as petrological thin sections. Their mineral components differ in their relative hardness, some are only a few microns in diameter, while others are water soluble or may be damaged by water, and the material itself may be fragile due to pre-existing cracks, voids and porous areas. As already noted, some of these problems may be overcome by the use of non-aqueous grinding fluids, epoxy resin impregnation and adhesives which effectively stabilise the fabric of the concrete. However, the proper examination of a thin section of concrete requires larger area and reduced thickness sections compared to those used for rocks. Also retention of the original section edges is essential for an adequate examination. Consequently, the finished thin sections typically 50 × 75 mm or larger with thicknesses of 20–25 μm, are generally considered more appropriate for concretes and related materials.

The first semi-routine preparation of large-area thin sections of concrete was described by Poulsen (1958) and Andersen and Petersen (1961). By a combination of hand and mechanical techniques, they prepared a wide range of large-area thin sections for the investigation of the alkali–aggregate reaction in Denmark. Jones et al. (1966) used a modified surface grinder with a vertical spindle and diamond cup wheel to produce large-area thin sections of concrete on a semi-automatic basis. Further development of this technique was made by

1   2   3   4

Equipment for automated thin section preparation:
- Diamond rollers flanked by eccentric ball bearing system for accurate and automated thickness control (#1)
- Work-piece placed on the two-sided vacuum chuck (#2)
- Grinding the work-piece on various diamond roller sizes (#3 and #4)

5   6   7   8

Preparation of fluorescence impregnated thin sections comprises the following steps:
- Pre-impregnate fragile samples
- Mount a reference glass on back side of sample (#5; the final thin section will be placed at the white hatched line)
- Cut to ~10 mm thickness incl. reference glass
- Grind with diamond rollers nos. 1–3 (#3 and #4)
- Vacuum impregnate with fluorescent epoxy
- Re-grind ~5 μm off the impregnated surface
- (default: Polish the surface)
- Mount object glass with UV-hardening glue (#6) and cure in UV light (#7)
- Cut to ~0.4 mm thickness using the diamond saw
- Pre-set the desired grinding thickness
- Grind with rollers nos. 1–3, thin section is ready
- Mount cover glass using UV-hardening glue to get the final thin section (#8)
- (Default: Polish the surface for e.g. SEM analysis)

Figure 3.17 Example of a system for the semi-automated preparation of thin sections, together with a summary checklist of the steps required for preparing thin sections from samples integrated with flourescent resin. (Courtesy of Peter Laugesen of Pelcon Materials & Testing ApS, Ballerup, Denmark.)

St John and Abbott (1983) to enable it to be used to prepare 25 μm thick sections of concrete up to 150 × 125 mm in size on a routine basis. Concurrently with these developments, Logitech Ltd. mechanised the traditional lapping techniques for the preparation of thin sections (Wilson 1973). Initially the Logitech machines were intended for multiple preparation of routine rock thin sections, but with further development (Wilson and Milburn 1978), sections of up to 100 × 75 mm may be prepared with full edge retention so the technique is now applicable to concrete. This size can now be extended to 150 × 100 mm (see Figure 3.12).

Lapping sections with alumina grinding powder as opposed to grinding with fine diamonds embedded in a bronze cup wheel produces thin sections with differing characteristics. The diamonds embedded in the cup wheels are usually about 60 μm (260–300 mesh) in size, and while this produces a reasonably smooth surface finish, it is not as fine as that produced by lapping with 5 μm alumina. However, lapping frosts the surface of the mounting resin and can obscure fine detail. Also it can leave alumina particles embedded in the sample unless special care is taken to clean the section surfaces before mounting. These effects are absent where a diamond cup wheel is used, but unless the diamond cup wheel is in good condition, some crystal shattering may occur especially where the section thickness is reduced to 25 μm. Consequently, unless great care and skill are used in the manufacture of thin sections of concrete and related materials, a number of artefacts can be introduced as a result of careless manufacture, some of the common faults are as follows:

- Small particles of grinding medium may become trapped and incorporated into the surface of the finished thin section.
- Grinding debris may infill cracks and small voids and can lead to difficulties in interpretation.
- Some voids and fine cracks may not become infilled during the resin impregnation making it difficult to be certain whether they are original or manufactured features.
- Some particles may be 'plucked' from the section during grinding leaving voids that are not original features of the fabric.
- Cracking may also be introduced during grinding, or from 'plucked' particles being dragged across the surface during grinding.
- Cracking marginal to aggregate particles can be introduced particularly where very hard aggregate particles are being stressed by being ground in a softer matrix.
- Mineralogical changes and/or shrinkage microcracks can develop in the cementitious matrix as a result of overheating during initial curing, or uneven friction heating effects resulting from the grinding processes.
- Carbonation may develop if the thin section is not adequately protected from atmospheric CO_2 during a prolonged preparation or storage.
- Entrapped micro-bubbles of air can be introduced during the resin impregnation and mounting processes.
- Uneven thickness, particularly thinning near margins, can occur if the finishing stages of grinding are not carried out with sufficient skill, or the adhesives used are applied unevenly.

The diamond grinding and alumina lapping techniques can be combined to provide a ground surface. The thin section is first cut on a surface grinder and then the upper surface lapped on a Logitech or similar lapping machine to produce a surface which is ground finely enough to be used for a general examination by SEM. Using a removable cover slip, the section is examined using a petrological microscope, and then after removal of the cover glass, further polishing and application of a suitable conductive coating, a more detailed examination under the SEM is possible.

The concrete petrographer must be familiar with the preparation techniques that are used as these affect the petrographic examination. As an example, the effective use of epoxy resin is dependent on drying the concrete of free moisture, the temperature of drying and the polymerisation of the resin. Temperatures must not exceed 60°C as dehydration of gypsum and ettringite can occur, as can microcracking due to temperature gradients. Thus, drying of a 10–15 mm thick slice of concrete will require at least 16 h, and if lower temperatures are used, even longer drying times may be required.

The use of a dye in the impregnating resin, as described in Section 3.5.4, also needs to be specified as appropriate to allow comparison to be drawn between specimens. The routine use of a dye in the impregnating resin to outline cracks and pores is controversial as it can obscure detail and ideally should not be required if sectioning is carried out correctly. However, the yellow fluorescent dyes in general use are effective in highlighting areas of porosity and determining water/cement ratio (see Section 4.4 and Jensen et al. 1985). One compromise is to select and cut additional smaller sections, to include the fluorescent dye so that the large sections are retained for general examination.

The types of resins used are critical to the sectioning process. Impregnating, casting and adhesive types of resins are required, each having specific properties appropriate to the application (see Table 3.4). The problem in specifying particular resins is that formulations are usually confidential to the manufacturer and code designations change with time. The casting resin is the easiest to specify, is usually an epoxy and should be readily available from all manufacturers as it is one of the basic resins used for the manufacture of a range of products. It is an unmodified liquid resin derived from epichlorohydrin and bisphenol. A typical epoxy formulation will have viscosities that can range from 90 to 1300 mPa s at 25°C. When used with an appropriate amine type hardener followed by hardening at room temperature, then finally cured at 30°C–60°C, a hard cast results that is excellent for embedding slices of concrete in preparation for machining.

The UV-hardened urethane/acrylic is now widely used as an adhesive, but the adhesive used previously was also an unmodified epoxy. If used, it should not contain solvents but may contain a wetting agent. It needs to be less viscous than the casting resin and its setting shrinkage should not be excessive. The choice of a resin suitable for impregnation presents many problems not the least being the fact that concrete is a difficult material to impregnate unless the viscosity of the resin is close to that of water. The depth of impregnation of the fabric of even highly porous concretes is very limited even when using an epoxy intended for crack injection. When concretes impregnated with these resins are examined in thin section, the resin often appears as crumpled, heavily shrunk coatings inside pores and microcracks, and it is questionable whether more than superficial impregnation of the fabric penetrates beyond a depth of 0.5–1 mm. Because impregnation of concrete is difficult, only highly porous and friable specimens should be impregnated before casting and this should be regarded as a consolidation process rather than impregnation. In practice, impregnation to consolidate the fabric of the layer of concrete in the thin section is usually undertaken when the ground slice is glued to the glass mounting plate. While this impregnation is superficial, this does not matter as the final section is only 25–30 μm thick.

The types of epoxy resin discussed earlier and shown in Table 3.4 have refractive indices typically around 1.58 which are higher than the 1.54 of the traditional mountants for rock thin sections. Special epoxy resins with a refractive index of ~1.54 are available but these are not necessarily suitable for large-area thin sections because of high setting shrinkage. As already noted, UV curing cyanoacrylate types are also now in common use. A well-equipped laboratory will have a number of resins and adhesives available for use, and the particular choice will depend on the application.

Whatever resins are used, the requirements for thin sectioning are stringent. The resins must be able to be used at temperatures up to a 60°C maximum, not have excessive setting shrinkage, be hard and have good adhesive properties and also be resistant to the oils and solvents used during sectioning. Experience to date indicates that epoxy-based resins are still generally superior for this purpose. Additional difficulties arise when sections exceed 75 × 50 mm in size so that thicker glass mounting plates must be used. Dependent on size, a 3 or 4 mm float glass is used and only one side is ground or lapped to provide the essential flat, parallel surface for mounting the concrete slice. If both sides of the glass mounting plate are ground the glass plate, will lose strength and tend to break easily. The thicker glass mounting plates will also require the use of a longer focus condenser cap on the substage assembly of the petrographic microscope.

The choice of large-area thin-sectioning technique may also be dictated by the nature of the investigation. While the production of sections using surface grinders with diamond cup wheels has some advantages, the method is not applicable to the examination of steel reinforcement embedded in concrete. If the steel/paste interface and associated alteration products are to be retained for examination, the section must be prepared by a lapping technique. This is because the type of diamond and bond used in the cup wheels will not grind steel without smearing and overheating occurring. A method for overcoming this problem has been proposed by Garrett and Beaman (1985).

3.5 SPECIMEN PREPARATION FOR SPECIAL PURPOSES

Concrete and its related materials are complex assemblages of mineral materials ranging from millimetres to sub-micron in sizes. Many of the minerals are themselves chemically complex, present as solid solution or amorphous materials which may not be resolved under the microscope. Increasingly in many investigations, the petrographer finds it preferable and more economic to use specialised techniques such as staining or etching, chemical analysis, x-ray diffraction, infrared spectroscopy, thermal analysis, electron-microscopy and microanalytical techniques in addition to, or instead of, extended optical microscopy (see Section 2.3.2). In particular, the SEM is an invaluable and frequently used investigative tool. Such microscopes are usually fitted with an x-ray microanalytical facility which allows at least semi-quantitative elemental analysis of material in the specimen, thus greatly enhancing its diagnostic and investigative value (see Section 2.5.1). Modern electron microscopes with windowless detectors allow light elements such as sodium, oxygen and carbon to be detected (see Section 2.5.2).

The SEM is not usually appropriate for the evaluation of the composition of a 'representative' sample because the specimens prepared for examination need to be small, and if quantitative analyses are also required, they need to be polished to at least a 1 μm diamond polish. The problems of the degassing of large specimens in the SEM and the 30 mm diameter practical limit to specimen polishing necessitate careful choice of specimen by the petrographer. If statistically representative analyses are required, optical microscopy, or other means such as x-ray diffraction, or chemical techniques are generally more suitable.

A method for the preparation of representative test specimens for the analysis of concrete by chemical methods is given in BS 1881: Pt 124 (1988). The specific methods for preparing samples for x-ray diffraction and x-ray fluorescence analysis are described in detail by Burhke et al. (1998). The heterogeneity of concrete and aggregates requires that several kilograms be crushed in order to obtain a representative subsample. A single sample portion requires a final subsample size of 100–200 g passing a 150 μm sieve. BS 1881 also requires the minimum linear dimension of the sample to be at least five times that of the nominal maximum dimension of the aggregate particles and also to be not less than 1 kg for the analysis of concrete. It also points out that a major source of error in the analysis of hardened concrete is inadequate sample preparation.

3.5.1 Small selected specimens

During the inspection of structures, or the examination of cores and other laboratory samples, any samples of efflorescence, exudations and crack infillings should be collected and retained for later detailed study. These samples are often only a few milligrams in size and should first be examined without pretreatment. The initial steps are examination using a stereo-binocular microscope followed in some circumstances by an examination of part of the sample in immersion oil using a petrographic microscope. These initial steps will often provide valuable information about the number and type of constituents present and will indicate whether further detailed analysis is required.

Examination of core, broken and cut surfaces using the binocular microscope may reveal crack or void infillings that can be sampled. The examination may also provide sufficient information to allow particular small segments to be cut from the larger sample for preparation for investigation using the specialised equipment referred to earlier and described more fully in Chapter 2.

SEM and microanalytical investigations can be carried out on the original unmodified material, and with some modern equipment, an electrical surface coating may not be essential for a first examination, though surface charge may build up and cause fluorescence. However, where reliable, quantitative chemical data are required, the sample must be prepared as a polished mount with an electrically conducting carbon surface film so that its x-ray emissions may be directly compared against the elemental standards used for calibrating the instrument.

3.5.2 Powdered specimen

Many specialised techniques, for example, x-ray diffraction, require the specimen to be in the form of a fine powder (Klug and Alexander 1974). If the sample is already available as a powder, consideration only needs to be given to the particle size and to obtaining a representative subsample. Subsamples required for these techniques are typically less than 10 g and if necessary can be crushed by hand in a small percussion mortar and then ground by hand in a small agate pestle and mortar. Consideration should be given to possible contamination from the mortar and also whether the sample may be sensitive to the local heating effects resulting from grinding. Contamination can mask elements/ compounds of interest in an analysis and can be significant, particularly if the element sort is only present in low concentrations.

To summarise, x-ray diffraction analysis in general requires fine powder with a maximum particle size of less than 45 μm. X-ray fluorescence analysis requires particles less than 100 μm. It is possible to enrich the hardened paste fraction of concrete by at least 100% by hand selection and sieving with minimal contamination from the crushing apparatus or degradation of phases due to heating. This method is most appropriate for the identification of phases within cement pastes and for minor and trace analysis. It has been used successfully for this purpose by Goguel and St John (1993a,b). Selective sampling may also be used for the detection of organic admixtures by Fourier transform infrared spectroscopy in contrast with the laborious chemical extraction of large powdered samples used by Hime (1974) and Rixom (1978) and proposed by the Concrete Society (1989).

ELECTRICALLY CONDUCTING THIN FILMS

The electrically conducting film typically of the order of 20 nm in thickness is usually applied using a low-vacuum 'sputter coating' unit, or in a high-vacuum evaporative coater. If x-ray micro-analysis is envisaged, the coating material used is normally carbon since it absorbs less of the x-ray energies generated from the specimen than a heavier element such as gold, though gold with its high back scattering potential and fine crystallite size in the coated film gives a clearer more detailed backscattered image of the specimen surface. However, a range of other conducting coatings including gold/palladium, aluminium, chromium, iridium and silver are sometimes used. One development uses osmium as a sputter coating material which, though almost as heavy as gold ($Z = 190$, gold $Z = 196$), is claimed to give more uniform coating with finer texture and to only need a 1 nm thickness to provide adequate electrical conductivity.

REMOVAL OF CONDUCTING COATINGS

Removal of coatings from specimens may best be accomplished by polishing with 0.25 µm diamond. Metal coatings can sometimes be removed by wiping the polished surface, though traces are likely to remain in cracks. Rough surface gold-coated specimens require a different approach. Gold can be removed by treatment with a 10% aqueous solution sodium cyanide (for toxic solutions, precautions are essential) (Sela and Boyde 1977). Silver can be removed using photographic 'Farmers reducer' (Mills 1988).

When very fine powders are to be examined, a method referred to as dispersion staining can be used for identifying fine particles such as dusts, asbestos, industrial materials and cases where it is necessary to identify a very small amount of a particular mineral or material, even as little as one particle, among a large number of other particles. Details of the method, determinative tables and graphs are given by McCrone and Delly (1973). A special objective with a central stop and annulus is required which may be purchased from the McCrone Institute. The technique consists of placing the fine particles in immersion liquids of high dispersion which are commercially available from R.P. Cargille Laboratories Inc., New Jersey, United States, and observing the particles under the microscope using the special objective. Differences between the refractive indices of the immersion liquid and the particles produce characteristic colour fringes around the margins of the particles. These fringes can be used for identification. If one particle with a different refractive index is present, it will have different colour fringes and will stand out in the field of view. In this respect, dispersion staining is a powerful technique and should always be kept in mind by the petrographer especially when dealing with fine particulate, heterogeneous materials.

3.5.3 Preparations involving etches, stains and dyes

Techniques using chemical solutions or vapours to selectively stain or etch minerals as an aid to their identification have been used for many years. Geologists have made use of such techniques to identify a variety of rock-forming minerals and clays (Reid 1969; Allman and Lawrence 1972), while cement chemists have used etching techniques to identify phases present in polished sections of clinkers and slags (Campbell 1986). As outlined

Figure 3.18 Voids and fine cracks picked out by resin containing yellow fluorescent dye on a cut and ground concrete surface. Width of field 100 mm.

in Section 3.5.4, dyes have been routinely incorporated into impregnating resins to enable small cracks, voids and porous areas of a concrete surface or thin section to show up clearly (Figure 3.18), but this does not identify the mineral constituents present.

The basic mechanism for both stains and etches is the same in that a chemical reagent reacts selectively with specific minerals. In the case of etches, the surface layers of the minerals are removed in solution, and in the case of stains, a reaction product remains on the surface and is either coloured or a colour can be developed by further treatment. The complexities of the individual recipes arise from the need to select appropriate reactant concentration, reaction time and temperature conditions so that the desired effect is produced. Some skill and experience are usually necessary to ensure that the mineral of interest is neither under- or overetched nor stained.

3.5.4 Dyes for use with resins

There are two types of colourants suitable for incorporation into the epoxy resins which are used for specimen impregnation. The first type consists of dry pigments in a form which may be mixed directly with the resin prior to impregnation, but unless the mixing is very thorough, a patchy uneven colouring is produced and the coloured pigment particles may not fully permeate the finer pores in the cement paste. An alternative approach is to use premixed colouring pastes marketed by major producers of epoxy resin which dissolve in the resin when heated to 100°C. These pastes are available in red, green and blue and provided they are thoroughly intermixed with the resin at 100°C produce an even coloration.

The second type of colouring agents consists of dyes that are soluble in the resin. These types are to be preferred in that they readily impart a uniform colour to the resin. The colourants of greatest value to concrete petrographers are dyes both for highlighting cracks and voids in white light as is illustrated in Figure 3.18 and for their fluorescent properties in ultraviolet light which can be used to detect porosity (see Section 2.5.1). These dyes need to be reasonably stable to resist fading. A yellow dye known as Albisol Brilliant Yellow R has been found particularly suitable and is manufactured under a number of names that will be found listed in the *Colour Index* (2009) available online from the Society of Dyers and Colourists.

3.5.5 Stains for identification of minerals

Most aggregates in concrete are usually easily identified without recourse to staining. However, some carbonate aggregates do present problems because distinctions between limestones, dolomites and dolomitic limestones are not always clear, particularly if the rock is fine grained; staining can provide a simple means of identification (see Figure 3.19). Identification of carbonates is important when investigating the alkali–carbonate reaction (see Chapter 6). Allman and Lawrence (1972) detail several methods for separating these rock types using selective stains. One of the best of these techniques published by Dickson (1965, 1966) allows differentiation of all the common carbonate rocks and if used carefully will pick out zoning and overgrowths on individual particles. Sulphate minerals are also sometimes difficult to identify when they are present as small particles mixed with a fine aggregate and also as fine-grained sulphate phases such as ettringite in cement paste. An effective stain which may be adapted for use on ground or polished surfaces and on thin-sections is described by Poole and Thomas (1975) and has been used to confirm the presence of sulphate bearing phases in studies of sulphate attack on concrete (Harrison 1992a,b). The development of alkali–silica gel in concrete is discussed in Section 6.7.

This gel is quite difficult to identify unequivocally on broken and sawn concrete surfaces but readily absorbs the uranyl ion which fluoresces strongly in ultraviolet light (see Section 2.3.2). The method described by Natesaiyer and Hover (1988, 1989) is useful as a method for detecting the presence of alkali–silica gel when in the field. A summary outline of a number of these methods is given in Table 3.5, although the original references should be consulted before they are attempted for the first time. These staining techniques can be applied to specimens before or after resin impregnation. All that is required is that the exposed surface under test is chemically clean.

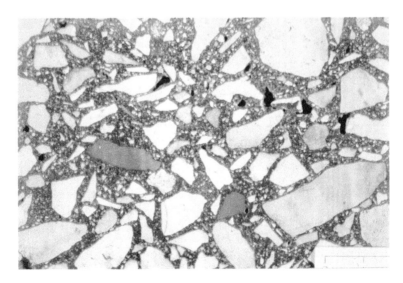

Figure 3.19 A cut and ground concrete surface stained with acidified alizarin red S differentiating coarse aggregate limestone stained red from dolomite that remains white. Width of field 110 mm.

Table 3.5 Summary of some useful staining techniques for concrete petrography

Specimen	Method outline	Observation	Reference
Sulphates			
Broken, polished or ground surfaces Thin sections Particulates	2 min immersion in 2:1 mixture of $BaCl_2$:$KMnO_4$ 6% solution. Wash first in water and then in saturated oxalic acid solution.	Ettringite, gypsum anhydrite stain, pink to purple.	Poole and Thomas (1975)
Broken, polished or ground surfaces Thin sections Particulates	A few seconds immersion in 10% mercuric nitrate solution. Acidified with 1% nitric acid.	Gypsum and anhydrite stain yellow.	Hounslow (1979)
Broken, polished or ground surfaces Particulates *Alkali–Silica Gel*	2–3 min immersion in a cold solution of 0.1–0.2 g Alizarin Red S in 25 mL methanol boiled with 50 mL of 5% NaOH solution.	Gypsum stains purple; anhydrite remains white.	Friedman (1959)
Broken, polished or ground surfaces Thin sections	15 min immersion in 10% uranyl acetate and 1.5% acetic acid solution; washed in water.	240 nm wavelength. UV light produces a green fluorescence.	Natesaiyer and Hover (1988, 1989)
Broken, polished or ground surfaces Thin sections	72 h immersion in 4 M cuprammonium sulphate; washed with water.	ASR gels in cracks and voids stain sky blue.	Poole, McLachan and Ellis (1988)
Broken, polished or ground surfaces Thin sections	Immersion in saturated sodium cobaltinitrate, then Rhodamine-B base solution; washed in water.	Alkali-rich gel stains bright yellow; calcium-rich gels stain pink.	Guthrie and Carey (1997, 1998)
Carbonates			
Broken, polished or ground surfaces Thin sections Particulates	10 s etch in 10% HCl, 10 s immersion in 1:1 Alizarin. Red S + 0.9 g Potassium ferri-cyanide in 100 mL 1.5% HCl; washed with water.	Calcite stains pink. Ferroan calcite stains purple/blue; brucite stains purple. Dolomite remains unaffected.	Dickson (1965, 1966) Haines (1968)
Broken, polished or ground surfaces Thin sections Particulates	5 min immersion in boiling solution of titan yellow (0.2 g) in 25 mL methanol + 30% NaOH solution.	Dolomite and magnesite stain deep orange/red. Calcite and aragonite remain unaffected.	Tucker (1988)
Broken, polished or ground surfaces Thin sections Particulates	10 min immersion in Fiegl's solution (1 g silver sulphate boiled with 11.8 g manganese sulphate in 100 mL water; 2 drops in 10% NaOH solution).	Aragonite stains black; calcite and dolomite remain unaffected.	Fiegl (1937)

3.5.6 Etching procedures for cement clinkers

The identification of the mineral phases in cement clinkers using reflected light (incident light) microscopy requires the use of selective etching techniques and will also reveal structural features such as twinning and zoning. A comprehensive review of these techniques is given by Campbell (1986), together with colour photomicrographs illustrating the results obtained. Many of the techniques require very brief exposure of the specimen to the etching solution or vapour. The common method used is to invert the section and dip the polished face into a thin layer of the etchant contained in a shallow Petri dish. In the case of hydrofluoric acid vapour, the hydrofluoric acid, often diluted to slow the etching down, is placed in a platinum crucible and the inverted polished face is held above it. The etched surface is washed with alcohol to stop the reaction and the surface quickly dried in a current of warm air. Figures 3.20 through 3.22 illustrate the differentiation of anhydrous cement phases obtained using etching techniques summarised in Table 3.6. Skill, practice and experience are necessary to obtain clear results from such procedures. Overetching is a common cause of unsatisfactory results, but lowering the temperature of etching allows better control and gives better results for some etches (Fundal 1980). Some additional etching techniques are summarised by Ingham (2011) in his Appendix B.

In all staining and etching procedures, a clean, high-quality polished specimen surface is essential if unambiguous results are to be obtained, but in each case, reference should be made to the original published method before any particular procedure is attempted for the first time.

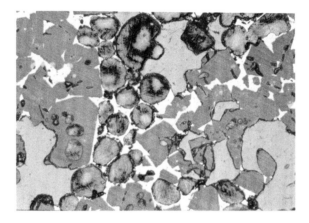

Figure 3.20 Anhydrous OPC clinker etched with hydrofluoric vapour. 'Alite' (C_3S) brown, 'Belite' (C_2S) blue, tricalcium aluminate (C_3A) grey and ferrite white. Specimen etched with 10% KOH solution at 30°C. Width of field 180 μm reflected light. (Courtesy of M. Grove.)

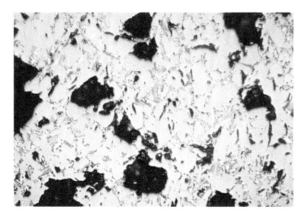

Figure 3.21 Polished anhydrous OPC clinker etched with a 10% KOH solution at 30°C. The addition of alkali changes the C_3A lattice from cubic to orthorhombic. Small elongate interstitial prismatic pale grey/blue C_3A crystals are clearly defined, the black areas are voids. Width of field 100 μm, reflected light. (Courtesy of M. Grove.)

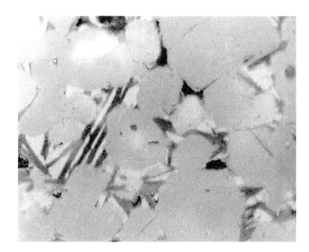

Figure 3.22 High-alkali cement clinker etched with 10% KOH at 30°C. The rounded C_2S has etched to a grey colour. The C_3A has absorbed alkali ions into the cubic lattice structure to an orthorhombic form, the 'dark prismatic interstitial phase'. Between the lamellae, the light-coloured infill is C_4AF. Width of field 60 μm reflected light. (Courtesy of M. Grove.)

Table 3.6 Appearance of clinker phases in reflected light using various etchants

Clinker phase	Etchant or stain and etchant	Immersion/exposure time (s)	Appearance in reflected light
Free lime, CaO	Distilled water	3–5	Etches rapidly
Tricalcium silicate	10% KOH		Does not etch
C_3S (alite)	HF vapour	2–5	Buff brown
	HNO_3 in alcohol[a]	2–15	Greyish brown
	NH_4NO_3/salicylic acid[b]	25–30/30	Yellow green[c]
Dicalcium silicate	10% KOH		Does not etch
C_2S (belite)	HF vapour	2–5	Blue
	HNO_3 in alcohol[a]	2–15	Blue
	NH_4NO_3/salicylic acid[a]	25–30/30	Brown (when C_3S is yellow green)
Tricalcium aluminate	Water	3–5	Etches
C_3A	10% KOH	10–20	Bluish grey
	HF vapour	2–5	Light grey
	HNO_3 in alcohol[a]	2–15	Light grey
Ferrite	10% KOH	10–20	White
C_4AF	HF vapour	2–5	White
	HNO_3 in alcohol[a]	2–15	White

[a] Gille et al. (1965). Nitric acid in alcohol, using 1:100 or 1:1000 solutions, prepared at least a fortnight prior to use.
[b] Ahmed (1994). Stain using 1 g NH_4NO_3, 150 mL isopropyl alcohol, 20 mL water, 20 mL ethyl alcohol and 20 mL acetone; then etch using 0.2 g salicylic acid, 25 mL isopropyl alcohol and 25 mL water.
[c] Depending on the time, the colour ranges from light brown to brown to purplish brown to blue to blue-green to yellow green.

REFERENCES

Ahmed, W.D. 1991: Advances in sample preparation for clinker and concrete microscopy. In *Proceedings of the 13th International Conference on Cement Microscopy*, Tamba, FL, Int'L Cement Microscopy, pp. 17–29.

Ahmed, W.U. 1994: Petrographic methods of analysis of cement clinker and concrete microstructure. In *Petrography of Cementitious Materials*, DeHayes, S.M. and Stark, D. (eds.), ASTM STP 1215, Philadelphia, PA.

Allman, M. and Lawrence, D.F. 1972: *Geological Laboratory Techniques*. Blandford Press, London, U.K.

Andersen, P.E. and Petersen, RH. 1961: Drilling of concrete cores and preparation of thin sections. *Rilem Bulletin* 11, 94–106.

ASTM C42/C42M-04. 2004: *Standard Test Method for Obtaining and Testing Drilled Cores and sawed Beams of Concrete*. ASTM, Philadelphia, PA.

ASTM C183-08. 2008: *Standard Practice for Sampling and the Amount of Testing of Hydraulic Cement*. ASTM, Philadelphia, PA.

ASTM C204-07. 2007: *Standard Methods for Fineness of Hydraulic Cement by Air-Permeability Apparatus*. ASTM, Philadelphia, PA.

ASTM C295–09. 2009: *Standard Guide for Petrographic Examination of Aggregates for Concrete*. ASTM, Philadelphia, PA.

ASTM C457/C457M. 2010: *Standard Test Method for the Microscopical Determination of Parameters of the Air Void System in Hardened Concrete*. ASTM, Philadelphia, PA.

ASTM C786/C786M. 2007: *Standard Test Methods for Fineness of Hydraulic Cement and Raw Materials by the 300-mue (No.50), 150-mue (No.100), and 75-mue (No.200) Sieves by Wet Methods*. ASTM, Philadelphia, PA.

ASTM C823/C823M–07. 2007: *Standard Practice for Examination and Sampling of Hardened Concrete in Construction*. ASTM, Philadelphia, PA.

ASTM C856–04. 2004: *Standard Practice for Petrographic Examination of Hardened Concrete*. ASTM, Philadelphia, PA.

ASTM D75/D75M. 2009: *Standard Practice for Sampling Aggregates*. ASTM, Philadelphia, PA.

ASTM E20-85(94). 1985: *Standard Practice for Particle Analysis of Particulate Substances in the Range of 0.2 to 75 Micrometers by Optical Microscopy*. ASTM, Philadelphia, PA.

ASTM E105-04. 2004: *Standard Practice for Probability Sampling of Materials*. ASTM, Philadelphia, PA.

BS 812-102. 1989: *Testing Aggregates. Methods for Sampling*. This has been partly replaced by BS EN 932-1 (1997). BSI, London, U.K.

BS 812-104. 1994: *Testing Aggregates. Procedure for Qualitative and Quantitative Petrographic Examination of Aggregates*. BSI, London, U.K.

BS 1881-124: 1988: *Testing Concrete Part 124. Methods of Analysis of Hardened Concrete*. BSI, London, U.K.

BS EN 196-6, 2010: *Methods of Testing Cement. Determination of Fineness*. BSI, London, U.K.

BS EN 196-7, 2007: *Methods of Testing Cement. Methods of Taking and Preparing Samples of Cement*. BSI, London, U.K.

BS EN 932-1, 1997: *Tests for the General Properties of Aggregates. Methods of Sampling*. BSI, London, U.K.

Buhrke, V.E. Jenkins, R., and Smith, D.K. 1998: *Preparation of Specimens for X-Ray Fluorescence and X-Ray Diffraction Analysis*. Wiley, New York.

Campbell, D.H. 1986: *Microscopical Examination of Interpretation of Portland Cement and Clinker*. Construction Technology Laboratories, Portland Cement Association, Skokie, IL.

Concrete Society. 1989: Analysis of hardened concrete. Report of a Joint Working Party of the Concrete Society and Society of Chemical Industry. Concrete Society Technical Report No. 32, 117.

Concrete Society. 1992: *Non-Structural Cracking in Concrete*, TR. 22. 3rd ed. Concrete Society, Camberley, U.K.

Concrete Society. 1999: *Alkali-Silica Reaction: Minimising the Risk of Damage to Concrete. Guidance Notes and Model Clauses for Specifications*, TR.30. 3rd ed. Concrete Society, Camberley, U.K.

Concrete Society. 2000: *Diagnosis of Deterioration in Concrete Structures*, TR 54. Concrete Society, Camberley, U.K.

Concrete Society. 2008: *Assessment, Design and Repair of Fire-Damaged Concrete Structures*, TR.68. Concrete Society, Camberley, U.K.

Concrete Society. 2010: *Concrete Petrography: An Introductory Guide for the Non-Specialist*, TR 71. Concrete Society, Camberley, U.K.

CSA A23.2 – 09. 2009: *Test Methods and Standard Practices for Concrete*. Canadian Standards Association, Mississauga, Ontario, Canada.

CUR (Centre for Civil Engineering Research and Codes). 2008: CUR-Recommendation 89. *Measures to Prevent Concrete Damage by the Alkali-Silica Reaction*. Official English translation, 2nd rev. ed. Appendices A, B and C. Gouda (Board of CURNET), pp. 15–23.

Dickson, J.A.D. 1965: A modified staining technique for carbonates in thin section. *Nature* 205, 587–589.

Dickson, J.A.D. 1966: Carbonate identification and genesis as revealed by staining. *Journal of Sedimentary Petrology* 361 (2), 491–505.

Dolar-Mantuani, L. 1983: *Handbook of Concrete Aggregates: A Petrographic and Technological Evaluation*, Noyes Publications, Park Ridge, NJ.

Fiegl, F. 1937: *Qualitative Analysis by Spot Tests*. Nordemann, New York.

Fookes, P.G. 1981: Concrete deterioration and repair–Discussion. *Proceedings of the Institution of Civil Engineers* 70, Part1, 330–331.

Fookes, P.G. Pollack, D.J., and Kay, E.A. 1982: Concrete in the Middle East, Part 2, Reprints of three articles from *Concrete*. *The Concrete Society, UK*.

Francois-Bongarcon, D. 1998: Extensions to the demonstration of Gy's Formula. *Proceedings of Annual Conference*, Montreal 98, May 98. (*Also down-loadable html from the world-wide web*).

Friedman, G.M. 1959: Identification of carbonate minerals by staining methods *Journal of Sedimentary Petrology* 49, 636–637.

Fundal, E. 1980: Microscopy of cement raw mix and clinker. *FLS-Review* 25, F.L. Smidth Laboratories, Copenhagen, Denmark.

Garrett, H.L. and Beaman, D.R. 1985: A method for preparing steel reinforced mortar or concrete for examination by transmitted light microscopy. *Cement and Concrete Research* 15, 917–920.

Gille, F., Dreizler, I., Grade, K., Kramer, H., and Woermann, E. 1965: *Microscopy of Cement Clinker—Picture Atlas*. Benton-Verlag GmbH, English translation by P. Schmid 1980.

Goguel, R.L. and St. John, D.A. 1993a: Chemical identification of Portland cements in New Zealand concretes–I. Characteristic differences among New Zealand cements in minor and trace element chemistry. *Cement and Concrete Research* 23 (1), 59–68.

Goguel, R.L. and St. John, D.A. 1993b: Chemical identification of Portland cements in New Zealand concretes. 11. The Ca-Sr-Mn plot in cement identification and the effect of aggregates. *Cement and Concrete Research* 23 (2), 283–293.

Guthrie, G.D. and Carey, J.W. 1997: A simple environmentally friendly, and chemically specific method for the identification and evaluation of the alkali-silica reaction. *Cement and Concrete Research* 27 (9) 1407–1417.

Guthrie, G.D. and Carey, J.W. 1998: A geochemical method for the identification of ASR gel. TRB Paper No. 991261, Transportation Research Board, Washington, DC.

Green, J.K. 1976: Some aids to the assessment of fire damage. *Concrete* 10(1), 14–18.

Gy, P.M. 1982: *Sampling of Particulate Materials Theory and Practice*. Elsevier, Amsterdam, the Netherlands.

Gy, P.M. 1998: *Sampling for Analytical Purposes*. John Wiley & Sons, U.K., pp. 1–150.

Haines, M. 1968: Two staining tests for brucite in marble. *Mineralogical Magazine* 36, 886–888.

Harris, P.M. and Sym, R. 1990: Sampling of aggregates and precision test standards. In *Standards for Aggregates*, Pike, D.C. (ed.). Ellis Horwood, New York, pp. 19–63.

Harrison, W.H. 1992a: Assessing the risk of sulphate attack on concrete in the ground. BRE Information Paper, P 15/92. BRE, Watford, U.K.

Harrison, W.H. 1992b: Sulphate resistance of buried concrete. The third report on long-term investigations at Northwick Park and on similar concretes in sulphate solutions at BRE. Building Establishment Report. BRE, Watford, U.K.

Hime, W.G. 1974: Multitechnique approach solves construction materials failure problems. *Analytical Chemistry* 46, 1230A.

Hornain, H., Marchand, J., Ammouche, A., Commene, J.P., and Moranville, M. 1995: Microscopic observations of cracks in concrete–A new sample preparation technique using dye impregnation. In *Proceedings of the 17th International Conference on Cement Microscopy*, Alberta, Canada, pp. 271–282.

Houndslow, A.W. 1979: Modified gypsum/anhydrite stain. *Journal of Sedimentary Petrology* 49, 636–637.

Hutchinson, C.S. 1974: *Laboratory Handbook of Petrographic Techniques*. John Wiley & Sons, New York.

Ingham, J.P. 2011: *Geomaterials under the Microscope*. Manson Publishing Ltd., London, UK.

Institution of Structural Engineers. 1992: Structural Effects of Alkali-Silica Reaction. Technical Guidance on the Appraisal of Existing Structures. Published by SETO Ltd., London, UK.

Jensen, A.D., Eriksen, K., Chatterji, S., Thaulow, N., and Brandt, I. 1985: Petrographic analysis of concrete. *Danish Building Export Council 12*.

Jones, J.C., Hawes, R.W.M., and Dyson, J.R. 1966: Semi-automatic preparation of ultra-thin and large-area thin sections. *Transactions of the British Ceramic Society* 65, 603–612.

Klein, C. and Hurlburt, C.S. 1993: *Manual of Mineralogy* (after James D. Dana), 21st ed. John Wiley & Sons Inc., New York.

Klug, H.P. and Alexander, L.E. 1974: *X-Ray Diffraction Procedures for Polycrystalline and Amorphous Material*. Wiley Interscience, New York.

Larsen, E.S. and Berman, H. 1934: *Microscopic Determination of the Monopaque Minerals*, 2nd ed. USGS, Bulletin 848. Washington, DC.

McCrone, W.C. and Delly, J.G. 1973: *The Particle Atlas, l &2*. Ann Arbor Science Publishers.

Mills, A.A. 1988: Silver as a removable conductive coating for scanning electron microscopy. *Scanning Microscopy*, 2 (3). *Electron Microscopy Sciences*, 1265–1271.

Moreland, G.C. 1968: Preparation of polished thin sections. *American Mineralogist* 53, 2070–2074.

Natesaiyer, K.C. and Hover, KC. 1988: In situ identification of ASR products in concrete. *Cement and Concrete Research* 18, 455–463.

Natesaiyer, KC. and Hover, KC. 1989: Further study of an in-situ identification method for alkali-silica reaction products in concrete. *Cement and Concrete Research* 19, 770–778.

Pitard, F.F. 1989: *Pierre Gy's Sampling Theory and Sampling Practice*. Vol. 2. CRC Press, Boca Raton, FL.

Poole, A.B., McLachlan, A., and Ellis, DJ. 1988: A simple staining technique for the identification of alkali-silica gel in concrete and aggregate. *Cement and Concrete Research* 18, 116–120.

Poole, A.B. and Thomas, A. 1975: A staining technique for the identification of sulphates in aggregates and concrete. *Mineralogical Magazine* 40, 315–316.

Poulson, E. 1958: *Perforation of Samples for Microscopic Investigation*. NIBR and Academy of Technical Sciences, Copenhagen, Progress Report M1, pp. 42–45.

Reed, F.S. and Mergner, J.L. 1953: Preparation of rock thin sections. *American Mineralogist* 53, 1184–1203.

Reid, W.P. 1969: Mineral staining tests. *Colorado School of Mines Mineral Industries Bulletin* 12 (3), 1–20.

RILEM AAR-6.1 2013: Guide to diagnosis and appraisal of AAR damage to concrete in structures. Part 1: Diagnosis. RILEM/TC-ACS/08/14. Published by Springer 2013. ISBN 978-94-007-6566-5.

Rixom, M.R. 1978: *Chemical Admixtures for Concrete 234*. E. & F.N. Spon, London, U.K.

Roberts, M.H. 1986: Determination of the chloride and cement contents of hardened concrete. *BRE Information Paper*, IP 21/86, 4.

Sela, J. and Boyde, A. 1977: Cyanide removal of gold from SEM specimens. *Journal of Microscopy* 111, 229–231.

Smith, M.R. and Collis, L. (eds.) 2001: Aggregates: Sand, gravel ad crushed rock aggregates for construction purposes. In *Geological Society Engineering Geology Special Publication, No.9* 3rd ed. The Geological Society of London, U.K.

Society of Dyers and Colourists 2009: *Colour Index, 4th Edition Part 1. Pigments and solvent dyes*. SDC, Bradford, U.K.

St. John, D.A. and Abbott, J.H. 1983: Semi-automatic production of concrete thin-sections. *Industrial Diamond Review* 43 (494), 13–16.

Tucker, M. (ed.) 1988: *Techniques in Sedimentology*. Blackwell Science Ltd., London, U.K.

US Department of the Interior. 1981: *Concrete Manual*. Water Resources Technical Publication, 8th rev. ed.. US Department of the Interior, Denver, CO.

Walker, H.N. April 1979: Evaluation and adaptation of the Dobrolubov and Romer method of microscope examination of hardened concrete. *Virginia Highway and Transportation Research Council* VHTRC 79-R42.

Walker, H.N. 1980: Formula for calculating spacing factor for entrained air voids. *Cement, Concrete, and Aggregates CCAGDP* 2 (2), 63–66.

Wilson, R.B. 1973: *Preparation of Microscope Slides of Rocks, Minerals and Other Research Materials*. Logitech Ltd. Glasgow, Glasgow City, U.K., pp. 1–12.

Wilson, R.B. and Milburn, G.T. 1978: The preparation of microscope slides of rocks, minerals and other research materials. In *Third Australian Conference on Science Technology*, Australian National University, Canberra, Australian Capital Territory, Australia.

Chapter 4

Composition of concrete

4.1 SCOPE

There is frequently the need to establish the identity of the mix constituents within hardened concrete and sometimes also the mix proportions. Petrographic examination is a direct and effective method of achieving these objectives. Typical causes of the need for such analyses include disputes during construction, materials assessments during the life of a structure for condition or for predicting performance in respect of a change of use and matching constituents for repair concretes. In one region of the United Kingdom, petrographic examination has been adopted as the method for classifying concrete materials during routine building surveys (Stimson 1997; Royal Institution of Chartered Surveyors 2015). A recent introductory guide to the usefulness of concrete petrography was published by the Concrete Society (2010) and Winter (2012a) gives a good summary of the allied technique of scanning electron microscopy (SEM) of cement and concrete.

Cement types, aggregates, mineral additions or 'supplementary cementitious materials' (SCMs), pigments and fibres are usually identifiable using a range of petrographic techniques. However, the many admixtures, which are increasingly used in small dosages to modify the properties of concrete, are more difficult to characterise by either petrographic or chemical techniques. The principal mix parameters may each be readily evaluated by petrographic methods, especially water/cement (or water/binder) ratio, aggregate/cement ratio, coarse–fine aggregate ratio and void content (including entrapped and entrained air voids), provided that adequately large and representative samples are available.

4.2 CEMENT TYPES AND BINDER CONTENT

4.2.1 Anhydrous Portland cement phases and clinker

The sintered product of burning limestone and an aluminosilicate rock (clay or shale) at temperatures of up to 1500°C is known as Portland cement 'clinker' (or up to nearly 2000°C in the case of white Portland cement: see Section 4.2.2). The grey powder generally called 'cement' is produced by intergrinding the clinker and ~5% gypsum to act as an early set retarder. Lea (1970) and Hewlett (1998) give detailed accounts of the history of the development of Portland cement. Good explanations of cement production are given by Bye (1983, 1999) and Locher (2006), with an update on modern processes by Bye (2011). An excellent introduction is provided by Winter (2012c).

The properties of Portland cement can be modified by altering its chemical composition by varying the raw feed materials and mix proportions, also the rate of cooling of the clinker and other factors, in order to produce different clinker mineralogies and hence different Portland cement types (Table 4.1a and b). Fineness (or specific surface) can also be varied,

Table 4.1a Principal Portland cement types

Cement type	CEM	Product	Clinker mass (%)	Other mass (%)a	Other, constituenta	Previous BS equivalent
Portland	I	I	95–100	—	—	BS 12
Sulphate-resisting Portland		I-SR 0	95–100[b]			BS 4027
		I-SR 3	95–100[b]			
		I-SR 5	95–100[b]			
Portland–slag	II	II/A-S	80–94	6–20	Blastfurnace slag	BS 146
		II/B-S	65–79	21–35		
Portland–silica fume		II/A-D	90–94	6–10	Silica fume	—
Portland pozzolana		II/A-P	80–94	6–20	Natural pozzolana	—
		II/B-P	65–79	21–35		
		II/A-Q	80–94	6–20	Natural calcined pozzolana	
		II/B-Q	65–79	21–35		
Portland–fly ash		II/A-V	80–94	6–20	Siliceous fly ash	BS 6588 (pfa)
		II/B-V	65–79	21–35		
		II/A-W	80–94	6–20	Calcareous fly ash	—
		II/B-W	65–79	21–35		
Portland–burnt shale		II/A-T	80–94	6–20	Burnt shale	—
		II/B-T	65–79	21–35		
Portland–limestone		II/A-L	80–94	6–20	Limestone[c]	BS 7583
		II/B-L	65–79	21–35		
		II/A-LL	80–94	6–20	Limestone[c]	
		II/B-LL	65–79	21–35		
Portland composite		II/A-M	80–88	12–20	Combination	—
		II/B-M	65–79	21–35		
Blastfurnace	III	III/A	35–64	36–65	Blastfurnace slag	BS 146 and BS 4246[d]
		III/B	20–34	66–80		
		III/C	5–19	81–95		
Sulphate-resisting blastfurnace		III/B-SR	20–34	66–80	—	
		III/C-SR	5–19	81–95		
Pozzolanic	IV	IV/A	65–89	11–35	Silica fume, pozzolana or fly ash	—
		IV/B	45–64	36–55		BS 6610
Sulphate-resisting pozzolanic		IV/A-SR	65–79[e]	21–35		—
		IV/B-SR	45–64[e]	36–55		—
Composite	V	V/A	40–64	18–30 + 18–30	Blastfurnace slag + pozzolana or siliceous fly ash	—
		V/B	20–38	31–49 + 31–49		

Source: As specified in BS EN 197-1, *Cement. Composition, Specifications and Conformity Criteria for Common Cements*, British Standards Institution, London, U.K., 2011.

[a] Up to 5% 'minor additional constituents' permitted in all cases.
[b] C_3A content = 0% (I-SR 0), ≤ 3% (I-SR 3), or ≤ 5% (I-SR 5).
[c] L = total organic carbon not >0.50%, LL = total organic carbon not >0.20%.
[d] BS 4246 covered cement with 50%–85% blastfurnace slag.
[e] C_3A content = ≤ 9%.

Table 4.1b Principal Portland cement types

ASTM C150: Portland cement		
Type	Description of application	Key characteristics, cf. type I
I	When the special properties specified for any other type are not required	
IA	Air-entraining cement for the same uses as Type I, where air entrainment is desired	Type I containing an interground air-entraining addition
II	General use, more especially when moderate sulphate resistance is desired	8% max. C_3A
IIA	Air-entraining cement for the same uses as Type II, where air entrainment is desired	Type II containing an interground air-entraining addition
II(MH)	General use, more especially when moderate heat of hydration and moderate sulphate resistance are desired	Sum of C_3S and 4.75 C_3A limited to 10 or tested heat of hydration at 7 days that is 290 kJ/kg max
II(MH)A	Air-entraining cement for the same uses as Type II(MH), where air entrainment is desired	Type II(MH) containing an interground air-entraining addition
III	When high early strength is desired	Strength, 12 MPa min @ 1 day and 24 MPa min @ 3 days; plus 15% max C_3A
IIIA	Air-entraining cement for the same uses as Type III, where air entrainment is desired	Type III containing an interground air-entraining addition, plus strength criteria slightly lower
IV	When a low heat of hydration is desired	35% max C_3S and 40% max C_2S and 7% max C_3A
V	When high sulphate resistance is desired	5% max C_3A and 25% max $C_4AF + 2(C_3A)$

ASTM C595: blended hydraulic cements		
Type	Description of composition	Characteristics
IS(X)[a]	Portland blastfurnace slag cement	Where X is the % by mass of slag
IP(X)[a]	Portland–pozzolan cement	Where X is the % by mass of pozzolan
IL(X)[a]	Portland–limestone cement	Where X is the % by mass of limestone
IT(AX)(BY)[a]	Ternary blended cement	Where AX is the type and % by mass of constituent A (S = slag, P = pozzolan, L = limestone) and where BY is the type and % by mass of constituent B (S = slag, P = pozzolan, L = limestone)

Source: As specified in ASTM C150, *Specification for Portland Cement*, American Society for Testing and Materials, West Conshohocken, PA, 2012; ASTM C595, *Standard Specification for Blended Hydraulic Cements*, American Society for Testing and Materials, West Conshohocken, PA, 2012.

[a] Special properties indicated by the following additional suffices: (A), air entraining; (MH), moderate heat of hydration; (MS), moderate sulphate resistance; (HS), high sulphate resistance; (LH), low heat of hydration.

with effects on the rates of hardening and strength gain, and there has been a tendency over time for cements to be more finely ground: generally, pre-1950s Portland cements will be found to be more coarsely ground than their modern equivalents.

Nearly all hydrated cement paste within concrete contains residual kernels of unhydrated cement or clinker, although the proportion and grain size vary greatly according to a number of factors, and these pieces of unhydrated clinker can be used to characterise the cement type by mineralogical analysis.

The microscopical study of Portland cement clinker to determine mineralogy commenced in the last years of the nineteenth century (Le Chatelier 1905). Törnebohm (1897) gave the names alite, belite, celite and felite to four distinctive crystalline components that he and Le Chatelier had observed within Portland cement clinker, plus an isotropic residue phase. Insley (1936, 1940) later demonstrated that Törnebohm's 'alite' was tricalcium silicate (C_3S), that 'belite' and 'felite' were two different habits of (usually beta) dicalcium silicate (βC_2S), that 'celite' was a calcium aluminoferrite phase and that the isotropic residue contains calcium aluminates (mainly C_3A) and glass. The calcium aluminoferrite phase was first thought to be C_4AF (brownmillerite) but is now known to be a solid solution series ranging from C_2F to just beyond C_6A_2F (Midgley 1964), although according to Lea (1970) the median value is close to C_4AF, which is the compound assumed by Bogue (1955) for his normative calculations from chemical analyses of cement.

The identification of clinker mineralogy has been described in detail elsewhere (Insley and Fréchette 1955; Midgley 1964; Gille et al. 1965; Lea 1970; Bye 1983, 1999; Campbell 1986). In general, petrographic examination of Portland cement clinker reveals a texture of relatively large crystals of alite (C_3S) and belite (C_2S) set in a matrix mainly comprising C_3A and calcium aluminoferrite. The principal characteristics of the Portland cement minerals are detailed in the 'Glossary of minerals'.

Briefly, the texture of a cement clinker is dominated by pseudo-hexagonal crystals of C_3S, varying from 25 to 65 µm in size, together with less frequent rounded crystals of C_2S usually with polysynthetic twinning, varying from 10 to 40 µm in size, set in a poorly differentiated matrix. The crystals of C_3S are usually well formed and dispersed, but C_2S can vary from well-scattered crystals to clusters or nests and is nearly always present as small relict inclusions in C_3S crystals. Modern cements are liable to be finer than the ranges shown earlier, possibly with median particle size in the range from 10 to 20 µm.

Differentiation of the aluminate and ferrite phases in the matrix is primarily a function of the rate at which the clinker has been cooled below about 1250°C, which is approximately the liquidus temperature. In modern cement manufacture, clinker is cooled moderately fast, so that even with adequate etching, details of the matrix may not be well differentiated. It is only when the clinker has been more slowly cooled in the manufacture of older cements that the aluminate and ferrite phases stand out clearly.

Microscopical examination of polished and etched surfaces in reflected light is the most useful method, but thin-section examination can also be helpful (Campbell 1986). The techniques are equally applicable to portions of unground clinker, samples of cement or pieces of concrete matrix containing residual kernels of clinker. Table 4.2 summarises the appearance of the main cement minerals in reflected light using various etchants. Fundal (1980) has demonstrated that clinker microstructures can also be used to assess clinker properties and quality.

4.2.1.1 Calcium silicates (C_3S, C_2S)

Accounting for nearly 80% of a typical ordinary Portland cement clinker, C_3S and C_2S form crystals that appear as agglomerated phenocrysts, with the other cement constituents forming the fine-grained interstitial component. The dominant C_3S (alite) usually forms large euhedral pseudo-hexagonal crystals, which are colourless in thin section but appear brown in reflected light after HF etching (Figure 4.1). Inclusions of C_2S, lime (CaO) or periclase (MgO) are not uncommon within C_3S crystals; also alite is typically impure, with small proportions of magnesium and aluminium substituting for silica.

The C_2S (belite and felite, but now usually just referred to as belite), by contrast, usually forms anhedral or sub-hedral crystals, and in some cases, these appear well rounded in cross section. It is believed that C_2S in Portland cement clinker is usually the βC_2S polymorph, although Insley

Table 4.2 Appearance of clinker phases in reflected light using various etchants

Clinker phase	Etchant or stain and etchant	Immersion/exposure time(s)	Appearance in reflected light
Free lime, CaO	Distilled water	3–5	Etches rapidly
Tricalcium silicate	10% KOH		Does not etch
C_3S (alite)	HF vapour	2–5	Buff brown
	HNO_3 in alcohol[a]	2–15	Greyish brown
	NH_4NO_3/salicylic acid[b]	25–30/30	Yellow-green[c]
Dicalcium silicate	10% KOH		Does not etch
C_2S (belite)	HF vapour	2–5	Blue
	HNO_3 in alcohol[l]	2–5	Blue
	NH_4NO_3/salicylic acid[a]	25–30/3 0	Brown (when C_3S is yellow-green)
Tricalcium aluminate	Water	3–5	Etches
C_3A	10% KOH	10–20	Bluish grey
	HF vapour	2–5	Light grey
	HNO_3 in alcohol[l]	2–15	Light grey
Ferrite	10% KOH	10–20	White
C_4AF	HF vapour	2–5	White
	HNO_3 in alcohol[a]	2–15	White

[a] Gille et al. (1965). Nitric acid in alcohol, using 1:100 or 1:1000 solutions, prepared at least a fortnight prior to use.
[b] Ahmed (1994). Stain using 1 g NH_4NO_3, 150 mL isopropyl alcohol, 20 mL water, 20 mL ethyl alcohol and 20 mL acetone, then etch using 0.2 g salicylic acid, 25 mL isopropyl alcohol and 25 mL water.
[c] Depending on time, the colour ranges from light brown to brown to purplish brown to blue to blue-green to yellow-green.

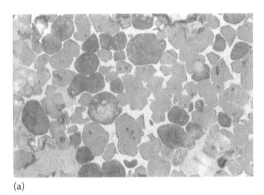

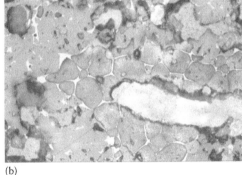

(a)   (b)

Figure 4.1 (a) Photomicrograph of unhydrated Portland cement clinker in reflected light after etching with HF vapour. Brown is C_3S, blue is C_2S and the matrix is differentiated into C_3A (light grey) and ferrite (white). Width of field 0.6 mm. (b) Photomicrograph of nest of C_2S crystals (blue) in cement clinker in reflected light after etching with HF vapour (a resin-filled hole in the centre). Width of field 1 mm.

(1936) demonstrated at least three different types of βC_2S according to crystal twinning styles. French (1991a) has referred to four types of belite, according to the number of sets of inversion lamellae and the presence of inclusions. C_2S is typically slightly coloured in thin section (yellow, brown or green) and generally appears blue in reflected light after HF etching (Figure 4.1).

In slowly cooled clinker, the C_2S can occur also as very small rounded particles dispersed in the interstitial component and/or as 'rims' of small particles around the edges of the large C_3S crystals. Clusters of rounded C_2S crystals have sometimes been termed 'bunch of grapes'

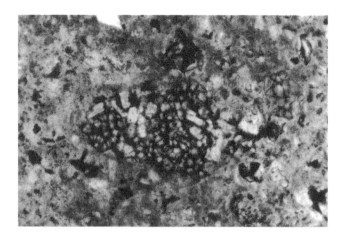

Figure 4.2 Photomicrograph of cluster of C$_2$S crystals in concrete in thin section, exhibiting the 'bunch of grapes' appearance. Width of field 0.5 mm. Plane polars. (Photograph courtesy of RSK Environment Ltd., U.K.)

or 'basket of eggs' texture because of their appearance in thin section (Figure 4.2), and when large, this texture can indicate an unacceptable degree of inhomogeneity in the kiln raw feed (Fundal 1980; Bye 1983, 1999).

4.2.1.2 Calcium aluminates and calcium aluminoferrites (C$_3$A, C$_4$AF)

Forming in the order of 10% of ordinary Portland cement, C$_3$A typically forms one component of the interstitial material between the C$_3$S and C$_2$S crystals. This 'dark' interstitial material has been subdivided into three types: small rectangular outlined crystals believed to be C$_3$A, long prismatic crystals (said by Insley and Fréchette [1955] to be 'the more usual form') thought to be a C$_3$A polymorph modified by alkali in solid solution and an amorphous (or glass) phase. The precise nature of the amorphous interstitial material appears uncertain, but it appears to be logical and agreed that the amount of glass increases for rapidly cooled clinker.

In thin section, the C$_3$A crystals are colourless to brown and frequently isotropic. In reflected light after HF etching, the C$_3$A material is distinguished by being light grey in appearance and clearly interstitial in occurrence.

The other element of the interstitial component (the 'light' interstitial material) is the calcium aluminoferrite (or simply 'ferrite') phase: the old term 'celite' has fallen out of use. The ferrite material is distinctive in both thin section and reflected light. Ferrite has a much higher refractive index than the other clinker constituents (1.9–2.0 compared with about 1.7 for C$_3$S, C$_2$S and C$_3$A), is highly birefringent and is pleochroic, typically ranging from pale amber to dark reddish brown (French [1991a] reports a colour range from dark green to almost opaque). Apart from being truly interstitial, thus taking the form imposed upon it by the neighbouring constituents, ferrite can occur as small prismatic or dendritic crystals or as fibrous material. In reflected light, the ferrite phase is unaffected by most etching agents and appears nearly white and highly reflectant.

4.2.1.3 Lime and periclase (CaO, MgO)

Significant amounts of 'free' lime (CaO) may occur in Portland cement clinker if the raw feed mixture was overloaded with limestone or if that mixture was incompletely burned in the kiln, but a normal range would be from 0.5% to 1.5%. Excessive proportions of

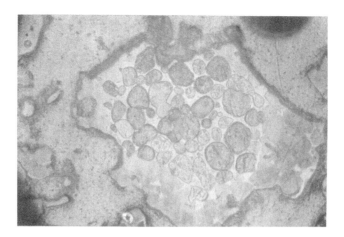

Figure 4.3 Photomicrograph of free lime in Portland cement clinker in reflected light after etching with light HF vapour. Width of field 0.6 mm.

CaO lead cements to be 'unsound', and standard physical tests are routinely carried out on cement to avoid this problem (BS EN 196-3, 2005), but Vivian (1987) has reported delayed expansion in concrete caused by 'hard-burnt' uncombined free lime which was not detected by routine testing.

When present in clinker, CaO is almost always found as distinctive spheroidal particles and never exhibits its cubic crystal faces (Figure 4.3); CaO might also occur as small inclusions within C_3S or C_3A crystals. CaO grains are often very small but occasionally form rounded crystals, or clusters of small crystals, which might be as large as the C_3S crystals. In reflected light, CaO particles are distinguished by their rapid etching with water (whereas etching to distinguish other clinker constituents typically involves KOH or HF reagents) and their spherical shape.

Periclase (MgO) can form in clinkers in which the magnesium content is higher than that which can be taken into solid solution (about 1.5% according to Bye 1983, 1999) and if the clinker is not cooled too rapidly. Usually, such discrete MgO averages only around 1% of the clinker, but occasionally might amount to 5% or more. Although MgO appears to have some value as a flux in cement making, excessive proportions of free MgO can sometimes lead to long-term unsoundness caused by expansion resulting from very slow hydration (Lea 1970): this led the Americans to introduce an autoclave test for assessing cement soundness (ASTM C151 2009).

MgO forms in association with the other interstitial components (e.g. C_3A and ferrite), as small angular crystals or clusters of crystals. In reflected light, these crystals are readily identifiable as highly reflecting grains on a polished but unetched surface, when they are emphasised by the relief caused by their relative hardness and by distinctive dark borders (Figure 4.4).

4.2.1.4 Other phases and gypsum

The other significant but minor constituents of Portland cement clinker are the alkali sulphates and double sulphates, which are the source of most of the alkalis associated with Portland cement. Excess sodium and potassium in the clinker mix appear to combine with calcium silicates and calcium aluminates in a complex manner (Lea 1970). According to Bye (1983, 1999),

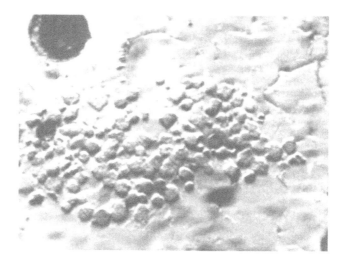

Figure 4.4 Photomicrograph of periclase in Portland cement clinker in reflected light, unetched. (From Fundal, E., Microscopy of cement raw mix and clinker, *FLS-Review 25*, F.L. Smidth Laboratories, Copenhagen, Denmark, 1980.) Width of field 0.150 μm.

water etching of clinker specimens removes the alkali sulphates and creates pores, suggesting that these alkali sulphates might be observed in clinker as later-stage pore infillings.

Gypsum (or occasionally anhydrite) is not a component of the clinker but is instead added to the clinker at the grinding stage and is thus an essential but discrete constituent of the cement power produced. If milling temperatures are too high, the gypsum ($CaSO_4 \cdot 2H_2O$) can become dehydrated to form the 'hemihydrate' ($CaSO_4 \cdot 0.5H_2O$), giving rise to problems of 'false setting' (i.e. apparent setting of cement which can be removed by remixing) caused by rapid rehydration on exposure to moisture (cf. setting of 'plaster of Paris' which is hemihydrate, $CaSO_4 \cdot 0.5H_2O$). The gypsum reacts more or less immediately with C_3A during Portland cement hydration (Lea 1970) and is rarely observed as a residual unreacted cement constituent within hardened concrete or mortar.

Limestone (calcite) fines are an increasingly common addition to many Portland cements as a component added during manufacture at levels ranging between 5% and 30%. These additions were initially considered as primarily inert filler material. However, when added as nano-sized particles, they are now thought to contribute to the formation of additional calcium aluminate and hemicarboaluminates within the hydrate phases and also act as seeding crystallisation points for the formation of additional hydration products including calcium-silicate-hydrate (CSH) gel (Sato and Beaudoin 2007). It has also now become commonplace for cement manufacturers to pre-blend their basic Portland cement product with a range of mineral additions and in various proportions, as indicated in the classification shown in Table 4.1.

4.2.2 Identification of cement type in concrete

As the type of cement (Table 4.1) is often specified for concrete, disputes can arise concerning whether or not the correct cement was used. More importantly, the type of cement might have an important influence over the durability of a concrete being assessed for condition or for future use. In the past, this distinction has most often concerned the difference between ordinary and sulphate-resisting Portland cements (OPCs and SRPCs), but increasingly, questions arise concerning blended cements (see Section 4.2.4) or the presence and quantity of separately batched mineral additives (such as ground-granulated blastfurnace slag [ggbs] or pulverised-fuel ash [pfa], see Section 4.6). These cement types or blends of materials can be traditionally identified from their mineralogical characteristics, but increasingly, investigators use SEM with an in-built energy-dispersive X-ray (EDX) microanalysis facility.

4.2.2.1 General principle of optical microscopical methods

Hardened concrete contains residual unhydrated particles of cement (or clinker), although the quantity and the particle size both vary widely, depending upon composition, curing and exposure. Concretes in which mineral additives (e.g. ggbs, pfa, natural pozzolanas) have been deliberately used will always contain, often substantial, proportions of unreacted mineral additive material. In most cases, microscopical examination of these various residual particles enables the cements and additives to be identified from their mineralogy, and statistical analysis offers prospects for quantifying blend proportions. However, in a hardened concrete, owing to varying reactivity of the binder constituents, the relative proportions of various residual particles will not necessarily be a reliable indicator of their original proportions.

A discussion of the commonly employed method used for distinguishing between ordinary Portland cement (OPC) and sulphate-resisting Portland cement (SRPC) provides an effective introduction to the approach required for identifying any cement and/or mineral additive contents of hardened concrete (these abbreviations are now discouraged, as modern cement composition is more complicated: see Tables 4.1a and 4.1b).

When ground Portland cement is examined in transmitted light, the majority of the particles appear monomineralic. The most obvious particles are crystals of C_3S, which are often broken and rarely have any adherent matrix. Discrete crystals of C_2S are much less common and are more likely to have some matrix material attached to them. The largely monomineralic character of most of the cement particles is not surprising, considering that 80%–90% of the particles are less than 45 µm in size in a typical modern cement. As discussed further in Chapter 5, the residues on a 90 µm sieve vary from as little as 2% for modern cements to as much as 10% for older cements. These oversize particles, which range from 90 to 300 µm or greater in size, consist of matrix in which are commonly embedded C_2S and less commonly C_3S crystals. In hardened cement paste, these remnant grains are surprisingly resistant to hydration, and their examination in polished section can be used to differentiate some cement types.

Traditionally, some types of Portland cement were ground finer to achieve a 'rapid-hardening' effect, and when examining cements, clear observation of smaller mean and maximum particle sizes permits identification of such rapid-hardening Portland cement (RHPC, ASTM Type III) (Sibbick 2011). However, in many parts of the world, the main product is now finely ground and this distinction is less useful. Also, the more finely ground nature of modern cement reduces the frequency of residual unhydrated cement particles within concrete that are large enough to be mineralogically assessed using optical microscopy.

4.2.2.2 Microscopical procedure

A method for distinguishing OPC from SRPC is given in BS 1881-124 (1988), and a good account of the procedure is given by Grove (1968). In considering the reflected light techniques, it is important to realise that different effects are produced by different etchants and, commonly, these are selected to highlight a constituent of particular interest (Table 4.2). In Section 4.2.1, for example, reference has been made to the appearance of various clinker phases when viewed after HF vapour etching, which is frequently used today as a general etchant; it was also explained that CaO was best detected by water etching and that

MgO was most obvious on the polished surface prior to etching. The technique described by Grove (1968) employs a 100 g/L aqueous solution of KOH as the etching agent, which highlights the aluminate phase and is thus particularly applicable to the differentiation of OPC and SRPC.

Conventional sulphate attack on concrete (see Chapter 6) is most usually associated with the expansive secondary formation of ettringite within hardened concrete as the result of reaction between externally derived sulphates in solution and the calcium aluminate and calcium monosulphoaluminate hydrate phases in the hydrated cement paste. SRPC is based on the principle of reducing the C_3A content of the interstitial clinker phase in manufacture by correspondingly increasing the content of ferrite (notionally C_4AF), although it remains uncertain why the calcium aluminoferrite is not also susceptible to sulphate attack (Lea 1970). A review of sulphate attack and the effectiveness of SRPC was carried out by Lawrence (1990).

SRPC was thus defined by limiting the permitted content of C_3A to 3.5% in the previous BS 4027 (1996), which is now superseded by BS EN 197-1 (2011). The sulphate-resisting cement products in BS EN 197-1 (see Table 4.1a) are primarily CEM I-SR 0, CEM I-SR 3 and CEM I-SR 5 (with C_3A limits of 0%, ≤3% and ≤5%, respectively), which roughly equate to the former BS 4027. BS EN 197-1 also includes CEM III/B-SR, CEM III/C-SR, CEM IV/A-SR and CEM IV/B-SR (with a C_3A limit of ≤9% on the Portland cement clinker component of the CEM IV products). In ASTM C150 (ASTM C150 2012), the permitted content of C_3A is limited to 5%. As research suggested that the effectiveness of low-C_3A cements was reduced at higher C_4AF contents, ASTM C150 also places a 25% upper limit on the total content of C_4AF plus twice the C_3A.

In distinguishing SRPC from OPC, therefore, the ratio of C_3A to C_4AF (ferrite) is critical. In Grove's method, a highly polished surface of the concrete matrix (see Chapter 3) is treated with the KOH solution for 10 or 20 s at room temperature, then washed in ethanol. It is important to avoid over-etching the specimen. When the residual unhydrated cement particles in the etched surface are examined under a reflected light microscope, the C_3A exhibits a distinctive blue or blue-grey colour, while the ferrite is unetched and appears bright and highly reflectant; the other clinker phases are dull by comparison.

In most cases, the difference between representative clinker grains of OPC or SRPC is visually apparent (Figure 4.5), with the volume ratio of ferrite to C_3A rarely exceeding 2:1 for OPC, but being more than 5:1 for SRPC. In fact, the C_3A can be difficult to find in many SRPCs, such is the dominance of ferrite: Grove (1968) states it is usually necessary 'to search' for C_3A, and

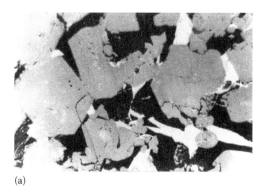
(a)

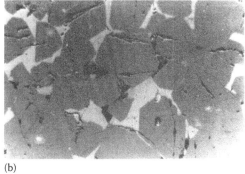
(b)

Figure 4.5 Photomicrographs: comparison of (a) OPC and (b) SRPC clinkers in reflected light after etching with 10% KOH. Width of fields 0.250 μm. (From Grove, R.M., *Silicates Indust.*, 10, 317, 1968.)

French (1991a) agrees that ferrite 'makes a very large part' of the matrix in SRPC. Therefore, although the relative proportions of C_3A and ferrite could be quantified by point counting in Grove's method (Weigand 1994), in practice, this is rarely necessary for the experienced cement microscopist. De Rooij (2011) warns that, in a hardened concrete, much of the original C_3A might have been consumed by early-age hydration reactions, such that an apparent paucity of residual C_3A might be misleading; partly for this reason, the authors recommend weighting the findings towards the coarser residual cement grains (see Section 4.2.2.3).

4.2.2.3 Interpretation of the findings and some difficulties

In many cases, the KOH etching method is a reliable and comparatively straightforward means of confirming whether the cement used in a concrete was OPC or SRPC. The optional method using HF vapour is usually also effective, except that the visual contrast between C_3A (light grey) and ferrite (white) in the interstitial component is rather less obvious; however, the HF method also provides information about the other clinker minerals, notably C_3S and C_2S.

Optical studies of Portland cement clinker prior to grinding illustrate its often variable texture and composition, and analysis of these textural variations can assist cement technologists in perfecting manufacturing techniques (Fundal 1980). It therefore follows that, after grinding, the individual cement grains will be variable in composition and only the exceptionally large grains are likely to approach being representative of the original overall clinker. When examining specimens of hardened concrete matrix, the cement-type determination has to be based upon whatever residual unhydrated particles of cement remain, and the limitation on representativeness of individual grains must be taken into account. At least twenty such residual particles should be examined and the mineral composition of each assessed.

The authors also recommend that the approximate size of each particle should be recorded and the compositional findings classified according to size band (<20, 20–40 and >40 µm bands have been found satisfactory). Then, in making final judgement in apparently marginal cases, reliance weighting should be placed upon the findings for the coarser particles (>40 µm), and, if necessary, further analyses should be carried out, perhaps using additional specimens of the concrete matrix in question. One form of practical reporting of results is shown, with an example in Table 4.3.

Mixtures of Portland cement types in concrete are uncommon and are very difficult for the microscopist to detect with certainty. Such mixtures may result from contamination, for example, within cement silos or from inadvertent use of different types of cement (or different sources of the same type) at the concrete-making stage. However, clinkers of either the same or different cement types from different sources are sometimes deliberately blended. One U.K. manufacturer has employed blends of OPC and SRPC in order to create a 'reduced alkali' cement (the alkali content of SRPC is typically lower than that of the related OPC), for possible use with aggregates considered to be potentially alkali reactive (see Chapter 6). Such mixtures of cement types can sometimes be identified and possibly even quantified by statistical treatment of the microscopical findings, although a comparatively large number of residual unhydrated particles would need to be found in the concrete matrix and examined.

The microscopical method obviously depends upon the hardened concrete matrix containing a sufficient number of residual unhydrated cement particles of microscopically resolvable size, and this is not always the case. In some modern very finely milled cements (e.g. RHPC), such unhydrated particles might sometimes be hard to find.

French (1991a) has claimed that the content of residual unhydrated cement in concrete is related to the water/cement ratio and the temperature during curing. In concretes made at 20°C, for example, he found that the cement powder was 'virtually completely hydrated' for water/cement ratios of 0.6 or more. Similarly, for concrete cured at between 40°C and

Table 4.3 Determination of Portland cement type by reflected light microscopy – the suggested form and a typical example

Sample details and methods of treatment			
Laboratory ref:	11525	Site ref:	50
Location:		Block A, south elevation	
Type of material:		Brickwork jointing mortar	
Preparation and procedure:		Polished specimen etched for 3 s in HF vapour	
Test results – apparent cement type			
Unhydrated grain size:	>40 μm	40–20 μm	<20 μm
Number of grains:			
OPC appearance	—	2	3
SRPC appearance	3	12	10
HAC appearance	—	—	—
WPC appearance	—	—	—
Total:	3	14	13
Percentage of grains:			
OPC appearance	—	14	23
SRPC appearance	100	86	77
HAC appearance	—	—	—
WPC appearance	—	—	—
Total:	100	100	100
Test results – other observations			
Mineral additions:		Building lime	
Other constituents or features:		Yellow pigment	
Conclusions			
Apparent cement type:		SRPC	
Other comments:		Mortar matrix that was soft	

Note: SRPC in a mortar with building lime.

50°C, French found that hydration was 'generally complete', although others have reported a 'limited' presence of residual clinker for concretes cured up to ~80°C (Patel et al. 1995). In the authors' experience, for the bulk of concretes made at temperatures below 20°C and/or at water/cement ratios of less than 0.6, there is usually a sufficiency of unhydrated cement particles in the matrix for the cement type to be identified microscopically.

Although OPC and SRPC have traditionally been by far the most commonly used cements, there are other specialised varieties of Portland cement, increasingly blended cements (see Section 4.2.4) and non-Portland cements such as aluminous cements. Good summaries of the range of modern cement types are given by Neville and Brooks (1987, 2010) and Moir (2003). The possibility of these other cement types, or the possible presence of separately batched mineral additives, must not be forgotten by the microscopist when examining concrete.

Microscopy is particularly useful, for example, for confirming the presence of white Portland cement (WPC), which is frequently employed for concrete for architectural usage, sometimes with various pigments that mask the white cement coloration. White Portland cement is made by using limestone and aluminosilicate raw materials, such as bauxite, chalk and china clay, which are low in the iron which usually gives OPC its distinctive grey colour. The resultant white clinker has less than 1% of ferrite and a substantially higher C_2S content than OPC, while the aluminate content is comparable with OPC or even a little higher than average (Bye 1983, 1999). Thus, the near absence of ferrite and the high content of C_2S make

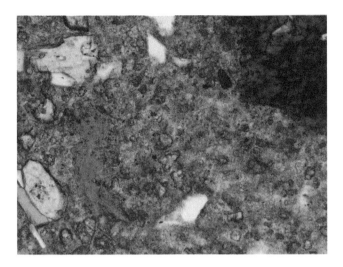

Figure 4.6 Typical appearance of white Portland cement in concrete, showing a near absence of ferrite and a substantially higher content of C_2S than for ordinary Portland cement. (Photomicrograph courtesy of Paul Bennett-Hughes, RSK Environment Ltd., U.K.)

the cement relatively easy to identify microscopically (Figure 4.6). White Portland cement has a lower alkali content and typically also lower strength properties than conventional OPC.

Unhydrated cement particles are visible in thin sections of concrete, especially if the sections are ground thinner than usual for petrographic purposes (say to 10 or 15 μm, rather than the standard 25 μm for concrete and related materials), but it is extremely difficult to distinguish between types on a mineralogical basis, mainly because the grain size of the critical interstitial components is close to the practical resolution limits of the optical system. In most Portland cement concretes, it is the distinctive clusters of round C_2S (belite) crystals set in the clinker matrix (Figure 4.2) that are most readily noticed as residual cement material in thin section (Parker and Hirst 1935).

Extreme cases might be discernible in thin section when present as comparatively large residual grains. Examples would include SRPC which is notably ferrite rich or white Portland cement which is equally notably ferrite poor and also high in C_2S or a non-Portland type of cement (such as high-alumina cement [HAC], see Section 4.2.4). The presence of mineral additions, especially ggbs or pfa, either as a blended cement or separately added, is usually detectable in thin section, but silica fume might be difficult to resolve unless present in agglomerations (see Section 4.6.3).

It is theoretically possible to judge the likelihood of RHPC being present in a concrete from thin sections, by studying the maximum size of the unhydrated cement particles. The bulk of remnant cement grains in RHPC will be found to be less than 20 μm in size and difficult to find in older concretes. However, as mentioned earlier, in more recent concrete, it will not be possible reliably to discern RHPC from a standard production Portland cement that has been routinely ground to RHPC fineness.

4.2.2.4 SEM and other methods

An effective modern alternative or extension to the range of optical techniques for identifying cement type in concrete is characterisation and EDX microanalysis of residual cement particles during examination using SEM. A recent specialised treatment may be found in Winter (2012a). French (1991a) described three approaches to the determination of 'apparent' cement type during SEM examination of polished sections of concrete:

1. Identification of the interstitial phases in residual cement particles (this is comparable with the optical methods)
2. Overall chemical analysis of the larger residual cement particles
3. Chemical analysis of the hydrated paste

In the first approach, analysis of a 'spot' 5 μm in diameter enables the proportions of C_3A and C_4AF to be determined using the Bogue (1955) equations, but the analyses can be unreliable owing to the small size of the interstitial areas and consequent problems of interference from surrounding phases. Clearly, the overall microanalysis of residual cement particles, the second approach, ought to be effective at characterising the cement, provided, as with the optical methods, that a sufficiently large number of particles are analysed. The method also offers a prospect of identifying different sources of the same type of cement, as each cement source might be expected to be chemically distinctive. The third approach, area microanalysis of the hydrated cement paste, is complicated by the additional presence of aggregate dust. French (1991a) states that the method yields one of the three findings: high C_3A implying OPC, low C_3A implying SRPC or ambiguous apparently intermediate C_3A content which a statistical analysis of the data might help to clarify.

An excellent account of using modern SEM–EDX techniques for studying the composition of cementitious materials is to be found in Stutzman (2007, 2012), and general guidance on the SEM technique is now available in an ASTM standard (ASTM C1723 2010), which supplements the long-standing practice for concrete petrography using optical microscopy (ASTM C856 2011).

Hammersley (1980) found few other techniques helpful in identifying cement type. X-ray diffraction (XRD), infrared absorption spectroscopy, differential thermal analysis (DTA) and chemical analysis were each found to be useful in studying hydration or in distinguishing between Portland and non-Portland types of cement in concrete but were of limited application in identifying the type of Portland cement. One of these, overall chemical analysis of concrete, is frequently used in commercial practice to suggest the probable cement type present, but where the aspect of cement type is important, confirmatory optical or electron microprobe identification is indispensable.

One clear recent development is the increasing use of quantitative XRD–Rietveld analysis by cement companies and some analytical companies. Increasing expertise in its use with cement compositional analysis has shown it to be an increasingly reliable method. Comparison with the 'Bogue' analysis and clinker microscopy shows it to be at least comparable with microscopy and considerably better than Bogue (Stutzman 2004).

Where the ranges of possible cement and aggregate sources are limited and known, it can be possible to 'fingerprint' the cement present by selective chemical analysis of the concrete material. Goguel and St John (1993a,b), for example, devised a simple procedure for New Zealand concretes whereby analysis for Ca, Sr and Mn enabled the cement source to be identified. In modern practice, as the aforementioned, direct SEM–EDX microanalysis of a representative number of residual cement grains can enable a cement source to be confirmed.

4.2.3 Hydrated cement phases

The four principal cement constituents (C_3S, C_2S, C_3A and C_4AF; see Section 4.2.1) are each hydraulic and thus react with water on mixing to form a range of hydrated phases. Detailed accounts of the complex chemistry and mineralogy of cement hydration are given elsewhere

(Copeland and Kantro 1964, Lea 1970, Czernin 1980, Bye 1983, 1999, 2011, Taylor 1990, 1997, Hewlett 1998, Bensted and Barnes 2002, Moir 2003 and Locher 2006). The initial post-hardening result of hydrating Portland cement at normal temperatures is a groundmass of poorly crystalline and microporous hydrated compounds with scattered inclusions of crystallised portlandite (calcium hydroxide), other crystalline phases and residual kernels of unhydrated cement. A simplified illustration of cement hydration has been provided by Moir (2003) and is reproduced in Figure 4.7. Carbonation and general recrystallisation of this hardened 'cement paste' takes place gradually with time (see Chapter 5).

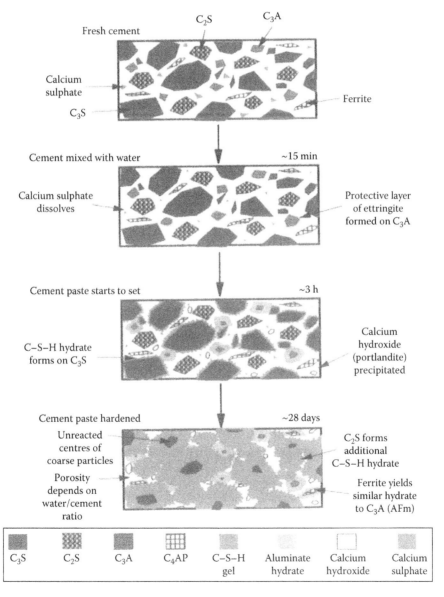

Figure 4.7 Simplified illustration of cement hydration. (From Moir, G., Cements, in *Advanced Concrete Technology – Constituent Materials*, Newman, J. and Choo, B.S. [eds], Chapter 1, pp. 1/3–1/45, Butterworth-Heinemann [Elsevier], Oxford, U.K., 2003.)

4.2.3.1 CSH and the microstructure of cement paste

As the calcium silicates (C_3S and C_2S) dominate the composition of Portland cement, it follows that the constitution of hydrated Portland cement paste is largely dictated by the reactions between these calcium silicates and water (idealised chemical equations are given in Table 4.4). Tricalcium silicate (C_3S) reacts with water and hydrated calcium silicate is quickly formed, together with a supersaturated solution of calcium hydroxide from which crystals of calcium hydroxide are subsequently precipitated, finally leaving a 'pore solution' which is saturated in respect of calcium hydroxide. Dicalcium silicate (βC_2S) exhibits a broadly similar hydration reaction, albeit much less rapidly and producing two-thirds less portlandite.

The hydrated calcium silicate, which is principally responsible for the binding properties of Portland cement, appears to exhibit a highly variable chemical composition (lime–silica ratios typically range from 1 to 2) and is practically amorphous. Two types of hydrated cement paste (or 'gel' as it is often called) have generally been recognised: C–S–H (I), which is structurally related to naturally occurring mineral tobermorite, and C–S–H (II), which is more related to another natural mineral jennite (Taylor 1997). Previously, some cement mineralogists have referred to the hydrated calcium silicate phase of cement paste as 'tobermorite gel' (Brunauer et al. 1958). In reality, it now seems that C–S–H in concrete is usually a mixture of phases related to tobermorite and jennite (Richardson 2004, 2008). Richardson (2004) is well illustrated by helpful diagrams and transmission electron microscopy micrographs. SEM micrographs of Portland cement hydration processes and products have been provided by Möser and Stark (2007) and Möser (2009).

Hydration of the aluminate and ferrite phases (C_3A and C_4AF) also involves the alkali sulphate and gypsum components of Portland cement (in the absence of gypsum, the extremely rapid hydration of C_3A would lead to 'flash setting'). The precise series of reactions which occur are seemingly dependent upon a number of factors, including the relative proportions present of C_3A, sulphates and calcium hydroxide, but generalised chemical equations are included in Table 4.4.

Initially, C_3A and calcium sulphate react rapidly to form ettringite, hydrated calcium sulphoaluminate. The ettringite envelopes the unhydrated C_3A and effectively prevents the cement from setting too quickly. With time the ettringite is replaced by calcium

Table 4.4 Idealised hydration reactions for Portland cement mineral phases

Tricalcium silicate:

$$2(3CaO \cdot SiO_2) + 6H_2O \rightarrow 3CaO \cdot 2SiO_2 \cdot 3H_2O + 3Ca(OH)_2$$

Dicalcium silicate:

$$2(2CaO \cdot SiO_2) + 4H_2O \rightarrow 3CaO \cdot 2SiO_2 \cdot 3H_2O + Ca(OH)_2$$

Tricalcium aluminate and gypsum:

$$3CaO \cdot Al_2O_3 + 3\left(CaSO_4 \cdot 2H_2O\right) + 26H_2O \rightarrow \underset{\text{ettringite}}{3CaO \cdot Al_2O_3 \cdot 3CaSO_4 \cdot 32H_2O}$$

then (partially)

$$2\left(3CaO \cdot Al_2O_3\right) + 3CaO \cdot Al_2O_3 \cdot 3CaSO_4 \cdot 32H_2O + 4H_2O \rightarrow 3\left(\underset{\text{monosulphate}}{3CaO \cdot Al_2O_3 \cdot CaSO_4 \cdot 12H_2O}\right)$$

then (partially)

$$3CaO \cdot Al_2O_3 + Ca(OH)_2 + 12H_2O \rightarrow 4CaO \cdot Al_2O_3 \cdot 13H_2O$$

Tetracalcium aluminoferrite:

$$4CaO \cdot Al_2O_3 \cdot Fe_2O_3 + 4Ca(OH)_2 + 22H_2O \rightarrow 4CaO \cdot Al_2O_3 \cdot 13H_2O + 4CaO \cdot Fe_2O_3 \cdot 13H_2O$$

Source: After Czernin, W., *Cement Chemistry and Physics for Civil Engineers*, translated by Amerongen, C. van (ed), 2nd English ed., George Godwin Ltd, London, U.K. for Bauverlag Gmbh, Wiesbaden, Germany, 1980.

monosulphoaluminate ('monosulphate'). As reaction occurs between residual C_3A and the initially formed ettringite coating layer, once the primary sulphates have been consumed, normal setting of the cement proceeds. If there is residual C_3A present, in the presence of calcium hydroxide, later-stage hydrated calcium aluminate can also be formed, usually of the form C_4AH_{13}, but the 'hydrogarnet' (C_3AH_6) might be formed under some circumstances. The much slower hydration of C_4AF is roughly analogous to that of C_3A after the consumption of the sulphates (Table 4.4), leading to the compound C_4FH_{13}.

The considerable intermixing of the various hydrated phases in the main period of hydration results in an extremely complicated mixture of phases. Its microstructure is difficult to resolve even using electron microscopy. Under the SEM, the C–S–H phases are often said to appear as filaments or tubular structures, although Jennings and Pratt (1980) have suggested that this appearance results from the rolling up of thin amorphous sheets caused by the drying involved in specimen preparation for electron microscopy. C–S–H also occurs pseudomorphing the original cement grains. The portlandite and hydrated calcium aluminate form hexagonal platelets, while crystalline ettringite and monosulphate form distinctive clusters of acicular crystals. Some typical SEM views of hydrated cement paste are shown in Figure 4.8. Diamond (1976) has estimated a typical composition of almost fully hydrated paste to be (see also Chapter 5):

CSH	70%
Portlandite	20%
Ettringite/monosulphate	7%
Minor phases	3%

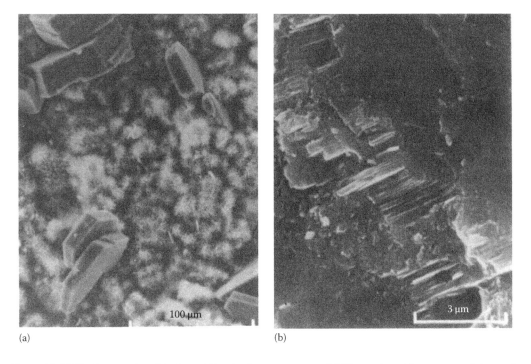

(a)   (b)

Figure 4.8 SEM micrographs of hydrated Portland cement paste: (a) Three-day-old paste containing well-defined portlandite crystals. (b) Fracture through dense 'groundmass' in 8-month-old paste. (From Bye, G.C., *Portland Cement – Composition, Production and Properties,* Institute of Ceramics/Pergamon Press, Oxford, U.K., 1983.)

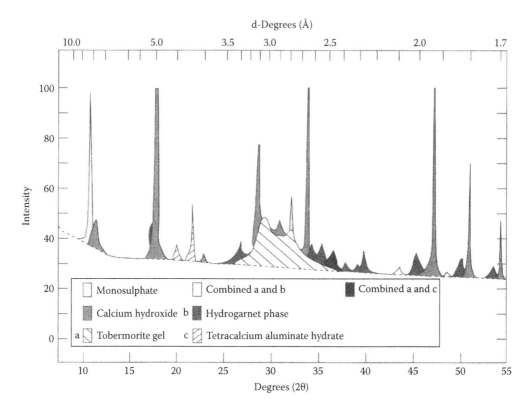

Figure 4.9 Schematic X-ray diffractogram of Portland cement paste. (From Copeland, L.E. and Kantro, D.L., Chemistry of hydration of Portland cement at ordinary temperature, in *The Chemistry of Cements*, Taylor, H.F.W. [ed], Vol. I, Chapter 8, pp. 313–370, Academic Press, London, U.K., 1964.)

Hydrated cement paste is porous, exhibiting a specific surface area in the region of 200 m²/g, implying pores of <10 nm diameter which are not easily observed by SEM (Bye 1983, 1999). Powers and Brownyard (1947/1948) recognised two types of cement paste porosity: gel pores contained within an area of CSH particles and relatively coarser capillary pores representing space originally occupied by air or by mix water not used in hydration. The latter are clearly related to the cement and water contents of the concrete mix and are used as the basis of several methods for estimating the original water/cement ratio of concrete (see Section 4.4).

The hydrated cement paste in samples of concrete is difficult to study by any optical system. As most of the cement 'gel' is virtually amorphous, it appears featureless and isotropic in thin section, although the crystallinity, grain size and distribution of portlandite are usually readily apparent under the polarising microscope (see Section 4.2.3.3). The composition of cement paste can be qualitatively determined by XRD (Figure 4.9), provided that interference from aggregate constituents can be avoided. Over an extended period of time, cement paste gradually alters and crystallises, and especially carbonation frequently occurs as a result of reaction with the atmosphere (see Chapter 5).

4.2.3.2 Degree of hydration

The extent to which the cement in concrete becomes hydrated principally depends upon the water/cement ratio and the degree of curing. The degree of hydration is important as the concrete strength is influenced not only by the content of cement in the mix but also by the degree of hydration of that cement. A low water/cement ratio is usually desirable, provided that the concrete can be compacted. However, exceptionally low water/cement ratios (say about 0.20) can also lead to under-hydration of the cement in some circumstances and, sometimes, consequently weak concrete. Conversely, an excessively high water/cement ratio can lead to almost complete hydration but cause a large capillary pore volume in the cement paste (even 'bleeding') and thus impair both strength and impermeability of the concrete.

Thus, for any combination of constituents, required mix properties and conditions, there is a critical degree of hydration associated with the maximum attainable strength and minimised permeability. Of course, except for concretes kept in completely dry conditions, the degree of hydration is liable to increase with time after the initial period of hardening and strength gain, as residual cement particles very gradually become hydrated.

Determination of the degree of hydration for a particular concrete, therefore, might sometimes assist in the assessment of concrete strength and permeability. Such estimates are feasible using chemical analysis, whereby the determined cement and CO_2 contents are compared with the content of combined water, but the method requires generalised assumptions and is complicated by the presence of aggregates. As the anhydrous cement compounds are crystalline, XRD analysis is potentially effective, but interference from the poorly crystalline hydrates and aggregate materials makes reliable interpretation difficult; moreover, determining a content of residual cement does not assist with finding the original cement content.

Microscopical examination in reflected light of polished specimens of concrete matrix (see Section 4.2.2) offers a direct method of determining degree of hydration. In concretes of unknown mix design, the volumetric ratio between the contents of hydrated and residual unhydrated materials in the cement paste is clearly related to the degree of hydration, although many factors affect the volume change accompanying hydration, so that such an approach can only be qualitative.

Ravenscroft (1982) has described a microscopical quality control method for determining the degree of hydration for concretes of known original mix proportions. In this method, the volume proportions of residual cement particles and fine aggregate in portions of the hardened cement matrix are compared with those of the original mix proportions. The volumetric determinations by point counting under the reflected light microscope are converted to gravimetric results by assigning appropriate specific gravity values to the cement and aggregate constituents.

In this way, the degree of hydration (Dh) is determined as follows:

$$Dh = \frac{(Cm/Sm - Cs/Ss) \times 100}{Cm/Sm}$$

where
   Cm/Sm is the cement to sand ratio by mass in original concrete mix
   Cs/Ss is the unhydrated cement to sand ratio by mass in the sample

An example of this determinative method is shown in Table 4.5.

4.2.3.3 Portlandite (Ca(OH)₂)

It was explained earlier that portlandite ($Ca(OH)_2$) is formed during the hydration of the calcium silicates in Portland cement and it is thus characteristic of uncarbonated cement paste. Typically, $Ca(OH)_2$ forms as hexagonal crystals with a perfect basal cleavage. In hardened cement paste, it appears as anhedral or euhedral crystallites and plates or as short prisms, up to about 100 μm in maximum crystallite length when precipitated during the

Table 4.5 Determination of the degree of hydration

The formula proposed by Ravenscroft (1982) is

$$Dh = \frac{(Cm/Sm - Cs/Ss) \times 100}{Cm/Sm}$$

where

Dh is the degree of hydration

Cm/Sm is the cement/sand ratio in mix

Cs/Ss is the unhydrated cement/sand ratio in sample

If a concrete had mix proportions of 1:2:4,

Cm/Sm = 1/2 = 0.5

If point counting of the mortar component had given

Unhydrated cement 105 points

Sand grains 895 points

Cs/Ss = 105/895 = 0.12

Thus, the degree of hydration (Dh) may be calculated from

$$Dh = \frac{(0.5 - 0.12) \times 100}{0.5} = 76\%$$

Note: A hypothetical example.

main hydration phase. The crystallites are distinctive in thin section under the polarising microscope, being colourless and highly birefringent (see the 'Glossary of minerals').

Christensen et al. (1979) reported that using optical microscopy, portlandite appears to occur in hardened concrete in three forms (Figure 4.10):

1. Small crystals uniformly distributed in the cement paste (Figure 4.10a)
2. Coarser crystals at aggregate particle boundaries and/or at the edge of voids
3. Recrystallised relatively very coarse crystals in cracks and cavities (Figure 4.10b)

Also, Berger and McGregor (1973), using electron microscopy, suggested that submicroscopic ($\sim$0.1 μm) globular particles scattered within the CSH were probably amorphous $Ca(OH)_2$. In comparing different techniques for determining the content of $Ca(OH)_2$ in set Portland cement, Midgley (1979) found that XRD, which only detects crystalline material,

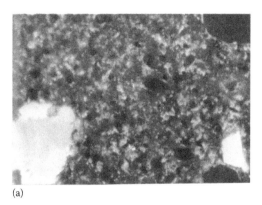
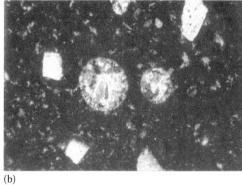

(a)   (b)

Figure 4.10 Photomicrographs of portlandite as initial hydrates in thin section: (a) Disseminated crystallites. (b) Infilled voids. Width of fields 0.4 mm. (Photographs courtesy of Mr Mark Knight, Sandberg LLP, London, U.K.)

underestimated the amount present. For example, in one sample containing ~16% $Ca(OH)_2$ determined using DTA, XRD indicated only ~13% and Midgley attributed this to the presence of ~3% amorphous (i.e. non-crystalline) portlandite in the sample.

Referring to the conspicuous portlandite that occurs at the interface between aggregate particles and the cement paste, French (1991a) noted that the quantity of such material appears to be dependent upon the aggregate type. He found that, for a given water/cement ratio, relatively thick layers of portlandite develop on siliceous particles, such as chert or quartz, but that the layers on carbonate rock surfaces were thin and impersistent. Also, the amount of this boundary portlandite evidently increased with aggregate surface area, being apparently more abundant in concretes made with finer sand.

The observed distribution and crystallite size of the portlandite in the cement paste can sometimes be used as indicators of the likely properties of the concrete mix or as evidence of the conditions to which the concrete has possibly been exposed, although any such interpretations should be treated with circumspection and must be corroborated by other factors. For example, portlandite crystals formed in concretes made with high water/cement ratios tend to be well defined, relatively coarse and predominantly situated along aggregate boundaries and void edges, while those formed at lower water/cement ratios are often irregular in shape, smaller and distributed uniformly through the paste (French 1991). According to Jepsen and Christensen (1989), the 'unusually large' $Ca(OH)_2$ crystals which form when 'the concrete is too wet' (i.e. high water/cement ratio) can be indicative of bleeding.

Portlandite size and form can also indicate frost or freeze–thaw damage to concrete. The solubility of portlandite decreases with rising temperature (Taylor 1964), and the corollary is that the solubility is increased at lower temperatures. For example, Bassett (1934) demonstrated that the solubility was 1.3 g/L as CaO at 0°C, dropping to 1.2 g/L at 18°C or 0.5 g/L at 100°C (see also Chapter 5). According to Idorn (1969), microcrystalline $Ca(OH)_2$ can thus redissolve and be re-precipitated during freezing and thawing cycles, allowing large crystalline accumulations of portlandite to form (Figure 4.11). The occurrence of such accumulations, plus evidence of chemically unaltered cement paste (Idorn cites the presence of unhydrated C_2S), 'is a good indication' that freeze–thaw action might have caused damage.

Again, however, caution must be exercised in making such interpretations, and any apparent freeze–thaw action indicated by portlandite appearance should be verified by the presence of other characteristic features, such as macro- and microcracking patterns. As correctly reported by French (1991a), very coarse portlandite crystals can occur in factory-made concrete subjected to steam curing, and again his is related to the reduced solubility of portlandite at elevated temperatures. In these cases, the portlandite might be intergrown with ettringite (see the next section). Patel et al. (1995) have similarly shown that relatively coarse portlandite can be characteristic of concretes cured at elevated temperatures, and, indeed, according to de Rooij (2011), the presence of relatively large portlandite developments can also be a telltale sign of high levels of heat generation within the concrete.

Calcium hydroxide can be dissolved by percolating water, often being redeposited in open spaces elsewhere within the concrete. Thus, often comparatively very coarse secondary development of portlandite within voids and open cracks within hardened concrete (i.e. Type 3 in Christensen et al., 1979) are fairly reliable indicators of such leaching by the migration of water through the concrete. Commonly such secondary portlandite appears either as a lining to the void or crack (Figure 4.12) or as a more or less complete infilling by coarse platy crystals. Such portlandite leaching products often occur in close association with secondary ettringite deposits. Portlandite is also subject to carbonation, but this is dealt with in detail in Chapter 5.

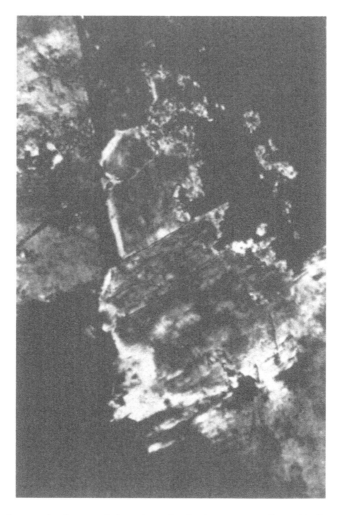

Figure 4.11 Photomicrograph of accumulation of portlandite in concrete affected by frost. The field of view is about 0.4 mm across. (From Idorn 1967; Idorn, G.M., *The Durability of Concrete,* Technical Paper PCS 46, The Concrete Society, London, U.K., 1969.)

4.2.3.4 Ettringite and some other complex phases

As already explained, ettringite $(3CaO \cdot Al_2O_3 \cdot 3CaSO_4 \cdot 26\text{–}32H_2O)$ is the initial cement hydration product of C_3A and water in the presence of calcium sulphate. When the hydration of C_3A continues after the supply of sulphate ions diminishes, then the first-formed ettringite decomposes and is progressively replaced by the 'monosulphate' form $(3CaO \cdot Al_2O_3 \cdot CaSO_4 \cdot 12H_2O)$. These are sometimes referred to as the 'high' (ettringite or tri-sulphate) and 'low' (monosulphate) forms of calcium sulphoaluminate hydrate.

However, these hydration products are generally submicroscopic, and it is usually secondary ettringite that is observed under the optical microscope during the examination of concrete. A particular case is the more coarsely crystalline ettringite that can eventually develop in some concretes cured at very high temperatures (French 1991a, see also 'Delayed ettringite formation' in Chapter 6). Otherwise, the secondary ettringite usually results either from one (the 'conventional') type of sulphate attack (see Chapter 6) or, perhaps more commonly, from leaching and redeposition through the percolation of water.

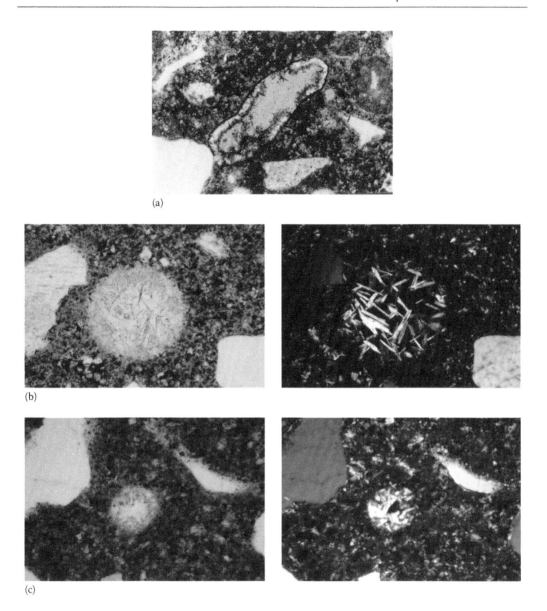

Figure 4.12 Photomicrographs of secondary portlandite deposits in voids: (a) Elongate void lined with port-landite, ppl, width of field ~1.5 mm. (b) Spherical void filled with portlandite, left ppl and right xpl, widths of field ~0.6 mm. (c) Spherical void lined with portlandite, left ppl and right xpl, widths of field ~0.3 mm. (Photographs in [b] and [c] courtesy of Dr Ted Sibbick.)

In the late nineteenth century, the needle-like crystals found to be associated with the type of deterioration now recognised as (conventional) sulphate attack were termed 'cement bacillus', and Candlot (1890) identified these 'bacilli' as ettringite. Acicular crystals of ett-ringite are rather common in concrete, often lining or filling voids or cracks (Figure 4.13), and it is important to realise that these have most usually resulted from the redistribution of sulphates within the concrete by migrating water and thus might indicate a degree of leaching: expansive sulphate attack is comparatively rare and, if suspected, should be corroborated by other observations.

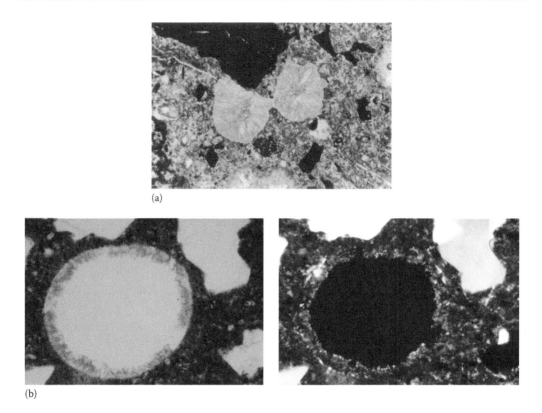

(a)

(b)

Figure 4.13 Photomicrographs of ettringite in voids in concrete in thin section: (a) Voids filled with ettringite, ppl, width of field ~0.9 mm. (Photograph courtesy of RSK Environment Ltd., U.K.) (b) Spherical void lined with ettringite, left ppl and right xpl, widths of field ~0.7 mm. (Photographs courtesy of Dr Ted Sibbick.)

Detailed descriptions of the complex compositions, structures and properties of ettringite and monosulphate are given by Turriziani (1964) and Lea (1970). Typically, ettringite forms long, thin, colourless 'needles', of low to medium birefringence, commonly occurring in radiating clusters or sometimes as crustiform linings. The optical properties are detailed in the 'Glossary of minerals' and summarised in Table 4.6. Ettringite is relatively easy to detect by DTA (Midgley and Rosaman 1960), when it is characterised by a large endotherm at low temperature (110°C–150°C) or by XRD (Midgley 1957; Crammond 1985a).

In recent years, it has been discovered that early prolonged high temperature (say >65°C–75°C) curing of precast concrete can interfere with the normal hydration reactions of C_3A and sulphates, initially suppressing the formation of calcium sulphoaluminate hydrate, so that later expansion and damage can be caused by 'delayed ettringite formation' (DEF) when the concrete is subsequently exposed to moist conditions or wetting (Heinz and Ludwig 1987; Lawrence et al. 1990). Heinz and Ludwig (1987) reported that the expansive secondary ettringite formation occurred mainly at the interfacial zone between cement paste and aggregate particles (Figure 4.14).

Lawrence et al. (1990) gave some preliminary recommendations for minimising the risk of DEF, variously controlling the premature application of high temperature (i.e. not before 3 or 4 h after casting), the rates of both raising and later reducing the curing temperatures

Table 4.6 Some secondary deposits in concrete

Compound/ mineral	Chemical formula	Occurrence frequency	Refraction indices	Form
Calcium carbonate/ calcite	$CaCO_3$	Very common	$\omega = 1.658$ $\varepsilon = 1.486$	Fine-grained, white or grey masses or coatings in the cement paste, in voids, along cracks or on exposed surfaces
Calcium carbonate/ aragonite	$CaCO_3$	Rare	$\alpha = 1.530$ $\beta = 1.680$ $\chi = 1.685$	Minute white prisms or needles in voids or cracks
Calcium carbonate/ vaterite	$CaCO_3$	Common	$o = 1.544–1.550$ $E = 1.640–1.650$	Spherulitic, form birefringent, white encrustations on moist-stored laboratory specimens; also identified in sound concrete by XRD
Calcium sulphoaluminate hydrate/ ettringite	$3CaO \cdot Al_2O_3 \cdot 3CaSO_4 \cdot 32H_2O$	Very common	$\omega = 1.464–1.469$ $\varepsilon = 1.458–1.462$	Fine white fibres or needles or spherulitic growths in voids or cracks or in the cement paste
Calcium sulphoaluminate hydrate/ monosulphate	$3CaO \cdot Al_2O_3 \cdot CaSO_4 \cdot 12H_2O$	Very rare	$\omega = 1.504$ $\varepsilon = 1.49$	Minute white to colourless hexagonal plates in voids and cracks
Calcium aluminate hydrate	$4CaO \cdot Al_2O_3 \cdot 13H_2O$	Very rare	$\omega = 1.53$ $\varepsilon = 1.52$	Mica-like, colourless, pseudo-hexagonal twinned crystals in voids
Hydrous sodium carbonate/ thermonatrite	$Na_2O \cdot CO_2 \cdot H_2O$	Rare	$\alpha = 1.420$ $\beta = 1.506$ $\chi = 1.524$	Minute inclusions in alkali–silica gel
Hydrated aluminium silicate/ paraluminite	$2Al_2O_3 \cdot SO_3 \cdot 15H_2O$	Very rare	$\alpha = 1.463 \pm 0.003$ $\beta = 1.471$ $\chi = 1.471$	In cavities in intensely altered concrete
Calcium sulphate dihydrate/ gypsum	$CaSO_4 \cdot 2H_2O$	Unusual	$\alpha = 1.521$ $\beta = 1.523$ $\chi = 1.530$	White to colourless crystals in paste, in voids or along aggregate surfaces in concrete or mortar affected by sulphate or seawater attack
Calcium hydroxide/ portlandite	$Ca(OH)_2$	Ubiquitous	$\omega = 1.574$ $\varepsilon = 1.547$	White to colourless hexagonal plates in paste, voids and cracks; reduced frequency in the presence of pozzolanic constituents
Magnesium hydroxide/ brucite	$Mg(OH)_2$	Unusual	$\omega = 1.559$ $\varepsilon = 1.580$	White to yellow fine-grained fillings and encrustations in concrete attacked by seawater
Hydrous silica/ opaline silica	$SiO_2 \cdot nH_2O$	Unusual	$\eta = 1.43$ (varies with water content)	White to colourless, fine-grained amorphous; resulting from intense leaching or carbonation of paste

(Continued)

Table 4.6 (Continued) Some secondary deposits in concrete

Compound/ mineral	Chemical formula	Occurrence frequency	Refraction indices	Form
Alkali–silica gel	$Na_2O \cdot K_2O \cdot CaO \cdot SiO_2 \cdot nH_2O$	Quite common	$\eta = 1.46–1.53$	White, yellowish or colourless; viscous, fluid, waxy, rubbery or hard, in voids and cracks (See Chapter 6.)
Hydrated iron oxides/limonite	$Fe_2O_3 \cdot nH_2O$	Common	Opaque or nearly so	Brown deposits and stains in cracks and in paste around iron-bearing aggregate grains (e.g. pyrite) or associated with nearby reinforcement corrosion
Calcium carbo-sulpho-silicate hydrate/ thaumasite	$CaSiO_3 \cdot CaCO_3 \cdot CaSO_4 \cdot 15H_2O$	Quite common	$\omega = 1.504$ $\varepsilon = 1.468 \pm 0.002$	Needles of similar general appearance to ettringite, with which it can occur; forms in concrete exposed to damp cold conditions, sewer pipes subject to sulphate attack (See Chapter 6.)
Potassium–calcium sulphate hydrate/ syngenite	$(K_2Ca(SO_4)_2) \cdot H_2O$	Very rare	$\alpha = 1.501$ $\beta = 1.51$ $\chi = 1.51$	Fibrous material found in cavities and zones peripheral to slate aggregate particles
Hydrated magneso-alumino carbonate/ hydrotalcite	$Mg_6Al_2(OH)_{16}$ $CO_3(H_2O)_4$	Very rare	$\omega = 1.510 \pm 0.003$ $\varepsilon = 1.495 \pm 0.003$	Foliated platy to fibrous masses

Source: Adapted from ASTM C856, *Standard Practice for Petrographic Examination of Hardened Concrete,* American Society for Testing and Materials, West Conshohocken, PA, 2011.

Note: Also, see Chapters 5 and 6 and the 'Glossary of minerals' for further details.

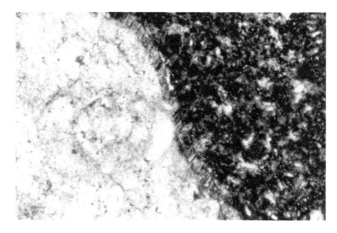

Figure 4.14 Photomicrograph of ettringite formed in a parting at a cement/aggregate boundary in DEF, xpl, width of field ~0.4 mm. (Photograph courtesy of RSK Environment Ltd., U.K.)

(i.e. not greater than 20°C/h) and the maximum level of curing temperature (i.e. not greater than 60°C or 70°C). Cement chemistry also appears to be a critical factor (Lawrence 1995).

Although the occurrence of secondary ettringite observable to optical microscopy, particularly when infilling peripheral cracks around aggregates, is often a feature of concrete damaged by DEF, caution should be exercised in attributing damage to this cause on the sole basis of ettringite presence, unless substantiating evidence is available including information relating to the conditions of manufacture. Deng and Tang (1994) suggested that only ettringite which forms in certain ways within the cement paste can cause expansion, while that precipitating within existing voids or cracks exerts little or no expansive stress. Similarly, Poole et al. (1996) concluded from their experiments that the growth of ettringite within the cement paste was the principal cause of expansion in cases of DEF and that ettringite rims around aggregate particles were not always formed. A full and up-to-date account of DEF is given in Chapter 6.

In 1965/1966, Erlin and Stark reported on concrete deterioration caused by the secondary formation of the complex compound thaumasite ($CaSiO_3 \cdot CaCO_3 \cdot CaSO_4 \cdot 15H_2O$). An early review of the thaumasite literature is given by Van Aardt and Visser (1975), who demonstrated that thaumasite forms preferentially at lower temperatures (i.e. 5°C rather than 25°C). A full and up-to-date account of thaumasite and its effects is given in Chapter 6.

According to Crammond (1985b), thaumasite forms in conditions that are very damp as well as cold and also when there are abundant proportions of sulphate and carbonate ions available. Typically, thaumasite and ettringite form together and both can contribute to expansion and degradation of the matrix (see Chapter 6). Berra and Baronio (1987) described a case in which portions of a concrete tunnel lining had been 'transformed into an incoherent whitish mass' consisting chiefly of thaumasite resulting from attack by water containing both sulphates and aggressive CO_2. Crammond and Halliwell (1995) have described cases of the thaumasite form of sulphate attack (TSA) in which the required carbonate ions were derived from limestone and dolomite dust within the concrete.

Thaumasite is superficially rather similar in appearance to ettringite under the microscope and indeed often occurs as intimate mixtures with ettringite, but thaumasite usually has strong birefringence in contrast to the weak birefringence of ettringite (Varma and Bensted 1973). Although the XRD patterns for thaumasite and ettringite are quite similar, Crammond (1985a) has demonstrated a quantitative procedure for analysing mixtures of thaumasite, ettringite and gypsum in concretes and mortars. French (1991a) has suggested that there might be solid solutions between thaumasite and ettringite.

Sibbick and Crammond (2001) have described examples of the coexistence of thaumasite and ettringite in mortars affected by 'thaumasite formation' and the 'TSA'. Well away from the degraded areas, they identified air voids containing clear secondary ettringite linings. The ettringite appeared as fine needle-like, but well-formed preferentially orientated (rigid-harder) deposits. By contrast, the nearby thaumasite deposits varied from poorly structured mush, lower-birefringence deposits through increasing density and higher-birefringence material with a more highly structured development. It is important to recognise that thaumasite can occur in this low-birefringence form that is not compositionally ettringite (Sibbick et al. 2002).

Chlorides added to a concrete mix or entering hardened concrete as a contamination can combine with C_2A or C_4AH_{13} to form complex calcium chloroaluminate hydrates (e.g. $3CaO \cdot Al_2O_3 \cdot CaCl_2 \cdot 10H_2O$ – Friedel's salt). According to Figg (1983), virtually all of the calcium chloride ($CaCl_2$) added to concrete as an accelerator usually becomes combined into this low-chloride chloroaluminate, although the high-chloride form has been reported from the USSR when exceptionally large amounts of $CaCl_2$ were used for placing concrete at sub-zero temperatures. Damage to concrete arising from the formation of calcium chloroaluminate hydrates appears to be uncommon.

Similar chloroaluminate compounds can form when sodium chloride is present, either in the mix water (e.g. seawater or saline mix water) or as an externally derived contaminant (e.g. seawater, de-icing chemicals, sabkha ground conditions in some arid desert regions). Free chloride in concrete is a major cause of reinforcement corrosion (see Chapter 6), so that the amount of calcium aluminate available to react with chloride salts is an important consideration when setting tolerance limits for chloride content in reinforced concrete. For example, former BS 8110-1 (1997) set a limit of 0.4% chloride ion for reinforced OPC concrete with a relatively high C_3A content, but a limit of only 0.2% chloride ion for reinforced SRPC concrete with a restricted C_3A content. Similar criteria are given in BS EN 206-1 (2000) and BS 8500-1 and BS 8500-2 (2006), which together replaced BS 8110. In reinforced concretes exposed to conditions of both sulphate and chloride, therefore, it may sometimes be deemed counterproductive to specify a type of Portland cement to resist the sulphates, which might be less able to inhibit the effect of the chlorides on the embedded steel.

The calcium chloroaluminates also form needle-like crystals and under the microscope are difficult to distinguish from ettringite. However, the XRD pattern permits distinction (Crammond 1985a) and, of course, EDX microanalysis can be carried out under the SEM (Figure 4.15). Reaction between sulphates or chlorides and the aluminoferrite compounds of cement can also give rise, respectively, to various sulphoferrite and chloroferrite analogues (Lea 1970).

Figure 4.15 SEM micrograph of hexagonal plates of calcium chloroaluminate hydrate. (From French, W.J., *Quarter. J. Eng. Geol.*, 24(1) 17, 1991a.)

4.2.4 Blended and special cements

It has now become common worldwide for the cement in concrete mixes to be partially replaced by mineral additions or supplementary cementitious materials ('SCMs') of cement fineness or finer in the more recent cases of microsilica and metakaolin (Malhotra 1987; Lewis et al. 2003). Indeed, some European countries have been using blended cements for much more than half a century. These supplementary materials can modify the properties of both fresh and hardened concretes. The uses and potential benefits of using either ggbs or pfa, which are the predominant materials used, are well documented, and one useful brief resume is given by Dewar and Anderson (1992).

Among other advantages, subject to good concrete-making practice, either ggbs or pfa might be expected to reduce costs (as a cement replacement), to reduce the heat of hydration which is especially important in large concrete pours and usually to improve resistance to a range of durability threats, such as sulphate attack or alkali–aggregate reactivity (see Chapter 6).

On the mainland of Europe and increasingly also in the United Kingdom, it has been a traditional practice for such mineral additives to be blended with the Portland cement component during the cement manufacturing process (Corish 1989). The use of such blended cements obviously simplifies the site batching of concrete but permits less flexibility of mix design. Apart from rare cases of inadequate concrete mixing leading to a markedly uneven distribution when the additive has been added at the mixer, it is usually not possible for petrographic examination to differentiate between concretes made using separate additive and cement components and those made using factory-blended cements.

4.2.4.1 Portland–limestone cements

Incorporation of up to 5% finely ground limestone filler is now permitted in Portland cement (e.g. BS EN 197-1 [2011] for CEM I: see Table 4.1a). However, varieties of CEM II are also recognised (see Table 4.1a) in which proportions of limestone from 6% to 35% are interground with Portland cement clinker to produce 'Portland–limestone cement' (PLC). The BS EN 197-1 (2011) products are CEM II/A-L or LL (6%–20% limestone) and CEM II/B-L or LL (21%–35% limestone), 'L' being a limestone with not more than 0.5% organic carbon, while 'LL' being a limestone with not more than 0.2% organic carbon. According to the review by Price (2004), by comparison with CEM I concrete, PLC can enhance concrete workability and reduce bleeding, and also the hydration reactions are modified, typically leading to earlier setting, reduced porosity and higher early strength. Durability appears generally comparable with CEM I concrete, although there is some evidence of increased vulnerability to sulphate action (see Chapter 6). Dilution of the Portland cement clinker component suggests some environmental benefits from the use of PLC.

At present, petrographers will perhaps mainly encounter PLC in precast and decorative concrete products in the United Kingdom, with users taking advantage of its typically lighter coloration and smoother surface finish, but it has been more widely used for more than 25 years on the European mainland, and the authors have recently experienced its routine application for general concrete in some other parts of the world. Concrete with PLC varieties containing higher limestone contents will be distinguished by the abundance of finely divided limestone/calcite included within the cement paste, although this might only be obvious at higher magnifications and/or using SEM (see Figure 4.16).

4.2.4.2 Ggbs and Portland blastfurnace cements

The nature of ggbs is described in more detail in Section 4.6.2. A by-product of the iron-making industry, granulated or pelletised blastfurnace slag is produced by the rapid quenching of molten blastfurnace slag as it passes through water sprays, followed by either water

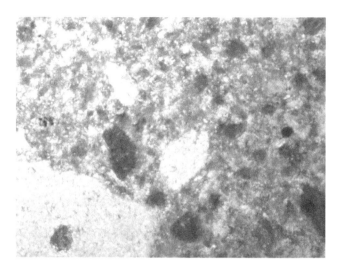

Figure 4.16 Appearance of PLC in hardened cement paste in thin section visible as light/dark brown flecks xpl, field of view 0.6 m. (Photomicrograph courtesy of Paul Bennett-Hughes, RSK Environment Ltd., U.K.)

granulation or pelletisation on a spinning drum (Hooton 1987). The granulated material produced is largely glassy, typically at least 95%, being glass with a chemical composition close to that of the bulk composition: around 40% lime, 30% silica, 15% alumina and variable proportions of magnesia and other constituents (Lee 1974). The minority crystalline components are dominated by melilite or merwinite (see Section 4.6.2) with some oldhamite and residual native iron (Scott et al. 1986). Ground pelletised slag is essentially similar but tends to contain more vesicles (gas bubbles).

Hydration products are slow to form when ggbs is exposed alone to water, but the 'latent hydraulicity' of ggbs is activated by the calcium hydroxide and alkalis liberated during the hydration of Portland cement (Lea 1970; Hooton 1987). This activation can be enhanced by the addition of lime, which accelerates the breakdown of the glassy ggbs structure. It seems that the hydration products of a binder containing both Portland cement and ggbs are broadly similar to those of Portland cement alone, except that the nature of the C–S–H produced is slightly modified (Richardson 2004, 2008) and the content of portlandite ($Ca(OH)_2$) is reduced because of the reduced lime/silica ratio. Small and infrequent proportions of hydrated gehlenite, C_2ASH_8 (gehlenite is an endmember of the melilite solid solution series), and hydrogarnets, $C_4(A,F)H_{13}$, have also been reported. However, it is the usually considerable quantity of residual unreacted ggbs in hardened concrete that will be apparent to the petrographer and which enables the presence of ggbs to be identified.

Fresh concrete made using Portland blast-furnace slag cement has a fairly distinctive dark bluish-green appearance, but with time following oxidation, the colour becomes similar to that of normal concrete. Unreacted remnants of ggbs appear in the cement paste matrix as angular, even shard-like, particles of silicate glass. They are clearly visible either in thin section under the petrological microscope (Figure 4.17) or in highly polished surfaces, etched with HF vapour, under the reflected light microscope (Figure 4.18). It is possible that hydraulic reactions only affect the smaller, submicroscopic ggbs particles as the coarser pieces observable under the microscope typically exhibit sharply defined boundaries and yield no apparent evidence of any interfacial reaction, even after many years. French (1991b) has suggested that because ggbs particles as small as 1 μm can be seen by SEM to be completely unreacted (Figure 4.19), ggbs particles might be divided into those which react entirely to produce hydrates and those which are essentially inert.

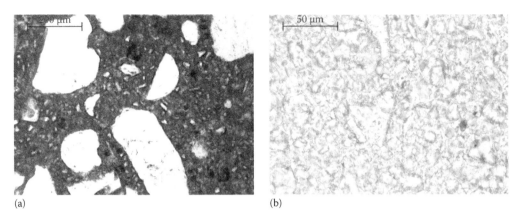

Figure 4.17 Photomicrographs of unreacted ggbs grains in concrete in thin section, visible as greyish angular particles often with dark edges: (a) Scattered ggbs grains in a concrete matrix. (b) Closer view of ggbs powder. Scale bars on images. (Photographs courtesy of Paul Bennett-Hughes, RSK Environment Ltd., U.K.)

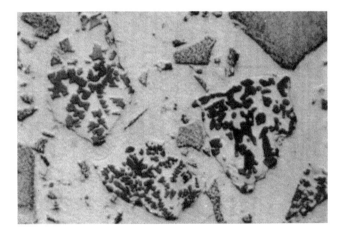

Figure 4.18 Photomicrograph of unreacted ggbs grains in concrete in reflected light after etching with HF vapour, width of field 0.3 mm.

In thin sections and polished specimens, the bluish-green coloration associated with hydrated slag mixes is typically seen to be heavily concentrated in and around the residual Portland cement grains (Figure 4.20), rather than within the CSH product or associated with residual slag grains (Sibbick 2011). Possibly, this results from migration of the sulphide phases from the slag during the hydration process.

While identifying the presence of significant ggbs in a hardened concrete presents few difficulties, quantifying the relative proportions of Portland cement and ggbs is much more problematic. The chemical procedure given in BS 1881-124 (1988) relies upon the determination of sulphide and is thus inappropriate if any other of the concrete constituents contains sulphide, and is also ineffective if all or part of the sulphide has been oxidised to form sulphate. It is possible microscopically to point-count the residual unreacted ggbs and Portland cement clinker grains within the cement paste matrix, but the result needs to assume a relationship with the original cement–ggbs ratio as represented in the hydrated groundmass, which will not be known and might not be consistent in any case. If French (1991b) is correct, the relationship between the hydrated ggbs and the unreacted ggbs could be highly variable. If a reference concrete with

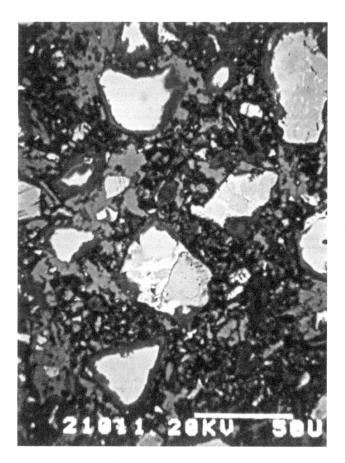

Figure 4.19 SEM micrograph of unreacted ggbs grains in concrete visible as grey/white particles in the hardened matrix.

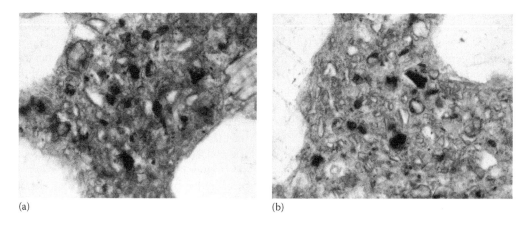

(a)   (b)

Figure 4.20 Concentration of bluish-green coloration around Portland cement grains: (both a and b) ppl, width of field 285 μm. (Photomicrographs courtesy of Dr Ted Sibbick.)

known mix proportions using the same constituents as the sample being analysed is available, this point-counting technique might be feasible (see the example in subsection 4.9.3).

French (1991b) advocated a method using EDX-type microanalysis, whereby the cement–ggbs mixture in the hydrated paste is estimated from chemical composition data separately determined by spot analyses of unreacted ggbs, residual cement clinker and the hydrated groundmass. However, although this type of analysis is now readily available, even this could be complicated by his postulated division between inert and reactive types of ggbs particle. It is possible that a combination of the microscopic point-counting and micro-analytical methods could enable a reasonably dependable estimate of cement and ggbs mix proportions to be determined (see Section 4.9).

4.2.4.3 Pfa, fly ash and Portland pozzolanic cements

Pulverised bituminous coal (pulverised-fuel) is burnt at some electricity power stations, producing 'pfa' as a by-product, about 20%–25% of which becomes fused together like clinker and is known as 'furnace bottom ash', while the remaining 75%–80% is a fine pfa dust or 'fly ash' which may be collected from the combustion gases (Owens 1980-2, Berry and Malhotra 1987, Owens 1999, Sear 2001, Sear and Owens 2006). Note that the term 'fly ash' has a wider application than pfa, sometimes being used for the product of burning a variety of other materials, such as rice husk (see Section 4.6.4). This discussion focuses on coal fly ash.

The chemical composition of pfa or fly ash is dependent upon the rank of coal burnt and also the nature of the clay impurities in the coal, but all are dominated by silica (45%–50%) and alumina (25%–30%). Two main types are recognised (ASTM C618 2008): 'Class C' types are produced from sub-bituminous coal (or even lignite) and are relatively high in calcium content (>10%), whereas 'Class F' types are produced from bituminous coals and are lower in calcium content (<10%). U.K. power stations produce only 'Class F' pfa. In the United States and other parts of the world, substantial volumes of 'Class C' ash are employed as SCM additions to concrete, and these 'Class C' ashes tend to be more reactive and therefore often leave little visible residual material when well activated (Sibbick 2011). Age and degree of activation also need to be considered.

It seems likely that analysis using SEM-EDX offers the best means of characterising fly ash type and assessing its activation potential (Chancey et al. 2010; Sibbick 2012). In this technique, polished surfaces are examined using SEM-EDX, with high-quality elemental maps being produced and converted into phase maps, including glassy phases exhibiting various ranges in composition; image analysis then allows these phases to be quantified. In principle, comparison of a fly ash material before and after incorporation into a cementitious system enables its reactive and residual inert phases to be determined.

Mineralogically, most pfa-type materials are dominated (45%–70%) by aluminosilicate glass (Hubbard et al. 1985), principally in the form of solid spheroidal particles less than 20 μm in diameter (RILEM 1988) and occasionally forming hollow 'cenospheres' (Raask 1968). Hubbard et al. (1985) found the crystalline component of U.K. pfa materials variably to consist of mullite ($Al_6Si_2O_{13}$), haematite, quartz and magnetite, with about 5%–10% unburnt coal. The product from any one power station was found to be rather consistent, and the variations between stations were largely correlated to the clay impurities in the coal used, aluminosilicate glass being generated by the fusion of illite and mullite being formed by the recrystallisation of kaolinite. However, it is known that the furnace design and parameters can also influence the character of the fly ash produced. McCarthy et al. (1984) reported a different and wider variety of crystalline phases for 'Class C' pfa materials from western United States, perhaps accounting for 'Class C' materials being cementitious as well as pozzolanic, whereas 'Class F' ashes are only normally pozzolanic (RILEM 1988).

The presence of pfa is readily identified in thin sections of concrete under the petrological microscope because of the distinctive spheroidal or ellipsoidal glassy particles or even relatively coarser hollow 'cenospheres' (Figure 4.21). In their studies of pfa concrete structures, Thomas and Matthews (1991) routinely used SEM of broken surfaces to confirm the presence of pfa, as betrayed by either unreacted spherical particles or relicts of mullite crystals (Figure 4.22). Residual carbon particles and clusters of, or cenospheres filled with, still non-reacted spheres and other lesser crystalline, often non-spherical, 'contaminants' (quartz, calcite, lime, calcium sulphides, etc.) are also typically present in varying degrees (Sibbick 2011). A comparison of the SEM-EDX appearance of typical examples of 'Class C' and 'Class F' fly ash in concrete is shown in Figure 4.80 (in Section 4.6.1). However, like ggbs, the quantitative determination of pfa in concrete is complicated by the small particle size, which ranges down to submicroscopic and the uncertain extent to which some proportion of the pfa in hardened concrete has possibly been consumed in the formation of hydrates (see Section 4.9.2).

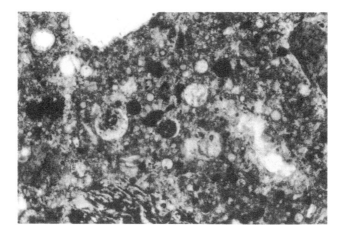

Figure 4.21 Photomicrograph of pfa cenospheres in concrete in thin section, ppl, width of field ~0.5 mm. (Photograph courtesy of RSK Environment Ltd., U.K.)

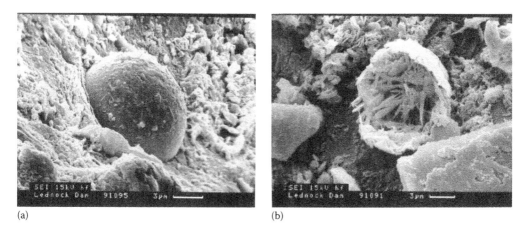

(a)   (b)

Figure 4.22 SEM photomicrographs of pfa in concrete: (a) Unreacted cenosphere. (b) Relicts of mullite crystals present as needles in the corroded sphere. (From Thomas, M.D.A. and Matthews, J.D., *Durability Studies of Pfa Concrete Structures*, BRE Information Paper IP 11/91, Building Research Establishment, Watford, U.K., 1991.)

The aluminosilicate glass phase in pfa is considered to be genuinely pozzolanic and may react in the presence of water with calcium hydroxide liberated by the normal hydration of Portland cement to form stable hydrated cementitious compounds (Eglinton 1987). Such pozzolanicity is likely to vary with both chemical composition and particle size, but the extent to which pozzolanically derived hydrates exist in concretes made using pfa remains uncertain. Abundant residual and apparently completely unreacted pfa particles are a characteristic feature of such concretes. Semi-quantitative techniques, including the point counting of residual pfa and cement clinker particles in optical microscopy or the visual assessment of those residual particles in SEM, enable the original cement and pfa proportions to be estimated.

Chemical techniques are fraught with difficulty, especially in the absence of reliable reference constituent samples, although some success has been claimed for complicated procedures using differential solubility rates (e.g. Gomà 1989). As it seems likely that the hydrates produced by any pozzolanic activity of pfa will be different in composition and character from those generated by Portland cement hydration, it is possible that an EDX-type microanalytical technique could be developed (see earlier discussion on SEM-EDX for characterising fly ash). However, de Rooij (2011) advises that the C–S–H arising from the interaction of cement and fly ash is both complex and variable, presenting a challenge to the EDX-type approach.

Sibbick (2011) has found the degree of 'opalescence' or milkiness of the resultant hydrated matrix to be a useful indication of the degree of hydration and activation the pfa/fly ash has achieved (see Figure 4.23). Opalescence differences occur with other SCMs in a similar

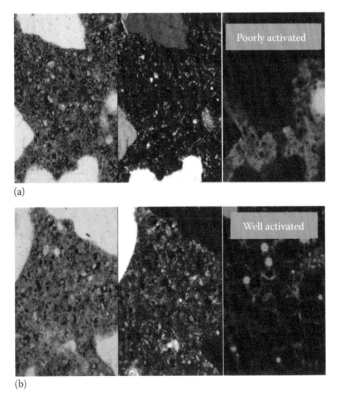

Figure 4.23 Opalescence of cement–pfa pastes of similar age with differing degrees of hydration and activation: (a) Poorly activated, ~0.6 mm, left to right, ppl, xpl, uvl. (b) Well activated, ~0.6 mm, left to right, ppl, xpl, uvl. (Photomicrographs courtesy of Dr Ted Sibbick.)

Table 4.7 Some natural and artificial pozzolanas used in concrete

Type of source	Type of pozzolana	Pozzolanic classification after RILEM (1988)
Natural	Volcanic tuff (e.g. trass)	—
	Volcanic ash/earth	—
	Diatomaceous earth/diatomite	—
	Bauxite (e.g. moler in Denmark)	—
Artificial – waste/by-product	Fly ash/pfa – high calcium[a]	Cementitious and pozzolanic (II)
	Fly ash/pfa – low calcium[a]	Normal pozzolanic (IV)
	ggbs	Cementitious (I)
	csf/microsilica	Highly pozzolanic (III)
Artificial – synthetic	Burnt clays and shales	—
	Burnt moler/diatomite	—
	Rice husk ash – controlled incineration	Highly pozzolanic (III)
	Metakaolin	Highly pozzolanic (III)[b]

[a] High calcium > 10% CaO, low calcium < 10% CaO, as defined in RILEM (1988).
[b] Not included in RILEM (1988), but here allocated to Class III.

manner. It is thought to be the result of consumption of finer SCM material, the development of a denser resultant hydrate product and obviously higher or complete consumption of the calcium hydroxide. Otherwise, similar paste opalescence only appears to occur in cement pastes 'contaminated' with various external salts (Sibbick 1993, 2010).

Assessing a cementitious matrix in terms of hydration and activation can highlight reasons for unexpected performance, such as low strength, setting time issues or retardation in a particular concrete mix. If one or other cementitious addition appears not to have 'reacted', this might suggest a problem with that particular material or perhaps an incompatibility in the concrete mix combination (Sibbick 2011).

Pfa has long been the main pozzolana used in the United Kingdom, previously either as a 15%–35% Portland cement replacement in 'Portland pfa cement' (BS 6588 1996) or as a 35%–40% replacement in 'Pozzolanic cement with pfa as pozzolana' (BS 6110 1985). These blends are essentially now accommodated as CEM II/A-V (6%–20% fly ash), CEM II/B-V (21%–35% fly ash) and CEM IV-A (11%–35% fly ash) products within the classification of BS EN 197-1 (2011); see Table 4.1a.

However, on the mainland of Europe and elsewhere, a range of other natural and artificial pozzolanas are available and sometimes used in concrete (Table 4.7). The natural pozzolanas are mainly volcanic ash or tuff deposits, including the German ground 'trass' materials, the 'Santorin earth' from Greece and of course the Roman source of zeolitic tuff from Pozzuoli near Mount Vesuvius in Italy which gives pozzolana its name. The diatomaceous earths are the other main group of natural pozzolanas. Apart from pfa, the artificial pozzolanas mainly include burnt clay or shale, but other materials have been used, for example, 'rice husk ash' (RHA). Further information on pozzolanas may be obtained from Lea (1970), Mehta (1985), RILEM (1988) and Lewis et al. (2003).

4.2.4.4 High-alumina cement

HAC, or calcium aluminate cement (CAC), differs from Portland cement in being manufactured by the fusion of limestone and bauxite, instead of limestone and clay or shale (Robson 1962, 1964). As a consequence, HAC clinker consists chiefly of calcium aluminates with much smaller amounts of ferrite and silicate phases. Patented in 1908, the production of

HAC commenced in France in 1913 and in Britain in 1925. In addition to its rapid strength gain properties, Bied (1926) discovered the superior sulphate-resistance properties of HAC. Following some structural failures in the United Kingdom in the 1970s, the use of HAC declined, but the material has continued to be used for specialist purposes and has now undergone something of a reassessment, including the introduction of the alternative term 'calcium aluminate cement'. Recent accounts will be found in Cather (1997), Scrivener and Capmas (1998), Neville (2006) and Fentiman et al. (2008).

Reference to the phase equilibrium diagram for the $CaO-Al_2O_3-SiO_2$ system shows that HAC falls almost entirely within the stability field of monocalcium aluminate (CA), which is found to be the main phase present and the principal cementitious material (Table 4.8). Other aluminate compounds include $C_{12}A_7$, CA_2 and C_3A (rare), while larnite (βC_2S) and/or gehlenite (C_2AS) may also be present depending upon silica content. Calcium alumino-ferrites form with compositions in the range $C_4AF-C_6AF_2$, while most of the ferrous iron appears to be contained within wüstite (FeO)-bearing glass.

One characteristic constituent of HAC, which is fibrous and pleochroic and variously known as 'pleochroite' or 'Q', has proved difficult to identify chemically (Sourie and Glasser 1991). However, Kapralic and Hanic (1980) have established the general formula for 'Q' as $Ca_{20}Al_{32-2x}Mg_xSi_xO_{68}$ (where x can vary from 2.5 to 3.5), which is believed to be isostructural with 'pleochroite' wherein FeO substitutes for MgO and Fe_2O_3 substitutes for some Al_2O_3.

HAC clinker is generally dominated by short prismatic crystals of CA, mixed with ferrite, other calcium aluminates and sometimes gehlenite. The crystal sizes are very dependent upon cooling rate (Figure 4.24). The more slowly cooled (and hence coarser textured) HAC clinker can be studied using optical microscopy, when the cement type is easily distinguishable

Table 4.8 Anhydrous mineralogy of high-alumina (calcium aluminate) cements

Phase	Mineralogical composition, % by mass[a]										
	1	*2*	*3*	*4*	*5*	*6*	*7*	*8*	*9*	*10*	*11*
CA	60	60–65	58	59	46	57	36	50	26	29	45
Grossite, CA_2	—	—	—	—	4	11	5	15	—	—	5
Mayenite, $C_{12}A_7$	—	—	—	4	2	2	2	—	31	38	13
Gehlenite, C_2AS	15–20	2	—	2	18	17	47	28	18	15	21
C_2S	10–15	10	—	—	—	—	—	—	—	—	—
Pleochroite	—	2	20	8	5	3	—	1	11	13	7
Spinel	—	—	—	2	1	<1	3	—	—	—	—
Iron-bearing glass	10	5	—	—	—	—	—	—	—	—	—
Wüstite, FeO	—	—	12	3	—	—	—	—	—	—	—
Maghemite, γFe_2O_3	—	—	—	4	6	—	—	—	—	—	—
Brownmillerite, C_4AF	—	—	—	7	9	—	—	—	—	—	—
C_6AF_2	—	—	10	—	—	—	—	—	—	—	—
Other iron compounds[b]	vl	h	—	—	—	—	—	—	—	—	—
C_3FT	—	—	—	10	8	1	—	—	—	—	—
Perovskite, CT	—	—	—	—	—	5	4	4	5	—	5
Anhydrite, $CaSO_4$	—	—	—	—	—	—	—	—	2	2	2
Yeelimite, $Ca_4Al_6O_{12}SO_4$	—	—	—	—	—	4	2	1	<1	—	2
Kuzelite, $Ca_4Al_2(OH)_{12}(SO_4)\cdot 6(H_2O)$	—	—	—	—	—	—	—	—	6	—	—
Oldhamite, CaS	—	—	—	—	—	—	—	—	—	3	—

[a] Cements 1 and 2 from Robson (1964), 3 from Midgley (1967) and 4–11 from Pöllman et al. (2008).
[b] vl, very low; h, high.

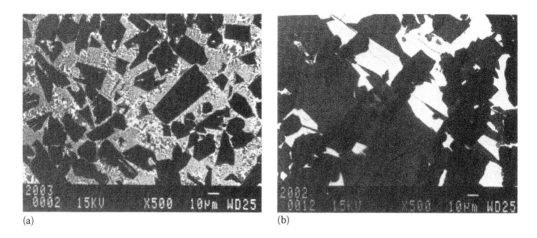

(a)   (b)

Figure 4.24 SEM micrographs of a HAC clinker sample: (a) Fast cooling (~300°C/min), crystals of CA in a dendritic matrix including C_2AS. (b) Slow cooling (~5°C/min), coarser dark CA crystals, plus grey C_2AS and white ferrite. (From Sourie, A. and Glasser, F.P., *Trans. J. Br. Ceram. Soc.*, 90(3), 71, 1991.)

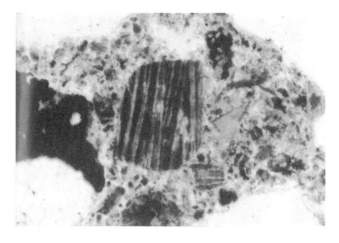

Figure 4.25 Photomicrograph of pleochroite in HAC clinker in concrete in thin section, ppl, width of field ~0.2 mm. (Photograph courtesy of RSK Environment Ltd., U.K.)

using either etched polished specimens or thin sections (Figures 4.22 and 4.25). Pleochroite distinctively occurs as long lathlike crystals with a striking pleochroism (Figure 4.21).

Residual pieces of such clinker are nearly always observable in the matrix of HAC concrete, especially since, at the recommended low water/cement ratios, typically less than half of the cement hydrates. Figure 4.26 illustrates the typical appearance of HAC concrete matrix in thin section. Chemical tests have also been described for the rapid differentiation of HAC from Portland cement in concretes (Roberts and Jaffrey 1974).

The hydration and ageing of HAC is summarised by Midgley and Midgley (1975) and Sims (1977) (Figure 4.27). The dominant CA reacts rapidly with water to produce the components that give HAC concrete its high early strength. Pleochroite and βC_2S react more slowly but appear to contribute to strength gain in due course. The iron compounds and any C_2AS are either inert or only slowly reacting. The initial hydration products at normal temperatures are principally hexagonal CAH_{10}, with some C_2AH_8, depending upon cement chemical composition and the temperature, and alumina gel. It appears that mineralogical

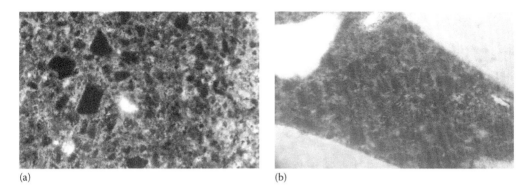

(a)   (b)

Figure 4.26 Photomicrographs of HAC concrete matrix in thin section; (a) the dark crystals are mainly CA. xpl, width of field ~0.3 mm, (b) showing pleochroite (black/purple), ppl, width of field 20.3 mm. (Photographs courtesy of Mr Mark Knight, Sandberg LLP, London, U.K., and Paul Bennett-Hughes, RSK Environment Ltd., U.K., respectively.)

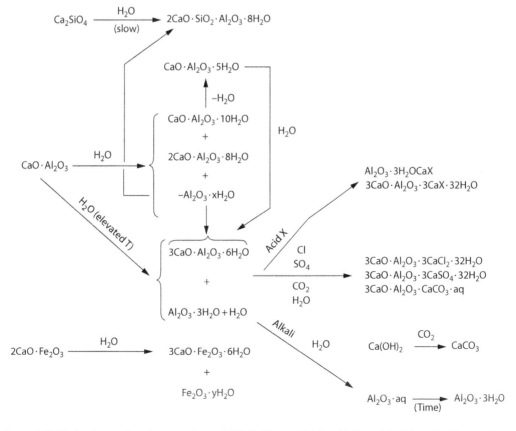

Figure 4.27 Hydration and ageing reactions of HAC. (From Midgley, H.G. and Midgley, A., *Magaz. Concr. Res.*, 27(91), 59, 1975.)

changes might occur at or near concrete surfaces exposed to the atmosphere, such that CAH_{10} dehydrates to CAH_5 (Midgley 1967) or even perhaps decomposes to alumina gel and calcite (Robson 1964).

These early-formed hexagonal calcium aluminate hydrates are metastable and spontaneously invert (or 'convert') with time to the cubic hydrogarnet (C_3AH_6) and monoclinic gibbsite (AH_3). When HAC is hydrated at an elevated temperature (40°C), Scrivener and Taylor (1990) found that CAH_{10} and an 'inner product' were still the initially formed phases, which within a matter of hours converted into C_2AH_8 and AH_3, then into C_3AH_6 and more AH_3. The 'conversion' reaction may be idealised as

$$3CAH_{10} \rightarrow C_3AH_6 + 2AH_3 + 18H$$

$$\text{or } 3C_2AH_8 \rightarrow 2C_3AH_6 + AH_3 + 9H$$

The factors controlling the rate of conversion and the magnitude of its effect on the properties of HAC concrete are most importantly temperature and water/cement ratio: other factors include humidity, possibly stress and notably alkalis (e.g. possibly deriving from adjacent Portland cement products, some aggregates and even mould release agents). These controlling factors actually influence the nature of the mineralogical phase change, including the higher mineral density of C_3AH_6 as compared with that of CAH_{10}, the increase in crystallite size and the complete change in habit on conversion, which effects crystal packing (Sims 1977).

The progress of the conversion reaction is usually assessed by measuring the 'degree of conversion', whereby the relative proportions of CAH_{10} and C_3AH_6 (or AH_3) are determined by mineralogical analysis, most commonly using DTA (Chemical Society 1975) (Figure 4.28). However, it should be appreciated that this concept of 'degree of conversion' is an oversimplification and that the use of DTA for this purpose is subject to considerable imprecision.

Generally, the conversion of HAC in concrete leads to a progressive loss of strength, affecting structural integrity, and a simultaneous increase of porosity. The relationship between conversion and strength loss is complex and practically unpredictable: in any case, in practice, strength loss frequently continues even after the attainment of full conversion, especially when the concrete is not constantly dry. Although some structural collapses occurred in the United Kingdom (Building Research Establishment 1973, Bate 1974, Neville 1975) and later also in Spain (Figure 4.29), assuming no building faults irrespective of concrete type, a majority of structures containing HAC concrete survive the detrimental effects of conversion alone. The structural use of HAC concrete ceased in the United Kingdom in the mid-1970s.

Converted and hence more permeable HAC concretes become vulnerable to further decay of the concrete itself and reduced protection for any embedded steel. Further deterioration of converted HAC may variously result from carbonation, alkaline hydrolysis, sulphate attack, chloride attack or even combinations of these agencies (BRE 1981; Collins and Gutt 1988; BRE 1994). HAC conversion is discussed further in Chapter 5.

The petrographic examination of any HAC concrete must therefore address the conversion state of the cement matrix, as well as identifying the presence of HAC, and the extent to which any secondary chemical actions might have further impaired the strength and integrity of the material. In particular, now that most HAC concrete in existing structures in the United Kingdom can be assumed to be highly converted, concern has been expressed over the possibly detrimental effects of carbonation of the HAC concrete cover to the depth of steel reinforcement or pre-stressing wires (Crammond and Currie 1993; Currie and

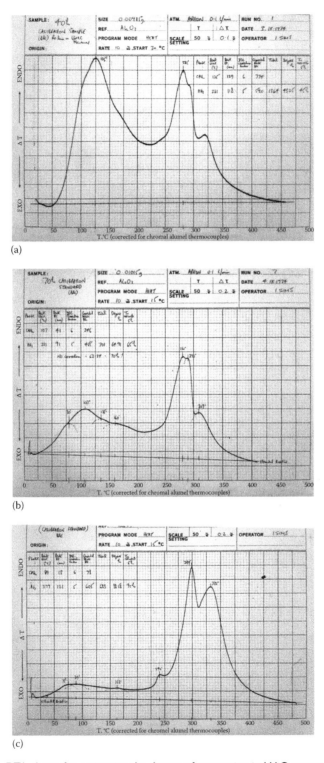

Figure 4.28 Example DTA charts for estimating the degree of conversion in HAC concrete: (a) 40%/45% converted, with similar endothermic peaks for CAH_{10} and AH_3; (b) 65%/70% converted, with AH_3 peak dominant over that for CAH_{10} and (c) 90% converted, with dominant peaks for AH_3 and C_3AH_6.

Figure 4.29 Collapse of floors in an apartment block caused by conversion of HAC concrete beams, Barcelona, Spain: (a) Exterior of affected block. (b) Interior view from the roof, showing the progressive floor collapse. (c) Detail showing failed beams. (d) Detail showing a non-failed beam with temporary support.

Crammond 1994). It is feared that such steel is at greater risk of corrosion. According to Building Research Establishment Digest 392 (1994), 'petrography (optical microscopy) is the only cost-effective way to establish the depth of carbonation' of HAC concrete.

Carbonation of HAC is mineralogically quite different from that of Portland cement (see Chapter 5). The mechanism may be envisaged as C_3AH_6 reacting with CO_2 from the atmosphere to form an intermixture of calcium carbonate (mainly calcite) and aluminium hydroxide (mainly gibbsite). It seems that damp concrete is particularly likely to undergo this carbonation reaction, so that particularly samples from continuously or intermittently

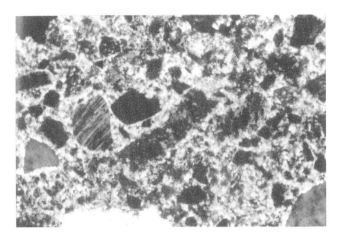

Figure 4.30 Photomicrograph of carbonated HAC concrete in thin section, xpl, width of field ~0.4 mm. (Photograph courtesy of RSK Environment Ltd., U.K.)

damp locations should be examined. Wholly or partially carbonated HAC is easily identified in thin section (Figure 4.30).

A particularly damaging form of HAC carbonation is associated with a process known as 'alkaline hydrolysis', whereby alkaline solutions ingress HAC concrete as a result of water leakage via adjacent Portland cement concrete materials (e.g. concretes or screeds overlying HAC concrete beams). The alkaline solutions dissolve and decompose the calcium aluminate hydrates, to form alkali aluminate and calcium hydroxide, which can then be carbonated by atmospheric CO_2 to form calcite and gibbsite, so regenerating the alkali hydroxide to react again. A case study, concerning alkaline hydrolysis, possibly combined with sulphate attack, affecting HAC beams in a school in North West England built in the 1960s, has been described by Dunster and Weir (1996).

4.2.4.5 Supersulphated and other special cements

A great majority of the concretes and mortars encountered by the petrographer will be found to be made using one of the variants of Portland cement and most commonly OPC. It has become quite common for ggbs, pfa or another pozzolanic material to be present as a binding material in combination with Portland cement, and these have been discussed in earlier sections. HAC is occasionally encountered and is quite distinctive as described in the previous section. Comparatively rarely, concretes or mortars will be found to contain an atypical binder, sometimes based upon or containing Portland cements and sometimes being quite different. While the petrographer obviously must be aware of such special or proprietary cements, apparently unusual binders in concrete are, at present, always more likely to be a modified or altered form of Portland cement than a completely different type of material.

'Supersulphated cement' is the name given to a form of cement based upon ggbs (80%–85%) activated by calcium sulphate (i.e. 10%–15% anhydrite or burnt gypsum) and a small proportion of Portland cement (5%). The cement develops only a low heat of hydration and is particularly resistant to sulphates and other aggressive chemicals. It has been used to only a limited extent in the United Kingdom but has been utilised more widely elsewhere, notably in Belgium. Apart from the high content of residual ggbs (see Section 4.6), the hydration products are said to be mainly ettringite and a tobermorite-like phase (Lea 1970).

Table 4.9 Some special non-Portland types of cement and their principal characteristics

Cement type	Manufacture/composition	Advantageous properties
HAC	Fusing limestone and bauxite/mainly CA	Rapid strength development, good refractory and sulphate resistance
Slag cement	Ggbs activated by hydrated lime	Chemical resistance
Supersulphated cement	80%–85% ggbs activated by 10%–15% interground anhydrite (calcium sulphate) and with a small proportion of Portland cement or lime	Chemical resistance
BRECEM	Blend of HAC with ggbs	Benefits of HAC without the problems or conversion
Masonry cement	Finely interground Portland cement with limestone, or another inert filler, sometimes plus a plasticising and/or waterproofing admixture	Intended for use in mortars for masonry jointing, producing a more plastic mortar than that using ordinary Portland cement
Magnesium oxychloride or Sorel cement	Magnesite (magnesium carbonate) is calcined, ground and mixed with a magnesium chloride solution/brucite (magnesium hydroxide) bound by magnesium oxychloride	Produces a hard strong material, used for flooring when mixed with inert filler and a pigment

A list of some special cements and their principal characteristics is given in Table 4.9. Of these, the petrographer is liable to encounter masonry cement, particularly when examining mortar samples. Masonry cements based on mixtures of building lime and OPC are discussed in Section 4.2.5, but other types of 'masonry cement' are encountered which comprise OPC with a finely ground inert material and an air-entraining agent. In the United Kingdom, the inert filler is usually ground limestone, but at least one variety is made using a finely ground quartz filler.

Experimental programmes have indicated promising results for blends of HAC with ggbs and other materials, but it is not believed that such cements have yet been widely used in practice. The addition of ggbs to HAC has been reported to inhibit conversion by the formation of C_2ASH_8 (strätlingite) instead of C_3AH_6 as the main product of hydration (Majumdar et al. 1990; Richardson and Groves 1990; Majumdar and Singh 1992). Other additives might also be effective in preventing or inhibiting HAC conversion, including microsilica and metakaolin (Majumdar and Singh 1992) or even powdered limestone or quartz (Piasta et al. 1989). Some discussion of blends and modifications of cements for special purposes will be found in Chapter 8.

4.2.5 Building lime and cement/lime mixtures

Prior to the invention of Portland cement, most mortar and concrete materials were made using lime as the binder. The use of lime has been traced to the early Egyptian and Greek civilisations, and the Romans became technically sophisticated in the manufacture of lime concretes (Granger 1962; Davey 1965; Malinowski 1982). Lime mortars are still used today, especially with stone masonry; indeed recently, there has been something of a resurgence in the use of lime-based materials (Livesey 2011). Frequently, lime is used in tripartite mixtures with sand and Portland cement, particularly for brickwork jointing mortar, and the petrographer must be aware of this possibility, especially when examining samples of mortar. Further detailed consideration of lime mortars and renders will be found in Chapter 9.

Lime is manufactured by the burning of limestone, when, at a temperature of around 900°C, the calcium carbonate decomposes, so that CO_2 is driven off leaving only calcium oxide (known as 'quicklime'). This process is known as 'calcination'. For building purposes, the lime is then hydrated by slaking to produce calcium hydroxide (known as 'slaked lime'). Hardening and strength development is achieved by carbonation to reform calcium carbonate when slaked lime is exposed to the atmosphere. This process has been termed 'the lime triangle' (de Rooij 2009) and is illustrated by the following chemical equations:

$$CaCO_3 \rightarrow CaO + CO_2$$

Calcite   Lime

(4.1)

$$CaO + H_2O \rightarrow Ca(OH)_2$$

Lime   Calcium hydroxide

(4.2)

$$Ca(OH)_2 + CO_2 \rightarrow CaCO_3 + H_2O$$

Calcium hydroxide   Calcite

(4.3)

The burning of limestones containing a proportion of clay leads to the production of 'hydraulic' limes, which contain calcium silicates, calcium aluminates and other phases in addition to calcium oxide. Depending on the proportion of clay in the original limestone, these hydraulic limes range in silica and alumina content from as low as 1%–2% ('fat' limes) up to 50% ('eminently hydraulic' limes), with intermediate materials being variously termed 'lean' or 'semi-hydraulic' limes. The fat limes harden only by the carbonation process shown in the preceding equations, whereas the hydraulic and semi-hydraulic limes harden wholly or additionally by the hydration of the silicate and aluminate compounds.

4.2.5.1 Building lime

Building limes are classified according to their composition and whether or not they exhibit any hydraulic properties (i.e. an ability to set under water). BS 890 (1972/1995), which is described first as follows, was replaced by BS EN 459-1 (2001/2010).

BS 890 (1972/1995) gave four types of hydrated lime (either as fine dry powder or as lime putty):

1. Hydrated high-calcium lime (CL) (white lime)
2. Hydrated high-calcium by-product lime
3. Hydrated semi-hydraulic lime (grey lime)
4. Hydrated magnesian lime

BS 890 excluded the class of 'eminently hydraulic' limes, which are stated to be 'not widely available in the United Kingdom'. The by-product lime (2) is derived from the wet-process manufacture of acetylene from calcium carbide, and, in BS 890, the only specified difference from white lime (1) was an additional limitation on soluble salts.

In the United Kingdom, 'white limes' (or fat limes) are manufactured from Chalk, Jurassic, oolitic limestone and from some of the purer Carboniferous limestones. Semi- (or moderately) hydraulic limes are made by burning limestones which contain a proportion of clay or shale, such as the greystone from the Lower Cretaceous of the U.K. Thames Basin. According to Roberts (1956), after burning an argillaceous limestone at 1000°C, the products include $\beta C_2 S$, $C_2 AS$ and $C_4 AF$ plus other silicates and aluminates. The hydraulic properties are mainly attributed to the hydration of $ßC_2S$. White lime can be converted into hydraulic lime by the addition of pozzolanic material at the mortar or concrete mixing stage, and the Romans made particular use of this technique, employing various volcanic ash materials or even ground brick (Davey 1965).

The eminently hydraulic limes can contain up to 50% silica and alumina (Lea 1970) and have been made in the United Kingdom from burning Blue Lias, Chalk Marl and the shaley limestones from the Carboniferous in Scotland (the aptly termed 'cementitious group'). There is an uncertain boundary between these eminently hydraulic limes and the so-called natural cements, made historically by calcining natural mixtures of calcareous and argillaceous material, for example, the 'Roman cement' made from septarian nodules occurring in the London clay.

The hydrated magnesian limes are broadly similar to white lime, except that the carbonate rock parent contained the mineral dolomite $(CaMg(CO_3)_2)$ in analogous addition to calcite $(CaCO_3)$. An amount of magnesium oxide (MgO or periclase) is formed on calcination, subsequently slaked to magnesium hydroxide $(Mg(OH)_2$ or brucite) and finally carbonated to $MgCO_3$ or magnesite.

BS EN 459-1 (2010), which replaced BS 890 (2001), essentially recognises two types of building lime: 'air lime' (translated from a term used in most European countries) in main clause 4 and 'lime with hydraulic properties' in main clause 5 (see Table 4.10).

Air lime, which has no hydraulic properties, is subdivided into two 'families' (CL and dolomitic lime [DL]) and into two 'forms' (quicklime and hydrated lime, of which the latter is available as powder, putty or slurry). CL is further classified in to three designations on the basis of its content of lime + magnesia (CaO + MgO, with the overall MgO content ≤5%): CL 90, CL 80 and CL 70 (for ≥90%, ≥80% and ≥70%, respectively). DL is classified on a similar basis, but into four designations depending upon the contents of CaO and MgO: DL 90-30, DL 90-5, DL 85-30 and DL 80-5 (for ≥90% CaO - ≥30% MgO, ≥90% CaO - >5% MgO, etc., respectively).

Lime with hydraulic properties, which can set and harden when mixed with water and/or under water, but for which reaction with atmospheric carbon dioxide also remains part of the hardening process, is subdivided into three 'families': natural hydraulic lime (NHL) (produced from argillaceous or siliceous limestones), formulated lime (air lime and/or NHL with declared hydraulic or pozzolanic additions) and hydraulic lime (nominal lime content with other non-declared constituents). NHL is further classified into three designations on the basis of compressive strength (minimum to maximum) at 28 days: NHL 2, NHL 3.5 and NHL 5 (for ≥2 to ≤7 MPa, ≥3.5 to ≤10 MPa and ≥5 to ≤15 MPa, respectively). These designations also have required minimum characteristic 'available lime' (as $Ca(OH)_2$) values, which obviously decrease with strength (see Table 4.10). Lime-based binders from the 'formulated' and 'hydraulic' families, as defined in BS EN 459-1 (2010), are difficult for petrographers to recognise as such in hardened mortar or concrete, except perhaps when reference samples of particular products are available for comparative examination.

In contrast to the wealth of literature on the texture and composition of hydrated Portland cement, there has been little study in recent times of the nature of set lime binders. The cementing mechanism has been studied by Moorehead and Morand (1975) and Moorehead (1986). The rate and extent of hardening is governed by the diffusion of CO_2 from the atmosphere to the hydrated lime reaction site, where carbonation occurs by a solution mechanism, leading to a 35% increase in mass and a 12% increase in solids volume.

Significant expansion does not occur on carbonation, however, as the internal pores accommodate the increase in volume and the whole composite becomes denser and less permeable. The heat generated by the carbonation reaction, 74 kJ/mol according to Moorehead (1986), is sufficient to evaporate the water liberated by the reaction (see Equation 4.3) and

Table 4.10 Summary of lime classification in BS EN 459–1:2010

Air lime							
Family	Type	% CaO + MgO	% MgO[a]	% CO$_2$[b]	% SO$_3$	Form[c]	Properties
Calcium lime	CL 90	≥90	≤5	≤4	≤2	Q S[d]	R and P[e]
	CL 80	≥80	≤5	≤7	≤2	Q S[d]	R and P[e]
	CL 70	≥70	≤5	≤12	≤2	Q S[d]	R and P[e]
Dolomitic lime	DL 90–30	≥90	≥30	≤6	≤2	Q S[d]	R and P[f]
	DL 90–5	≥90	≥5	≤6	≤2	Q S[d]	R and P[f]
	DL 85–30	≥85	≥30	≤9	≤2	Q S[d]	R and P[f]
	DL 80–5	≥80	≥5	≤9	≤2	Q S[d]	R and P[f]

Lime with hydraulic properties					
Family	Type	Compressive strength, MP[a]		% Available lime	% SO$_3$
		7 days	28 days		
Natural hydraulic lime	NHL 2	—	≥2 to ≤7	≥35	≤2
	NHL 3.5	—	≥3.5 to ≤10	≥25	≤2
	NHL 5	≥2	≥5 to ≤15	≥15	≤2
Formulated lime	FL A 2	—	≥2 to ≤7	≥40 to <80	≤2
	FL B 2			≥25 to <50	≤2
	FL C 2			≥15 to <40	≤2
	FL A 3.5	—	≥3.5 to ≤10	≥40 to <80	≤2
	FL B 3.5			≥25 to <50	≤2
	FL C 3.5			≥15 to <40	≤2
	FL A 5	≥2	≥5 to ≤15	≥40 to <80	≤2
	FL B 5			≥25 to <50	≤2
	FL C 5			≥15 to <40	≤2
Hydraulic lime	HL 2	—	≥2 to ≤7	≥10	≤3[g]
	HL 3.5	—	≥3.5 to ≤10	≥8	≤3[g]
	HL 5	≥2	≥5 to ≤15	≥4	≤3[g]

Note: The BS EN should be consulted for full details of the requirements. All percentages are by mass.

[a] Up to 7% MgO permitted if soundness test is passed.
[b] Higher values permitted if all other chemical requirements and the testing frequency are satisfied.
[c] Q, quicklime; S, hydrated lime.
[d] S is subdivided: S, powder; S PL, putty; and S ML, slurry or 'milk of lime'.
[e] Properties used in designation of CL Q: R, reactivity (R5–R2 or R$_{sv}$), and P, particle size distribution (P4–P1 or P$_{sv}$).
[f] Properties used in designation of DL Q: R, reactivity (R5, R2, R1 or R$_{sv}$), and P, particle size distribution (P4–P1 or P$_{sv}$).
[g] Up to 7% SO$_3$ permitted if soundness test is passed.

sometimes also the residual capillary water, so that in some circumstances carbonation may be terminated prematurely.

Ancient lime mortars and concretes are quite distinctive in thin section under the petrological microscope, with the matrix being predominantly finely crystalline calcite (Figure 4.31), together with residual particles of partially burnt limestone, charcoal contamination from the original lime kiln and patches of more coarsely crystallised calcite (Figure 4.32) deriving from the slow carbonation of pockets of trapped slaked lime (Sims 1975). Malinowski and Garfinkel (1991) have likened the textural appearance of such material to the natural rock travertine.

Old mortars and concretes originally made using a hydraulic type of lime might initially appear similar to those containing non-hydraulic lime, because of the ultimate carbonation of calcium silicate hydrates, but relict particles of unhydrated dicalcium silicate and other potentially hydraulic constituents are liable to be observable. Sims (1975) used XRD to search for evidence of the presence of calcium silicates or the formation of calcium silicate

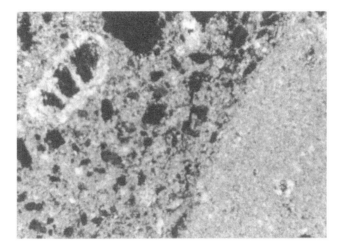

Figure 4.31 Photomicrograph of Roman lime mortar in thin section, the matrix comprising finely crystalline calcite. The aggregate particle on the right of the view is chalk, xpl, width of field ~0.9 mm.

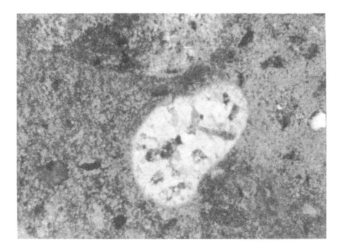

Figure 4.32 Photomicrograph of Roman lime mortar in thin section, exhibiting a pocket of coarsely crystallised calcite, xpl, width of field ~0.5 mm.

hydrate phases in some Roman concretes made using ground brick, but in those particular samples, the brick did not appear to have taken part in a pozzolanic reaction with the lime.

An excellent account of the modern investigation of lime-bound mortars and related materials will be found in Ingham (2003). He especially describes and illustrates (see Figure 4.33) microscopical examinations to determine the binder and sand constituents and the often characteristic texture of a lime-based binder, including pockets of coarser calcite deriving

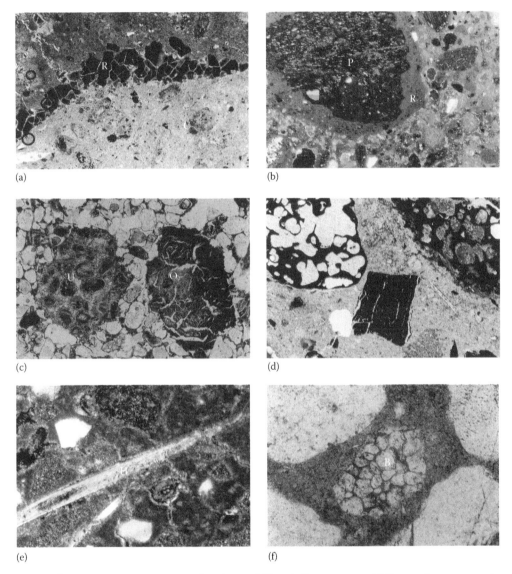

Figure 4.33 Some microscopic textures of ancient and historic lime mortars: (a) Roman lime mortar, showing a remnant of only partially burnt limestone (R = burnt rim, C = unburnt core); (b) Roman lime concrete, showing pozzolanic reaction (R) between lime and pumice (P) aggregate; (c) late medieval lime mortar, showing inclusions of unburnt (U) and overburnt (O) limestone; (d) late eighteenth century lime mortar, showing inclusions of charcoal (black) from the lime manufacture, (e) nineteenth century lime plaster, showing animal hair (H) reinforcement; and (f) hydraulic lime render from the 1930s, showing a relict grain of dicalcium silicate (B, belite). Selected photomicrographs from Ingham (2003), which is also reproduced in its entirety in Brocklebank (2012): refer to the paper for full details. (Colour images courtesy of Jeremy Ingham, Mott MacDonald Ltd., Croydon, U.K.)

from the slow carbonation of trapped slaked lime, variously underburnt and/or overburnt lime inclusions (usually abundant in historic materials) and the possible presence of relict grains of dicalcium silicate ('belite' or $\beta C_2 S$), the frequency of which might be expected to increase with hydraulicity of the lime.

The authors find that these distinctive features are often less apparent in modern lime-based products, when lime manufacture has been better controlled, the raw constituents are more thoroughly and finely ground and there is the possibility of small and variable proportions of similarly finely ground Portland cement having been added to the mix (see Section 4.2.5.2). One common textural feature, as the traditional disciplines of using, protecting and maturing lime mortar and render work can no longer be assumed, is the characteristic network of voidage and microcracks indicative of excessively rapid drying (Figure 4.34).

Moorhead (1986) has studied the microstructures of laboratory-prepared lime-cemented materials using electron microscopy. In the mixture prior to carbonation, the slaked lime binder was visible as distinctive platy crystals of calcium hydroxide (Figure 4.35a). After

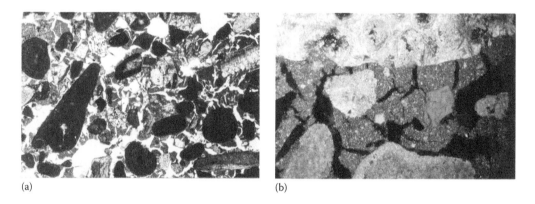

(a)   (b)

Figure 4.34 Characteristic microtextures of lime mortars subjected to excessively rapid drying (a) ppl and (b) xpl: field of view for photomicrograph (a) is 1.4 m; for photomicrograph (b), it is 0.6 m. (Photomicrographs courtesy of Paul Bennett-Hughes, RSK Environment Ltd., U.K.)

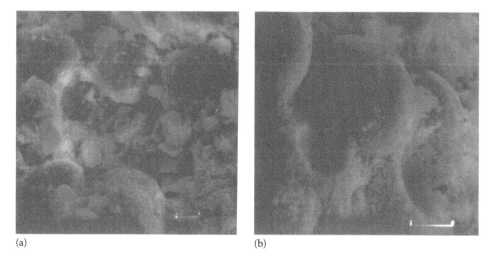

(a)   (b)

Figure 4.35 SEM micrographs of building lime (a) before and (b) after carbonation. (From Moorhead, D.R., *Cement Concr. Res.*, 16(5), 700, 1986.) The white bars are equal to 10 μm.

carbonation, cryptocrystalline and amorphous calcite were found to have formed and to have infilled the larger voids and spaces between aggregate particles (Figure 4.35b). The calcite grain size was generally less than 1 μm, with a significant proportion less than 0.1 μm and sometimes with more than 50% of the calcite being amorphous or nearly so in respect of XRD detection. Moorhead also found up to 8% residual uncarbonated calcium hydroxide.

It is expected that, with time, such a texture will recrystallise to form the type of microcrystalline groundmass apparent in ancient lime mortar and any residual calcium hydroxide is likely to become carbonated. However, such a maturation time will vary very widely, depending upon the compactness of the lime mortar material itself and the environmental conditions to which it is exposed. The authors have encountered old lime mortars that exhibit a fully carbonated hard outer crust and a softer interior of incompletely carbonated material. Building lime is discussed further in Chapter 9.

4.2.5.2 Cement/lime mixtures

Portland cement mortars for masonry work are today generally preferred over lime mortars for their much more rapid strength gain characteristics. However, masonry mortars must not be over-strong and a reduction in Portland cement content to control the ultimate strength typically leads to the production of 'harsh' and unworkable mixes. Combined lime–cement mortars overcome this problem and are widely used, often by adding Portland cement to a pre-mixture of sand and lime. It is common for mortars to be gauged such that the lime and Portland cement components are present in equal volumes: traditional mix proportions of 1:1:6 by volume are most frequently encountered.

In the chemical analysis of hardened mortars, the presence of building lime will be suggested by the significant excess of calcium over that required for combination with soluble silica in calculating the Portland cement content. However, such calcium could also derive from calcium carbonate in the aggregate (i.e. particles of limestone, chalk or shell fragments), and therefore only microscopy can unequivocally identify the presence or absence of a substantial content of building lime. Once microscopic examination has established the nature of the non-Portland cement source or sources of calcium, the chemical analysis can be used to estimate the original mix proportions.

Identifying the presence of building lime in a cement/lime/sand mortar by microscopy is not always straightforward, especially if it is likely that the hydrated Portland cement has been subjected to carbonation. The absence of calcareous aggregate material in a sample found to contain significantly excess calcium by chemical analysis is clearly indicative of the presence of building lime. Pockets of trapped calcium hydroxide, whether or not subsequently carbonated, can also indicate the presence of building lime, provided that it can be shown that the 'pocket' is not instead a void secondarily infilled with portlandite as a result of leaching.

The most positive evidence of building lime is the identification of relict burnt or partially burnt limestone particles (Figure 4.36). Spheroidal pfa particles might be observed sporadically in the matrix, but these could be derived either from building lime or from the Portland cement (modern kilns for both lime and Portland cement manufacture frequently use pfa and traces of residual fuel occur in the products).

Microscopy is sometimes required to distinguish materials, usually mortars or renders, made using hydraulic lime from those made from or including Portland cement; the presence of soluble silica in hydraulic lime can suggest the presence of Portland cement in the chemical analysis method (see Section 4.2.6). Theoretically, the difference should be determinable by microscopy, because Portland cement critically comprises C_3S (alite) as the main

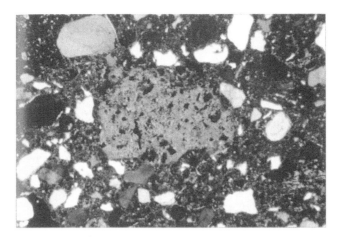

Figure 4.36 Photomicrograph of lime-bound mortar in thin section, containing a relict particle of burnt limestone. Width of field 4 mm. Crossed polars. (Photograph courtesy of RSK Environment Ltd., U.K.)

component, whereas hydraulic lime will contain βC_2S (belite) as the main active siliceous constituent and no alite. However, this distinction is complicated by the finely ground nature of modern Portland cement, thus restricting the frequency of resolvable unhydrated relicts of its C_3S component, which is in any case more likely to have been consumed than the C_2S component, making it difficult to be sure whether or not an apparent absence of alite confirms an original absence of Portland cement.

In such cases, it is usually possible to resolve the problem by thorough inspection using SEM, with the assistance of EDX microanalysis, to determine whether unhydrated alite or relict structures of hydrated alite are present to indicate some Portland cement (Winter 2012b). Alternatively, relatively large belite crystals (say >10 µm), either as unhydrated belite or relict hydrated structures, may be indicative of Portland cement rather than lime. Semi-quantitative EDX microanalysis of scanned regions of paste may also be helpful, although caution is required wherever that paste is secondarily altered. The original textures might have been altered beyond recognition by subsequent carbonation and/or leaching of the cementitious material, so that, whenever possible, samples for determination of original composition should be sought that have been least exposed in service (e.g. bedding mortar from the wall interior, rather than its surface edge).

According to Winter (2012b), using SEM and EDX microanalysis, there are four criteria that are helpful in establishing whether or not Portland cement was part of the original mix (see the example in Figure 4.37):

- Is there any residual alite (C_3S), confirmed using EDX?
- Are there any hydrated relicts showing unambiguously the characteristic hexagonal shape of alite, either as C–S–H or carbonated C–S–H or by delineation of the alite crystal shape by residual ferrite?
- Are individual belite (βC_2S) crystals more than about 10 µm across? (Belite in lime is typically much smaller than belite in Portland cement, owing to the lower burning temperature of lime.)
- By scanning areas of paste (avoiding any aggregates and bearing in mind that the sample will not be ideally homogenous), is the Si/Ca ratio typically in excess of that expected in a lime mortar?

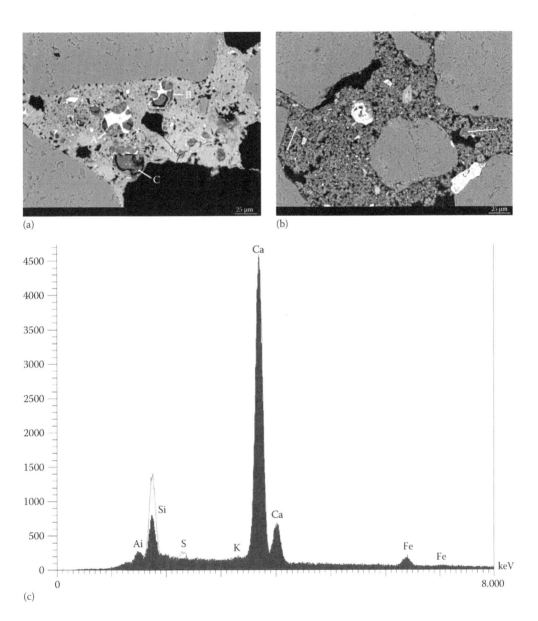

Figure 4.37 Differentiation of cement/lime and lime mixtures using SEM: (a) A typical paste region in a cement/lime mortar mixture, showing (labelled A, B and C) relict structures of Portland cement grains – the bright white areas within these structures are ferrites, whereas the darker features are hydrated and carbonated C–S–H from alite and belite hydration (which in grain C is probably an alite relict); smaller rounded dark features (not labelled) are due to belite. (b) A typical paste region in a mortar made with only hydraulic lime, at the same magnification as (a), showing a carbonated lime texture with some notably small grains (arrowed) of partly carbonated C–S–H, which could be relicts of small belite crystals or clusters of even smaller belite crystals; there is no alite and, in this case, no unhydrated belite relicts – the bright white features are iron-rich particles, with the larger ones being from the sand and the smaller ones from the lime. (c) Example EDX spectra, overlain and standardised (to the intensity of the Ca K-alpha peak), from the aggregate-free paste areas of the cement/lime and hydraulic lime mixtures in (a, blue line) and (b, red infill), respectively, showing that the Si K-alpha peak is more intense (a higher Si/Ca ratio) for the cement-containing material. (Material courtesy of Nick Winter of WHD Microanalysis Consultants Ltd., Woodbridge, U.K.)

4.2.6 Cement or binder contents in concrete

Microscopy is effective at characterising cement types and the presence within concrete of mineral additions (or 'SCMs') and various aggregate and filler constituents, but quantification is more complicated (see subsection 4.9 for a discussion on modal analysis). It is common for chemical analysis to be used to estimate the original cement content, and standard methods are available: BS 1881-124 (1988). The application and reliability of these methods have been discussed by Figg (1989), and more recently by Crofts (2006), and a new Concrete Society Working Party has recently reconsidered the matter (see later in this section). As these chemical techniques are commonly used alongside petrography in the assessment of hardened concrete, it is appropriate to provide a brief introduction.

While petrographic methods can be used to assess the cement or binder contents in concrete, it is customary to use chemical methods of analysis in addition. The current BS 1881-124 (1988) method principally employs the determination of soluble silica and calcium oxide contents to estimate the cement content of the concrete. Ideally, reference samples of the known mix constituents are analysed in parallel with the concrete, but frequently such samples are not available, with varying effects on the reliability of the outcome.

According to Crofts (2011), while some of the methods have developed since 1988 (such as soluble silica now being determined typically using atomic absorbance spectrophotometry rather than gravimetry), the principle has not changed: soluble silica and calcium oxide contents for the hardened concrete are compared with determined or assumed values for the cement/binder, then the approximate cement or binder contents are calculated accordingly (additionally, sulphide content can be used to estimate any content of ggbs). Thus, uncertainty in the method is derived from the analysis itself and the validity of the assumed soluble silica and calcium oxide contents used in the calculations. BS 1881-124 includes some precision data obtained from trials using concrete made with Portland-type cement and either siliceous or calcareous aggregates, and the limits on the accuracy of the results should be considered and reported (Table 4.11).

BS 1881-124 states the following:

1. The procedures within are suitable for concrete made using Portland-type cement and, 'in favourable circumstances', containing ggbs.
2. The analysis of concrete prepared using other cements and the determination of pfa content are 'outside the scope' of BS 1881-124.

Table 4.11 Precision data cited for cement content determinations

Type of aggregate	Range of cement contents investigated kg/m³	Oxide used to calculate cement content	Precision estimates, kg/m³ (see Note 1)			
			Repeatability r	Reproducibility R	Within-laboratory standard deviation σ_r	Between-laboratory standard deviation σ_L
Flint	240–425	CaO	40	60	15	15
		Soluble SiO₂	40	60	15	15
Limestone	300–350	Soluble SiO₂	40	50	15	10

Source: After BS 1881-124, *Testing Concrete, Part 124: Methods for Analysis of Hardened Concrete*, British Standards Institution, London, U.K., 1988, Appendix A.

Note 1: These precision data were determined from an experiment conducted in 1983, when each of the 18 laboratories analysed split level pairs of samples of three different concretes. The results from two laboratories were rejected as outliers using the criteria of BS 5497-1. If the outlier results had been included in the statistical analysis, the effect would have been to increase repeatability by about 50% and reproducibility by about 100%.

As such, the majority of modern concrete mixes potentially lie outside the scope of the methods given in BS 1881-124:1988. De Rooij (2011) has noted that BS 1881-124 relies on bulk analysis of a concrete sample, which will be complicated by the multi-constituent complexity of modern concretes, and also that the breadth of permitted variants in manufactured cements (e.g. see Table 4.1a) can invalidate assumptions on which the final calculations might be based.

Concrete Society Technical Report No. 32 (1989) and (Figg 1989) on the analysis of hardened concrete was published in order to aid interpretation and understanding, as well as to provide guidance on selection and use of the methods available. The methods presented and discussed therein are generally those in BS 1881-124:1988 and include some microscopical techniques. A new working group was convened in 2010 to review and revise TR 32 and a second edition was published recently (Concrete Society 2014). Although much of the 1989 edition was found to remain correct, the edition updated the standards and aimed to take account of changes in concrete technology, especially the increasing use of mineral additions or SCMs. Importantly, the working group arranged precision trials, which confirmed '*that there is significant doubt regarding the accuracy of the BS 1881-124 (1988) analysis procedures when applied to contemporary concrete,*' but they also recognised the limitations of their own trials and recommended '*a much larger research trial.*' Crofts (2014) has provided a helpful summary of the precision trial findings.

4.3 AGGREGATE TYPES AND CHARACTERISTICS

Several general treatments of aggregates are available, including Fookes (1997), Sims and Brown (1998), Smith and Collis (1993, 2001), Poole and Sims (2003), Alexander and Mindess (2005) and, as a particularly helpful introduction, Fookes and Walker (2011, 2012).

4.3.1 Petrographic identity of aggregate

Aggregate particles typically comprise about three-quarters of the volume of a concrete, and their petrological type, maximum size, size grading, surface texture and shape have influences on both the engineering and architectural properties of the concrete. The petrographic identity of the aggregate (i.e. the rock or mineral type or types of which it is composed) is an important consideration in establishing the composition of any concrete. In many cases, the petrographic identity of the aggregate will indicate provenance, at least to within particular regions, and in rare cases, it is possible to pinpoint specific quarry or pit sources.

However, the petrographer should concentrate aggregate description on to those features which have a potential bearing upon concrete properties: for instance, a petrogenetic analysis of a granite is not relevant to concrete performance, whereas assessing the extent of secondary alteration of the granite could provide valuable information pertinent to concrete durability.

Various undesirable or 'potentially deleterious' constituents might also be identified, including inter alia clay lumps, discrete (or 'free') mica, gypsum, pyrite and alkali-reactive materials. The presence of such constituents in the concrete will frequently need to be quantified and any evidence of detrimental activity recorded (see Chapters 5 and 6). Further information on undesirable aggregate constituents may be found in Sims and Brown (1998), and a list of some of the more common is given in Table 4.12.

The aggregates can be studied on broken concrete surfaces or in the sides of drilled cores. However, the aggregates are most easily identified and described in finely ground large-area surfaces, using unaided visual and low-power microscopical procedures. Fine-grained coarse aggregates and the fine aggregates are most easily examined in thin section under a petrological microscope.

Table 4.12 Some potentially deleterious constituents found in aggregates

Potentially deleterious constituent	Possible adverse effect in concrete[a]				
	i	*ii*	*iii*	*iv*	*v*
Clay coatings on aggregate particles	—	✓	–	–	–
Clay lumps and altered rock particles	—	—	✓	✓	✓✓
Absorptive and microporous particles	—	—	✓	✓	✓✓
Coal and lightweight particles	—	—	—	—	✓✓
Weak or soft particles and coatings	—	✓✓	✓	✓	✓✓
Organic matter	✓✓	—	✓	—	—
Mica	—	—	✓✓	—	✓
Chlorides[b]	✓	—	—	✓	—
Sulphates	✓	—	—	✓✓	✓
Pyrite (iron disulphide)	—	—	—	✓✓	✓✓
Soluble lead, zinc or cadmium	✓✓	—	—	—	—
Alkali-reactive constituents	—	✓	—	✓✓	—
Releasable alkalis	—	—	—	✓✓	—

Source: Compiled from information in Sims, I. and Brown, B.V., Concrete aggregates, in *Lea's the Chemistry of Cement and Concrete*, 4th ed., Hewlett, P.C. (ed), Chapter 16, pp. 903–1011, Arnold, London, U.K., 1998.

[a] i, chemical interference with the setting of cement; ii, physical prevention or good bond between the aggregate and the cement paste; iii, modification of the properties of the fresh concrete to the detriment of the durability and strength of the hardened material; iv, interaction between the cement paste and the aggregate which continues after hardening, sometimes causing expansion and cracking of the concrete; and v, weakness and poor durability of the aggregate particles themselves.

[b] The main problem with chlorides in concrete is associated with the corrosion of embedded steel.

✓✓, main effect; ✓, additional effect.

4.3.1.1 Typical aggregate combinations

A concrete mix design usually aims to achieve a reasonably continuous aggregate particle size distribution (or grading), so that inter-particulate space, to be occupied by cement paste, is minimised after compaction. This is typically achieved by blending a coarse aggregate (say >4 or 5 mm particle size) with a sand or fine aggregate (say <4 or 5 mm particle size). Occasionally, a natural sand and gravel material will exhibit a suitably continuous grading and may be used as a single 'all-in' aggregate. It is more common for the sand and gravel components of a natural material to be processed separately and then recombined in a controlled way to achieve a continuous grading. Crushed rock coarse aggregates are frequently combined either with natural sand fine aggregates or with a controlled blend of the crushed rock fines produced as a by-product (i.e. 'crusher-run fines') with a natural sand.

It is therefore necessary to recognise that the aggregates within a hardened concrete sample might represent a combination of materials from various different sources. It is usually convenient separately to describe the coarse and fine fractions, with 4 or 5 mm being a common threshold particle size between the two, unless another demarcation is made apparent by a clear compositional change.

4.3.1.2 Crushed rock coarse aggregates

The presence of crushed rock coarse aggregates in concrete is usually indicated by a relatively angular particle shape (see Section 4.3.3) and an essentially single-component (i.e. 'monomictic') composition. The rock type is usually straightforward to classify within broad generic groupings, such as those that were listed in BS 812-102 (1989) (Table 4.13)

Table 4.13 Names and simple definitions for rock types commonly used for aggregates

Petrological term	Description
Agglomerate	Pyroclastic rock, comprising coarse to very coarse, mainly rounded, volcanic fragments; see also volcanic breccia.
Amphibolite	Metamorphic rock consisting mainly of amphibole and plagioclase.
Andesite	Fine-grained, usually volcanic, variety of diorite.
Arenite	Essentially synonymous with sandstone (e.g. quartz arenite).
Argillite	Sedimentary rock comprising clay-sized grains (see mudstone); term often restricted to slightly metamorphosed mudstone varieties.
Arkose	Type of sandstone or gritstone containing over 25% feldspar.
Basalt	Fine-grained basic rock, similar in composition to Gabbro, usually volcanic.
Breccia	Rock consisting of angular, unworn rock fragments, bonded by natural cement; sedimentary, volcanic and fault (see cataclasite) varieties.
Cataclasite	Metamorphic rock formed by brittle fragmentation during shearing, consisting of angular fragments or 'clasts' in a fine-grained matrix; varieties include fault breccia.
Chalk	Very fine-grained limestone of Cretaceous age, usually white.
Chert	Rock comprising cryptocrystalline (resolved only with a high-power microscope) silica.
Claystone	Sedimentary rock containing more than 67% clay-sized particles.
Conglomerate	Rock consisting of rounded pebbles bonded by natural cement.
Diorite	An intermediate plutonic igneous rock, consisting mainly of plagioclase feldspar, with hornblende, augite or biotite.
Dolerite	Basic igneous rock, with grain size intermediate between that of gabbro and basalt; the term diabase is almost synonymous but often applied to highly altered dolerite.
Dolomite	Rock (or mineral) composed of calcium magnesium carbonate; the dolomite term is usually applied to carbonate rock with >50% mineral dolomite.
Dolostone	Term used by some as an alternative to dolomite rock, especially when the rock almost completely comprises mineral dolomite.
Flint	Cryptocrystalline silica originating as nodules or layers in chalk; variety of chert.
Gabbro	Coarse-grained basic plutonic igneous rock, consisting essentially of calcic plagioclase feldspar and pyroxene, sometimes with olivine.
Gneiss	Foliated and banded rock, produced by intense metamorphic conditions.
Granite	Coarse-grained acidic plutonic igneous rock, consisting essentially of quartz and alkali feldspar, typically with mica.
Granodiorite	Coarse-grained igneous rock between granite and diorite in composition, similar to granite but with less alkali feldspar and more plagioclase feldspar.
Granulite	Metamorphic rock with granular texture and no preferred orientation of minerals.
Greywacke	Dark-coloured, impure type of sandstone or gritstone, composed of poorly sorted fragments of rock, plus quartz and other minerals; the coarser grains are usually strongly cemented in a fine matrix; some classifications stipulate >15% fine matrix.
Gritstone	Sandstone, with coarse and usually angular grains.
Hornfels	Thermally metamorphosed non-foliated rock, containing substantial amounts of rock-forming silicate minerals.
Ironstone	Imprecise term, typically used for compact sedimentary rock, sometimes clayey and/or calcareous, containing large amounts of iron minerals; variously beds or nodules.
Limestone	Sedimentary rock, consisting predominantly of calcium carbonate.
Marble	Metamorphosed limestone.
Microdiorite	Igneous rock with grain size intermediate between that of diorite and andesite.
Microgranite	Acid igneous rock with grain size intermediate between that of granite and rhyolite.

(Continued)

Table 4.13 (Continued) Names and simple definitions for rock types commonly used for aggregates

Petrological term	Description
Mudstone	Argillaceous (clay-sized grains) sedimentary rock, distinguished from shale by being massive, not fissile; the term 'mudrock' includes clay- and silt-sized particles; calcareous mudstone contains up to 50% calcite and can have properties similar to limestone.
Mylonite	Fine-grained rocks formed by brecciation, milling and sometimes recrystallisation of rocks during deformation.
Obsidian	Rock comprising dark, often black, volcanic glass of rhyolitic composition.
Pegmatite	Very coarse-grained igneous rock, forming veins and dykes associated with plutonic rocks.
Peridotite	Ultrabasic plutonic igneous rock, consisting of 40%–90% olivine, with pyroxene, hornblende and other minerals; see an altered variety serpentinite.
Porphyry	Largely disused term for various igneous rocks with a 'porphyritic' texture.
Quartzite	Metamorphic (meta-quartzite) or sedimentary (ortho-quartzite) rock, composed almost entirely of quartz grains.
Rhyolite	Fine-grained or glassy acid igneous rock, usually volcanic.
Sandstone	Sedimentary rock, composed of sand grains naturally cemented together.
Schist	Metamorphic rock in which the minerals are arranged in nearly parallel bands or layers; platy or elongate minerals, such as mica or hornblende, cause fissility that distinguishes it from gneiss.
Serpentinite	Metamorphosed basic or ultrabasic igneous rock, composed of serpentine with relicts of olivine and other primary minerals.
Shale	Argillaceous (clay-sized grains) sedimentary rock, characterised by fissility parallel to bedding.
Siltstone	Fine-grained sandstone, comprising silt-sized particles naturally cemented together.
Slate	Rock derived from argillaceous sediments or volcanic ash by metamorphism, characterised by cleavage planes independent of the original stratification.
Syenite	Intermediate plutonic igneous rock, consisting mainly of alkali feldspar with plagioclase, hornblende, biotite or augite.
Trachyte	Fine-grained, usually volcanic, variety of syenite.
Tuff	Consolidated volcanic ash; pyroclastic rock.

Source: Originally after BS 812-102, *Testing Aggregates, Part 102, Methods for Sampling*, British Standards Institution, London, U.K., 1989; but now extended with reference to BS EN 12670, *Natural Stone, Terminology*, British Standards Institution, London, U.K., 2002; and Applied Petrography Group, *A Glossary of Terms and Definitions, Part 1. Aggregate Terminology*, Special Report 3, Part 1 (APG SR3 Part 1). APG, Engineering Group, Geological Society, London, U.K., 21pp (edited by Poole, A.B., ppAPG) 2010.

or defined in ASTM C294 (2005) or BS EN 12670 (2002). The Applied Petrography Group (APG, affiliated to the Engineering Group of the Geological Society) is developing recommended terminology for aggregates (APG-SR3 2010).

French (1991a) has highlighted the classification difficulties that can arise because the rocks seen as aggregate particles within concrete are separated from their field association and thus have to be identified from their mineralogies and textures alone. This maximises the possibilities of different interpretations by different petrographers and sometimes leads to a lack of confidence in petrography by engineers who do not always appreciate the imprecise and widely overlapping nature of petrological nomenclature.

There would be much to be gained from a universally standardised and unambiguous petrological classification scheme for engineering purposes, but this is easier stated than achieved: this problem is discussed in Sims and Brown (1998) and more fully in Chapter 6 of Smith and Collis (1993, 2001). French (1991a) advocates 'full descriptions of the rocks…, illustrated with photomicrographs, rather than the simple use of "standard" terminology'. However, for purely compositional analyses, this might lead to unnecessarily detailed work and clearly the descriptive detail justified must be tailored to the objectives of the petrographic examination (see also Chapters 5 and 6).

Even a crushed rock aggregate will not necessarily exhibit an entirely uniform petrographic composition. A crushed sedimentary rock, for example, might comprise a variety of types reflecting the bedded sequence within the source quarry. Crushed limestones, for another example, might exhibit a range of carbonate rock types, including dolomitic varieties and silicified material, and sporadic chert particles are very common. Crushed greywacke might comprise a complete range from grain-dominant to matrix-dominant, virtually mudstone, types.

Igneous rock sources can also give rise to mixtures, for example, as the result of vein swarms, or small dykes within the main rock body, or, in the case of volcanic rocks, variations within or between successive lava flows or tuff deposits. All types of rock might exhibit additional variations as the result of patchy or zonal alteration and weathering. Variants considered potentially to be of engineering significance should be separately identified and semi-quantified.

The petrographer must be aware of the heterogeneity likely to arise from these variations in aggregate composition and, whenever practicable, should ensure that adequately large sample areas are examined for the results to be sensibly representative.

4.3.1.3 Natural gravel coarse aggregates

Natural gravel coarse aggregate in concrete is usually indicated by rounded or sub-rounded particle shape (see Section 4.3.3) and a mixed or multi-component (i.e. 'polymictic') composition. Oversized gravel particles will be crushed, giving rise to the added presence of angular particles and also particles which are partly well rounded and partly angular. Determination of petrographic composition should follow the same principles as those for crushed rock aggregate, except that a variety of completely different rock types might be encountered and each should be separately identified and semi-quantified.

Marine gravel coarse aggregate can usually be distinguished from an equivalent terrestrial material, by the presence in the former of recent calcareous shell debris, although in coarse aggregate this might form only a small proportion and then must be distinguished from fossil shell debris that is sometimes present in terrestrial sand and gravel deposits.

The compositional presentation of the aggregate observations will depend upon the rock constituents identified and their relative proportions. For example, a gravel aggregate that is found to comprise 80%–90% flint and 10%–20% varied assortment of other materials might reasonably be labelled a 'flint gravel'. Another gravel aggregate which is found to comprise 20% greywacke, 20% chert, 20% limestone, 20% quartzite and 20% varied others might better be described as a mixed or 'polymictic' gravel mainly containing greywacke, chert, limestone and quartzite.

Visual estimation of relative proportions is usually adequate for concrete compositional analysis purposes, when the percentage estimate charts devised for sediments are usually helpful (Terry and Chilingar 1955) (Figure 4.38). Quantitative determinations by point counting are arduous and time consuming and will not always be necessary; certainly, they will not be justified unless the concrete sample area is large enough to be adequately representative.

4.3.1.4 Crushed rock and natural sand fine aggregates

The identification and description principles for coarse aggregates largely apply similarly to fine aggregates, except examination is usually carried out under a microscope and monomineralic grains will often be present in addition to rock particles. Particle shape and uniformity of composition are again the main features that distinguish crushed rock and natural sand materials, although blends can be difficult to differentiate if the natural sand particles are

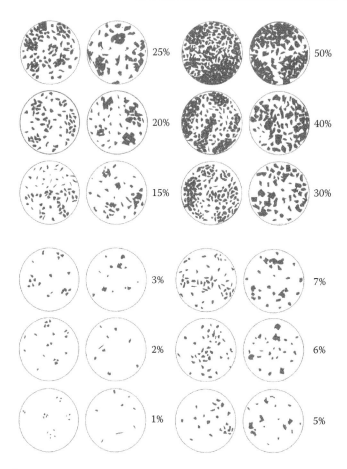

Figure 4.38 Charts for estimating percentage composition of rocks and sediments. (From Terry, R.D. and Chilingar, G.Y., *J. Sediment. Petrol.*, 25(3), 229, 1955.)

not well rounded. Natural sand particles can sometimes exhibit secondary surface growths, or superficial erosion or weathering zones, when microscopically examined in thin section.

Mixed or polymictic sand compositions are readily quantified by point counting in thin section under the microscope. Particle size differentiation of various constituents is quite common with natural sands, and any quantified analysis must take this into account. For example, the flint-bearing natural sands of south-eastern England typically contain their flint within the coarser particle size range (say 1–5 mm), whereas the <1 mm fractions are dominated by quartz. Marine sands will usually be distinguished from terrestrial sands by the presence in the former of recent shell debris, sometimes in considerable quantity.

4.3.1.5 Recycled aggregates

Although they are used increasingly for hardcore and fill and in road making, to date, there has been only limited use of recycled aggregates in new concrete (Hack and Bryan 2006; Dhir and Paine 2007), but modern awareness of environmental and sustainability issues makes it likely that there will be an increasing use of such materials in the future (Langer 2009). Potential use of coarse 'recycled concrete aggregate' (RCA, principally comprising crushed concrete) is recognised in BS EN 8500-2 (2006), where it is permitted for use in concrete up to strength Class C40/50 and with certain durability classes.

However, the more common coarse recycled aggregate principally comprising crushed masonry (i.e. brickwork and blockwork), designated 'RA' in BS EN 8500-2, is only permitted provided that there are additional provisions in the project specification to allow for variability. BS EN 8500-2 is also more cautious about the suitability of fine RCA or RA for use in concrete, mainly owing to concerns over sulphates deriving from the concentration of plaster dust and also general variability of composition in the case of fine RA. Dhir and Paine (2007), in conducting a research programme for WRAP (Waste and Resources Action Programme) in the United Kingdom, found that 20% RCA or RA had little effect on concrete performance and suggested a performance-related classification for aggregate combinations containing RCA and/or RA: Class A would be suitable for a wide range of exposure conditions, Class B would probably include the majority of such combinations and would be suitable for 'moderate' exposure conditions (up to XC-4 and XF-2 as defined in BS EN 206-1 (2000)) and Class C would only be usable in the mildest conditions (say XO and DC-1).

BS EN 8500-2 (2006) includes a method for determining the pre-use composition of RCA or RA, in which a representative sample is sieved and then each size fraction hand separated into (1) concrete and normal-weight aggregate, (2) masonry (brick and block), (3) lightweight block materials (apparently <1000 kg/m^3), (4) asphalt and (5) other foreign matters (e.g. wood, glass, plastics). The separated components are weighed and expressed as percentages. An analogous approach could clearly be employed by a petrographer examining a hardened concrete containing recycled aggregate, quantifying observations and using a similar classification scheme.

Particular concern has usually focused on the potential for unpredictable variation, although WRAP (2005) has endeavoured to make some allowance for this issue in their quality protocol for producers. As well as a potential problem with included sulphates (especially in fine RCA and RA, see this section above), it has also been anticipated that recycled aggregates might present an increased risk of alkali–aggregate reaction (AAR, see Chapter 6), owing to the possibility of introducing additional alkalis and/or unforeseen reactive constituents into the new concrete or, in the case of RCA, perhaps even including some already active AAR. However, in the United Kingdom, RCA is now accepted as having 'normal reactivity' within BRE Digest 330 (2004), with specified conditions, whereas it was previously assumed to have 'high reactivity'. Also, research by McCarthy et al. (2009), again for WRAP, has reported encouragingly low expansions in an extensive programme of AAR tests.

4.3.2 Particle size and aggregate size grading

The overall particle size distribution (or 'grading') of aggregate exerts an important influence over the properties of concrete, particularly by controlling with other factors the mix workability and the feasibility of achieving a satisfactorily high density by normal compaction methods. The aggregate maximum particle size and overall gradings are therefore necessary features to be established as part of any petrographic assessment of concrete composition.

In practice, modern design mix techniques and the availability of a range of admixtures for modifying concrete properties can enable successful concrete to be made using virtually any grading, provided that grading remains reasonably consistent during a period of supply and use. The petrographic analysis of a population of samples from the same or supposedly similar mixes must therefore address the aspect of aggregate grading consistency.

Aggregate particle size and grading are traditionally characterised and quantified by sieve analysis (e.g. BS 812-103:1 1985 and BS EN 933-3 1997), wherein the proportions by mass passing through each of a series of sieve sizes are computed. The set of sieve sizes chosen for this purpose varies between different methods; those specified in British (now superseded) and American Standards are shown in Table 4.14 for (a) coarse and (b) fine aggregates,

Table 4.14a Coarse aggregate grading limits specified by BS and ASTM standards

BS Type

BS 882:1992 Sieve (mm)	Graded aggregates (mm)			Single-sided aggregates (mm)				
	40–5	20–5	14–5	40	20	14	10	5
	% by Mass passing							
50.0	100	—	—	100	—	—	—	—
37.5	90–100	100	—	85–100	100	—	—	—
20.0	35–70	90–100	100	0–25	85–100	100	—	—
14.0	25–55	40–80	90–100	—	0–70	85–100	100	—
10.0	10–40	30–60	50–85	0–5	0–25	0–50	85–100	100
5.0	0–5	0–10	0–10	—	0–5	0–10	0–25	45–100
2.36	—	—	—	—	—	—	0–5	0–30

ASTM Size (mm)

ASTM C33 Sieve (mm)	90–37.5	63–37	50–25	50–4.75	37.5–19	37.5–4.75	25–9.5	25–12.5	25–4.75	19–9.5	19–4.75	12.5–4.75	9.5–2.36	9.5–1.18	4.75–1.18
	% by Mass passing														
100	100	—	—	—	—	—	—	—	—	—	—	—	—	—	—
90	90–100	100	—	—	—	—	—	—	—	—	—	—	—	—	—
75	—	100	—	—	—	—	—	—	—	—	—	—	—	—	—
63	25–60	90–100	100	100	—	—	—	—	—	—	—	—	—	—	—
50	—	35–70	90–100	95–100	100	100	—	—	—	—	—	—	—	—	—
37.5	0–15	0–15	35–70	—	90–100	95–100	100	100	100	—	—	—	—	—	—
25	—	—	0–15	35–70	20–55	—	90–100	90–100	95–100	100	100	—	—	—	—
19	0–5	0–5	—	—	0–15	35–70	40–85	20–55	—	90–100	90–100	100	—	—	—
12.5	—	—	0–5	10–30	—	—	10–40	0–10	25–60	20–55	—	90–100	100	100	—
9.5	—	—	—	—	0–5	10–30	0–15	0–5	—	0–15	20–55	40–70	85–100	90–100	100
4.75	—	—	—	0–5	—	0–5	0–5	—	0–10	0–5	0–10	0–15	10–30	20–55	85–100
2.36	—	—	—	—	—	—	—	—	0–5	—	0–5	0–5	0–10	5–30	10–40
1.18	—	—	—	—	—	—	—	—	—	—	—	—	0–5	0–10	0–10
0.30	—	—	—	—	—	—	—	—	—	—	—	—	—	0–5	0–5

Source: Adapted from superseded BS 882, Specification for Aggregates from Natural Sources for Concrete, British Standards Institution, London, U.K., 1992 and ASTM C33, Standard Specification for Concrete Aggregates, American Society for Testing and Materials, West Conshohocken, PA, 2011.

Table 4.14b Fine aggregate grading limits specified by BS and ASTM standards

| Type Sieve size | BS 882 | | | | ASTM C33 |
	Overall limits	Coarse (C)	Medium (M)	Fine (F)	General[a]
	% by mass passing a sieve				
10 mm	100	100	100	100	—
9.5	—	—	—	—	100
5	89–100	89–100	89–100	89–100	95–100
4.75	—	—	—	—	80–100
2.36	60–100	60–100	65–100	80–100	80–100
1.18	30–100	30–90	45–100	70–100	50–85
600 μm	15–100	15–54	25–80	55–100	25–60
300	5–70	5–40	5–48	5–70	5–30
150	0–15/20[b]	0–15/20[b]	0–15/20[b]	0–15/20[b]	0–10
75					0–3.0[c]

Source: Adapted from superseded BS 882, Specification for Aggregates from Natural Sources for Concrete, British Standards Institution, London, U.K., 1992 and ASTM C33, Standard Specification for Concrete Aggregates, American Society for Testing and Materials, West Conshohocken, PA, 2011.

a Additional criteria are given in ASTM, depending on the concrete mix, and ensure evenness of grading.
b The upper limit is increased from 15% to 20% for crushed rock fines, except for heavy-duty floors.
c Increased to 5.0% for concrete not subjected to abrasion or when <75 μm material consists of 'dust of fracture' rather than clay or to 7.0% for concrete not subject to abrasion and where the <75 μm material is 'dust of fracture' and not clay.

although graphically presented results ought to be similar. The nominal particle sizes used to describe standardised aggregate gradings similarly vary (Table 4.14a,b).

Former specification BS 882 (1992) has now been superseded and replaced by the European Standard, BS EN 12620 (2002/2008), wherein the grading compliance criteria are more complicated, in the laudable interests of both maximising materials usage by not imposing artificial limits on natural materials and seeking to satisfy the important parameter of grading consistency. As shown in Table 4.15, BS EN 12620 provides some general requirements for the minimum (d) and maximum (D) particle sizes, but otherwise imposes permitted tolerances on the midsize sieve relative to the grading declared for the aggregate by the supplier. Aggregate sizes are described using the designation d/D, in millimetres. In the United Kingdom, guidance on using BS EN 12620 is provided in PD 6682-1 (2009), wherein a table relates the traditional U.K. aggregate sizes to the new BS EN designations (see Table 4.16). It can be seen, for example, that a former '20-5 mm graded' coarse aggregate becomes '4/20' in the new parlance (the fine/coarse threshold is 4 mm rather than 5 mm), while a former 'M (medium)' sand becomes '0/2'.

Sieve analysis results are influenced by a number of factors (Smith and Collis 1993, 2001), including inter alia sieve aperture shape, aggregate particle shape (see Section 4.3.3), whether sieving is carried out dry or wet and the energy and duration of the sieve shaking. Broadly speaking, however, sieved size fractions are determined by the largest dimension of a cuboidal particle or the larger width dimension of an elongated particle. It is therefore

Table 4.15 Aggregate grading requirements of BS EN 12620 (2002/2008), for coarse and fine aggregates. Separate requirements are also given in BS EN 12620 for natural graded 0/8 mm and all-in materials. Consult BS EN 12620 for full details and requirements.

Aggregate	Size, mm	% Passing by mass					Grading category
		2 D	1.4 D	D	d	d/2	
Coarse (general and single sized)[a]	$D/d ≤ 2$ or $D ≤ 11.2$	100	98–100	85–99	0–20	0–5	G_C 85/20
				80–99			G_C 80/20
	$D/d > 2$ and $D > 11.2$			90–99	0–15		G_C 90/15
Coarse (additional for graded)[b]		% Passing by mass – midsize sieves					
	D/d Midsize	Overall limits		Tolerances on declared grading			
	<4 $D/1.4$	25–70		±15		G_T 15	
	≥4 $D/2$			±17.5		G_T 17.5	
Fine (general)[a]		% Passing by mass					Grading Category
	Size, mm	2 D	1.4 D	D	d	d/2	
	$D ≤ 4$ and $d = 0$	100	95–100	85–99	—	—	G_F 85
Fine (additional tolerances on declared grading for all fine aggregates)[c]		Tolerances on declared grading, % passing by mass					
	Sieve size, mm	0/4		0/2		0/1	
	4	±5[d]		—		—	
	2	—		±5[d]		—	
	1	±20		±20		±5[d]	
	0.250	±20		±25		±25	
	0.063[e]	±3		±5		±5	

[a] From Table 4.2 in BS EN 12620 (2002/2008).
[b] From Table 4.3 in BS EN 12620 (2002/2008).
[c] From Table 4.4 in BS EN 12620 (2002/2008).
[d] Further limited by the general requirement for % passing D.
[e] In addition, separate % passing 0.063 mm limits apply (Table 4.11 in BS EN 12620 (2002/2008)).

Table 4.16 Overall grading limits complying with BS EN 12620 but relating to traditional aggregate sizes used in the United Kingdom. Based on Tables C.1 and D.1 in BS PD 6682–1 (2009). Refer also to Table 4.15. Consult BS PD 6682–1 (2009) for full details and guidance.

	Coarse aggregates							
	% by mass passing for d/D sizes							
	Graded aggregates[a]			Single-sized aggregates				
Sieve size, mm	4/40[b]	4/20[b]	2/14[b]	20/40[c]	10/20[c]	6.3/14[c]	4/10[c]	2/6.3[d]
80	100	—	—	100	—	—	—	—
63	98–100	—	—	98–100	—	—	—	—
40	90–99	100	—	85–99	100	—	—	—
31.5	—	98–100	100	—	98–100	100	—	—
20	25–70	90–99	98–100	0–20	85–99	98–100	100	—
16	—	—	—	—	—	—	—	—
14	—	—	90–99	—	—	85–99	98–100	100
10	—	25–70	—	0–5	0–20	—	85–99	98–100
8	—	—	—	—	—	—	—	—
6.3	—	—	25–70	—	—	0–20	—	80–99
4	0–15	0–15	—	—	0–5	—	0–20	—
2.8	—	—	—	—	—	0–5	—	—
2	0–5	0–5	0–15	—	—	—	0–5	0–20
1	—	—	0–5	—	—	—	—	0–5

	Fine aggregates				
	% by mass passing for d/D sizes[e–g]				
	0/4	0/4	0/2	0/2	0/1
Sieve size, mm	CP	MP	MP	FP	FP
8	100	100	—	—	—
6.3	95–100	95–100	—	—	—
4	85–99 (±5)	85–99 (±5)	100	100	—
2.8	—	—	95–100	95–100	—
2	—	—	85–99	85–99	100
1	(±20)	(±20)	(±20)	(±20)	85–99 (±5)
0.5[f]	5–45	30–70	30–70	55–100	55–100
0.25	(±20)	(±20)	(±25)	(±25)	(±25)
0.063	(±3)	(±3)	(±5)	(±5)	(±5)

[a] Category $G_T17.5$ (see Table 4.15).
[b] Category $G_C90/15$ (see Table 4.15).
[c] Category $G_C85/20$ (see Table 4.15).
[d] Category $G_C80/20$ (see Table 4.15).
[e] Category G_F85 (see Table 4.15).
[f] Coarseness/fineness (CP, MP or FP) is described in Table B.1 of BS PD 6682–1 (2009), according to % passing 0.5 mm.
[g] Tolerances are shown in parentheses for those fractions required to be declared.

these dimensions which must be measured and recorded when evaluating aggregate grading in hardened concrete.

The nominal size and grading of aggregates in concrete can be determined by petrography, preferably from sawn and finely ground surfaces, or using large-area thin sections (St John 1990). It is usually more helpful to attribute the maximum particle size observed to the nearest standard size (i.e. measured 22 mm might be better reported as nominal 20 mm), except that anomalously large particle (e.g. 'plums') should be additionally recorded.

It is important to recognise that random cross sections through the centres of aggregate particles almost always exaggerate the dimensions (see Figure 4.39): only those rare cross sections exactly perpendicular to the dimension concerned will provide an accurate measurement; all otherwise oblique sections are likely to suggest a larger than actual dimension. Conversely, sections which transect only the tip of a particle might *under*state the size of that particle.

Figure 4.39 Sketch illustrating that particle dimensions are usually misrepresented by cross sections. A variety of particle shapes are overlaid by a grid of traverse lines. Only the solid intercepts in the 2D sketch would accurately reflect the particle size as measured by sieve analysis, while the dashed intercepts would all either underestimate or overestimate particle size. Even the solid intercepts in the sketch might not be accurate in three dimensions.

Thus, in determining aggregate particle sizes, the petrographer must make intuitive allowance for these 'apparent' particle dimensions in all sectional specimen surfaces. These inherent problems of viewing 3D materials in two dimensions are addressed by stereology, but these specialised techniques will not assist with routine examinations (Russ and Dehoff 2000).

Grading analyses can be quantified from sawn and ground surfaces or large-area thin-section specimens, using linear-traverse or particle counting techniques. However, this is exceptionally arduous, in view of the large number of measurements needed to satisfy the demands for adequate precision, and rarely justifiable for normal concrete petrography. French (1993) described a procedure based upon the geometrical relationship between the 'grading curve' of chord intersects along lines of traverse and the sieved grading curve for an idealised aggregate of spherical particles.

In this chord method, lines of traverse are described across a large sample area, spaced at 10 mm intervals and totalling ~4 m of traverse. By examining these lines of traverse under a microscope, the total length of traverse, the number of aggregate particles encountered and the individual chord intersect lengths are recorded. The chord lengths are corrected to compensate for their typical underestimation of particle diameter (with spherical particles, the correction factor is suggested by French to be ~1.5). A chord correction factor can be estimated for a particular concrete by comparing the average chord length against the average particle size, calculated from the total length of traverse and the total number of particles encountered, assuming an overall aggregate volume for the concrete (say 70%). Corrected chord lengths are then allocated into convenient and appropriate size fractions (e.g. 20–14 mm, 14–10 mm and 10–5 mm), and a grading curve can be obtained from the relative proportions of these size fractions.

The method requires the painstaking examination of large areas of ground concrete surface and is potentially subject to errors arising from its assumptions: the aggregate particles will not be spherical and in many cases will be strongly asymmetrical, the distribution of aggregate particles may be uneven, the particle densities may vary (affecting the relationship between the volumetric data and the originally gravimetric grading) and the total aggregate volume may be unknown and/or atypical. Although the procedure may have merit in critical investigations, once its accuracy and precision have been demonstrated, the authors do not consider that such quantified aggregate gradings should be undertaken as part of routine concrete petrography. In time, image analysis linked to stereological principles might enable aggregate grading to be routinely quantified, in similar ways to those currently used for air voids (see Section 4.5.2).

However, with experience, petrographers will be able subjectively to assess the evenness of the overall aggregate grading (see Figure 4.40). In this way, it should be possible to identify any particle size excesses or deficiencies in the overall grading which might have occurred either as an unintended result of blending aggregates with slightly incompatible gradings or as a deliberately designed effect (e.g. 'gap-graded' concrete). Often, it will be appropriate to describe the maximum particle size and grading separately for the coarse and fine aggregates, but it will not usually be possible for the petrographer to differentiate an overall coarse aggregate grading which was originally compiled using 'single sizes' processed from the same aggregate source.

The grading of the fine aggregate (i.e. particles mainly passing a 4 or 5 mm sieve or 'sand') is particularly critical, as this can have significantly influenced the strength and durability of the hardened concrete. The former BS 882 (1992) and essentially also BS EN 12620 (2002/2008) stipulate a fine aggregate grading system which comprises a wide overall grading envelope, plus three overlapping additional sets of grading limits formerly designated C, M and F (i.e. coarse, medium and fine) and now basically 0/4, 0/2 and 0/1, respectively (see Tables 4.14b and 4.16).

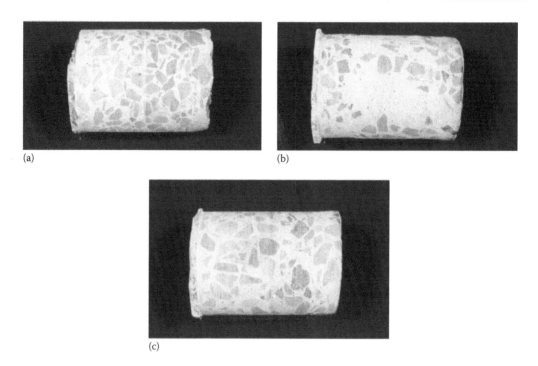

(a)

(b)

(c)

Figure 4.40 Photographs of concrete cores, illustrating three different appearances: (a) Evenly graded and distributed coarse aggregate. (b) Unevenly distributed coarse aggregate. (c) Apparently gap-graded coarse aggregate (although this concrete has the appearance of being gap graded, the coarser sand sizes are similar in colour to the cement matrix and thus not easily discernible).

Although it is less arduous to quantify the grading of a fine aggregate using a counting technique in thin section (this is standard practice for sedimentary petrographers in analysing the particle sorting of natural sandstones), it is nevertheless rarely considered essential in concrete examination. However, the nearest appropriate BS 882 or BS EN 12620 (or others as appropriate) sand grading designation should be determinable, when it should be appreciated that many aggregates will comply with two of the three overlapping limit sets.

In U.K. concretes made before 1983, the fine aggregate grading should be assessed with reference to the now long superseded four-zone system first introduced in the 1954 edition of BS 882. This system, which was based on a survey of available U.K. sand gradings and was not intended to imply any relationship with quality or performance, described four zones from the coarsest (Zone 1) to the finest (Zone 4) in which the proportional ranges of material passing 600 µm did not overlap (Table 4.17). This offers a prospect to the petrographer of identifying the sand grading zone with some reliability in critical cases, by preferential particle counting in thin section of the 600–300 µm size fraction.

The quantities in aggregates of very fine and ultra-fine material (i.e. those passing the 75 or 63 µm sieves according to different standards authorities) are particularly critical to the properties of concrete, but these constituents are usually difficult for the petrographer to assess using only optical microscopy. However, the actual particulate size of this material, especially whether it is mainly silt sized (i.e. 2–63/75 µm) or mainly clay sized (<2 µm), and the mineralogical composition, especially whether or not clay minerals are present, are important considerations, and these aspects can sometimes be established by optical petrography. Clay fines will sometimes occur coagulated as coarser clay aggregations or

Table 4.17 Superseded British Standard fine aggregate grading limits (note the non-overlapping 600 μm limits)

Type	Zone 1	Zone 2	Zone 3	Zone 4
Sieve Size	% by Mass Passing[a]			
10.0 mm	100	100	100	100
5.0	90–100	90–100	90–100	95–100
2.36	60–95	75–100	85–100	95–100
1.18	30–70	55–90	75–100	90–100
600 μm	15–34	35–59	60–79	80–100
300	5–20	8–30	12–40	15–50
150	0–10[b]	0–10[b]	0–10[b]	0–15[b]

Source: After BS 882–2: 1973.

[a] A cumulative total tolerance of 5% is permitted on the limits not given in bold type.
[b] For crushed stone (rock) sands, the permissible limit is increased to 20%.

'clay lumps'. A more detailed discussion of the factors influencing the effect of fines on concrete properties is included in Sims and Brown (1998).

SEM and/or XRD is increasingly available to practising concrete petrographers and can be considered, as extensions to the optical examination, to study inter alia the content, characteristics and distribution of these very fine and ultra-fine aggregate constituents.

BS EN 12620 (2002/2008) includes a classification scheme for this 'fines content' (based on material passing the 63 μm sieve), although there are several options for each type of aggregate, with no guidance on which to specify in a particular case; for example, for a coarse aggregate, the optional fines content categories are ≤1.5%, ≤4%, >4% and 'no requirement'. However, some guidance is provided in an annex for assessing 'fines quality' (or 'harmfulness of the fines in finer aggregate'), which includes possible applications of the 'sand equivalent' or 'methylene blue' tests, although again no interpretation criteria are suggested.

As an aid to the petrographer, destructive chemical techniques can sometimes be employed to determine aggregate grading, using representative but disposable samples. Figg (1989) describes a 'mechanical' technique (to distinguish it from the 'microscopic' methods), whereby hardened concrete is first carefully disaggregated by physical means, then the residual cementitious materials are removed mainly using acid treatment and finally, the resultant loose aggregate is subjected to a conventional sieve analysis. This procedure is potentially quite accurate but somewhat operator dependent in respect of reliability and of limited usefulness when the aggregate itself includes or comprises acid-soluble components.

4.3.3 Particle shape

Aggregate particle shapes and their variations can influence the concrete mix characteristics and hence the properties of the resultant hardened concrete. Particle shape also affects the achievable degree of cement–aggregate bonding and component interlock within concrete, which are each of critical importance to the mechanical properties. A careful description of aggregate particle shape should, therefore, be included in any petrographic determination of concrete composition.

The shapes of aggregate particles are typically described using the two parameters 'sphericity' and 'roundness', long established for use in characterising the grains in clastic sedimentary rocks (Figure 4.41). A simplified version of such a scheme was included in BS 812 (1975), except that the low-sphericity particles were differentiated into 'flaky' (i.e. thickness small relative to other dimensions) and 'elongated' (i.e. length large relative to other dimensions).

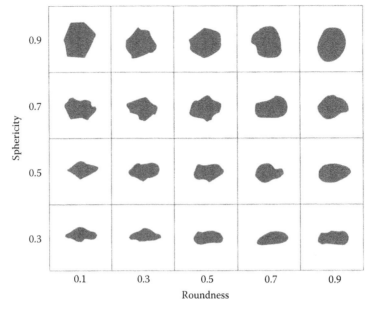

Sphericity = nominal diameter/maximum intercept

Roundness = average radius of corners and edges/
radius of maximum inscribed circle

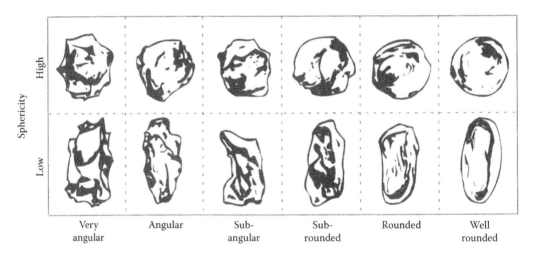

Figure 4.41 Two comparable charts for assessment of aggregate particle shape. (From Sims, I. and Brown, B.V.,
Concrete aggregates, in *Lea's the Chemistry of Cement and Concrete*, 4th ed., Hewlett, P.C. [ed],
Chapter 16, pp. 903–1011, Arnold, London, U.K., 1998; Derived from Krumbein, W.C. and
Sloss, L.L., *Stratigraphy and Sedimentation*, 2nd ed., Freeman, San Francisco, CA, 1963; Powers, M.C.,
J. Sediment. Petrol., 23(2), 117, 1953.)

BS 812-105 (1989) includes methods for quantifying the contents of flaky and elongated particles as flakiness and elongation indices.

Generally speaking, it is considered that, in terms of shape, 'cuboidal' aggregate particles characterised by high sphericity and low roundness (i.e. very angular) are to be
preferred, with particles which are both flaky and elongated being the least desirable.

Although this is a fair overview, many aggregates which are far from ideal in terms of particle shape are used in concrete without significant impairment to its properties, especially if the shape characteristics are consistent and compensatory measures are taken in the mix design.

The aggregate particle shapes can be described, using the aforementioned morphological terms, as part of a petrographic examination of hardened concrete. The sphericity aspects are liable to relate to rock type: for instance, closely cleaved or bedded rocks tend to yield flaky and sometimes also elongated particles. Roundness, by contrast, is more usually related to the type of source: for example, crushed rocks tend to form angular particles, whereas natural gravels are more likely to contain variably eroded and thus rounded particles.

Mixtures of aggregate particle shapes are quite common, as shapes may be modified by processing, such as crushed oversized gravel particles, which can be in part rounded and in part angular. Particle shapes in some crushed aggregates are also dependent to an extent upon the crusher type: for example, jaw and cone crushers can exacerbate a tendency of some rocks to form flaky particles. When a degree of quantification of aggregate particle shape in concrete is required, a visual estimate rather than a time-consuming particle count is usually adequate.

When required, schemes exist for a more critical consideration of particle shape parameters which are most likely to have an effect on the aggregate performance. Lees (1964), for example, devised a system for categorising aggregate particles in terms of axial ratios obtained from the longest, intermediate and shortest orthogonal dimensional axes. This gives a plot of flatness and elongation ratios divided into the four fields: cuboidal (equidimensional), elongated, flaky and flaky elongated.

Alternatively, French (1991a) proposed the measurement of 'aspect ratios' (i.e. longer dimension of the particle divided by the shortest), which range from unity for highly spherical particles to about 1.7 for regular cubic grains. He considered that when all of the particles have aspect ratios of 3 or less, the shape factor has little influence on the quality of the concrete.

4.3.4 Particle shape recognition

This subsection has been compiled with the assistance of Graham True, who has been pioneering the use of desktop flat-bed scanners for analysing voidage and aggregate particle shape in concrete: also see Section 4.5.2 (air-void content) for a further description of the flat-bed scanning technique.

Particle shape assessment may be obtained by using standard morphological shape descriptors, such as *form, sphericity, roundness/angularity* and *irregularity*.

4.3.4.1 Form and sphericity

It is possible to consider aggregate particles in three perpendicular dimensions, length, breadth and thickness (L, I and S), in order to characterise *form*. Form is independent of whether the edges or corners are sharp or round and was defined by Wadell (1932) as a measure of the ratio of the surface area of a particle to its volume. For practical purposes, this ratio is difficult to measure and has been interpreted by Krumbein (1941) as being the ratio of the volume of the particle to the volume of a circumscribing sphere. The cube root of this is termed the *sphericity* of the particle.

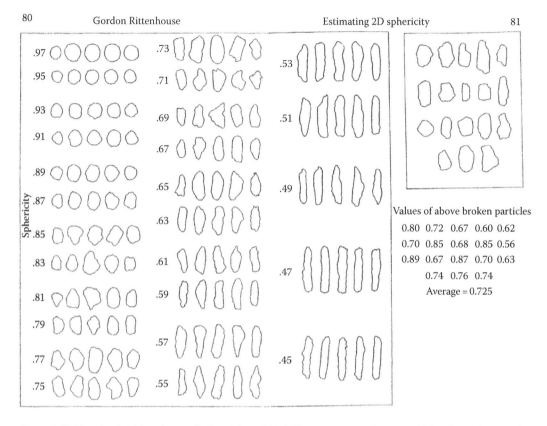

Figure 4.42 Visual sphericity chart calculated from Wadell's projection sphericity. (After Rittenhouse, G., *J. Sediment. Petrol.*, 13(2), 79, 1943.)

Rittenhouse (1943) developed a visual sphericity comparison chart (Figure 4.42), based on a practical 2D method of determining sphericity proposed by Wadell (1933) in which he describes the degree of sphericity as

$$Degree\ of\ sphericity,\ \emptyset = \frac{d}{D}$$

where
    d is the diameter of a circle of an area equal to the area obtained by projection of the surface containing the largest and intermediate diameters of the particle
    D is the diameter of the smallest circle that will circumscribe the particle

4.3.4.2 Roundness and angularity

Most authors consider roundness and angularity to be opposites, and some use these descriptions to define precise concepts of particle shape. In most cases, it is recognised that roundness is a measure of surface projections rather than a consideration of the overall particle outline and resemblance to circularity.

Wadell's method of assessing roundness is to assess the radii of the inscribed circle, edges and corners, all from a drawing made of the particle. From these, a ratio is calculated that

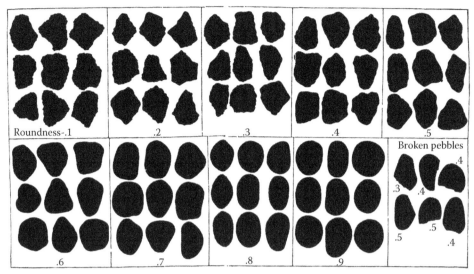

Pebble images or visual roundness

Figure 4.43 The Krumbein roundness images, in groupings, derived from Wadell's definition of roundness. (After Krumbein, W.C., *J. Sediment. Petrol.*, 11(2), 64 1941.)

indicates the relative size of the average radius of curvature to the radius of the inscribed circle. This is the *roundness* of the particle.

Krumbein (1941) developed a visual method of assessing roundness of aggregate particles (Figure 4.43), by comparing them individually against standard images of known roundness. The images were redrawn from pebbles measured by Wadell's method with nine pebbles in each grouping that produced an average value that agrees well with Wadell's derived values. A section of broken pebbles is also included.

4.3.4.3 Irregularity

Irregularity is defined by Blott and Pye (2008) as the significant indentations (concavities) and projections (convexities) on the surface of a particle. They proposed a measure of this irregularity in a 2D image of a particle to the depth of the concavities (measured from the centre of the largest inscribed circle) and the length of the projections (measured from the convex hull, a line of the shortest distance that connects the maximum projections on a particle). An irregularity index can then be calculated as follows:

$$I_{(2D)} = \sum \frac{y - x}{y}$$

where

x is the distance from the centre of the largest inscribed circle to the nearest point of any concavity

y is the distance from the centre of the largest inscribed circle to the convex hull, measured in the same direction as x

The sum of all the values of the concavities is totalled. Where the distance to the convex hull is not easy to measure, this can be estimated by measuring the distance to the projections adjoining any cavity.

4.3.4.4 Use of desktop flat-bed scanners

Blott and Pye (2008) suggest that analysis of the behaviour of aggregates in concrete may be undertaken in terms of simple descriptors of *form, sphericity, roundness/angularity* and *irregularity*. The Québec Ministry of Transportation (QMOT) uses image analysis on a regular basis to verify the angularity of hot-mix aggregates as reported by Janoo (1998). Uthus et al. (2005) state that digital computer analysis enables almost any geometrical parameter to be calculated in two dimensions.

Desktop flat-bed scanners are now readily available with suitable resolution (i.e. >600 dpi) to scan both aggregate particles (backlit scanning) and also concrete core samples (reflective scanning) to obtain outline images of aggregate profiles. A scanning method for core samples is included in Section 4.5.2 for assessing voids, as described by True et al. (2010).

It can be shown that the image of aggregate obtained from the surface of a concrete core has similar characteristics to that of the same aggregate source obtained by backlit scanning of loose aggregate particles. This is not unduly surprising since both aggregate profiles are from the same aggregate source and profile characteristics are therefore significantly similar.

Uthus et al. (2005) provide an illustration of the basic measurements that can be made from a binary digital image of a particle (Figure 4.44). It can be seen that particle area and

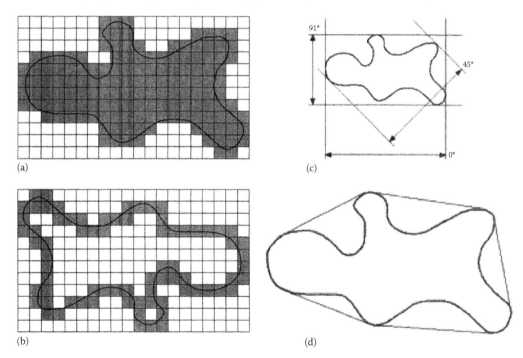

Material characterisation using image analysis

Figure 4.44 Illustration of the basic measurements obtained by digital imaging. (a) Area of an aggregate calculated through image analysis, (b) perimeter of an aggregate calculated through image analysis, (c) typical Feret measurement and (d) illustration of convex perimeter. (After Uthus et al. 2005.)

profile accuracy are dependent on the resolution of the image. For scans of aggregate profile, a resolution setting of 800 dpi is advisable. Figure 4.44 shows that estimates of (a) particle area, (b) perimeter, (c) aspect ratio (length/breadth) (d) and convex perimeter can be obtained by simple 2D surface scans.

The opportunity is therefore available to scan concrete core samples and, from the profile of aggregates on the surface of the cores, obtain an objective profile characterisation of the aggregate. This may prove useful, in conjunction with an estimate of entrapped air voidage, in understanding, for example, why a particular concrete has provided relatively low compressive strength results. It is recognised that if the particle shape of an aggregate changes, such that more *angular–irregular* or *elongated* material is included, without compensating for the voidage change, this introduces in the mix by increasing the fines content (sand and cement), then voidage such as entrapped air voidage is likely to increase at the cost of a reduction in strength.

Figure 4.45, taken from Janoo (1998), quoting the work of Ishai and Tons (1977), shows how aggregate profile characteristics generate and can influence inter-particle packing porosity.

Packing porosity and macro-surface voids together provide a measure of the total void content of an aggregate. This can range between 33% for perfectly spherical particles and 50%, and more, for concrete containing angular aggregate. Total voidage content of an aggregate source can be determined by the use of a voidmeter or by water displacement.

Figures 4.41 and 4.42 provide aggregate profile charts that represent measures of *sphericity* and *roundness* as used in morphological field studies. The originals of these can be scanned and used to provide a digital means of assessing and classifying aggregate source material as well as aggregate profiles scanned from concrete core samples. Another simple property, *aspect ratio*, can be obtained from Feret measurement as shown in Figure 4.44c, obtained from suitable imaging software.

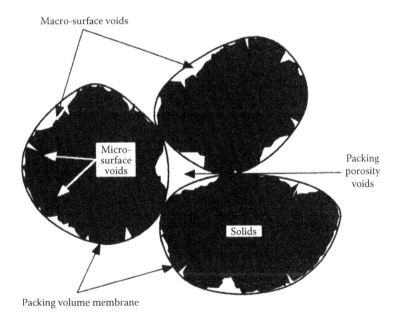

Figure 4.45 Inter-particle packing. (After Ishai, I. and Tons, E., *J. Test. Eval*, 5(1), 3, 1977; from Janoo 1998.)

Table 4.18 Profile tools for aggregate morphology in concrete

Morphological profile tool	Definition, description and property measurement
Percentage concavity (% concavity)	$\% \ Concavity = 100 \times \left[\dfrac{convex\ area - area}{area} \right]$ Assessment of *irregularity, roundness/angularity* and *macro-surface voids*
Riley inscribed circle sphericity (*Riley ICS*)	$Riley\ ICS = \sqrt{\dfrac{Diam\ of\ largest\ inscribed\ circle}{Diam\ of\ smallest\ circumscribed\ circle}}$ Measurement of *circularity*
Aspect ratio (*Aspect Ratio*)	$Aspect\ Ratio = \dfrac{Length}{Breadth} = \dfrac{Largest\ Feret\ length}{Shortest\ Feret\ length}$ *Elongation*

Three aggregate morphological tools can be used to investigate *sphericity, roundness/ angularity, circularity* and *elongation* as well as *irregularity* of aggregate profile scans, as shown in Table 4.18.

These three morphological profile tools, *% concavity, Riley inscribed circle sphericity (ICS)* and *aspect ratio*, can be calibrated against the Krumbein and Rittenhouse visual comparison charts, and Figures 4.46 and 4.47 show good correlation of the trend lines, except for *aspect ratio* and *Riley ICS* against the Krumbein visual comparison chart. However, both these profile tools have relevance as a means of identifying particle shape characteristics, and Figure 4.48 shows the output obtained from scanning and analysis using ImageJ software of two sets of 120 aggregate particles. Both are nominally 20 mm particle size: one is a crushed rock and the other selected rounded natural gravel.

The graphs show the results presented as a frequency distribution and indicate a significant difference in the *% concavity*, whereas the *aspect ratio* and *Riley ICS* are similarly distributed.

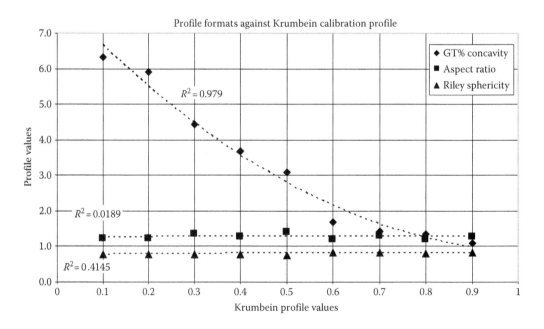

Figure 4.46 Calibration against the Krumbein chart.

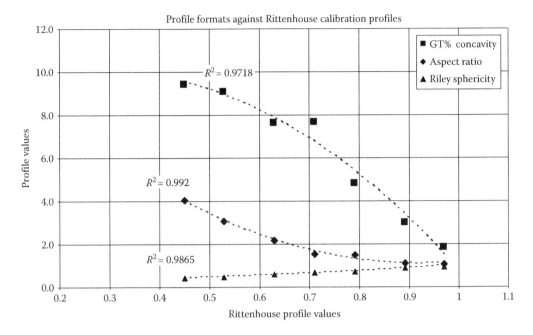

Figure 4.47 Calibration against the Rittenhouse chart.

Another illustration of how particle shape characterisation from scanned digital images can help interpret the particle profile of an aggregate source can be seen in Figure 4.49, where *roundness* and *irregularity* can be seen presented in the form of % *concavity*. The data shown have been taken from two aggregate sources (as shown in Figure 4.48), one a crushed rock and the other a natural rounded gravel.

The natural rounded gravel was found to contain a significant amount of crushed or broken particles, and this is illustrated by the three sets of data: one contains two samples of the crushed rock aggregate, the second two samples of the natural rounded aggregate with just selected rounded particles and the third a pair of samples, natural rounded, selected as crushed or broken particles.

The three sets of profiles show the crushed aggregate and the selected broken natural aggregate to overlap in terms of the % concavity, whereas the selected rounded natural aggregate was distinctly separate with a % *concavity* of 1% and less, except for particle 11, as shown in Figure 4.50, which indicated a % *concavity* of 4.44%. When core samples were taken from concretes containing the crushed aggregate and then the nominally rounded aggregate, it was found difficult to distinguish between the two aggregate sources until it was noted that a significant proportion of the apparently natural rounded aggregate was indeed broken and therefore more angular and irregular and this significantly influenced the % *concavity*, but not the *Riley ICS* or the a*spect ratio*.

Desktop flat-bed scanners are therefore useful in morphological classification of concrete aggregate profiles and can be used to assess aggregate sources as well as to classify and analyse the aggregate profiles found on the surface of concrete core samples. The data obtained from classification can be used to assist an investigation of questionable concrete strength, in conjunction with an assessment of entrapped air voidage. Broadly similar approaches, using analysis of microscopical (optical and/or SEM) images, can be applied for the characterisation of fine aggregates.

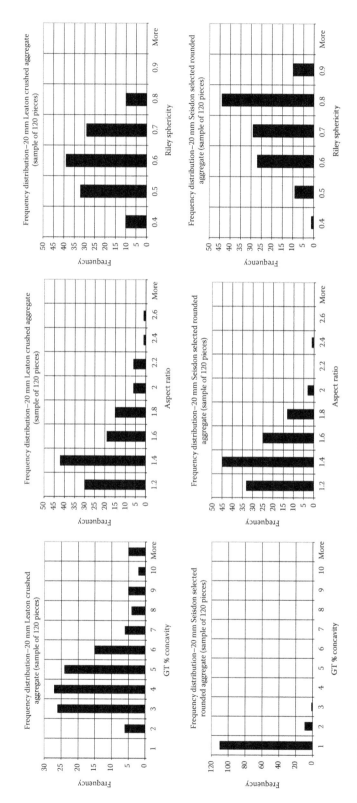

Figure 4.48 Comparison of crushed olivine dolerite with an irregular–subangular/subrounded quartzite gravel.

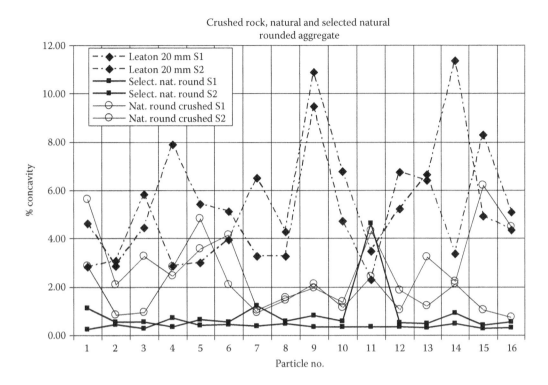

Figure 4.49 Illustration of how % concavity can identify the shape characteristics of aggregate.

Figure 4.50 Particle No. 11 in Figure 4.49.

A further application of the use of desktop scanners has been as a site tool, whereby if placed inverted on a exposed aggregate floor or precast unit, it is possible to obtain an image of the surface and then to analyse the aggregate particle size distribution and compare this with a sample panel as a check on apparent compliance.

4.3.5 Other particle characteristics

When examining samples of hardened concrete, the petrographic composition and the aggregate particle sizes and shapes are readily described. Other factors relating to the particle surfaces or their internal microstructures are less straightforward but can often have a significant effect on concrete properties.

The interface between aggregate particle surfaces and cement paste is one of the most important parameters in determining the mechanical performance of concrete. Although aggregate–paste interfacial zones are often studied in some detail during failure investigations (see Chapters 5 and 6), their potential importance is frequently overlooked during routine assessments of composition and quality.

It is, therefore, recommended that the aggregate–paste interface should be regarded and described as if it was an additional *component* of a concrete. According to Sarkar and Aïtcin (1990), theoretical models for the elastic behaviour and mechanical properties of concrete prove inadequate when they fail to address the aggregate–cement interface.

Physical strength and integrity of the bond achieved at the aggregate–paste interface variously depend upon the character of the aggregate particle surface and the nature of the immediately adjacent hydrated cement. The micro-topography of aggregate particle surfaces should be examined for 'texture' and for coatings or encrustations. The immediately adjacent cement paste zone should be examined for entrapped voids, microporosity and the concentrated development of mineral phases, such as portlandite or ettringite.

Many attempts have been made over the years to find an objective method for measuring the roughness of aggregate particle surfaces, most recently using Fourier and fractal techniques, but the simple descriptive schemes, such as that included in BS 812-105 (1989), remain the most widely used. However, it is difficult to apply these textural concepts to aggregates in hardened concrete, particularly when viewed in section, when only the general roughness or otherwise of the aggregate–cement interface will be readily apparent.

It is reasonable to assume, within limits and with other factors held constant, that relatively rougher or more rugose particle surfaces will facilitate stronger aggregate–paste bonds, because of improved physical keying and enlarged contact areas. However, other factors are not held constant in practice, and extremes of surface texture are likely to be disadvantageous: for example, unduly rough aggregates will adversely influence workability and might also consume quantities of cement paste without any commensurate binding benefit.

Dolar-Mantuani (1983) considered that there is a complex interrelationship between the main textural features of aggregate surfaces and the quality of the bond to cement paste. In particular, she stressed the importance of surface absorption, in comparison with surface roughness, with the strength of bond being potentially increased by the penetration of cement slurry into permeable aggregate surfaces.

One of the fundamental activities in placing in situ concrete, or in moulding precast concrete products, is some form of compactive effort, such as internal or external vibration, whereby the constituents are consolidated and entrapped air is reduced to a minimum. This compactive activity tends to promote segregation and bleeding. Excessive compaction can lead to impaired concrete quality, although the inherent susceptibility of concretes to segregation and bleeding also depends upon a variety of mix factors.

The most familiar evidence of bleeding is the formation of a layer of water on the upper surface of a freshly emplaced concrete and within normal limits, this is a feature which helps to prevent premature surface drying and consequential cracking. However, localised internal bleeding also occurs throughout the body of a concrete, typically manifesting itself by the concentration of water at the aggregate–paste interfaces.

This can lead to the formation of unusually porous hardened paste at the transition zone, because of the locally high water/cement ratio. Otherwise, if water has completely segregated to form either a pocket or a continuous film at the aggregate particle surface, voids or spaces can be created if the water remains trapped during the plastic phase and only evaporates from the concrete after hardening.

Such evidence of internal bleeding is often apparent at the aggregate–paste interface on the underside (as cast) of an aggregate particle (see Figure 4.51). Entrapped air voids might also be found in such locations, which tend to occur most readily with elongated aggregate particles that have become orientated parallel to the concrete upper surface.

The film of fluid which can be formed at the aggregate/paste interface as the result of internal bleeding will naturally have the chemical composition which is characteristic of the mix water in freshly placed concrete, being rich in calcium hydroxide and sulphates (Lea 1970). It is therefore quite common for layers of portlandite and/or ettringite to be deposited at the aggregate–paste interface as the water evaporates (Figure 4.52). Sarkar and Aïtcin (1990) describe the well-crystallised nature of these developments. French (1991a) suggests that these layers of portlandite particularly form at the surfaces of siliceous aggregate particles and the amount of portlandite precipitated usually increases as the particle size of the aggregate decreases.

The efficacy of the bond between aggregates and hardened cement paste can obviously be influenced by any material which interposes between these two components, including the layers of water and crystalline deposits discussed earlier. Dust coatings of aggregates particles are liable severely to reduce the aggregate–paste bond, although these might be difficult to detect during the examination of concrete using purely optical techniques.

Thicker encrustations on the surfaces of aggregate particles will be more visible to the petrographer, but need not be detrimental. One of the most common encrustations will be calcite, which will sometimes have been formed in situ within sand and gravel deposits as the result of precipitation from percolating groundwaters or might be the residual natural

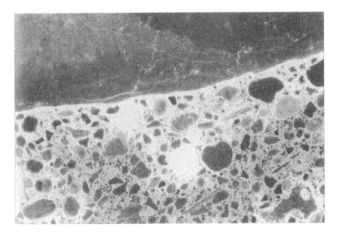

Figure 4.51 Photomicrograph of internal bleeding in concrete in thin section, evidenced by voids and porous cement paste beneath an aggregate particle, uvl, width of field ~4 mm. (Photograph courtesy of RSK Environment Ltd., U.K.)

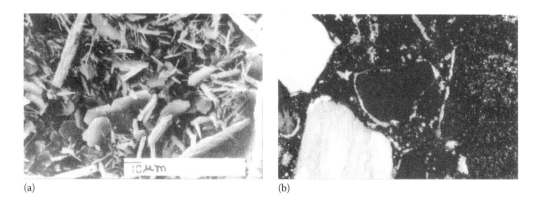

(a)   (b)

Figure 4.52 Ettringite and/or portlandite at the aggregate/cement interface in concrete: (a) SEM micrograph showing ettringite and portlandite in the transition zone. (b) Thin-section photomicrograph showing portlandite around an aggregate boundary. xpl, width of field ~0.9 mm. (Photograph [a] from Sarkar, S.L. and Aïtcin, P.-C., The importance of petrological, petrographical and mineralogical characteristics of aggregates in very high strength concrete, in *Petrography Applied to Concrete and Concrete Aggregates*, Erlin, B. and Stark, D. [eds], pp. 129–144, ASTM STP-1061, Philadelphia, PA, 1990, and [b] courtesy of RSK Environment Ltd., U.K.)

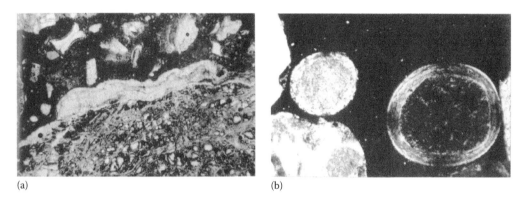

(a)   (b)

Figure 4.53 Photomicrographs of thin sections showing aggregate particles encrusted with (a) calcite (encrusting serpentinite wadi gravel in concrete), ppl, width of field ~1.5 mm, and (b) gypsum (encrusting dune sand grain), xpl, width of field ~0.5 mm.

cement from an eroded conglomeritic sedimentary rock. Such partial calcitic encrustations are quite common with the wadi gravels widely used for aggregate in the Arabian Gulf (see Figure 4.53a), and these have not usually been considered to be detrimental, unless contaminated by potentially deleterious salts.

In some parts of the Arabian Gulf region, prospective aggregates can be seriously contaminated by potentially detrimental salts (Fookes and Collis 1975; Sims 2006). Although it is feasible for water-soluble salts to be removed by washing, a greater long-term threat to concrete durability is posed by gypsum, which is only very slowly soluble in water and might well survive aggregate processing either as discrete particles or as surface coatings (see Figure 4.53b). Crammond (1984, 1990) has demonstrated that expansion approaching 1% over about 5 years can be caused in mortar-bar tests at room temperature by as little as 5% crystalline gypsum in an aggregate.

Some sand and gravel aggregates can be contaminated with clay, such that particles are coated with thin layers or even films of clay. This will inhibit the formation of an effective bond and might even cause disruption of an initial bond as the result of moisture movements affecting the clay material.

Although the factors affecting the aggregate–paste interface are usually the more critical, the internal characteristics of aggregate particles are occasionally important, especially in the case of high- and very-high-strength concretes. There are obviously certain constituents that are themselves usually regarded as undesirable at best and deleterious at worst. These are fully described by Sims and Brown (1998) and summarised here in Table 4.12. Otherwise, this discussion refers to aggregate particles which are compositionally acceptable, but which might exhibit questionable microstructural features.

It is usually a reasonable assumption that rock particles that survive the rigours of aggregate processing will possess adequate strength to perform their role as a constituent of concrete. However, this might not apply to higher-strength concretes (Sarker and Aïtcin 1990) or to concretes which are subjected in service to exceptional shear stresses, when otherwise stable aggregate particles might prove to have potentially vulnerable flaws. Crushed rock aggregate particles, for example, can exhibit potentially vulnerable microfractured surface zones.

Sometimes, rocks will have been fragmented or become micro-brecciated as part of their geological history or as a result of blasting in the quarry. Aggregates produced from such materials might comprise particles that contain potentially numerous weak points. The trachyte shown in Figure 4.54, for example, was found to exhibit a limonitised natural micro-brecciation, and while the rock performed admirably in a range of aggregate strength tests prior to its use, it failed seriously in service and in subsequent investigation using the magnesium sulphate soundness test method (ASTM C88 2005, BS 812-121 1989).

Rocks can also exhibit natural fabrics or textures that might adversely influence their performance as aggregates. Metamorphic rocks, for instance, frequently display a schistose or gneissose fabric, often causing aggregate particles to be strongly anisotropic in their mechanical behaviour. Because such rock fabrics tend also to create flaky and/or elongated aggregate particle shapes, which can then adopt a preferred orientation, effectively, 'planes' of potential weakness can sometimes be created within a concrete containing these types of aggregates.

The integrity of rock aggregate particles might also be impaired where crystalline textures have been 'loosened' by weathering or by stress relaxation following quarrying. Certain types of rocks, particularly those comprising crystalline calcite, such as some marbles and limestones, are occasionally also liable to thermal hysteresis, whereby thermal cycling can

Figure 4.54 Photomicrograph of trachyte aggregate particle in thin section, showing limonitised micro-brecciation, ppl, width of field ~0.5 mm.

lead to irreversible expansions and deformations of aggregates particles within hardened concrete (Senior and Franklin 1987).

When hardened concrete is being investigated for its condition and/or future performance, it is sometimes appropriate to include an assessment of the various characters of the aggregate particles and their aggregate–paste interfacial zones. One example scheme is the Damage Rating Index (DRI), developed in Canada (Grattan-Bellew 1995, Grattan-Bellew and Mitchell 2006), which is fully described in Chapter 6.

4.4 WATER/CEMENT RATIO

Water/cement ratio is a key criterion in determining concrete behaviour and performance. Already a complex aspect that is difficult reliably to establish retrospectively for hardened concrete samples, the matter is now frequently further complicated by the presence of mineral additions or 'SCMs', when the critical relationship is more properly termed 'water/binder ratio' or even 'water/cementitious materials ratio'. It is always the 'water/binder ratio' that is important, whether the 'binder' is just ordinary Portland cement or some interground or separately blended mixture of cement and additions, but inevitably casual use of the established term 'water/cement ratio' will persist.

It is recommended that petrographers should use the term 'water/cement ratio' when the binder consists of *only* ordinary Portland cement and the term 'water/binder ratio' whenever the concrete contains mineral additions or SCMs in addition to Portland cement. De Rooij (2011) has sensibly suggested that, in Europe, the term 'water/cement ratio' might be widened to include any cement known to relate to one of the BS EN 197-1 (2011) products (see Table 4.1a). He has also noted the complicating factor that, in some concrete mixes, only a proportion of an added SCM is considered to be part of the 'binder'.

4.4.1 Definitions and relationship to concrete properties

Although many factors contribute to the engineering properties of concrete, the relationship between cement (or binder) and water is typically a matter of crucial importance, because it is the reactions within and hydration of this binder during a plastic phase, and its consequent hardening into a strong composite with aggregates, which affords the material most of its advantageous properties. The properties and hydration processes of Portland cement binders are complicated in detail, and the resultant microstructures, particularly the pore structures, are related to the strengths of hardened cement paste and concrete.

Aggregate–paste bond is, for normal-range concrete, generally the limiting factor regarding the mechanical behaviour of most concretes. Yet, even if the aggregate is disregarded, the strength typically achieved by hardened cement paste is up to 1000 times lower than the 'theoretical strength' which may be calculated, according to Neville and Brooks (1987, 2010). This difference is accounted for by Griffiths' theory of fracture mechanics, whereby minute cracks or other flaws in brittle materials act as stress raisers by concentrating stress at their tips. There are various micro-defects and discontinuities within the complicated structure of cement paste, but around half of the volume consists of micropores and these are the principal 'flaws' which critically control the strength potential.

In most cases, the content of micropores in the cement paste is determined by the quantity of water used in relation to the amount of cement, traditionally expressed as the 'water/cement ratio'. Thus, Neville and Brooks (1987, 2010), discussing the factors influencing concrete strength, stated: 'The most important practical factor is the water/cement ratio, but the underlying parameter is the number and size of pores in the hardened cement paste'. De Rooij (2011) reminds us that porosity in the paste is also controlled by particle packing.

The hardened cement paste comprises the solid hydrates plus the adsorbed water held in very tiny 'gel pores' (around 2 nm in diameter). Because this hydrated cement occupies a smaller volume than that of the original dry cement and water, there are residual spaces or 'capillary pores' (much larger than gel pores, at around 1 µm or 1000 nm in diameter). If the cement was fully hydrated and there was no excess of water over that required for cement hydration, these capillary voids would amount to about 18% of the original volume of dry cement (Neville and Brooks (2010). Detailed up-to-date information on the pore structure of cementitious materials may be found in Aligizaki (2005).

In practice, there is always an excess of water over that required for cement hydration, essentially acting as a lubricant to enable concrete to be mixed and placed. As the amount of excess water increases, so the volume of capillary voids, initially filled with water, increases. Powers (1949a) developed the diagram shown in Figure 4.55 to illustrate the relationships between the water/cement ratio, the degree of hydration and the relative proportions of unhydrated cement, hydrated cement and water-filled capillary voids. The role of water in concrete and its influence on key properties was the subject of an important Concrete

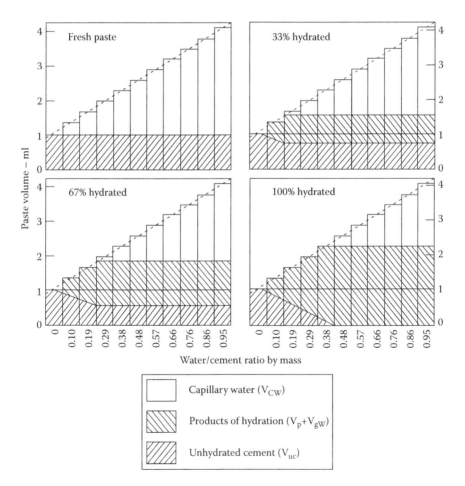

Figure 4.55 Composition of cement paste at various stages of hydration. The percentage indicated applies only to pastes with enough water-filled space to accommodate the products of hydration at the degree of hydration indicated. (From Powers, T.C., *ASTM Bull.*,158, 68, 1949a.)

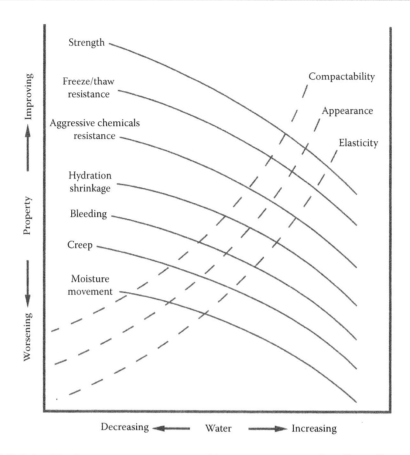

Figure 4.56 Relationships between water content and key concrete properties. (From Concrete Society, The role of water in concrete and its influence on properties [Concrete Society discussion document, prepared by a working party of the Concrete Society], Ref. CS 156, The Concrete Society, Camberley, U.K., 2005.)

Society discussion document (Concrete Society 2005), which includes the schematic diagram reproduced in Figure 4.56.

It is apparent from the preceding discussion that the water/cement ratio is an important criterion in the composition of normal-range concrete, controlling the microporosity of the cement paste and hence having a critical influence on concrete strength properties. Furthermore, because durability is generally related to the water permeability of concrete, which is in turn largely dependent upon cement paste porosity, water/cement ratio also has an important bearing upon durability.

4.4.2 Indicators of water/cement ratio

Variations in the relative proportions of cement and water in the mix have significant influences on the physical and mineralogical characteristics of the resultant concrete. It therefore follows that study of the relevant features of hardened concrete can provide an insight into the original water/cement ratio of the mix. Although procedures have been devised for using these factors to quantify the original water/cement ratio for a given concrete sample, in truth, the result can only ever be *indicative*, because the determinable parameters which are influenced by water/cement ratio are all also affected by various other factors in complex and practically unpredictable ways.

In practice, accurate estimation of the water/cement ratio in a concrete is rarely necessary. Often, the petrographer will only be required to establish whether or not a particular concrete had a water/cement ratio which significantly exceeded a given value. Also, many structural concrete mixes will have been made using water/cement ratios within a narrow range, and the imprecision of the determinative methods would not allow these mixes to be differentiated. As cement paste volume increases with water/cement ratio, for a given cement content, it is often possible to assess the likely level of water/cement ratio from the paste volume, and this will frequently suffice (Figure 4.57).

4.4.2.1 Water voids and bleeding

During and after the placement of concrete, there is a tendency in the plastic phase for the constituents to segregate, with coarse aggregate particles settling and excess water and entrapped air rising towards the upper surface. This effect is minimised in good-quality cohesive mixes, but in poor concrete leads to materials segregation. The segregation of water from the other constituents within concrete is known as *bleeding* and, within limits, is a typical feature of all normal concrete mixes. Excessive bleeding, however, can detrimentally affect both the strength and durability of concrete and has a variety of causes.

The likelihood of excessive bleeding increases with water content, so that the occurrence of excessive bleeding is one sign of a possibly high water/cement ratio. Bleed water either reaches the surface of the concrete, where it evaporates, or becomes trapped within the concrete interior at the time of setting. It is common for such entrapped bleed water to accumulate beneath aggregate particles (Ash 1972), leading either to water voids or to local zones of conspicuously porous cement paste (Figure 4.58). Bleed water may also occupy cavities and voids left by incomplete compaction, leading to smooth internal coatings of secondary deposits within the void which help to distinguish water voids from air voids. In some cases, bleed water can migrate along particular channels within the concrete, which later manifest themselves as conspicuously porous zones or tortuous channels within the cement paste (Figure 4.59).

The presence of water voids beneath aggregate particles in hardened concrete is readily observable during the macroscopical examination of the sides of drilled cores. Even zones of conspicuously porous cement paste beneath aggregate particles are frequently betrayed by relatively greater erosion during drilling, producing slight grooves under the aggregates in the core side surfaces. Otherwise, the more porous zones beneath aggregates or along bleed channels can be detected with the aid of differential absorption techniques. The simplest method is to wet the drilled or sawn concrete surface and to make observations during drying: the more porous zones are the last to dry. Ash (1972) described a method using red ink, but today, impregnation with a low-viscosity resin containing a fluorescent dye is probably the most efficient technique, whereby porous zones absorb more resin and thus subsequently fluoresce more in ultraviolet (UV) light.

4.4.2.2 Capillary porosity

It was explained in Section 4.4.1 that hydrated cement paste inevitably contains capillary pores around 1 μm in size and that the total volume of capillary pores increases with increasing mix water. The correlation shown in Table 4.19 indicates that the relationship between the capillary porosity of cement paste and the water/cement ratio is particularly pronounced at the critical lower end of the range, the capillary porosity trebling between water/cement ratios of 0.40 and 0.55 (Christensen et al. 1979). These complex interrelationships have been reworked by Hansen (1986) and Jensen and Hansen (2001).

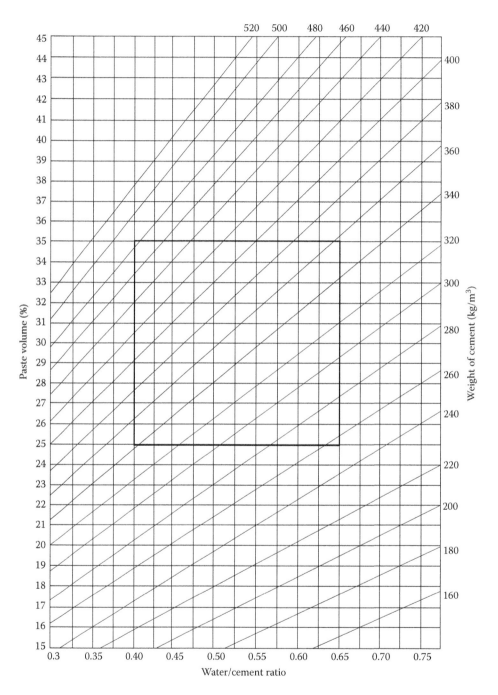

Figure 4.57 Relationship between water/cement ratio, hardened cement paste volume and cement content, assuming the cement specific gravity to be 3.12. The values in the large square represent a typical range of concretes.

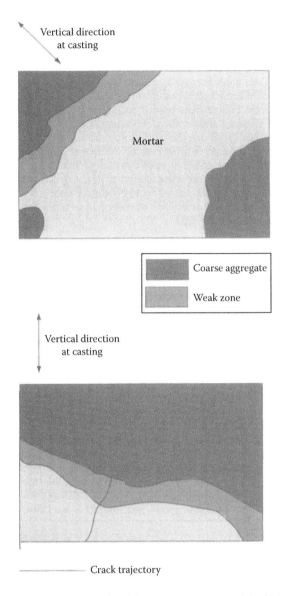

Figure 4.58 Sketches to illustrate water voids and porous paste caused by bleeding beneath aggregate particles. (From Ash, J.E., *ACI J.*, 69(4), 209, 1972.)

The measurement of capillary porosity is a crucial element in the physico-chemical determination of original water/cement ratio (see subsection 4.4.3), but, until recently, this important indicator of water/cement ratio could not be assessed by petrography because the capillary pores at around 1 μm in size are too small for observation by optical microscopy. Mielenz (1962), for example, one of the founders of modern concrete petrography, stated that 'the cement paste of hardened concrete includes ... a system of minute capillaries and pores not visible under the optical microscope'. However, the technique described by Christensen et al. (1979), involving impregnation of the concrete sample by low-viscosity resin containing a fluorescent dye and subsequent examination in UV light, has enabled the capillary porosity to be assessed directly by microscopy. This is the basis of a microscopical

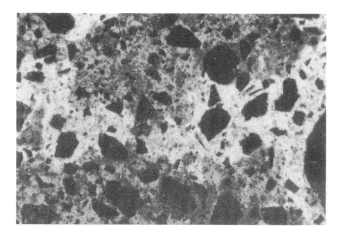

Figure 4.59 Photomicrograph of bleeding channel in concrete, uvl, width of field ~0.8 mm.

Table 4.19 Relationship between the capillary porosity of hardened cement paste and the original water/cement ratio

Water/cement ratio by mass	Capillary porosity of cement paste % by volume
0.40	8
0.45	14
0.50	19
0.55	24
0.60	28
0.65	32
0.70	35
0.75	38
0.80	41

Source: After Christensen, P. et al., *Nordisk Betong*, 3, 4, 1979.

method for the estimation of original water/cement ratio (see subsection 4.4.4). Liu and Khan (2000) recommended a combination of petrographic techniques for studying the capillary porosity of cement paste.

4.4.2.3 Mineralogical features

Portlandite is a typical product of the hydration of Portland cement (see Section 4.2.3), but the amount, crystallite size and disposition of portlandite vary with the original water/cement ratio and compaction. In all Portland cement concretes, unless a mineral additive is present which can react with and remove it, the portlandite forms within the cement paste and at aggregate–cement interfacial zones. When the water/cement ratio is high, these portlandite crystallites become unusually large and more clearly defined (Jepsen and Christensen 1989; French 1991a), although the size of portlandite crystallites is also temperature dependent (see Section 4.2.3). French (1991a) has also observed that the coarser portlandite crystallites formed with high water/cement ratios tend to occur as clusters, whereas the smaller crystallites formed at lower water/cement ratios are more uniformly distributed. French (1991a) further suggested that 'there is probably slightly more [portlandite] in concretes made with higher water/cement ratios', but this could be an impression created by the coarser and therefore more visually apparent crystallites.

It seems obvious that higher degrees of hydration (see Subsection 4.2.3) must indicate higher water/cement ratios, but such would be an oversimplification. Although there is a relationship between the proportion of residual unhydrated cement particles and water/cement ratio, the degree of hydration is also strongly dependent upon curing conditions and the properties of the particular cement involved.

According to French (1991a), there is a direct correlation between the degree of hydration and original water/cement ratio for concretes made at around 20°C and cured for just a few days: he found cement paste to contain 7% or 8% unhydrated particles at 0.4 water/cement ratio, reducing to 3% or 4% at 0.5 water/cement ratio and virtually none (i.e. completely hydrated) at 0.6 water/cement ratio. However, these guidance values change at significantly lower or higher temperatures during setting and hardening, and also relict cement clinker grains are able progressively to hydrate over a period after hardening when there is continuous or intermittent exposure to moisture. Also, as de Rooij (2011) has noted, French's observations were probably obtained at a similar age (say 28 days) for each mix, and the apparent relationship between water/cement ratio and the content of residual unhydrated cement particles would become less pronounced over a longer period.

Erlin and Campbell (2000) reported a promising relationship, in experiments, between the microhardness of cement paste in concrete, as measured using Knoop or Rockwell indentation methods, and the known water/cement ratio of that matrix. Such a relationship might be expected, given that increasing water content leads to greater capillary porosity and thus weaker paste, but devising an objective technique that does not need to be separately calibrated for each cement or binder type could challenge its practicality. Moreover, obtaining reliable paste microhardness values for concrete is challenged by underlying conditions at the test position (e.g. hard aggregate or an air void); also any relationship with water/cement ratio will vary with time, as paste strengthens with continuing hydration or is later altered by carbonation or weathering.

The practical usefulness of these mineralogical factors, portlandite and degree of hydration, for indicating original water/cement ratio is therefore limited, although they might be helpful indicators when taken together with other features, such as the amount of bleeding and the level of capillary porosity. Original water/cement ratio is most frequently estimated for hardened concrete using the physico-chemical method (see Section 4.4.3), but fluorescence microscopy (see Section 4.4.4) has some advantages, as well as some pitfalls, and is a useful additional aid to the concrete petrographer.

4.4.3 Determination of water/cement ratio by the physico-chemical method

As with cement content determination, the physico-chemical method was reviewed by a new Concrete Society working group, in preparation of the recently published second edition of TR 32 (Concrete Society 2014) (see Section 4.2.6).

To establish the original water/cement ratio, it is clearly necessary to determine the respective proportions of cement and water in the original mix. The cement content may be estimated from a partial chemical analysis, particularly using the values obtained for soluble silica and acid-soluble calcium, provided the aggregates do not contain acid-soluble calcium: the method is detailed in BS 1881-124 (1988) and was earlier discussed in subsection 4.2.6. The original water content is estimated from determined values of combined water of hydration and water in excess of that required for hydration. Combined water is determined by measuring the amount of water driven from a pre-dried powdered sample of the concrete by ignition. Excess water is calculated from the volume of capillary pores, which is in turn determined by the vacuum saturation of a dried slice specimen with a liquid of known density. Historically, 1,1,1 trichlorethane has been used for this purpose, as directed by BS 1881-124, but environmental concerns have led to the cessation of production of this solvent; tetrachloroethylene has been found to be a suitable replacement. Table 4.20 gives comparison data obtained by Crofts for 1,1,1 trichlorethane and tetrachloroethylene, showing that the two solvents produce closely similar findings.

According to Figg (1989), in the case of sound, uncarbonated portions of concrete, within the normal range of mixes containing 200–500 kg/m³ of cement, with low-porosity aggregates and no cement replacement materials, the accuracy of the result is likely to be within ±0.05 of the actual water/cement ratio, and the reproducibility (R) is of the order of ±0.05 for concretes with water/cement ratios in the range 0.4–0.8. However, these criteria may not always apply and, in practice, the determined values of original water/cement ratio by this method should be treated with considerable circumspection, even when obtained by experienced analysts.

French (1994) considered the BS 1881-124 procedure for the measurement of water/cement ratio and concluded: *'According to BS 1881 the determination of cement content has a repeatability of about 40 kg/m³ where the cement content is of the order of 400 kg/m³.*

Table 4.20 Comparison data for BS 1881–124 tests for original water/cement ratio

a) Comparison of results using 1,1,1-trichlorethane (CH₃CCl₃) and tetrachloroethene (C₂Cl₄)

| Sample | Capillary porosity, % by mass | | Original total water/cement ratio | | |
	CH_3CCl_3	C_2Cl_4	CH_3CCl_3	C_2Cl_4	Difference
1	4.99	5.14	0.52	0.53	+0.01
2	5.12	5.06	0.55	0.55	0
3	5.46	5.37	0.55	0.55	0
4	5.72	5.64	0.66	0.65	−0.01
5	5.65	5.67	0.66	0.66	0
6	5.87	5.78	0.71	0.71	0
7	6.24	6.19	0.92	0.92	0
8	5.20	5.10	0.60	0.59	−0.01
9	5.26	5.32	0.62	0.62	0

b) Duplicate testing using tetrachloroethene (C₂Cl₄)

Sample	Capillary Porosity, % by Mass	Difference
A	5.05	0.04
	5.09	
B	6.06	0.07
	5.98	
C	5.03	0.06
	4.97	
D	5.38	0.07
	5.31	
E	5.29	0.06
	5.23	

Source: Data courtesy of Dr David Crofts, RSK Environment Ltd, U.K.

Note: (a) Using the previous (but now withdrawn) standard solvent 1,1,1-trichlorethane and the new replacement solvent tetrachloroethene, for cement contents ranging from 9% to 17% and water/cement ratios ranging from 0.50 to 0.90, and (b) duplicate testing to validate the BS 1881–124 method using tetrachloroethene.

The interlaboratory reproducibility is 60 kg/m³. The corresponding figures are not given for the water content measurement but examination of the method suggests that the relative error is unlikely to be better than that for the determination of cement content and a reproducibility of 30 kg/m³ in 200 kg/m³ may be of the right order. Because of the summation of errors, if a water/cement ratio of 0.5 is considered, the reproducibility of the determination is 0.10 and the result must be quoted as 0.5 ± 0.10'. These comments do not take into consideration the uncertainties introduced by the commonly required assumptions regarding the water absorption of the aggregate and the effects of voidage on the measurement of capillary porosity (Eden 2011).

The outcome of the recent TR 32 precision trial for the physico-chemical method (Concrete Society 2014) was especially disappointing, leading the working group to advise that the method *'did not appear to be sufficiently accurate to provide useful data to the end user.'* However, the group recognised the limitations of the trial and recommended *'a much larger research trial'* (see also Crofts 2014).

Moreover, the method produces an averaged finding for a relatively large slice specimen (typically a 20 mm thick slice across a 150 mm diameter core sample) and is thus insensitive to small-scale variations. The determined water/cement ratio will also underestimate the original water content for concretes affected by excessive bleeding or extensive carbonation but overestimate the original water content for concretes which have been physically damaged or chemically attacked.

4.4.4 Determination of water/cement ratio by fluorescence microscopy

4.4.4.1 *Principle of the method*

The capillary porosity of the hardened cement paste in concrete increases as the water/cement ratio of the concrete mix is increased (see Section 4.4.2). Christensen et al. (1979) and Thaulow et al. (1982) demonstrated that the capillary porosity of a cement paste could be determined, on a comparative basis, by impregnating a suitable sample with low-viscosity resin containing a fluorescent dye and then examining a thin section under a microscope using UV illumination. The greater the capillary porosity, the more fluorescent resin is absorbed and the brighter is the fluorescence in UV light. If reference samples of similar concrete and known water/cement ratio are prepared or available, it is possible, by visual comparison, to allocate an equivalent water/cement ratio to the sample in question. Unlike the physico-chemical method (see Section 4.4.3), which is applied to a bulk sample of the concrete, this microscopical technique also provides localised information.

The fluorescence technique has been shown potentially to provide a reliable determination of the water to cement ratios of hardened concrete over a range from 0.35 to 0.70 (Sibbick et al. 2007). An early review was conducted by St John (1994a), and a recent critical review of the method of optical fluorescence microscopy is given by Neville (2003, 2006). The recently published second edition of Concrete Society TR 32 (2014) includes a good description of the fluorescence microscopy technique. Given the poor outcome for the physico-chemical method in their precision trial (see Section 4.4.3), it might have been expected that the fluorescence microscopy technique would have been evaluated as an alternative, but unfortunately it was not included in the TR 32 precision trial programme.

Neville (2003, 2006) has been especially critical of accuracy claims made for the fluorescence technique by Jakobsen et al. (2000), who had suggested that the procedure could determine water/cement ratio to 'an accuracy of about 0.02'. The issue had arisen during a litigation in the United States, and Neville (2006) reproduces a 2005 judge's ruling, excluding evidence from the fluorescence method on the basis that it had not achieved international consensus. Technically, Neville stressed the importance of having suitable reference samples and demonstrated that the accuracy claims made by Jakobsen et al. were statistically untenable. The present authors agree with these conclusions by Neville but suggest that the procedure nevertheless has merit when used correctly and cautiously by experienced petrographers. De Rooij (2011) stresses the difference between the relatively good accuracy that might be achievable for water/cement ratio in a localised portion of paste under the microscope and the relatively poorer accuracy that will pertain to a representative sample of concrete for which the overall water/cement ratio is estimated from multiple localised assessments.

4.4.4.2 *Transmitted and reflected light procedures for determination of equivalent water/cement ratio using thin sections*

A standard test procedure for the determination of water/cement ratio by fluorescence microscopy is available and published as NT BUILD 361 (Nordtest 1999), and this technique has become widely used in Scandinavian and Nordic countries as well as in North America. Similar procedures have been commercially applied in the United Kingdom by a number of laboratories, and the following description is based upon the Sandberg technical procedure (Sandberg 1987, compiled by Ian Sims, and Sandberg 2010, as updated by Mike Eden).

A representative slice from the concrete specimen in question is vacuum impregnated with a low-viscosity epoxy resin containing a fluorescent dye, such as Magnaflux ZL7 or Dayglo Hudson Yellow, and then used to prepare a petrographic thin section. In some cases, the concrete will have needed such vacuum impregnation prior to slicing. The thin section is then examined with a petrological or dedicated fluorescence microscope fitted with the appropriate filters to provide UV illumination under either transmitted or reflected light conditions.

The intensity of fluorescence of the concrete matrix under the microscope is proportional to the concentration of dye in the resin and then the quantity of fluorescent resin that has permeated into the matrix, which in turn is dependent upon the microporosity. As the microporosity of the matrix is associated with the original water/cement ratio of the concrete, in the absence of other significant factors, the intensity of fluorescence is an indirect measure of the original water/cement ratio. By comparison of the test thin section (or thin sections) with suitable reference specimens of known water/cement ratio, the *equivalent* water/cement ratio of the concrete sample may be estimated. Parameters such as the amounts of unhydrated cement and the size and abundance of portlandite (see Section 4.4.2) can also be usefully compared between reference thin sections and test sample to assist in the overall assessment of water/cement ratio.

The procedures followed in the preparation of the thin sections are critical to effectiveness of the method, and it is essential that this preparation is only carried out by a technician experienced in the preparation of concrete thin sections. There are several avoidable pitfalls in preparing thin sections for use in water/cement ratio determination (Eden 2011):

- It is particularly critical to ensure that the specimen is impregnated by fluorescent resin to the maximum extent permitted by the microporosity. St John (1994a) found incomplete impregnation to be a potential limitation of the method, especially in the case of the denser cement pastes which derive from lower water/cement ratios (say less than 0.4) or blends of cement with mineral additives (e.g. ggbs, pfa, microsilica). Mayfield (1990) overcame this difficulty for the reflected light method (see Section 4.4.4.3) by using an alcohol/dye mixture, but this cannot be used in a concrete thin section.

- Concrete thin sections are usually prepared to a thickness in the range 20–30 μm, with 25 μm being optimum (but geological laboratories are typically standardised on 30 μm). St John (1994a) found that small variations in thin section thickness had a roughly proportionate influence on the intensity of fluorescence under the microscope (i.e. a 3 μm variation in thickness changed the intensity by around 10%), so that consistency in thin-section thickness was essential, both within and between specimens. Adjustments to the UV light intensity, if available, can help to compensate for any such variations in thin-section thickness and/or in fluorescence dye concentration in the resin, but the settings must be the same for the unknown and reference thin sections (see Figure 4.60).
- It is likely that fluorescent light examination in reflected light (epifluorescence) would be less sensitive to variation in thin section thickness than transmitted light fluorescence.

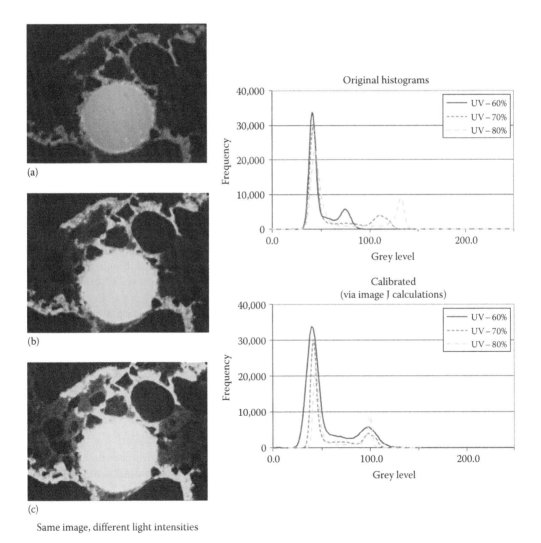

Same image, different light intensities

Figure 4.60 Compensating for fluorescence intensity caused by dye concentration in the resin and/or thin-section thickness: the same view of void and paste using (a) 60%, (b) 70% and (c) 80% UV light intensity. (Images courtesy of Mario de Rooij, TNO, Delft, the Netherlands.)

- The temperatures reached during the curing of the impregnating resin and drying of the concrete specimen should not be excessive (Applied Petrography Group 2008: APG-SR2), as excessive heating during the preparation of the thin section can exaggerate the microporosity of the cement paste. De Rooij (2011) advises that such problems are avoided if temperatures are kept below 40°C.
- The concrete specimen should not be excessively exposed to water or air during the final stages of preparation of the thin section, as any secondary hydration, leaching and carbonation all have the potential to modify the apparent porosity of the cement paste.
- There is much microscopic damage associated with diamond sawing, and unless this damage is removed by grinding and polishing, there is the potential for microporosity to be exaggerated. However, the grinding necessary to obtain a 25 μm thick thin section will usually ensure that such sawing damage is eliminated.
- Unless the concrete surfaces are flat prior to impregnation, there is the risk that the impregnation may be partially removed from the high areas of the surface during the grinding and polishing of the concrete specimen. This has the potential to lead to an underestimate of the microporosity of the cement paste. It is therefore sometimes necessary for impregnation with fluorescent resin to be repeated at stages during the thin-sectioning process.

With a suitable set of reference concrete thin sections and under the following conditions, it should be possible to measure water/cement ratio with a precision of ±0.05 (Eden 2011):

1. The analysis is carried out by an experienced concrete petrographer.
2. The unknown concrete is of very similar composition to the available reference concretes. The presence of cement replacement materials (SCMs) and organic admixtures such as plasticisers will greatly add to the uncertainty in the measurement of water/cement ratio unless the reference concretes contain the same additives or admixtures as the unknown concrete.
3. The unknown concrete is either of recent origin or in a non-aggressive environment where it is unlikely to have undergone secondary hydration or leaching due to moisture ingress and has not been affected by processes of deterioration such as AAR or DEF (see Chapter 6).
4. The area of concrete being assessed for water/cement ratio is not carbonated. In some circumstances, carbonation can greatly reduce microporosity, and in damp environments, it may lead to an increase in microporosity. Of course, the technique can still provide information on the microporosity and other features of any carbonated regions.
5. The water/cement ratio is in the range of 0.40–0.60. Sibbick et al. (2007, 2012) have cited a range of 0.35–0.70.

In many situations, these rather demanding conditions cannot be wholly met, and in such situations, the precision of the method for determining overall water/cement ratio will be much reduced and may be of the order of ±0.10 at best. The thin-section fluorescence method however has many advantages over the physico-chemical method of water/cement ratio measurement for bulk concrete samples. The thin-section fluorescence method is unaffected by void content, aggregate porosity and cracking and does not rely on assumptions about the binder type and composition. In all cases, petrography can provide an indication of the likely precision of the measurement and can be used to evaluate other water/cement ratio indicators such as the quantity of unhydrated cement and size and abundance of the portlandite crystals.

If the main uncertainty concerns the *overall* water/cement ratio, to ensure representativeness, it will be necessary to examine a range of specimens, taken from different parts of the concrete sample or unit under investigation and with various orientations to the original direction of placement. Otherwise, the technique is particularly useful for studying local variations in water/cement ratio, and, for example, often the differences between the near-surface zones and the interior of the concrete will be especially relevant to durability. However, the method will yield misleading equivalent water/cement ratio values for concrete areas that have been carbonated or otherwise altered in a manner likely to have affected the microporosity.

Unless a special microscope is available, most concrete petrographers will find it convenient to make temporary adjustments to the standard petrological microscope. The microscope must be fitted with a sufficiently powerful light source, such as a 100 W halogen light source or a high-intensity LED illumination.

For transmitted light fluorescence microscopy, a blue excitation filter, such as a Leitz Type BG12, is installed to cover the emergent aperture of the light source, and a yellow barrier filter, such as a Leitz Type K510, is inserted as an accessory plate, to detect the broad peak of yellow fluorescence of around 550 nm emitted by the dye. Both the polariser and analyser are removed from the light path and the condenser adjusted to give Kohler illumination. It is usually necessary to carry out this microscopy in a semi-darkened room. The setting of the aperture diaphragm is critical and preferably should not be changed during measurements; certainly unknown and reference specimens must always be examined under the same conditions and apparatus settings. For reflected light fluorescence microscopy or epifluorescence microscopy, an excitation filter such as Zeiss BP 450–490 and an emission filter such as Zeiss BP 515–565 are used in conjunction with a beam splitter such as Zeiss FT 510 (Eden 2011).

On examining the thin sections using this system, all regions invaded by fluorescent resin will show a yellow colour, the brightness or intensity being controlled by the quantity of resin absorbed by that part of the specimen. Non-opaque materials which have low microporosity will show as blue in colour. Cracks, microcracks and voids will exhibit maximum intensity of fluorescence. By concentrating attention on the cement paste regions, it will be possible to identify areas relatively exhibiting the maximum, minimum and modal average (i.e. most frequent) degrees of fluorescence. If suitable reference sections are available (see below in this section), it will be possible, on a comparative basis, to allocate maximum, minimum and overall *equivalent* water/cement ratios to these areas. However, Sibbick (2011) warns that light glare from numerous smaller-sized voids in highly air-entrained concrete can make this assessment considerably more difficult.

One major feature that is also highlighted well in the fluorescence mode, but not always highlighted in thin sections produced without the fluorescence dye, is the degree of paste homogeneity (Sibbick 2011). This is often an important factor in the strength of a concrete and is particularly indicative of when extra water has been added to a mix at a later stage (tempering/retempering) or when a concrete has been poorly mixed. These features are probably more common in areas of the world where mix water is withheld and/or added at site, to improve later workability and to compensate for potentially significant water loss from high environmental temperatures in which concrete is being placed. Highly heterogeneous paste may have strength issues, but also estimations of the water/cement ratio by this technique become more variable. Of course, in some cases, the localised information that is obtainable using fluorescence microscopy may assist in explaining such heterogeneity.

The quantified estimation of water/cement ratios should only be attempted when a full range of relevant reference thin sections is available for the concrete under investigation. Such reference sections should be prepared from laboratory-prepared concrete castings, wherein the type of cement, any mineral addition(s), aggregates and any admixtures are essentially similar to the concrete in question. A sequential range of mixes should be prepared in which the water/cement ratios are strictly controlled in the range from 0.35 to 0.75. Some relative information on the original water/cement ratio may be obtainable using a limited range of relevant reference thin sections (say variously higher or lower than the available references).

Sibbick et al. (2007) and Sibbick (2012) advise that these carefully prepared reference specimens need to be stored and used out of strong light sources and also recommend that these are replaced with equivalent new specimens every few years.

The relationship between original water/cement ratio and microporosity of the paste, and hence fluorescence exhibited by that paste, potentially might also change as the concrete matures, especially for some slower-reacting binders. Accordingly, some petrographers maintain sets of reference samples that include similar compositions at different ages, say 28 days, 1 year and 3 or more years (de Rooij 2011); this enables samples for assessment to be compared with references that can be expected to have a similar degree of maturity. An example that exhibits an unusually pronounced difference in fluorescence between 28 days and 5 years for concrete made using a CEM III binder is shown in Figure 4.61.

By contrast, Sibbick (2012) regards such a difference with time as being exceptional, with the microporosity depicted by fluorescence typically being similar for a given concrete at 28 days and at later ages, providing similar representativeness of samples in terms of hydration and SCM activation and similarly consistent qualities of impregnation and specimen preparation. It is clearly very important to ensure that the more mature materials are fully impregnated with fluorescent resin, as any incomplete impregnation would give a false impression of reduced microporosity. Recent work by Sibbick et al. (2013) has confirmed the usefulness of the fluorescence technique for concretes made using blended cements or SCMs but also highlighted the key importance of using *relevant* reference sets, because of the ways in which the various types of addition influence the light intensities in different ways. Research is continuing on these issues and the petrographer will need to refer to later reports in due course.

Visual comparisons of these reference thin sections with those prepared from the concrete sample, using the same system at the same time, enable the closest similarities of fluorescence to be established, including intermediate values and, in the best circumstances (see conditions listed earlier), yielding *equivalent* water/cement ratios to the nearest 0.05. Jakobsen et al. (2000) are very supportive of the fluorescence technique and claim that

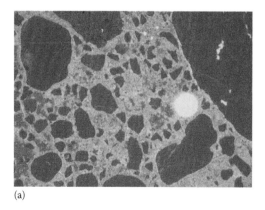
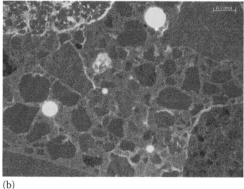

(a)   (b)

Figure 4.61 Potential influence of time on degree of hydration and microstructure of concrete, which influences its appearance in fluorescence microscopy: the same CEM III-B concrete with water/binder ratio of 0.45 at (a) 28 days and (b) 5 years. (Images courtesy of TNO, Delft, the Netherlands.)

experienced petrographers using good-quality thin sections can achieve water/cement ratio determinations as close as ±0.02. Jakobsen and Brown (2006) stressed the criticality of the thin-section preparation and reported a 'round-robin' trial showing remarkably low standard deviations for water/cement ratio determinations on six concretes, involving seven petrographers (from three laboratories) and five groups of specimens. De Rooij (2011) considers that such accuracy is potentially achievable for each individual estimate under the microscope, but not for the overall estimation of water/cement ratio based on collective overall assessment from multiple individual estimates.

The authors, supported by Neville (2003) and de Rooij (2011), would not agree that the accuracy apparently suggested by Jakobsen et al. (2000) and Jakobsen and Brown (2006) is realistically feasible on a routine basis for the determination of overall water/cement ratios. In all the best circumstances, which are rarely achieved in commercial concrete petrography, the accuracy for such overall estimates could reach ±0.05, but in reality, ±0.10 is a more credible best-achievable confidence limit.

Photomicrographs illustrating some typical appearances of concrete matrices under fluorescence microscopy are shown in Figure 4.62. In well-mixed concrete, the degree of fluorescence should be reasonably uniform within the cement paste regions, but a patchy appearance is quite common, indicative of an uneven distribution of cement and water in the mix, provided that the specimen is fully impregnated and of consistent thickness

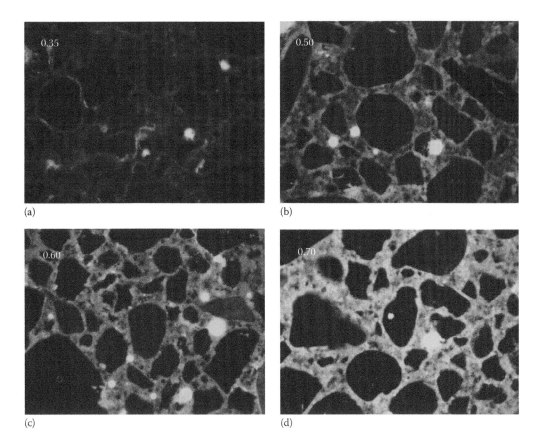

(a) (b) (c) (d)

Figure 4.62 Photomicrographs of fluorescence microscopy in thin section for concretes of different water/cement ratio: (a) 0.35, (b) 0.50, (c) 0.60 and (d) 0.70, uvl, widths of field 2.25 mm. (Photographs courtesy of Dr Ted Sibbick.)

(each of which can give rise to a similar patchiness). Local areas of apparently higher water/cement ratio may be associated with bleeding, and near-surface areas will frequently exhibit enhanced microporosity when not carbonated.

St John (1994a) confirmed the usual presence of a gradation of fluorescence that could be related to water/cement ratio but considered that estimates outside the range 0.4–0.6 were less reliable. Even within the 0.4–0.6 range, he thought that values were more realistically gauged to the nearest 0.1, rather than 0.05. At water/cement ratios exceeding 0.6, St John found that the fluorescence emission reached saturation unless the concentrations of dye in the resin had been reduced accordingly (which would also be required in the reference specimens), whereas at ratios below 0.4, it was difficult to be confident that complete impregnation had been achieved. It was also noticed that the proportions of portlandite and unhydrated clinker grains influenced the appearance of the fluorescence, although these factors can also be used to corroborate the water/cement or water/binder ratio determined by fluorescence.

Recently, Sibbick et al. (2012), developing the work by Sibbick et al. (2007), have found that a similar relationship potentially exists between fluorescence light intensity and water content (i.e. water/cement or water/binder ratios) for concrete materials containing both Portland cement and various additions or SCMs. They concluded that the method 'remains a valid and reliable tool for the determination of water to binder ratio of hardened concrete over the [water/binder ratio] range 0.35–0.70, regardless of mix proportions, providing proper references are employed'. Also, no 'rapid ageing effect' was found for fluorescent resin, with comparable mixes yielding similar fluorescent light intensities for reference thin-section sets made over a 10-year period (2001, 2006 and 2011). They also noted that the various SCM mixes tended to exhibit brighter fluorescence light intensities than the equivalent Portland cement-only mixes, which they suggested might be caused by the differing levels of translucency of the various residual SCMs.

In summary, the fluorescence microscopy method using transmitted light is potentially usable for estimating the original water/cement ratio, but its reliability depends critically on being able to make comparisons with entirely suitable reference samples and other factors, when accuracy in practical cases (assessing unknown field samples, rather than laboratory-prepared materials) is unlikely to be confidently better than ±0.1. In the absence of reference samples, the authors have found that the degree of fluorescence does not necessarily provide a dependable indication of original water/cement ratio, even on a broad range relative basis (say, low ≤0.35, normal 0.40–0.65 or high ≥0.70), and correlation of such an approach with the physico-chemical method (see Section 4.4.3) can be poor (Table 4.21).

The fluorescence microscopy method is particularly useful for detecting small-scale variations in microporosity and/or water/cement ratio and has the advantage of being completely complementary to the routine petrographic examination of concrete (using the same thin section if prepared appropriately and the same microscope assembly).

Systems have been developed (Elsen et al. 1995; Jakobsen et al. 1995) for conducting the examination phase of this method using automated image analysis techniques, and according to Sibbick (2011), these can be useful, especially for more heterogeneous mixes, but are not always applicable, for example, when patchy areas of the concrete are carbonated or exhibit variable degrees of hydration or activation. De Rooij et al. (2011) provide a very useful summary of the challenges encountered in using image analysis in the estimation of water/cement ratio by fluorescence microscopy. In advice to the present authors, however, de Rooij (2011) has expressed some confidence that, with the latest equipment and armed with suitable reference specimens, reflecting both appropriate composition and microstructural maturity, it will be possible successfully to develop this image analysis technique.

Bromley (2011) has investigated the use of image analysis to implement the method described by French (1991), for estimating apparent water/cement ratio from the determination of the

Table 4.21 Example comparison of original water/cement ratios separately determined using the fluorescent microscopy (transmitted light, without references) and physico-chemical methods

Type of concrete construction	Type of sample	Notional location number	Water/cement ratio range estimated using fluorescence microscopy		Water/cement ratio estimated using the BS 1881–124 method (mean)	Degree of agreement
			Overall relative	Range		
Existing floors	Cores	C1	Normal	0.40–0.65	0.7–0.9 (0.8)	Poor
	Cores	C2	Normal	0.40–0.65	0.8–1.1 (0.9)	Very poor
	Core	C3	Normal	0.40–0.65	0.8 and 0.9 (0.8)	Poor
	Core	C4	Normal	0.40–0.65	0.6–0.9 (0.8)	Poor
	Core	C5	Normal	0.40–0.65	0.8	Poor
	Core	C6	Low	≤0.35	0.7	Very poor
	Core	C7	Low	≤0.35	0.8	Very poor
New floors	Cubes	1	Normal	0.40–0.65	0.55–0.6 (0.6)	Good
	Cubes	2	Normal	0.40–0.65	0.5–0.55 (0.5)	Good
Road and pavement	Core	R1	Normal	0.40–0.65	0.6	Quite good
	Core	R2	Normal	0.40–0.65	0.6 and 0.6 (0.6)	Quite good
	Core	R3	Normal	0.40–0.65	0.9	Very poor
	Core	R4	Normal	0.40–0.65	0.7 and 0.7 (0.7)	Marginal
	Core	R5	Normal	0.40–0.65	0.6 and 0.6 (0.6)	Quite good
	Core	R6	High	>0.65	0.6	Poor
	Core	R7	High normal	Say 0.55–0.65	0.6	Good
	Core	R8	High normal	Say 0.55–0.65	0.7	Quite good
	Core	R9	High normal	Say 0.55–0.65	0.8	Quite poor
	Core	R10	High normal	Say 0.55–0.65	0.6	Good
	Core	R11	High normal	Say 0.55–0.65	0.7	Quite good
	Core	R12	High normal	Say 0.55–0.65	0.6	Good

Note: The data were obtained during an actual investigation in the British Isles, but the site identity must remain confidential; the concrete samples all comprised plain Portland cement–aggregate mixtures, with neither mineral additions nor admixtures, exhibiting thorough mixing and good compaction, with no deterioration and limited (some cracking in vicinity of location) damage.

number of unhydrated clinker grains per millimetre traverse of the paste, over a traverse length of at least 120 mm. Though the method is claimed by French to be accurate (up to a water/cement ratio of 0.6), it is very time consuming and tedious. Image analysis, using the reflected light fluorescent image of dye-impregnated cement paste in thin section, provides a modification of French's method, which does not sacrifice accuracy but is extremely rapid. It can be based on the number or total area of unhydrated cement clinker grains per unit area of paste. A polished thin section of standard mortar with water/cement ratio of 0.4 is used in the example shown in Figure 4.63.

Bromley (2011) has also developed image analysis techniques for estimating water/cement ratio by measurement of the fluorescent intensity of dye-impregnated cement paste (Christensen et al. 1979; Jensen et al. 1985). Direct measurement of the optical density or integrated brightness of the paste is both rapid and accurate, provided that suitable reference thin sections are available. Furthermore, he found that the relationship between water/cement ratio and integrated brightness is linear up to water/cement ratio of 0.8 at least. The method offers improved accuracy for the investigation of water-rich mixes. Bromley provides an example in Figure 4.64.

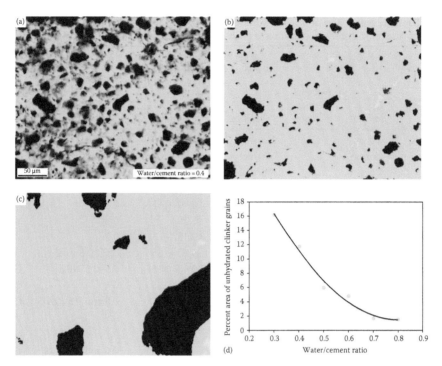

Figure 4.63 Image analysis of a polished thin section of mortar for water/cement ratio, using frequency of residual unhydrated cement particles (see also (c) in Figure 4.64): (a) Dye-impregnated polished thin section of standard mortar (water/cement ratio = 0.4) viewed under reflected UV illumination. (b) Application of extreme contrast enhancement to separate unhydrated crystalline cement clinker grains from hydrated paste and all other phases. (c) Isolation of unhydrated clinker grains using brightness histogram before counting or measuring total area. (d) Calibration graph derived from thin sections of standard mortars having water/cement ratios between 0.3 and 0.8 using percentage area of binder.

4.4.4.3 Reflected light procedure for the determination of equivalent water/cement ratio using ground polished specimens

Mayfield (1990) and his research team partially developed an alternative fluorescence microscopic method using reflected light with ground and polished specimens. In this technique, the concrete specimen, after grinding and polishing, is vacuum impregnated with 10^{-4} M fluorescein sodium in alcohol, which supposedly permeates the micropore system more easily and thus more reliably than low-viscosity resin. The specimen is then examined using a reflected light (i.e. metallurgical) microscope, modified in a similar way to the petrological microscope used in the thin-section technique, and estimates of *equivalent* water/cement ratio are similarly determined by comparison with reference specimens.

Instead of depending upon subjective visual appraisal, Mayfield (1990) demonstrated that it was feasible to quantify the emitted fluorescent light using a photodiode attached to the microscope, an A–D converter and a suitably programmed computer. In this way, he was able to show clear separations in the histograms produced for three different water/cement ratios (0.30, 0.35 and 0.40) when using hardened cement paste specimens (Figure 4.65).

Mayfield encountered some problems with mortar specimens, because of the dark areas caused by sand grains and unhydrated cement clinker particles: there was a need to correct for different aggregate/cement ratios. Once the sand aggregate content was taken into account, he recorded a relationship between the voltage of emitted light and water/cement ratio (Figure 4.66). It seems quite likely that the technique could be further refined and

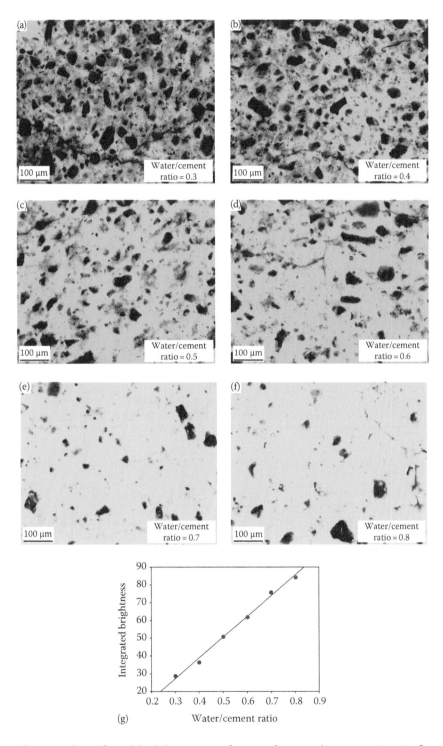

Figure 4.64 Image analysis of a polished thin section of mortar for water/cement ratio, using fluorescence of cement paste: (a to f) Polished thin sections of reference mortars having water/cement ratios between 0.3 and 0.8 viewed under reflected UV illumination. (g) Calibration graph using integrated brightness of standard mortars shown in Figures 4.64a–f.

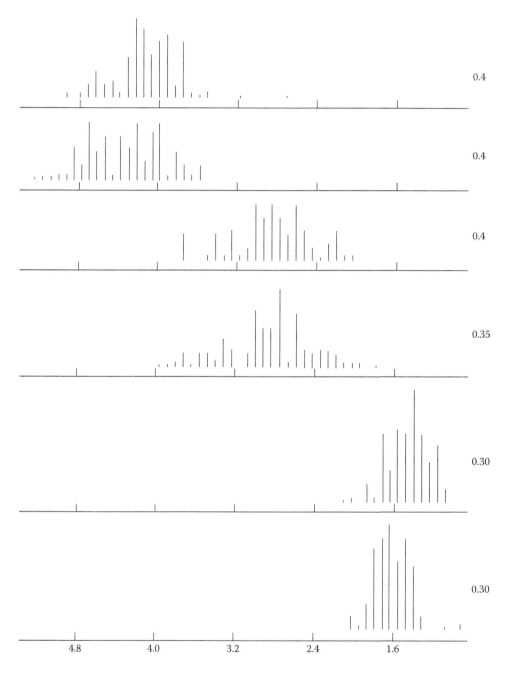

Figure 4.65 Histograms of emitted fluorescent light in reflected light for cement pastes of various water/cement ratios. (From Mayfield, B., *Magaz. Concr. Res.*, 42(150), 45, 1990.)

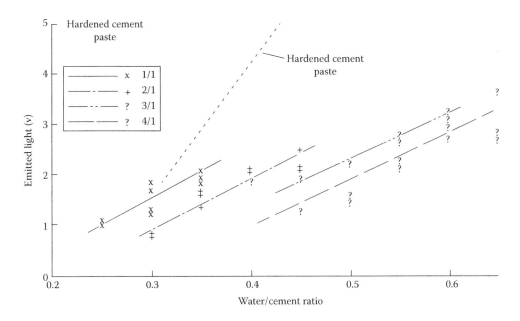

Figure 4.66 Relationship between emitted fluorescent light in reflected light, aggregate/cement ratio and water/cement ratio for mortars. (From Mayfield, B., *Mag. Concr. Res.*, 42(150), 45, 1990.)

perhaps successfully applied to concretes. Wirgot and Van Cauwelaert (1994) have developed a method for measuring fluorescence in the reflected light procedure using image analysis coupled with automatic statistical interpretation.

4.4.5 Determination of water/cement ratio using scanning electron microscopy

An apparently elegant, if somewhat arduous, procedure using field emission SEM (FE-SEM) in the backscattered mode has been proposed by Wong and Buenfeld (2007, 2009), developed from earlier work by Sahu et al. (2004) and Scrivener (2004). Working with polished, resin-impregnated specimens of cement paste, Wong and Buenfeld used the superior resolution of FE-SEM (over conventional SEM) and image analysis to quantify the four main components of hardened cement paste: (1) residual unhydrated cement, (2) crystalline and semi-crystalline hydration products, (3) capillary pores and (4) air voids from incomplete compaction and/or air entrainment (Figure 4.67). In order to relate these determined proportions back to the original ratio of water and cement, prior to hydration and shrinkage, the 'expansion coefficient of hydration' (δ_v) is derived, using the model of Powers and Brownyard (1947/1948). According to Wong and Buenfeld, this approach enables the proposed FE-SEM method to be applied to samples of unknown history without the use of laboratory reference specimens, which are essential for the fluorescence microscopy method (see Section 4.4.4). Their initial findings, for just cement paste, were certainly encouraging, as shown in Figure 4.68.

Although the reliability of the Powers and Brownyard (1947/1948) model for modern cements may be questionable (see Brouwers [2004, 2005] for a review of the pioneering work of Powers and Brownyard), the independence of the FE-SEM technique from reference specimens is a desirable aspect, and the initial findings for cement pastes suggested potentially good correlation between actual and estimated values of free water/cement ratio. However, subsequent

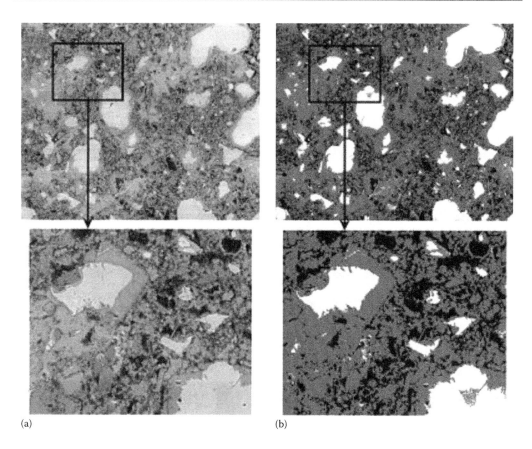

(a)   (b)

Figure 4.67 Example views of resin-impregnated cement paste under FE-SEM in backscattered mode, (a) before and (b) after 'segmentation' to enhance clarity and contrast for the image analysis. Each field of view is 240 × 192 µm or 67 × 54 µm for the insets. (From Wong, H.S. and Buenfeld, N.R., Estimating the water/cement [w/c] ratio from the phase composition of hardened cement paste, in *11th Euroseminar on Microscopy Applied to Building Materials*, 5–9 June 2007, Porto, Portugal, 11pp, 2007; Wong, H.S. and Buenfeld, N.R., *Cement Concr. Res.*, 39(10), 957, 2009.)

development of the method for application to mortars and concretes, while indicating that the method is potentially usable, has identified some problems (Wong et al. 2009).

First, the presence of aggregate particles as the major component, thus greatly increasing the heterogeneity of the microstructure, demands the cumulative analysis of many more views or 'frames' for each sample; it is perhaps surprising that Wong et al. only increased this number from 30 (for the cement paste work) to 50 (for the mortar and concrete). Second, given that it is obviously only the cement paste component that is needed for the image analysis, with mortar and concrete, there is considerable and time-consuming processing required completely to 'remove' the aggregate particles from the images to be analysed (Figure 4.69).

Even then, Wong et al. found that bleeding in the mortar and concrete laboratory samples significantly influenced the findings, with higher water/cement ratio mixes predictably tending to bleed the most and thus create larger underestimates of water/cement ratio, although good correlations between actual and estimated values were obtained in their experiments after making corrections for the bleeding. Overall, although further development might enable this method to become a reliable routine procedure in the future, at present, it appears impracticable for mortars and concretes of unknown history.

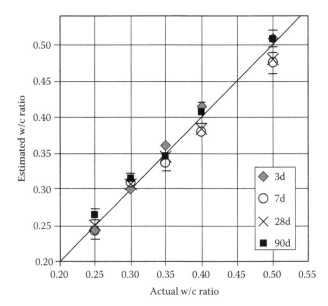

Figure 4.68 Correlations at 3, 7, 28 and 90 days between water/cement ratios determined using the FE-SEM method for cement paste and the actual values; error bars represent the 95% confidence limits. (From Wong, H.S. and Buenfeld, N.R., *Cement Concr. Res.,* 39(10), 957, 2009.)

4.5 AIR-VOID CONTENT AND AIR ENTRAINMENT

4.5.1 Types of voids in concrete

A fresh concrete mix is typically compacted, by vibration or ramming, in order to improve the packing of the aggregate and to minimise the residual content of *entrapped* air. Yet, however, efficiently the material has been compacted, most concrete inevitably retains some entrapped air. In fresh concrete, such entrapped air forms bubbles which range in size from microscopic spheres the size of cement clinker grains to irregularly shaped gas pockets the size of coarse aggregate particles or larger. Surface-active admixtures may be included in the mix to facilitate the formation of *entrained* air bubbles. After hardening of the concrete, these entrapped and entrained bubbles are fixed in place and become air voids. If a concrete has been under-compacted or the proportion of matrix is low relative to that of the coarse aggregate, the larger entrapped air voids may be abundant and interconnected, when the concrete is said to be *honeycombed*: this tends to occur in localised areas. Air-void definitions are given in Table 4.22.

Hover (1993a) has shown that the air-void system in a concrete can be represented as a void size distribution in a manner analogous to aggregate gradings (Figure 4.70). In this way, entrapped air voids might exhibit a grading intermediate to the coarse and fine aggregates, whereas entrained air voids will typically produce a grading equivalent to very fine sand.

Regardless of the size or the presence of admixtures, all air bubbles get into fresh concrete by the tumbling and folding actions of mixing and placing (Hover 1993a). The surface-active admixtures serve to stabilise and retain more of the smaller bubbles trapped during mixing and agitating. Thus, the so-called *entrained* air voids are actually only an enhanced variety of *entrapped* air void. Nevertheless, it is convenient, from a practical performance standpoint, to distinguish between entrapped and entrained air voids in concrete. Aerated (or foam or foamed) concrete is a special case in which abundant gas voids are generated either by chemical reactions between additives and the cement or by the use of foaming agents.

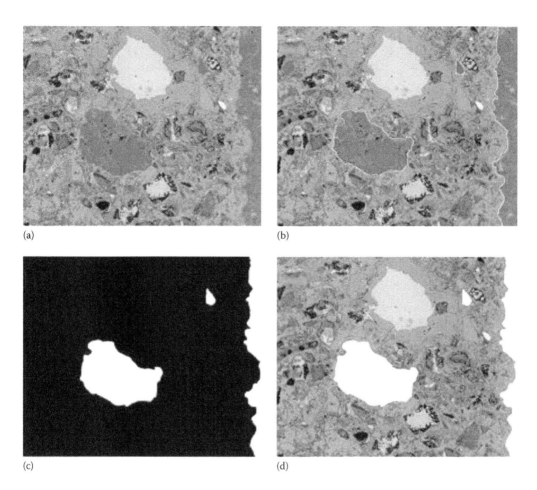

Figure 4.69 An example view of aggregate segmentation in the FE-SEM method for estimating original water/cement ratio of mortar or concrete. An original backscattered image at 500× magnification (a) has the aggregates identified and manually outlined (b), and then thresholding and particle detection are used to produce an aggregate binary mask (c), after which the aggregates can be 'removed' from the final image for analysis (d). Each field of view is 240 × 192 μm. (From Wong, H.S. et al., Estimating the water/cement [w/c] ratio of hardened mortar and concrete using backscattered electron microscopy, in *12th Euroseminar on Microscopy Applied to Building Materials*, 15–19 September 2009, Dortmund, Germany, 11pp, 2009.)

4.5.1.1 Entrapped air voids

Entrapped air voids are typically irregular in shape and, as shown in Figure 4.45, mainly greater than 1 mm in diameter, although any irregularly shaped smaller air voids might also be taken to be entrapped rather than entrained. The size distribution and disposition of entrapped air voids is also frequently markedly irregular within the concrete, commonly with a tendency for increases in both size and number towards surfaces (especially horizontal surfaces, as cast).

The content of entrapped air voids should be reduced as the degree of compaction improves for any concrete mix, but virtually *all* normal concretes can be expected to contain at least a small amount of entrapped air. In a non-air-entrained concrete which is *very* well compacted under ideal conditions (as in a laboratory-prepared test cube, for example), the entrapped

Table 4.22 Air-void system definitions

Term	Definition
Air void	A small space enclosed by the cement paste in concrete and occupied by air. This term does not refer to capillary or other openings of submicroscopical dimensions or to voids within particles of aggregate.
Entrained air voids	Air voids characteristically spherical in shape between 10 μm and 1 mm in diameter and should have a regular distribution in the concrete.
Entrapped air voids	Air voids mostly over 1 mm in diameter, typically irregular in shape and usually having an irregular distribution in the concrete, often increasing in size and number towards the surface. They are sometimes referred to as compaction voids.
Water voids	Filled with water at the time of setting of the concrete, water voids are irregular in shape, generally elongated and usually very large (several millimetre in size). They are typically found beneath particles of coarse aggregate or reinforcing bars, and their interior surface has a granular appearance instead of the usually glazed lustre of air voids.
Gas voids	Formed by the liberation of gas from the reaction between special admixtures and the cement. Their size and shape are variable.
Honeycombing	Interconnecting large entrapped air voids arising from inadequate compaction or lack of mortar.
Air-void content (A)	The proportional volume of air voids in concrete expressed as a volume percentage of hardened concrete.
Specific surface (α)	The surface area of the air voids expressed as mm^2 per mm^3 of air-void volume.
Spacing factor (L)	An index related to the maximum distance of any point in the cement paste from the periphery of an air void, expressed in μm or mm. In practice, this is approximately equivalent to half the mean air-void spacing within the cement paste.
Air/paste ratio (A/p)	The ratio of the volume of air voids to the volume of the cement paste in the hardened concrete.
Paste content (p)	The proportional volume of the cement paste in concrete, expressed as a volume percentage of the hardened concrete.

Source: After Figg, J.W., *Analysis of Hardened Concrete – A Guide to Tests, Procedures and Interpretation of Results – Report of a Joint Working Party of the Concrete Society and Society of Chemical Industry*, Technical Report No. 32, The Concrete Society, London, U.K., 1989.

air-void content will typically be of the order of 0.5%. On this basis, any air content additional to 0.5% has been termed 'excess voidage' in the U.K. Concrete Society Technical Report on core testing (Dewar 1976/1987) and may be used as a comparative measure of concrete quality.

The amount of entrapped air that might be expected in a well-compacted *site* concrete depends on a number of factors, but in general contents up to around 3.0% are not unusual. Contents of entrapped air voids substantially in excess of 3.0%, in all or parts of a concrete, might well be symptomatic of important mixing and/or placing deficiencies. The U.S. Bureau of Reclamation (1955) has shown that the content of entrapped air voids in a well-compacted concrete varies with maximum aggregate size, being approximately 1.0%, 2.0% and 3.0% air for 40 mm (1½ in.), 20 mm (¾ in.) and 10 mm (⅜ in.) nominal aggregate sizes, respectively. In honeycombed areas of concrete, the air-void content will probably range between 10% and 30%.

4.5.1.2 Entrained air voids

Entrained air voids are characteristically spherical in shape and, as shown in Figure 4.45, mainly range in diameter between 10 μm and 1 mm. Also, in contrast to the entrapped air voids, the entrained air voids usually exhibit a sensibly uniform distribution throughout the concrete matrix, except near to concrete surfaces (see below in this section) and for poorly mixed concrete.

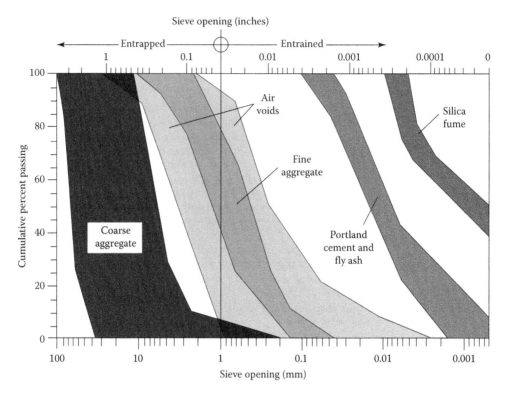

Figure 4.70 Air-void gradation, compared with aggregates, cement and some additions. (From Hover, K., *Concr. Const.*, 38(1), 11, 1993a.)

Air entrainment in concrete is achieved by the addition of organic surfactants, which, according to Usher (1980), 'lower the surface tension of water and facilitate bubble formation ... uniform dispersion and stability are achieved by the mutual repulsion of the negatively charged air-entrainer molecules and the attraction of the air-entrainer molecules for the positive charges on the cement particles'. These surfactants include neutralised wood resins, such as Vinsol resin, animal and vegetable fats and their salts and wetting agents such as alkyl aryl sulphonates, alkyl sulphates or phenol ethoxylates.

The factors affecting the success of air entrainment are examined in Brown (1983) and Neville and Brooks (1987, 2010). Air content can be increased or decreased by variations in inter alia temperature during mixing, mix workability, sand content and/or grading, cement fineness and/or alkali content, the presence of other organic materials and/or mineral additions (e.g. pfa) and even 'hardness' of the mix water (see Table 4.23). The influence of these factors varies for different types of air-entraining admixture: for example, the air-void system created by Vinsol resin is disrupted by high alkali contents, whereas that deriving from sulphonated hydrocarbon is unaffected (Pistilli 1983).

Controlled air entrainment of concrete has a number of potential benefits. One anonymous article in a concrete magazine perceived air entrainment as a 'shortcut' to durable concrete (Anon 1976), although that author was careful to add that it could not replace correct procedures in mix design or concreting. The beneficial and detrimental effects of entrained air, variously on fresh and hardened concrete, are thoroughly considered by Sutherland (1974) and reviewed in summary form by Usher (1980), Brown (1982) and Malisch (1990). Broadly, other things being unchanged, in fresh concrete, entrained air increases workability, improves cohesion and reduces bleeding. In hardened concrete, the prospects for

Table 4.23 Factors influencing the entrainment of air in concrete

Type of factor	Increasing air content	Decreasing air content	Example change	Estimate of effect (for 5% air content)[a]
Cement	Decrease in cement content[b]	Increase in cement content[b]	+50 kg/m³	0.5% reduction
	Decreased cement fineness	Increased cement fineness	—	—
	Increased alkali content (0.8%)	Decreased alkali content	—	—
Aggregate	Sand grading coarser	Sand grading finer	Zones 3 through 2 (see Table 4.12)	<0.5% increase
	Sand content increased	Sand content decreased	35%–45%	1%–1.5% increase
	Increased 600–300 μm fraction	Decreased 600–300 μm fraction	—	—
	Decreased – 150 μm fraction	Increased – 150 μm fraction	+50 kg/m³	0.5%
	Inclusion of organic impurity		Inclusion	Positive and negative effects
Additions	—	Inclusion of pfa	—	—
Admixtures	—	Greater than usual pfa carbon content	Inclusion	Significant
	Water-reducing admixture usage[c]	—	—	—
	Positive dispensing tolerance	Negative dispensing tolerance	±5%	±0.25%
Mix water	—	Increased hardness	—	—
Mix and mixing	More workable (i.e. higher slump)	Less workable (i.e. lower slump)	50–100 mm	1% increase
	Increased mixing efficiency[d]	Decreased mixing efficiency[d]	—	—
	Faster mixer rotation	Slower mixer rotation	—	—
	—	Prolonged mixing during transportation	1 h	Up to 0.25% reduction
			2 h	1% reduction
Environment	Lower temperature	Higher temperature	10°C–20°C	1%–1.25% reduction
	—	Steam curing	—	Incipient cracking caused by expansion of bubbles

Source: Modified from Brown, B.V., *Concrete*, 17(1), 45, 1983; Neville, A.M. and Brooks, J.J., *Concrete Technology*, Longman Group Limited, Harlow, U.K., 1987.

[a] These effects are only indicative and are not necessarily cumulative.
[b] Inclusive of sand content adjustment.
[c] Influence of superplasticisers is less clear.
[d] Optimum mixing time is required: too short a time causes a non-uniform distribution of the bubbles, while over-mixing gradually expels the bubbles.

durability are generally improved by a reduction in permeability, with resistance to freezing and thawing cycles and de-icing chemicals being notably enhanced. However, some long-term (currently 16 years and continuing) field exposure trials in the United Kingdom have indicated that, overall, entrained air has less influence than strength in relation to physical damage induced by freeze–thaw cycles (Troy 2011).

The principal disadvantage of air entrainment is the reduction in density and consequent proportional lowering of strength if there are no other changes to the mix design. However, in practice within the normal range of air contents used, the increased workability brought about by air entrainment usually allows the water/cement ratio to be reduced, which largely compensates for the potential loss of strength. The addition of a plasticising admixture to the concrete mix, in combination with the air entrainer, is also used to reduce further the water/cement ratio in order to combat strength loss (Hodgkinson and Rostam 1991).

Air-entrainment adequacy for various exposure conditions and with different coarse aggregate sizes is usually specified by placing limits on the *total* air content of the concrete (i.e. the sum of entrapped and entrained air voids). In the United Kingdom, total air content limits were formerly recommended for concrete exposure to freezing and thawing and de-icing salts in BS 5328 (1997). Guidance is now provided in BS EN 206-1 (2000) and complementary BS 8500-1 (2006), wherein minimum air contents are cited for various maximum aggregate sizes, to resist freezing and thawing ('XF' exposures). Hover (1993d) reviewed the specifications for concrete air content of a number of authorities in the United States. A summary of some standardised air content limits is given in Table 4.24. As a guide, most of these various limits indicate that, in a concrete made with nominal 20 mm aggregate, a total air content of around 5.5% or 6% should provide an appropriate level of protection, but the minimum air content criterion in BS 8500-1 is 3.5% for each exposure class, coupled with varying minimum binder contents and freeze–thaw-resisting aggregates for XF3 and XF4 exposure classes.

Although the addition of an air-entraining admixture will obviously produce air bubbles throughout the mix, in actuality, it is only the zone proximate to the exposed concrete surface where the air-void system is critical. Sandberg and Collis (1982) found this was the very zone in which air was most likely to be lost during compaction and finishing, in addition to any air lost during transportation of the concrete. In investigating some airfield and road pavement failures in which surface damage had occurred despite the apparent adequacy of total air contents measured on the fresh concrete, Sandberg and Collis found that up to 2% air could be lost in transit and up to a further 2% air was sometimes lost during compaction and finishing. Thus, the final air content in the crucial uppermost zone of concrete might be inadequate for the prevention of freeze–thaw damage.

Yingling et al. (1992) have investigated the loss of air in pumped concrete. It was found to be common to lose up to about 1% air and 25 mm of slump during pumping, but certain types of pump arrangement could occasionally lead to the loss of half the air content. The loss of air during handling and placing of concrete was reviewed by Hover (1993b), who makes a distinction between the relatively unimportant loss of larger voids, which might significantly affect the total air content, and the much more serious loss of the effective finer voids.

The use of some superplasticisers may destabilise the entrained air-void system. MacInnis and Racic (1986) determined that a nominal 5% air content was reduced on average to 3%, after the addition of a superplasticiser, and a 7% air content was reduced to 4%. Saucier et al. (1991), in summarising an extensive programme of investigation into air-void stability, confirmed inter alia the occasional incompatibility between superplasticising and air-entraining admixtures. Also, according to Gutmann (1988), some types of air-entraining agent are more prone to the coalescence of bubbles, creating larger air voids and poorer performance.

In addition to the proportional content of air voids by volume, it has long been recognised that the air-void *system*, especially the *spacing factor*, is also of critical importance to the

Table 4.24 Some European and American recommended air contents (and other limiting values) for Freeze–Thaw resistance

					Min. air content (%) and min. binder content (kg/m³) for max. aggregate size (mm)				
Exposure (EN 206–1)			Min. strength class	Max. water/ binder ratio	32 or 40	20	14	10	Notes[b]
Class	Conditions	Locations							
XF1	Moderate water saturation, no de-icing agent	Vertical surfaces exposed to rain and freezing	C25/30	0.60	**3.0** 260	**3.5** 280	**4.5** 300	**5.5** 320	1
			C28/35 or LC28/31	0.60	— 260	— 280	— 300	— 320	
XF2	Moderate water saturation, with de-icing agent	Vertical surfaces of road structures exposed to freezing and airborne de-icing agents	C25/30	0.60	**3.0** 260	**3.5** 280	**4.5** 300	**5.5** 320	1
			C32/40 or LC32/35	0.55	— 280	— 300	— 320	— 340	
XF3	High water saturation, no de-icing agent	Horizontal surfaces exposed to rain and freezing	C25/30	0.60	**3.0** 260	**3.5** 280	**4.5** 300	**5.5** 320	1, 2, 3
			C40/50 or LC40/44	0.45	— 320	— 340	— 360	— 360	
XF4	High water saturation, with de-icing agent or seawater	Road and bridge decks exposed to de-icing agents; surfaces exposed to direct spray containing de-icing agents and freezing; splash zones of marine structures exposed to freezing	C28/35	0.55	**3.0** 280	**3.5** 300	**4.5** 320	**5.5** 340	1, 2, 3, 4
			C40/50 or LC40/44	0.45	— 320	— 340	— 360	— 360	

European recommendations[a]

American recommendations

Mean total air content (% by volume) for nominal max aggregate size (mm)

Exposure	75	50	37.5	25	19	12.5	9.5
Mild[c,d]	1.5	2	2.5	3	3.5	4	4.5
Moderate[d,e]	3.5	4	4.5	4.5	5	5.5	6
Severe[d,e]	4.5	5	5.5	6	6	7	7.5
Destructive[f]	1.5–4.5	2.5–5.5	3–6	3.5–6.5	4–8	5–9	6–10

Source: Hover, K., *Concr. Const.*, 38(5), 361, 1993d.

Note: Minimum air content (%) values are given in boldface.

[a] Modified from BS 8500–1 (2006) and BS EN 206–1 (2000).

[b] Notes: 1, Where a concrete contains one or more admixtures in addition to an air-entraining agent, tests shall be carried out using BS EN 480–11 (2005), and the air-void spacing factor shall be not greater than 0.20 mm. 2, Cement designation CEM IVB-V is not suitable. 3, Freeze–thaw-resisting aggregates are required. 4, Cements or combinations containing more than 65% ggbs might not be suitable for wearing surfaces.

[c] ASTM C94 values shown based on 2% less than for moderate exposure with aggregates >25 mm and 1.5% less for aggregates up to 25 mm.

[d] Requirements of ACI 201.2R, ACI 211.1, ACI 318, ACI 345R and ASTM C94.

[e] A tolerance of ±1.5% air content is recommended or specified.

[f] Requirements of only ACI 301.

effectiveness of air entrainment in ensuring frost resistance. Powers (1949b) first proposed the significance of the average inter-bubble distance, or 'spacing factor', on the basis of his saturated flow hydraulic pressure hypothesis to explain the empirical observation that air entrainment of concrete improved the resistance to freezing and thawing. The spacing factor is defined in ASTM C457 (2010) as being an index related to the maximum distance of any point within the cement paste from the periphery of an air void, which is approximately half the average air-void spacing. A maximum spacing factor of 0.2 mm (0.008 in.) is often specified (Neville and Brooks 1987, 2010; Hover 1993d).

The Powers spacing factor has been frequently criticised but continues to be accepted as a useful parameter in assessing the likely effectiveness of an air-void system. The Powers spacing factor is calculated from the total air-void content, but Walker (1980) proposed the differentiation of entrapped and entrained voids and the calculation of a spacing factor for the small entrained air voids only. Philleo (1983) suggested a complicated alternative form of spacing factor, in an attempt to overcome variations in the air-void system within a concrete, which he suggested might be more important in determining performance. Chatterji (1984) has observed that while the Powers theory was based upon a model in which all the air voids are equal-sized spheres arranged in a simple cubic lattice, this is 'seldom satisfied in practice'. However, Chatterji also refuted the saturated flow hydraulic pressure hypothesis and challenged the 'physical and scientific relevance' of the spacing factor.

Routine monitoring of air content during construction is usually carried out on samples of fresh concrete, commonly using a pressure meter of the type described in BS EN 12350-7 (2009) or ASTM C231 (2010). However, such a method is incapable of distinguishing between entrapped and entrained air and obviously gives no information on either the size or spatial distributions of air voids in a concrete after placing and finishing. Moreover, it is quite likely that the air content of concrete will be significantly, perhaps critically, modified between sampling from the mixer or delivery vehicle and its final placement and consolidation (Sandberg and Collis 1982; Hover 1993c). For these reasons, the assessment of air content for suitable samples taken from the hardened concrete is essential to quantify the air-void system.

The addition of late-stage water (retempering) and reworking of a mix (probably more common in hotter climates) are known both to increase the total entrained air content and also to create enhanced clustering of air voids at aggregate boundaries (Kozikowski et al. 2005; Sibbick 2011). In extreme cases, this void clustering creates weaker bonds between the aggregate and cement paste, leading to reduced concrete strength and durability potential. In the example shown in Figure 4.71, the relative degrees of clustering were rated from 0 (none) to 3 (severe), using the system developed by Kozikowski et al. (2005).

4.5.2 Quantification of air-void content in hardened concrete

A number of methods have been devised for evaluating the air-void content of hardened concrete, but the direct microscopical measurement techniques, such as those described in ASTM C457 (2010) and BS EN 480-11 (2005), are the most precise and, given appropriate samples, arguably provide the most useful information. French (1991a) considered the ASTM C457-type methods to be 'far more reliable' than the other options. In the following account, these other options are briefly considered first, then the microscopical techniques are covered more fully in Section 4.5.3.

4.5.2.1 Visual assessment of excess voidage

Air voids are clearly visible in the sides of a concrete core sample, and any routine description should include reference to the apparent abundance and size of such voidage, as an indication

Rating 0 – No clustering   Rating 1 – Minor clustering

Rating 2 – Moderate clustering   Rating 3 – Severe clustering

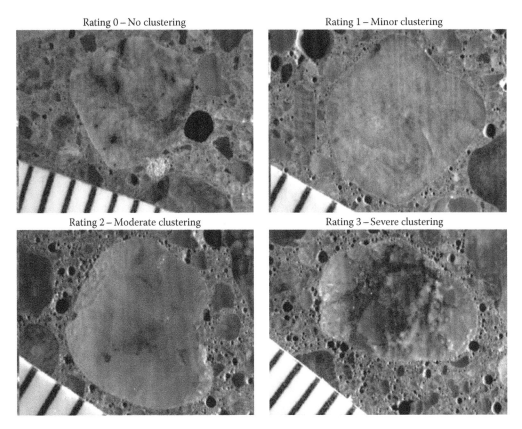

Figure 4.71 Clustering of entrained air voids at aggregate boundaries: showing examples of the relative severity rating (0–3) developed by Kozikowski et al. (2005). Scale shown is in millimetres. (Photographs by courtesy of Dr Ted Sibbick.)

of the success or otherwise of the compaction. In air-entrained concretes, the presence of microscopic bubbles in the concrete matrix might also be apparent when they are relatively abundant.

The Concrete Society Working Party convened by Dewar (1976/1987, Technical Report No. 11) developed standardised comparative techniques for evaluating the visible voids in concrete core samples, based on the notion that total air-void contents (for non-air-entrained concretes) greater than 0.5% were excessive. This 'excess voidage' is determined either by visual means or from the density test results. In the visual assessment, standard areas of core side surface (125 × 80 mm) are subjectively compared with actual-size photographs provided for five different levels of excess voidage (in the range 0%–13%). These are reproduced in Figure 4.72 (not at actual size). This approach has now been introduced into the U.K. National Annex of BS EN 12504-1 (2009), which is now referred to in BS 6089 (2010), but some issues with using core samples are discussed later (see Section 4.5.2.3.2).

In the density method, the actual density of the concrete, measured from core samples, is compared with a higher target density, with the difference being taken to indicate void content. In the Dewar (1976/1987) procedure, the target 'potential density' represents that of a properly compacted and cured test cube at 28 days when measured by water displacement; the 'actual density' is based upon water-soaked cores. A method previously favoured by the U.K. Department of Transport (1991) for pavement works was instead based upon a theoretical maximum *dry* density (TMDD) as the target, calculated from the mix proportions

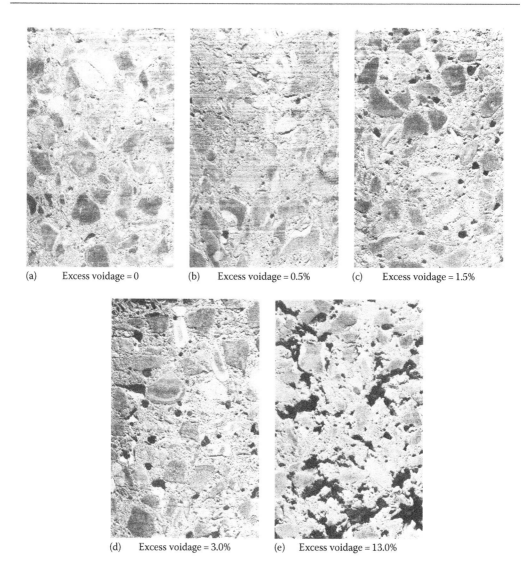

(a)   Excess voidage = 0   (b)   Excess voidage = 0.5%   (c)   Excess voidage = 1.5%

(d)   Excess voidage = 3.0%   (e)   Excess voidage = 13.0%

Figure 4.72 Photographs used for estimating 'excess voidage': (a) 0%, (b) 0.5%, (c) 1.5%, (d) 3% and (e) 13%. (From Dewar, J.D. [Convenor]. 1976 [1987]: *Concrete Core Testing for Strength – Report of a Concrete Society Working Party,* Concrete Society Technical Report No. 11, including Addendum [1987], Reprinted with addendum 1987, The Concrete Society, London, U.K.; BS EN 12504-1, *Testing Concrete in Structures, Part 1: Cored Specimens – Taking, Examining and Testing in Compression,* British Standards Institution, London, U.K., Includes National Annex NA, Guidance on the use of BS EN 12504-1., 2009.)

and constituent relative densities, and determining the dry density of cores. A worked example of this approach is given in Table 4.25.

Although the density difference methods might appear to produce precise and objective results, this is illusory even for a representative set of core samples, because of the likely errors in establishing the target density. Apart from the relative density values to be chosen for each of the various constituents, it is necessary to assume that the specified mix design has been precisely complied with throughout the concrete being evaluated. Figg (1989) suggested that the overall error may be 'unreasonably large' for non-trial concrete wherein there is uncertainty over composition.

Table 4.25 Void content estimation from density measurements

The formula given for the TMDD in the DoT (1991) is

$$TMDD = \frac{\left[(F \times W_1) + W_3 + W_4\right] \times 1000}{\dfrac{W_1}{P_1} + \dfrac{W_4}{P_4} + \dfrac{W_3}{P_3} + W_2}$$

where

F is the time factor for hydration of cement (given in the DoT [1991] table)
W_1 is the mass of cement, kg
W_2 is the mass of total water (in aggregate + additions), kg
W_3 is the mass of oven-dry fine aggregate, kg
W_4 is the mass of oven-dry coarse aggregate, kg
P_1 is the relative density of cement
P_3 is the apparent relative density of fine aggregate
P_4 is the apparent relative density of coarse aggregate

Using data obtained for the concrete in question, the TMDD was calculated as follows:

$$TMDD = \frac{\left[(1.19 \times 350) + 702 + 1162\right] \times 1000}{\dfrac{350}{3.12} + \dfrac{1162}{2.52} + \dfrac{702}{2.56} + 150.5} \qquad I$$

$$= \frac{\left[416.5 + 702 + 1162\right] \times 1000}{112.2 + 461.1 + 274.2 + 150.5} \qquad II$$

$$= \frac{2280.5 \times 1000}{998} \qquad III$$

$$= 2285 \text{ kg/m}^3$$

This TMDD is then compared against the measured dry density for samples of hardened concrete and the difference taken to be the air-void content, as follows:
Calculated TMDD = 2285 kg/m³
Less dry density measured from samples = 2184 kg/m³
Difference in kg/m³ = 101 kg/m³
Difference as % air voids = 4.4%

Note: Worked example of the method formerly given in the U.K. Department of Transport (1991).

4.5.2.2 Other methods for assessing air-void content

Erlin (1962) proposed a high-pressure method for the determination of total air content in hardened concrete, but despite being a direct method, this does not seem to have been adopted in practice. The procedure is particularly sensitive to changes in aggregate porosity and, in common with the pressure meter for fresh concrete, cannot differentiate between entrapped and entrained air.

The nature and objectives of an air-void system analysis of concrete have long seemed to be ideally suited for computerised image analysis. In practice, it has proved difficult to devise an entirely satisfactory specimen preparation technique. If a slice of concrete can be treated to create a sufficiently clear and invariable colour contrast between the air voids and all of the surrounding material, it is possible for a suitable image to be fully processed by a computer, effortlessly yielding all of the required parameters. Procedures have been described by Chatterji and Gudmundsson (1977) and MacInnis and Racic (1986): both consist of blackening the concrete and infilling the voids with bright white material. If satisfactory specimen preparation can be achieved, it is claimed that results can be obtained by image analysis which compare favourably with those found using the ASTM C457 microscopic method.

Jana (2007) has reported the findings of a 'round-robin' comparison of various image analysis methods with ASTM C457.

True et al. (2010) have recently developed new and practicable techniques for the assessment of the void content of concrete and concrete cores, employing a specially adapted desktop flat-bed scanner and image analysis (see also Section 4.3.4 for the earlier application of similar techniques in the characterisation of aggregate shape). A brief summary may be found in True and Searle (2012), and the following account has been prepared with the assistance of Graham True.

4.5.2.3 Use of desktop flat-bed scanners

4.5.2.3.1 Air-void content on flat concrete samples

Air-void content, as well as air bubble size and spacing, is currently tested to ASTM C457 and BS EN 480-11 (2005), both of which utilise microscopical determinations. The ASTM standard (also see Section 4.5.3) includes two methods, the linear-traverse (Rosiwal) method and the point-count method, both based on measurement of flat, sawn, ground and polished concrete samples. BS EN 480-11:2005 also requires samples, circa 100 mm by 150 mm by 20 mm thick, to be ground and polished, and voidage characteristics are determined from the frequency of intersections of air voids as well as air-void individual chord lengths.

Both these procedures may take many hours to perform, require expertise to carry out and are operator sensitive in the results obtained. However, once samples are sawn, ground and polished, they may then be scanned using a desktop scanner to obtain an image that can be analysed by a computer. The Michigan Technological University at Houghton has undertaken and published details of extensive studies (Peterson et al. 2001, 2009a,b; Carlson et al. 2006) including some undertaken for the U.S. Federal Highway Administration. These studies concluded the following:

- A flat-bed scanner is clearly a viable device for performing ASTM C457 (and therefore presumably also BS EN 480-11).
- In general, there is a good agreement between the scanner and other automated methods of point counting.
- The degree of correlation with manual results depends on the method used to establish thresholds for segmenting the greyscale image.
- Sample preparation is the key factor in obtaining quality results.

A procedure and software package for characterising air voids using a flat-bed scanner for performing ASTM C457 may be obtained from Michigan Technological University[1] (Peterson 2008; Peterson et al. 2009a,b). Sibbick (2012) has conducted some comparative trials between this flat-bed scanner option and the commercially available RapidAir method (see also Section 4.5.3), finding that while the findings were broadly similar (relatively higher and lower void contents were similarly identified), the flat-bed scanner in his trials was less sensitive to the extremely fine entrained voids.

4.5.2.3.2 Air-void content from concrete core samples

The aforementioned Concrete Society Technical Report No. 11 'excess voidage' method (Dewar 1976/1987 & BS EN 12504-1 (2009), U.K. NA) for determining the extent of

---

[1] At the time of writing, contact Jerry Anzalone at MTU at gcanzalo@mtu.edu. Link to current Adobe Photoshop script version at http://wiki.mtu.edu/index.php/BubbleCounter:Index.

entrapped air voids in concrete core samples has recently been reintroduced into BS 6089 (2010). It is stated in BS 6089 (2010) that in order to assess a structure correctly, an engineer should specify the reporting of 'excess voidage' and other properties. Additionally, before any test programme is commenced, BS 6089 (2010) suggests that there should be complete agreement between the interested parties on the validity of the proposed testing procedure and the criteria for acceptance. However, according to True (2011a,b), agreeing a procedure based on the 'excess voidage' may prove difficult for the following reasons:

- A single linear average relationship is assumed between 'excess voidage' and the reduction that has on in situ cube strength, that is, 6% strength reduction for each 1% of excess voidage up to 5% excess voidage. BS 6089 (2010) makes no reference to the source of data used to produce that average (i.e. Warren [1973]).
- The visual assessment of 'excess voidage', as stated in BS 6089 (2010), is subjective if it is done by comparison with reference samples and not by measurement, although the uncertainty can be reduced by having more than one determination.
- Visual assessment is currently performed in accordance with section NA.4.2 of the National Annex to BS EN 12504-1 (2009). This method requires test cores to be compared with reference photographs. The photographs are a flat image of a curved surface and therefore not a true representation of the voidage. Digital imaging scans of these photographs have suggested that they may not portray the percentage voidage they claim (True 2011a).
- BS EN 12504-1 (2009), NA.4.2, requires all cores under examination to be assessed by placing a cut-out card frame, 125 mm high × 80 mm wide (the size of the reference photographs given in the Standard: see Figure 4.46), over the core to assist the comparison. This negates the choice of using the preferred 100 mm diameter cores mentioned in BS 6089 (2010) to be used whenever possible to minimise corrections to core compressive strength.
- An alternative assessment of voidage based on comparison with the intended mix design density may not be viable since the mix density of the concrete in question may have undergone changes due to aggregate variations, cement content changes and especially water addition variations.

An objective assessment of voidage is required and several researchers have published the results of studies based on analysis of digital images obtained from prepared flat concrete samples (Peterson et al. 2001, 2009a,b; Peterson 2008; Carlson et al. 2006). According to True et al. (2010) and True (2011), similar image analyses may be carried out on digital images obtained from the curved surface of concrete core samples such as those prepared for compressive strength testing.

Figure 4.73 shows a desktop flat-bed scanner modified to allow a concrete core sample to be traversed over the glass platen to capture an image of the curved face. Connection is made to the scanner cross head within the case to drive an adjustable trolley over the flat glass platen that rolls cores over the traversing CCD. Adjustment of the trolley is required to suit the core diameter. Note the core has elastic bands attached to each end to act as tyres holding the core circa 1 mm above the glass platen.

A procedure has been developed to classify total voidage as well as voidage size distribution and aggregate shape characteristics (True et al. 2010). It was found necessary to prepare core samples to assist conversion into suitable binary images. A similar technique is used as suggested in BS EN 480-11 (2005), but in this case, blackening of the surface is achieved using felt tip markers and the voids whitened using hand zinc-based baby cream or hand cream. The resultant clear plain core image shown in Figure 4.74 was obtained after

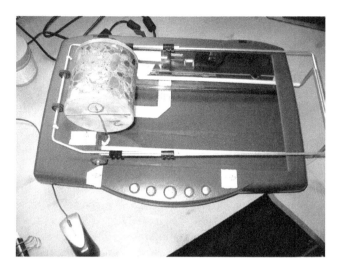

Figure 4.73 A typical flat-bed desktop scanner has been modified to drive a trolley over the glass face platen of the scanner and locate the lower part of the core over the traversing cross head so as to obtain a true, plain and undistorted image of the curved surface of the core.

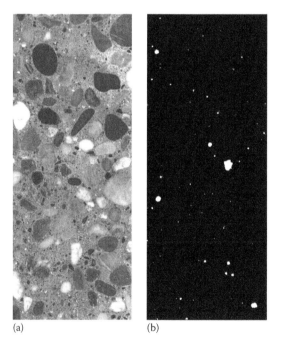

(a)   (b)

Figure 4.74 The left-hand image is as obtained (after only surface application of a white hand cream), and the right-hand image is after blackening solid areas and whitening voids.

applying a thin white hand cream over the surface of the core that converts any lime deposits to a transparent film, allowing a clear image to be obtained.

The black-and-white image on the right-hand side in Figure 4.74 has been analysed using ImageJ, a free download from the U.S. National Institutes of Health. Once the image size has been input and scaled into the software, the voidage may be determined relative to

Table 4.26 Voidage results from computer analysis of the right-hand black-and-white image in Figure 4.74

Particle size range (mm²)	Particle count	Particle area (mm²)	Average size (mm²)	Area fraction	Perim.	Circ.	Solidity
0 > max	554	104.62	0.19	0.7	0.84	0.79	0.86
0.1 > max	77	100.77	1.31	0.6	4.29	0.51	0.82
0.2 > max	50	96.68	1.93	0.6	5.51	0.51	0.84
0.3 > max	37	93.37	2.52	0.6	6.52	0.50	0.85
0.4 > max	29	90.51	3.12	0.6	7.51	0.49	0.86
0.5 > max	28	90.08	3.22	0.6	7.66	0.49	0.86
0.6 > max	24	87.93	3.66	0.5	8.28	0.48	0.86
0.7 > max	22	86.6	3.94	0.5	8.62	0.49	0.87
0.8 > max	21	85.86	4.09	0.5	8.85	0.48	0.87
0.9 > max	19	84.1	4.43	0.5	9.29	0.48	0.87
1.0 > max	16	81.27	5.08	0.5	10.09	0.48	0.87

particle size ranges as shown in Table 4.26. When a full range of data is considered, the number of voids noted is 554, provided that an area fraction or percentage voidage is 0.7%. Most of these are electrical noise, masquerading as tiny voids.

However, in the void size range (0.1–0.5 mm²) up to the maximum void size noted, the voidage was noted as 0.6% of the surface area. By considering voids in the size range 0.1–0.5 mm², errors due to inclusion of both electrical noise and inadequate resolution for the small size range being considered are removed. The optical sampling rate of the scanner used was set at 1200 dpi (not to be confused with resolution), this being adequate for void sizes circa 0.2 mm², which equates to a circular void 0.5 mm diameter.

When sampling for 'excess voidage', it has to be agreed what constitutes 'excess voidage', both in void size and amount. Currently, there is little agreement as to a precise definition. One possible approach is to consider all voidage over and above what may be taken as optimum entrained air voidage, as illustrated by CIRIA C559 (2001) and Pigeon and Pleau (1994). In those documents, a threshold appears between a maximum for entrained and a minimum for entrapped air voidage of void diameter 0.3 mm. With this assumption, the entrapped air voidage of the surface shown in Figure 4.74 and documented in Table 4.26, the entrapped air voidage appears to be 0.6%.

An estimated, voidage-free in situ compressive strength can be obtained by plotting the compressive strength results against void contents for a series of core samples taken from the same concrete batch in question, and from a best-fit trend line, the strength at zero voidage may be assessed.

Digital image scanning of the curved surfaces of core samples provides a significantly larger sample size or area than otherwise used for evaluation. Other options become available once a true image of the curved surface has been obtained, such as can be seen in Figure 4.75. It has been known for some time, as denoted in Concrete Society Technical Report No. 11 (Dewar 1976/1987), that aggregate shape can influence both the amount of entrapped air voidage and density, owing to a reduction in the packing of the aggregate.

Digital imaging using basic desktop flat-bed scanners now offers the ability to record and study the outer curved surfaces of concrete core samples, including estimates of voidage and aggregate shape characteristics; also, at higher resolution and after suitable sample preparation, it is possible to analyse entrained air characteristics.

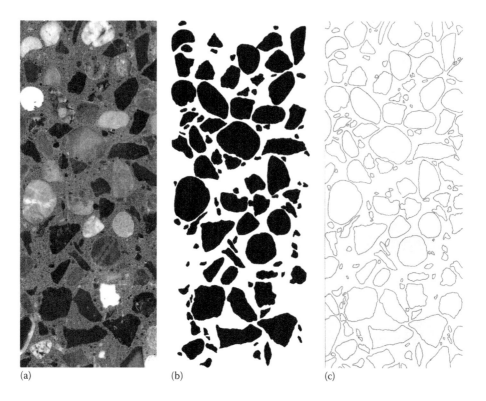

(a)   (b)   (c)

Figure 4.75 The left-hand scanned image of a concrete core surface (a) has been converted into the right-hand tracing of the aggregate perimeter profiles (c), and then all aggregate particles touching the perimeter of the frame have been removed and the remaining particles filled in the central representation (b). These types of images can also be used to classify aggregate morphological characteristics (i.e. roundness/sphericity/aspect ratio) for use in denoting the effect such aggregate shape changes may have on compressive strength.

4.5.3 Microscopical measurement of the air-void system

The procedures detailed in ASTM C457 (current version, 2011) are adopted internationally and form the basis of the following discussion, after which the more recent and broadly similar BS EN 480-11 (2005) will be addressed.

Specimens for ASTM C457 are prepared from the concrete sample by cutting a slice using a diamond saw and then progressively grinding one face of the slice to a smooth and flat surface. The finely ground specimen surface is then scanned under a good-quality optical travelling microscope, using one of the two techniques described: linear traverse or modified point count (see Figure 4.76). The observations collected during these scans are then used to calculate various parameters of the air-void system, including total air content and spacing factor.

Adequate preparation of the ground surface is critical, especially in the assessment of fine air systems (Sibbick 2011), where the saw abrasion and initial polish damage can also be mismeasured as finer bubbles. The precise details for grinding with progressively finer abrasives are given in ASTM C457, but the finished product must show 'excellent reflection' of a remote light source at low angle, be free of scratches and maintain sharp edges around the air voids. Consolidation might be necessary for weaker concretes, to avoid the plucking out of aggregate particles (creating false voids) and/or the crumbling of void edges, but this must not be of a sort that will infill the voids: ASTM C457 describes a molten wax technique. Figure 4.77 illustrates suitable surfaces when viewed through the microscope.

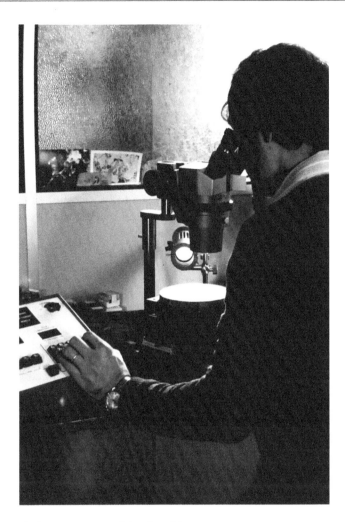

Figure 4.76 Microscopical measurement of the air-void system.

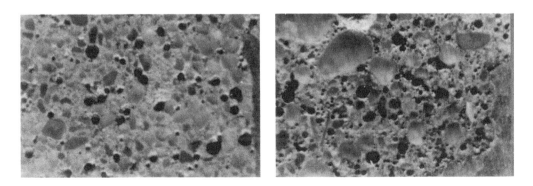

Figure 4.77 Two examples of suitable concrete surface preparation for air-void point counting to ASTM C457.

Both scanning methods envisage a series of parallel traverse lines across the specimen surface. In the linear-traverse option, the distance travelled across each component encountered as these lines are traversed is measured and cumulatively totalled; also the number of air voids intersected by the traverse line is recorded. The data collected are total length traversed (T_t), total length traversed across cement paste (T_p), total length traversed across air voids (T_a) and total number of air voids intersected by the traverse line (N). Aggregates are not measured as such but obviously could be obtained by deducting T_p plus T_a from T_t.

In the 'modified' point-count option, the traverses proceed step by step and the identity of the component under each stop or point reached is identified and added to the tally for that component. Again, the total number of air voids intersected by the traverse line is also recorded, and this represents the modification from conventional point counting. The linear distance between stops along the traverse (I) has to be determined (it is required to be between 0.64 and 5.1 mm). The data collected are total number of stops or points (S_t), total number of points over cement paste (S_p), total number of points over air voids (S_a) and total number of air voids intersected by the traverse line (N). The separate counting of other components, such as aggregates, or their differentiation, for example, between coarse and fine aggregates, can be achieved by adding to the range of tally counters.

Research has suggested that the two methods produce generally comparable results, but the modified point-counting system is preferred by many laboratories. Computer processors have been successfully linked to the counting devices, thus automating the various calculations given in ASTM C457 (2011) to establish the air-void system parameters (see the end of this subsection, including Figure 4.78).

The stipulated minimum area of specimen surface varies with the maximum aggregate size and the mode of calculation. For example, for a nominal 20 mm (19.0 mm) aggregate and employing the standard method of calculation, a minimum area of 7100 mm^2 is cited; the area of a diametral slice from a 150 mm diameter concrete core is thus more than sufficient. In the case of concretes containing an aggregate with particles in excess of 50 mm in size, when minimum specimen sizes can get impracticable, an alternative calculation procedure based upon the paste–air ratio is available, provided that the volumetric aggregate/paste ratio is known or has been determined. Table 4.27 gives the minimum lengths of traverse and minimum number of points specified in ASTM C457, which again vary according to maximum aggregate size. In the authors' experience, larger sample areas than those stipulated in ASTM C457 need to be examined when the concrete is unevenly mixed.

Formulae are provided in ASTM C457 for calculating air content (A), void frequency (n), paste–air ratio (p/A), specific surface (α) and spacing factor $(\bar{L})$. It has already been mentioned in Section 4.5.1 that Walker (1980) suggested an alternative formula for spacing factor, relating only to the smaller voids. Attiogbe (1993) has devised another formula for calculating 'mean spacing', which he suggests yields a better estimate of the actual spacing of the air voids in concrete than the standard spacing factor.

The statistical reliability of the results of the ASTM C457 test depends on many factors. Hover (1993c) contrasted the 1000 or so air voids evaluated in an average test, with the 10–15 billion air voids in a cubic yard of air-entrained concrete, and concluded, 'uncertainty or error is expected in such estimates'. The procedure may be expected to be subject to the reasonably predictable statistical errors associated with any counting technique but in addition will be affected inter alia by the quality of the specimen prepared, the apparatus, the magnification selected and the performance of the operator. A major inter-laboratory trial, involving twenty laboratories in ten countries, was reported by Sommer (1979), and the precision findings are summarised in Table 4.28.

The precision found in Sommer's trial was poor. Overall, the concrete prepared with 3.0% air-void content produced 95% confidence limits of ±1.5% for total air-void content,

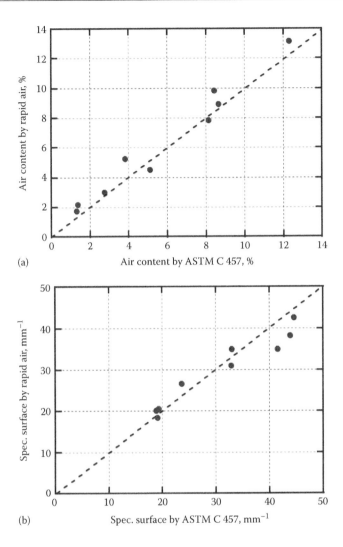

Figure 4.78 Relationship between air contents (a) and specific surface values (b) determined using the ASTM C457 method and the RapidAir image analysis system. (From Pade, C. et al., A new automatic analysis system for analyzing the air void system in hardened concrete, in *International Cement Microscopy Association Conference*, April 2002, San Diego, CA, 2002.)

when the specimens were prepared and measured by the participating (rather than orga-nising) laboratory. The average spacing factor was 170 μm, with 95% confidence limits of ±50 μm. Figg (1989) considered these values could be improved upon and recommended scanning the maximum possible surface area, using at least the minimum number of points recommended in ASTM C457, carrying out specimen preparation to the highest possible standards and using only a 50× magnification.

In ASTM C457, it is specifically stated that 'no provision is made for distinguishing among entrapped air-voids, entrained air-voids and water voids' and that 'any such distinction is arbitrary'. Nevertheless, the technique involves direct visual observation and differences between void size and morphology are clearly distinguishable. It is therefore common, and here recommended, for entrapped and entrained air voids to be separately counted, so that

Table 4.27 Minimum lengths of traverse and number of points for the ASTM C457 'modified' point-count option

Nominal or observed maximum size of aggregate in the concrete, mm	Minimum length of traverse for the determination of A, α or $\bar{L}$, mm	Minimum number of points for the determination of A, α or $\bar{L}$
150	4064	2400
75	3048	1800
37.5	2540	1500
25	2413	1425
19	2286	1350
12.5	2032	1200
9.5	1905	1125
4.75	1397	1000

Source: After ASTM C457, *Standard Test Method for Microscopical Determination of Parameters of the Air Void System in Hardened Concrete*, American Society for Testing and Materials, West Conshohocken, PA, 2011.

Note: According to ASTM C457, these limits should produce a standard deviation not greater than 0.5% for a 3% air content (a coefficient of variation of 17%); the coefficient of variation is reduced for air contents >3% and number of counts >1000.

Table 4.28 Precision of determinations of air-void contents and spacing factors

Concrete	I		II	
	Prepared and measured in Vienna	Prepared and measured by participant	Prepared and measured in Vienna	Prepared and measured by participant
Number of results	20	18	20	18
Average	2.92	2.73	6.48	5.87
Standard deviation, s	0.34	0.75	0.61	0.89
Lowest	2.52	1.77	5.44	4.39
Highest	3.58	3.90	7.71	7.74
Range	1.06	2.13	2.27	3.35
Precision, 2.8s	r = 0.95	R = 2.10	r = 1.71	R = 2.49
95% confidence limits (±1.96s)	—	±1.47	—	±1.74

Spacing factor, $\bar{L}$ (μm)

Concrete	I		II	
	Prepared and measured in Vienna	Prepared and measured by a participant[a]	Prepared and measured in Vienna	Prepared and measured by a participant[a]
Number of results	20	15	20	15
Average	168	172	87	98
Standard deviation, s	12	26	8	18
Lowest	148	136	75	66
Highest	192	215	99	132
Range	44	79	24	132
Precision, 2.8s	r = 34	R = 73	r = 22	R = 50
95% confidence limits (±1.96s)	—	±51	—	±35

Source: After Sommer, H., *Cem. Concr. Aggr. CCAGDP*, 1(2), 49, 1979 and Figg, J.W., *Analysis of Hardened Concrete – A Guide to Tests, Procedures and Interpretation of Results – Report of a Joint Working Party of the Concrete Society and Society of Chemical Industry*, Technical Report No. 32, The Concrete Society, London, U.K., 1989.

[a] All results from two laboratories excluded as 'outliers'.

the approximate relative proportions of those forms of air void, together comprising the total air content (A), can be calculated. Convenient identifying criteria may be derived from ASTM C125 (2010), wherein the following definitions are given: 'an entrapped air-void is characteristically 1 mm or more in width and irregular in shape; an entrained air-void is typically between 10 and 1000 µm in diameter and spherical or nearly so'. Some example ASTM C457 results, including this differentiation of air-void types, are given in Table 4.29.

In considering the interpretation of ASTM C457 results, it is important to recognise the objectives of the testing in question, particularly whether the issue is in compliance with a specified requirement or performance of a given concrete surface or conceivably both. Most, but not all, specification limits refer to pressure meter testing of fresh concrete prior to placement, so that the relevance in that respect of air contents determined later from samples of hardened concrete using a quite different technique is obviously questionable, particularly in marginal cases. It has already been explained in Section 4.5.1 that significant losses of air content can occur between original mixing and final placement.

Orientation of the specimen surface is a vital consideration. The standard requirement of ASTM C457 is for the slice to be sawn 'approximately perpendicular to the layers in which the concrete was placed', and this will clearly produce an overall average, which is best suited to any investigation which is aimed at checking the likely compliance with the original air content requirements. However, ASTM C457 is frequently used in practice for investigating failures, when the *overall* air content might be of limited relevance.

Following their experiences with some U.K. airfield pavements where surfaces had spalled in winter conditions despite compliant site test results for air content and satisfactory overall

Table 4.29 Microscopical determination of air-void content and parameters of the air-void system with sample data, including differentiation of entrapped and entrained air

Depth from surface, mm	20	40	60	80	100	Total/overall
Data determined – ASTM C457						
No. of points – aggregates	325	331	307	301	316	1580
Cement paste	135	123	124	149	126	657
Entrapped air	8	7	18	18	24	75
Entrained air	4	9	16	6	9	44
Total voids	12	16	34	24	33	119
Total points	472	470	465	474	475	2356
No. of air voids intersected	15	42	35	24	28	144
Calculations – ASTM C457						
Voids traversed, N	27	58	69	48	61	263
Traverse length l, mm	1.3	1.3	1.3	1.3	1.3	1.3
Total traverse T_t, mm	597	594	588	599	600	2978
Entrapped air content, %	**1.7**	**1.5**	**3.9**	**3.8**	**5.0**	**3.2**
Entrained air content, %	**0.8**	**1.9**	**3.4**	**1.3**	**1.9**	**1.9**
Total air content A, %	**2.5**	**3.4**	**7.3**	**5.1**	**6.9**	**5.1**
Void frequency, n	0.05	0.10	0.12	0.08	0.10	0.09
Paste content p, %	28.6	26.2	26.7	31.4	26.5	27.9
Paste–air ratio, p/A	11.2	7.7	36	6.2	3.8	5.5
Av. chord length, l	0.56	0.35	0.62	0.63	0.68	0.57
Specific surface, α	7.1	11.5	6.4	6.3	5.8	7.0
Spacing factor $\bar{L}$, mm	**0.94**	**0.49**	**0.57**	**0.81**	**0.65**	**0.69**

Note: The total air content is 5% but the entrained air content is only 2%.

findings in the ASTM C457 tests, Sandberg and Collis (1982) observed that 'the obvious and paramount requirement is that the entrained air is actually in that part of the hardened concrete which has to resist frost action'. They proposed that 'contractual acceptance should, ideally, be based on the entrapped and entrained air contents in the as-placed and hardened concrete'. This advice has been followed for some contracts, when periodic coring and application of ASTM C457 has been part of the specification requirements.

In such cases, when the air content and the parameters of the air-void system in the near-surface zone of concrete are under investigation, it will be more appropriate to prepare slice specimens which are orientated *parallel* to the exposed surface (i.e. parallel to the placement layering in a pavement concrete). Often, it will be appropriate to prepare several slices, providing specimen surfaces at various depths beneath the exposed concrete face. It is suggested that the 20 mm thick zone of concrete beneath the exposed surface is the most critical for freeze–thaw resistance, so that separate air-void measurements at levels of 5, 15 and 25 mm beneath the surface would provide suitable information about the adequacy of frost protection. Although this might be achieved, for indicative purposes, by separately counting traverses orientated parallel to the exposed surface on a slice sawn perpendicular to that surface, the limited number of counts at each depth level is unlikely to meet the statistical minimum requirement for reliability.

Of course, in many investigations, it might be appropriate to prepare specimens variously perpendicular and parallel to the exposed concrete surface, enabling thorough studies of both likely compliance and actual performance potential to be completed.

Sources of apparent contradiction between air content findings and actual performance include the type of air voids and the spacing factor. The pressure meter and density difference test methods only measure total air content, rather than entrained air, and the ASTM C457 procedure also determines only total air-void content in its unmodified form. Yet frost protection is only afforded by entrained air, so that an apparently suitably high total air content result that, in reality, comprises mainly entrapped air would provide misleading reassurance. It is primarily for this reason that the differentiation into entrapped and entrained air voids in the ASTM C457 test is being recommended, although the arbitrary and approximate nature of such a subdivision must not be overlooked.

As well as the presence of adequate entrained air voids, the spacing factor has been generally recognised as having a major influence on frost resistance (see Section 4.5.1), although it is less frequently included in specifications. Examples might therefore be encountered where freeze–thaw damage has occurred with concrete exhibiting an apparently adequate air content, even entrained air content, but an excessive spacing factor. A threshold spacing factor of 0.20 mm is commonly cited (Hover 1993d), below which concrete might be expected to exhibit satisfactory durability. Although this is lower than the 0.25 mm value originally envisaged by Powers (1949b), some research has indicated that even lower values might be required to ensure protection where de-icing chemicals are used (Klieger 1980).

BS EN 480-11 (2005) is similar in principle to ASTM C457 in linear-traverse mode. As the various traverse lines are scanned under the microscope at a magnification of 100× ± 10×, the linear values of solid portions (T_s) and air voids intercepted (T_a) are logged and the latter classified according to 28 standard size ranges. Total traverse length is the sum of the total lengths recorded for solid portions and air voids ($T_s + T_a = T_{tot}$). The data collected are subsequently used to calculate parameters including 'total air content' (A), 'specific surface of the air' (α) and 'spacing factor' ($\bar{L}$). There is an option for the separate recording of air voids that are 300 μm or less, so that the 'micro-air content' (A_{300}) can be calculated in addition to total air content. Because the T_a values are classified according to void size, it is also possible to assess the air void distribution, for which Annex B of BS EN 480-11 provides a helpful worked example.

Sibbick (2011) advises that computerised methods for measuring chord length, such as the RapidAir and equivalent systems, are essentially able to undertake linear-traverse air-void analysis in accordance with the ASTM C457 principles and produce completely reproducible results, whereas the manual linear-traverse method has been found to be extremely operator dependent, especially in terms of modal chord length distribution, even if void percentage and other parameters are similar.

In the RapidAir system, the surface of a ground slice specimen of concrete is coloured black using an ink pad, then white zinc paste is applied after the specimen has been heated to 55°C, so that the white paste melts and flows into the air voids. After cooling and removal of excess paste, the surface is inspected under a stereomicroscope and any non-air-void white paste (such as voids in aggregates and obvious cracks) coloured black using a marking pen. Analysis of the prepared surface is then carried out automatically by the RapidAir control unit. The pre-analysis surface preparation is both much more involved than that required for the conventional counting methods and obviously critical, but according to the Danish suppliers of RapidAir (Germann Instruments A/S 2011), the saving in time for analysis (claimed less than 17 min compared with 4–6 h) greatly exceeds the time required for pre-analysis preparation (claimed about 30 min). On the basis of a study involving 13 European laboratories, Pade et al. (2002) reported a good relationship between the air contents determined using conventional ASTM 457 measurement and the RapidAir system (see Figure 4.78). Jakobsen et al. (2006) have also reported on a 'round-robin' study of automated air-void analysis options. Carlson et al. (2006) found good correlation between the RapidAir system and the flat-bed scanning technique, but Sibbick (2012) thought the flat-bed scanner was less sensitive to very small voids (see Section 4.5.2).

Aarre (1995) has compared the use of thin-section specimens with the more conventional finely ground specimens described earlier in the ASTM C457 method. She found that the 'apparent' air content measured from thin sections was systematically lower than both that determined from comparable ground surfaces and the true air content. Lu et al. (2006) have applied X-ray microtomography to study pore structure and permeability in concrete. Air entrainment in concrete is further discussed in Chapter 5.

4.5.4 Aerated and foamed concrete

As we have seen, air-entrained concrete contains frequent, small to microscopic, spherical air voids or 'bubbles' that are unconnected and collectively amount to a few percent of the volume of the material. However, other concrete products and materials will be encountered in which the 'entrained' air content is deliberately much higher, commonly representing more or much more than 50% by volume. 'Controlled low-strength material' (CLSM) is a sand-rich concrete made using a powerful air-entraining admixture (Dransfield 2003), which has up to about 25% entrained air, with densities down to about 1800 kg/m³. In order to obtain higher entrained air contents and lower densities, the concrete has to be 'aerated' or 'foamed'.

In 'foamed' or arguably more accurately 'foam' concrete, the high air content is achieved by adding a foam, produced from a surfactant admixture and water in a foam generator, to cement grout or a mortar mixture of cement and filler. Any filler present is commonly fine sand and/or pfa, but other materials include limestone dust and even, occasionally, polystyrene beads. A useful introduction to foamed or foam concrete is given by Cox and van Dijk (2002), who explain that pore volume, overall density and filler type influence strength as much as cement content. CLSM and foamed/foam concrete have a variety of applications in which compressive strength is not a primary requirement and their low-density, flowing, self-levelling and stability (lack of bleeding or settlement) characteristics are beneficial. Thus, backfilling utility trenches and filling subterranean voids (such as redundant tunnels and sewers) are typical applications, when the inherently low strength makes it relatively easy to remove and thus also suitable for temporary works. There are also examples of foamed/foam concrete being used as a sub-base for roads and car parks (Cox and van Dijk 2002; CROW 2003), and increasingly, the material has a thermal insulating function (Jones et al. 2005).

Under the petrographic microscope, these aerated or foamed/foam materials are characterised by the exceptionally high content of entrained air voids and the lack (or only minimal content) of coarse aggregate, plus usually also the absence of normal-sized sand; some varieties comprise only hydrated cement and abundant entrained air voids or bubbles, but typically, the matrix between the voids contains both cement and a filler (Figure 4.79). Jones et al. (2005) have considered various ways in which the properties of foamed/foam concrete

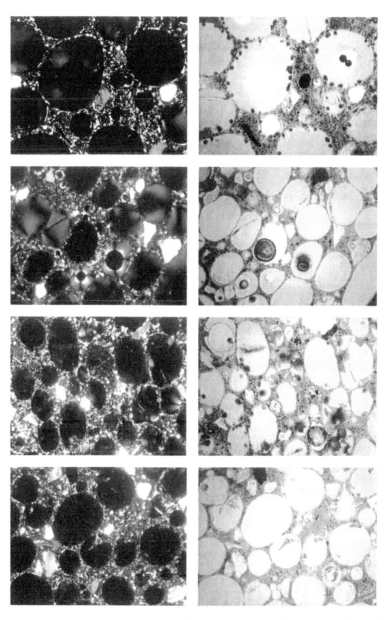

Figure 4.79 Some typical views of aerated or foamed/foam concrete in thin section: left-hand images are xpl and right-hand images are ppl. (Images copyright Building Research Establishment Ltd., Watford, U.K., reproduced with permission.)

could potentially be enhanced in the future, including the incorporation of recycled aggregate. Dransfield (2003) mentioned that heat of hydration can expand the air in foamed/foam concrete, 'tightening the mortar into the void so that no gaps remain'. However, the present authors were involved in investigating a major case of using large quantities of foamed/foam concrete for infilling an unstable system of historic mine workings in south-west England, when concern had been raised over heat evolution and the consequent risk of DEF (see Chapter 6). In that case, no evidence of DEF or its effects was identified. In another case, the authors investigated the 'failure' of foamed/foam concrete as a temporary filler, when injecting the concrete material into standing water had caused a catastrophic loss of the entrained air, leading to undesirable solidification into an exceptionally strong cement-rich mortar.

A familiar application for aerated concrete is in autoclaved precast concrete products, such as lightweight blocks (e.g. 'Thermalite' and similar). A comprehensive treatment of 'autoclaved aerated concrete' (AAC), based on the work of two RILEM technical committees, may be found in Aroni et al. (1993), including its definition: 'AAC is a lightweight cellular material which is formed by a chemical reaction between finely divided calcareous and siliceous materials. The calcareous component is usually lime and/or cement. The siliceous component usually consists of natural or ground sand and/or industrial by-products such as slag and pulverised-fuel ash (PFA), also known as fly ash. The cellular structure is achieved by either a chemical process causing aeration, or by introducing air voids by mechanical means into a slurry which contains no coarse material. Other processes may be used.' The present authors have encountered instability in blockwork walls made using AAC blocks of fly ash containing an excessive content of unburnt fuel. Further guidance on aerated blocks and other precast products is given in Chapter 7.

4.6 MINERAL ADDITIONS AND PIGMENTS

The use of mineral additions (or 'SCMs', as they are generally termed in North America, though not all additions are 'cementitious'), variously as cement replacement or filler materials, has already been discussed in Section 4.2.4 in the context of blended cements. In terms of identification and quantification, the mode of incorporation of the addition of SCM into the concrete, whether as an ingredient of the ex-factory cement or instead as an additional separate constituent, is immaterial. The following sections will concentrate on the methods for determining the presence, type (or types) and relative proportions of mineral additions or SCMs in hardened concrete.

4.6.1 Fly ash and pulverised-fuel ash

The terms *fly ash* and *pulverised-fuel ash* are commonly regarded as synonyms, but this is not strictly true. According to Cripwell (1992), 'fly ash is the generic term for all finely divided residues collected or precipitated from the exhaust gases of any industrial furnace', whereas pfa is 'that class of fly ash that is produced as a by-product specifically from the pulverisation of higher ranking coals'. This distinction can be important when considering the extensive international literature on 'fly ash' in concrete. Only 'Type F' fly ash material, as defined in ASTM C618 (2008), is used in the United Kingdom and would comply with the preceding definition for pfa, whereas in some parts of the world, 'Type C' fly ash, produced from low-ranking coals, is prevalent. A petrographer should thus take care in using the term 'pfa' unless the source and/or type of any fly ash present is known.

Books devoted to fly ash include those by Sear (2001) and Malhotra and Mehta (2005). Many reviews provide accounts of the properties and potentially beneficial influences of fly ash as a constituent within concrete (Owens 1980–1982, Berry and Malhotra 1987, Dewar and Anderson 1992, Owens 1999, Concrete Technology Unit, Dundee 2005, Sear and Owens 2006). One of the most recent and comprehensive reviews was compiled by the UK Concrete Society Working Party (Concrete Society 1991). There are rather fewer published studies of the *actual* performance of fly ash in structures. However, Cabrera and Woolley (1985) reported an appraisal of some power station foundations cast in 1957, in which most of the concretes contained a 20% cement replacement by pfa, but some sections lacked pfa but were otherwise similar.

They found that the pfa concretes had developed twice the strength of the plain concretes between 1 and 25 years and also that the porosity was much lower for the pfa concrete, which exhibited almost no carbonation. Thomas and Matthews (1993) investigated a number of structures and found similarly improved strength and reduced permeability for pfa concrete. They also identified significantly increased resistance to chloride penetration and alkali–silica reactivity (ASR).

Although fly ash is recognised to be a variably pozzolanic material, the precise mechanisms by which the material achieves its beneficial effects in concrete continue to be the subject of research. It is a matter of simple observation that, even in comparatively old concretes, much of the visible fly ash appears to be unreacted (French 1991a). When the concrete is young and still in its main strength gain phase, the fly ash behaves as an inert material, its principal effect being to provide nucleation centres for cement hydration (Montgomery et al. 1981; Fraay et al. 1989). At later periods, the fly ash participates in pozzolanic reactions, which account for the higher long-term strengths and improved impermeability. However, these reactions are complex and pozzolanicity appears to depend on a range of factors, including mineralogical composition and particle size range (Halse et al. 1984; Jun-yuan et al. 1984; Mehta 1985).

Modern concretes frequently contain 10%–35% fly ash by weight of the binder, although lesser and greater proportions may be encountered. Increasingly, blends of different additions, such as fly ash with silica fume (Ozyildirim and Halstead 1994), are being used. One recent major bridge and tunnel structure in Europe, crossing Storebælt in Denmark, included concrete designed to include proportions of pfa, ggbs and microsilica.

In the United Kingdom, pfa was specified for use with Portland cement by BS 3892-1 (1997), which included a requirement for particle size to be limited to a maximum of 12% retained on a 45 μm sieve, which in turn required careful selection and/or processing of candidate ashes. However, the European Standard BS EN 450 (1995 and now 2005) covered a wider range of ash products, including coarser materials with up to 40% retained on the 45 μm sieve. Among other things, this difference caused some concern over the relative effectiveness of BS 3892-1 and BS EN 450 additions as measures for counteracting AAR (Building Research Establishment 2002). Dhir and McCarthy (1998) have reported on the use of BS EN 450 fly ash in structural concrete.

4.6.1.1 Identifying fly ash or pfa

As fly ash, when used in normal amounts, does not substantially modify the colour of concrete, its presence will not usually be apparent to the unaided eye. However, microscopic examination of thin sections ground to 25 μm or less should enable the presence of any fly ash to be established with reasonable certainty, although quantification is much more difficult. The examination of concrete fracture surfaces by SEM is also effective for confirming the presence of fly ash (Thomas and Matthews 1991).

Fly ashes used for concrete additions are by-products of burning carbonaceous fuels (e.g. coal). Although it is common for fly ash to contain a few percent of residual unburnt fuel, which is a useful characteristic component when identifying the presence of fly ash, the majority of the material comprises the fused product of clay impurities in the original fuel. These clays, such as kaolinite and illite, give rise to the aluminosilicate glass and subordinate crystalline phases that dominate the mineral composition of fly ash. It is probable that the type and proportions of these clays contained as impurities in the original fuel dictate the final pozzolanicity of the fly ash (Hubbard et al. 1985).

Fly ash particles range from about 1 to 150 μm or so, with up to around 50% of those particles being smaller than the 20–30 μm thickness of the thin section, so that only the larger and largely unreacted particles are readily observable by optical microscopy. French (1991a) commented that 'in the case of pfa the iron-rich particles are the most conspicuous but the least representative of the globules present'.

Pfa mainly comprises solid and sometimes vesicular spherical glassy particles, although some other types of fly ash can include higher proportions of more irregularly shaped grains. A minority of these spherical particles consist of hollow 'cenospheres' (Raask 1968) which, because of their hollow lightweight nature, typically form a significant *volumetric* part of the material and represent a valuable characteristic indicator of the presence of pfa. The presence of *both* residual fuel particles and glassy cenospheres is virtually conclusive proof of the presence of some amount of fly ash. The cenospheres and other spherical grains can be discrete or agglomerated in clusters. Particle colour is very diverse, ranging from colourless to black and frequently being variously yellow, brown, red or grey. The typical appearance of these spherical particles, in thin section and under the SEM, is illustrated in Figures 4.21 and 4.22.

The mineralogical composition of fly ash is typically about two-thirds aluminosilicate glass. In pfa, the next most abundant constituent is usually crystalline mullite (aluminium silicate, $Al_6Si_2O_{13}$), with some proportions of quartz and iron-bearing phases such as magnetite and haematite. Hubbard et al. (1985) obtained modal compositions for a range of U.K. pfa materials, and these are reproduced in Table 4.30. Hubbard et al. (1984) found the cenospheres to be compounds of mainly glass with inclusions of mullite and quartz, frequently coated with the iron minerals derived from the furnace oxidation of pyrite. It is sometimes possible to observe mullite crystals in pfa in concrete thin sections, and using SEM examination, Thomas and Matthews (1991) found that relicts of mullite crystals could be identified (see Figure 4.22).

The 'Type C' fly ashes appear to exhibit higher proportions of crystalline material. McCarthy et al. (1984), for example, reported the presence of quartz, lime, periclase, anhydrite, spinel, calcium aluminate (C_3A), merwinite and melilite, plus occasionally alkali sulphates, sodalite and haematite, in a range of such ashes from the western United States. Mullite, a characteristic crystalline phase in 'Type F' fly ash, was found to be only a minor component and was not detected in half of their samples of 'Type C' material. Mehta (1985) similarly found no mullite in high-calcium fly ash, in which C_3A and anhydrite were identified. Harrison and Munday (1975) have reported findings of some anhydrite in U.K. pfa. In the authors' experience, very high-calcium types of fly ash can also contain some C_2S.

When examined under the SEM (see Figure 4.80), either as a powder or in a polished surface, each individual fly ash glassy particle will be compositionally different and thus of different potential reactivity. In Type C ashes, most of the calcium-rich particles react fairly effectively; thus, the residual material in a concrete is often atypical of the original material, being more silica and alumina rich. Compositionally, each individual particle can also contain localised porosity and compositional variations in a layered manner, such that a 'less reactive' outer surface may be surrounding a potentially more highly reactive inner core that cannot activate owing to its 'protected' position (Sibbick 2011).

4.6.1.2 Quantifying the content of fly ash or pfa

Although it is relatively straightforward to identify the presence of fly ash in concrete, by recognising under a microscope the characteristic composition and particle shapes described in the foregoing section, it is a more formidable task to quantify the amount present.

Table 4.30 Mineralogical composition of some U.K. pfa samples

U.K. region	Source	Aluminosilicate glass	Mullite $Al_6Si_2O_{13}$	Hematite Fe_2O_3	Magnetite Fe_3O_4	Quartz SiO_2	Coal
% by mass (for each value *n* = 15)							
SW	P13	62	14	6	4	2	12
	P14	65	13	7	6	5	6
SE	P15	50	19	9	9	4	9
	P16	43	30	9	6	8	3
	P17	52	22	10	5	5	6
Midlands	P18	67	11	6	8	5	3
	P19	55	16	11	8	5	4
	P20	63	18	7	8	3	1
	P21	66	13	7	6	3	5
	P22	53	12	10	7	8	10
	P23	63	12	9	8	5	2
	P24	62	16	6	7	5	4
NW	P25	70	12	5	6	2	4
NE	P26	45	28	9	7	7	4
	P27	70	9	7	7	2	3
	P28	70	12	6	6	3	2
	P29	64	11	11	5	2	6
	P30	62	11	11	7	6	3
Scotland	P31	40	41	6	3	9	5
	P32	29	43	10	4	8	7
P-Lytag	P33	48	37	3	2	7	3
	P28/P	72	12	5	6	3	3
	P25/P	73	9	6	6	2	3
	P20/P	74	10	7	7	1	1
	P31/P	45	35	7	4	5	3
	P24/P	67	13	6	7	4	3
Median values		62	13	7	6	5	4

Source: After Hubbard, F.H. et al., *Cem. Concr. Res.*, 15(1), 185, 1985.

Since compositional analysis by petrography is more and more frequently being carried out to check compliance with a specified mix design, such quantification is increasingly required. This is particularly the case since the conventional chemical analytical techniques, such as those described in BS 1881-124 (1988), are unreliable for concretes containing fly ash.

According to Figg (1989), pfa with a relatively high calcium content has a particularly significant effect on the BS 1881-124:1988 analysis for cement content: a pfa containing 12% CaO and present as 30% of the binder caused an 8% overestimate of cement content. French (1991a) claimed that the methods given for quantifying pfa and ggbs in BS 1881-124:1988 'are prone to some serious and often unknowable errors'. Such uncertainties worsen as the concrete ages, because variably greater amounts of pozzolanic activity will have taken place, thus varying the proportions of soluble silica in a largely unpredictable manner. Gomà (1989) has claimed success for a complicated chemical technique based upon differential solubility rates, and Figg (1989) maintained that an earlier technique (Figg and Bowden 1971) could be used if reference samples of *all* the concrete constituents were available.

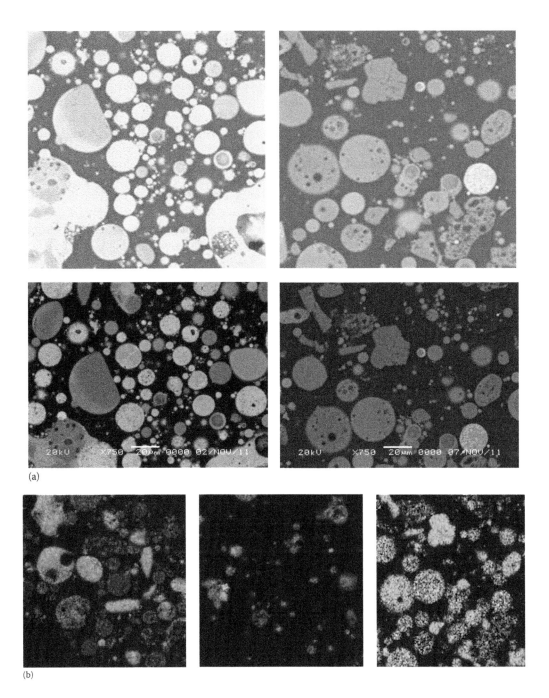

Figure 4.80 SEM comparison of Class C and Class F fly ash in concrete: (a) Backscattered electron images, Classes C left and F right, ~750×. (b) EDX chemical maps, ~750×, red, calcium; green, silica; blue, alumina; white, iron; and yellow, sodium. (Photographs courtesy of Ted Sibbick.)

Just as examination of concrete in thin section readily permits the presence of fly ash to be identified, so it is possible visually to estimate the relative proportions of fly ash and cement. However, this technique can only provide an extremely rough guide, because a majority of the fly ash particles will not be visible and the extent to which the fly ash and cement constituents have entered into pozzolanic or hydration reactions will not be known. The former objection is overcome in SEM examinations, where image analysis of highly polished surfaces is possible (French 1991a), but even this cannot take into account the uncertainties arising from the variable degrees of fly ash reaction.

Sibbick (2011) advises that while a reliably accurate determination of the original fly ash and/or slag contents, based on the residual particle components in a hydrated material, is difficult in thin section, identification of the relative amounts of residual fly ash and slag (plus residual cement) can help to establish if a significant overdose or under-dosing has occurred. In smaller batching plants, this can be a problem and leads to intermittent consequences with low strength and/or issues with such aspects as setting time, slump and surface finishing. Such an occurrence can be verified most effectively when a reference sample, known to have the correct mix, is compared directly with the sample under test. Wherever possible, both samples should be similar in terms of age, curing conditions and source location.

Figg (1989) mentioned the possible use of XRD to quantify the content of crystalline mullite, which is usually present in pfa, but this technique would only be feasible if the mullite content of the pfa used was known, was consistent and, when diluted within all the other materials present in the concrete, was satisfactorily detectable to be quantified.

In view of these difficulties with the various chemical and microscopical approaches to the quantification of fly ash content, it would appear that the only reasonably dependable technique involves the use of an SEM and associated EDX-type microanalysis. One procedure was developed by French (1991a, 1992), but it is important to recognise that techniques involving electron microscopy typically relate to very small specimens of concrete, so that care must be exercised to ensure that any findings can be taken as truly representative of the concrete in question.

French's approach involves collecting microanalyses from isolated particles of unreacted pfa and unhydrated cement under the SEM (at least 20 of each) and also carrying out area microanalyses of regions of cement paste. Then, using averaged compositions for the pfa, the cement and the hydrated paste, the ratio pfa/(pfa + cement) can be calculated from the triangular relationship of the three oxides which show the maximum contrast between pfa and Portland cement: CaO, Al_2O_3 and SiO_2. A precision for the instrumental analysis of ±0.03 at a ratio of about 0.50 is claimed, but the overall precision is likely to worsen by a variety of other potential sources of error.

Further guidance on the estimation of cement replacement by pfa is given in subsection 4.9.2.

4.6.2 Blastfurnace slag materials

Blastfurnace slag is a by-product of the iron-making process, and, in its air-cooled form, it is a dull grey, largely crystalline, rock-like material. This air-cooled slag, formed from the fusion of fluxing stone (limestone or dolomite) with 'gangue' residues (siliceous and aluminous) variously from the iron ore and the fuel, exhibits a chemical composition dominated by CaO, SiO_2 and Al_2O_3 and a mineralogical composition typically in the melilite field (Smolczyk 1980). Table 4.31 lists the main crystalline components of an air-cooled blastfurnace slag, and Figure 4.81 depicts the coarsely crystalline melilitic structure.

The large production of air-cooled blastfurnace slag in the United Kingdom is mainly used (>80%) for road aggregate (Gutt et al. 1974), with just a few percentage being employed as an aggregate or an inert filler for concrete. However, some blastfurnace slag is rapidly

Table 4.31 Mineralogical composition of an air-cooled blastfurnace slag

Main components	Solid solution of melilite, gehlenite and åkermanite	$2CaO \cdot Al_2O_3 \cdot SiO_2$ and $2CaO \cdot MgO \cdot 2SiO_2$
	Merwinite (in basic slags)	$3CaO \cdot MgO \cdot 2SiO_2$
	Diopside and other pyroxenes (in acid slags)	$CaO \cdot MgO \cdot 2SiO_2$
Minor components	Dicalcium silicate, $\alpha, \alpha', \beta, \gamma$ (in basic slags)	$2CaO \cdot SiO_2$
	Monticellite	$CaO \cdot MgO \cdot SiO_2$
	Rankinite	$3CaO \cdot 2SiO_2$
	Pseudo-wollastonite	$CaO \cdot SiO_2$
	Oldhamite	CaS
Seldom observed minor components	Anorthite (in acid slags)	$CaO \cdot Al_2O_3 \cdot 2SiO_2$
	Forsterite	$2MgO \cdot SiO_2$
	Enstatite (in acid slags)	$MgO \cdot SiO_2$
	Perovskite	$CaO \cdot TiO_2$
	Spinel	$MgO \cdot Al_2O_3$

Source: After Smolczyk, H.G., Slag structure and identification of slags, in *Seventh International Congress on the Chemistry of Cement*, Paris, France, Volume I, Principal Reports, Sub-Theme III, pp. I/3–I/17, 1980.

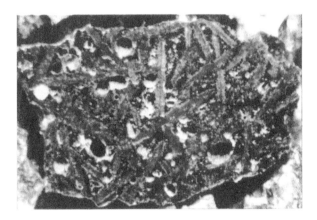

Figure 4.81 Section through a particle of air-cooled blastfurnace slag, showing large laths of melilite crystals. Width of field ~4 mm.

cooled in a granulation or pelletisation process, and the glassy product, which has latent hydraulic properties, is ground (or milled) to cement fineness or finer for use as an additive or cement replacement material in concrete. This ground granulated blastfurnace slag ('ggbs') now enjoys widespread and growing usage and will be frequently encountered by petrographers examining hardened concrete.

A number of reviews provide accounts of the properties and potentially beneficial influences of ggbs as a constituent within concrete (Smolczyk 1980; Hooton 1987; Werner 1987; Dewar and Anderson 1992). One of the most recent and comprehensive reviews was compiled by a U.K. Concrete Society Working Party (Concrete Society 1991). There is now appreciable practical experience with concrete containing ggbs, spanning more than 50 years worldwide, and it is evident that ggbs concretes are often notably durable. Figure 4.82 shows part of a wartime concrete submarine installation in Norway, wherein the sections made without ggbs are deteriorated by ASR and other factors, while the adjacent sections made with ggbs are in comparatively good condition (Jensen 1995).

Figure 4.82 Wartime submarine dock in Trondheim, Norway, showing cracked and uncracked portions for concrete, respectively, containing no ggbs (right) and ggbs cement replacement (left).

4.6.2.1 Identifying ggbs

It is common for ggbs to form up to 50% or more of the binder, so that the recognition of its presence should not pose a problem for the petrographer. Concretes made with ggbs exhibit a distinctive dark greenish-blue matrix coloration, although this changes to a creamy colour reminiscent of that displayed by non-ggbs concrete with time and exposure, apparently because of the oxidation of sulphides present in the ggbs. According to French (1991b), this colour change is not accompanied by any visually apparent features, even when examined under the scanning electron microscope (SEM). The lightening oxidation can penetrate deeply into ggbs concrete, and the absence of the dark greenish-blue colour from a sample should never be taken as an indication that ggbs is not present. It is possible to confirm that ggbs is present using either thin-section or reflected light optical microscopy or SEM, but, as with pfa discussed in Section 4.6.1, quantification is much more difficult. Sibbick (2011) advises that usually only occasional coarser-sized residual angular slag components remain largely intact and visible after the material is well hydrated.

The composition of ggbs is dominated by a calcium–aluminium–magnesium–silicate glass, which typically forms more than 70% of the material and frequently more than 90%. Crystalline phases are also found as inclusions within the glass, and these have been studied by Scott et al. (1986), who found that relatively small changes in the chemical composition of the slag melt dictated the principal crystalline material formed. The sections through the CaO–SiO_2–Al_2O_3–MgO quaternary system (at 10% MgO and 10% Al_2O_3) shown in Figure 4.83 illustrate how slag melts will crystallise either in the melilite or merwinite fields when quenched. Scott et al. (1986) reported that oldhamite (CaS) and native iron were the other main crystalline phases in ggbs.

The importance of glass content in the hydraulicity of ggbs has been the subject of considerable debate. It was initially assumed that reactivity would increase with glass content, but Demoulian et al. (1980) cast doubt on this notion, stating that 'for the same [chemical] composition, a perfect vitrification is not the criterion of an optimal reactivity'. Hooton and Emery (1983) similarly concluded, 'while an amorphous structure is fundamental to slag reactivity, the highest glass contents were not necessarily indicative of the highest strength development'. They suggested that the crystalline inclusions could act as hydration nuclei or could be associated with more disordered, and hence more reactive, glass than that present in more highly vitrified ggbs.

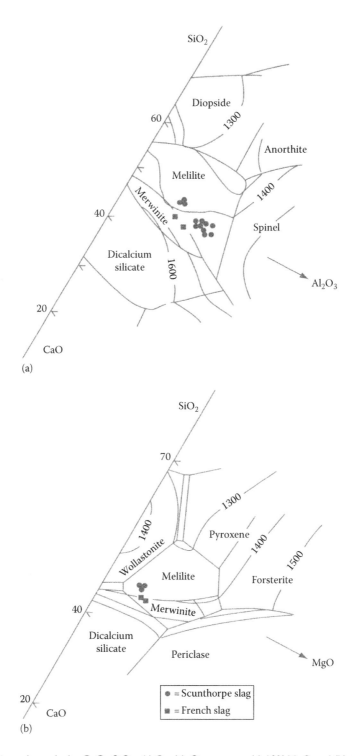

Figure 4.83 Sections through the $CaO–SiO_2–Al_2O_3–MgO$ system at (a) 10% MgO and (b) 10% Al_2O_3, showing the fields for melilite and merwinite. (From Scott, P.W. et al., *Mineral. Magaz.*, 50, 141, 1986.)

Frearson and Sims (1991) describe the investigations carried out during the preparation of BS 6699 (1986) and in particular the development of a test for checking that ggbs was adequately glassy. While melilitic ggbs formed mainly particles of transparent glass with occasional cloudy particles containing relatively large well-formed melilite crystallites, it was found that the merwinitic ggbs formed many more cloudy particles containing tiny rod-like or dendritic formations of merwinite crystallites (Figure 4.84). Despite the apparently lower glass content of the merwinitic ggbs, its actual performance was typically superior to that of the melilitic ggbs in their investigation. A test was therefore devised for BS 6699 (1986), based upon using reflected light microscopy to quantify the proportions of glass, glassy and crystalline particles, the 'glassy' particles being recognised at least as reactive as the pure glass particles and possibly more so. The later edition of BS 6699 (1992) instead opted for an XRD technique to quantify the overall glass (non-crystalline) content of ggbs (Gutt 1992), which is required to exceed 67%. BS 6699 (1992) was replaced by BS EN 15167-1 and BS EN 15167-2 (2006).

Drissen (1995) found that the estimation of glass content using transmitted light has advantages over other methods and described the techniques required. Taylor (1990) states that both microscopy and XRD methods will fail to predict reactivity if microcrystallinity is present. Another optical technique, which has been proposed for determining the hydraulic reactivity of granulated slags (Grade 1968), is based upon the fluorescence colours produced by UV light. The technique appears to be suitable for ranking slags from the same works, but it is uncertain whether it is applicable to slags from different sources.

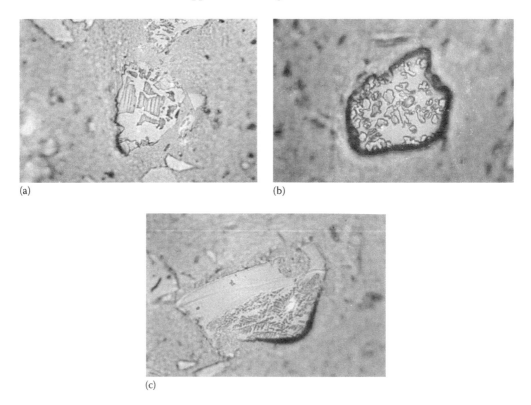

(a)

(b)

(c)

Figure 4.84 Photomicrographs of ggbs particles in reflected light after etching with nitric acid: (a) Melilite crystals within a glassy matrix. (b) Rod-like merwinite crystals within a glassy matrix. (c) Dendritic merwinite formations within a glassy matrix. (From Frearson, J.P.H. and Sims, I., *Concrete*, 25(6), 37, 1991.)

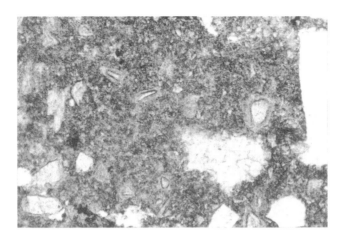

Figure 4.85 Photomicrograph of layers of hydrate around ggbs kernels in concrete in thin section, ppl, width of field ~0.8 mm.

French (1991b) suggested, from his observations of completely unreacted ggbs particles in concrete, even at sizes only resolvable using SEM (see earlier Figure 4.19), that ggbs particles are variously either inert or slowly reactive. If this is true, it is possible that the mineralogical character of the particle is the critical factor, perhaps with the 'glassy' particles and only some of the pure glass particles being the slowly reactive components.

Whatever the composition and microtexture of the particles, it is their generally glass-like nature which characterises ggbs in thin sections or polished sections of concrete (see Figures 4.17 and 4.18). The remnant particles are typically abundant, angular even shard-like and glassy in both fracture habit and isotropic properties. Ground particles of pelletised slag will be similar, but with an additional tendency to contain gas vesicles. Usually, ggbs particles are transparent or translucent and colourless or light brown, but occasionally may exhibit white, green or darker brown coloration. A few opaque discrete iron grains (or inclusions) might be apparent. Ggbs particles also exhibit higher relief than the enveloping cement paste and any fine quartz aggregate particles in the vicinity: the refractive index of ggbs is higher at ~1.63–1.64 than quartz at ~1.54–1.55 (McIver and Davis 1985).

In reflected light microscopy of concrete, any ggbs particles present usually yield a distinctive white to blue coloration after the polished specimen has been etched with HF vapour. SEM is also effective at confirming the presence of ggbs particles.

Although French (1991b) mentioned ggbs particles that exhibit no signs of reaction, it is common in older concretes to observe hydration zones surrounding kernels of many of the residual ggbs particles. Sometimes, these hydration zones are layered (Figure 4.85), and it has been suggested by Thaulow (1984) that, in some cases, these might be seasonal in a way analogous to annular rings in tree growth, perhaps offering a means of estimating the approximate age of such a concrete. It is interesting that the boundary between the latest hydration layer and the ggbs kernel is usually quite sharp rather than diffuse as might have been expected. Concrete matrices containing ggbs are also sometimes depleted in portlandite.

4.6.2.2 *Quantifying the content of ggbs*

Like pfa (see Section 4.6.1), while the presence of ggbs in a concrete is straightforward to detect, it is much more difficult to quantify. The chemical composition of ggbs is quite close to that of Portland cement, and *routine* chemical analysis, to BS 1881-124 (1988), for example, will not necessarily even indicate the presence of ggbs, let alone allow its quantification. Figg (1989) suggests that ggbs can be determined from the 'considerably higher' contents of sulphide and manganese, compared with Portland cement, although the contents of those constituents in the ggbs are required to be known. BS 1881-124:1988 provides a special chemical method for determining the percentage replacement of Portland cement by ggbs in concrete, based upon the sulphide content, but French (1991a,b) considers this method questionable because of the need to check for other sources of sulphide and also in view of the likelihood that an unknown proportion of the sulphide in the original ggbs will have oxidised by the time of the analysis.

Figg (1989) states unequivocally that 'determination of the relative proportions of ggbs and OPC in a hardened concrete is not possible by conventional microscopical procedures'. This is true, except that rough estimates are achievable from thin sections, for example, to assess whether or not the content of ggbs, estimated from the usually abundant remnant particles, is consistent with the specification. McIver and Davis (1985) reported a technique in which the thin section under consideration is compared with reference sections made from concretes of known ggbs contents. This approach might enable a reasonably reliable semi-quantitative estimate to be made, provided that the ggbs in the reference concretes is sufficiently similar to that in the concrete sample being analysed and the reference concretes are also comparable to the sample in terms of age and exposure history. Obviously, this is not practicable for concrete samples of unknown composition.

Sibbick (2011) agrees that this can be a useful process for indicating possible under- or overdosing with ggbs, by direct semi-quantitative comparison of residual slag components, between a reference sample of known correct original slag content and various test samples. Such a comparison exercise is also supported by the development of opalescent paste and the heavy consumption of calcium hydroxide, which is each characteristic of ggbs activity. He adds that the same general approach is possible for other types of mineral addition.

As with pfa (see Section 4.6.1), a quantitative procedure using EDX-type microanalysis offers the best chance of reliably determining the content of ggbs, although, as with any techniques involving use of an SEM and a microanalysis, it is important to ensure that the specimens analysed are adequately representative of the overall concrete concerned. Figg (1989) mentions a procedure in which spot microanalyses are made of remnant ggbs and unhydrated cement particles in the concrete, to establish the compositions of the original ggbs and cement materials, followed by multiple area analyses of uncontaminated hydrated paste; the relative proportions are then calculated from the data obtained. French (1991b) provides greater detail of this technique (which is broadly similar to that described in Section 4.6.1 for pfa) and also explains some of the probable sources of error, including variations in the water/binder ratio and alterations to the concrete such as the leaching of lime.

Further guidance on the estimation of cement replacement by ggbs is given in subsection 4.9.2.

4.6.3 Ultra-fine additions

Pfa and ggbs additions (see Sections 4.6.1 and 4.6.2) are each powders with a particle size range broadly similar to that of modern Portland cement. By contrast, the 'ultra-fine' additions, microsilica or silica fume and metakaolin, are each very much finer than Portland cement (Figure 4.86). According to ACI 234R (American Concrete Institute 2006), cited by Holland (2005), it has been estimated that, for a 15% silica fume replacement of cement, there are approximately two million particles of silica fume for each grain of Portland cement. In terms of nominal maximum particle, Holland (2005) cites 45 µm for Portland cement and just 0.5 µm for silica fume.

Figure 4.86 SEM comparison between Portland cement and silica fume. (From Holland, T.C., *Silica Fume User's Manual*, 183pp, Silica Fume Association technical report FHWA-IF-05-016 for the Federal Highway Administration, Washington, DC, 2005.)

4.6.3.1 Microsilica (condensed silica fume)

Silicon, ferrosilicon and other silicon alloys are produced by reducing quartz, with coal and iron or other ores, at very high temperatures (2000°C) in electric arc furnaces (Hjorth 1982). Some silicon gas (or 'fume') is produced in the process and reaches the top of the furnace with other combustion gases, where it becomes oxidised to silica by the air and then condenses as (<0.1–1 μm) spherical particles of amorphous silica. This material is usually known as 'silica fume' or, more properly, 'condensed silica fume' (csf); in the United Kingdom, the alternative term 'microsilica' has also been used (Concrete Society 1993), but the term 'silica fume' is preferred by BS EN 13263-1 (2005/2009). Microsilica is an ultra-fine powder (Figure 4.87a), with individual particle sizes between 50 and 100 times finer than cement or pfa, comprising solid spherical glassy particles of amorphous silica (85%–96% SiO_2) (Parker 1985, 1986). However, the spherical particles are usually agglomerated so that the effective particle size is much coarser. A recent overview on silica fume and its use in concrete was published as BRE IP5/09 (Dunster 2009).

Microsilica for use in concrete derives from the manufacture of ferrosilicon alloys and is modified, by densification, micropelletisation or slurrification, to facilitate transportation and handling (Male 1989). World supplies of microsilica are limited with total production probably being between 1 and 1.5 million tonnes. The United Kingdom currently imports only modest quantities of microsilica for use in concrete, mainly from Iceland, a country which itself has routinely used blended microsilica cement as a precaution against ASR for many years (Asgeirsson 1986). The potential benefits of using microsilica either as a

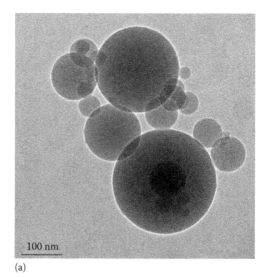

(a)

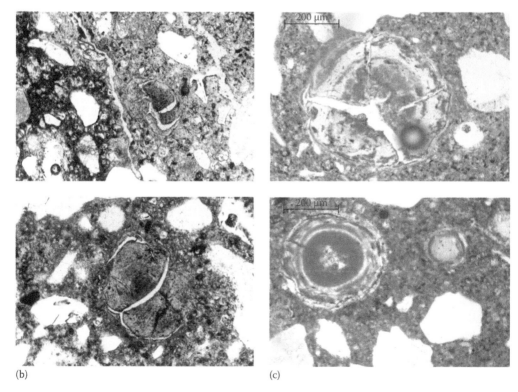

(b)   (c)

Figure 4.87 Microscopic appearances of silica fume: (a) Typical view of silica fume using SEM. (Micrograph courtesy of Robert Lewis of Elkem AS.) (b) Photomicrographs showing agglomeration of silica fume in concrete in thin-section optical microscopy, taken in 1994 from concrete derived from the Arabian Gulf, 200 μm magnification. (Courtesy of Sandberg LLP, London, U.K.) (c) More recent photomicrographs of silica fume agglomerations. (Courtesy of Paul Bennett-Hughes, RSK Environment Ltd., U.K.)

cement replacement material or as an addition to improve concrete properties have been well reviewed by Malhotra and Carette (1983), Sellevold and Nilsen (1987) and Holland (2005). Disadvantages would include the health hazards associated with fine silica materials and the increased cost of the concrete.

Although the use of microsilica in concrete has been limited to date, the petrographer may expect to encounter an increasing occurrence in the future and determining its presence or otherwise in a concrete sample is occasionally requested. Detecting microsilica in concrete is complicated by its usually small proportion (between 5% and 30% but typically around 10%) and its extreme fineness, with virtually all of the particles being substantially smaller than the 25–30 μm thickness of a conventional thin section. Moreover, because microsilica is an ultra-fine pozzolana, it is rapidly consumed in the initial hydration reactions, and, according to Figg (1989), after a month it is virtually undetectable even using SEM. However, the presence of microsilica is frequently betrayed to the petrographer by either or both of two factors: the occurrence of agglomerations and the modifying influence of microsilica on the cement paste.

The microsilica supplied for use in concrete has either been agglomerated by densification or micropelletisation or is suspended in a slurry with water (Male 1989). These agglomerations are supposedly redispersed during concrete mixing, and the slurries are often considered to redisperse most easily; superplasticisers are claimed to aid dispersion (Male 1989). However, St John (1994b) and St John et al. (1996) have challenged this supposition, finding that microsilica agglomerations typically survive and even that some represent fused microspheres, present in the microsilica before densification, which are incapable of dispersion.

St John (1994b) found that this agglomeration caused the mean particle size of microsilica to rise to between 1 and 50 μm, rather than the 0.1–0.2 μm frequently cited. The significance of this finding for the petrographer is that most microsilica concrete can be expected to contain some occasional agglomerations, which are coarse enough to survive pozzolanic reactivity and be visually detectable under the SEM and even in thin section (see Figure 4.87b and c). Sibbick (2011) clarifies that such agglomerations or clusters, which are not generally found when microsilica was added to the concrete mix as a slurry, typically exhibit an internal shrinkage crack pattern, and their silica-rich character can be confirmed by SEM-EDX. The conclusion that the presence of such microsilica agglomerations indicates poor dispersion and thus indicates a concrete mixing deficiency must be treated with caution, although an *unusual* abundance of coarse agglomerations might justify such an interpretation. Potentially, these microsilica agglomerations can also represent localised alkali–silica reaction sites within concrete.

The matrix of microsilica concrete is notably dense to thin section examination, and this optical denseness will frequently suggest the presence of microsilica, even if agglomerations are not apparent. This densifying effect has been attributed to the extreme fineness of microsilica, whereby 50,000–100,000 microspheres exist for every cement grain (Male 1989), allowing microsilica hydration products to infill the water spaces usually left within the cement hydrates (Figure 4.88). Overall coloration of microsilica concrete is often also slightly darker than a comparable material lacking microsilica, but this might not be apparent, especially for lower microsilica contents.

It has been found that calcium hydroxide (portlandite) is consumed by the pozzolanic reactions involving microsilica (Sellevold et al. 1982), the extent of portlandite depletion increasing linearly with microsilica content. Durning and Hicks (1991) considered this reduction in portlandite content to be a major factor in the effectiveness of microsilica concrete in resisting chemical attack and stated that the portlandite content is reduced to zero by a 30% microsilica replacement of ordinary Portland cement. In thin section examination, the combination of a notably dense paste and a complete or near absence of portlandite is a strong indication that microsilica might be present. More exhaustive inspection of the

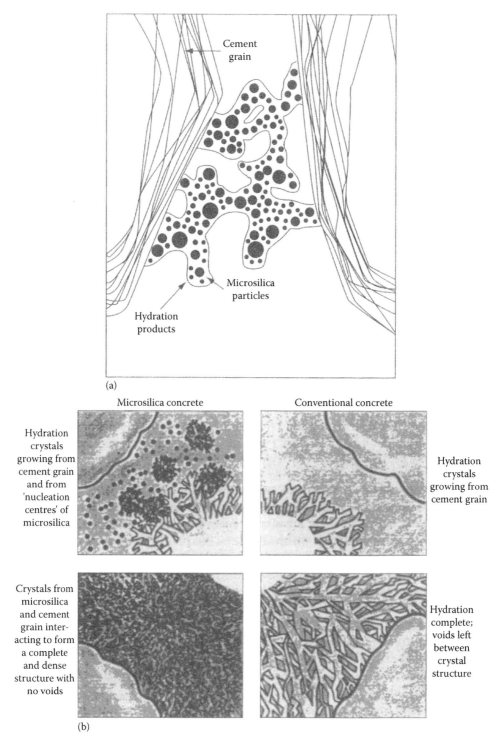

Figure 4.88 Sketches to illustrate the theoretical effect of microsilica in densifying the concrete matrix: (a) Microsilica subdividing the space between cement grains. (b) Comparison between conventional and microsilica concretes. (From Hjorth, L., *Nordic Concr. Res.*, 1, 9.1, 1982; Male, P., *Concrete*, 23(8), 31, 1989.)

thin section specimen might then reveal the sporadic and confirmatory occurrence of relatively coarse microsilica agglomerations.

Quantification of microsilica in concrete cannot be undertaken by any form of microscopy. Figg (1989) has suggested the use of electron probe microanalysis to establish the composition of the hydrated paste, presumably on the basis of an enhanced silica content relative to that to be expected from Portland cement alone. If the composition of the cement component is known, it might then be possible to calculate an approximate proportion for the microsilica.

4.6.3.2 Metakaolin

Metakaolin, a reactive pozzolanic material, is produced by the thermal activation of the mineral kaolinite, an hydrated aluminosilicate of formula $Al_2O_3 \cdot 2SiO_2 \cdot 2H_2O$. Thermal treatment of kaolinite initially drives off non-chemically bound water, at temperatures as low as 400°C, but more rapidly at 500°C or higher (He et al. 1995), when the highly endothermic formation of metakaolin by dehydroxylation occurs (Salvador 1995). The use of thermally activated clays for construction has a long history (Saad et al. 1982, Andriolo and Sgaraboza 1987, Kostuch et al. 1993). The use of metakaolin in modern engineering concretes has been reported by Asbridge et al. (1996) and Ryle (1999), and the role of metakaolin as a pozzolanic addition for cement-based systems has been more generally reviewed by Jones (2002).

Metakaolin has been widely discussed as a pozzolanic material (Ambroise et al. 1982, 1985a,b; Murat 1983a,b). Metakaolin also generally complies with the definition of a 'pozzolan' as provided by the American Society for Testing and Materials (ASTM C125 2007). Recently, the French standard NF P18-513 has been published for metakaolin as a pozzolanic addition for concrete (AFNOR 2010).

One challenge for the petrographer is that the reaction of metakaolin with calcium hydroxide gives products that are similar to those formed by the cement hydration itself, that is, hydrated silicates and aluminosilicates. However, Kostuch et al. (1993) reported that metakaolin refined the pore structure of the hydrated cement paste and it has been proposed that the products of the pozzolanic reaction give this refinement by reducing the mean pore size by depositing in the capillary pore structure of the hardened paste. This effect can also lead to a more tortuous pore network as suggested by Asbridge et al. (2001), which can ultimately reduce the effective permeability of the matrix. Poon et al. (2001) also observed that metakaolin reduced porosity and was more efficient than csf at refining the pore structure. The findings of Frias and Cabrera (2000) correlate with those of Bredy et al. (1989), who also found that metakaolin produced a finer pore structure. Khatib and Wild (1996a) and Chadbourn (1997) commented that metakaolin also decreased the threshold value (critical pore diameter).

Pettifer (1991) investigated metakaolin-containing cementitious products and found that metakaolin depletes calcium hydroxide from very early stages, within the first few days (see also Kostuch et al. [1993], Asbridge [1994] and Poon et al. [2001]). It has also been shown that calcium hydroxide can be completely removed from the hydrated Portland cement paste if sufficient metakaolin is present. Based on the reports of Kostuch et al. (1993) and Oriol and Pera (1995), this quantity seems to be between 20% and 40% metakaolin as a partial replacement for Portland cement.

The beneficial effect of metakaolin in refining the pore structure of Portland cement composites, coupled with potential reductions in calcium hydroxide levels, has been indicated to give specific performance advantages. These potential benefits include increased resistance to deterioration or expansion when subject to sulphate ingress (Singh and Osborne

1994; Khatib and Wild 1996b), reduced spalling when subject to freeze–thaw cycles as reported by Caldarone et al. (1994) and reduced concrete expansion attributable to alkali–silica reaction (Walters and Jones 1991; Jones et al. 1992; Verein Deutscher Zementwerke 1996; Ramlochan et al. 2000; Sibbick and Nixon 2000). Increased resistance to erosion by acids, increased resistance to water penetration and absorption and increased resistance to penetration by chlorides have been generally reviewed by Kostuch et al. (1993). Larbi (1991) specifically reported that metakaolin reduced water sorptivity in Portland cement mortars by around 56% and chloride diffusivity by about 98%.

The resistance of concretes incorporating metakaolin to chloride ion penetration has been investigated by Zhang and Malhotra (1995) and Caldarone et al. (1994), who reported that the resistance to chloride penetration was significantly higher than for control (Portland cement) concretes. Studies at the University of Dundee (Dhir et al. 1999) have also contributed to the knowledge of the effects of metakaolin on transport properties. They reported reductions in water sorptivity with increasing metakaolin partial Portland cement replacement levels up to approximately 45%. They also reported reductions in chloride diffusivity, resulting from the incorporation of metakaolin, of almost one order of magnitude at a partial Portland cement replacement level of 25%. Metakaolin will also have a beneficial effect on the strength of cement pastes, mortars and concretes through reductions in calcium hydroxide, pore size refinement and improved paste–aggregate bonding.

4.6.4 Natural pozzolanas and other additions

Although fly ash, ggbs and, to lesser extents, microsilica (or silica fume) and metakaolin (see the foregoing subsections) form the majority of the mineral additions used in concrete, natural pozzolanas have been used historically and continue to be added to concrete in some specific areas of the world. These natural pozzolanic materials sometimes exist in a natural form that facilitates their direct use in concrete after grinding, such as the volcanic or diatomaceous earths, while others need to be treated, for example, by heating or 'calcining', prior to grinding. Such materials include a range of silica-rich glassy volcanic rocks, such as pumicite, ignimbrites and other tuffs, opaline shales, other shales or clays and opaline cherts. A good review of the natural pozzolanas available in the United States is included in Mielenz (1983). Lewis et al. (2003) classified natural pozzolanas into four groups: (1) volcanic, (2) tuffs, (3) sedimentary (especially refers to diatomite) and (4) diagenetic (deriving from weathering of siliceous rocks). A good account of Turkish diatomite has been provided by Goren et al. (2001).

Generally, natural pozzolanic additives are used to replace between 10% and 30% of the cement in the binder, and, for an effective material, the beneficial effects appear broadly comparable with those of fly ash. In the United States, natural pozzolanas are specified, along with fly ash, in ASTM C618 (2008). In some cases, the presence of a finely ground natural pozzolanic additive could pose a problem for the unsuspecting petrographer, and care should be exercised when examining concrete from any areas in which such pozzolanas are traditionally used.

Some discussion of the problems in identifying pozzolanas in concrete will also be found in Chapter 5. The chemical method for identifying pozzolanas involves dissolution of the hardened cement paste and examination of the fine residue. In most cases, sufficient of the pozzolanic material is found in the fine residue to allow identification by optical means.

Generally, identification in thin section will depend on some unique characteristic of the pozzolana being visible either by optical or SEM. For instance, sponge spicules, often present in diatomite, may persist in concrete long after the diatom fragments have reacted. Where the presence of a pozzolana is known or suspected, the petrographer should examine samples of the likely pozzolanas for any unique characteristics before examining the concrete in thin section. This increases the chance of identification.

It must be accepted that in old concretes where the pozzolanic replacement is below 15%, identification may be difficult unless some unique characteristic of the pozzolana is present. Changes in the texture of the hardened cement paste and depletion of portlandite may be limited, subtle and insufficient to suggest the presence of a pozzolana. In such cases, determination of the Ca/Si ratio by analysis may suggest the presence of a pozzolana, but even this is not possible for wet chemical methods where calcareous aggregates are present.

Non-ferrous slags have not generally been considered suitable for use in concrete, either as aggregate or as addition in powdered form. However, some non-ferrous slags might be suitable as a pozzolana. Douglas et al. (1985) have suggested the potential usability of some Canadian non-ferrous slags. In a worldwide survey, a RILEM committee identified rice husk ash ('RHA') as one of the four principal pozzolanic/cementitious siliceous by-products used for concrete, alongside fly ash, ggbs and microsilica (RILEM 1988). In some rice-producing areas of the world, where a typical paddy comprises about 70% rice and 20% rice husk, incinerating the latter generates RHA, which contains up to 97% silica and can be highly pozzolanic, depending on the incinerating conditions (Muthadhi and Kothandaraman 2011).

4.6.5 Pigments

The typically white to dark grey colour of concrete is ordinarily determined mainly by the binder, with the fine and coarse aggregates having, respectively, much less influence. The shade of concrete surface coloration is also affected by other factors, including the type of formwork and the mode of curing. To obtain colours other than white to dark grey, the addition of a pigment is necessary. The use of such pigments is common in precast and 'architectural' concrete, such as paving, masonry, cladding units and 'cast stone'. Chemical stains and dyes are infrequently used, because in the longer term they fade in the alkaline environment within concrete.

Generally, the use of a pigment will be evident from the concrete colour, but a petrographer may be called upon to confirm the type of pigment used or even the presence of pigment in cases where only a tinting has been achieved. Most mineral pigments are ultra-fine powders, much finer than cement (mainly <1 μm particles), so that optical microscopy is of limited usefulness in identifying pigments and SEM will be required. However, the colour imparted to hydrated cement paste by the included pigment particles will be clearly apparent in thin section, as will its uniformity of dispersion.

Under the SEM, EDX-type microanalysis of pigment particles can be used to determine the basic type in the case of inorganic materials, although the various colours achieved using iron oxides might not be capable of differentiation. If the composition of the cement is known or can be established from microanalysis of representative unhydrated particles, area microanalyses of the pigmented cement paste might then enable the pigment content to be quantified. Sibbick (2011) confirms that the use of a reference specimen containing a known quantity of an Fe-based pigment can enable the probable presence or otherwise, or perhaps under- or overdosing, of that pigment to be assessed in a sample under investigation. SEM is unhelpful in the case of carbon black, which has a particle size of ~0.01 μm, which is much finer than most inorganic pigments and, as a result, tends to create murky thin sections.

Concrete pigment types, properties and effects have been reviewed several times (Coles 1978; Arnold 1980; Levitt 1985; Spence 1988). Good concise reviews were presented by Lynsdale and Cabrera (1989) and more recently by Levitt (1998) and Carter (1998). A summary of the main types of pigment likely to be found in concrete is given in Table 4.32, together with their colouring effect and some other characteristics. Some examples of pigmented concrete and cast stone products are given in Table 4.33 (see also Glossary of minerals).

Table 4.32 Colour and other properties of some types of pigment for concrete

Pigment type	Colour	Example (Arnold)	Particle size[a] μm	Bulk density[a] kg/m³	Water absorption[a] ml/100 g
Synthetic red iron oxide	Red	Pfizer	0.1	900	35
		R-1599	0.7	1500	20
Synthetic yellow iron oxide	Yellow	Mapico Yw	0.2 × 0.3[b]	800	50
		1000	0.2 × 0.8[b]	500	90
Synthetic black iron oxide	Black	Pfizer BK-5599	0.3	1100	33
Synthetic brown iron oxide	Brown	R-Coulston	0.3–0.6[c]	900	50
		Brown 537	0.1–0.2[c]	1000	30
Natural brown iron oxide	Brown	R-C Burnt Umber 15			
Chromium oxide	Green	Pfizer G-4099	0.3	1200	19
Carbon black (general purpose)	Black	Mapico Raven 410			
Carbon black (concrete grade)	Black	Mapico Raven 1040	0.01	500	100
Cobalt blue	Blue	Harshaw Blue 7540			
Ultramarine blue	Blue	Frank Davis Blue 410B			
Phthalocyanine	Green or blue	Mini Pigs Green 5069	0.01	500	0[d]
Toluidine red	Red	Du Pont RT-386-D			
Dalamar (hansa) yellow	Yellow	Du Pont YT-808-D			
Watchung (BON) red	Red	Du Pont RT-761-D			
Chrome yellow	Yellow	Du Pont Y-469-D			
Iron blue	Blue	Milori Blue 50–1750			
Titanium oxide	White		0.2	700	24

Source: Based on Arnold, D.R., *Cem. Concr. Aggr. CCAGDP*, 2(2), 74, 1980 and Levitt, M., *Concrete*, 19(3), 21, 1985.

[a] The physical property data, after Levitt (1985), do *not* relate to the examples cited by Arnold (1980).
[b] Needle-shaped crystals, hence the two particle dimensions.
[c] Brown iron oxides have a wider range of particle sizes than the others.
[d] Hydrophobic in their undiluted form and will not absorb water.

Most of the pigments in common usage are inorganic metallic oxides; the main exception is carbon black. Red, yellow, black and brown colorations are achieved with iron oxides, while green is usually derived from chromic oxide and blue from a copper complex with phthalocyanine. Bright white concrete can be achieved by using titanium oxide together with white cement. The most effective inorganic pigments are synthetic rather than natural, because the natural materials are typically less pure and hence exhibit a lower tinting strength. These inorganic pigments are essentially inert and stable, although it has been reported that, in certain conditions, the non-haematite iron oxides can alter to haematite (Koxholt 1985).

As pigment is an inert fine powder, excessive proportions have detrimental effects on the hydrated cement microstructure and on concrete strength. It is recommended in both BS 1014 (1975/1986) and ASTM C979 (2010) that the content of pigment is limited to a maximum of 10% by mass of cement. Lower dosages are envisaged by BS EN 12878 (2005), being 5% for inorganic solids and carbon black or 2% for organic pigments, but mixes with higher dosages can be subjected to tests. According to Lynsdale and Cabrera (1989), a

Table 4.33 Some pigmented concrete and cast stone products

Effect/colour required	Cement type	Aggregate examples	Admixture(s)	Pigment(s)
Brilliant white	White PC	Calcined flint, Carrara marble		None
	White PC	Calcined flint, quartz sand	Stearic acid	Titanium oxide
Light cream	White PC	Spanish dolomite		Yellow
Mid cream	White PC	Derbyshire limestone		Yellow
Buff	White PC	Grey or brown, with sand	Stearic acid	Yellow
Simulated bath stone	White PC	Crushed bath or Clipsham Stone	Aluminium stearate	Yellow or brown
Simulated york stone	White PC or OPC	Crushed sandstone, crushed limestone and crushed granite		Brown/yellow
Yellow	White PC	Crushed Cotswold limestone		Yellow
Green brick	White PC or WPC and OPC	Light buff limestone	Aluminium stearate	Phthalocyanine green[a]
Light pink	White PC	Derbyshire limestone		Light red
Deep pink	White PC	Mountsorrel granite		Light red
Red	White PC or OPC	Mountsorrel granite		Red
Red mortar	OPC and lime	Red, brown or buff sand	Stearic acid	Red
Brown	OPC	Mountsorrel granite, or quartzite gravel, or Cotswold limestone		Brown
Bronze	OPC	Criggion Green granite		Dark brown
Green	OPC	Criggion Green granite or green marble		Chromium oxide
Slate grey	OPC	Ingleton granite		Green/blue/black
Purple slate	OPC	Mountsorrel granite black marble		Red/black
Black	OPC	Clee Hill granite or Belgian black marble		Black

Source: Based on Levitt, M., *Concrete*, 19(3), 21, 1985 and Spence, F., *Concr. Forum*, 1, 15, 1988.

[a] 10/90 w/w fine silica dilution.

dosage in the range 3%–6% achieves a 'definite colour', while 'tinting' only requires ~1%, but a 'deeper shade' requires up to 10%. Each type of pigment has a colouring saturation level, and higher dosages do not necessarily produce even deeper colours.

Various disadvantages have been perceived for pigmented concrete, although in reality, these are often features common to all concrete which are just more noticeable in a coloured material (Spence 1988). Alleged 'fading' is frequently caused by a surficial overlay of lime bloom or efflorescence or simply grime from the atmosphere. Black or dark grey concretes pigmented with carbon black are said to be particularly vulnerable to loss of pigment from water-eroded surfaces (Lynsdale and Cabrera 1989). Colouring inconsistency can be overcome by using synthetic inorganic pigments, which achieve their colour saturation at relatively low dosage levels, and by ensuring quality control over concrete mix proportioning and the other variables such as curing.

4.7 CHEMICAL ADMIXTURES

Increasingly concrete mixes are designed to include one or more chemical admixtures, which are usually added at low dosage levels (mostly ~0.1% by weight of cement), to modify the properties of the fresh and/or eventually hardened concrete. Such modifications include workability improvement and water content reduction for a given workability, by plasticisers or superplasticisers, retardation or acceleration of set, air entrainment and waterproofing. Some admixtures achieve combinations of these modifications (e.g. retarder/plasticiser). A good account of the history of concrete admixtures has been provided by Jowett (2004).

Detailed accounts of chemical admixtures for concrete are to be found elsewhere (Rixom 1977; Neville and Brooks 1987/2010; Ramachandran 1995; Roberts 1998; Rixom and Mailvaganam 1999; Payne 2002; Roberts 2005; Dransfield 2003, 2006a; Spiratos et al. 2003). Roberts and Adderson (1983) review the benefits and some side effects of using superplasticisers. A list of the main admixture types and their chemical identities is given in Table 4.34.

These chemical admixtures cannot be directly detected by petrography, although in some cases, the effects created by an admixture might be observable, the most obvious example being air entrainment (see Section 4.5). Some plasticising and superplasticising admixtures can create a modicum of air-entrained voids (Roberts and Adderson 1983); Figure 4.89 shows clusters of air-entrained voids in blockwork jointing mortar, indicating the probable presence of a mortar plasticiser.

Chemical analytical techniques are required for detecting and quantifying the presence of an admixture, and various procedures are described by Figg (1989). However, except for chloride-based admixtures, which are now rarely used but will be encountered in many older concretes, most of these materials are organic in nature and present many difficulties to the analyst, particularly in view of the concentrations present as low as 0.02% by weight of cement and as many admixtures are not stable in the long term in the alkaline pore solution. St John (1992) presented chemical data on the analysis and stability of water-reducing agents in concrete: it was found that only the naphthalene sulphonates remained intact.

In some cases, it is possible to extract the organic admixture from a powdered concrete sample with solvents and then identify the nature of the substance using Fourier transform infrared (FTIR) spectroscopy or other specialised analytical techniques, but according to Sibbick (2011), the procedures are difficult, require assumptions on the equivalency of the concrete mix specifications and levels of hydration/activation and may be relatively expensive.

Table 4.34 Main types of chemical admixture for concrete and some of their varieties

Admixture type	Chemical type	Typical concentration mg/100 g concrete
Plasticisers	Lignosulphonates, hydroxycarboxcylic acid salts (e.g. gluconates, leptonates)	20
Pumpability aids (retarders)	Polyhydroxy compounds, tartrates/citrates	20
Superplasticisers	Naphthalene formaldehyde sulphonates	80
	Melamine formaldehyde sulphonates	
Air entrainers	Neutralised wood resins (Vinsol)	3
	Soaps of fatty acid–resin acid mix (tall oil soaps)	
	Surfactants (alkyl aryl sulphonates)	
Accelerators	Calcium chloride	300
	Calcium formate	300
	Triethanolamine	50
	Triethanolamine formate Morpholine derivatives	
Integral waterproofers	Fatty acids and salts (e.g. calcium stearate, oleic acid)	300
Mortar plasticisers	Neutralised wood resins, surfactants	50

Source: After Figg, J.W., *Analysis of Hardened Concrete – A Guide to Tests, Procedures and Interpretation of Results – Report of a Joint Working Party of the Concrete Society and Society of Chemical Industry*, Technical Report No. 32, The Concrete Society, London, U.K., 1989.

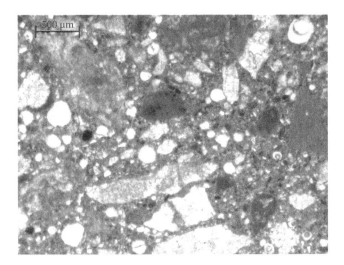

Figure 4.89 Photomicrograph showing clusters of entrained air voids in a blockwork jointing mortar, suggesting the use of a mortar plasticising admixture.

The potential effect of these variables on the resulting FTIR traces can limit the extent of any interpretations that might be made. There has been little published in this field since Hime et al. (1965), which contains the extraction techniques relevant to different groups of admixtures in use at that time.

More recently, direct FTIR analysis of a concrete specimen may be undertaken using a sampling technique, attenuated total reflectance (ATR), which can be used to probe a sample without extensive preparation or solvent extraction. In this technique, a beam of infrared radiation passes through and reflects internally within a crystal. Absorption in the IR region in the sample results in attenuation of an evanescent wave at the surface of the crystal in contact with the sample. This is then passed back to the IR beam and measured by a detector at the other end of the crystal. ATR allows direct investigation of a sample surface and, if the admixture(s) in question absorbs in a region of the IR spectrum not obscured by the bands attributable to the concrete, can provide a good indication of the presence or otherwise of small quantities of organic additives.

Simple qualitative tests for organic matter have sometimes been applied for indicative purposes, but the findings should be treated with circumspection as traces of organic material may well be derived from other concrete constituents (such as the aggregates) or as a minor contamination of the concrete in service.

One common component of many midrange water reducers, lignosulphonate, can be extracted from a hardened concrete in an alkali–carbonate solution. The concentrate derived from this dissolution can then be analysed using UV spectroscopy to determine the relative proportions of the lignosulphonate by comparing its detection at specific maxima. Such a technique can provide a comparative analysis with a sample known to contain the correct original admixture dosage. However, for this to be valid, all the other mix constituents have to be constant, with no additional contaminants incorporated for a reliable determination to be made (Sanders et al. 1995, Sibbick 2011).

Sibbick (2011) advises that although theoretically specific components of some modern admixtures, including nitrates, nitrites, glycols and various sucrose/gluconates, can also be analysed by various techniques, the efficient extraction of these components is problematic, because some of the substances can be lost (volatised) during the crushing, drying and extraction processes.

The recently published second edition of Concrete Society TR 32 (2014) includes a very useful account of the wide range of chemical admixtures available for use in concrete, including the various methods for preparing and analysing samples of hardened concrete to detect and characterise their presence. However, limited guidance is provided on their practical application during the investigation of concrete. In all cases, the availability of suitable reference materials is vital to the analyst charged with determining the presence and dosage of an admixture. Even so, the uncertainty of the modification of admixtures once incorporated into hardened concrete can hinder their successful identification. In the authors' experience, when the importance of the investigation justifies the amount of analysis and its possible cost, the identity and approximate dosage of an admixture can be established using specially prepared concrete reference samples (containing various known dosages of particular admixtures) and their analysis, alongside the samples of concrete in question, using semi-quantitative FTIR.

Examples of FTIR investigation into the presence or otherwise of admixtures in concrete are illustrated in Figures 4.90 and 4.91. In the case shown in Figure 4.90, four reference concretes were prepared, variously including no admixture, admixture A (a standard water-reducing plasticiser), admixture B (a proprietary 'new-generation' shrinkage-reducing special flooring admixture) and an equal mixture of A and B. These references (Figure 4.90a) and four samples from the floor in question (Figure 4.90b) were then solvent extracted, and the extracts were subjected to FTIR analysis. The FTIR spectra for the floor samples each showed a small absorbance peak (only weak in one of the four samples) that was only present in the reference spectra containing admixture A, demonstrating that the floor had been made using the standard admixture variety, rather than the special flooring admixture that had been specified.

In the case shown in Figure 4.91, two samples were derived from concrete that had been specified to contain a particular variety of lignosulphonate-based plasticiser, a reference sample of which was supplied by the manufacturer. Solvent extracts from the two concretes (C2 and C3), together with two samples (one allowed to dry) of the reference admixture, were subjected to FTIR analysis. The spectra for the two concrete samples were quite different. C2 contained a cellulosic substance, which was thus biochemically related to lignosulphonate (sulphonated lignin), but the spectrum was not a match for the particular plasticising admixture specified. C3 contained no cellulosic or other organic admixture, with the absorbance peaks being characteristic of a hydrocarbon, possibly a solvent.

4.8 FIBRE REINFORCEMENT

Conventional steel reinforcement (or non-corrodible alternatives for use in certain hostile environments) is outside the scope of this chapter. However, increasingly various types of fibre are added to concrete mixes, as mix constituents, and these will be encountered by petrographers. The potential benefits of adding fibres (especially animal hair and plant fibres) to mortars, plasters and concretes have been recognised since the earliest times (Davey 1961, 1965), and several types of building material were reliant on the properties of asbestos–cement mixtures for much of the twentieth century (see Chapter 7). Fibres have long been used in specialised or innovative types of modern cementitious material, such as glass fibre–reinforced cement/concrete (GRC) products, but their use as a designed constituent of structural, flooring, precast and even sprayed concretes is comparatively recent.

It seems that the main potential benefits of fibre reinforcement in concrete are reduced shrinkage cracking (fibres can modify the internal shrinkage stresses and thus control the formation and development of shrinkage cracks), increased impact resistance (perceived as a particular advantage for industrial floors) and improved fire performance. In some

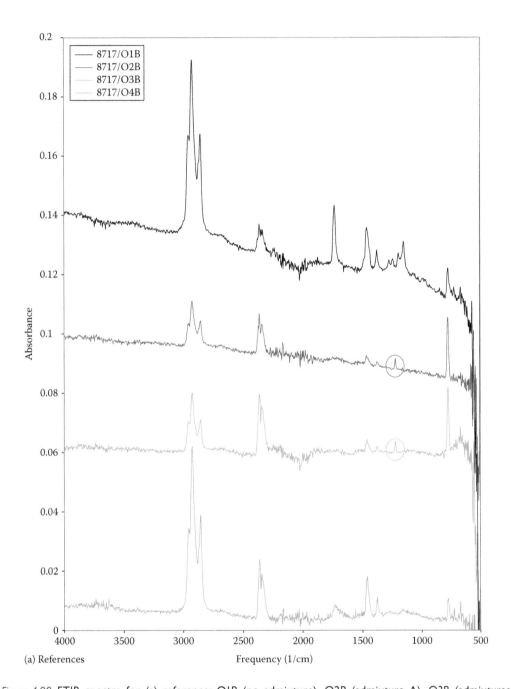

(a) References

Figure 4.90 FTIR spectra for (a) reference: O1B (no admixture), O2B (admixture A), O3B (admixtures A and B), O4B (admixture B); note highlighted peaks in O2B and O3B. *(Continued)*

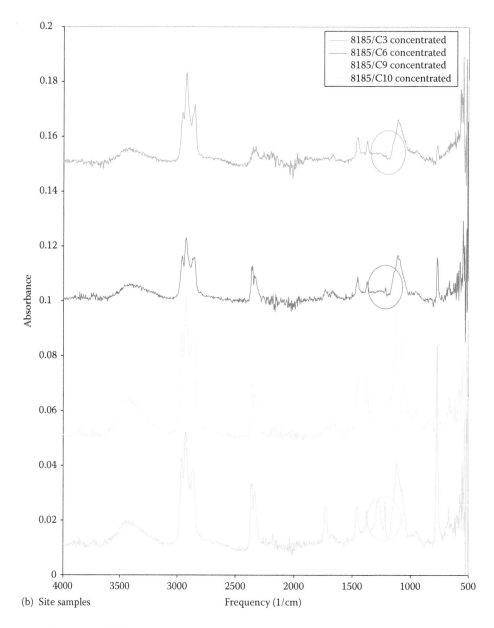

(b) Site samples

Figure 4.90 (Continued) FTIR spectra for (b) ex-floor concrete samples: note highlighted peaks consistent with admixture A. (Spectra courtesy of Dr David Crofts of RSK Environment Ltd., U.K.)

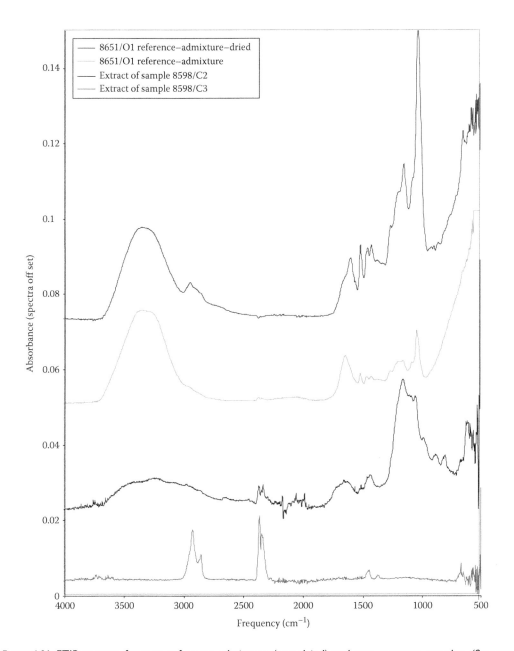

Figure 4.91 FTIR spectra for two reference admixture (one dried) and two concrete samples. (Spectra courtesy of Dr David Crofts of RSK Environment Ltd., U.K.)

circumstances, the use of fibre reinforcement within the concrete might enable the amount of conventional steel reinforcement to be reduced (or even occasionally and increasingly omitted) within the structural unit concerned. General treatments of fibre reinforcement will be found in Bentur and Mindess (1990), Sadegzadeh et al. (2001), Hannant (2003), Beddar (2004), Lambrechts (2005) and Purnell (2007). Brief guidance on recognising and assessing the main types of reinforcement fibre in current use is given in the following subsections. Petrographers will encounter combinations of different fibre types in some modern concretes.

4.8.1 Steel fibres

A most useful compilation on the use of steel fibres and the design of steel fibre–reinforced concrete (SFRC) has been published as Concrete Society Technical Report No. 63 (2007). It is emphasised in CSTR 63 that steel fibre reinforcement serves a distinctly different purpose to conventional reinforcement: 'Steel fibres are added to concrete mainly to influence the way in which concrete cracks as it fails. Microcracks form when concrete is loaded. Subsequently the microcracks coalesce to form macrocracks. Fibres can bridge cracks during loading and, hence, influence mechanical performance. Steel fibres do provide the concrete with a significant post-cracking strength (perhaps half the capacity of the uncracked cross-section) that can be taken into account in design.' Steel fibres are relatively coarse (diameters ≥0.4 mm, see below in this section) and mainly improve post-cracking toughness, but combinations with 'microfibres' (diameters <0.05 mm and typically polypropylene; see Section 4.8.3) are sometimes used to increase the elastic limit and strength of concrete by bridging microcracks.

Steel fibres for reinforcement are available in various sizes, but most have diameters between 0.4 and 1.3 mm, with lengths approximately ranging from 25 to 60 mm, and aspect ratio (ratio of length to diameter) is one of the factors considered to have a strong influence on performance. As well as size, steel fibres vary in their shape or form, being variously straight or deformed and sometimes with shaped ends (see Figure 4.92). These deformation and end features are thought to improve bond and anchorage and thus influence the behaviour of SFRC. Other factors that affect SFRC performance will be dosage

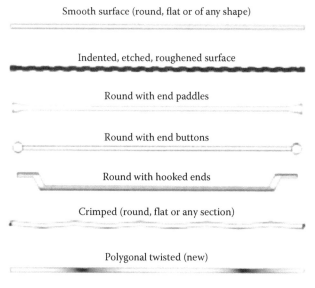

Smooth surface (round, flat or of any shape)

Indented, etched, roughened surface

Round with end paddles

Round with end buttons

Round with hooked ends

Crimped (round, flat or any section)

Polygonal twisted (new)

Figure 4.92 Diagram showing some types of steel fibre for reinforcement. (From Concrete Society, Guidance for the design of steel-fibre-reinforced concrete – Report of a Concrete Society Working Group, Technical Report No 63, The Concrete Society, Camberley, U.K., 2007.)

(in kg/m³), fibre count (number of fibres per unit volume of concrete) and fibre distribution (well dispersed or clustered into certain regions of the concrete or into random clumps), plus the steel material properties, such as tensile strength and elastic modulus. In addition to distribution homogeneity, Barnett et al. (2010) have demonstrated that fibre orientation, which is found to relate to the direction of flow of the fresh concrete, also has an important influence on performance. BS EN 14889-1 (2006) and ASTM A820 (2006) each classify steel fibres according to their method of manufacture (Groups I to V in BS EN 14889-1 and Types 1 to 4 in ASTM A820).

Steel fibre reinforcement is now quite commonly used for concrete floor slabs (especially for industrial applications) and external pavements (Williamson 2011), plus some precast concrete and sprayed concrete (shotcrete). According to CSTR 63 (Concrete Society 2007), steel fibre reinforcement facilitates major improvements in impact resistance and toughness, with various other potential benefits, some of which (such as abrasion resistance) derive indirectly from modification to the properties of the concrete material. Agostinacchio and Cuomo (2004) demonstrated that steel and polymer-modified fibres each also reduced plastic shrinkage movement, with the 60 mm long fibres being more effective than 30 mm long fibres in their trials.

The significant presence of any steel fibre reinforcement will usually be readily apparent, certainly from initial inspection of laboratory samples and often also from on-site surveys (see Figure 4.93). Petrographic examination will enable the shape and size range of any steel fibres to be determined, plus any other fibre types in combined or hybrid fibre–reinforced concrete, and especially the critical homogeneity of fibre distribution and fibre orientation can best be assessed by petrography. Typical steel fibre dosages may be in the range 20–50 kg/m³, with modern plant apparently being capable of batching to an accuracy of ±2 kg/m³, and determinations of fibre dosage and count in hardened concrete are best achieved by physical means. A suitable method is described in BS EN 14721 (2005/2007), whereby three or more concrete core samples are each tested for volume, then crushed such as to enable all the metal fibres to be collected (using a magnet where appropriate), after which the recovered fibres are weighed and expressed in terms of kg/m³ of concrete. In interpreting findings, it should be noted that a maximum batching tolerance of ±5% of the specified dosage is cited (for precast concrete) in BS EN 14650 (2005), though obviously the overall quantity and representativeness of the analysed samples must also be taken into account.

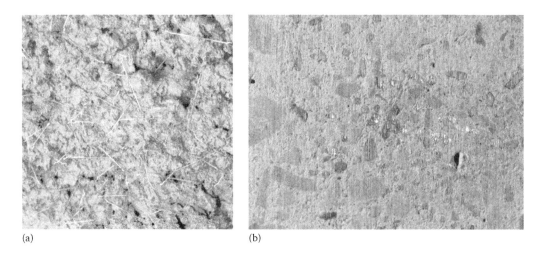

(a)   (b)

Figure 4.93 Steel fibres evident in concrete, (a) on site and (b) in a core sample.

4.8.2 Glass fibres

Glass fibres are rarely used in bulk concrete, owing to economic and mixing factors (Hannant 2003), but have been extensively used since about the 1970s in precast, thin-sheet, flat or shaped, decorative panels and facings for architectural or cladding applications. An early problem with these GRC products concerned disruptive inter-reaction between cement and the 'E glass' (an electrical grade of aluminosilicate glass) used to make the fibres, leading to corrosion and brittleness. The use of alkali-resistant glass fibres, containing zirconia, improves the performance by slowing the inter-reaction but does not necessarily prevent degradation (see Section 7.7.2).

Glass fibres are supplied in 'rovings' (continuous strands of fibres) of up to 64 strands or bundles, each strand or bundle comprising around 200 filaments of about 14 μm diameter, and today apparently with a fibre size holding the strands together that beneficially modifies hydration at the fibre–cement interface (Purnell 2007). Such rovings are chopped into short lengths for inclusion in mixes and products, those lengths varying with application (e.g. 12 mm for precast panels or longer, say up to about 35 mm, for sprayed panels). Typically, GRC contains 2.5%–5% chopped glass fibre strands, but smaller quantities (perhaps much less than 1% and especially when filaments can disperse from the strands) can be effective at inhibiting bleeding and plastic shrinkage cracking (Hannant 2003; Purnell 2007).

According to Muller and Morrison (2000), a degree of slip between filaments within the strands is a key mechanism in determining the mechanical behaviour of GRC. They also maintain that GRC can lose ductility owing to the bonding ('locking-up') effect of calcium hydroxide at the fibre–cement interface, which they say explains the beneficial effects of using metakaolin or other additions or a modified cement. Whatever the explanation, it is certainly the case that modern GRC frequently contains a blended binder and there is also an increasing use of glass fibre mats, meshes and unidirectional layers.

4.8.3 Polymer and other organic fibres

Apart from steel fibres (see Section 4.8.1), polymer fibres are most likely to be encountered during the petrographic investigation of modern concrete, although various other types of organic fibre, notably hair and cellulose (e.g. wood, cotton, paper, etc.), might be encountered with historic concrete and mortar materials (see Section 7.7.3 and Chapter 9). Polymer reinforced concrete is now often used for concrete floors and even external pavements, plus various other types of construction, many precast concrete products and also sprayed concrete.

Small quantities (say around 0.1% by volume) of short (<25 mm long) polymer fibres can modify the properties of fresh concrete, potentially reducing the extents of any bleeding and plastic shrinkage cracking. Agostinacchio and Cuomo (2004) demonstrated that both steel and polymer fibres reduced plastic shrinkage movement, with the 20 kg/m^3 dosage of polymer fibres being more effective than the 10 kg/m^3 dosage in their trials. The benefits for hardened concrete are more limited and less dependable, especially compared with SFRC (see Section 4.8.1), though some polymer fibre–reinforced concrete can exhibit improvements in post-cracking toughness and impact resistance. Grantham (2005) has correctly highlighted the erroneous substitution of polymer fibre reinforcement for air entrainment in the case of some concrete pavements subjected to freeze–thaw cycles. The present authors have similarly investigated an external non-air-entrained, but polymer fibre–reinforced, industrial concrete pavement in Northern England, which was damaged by cyclic freeze–thaw action exacerbated by heavy vehicular trafficking (Figure 4.94). It is possible for steel and polymer fibres to be used together in the same concrete, in order to benefit from both of their complementary but differing influences on performance.

(a)

(b)

(c)

Figure 4.94 Polymer fibre–reinforced concrete pavement damaged by freeze–thaw cycles – there was no air entrainment and the conditions were worsened by the heavy trafficking by manoeuvring heavy goods vehicles, but surface damage was not limited to areas subjected to the regular application of de-icing salt: (a) General view of damage in area subject to heavy trafficking and de-icing salts. (b) Surface deterioration in a non-salted zone. (c) As (b) but with 'inverted colour' image to highlight the surface crazing pattern.

Polymer fibres for concrete are defined and specified in BS EN 14889-2 (2006), which recognises 'microfibres' (<0.30 mm in diameter) as Class I, and the more recent 'macrofibres' (>0.30 mm in diameter) as Class II, for use when an increase in flexural strength is required. Microfibres are available in 'monofilamented' (Class Ia) or more commonly 'fibrillated' (an interconnected mass of fibres, Class Ib) forms (see Figure 4.95).

The fibrillated microfibres usually comprise polypropylene and are made by splitting and stretching sheets or extruded tapes of that polymer, creating individual tape-like fibrils or filaments that are about 15–100 μm thick and around 100–600 μm wide (Purnell 2007). Other than polypropylene, only polyvinyl alcohol (PVA) has been developed commercially for use in fibre-reinforced concrete, although aramid varieties (e.g. Kevlar) could potentially offer engineering benefits at a price. BS EN 14889-2 (2006) indicates that polymer fibres may be straight or deformed and that glued or wrapped fibre 'bundles' are also permitted. As polypropylene is hydrophobic, the fibres are usually surface coated ('spin finished'), to assist dispersion throughout and then to improve bond with the cementitious matrix. Thus, as well as identifying the presence and homogeneity of distribution of any polymer fibres within a concrete sample (Figure 4.96), petrographers should check for and record any

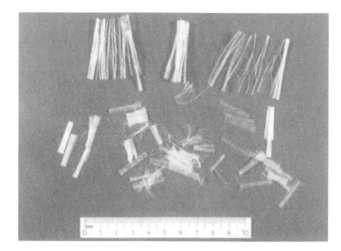

Figure 4.95 Some examples of monofilament and fibrillated polypropylene fibres. (From Hannant, D.J., Fibre-reinforced concrete, in *Advanced Concrete Technology – Processes*, Newman, J. and Choo B.S. [eds], Chapter 6, pp. 6/1–6/17. Elsevier, Oxford, U.K., 2003.)

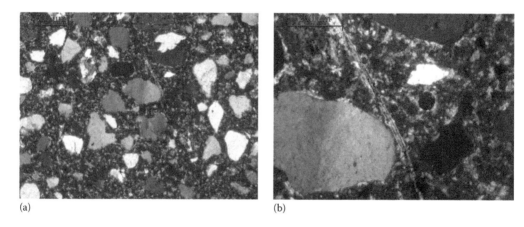

(a)   (b)

Figure 4.96 Appearance of polypropylene fibres in concrete: (a) General view of a concrete matrix. (b) Higher-magnification detail from the centre of (a), showing a portion of fibre.

tendency for those fibres to form clumps, which could possibly suggest defective fibres in the absence of any other evidence of poor mixing practice.

Thus, petrographers should be able to identify the presence of polymer fibres in hardened concrete and assess the success or otherwise of fibre dispersion, but more detailed study of the fibres is difficult on polished surfaces or in thin-section specimens, because typically only portions of individual fibres will be visible and even then at varying angles. In terms of fibre diameter, this can often be estimated with reasonable confidence from cross sections. However, assessments of fibre length and dosage will require the physical breakdown and/or chemical dissolution of a bulk concrete sample, to release all the fibres and allow them to be weighed collectively and have the dimensions of undamaged fibres measured.

This could be a broadly similar process to the BS EN 14721 procedure described for steel fibres (see Section 4.8.1), but obviously, polymer fibres are much more fragile, especially if bundles of filaments have become dispersed and can be damaged by physical breakdown of the concrete. One option is a relatively gentle physical breakdown initially, to enable most of the coarse aggregate to be removed, followed by chemical dissolution of the remainder, after which the polymer fibres can be separated from the insoluble residue (mainly sand and smaller coarse aggregate particles) by a flotation process. The liberated fibres can be subjected to measurements of dimensions, using the definitions and methods given in BS EN 14889-2 (2006) and also noting the permitted tolerances for the fibre manufacturer's declared values. Crofts (2011) advises that he has had some success with separating polymer fibres, for assessment, from 'gently' broken-down concrete using a plastics rod with a static charge.

Further discussion of fibre reinforcement in various products and mortars will be found in Chapters 7, 8 and 9. The authors have experience with the counterproductive use of polymer fibres in 'semi-dry' screed and also with moisture movement instability, leading to permanent deformation, arising from the use of cellulose fibres in some fibre–cement roofing products (Sims 2001).

4.9 ANALYSIS OF CONCRETE

4.9.1 Quantitative analysis of concrete composition

In the foregoing sections of this chapter, we have considered the various constitutional elements of concrete: cements or binders, aggregates, water, voids, additions and pigments, admixtures and fibres. Together these elements variously comprise the composition of a given concrete, with a particular set of relative (or mix) proportions. In this final subsection, we consider the analysis (sometimes previously termed the 'modal analysis') or assessment of a hardened concrete sample in order to endeavour to establish this composition and/or to quantify (or more realistically semi-quantify) these mix proportions. The most commonly employed procedures for the measurement of hardened concrete composition are based on standardised chemical analytical methods, such as BS 1881-124 (1988) (see Sections 4.2 and 4.4). However, there are circumstances in which the chemical procedures are inappropriate and/or unreliable.

Petrographic (or 'micrometric') determinations of cement/binder content can provide a reliable and cost-effective alternative procedure for the measurement of cement and aggregate contents in hardened concrete. The petrographic method has the advantage that it can be carried out as part of routine concrete petrography and can be used to identify factors that might be expected to interfere with standard chemical analytical methods, such as the unexpected presence of cement replacement materials, calcareous aggregates and/or aggregates with probable soluble silica contents.

Chemical determination of cement content in concrete is based on the soluble silica or the acid-soluble calcium values, or preferably both, determined for the concrete. However, this approach may be rendered unreliable when these constituents are also present in the aggregates or the original concentrations have been modified during the life of the concrete. Aggregates containing acid-soluble calcium compounds are common, such as limestone, and some rock materials will release silica during the extraction techniques employed in the chemical analysis of concrete. The unexpected presence of cement replacement materials, such as ggbs or pfa, can also lead to uncertainty in the measurement of cement content by chemical means (although BS 1881-124 [1988] provides a method for ggbs estimation using sulphide content: see also Section 4.2.6). Chemical alteration of concrete in service is commonplace, ranging from mild water leaching of exposed surfaces to severe chemical attack. In the case of alkali–silica reaction (ASR), the formation of soluble silica from reactive aggregates within the concrete occurs (see Chapter 6).

A procedure for the petrographic (or micrometric) measurement of concrete composition has recently been described by the APG (affiliated to the Engineering Group of the Geological Society, London), in APG-SR2 (2010). The first step in this process involves the measurement of volume proportions of aggregates, cement/binder paste (i.e. hydrated matrix of cement and any cementitious additions, plus residual unhydrated and/or unreacted particles) and voids by point-count analysis, using a method based on 'The Modified Point Count Procedure' given in ASTM C457 (2010) (see Section 4.5.3).

The point-count analysis is carried out using polished plate specimens with a surface area of at least 100 cm², but larger areas may be required for very coarse aggregates, and smaller areas may be acceptable for mortars or concretes without coarse aggregate. Figg (1989) suggested that the *minimum* surface area should be at least five times the nominal maximum aggregate size, which equates to 100 mm × 100 mm for 20 mm aggregate. In some cases, it may be appropriate to measure the ratio of fine aggregate to paste separately using a thin section and an automated point-counting stage fitted to a petrological microscope. Image analysis would be equally acceptable to point-count analysis, in principle, although careful calibration may be required for some concrete types – particularly those where there is low contrast between the aggregate and the cement paste (see Sections 4.3.4 and 4.5.2 for discussions on image analysis).

The measured volume proportions of aggregate and cement/binder paste can then be calculated as weight fractions as follows:

1. The aggregate content in kg/m³ can be simply calculated from the measured volume of aggregate and the aggregate density:

$$A\left(\mathrm{kg/m^3}\right) = D_\mathrm{A}\left(\mathrm{kg/m^3}\right) \times \left(V_\mathrm{A}/100\right)$$

where
   A is the aggregate content in kg/m³
   D_A is the density of the aggregate
   V_A is the volume percentage of aggregate

The aggregate density can generally be reliably assessed petrographically from the proportions of the various rock types present.

2. The cement/binder content in kg/m³ can be calculated from the measured volume of paste and the density of the cement/binder:

$$C\left(\mathrm{kg/m^3}\right) = \frac{\left(10 \times V_\mathrm{P}\right)}{\left(w/c + \left(1000/D_\mathrm{C}\right)\right)}$$

where
   C is the cement/binder content in kg/m³
   V_P is the volume of paste
   D_C is the density of the cement/binder in kg/m³
   w/c is the estimated or assumed original water/cement (or water/binder) ratio

Methods for estimating the original water/cement ratio were discussed in Sections 4.4.3 and 4.4.4. The value for D_C (assumed cement/binder density) is adjusted to take into consideration the presence of cement replacement materials, such as pfa or ggbs. The next subsection describes a procedure for the measurement of the level of cement replacement by pfa or ggbs. When there is uncertainty over the original water/cement ratio, the concrete composition can be calculated using several options (e.g. likely lowest, best estimate and likely highest water/cement ratios).

A worked example with 30% cement replacement by pfa provided the following results:

Volume of aggregate = 75%
Volume of paste = 25%
Petrographically measured water/binder ratio = 0.45
Binder composition = 70% by weight Portland cement and 30% by weight pfa
Assumed aggregate density = 2620 kg/m³
Assumed Portland cement density = 3140 kg/m³
Assumed pfa density = 2400 kg/m³
Calculated aggregate content = 0.75 × 2620 = 1965 kg/m³
Calculated cement density = (0.3 × 2400) + (0.7 × 3140) = 2918 kg/m³
Calculated binder content (cement + pfa) = (10 × 25)/(0.45 × (1000/2918)) = 315 kg/m³
Calculated water content = 0.45 × 315 = 142 kg/m³

It is very important to ensure that the ground surfaces used are adequately representative of the concrete under investigation. The slices of concrete must be derived from a location appropriate to the enquiry concerned, be correctly orientated to the placement layering of the concrete and must be large enough in area to overcome the inherent inhomogeneity of concrete (see the *minimum* size recommendations given earlier in this section). Slice orientation will usually be perpendicular to the placement layering, but other orientations might occasionally be appropriate. For example, a set of slices cut at selected levels parallel to the placement layering might be needed if possible segregation is being investigated.

Preparation of the concrete slice specimen surface is broadly similar to that for an air-void system analysis (see Section 4.5.3), except that the face must be very finely ground to enable the various concrete constituents to be clearly distinguished (see Figure 4.97).

Aggregate particle size will be under- or overestimated by this technique, because of oblique or non-central sectioning and depending upon particle shape (as explained in Section 4.3.2), so that the uncorrected content of fine aggregate will be under- or overestimated relative to the coarse aggregate content when the differentiation is only based on particle size (say taking 4 or 5 mm as the coarse/fine threshold). When the coarse and fine aggregates are petrographically distinctive, for example, in the case of a crushed limestone coarse aggregate and a natural quartz sand fine aggregate, this problem can be avoided.

No inter-laboratory trials have been published for the petrographic (or micrometric) method of determining concrete mix proportions. Axon (1962) reported a trial within one

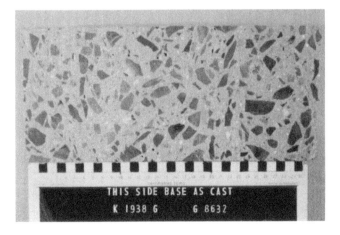

Figure 4.97 Large finely ground and polished slice of concrete suitable for micrometric analysis.

laboratory in the United States, wherein the actual and determined cement contents were remarkably close. In the best of seven mixes, the actual and determined values were the same, and in the worst, the determined value was no more than 29 kg/m³ different from the actual value of 280 kg/m³, which is a maximum error of about 10% relative. Such accuracy might be difficult to achieve for concrete samples lacking supporting information about their constituents. However, Figg (1989) states that 'research does suggest that the precision of micrometric analysis may be at least as good as, or possibly even better than, that attainable by chemical analysis'. This endorsement of the technique has not yet been tested by a statistically valid precision trial but is only thought likely to hold for analyses carried out carefully by a skilled and experienced concrete petrographer.

4.9.2 Estimation of cement replacement by pfa or ggbs

There are currently no standard procedures for the measurement of the level of cement replacement by pfa in hardened concrete. Standard chemical analytical techniques such as those described in BS 1881-124 (1988) are not applicable owing to the uncertainties in the quantities of acid-soluble calcium and soluble silica. BS 1881-124 does include provision for estimating ggbs content in concrete, using analysis for sulphide and assuming, in the absence of reference samples of the constituents, that any sulphide is derived from ggbs.

A method for determining the pfa content of hardened concrete based on SEM- and EDX-type microanalysis is described by French (1994), and the following description is based on a development of this test procedure by Eden (2011). A resin-impregnated polished surface is prepared from a representative portion of the concrete sample. The polished surface is examined with a SEM fitted with a calibrated EDX microanalysis system. Analyses are made of 15 or more 0.1 × 0.1 mm areas of cement/binder paste, excluding aggregate and (insofar as possible) aggregate dust. Spot chemical analyses of at least 15 pfa or ggbs particles are made. Ggbs particles are commonly of very uniform composition in a given concrete sample. With pfa, however, it is important to ensure that the spot analyses obtained for the pfa particles are representative, as pfa commonly contains small amounts of very iron-rich particles that are very conspicuous using backscattered electron imaging.

The mean chemical compositions of the paste (Portland cement and pfa/ggbs) and the pfa/ggbs particles are calculated. In many cases, it is appropriate to assume a composition for the Portland cement; otherwise, it may be possible to obtain a composition for the cement by making chemical analyses of the residual, more representative (larger) unhydrated cement grains.

From the mean compositions of the pfa/ggbs, Portland cement and paste areas (Portland cement and pfa/ggbs), the level of cement replacement by pfa or ggbs can be calculated. The following describes the measurement of the pfa content of concrete:

- The SiO_2 and CaO contents are calculated for the cement/binder paste with levels of cement replacement from 0% to 100% in 10% increments using the measured SiO_2 and CaO contents of the pfa and Portland cement (see Table 4.35, columns 1 through 3).
- The ratio $(SiO_2/(SiO_2 + CaO))$ is calculated for each 10% increment in cement replacement (see Table 4.35, column 4).
- A graph is plotted for the relationship between cement replacement level and the ratio $(SiO_2/(SiO_2 + CaO))$, and a curve is fitted as shown in Figure 4.98.
- The equation of this best-fit line can be used to calculate the pfa content of the binder from the mean SiO_2 and CaO contents of the binder measured with the SEM-EDX, and this is illustrated in the example given in Table 4.36.

Table 4.35 Microchemical data obtained for estimating the proportions of cement and pfa in a concrete binder

Ratio PFA/ (PFA + PC)	Calculated SiO₂ content of binder	Calculated CaO content of binder	Calculated ratio SiO₂/(SiO₂ + CaO) in binder
0	20.3	65.3	0.24
0.1	23.5	59.0	0.28
0.2	26.6	52.6	0.34
0.3	29.8	46.3	0.39
0.4	32.9	39.9	0.45
0.5	36.1	33.6	0.52
0.6	39.3	27.2	0.59
0.7	42.4	20.9	0.67
0.8	45.6	14.6	0.76
0.9	48.7	8.2	0.86
1	51.9	1.9	0.97

Note: These data are derived from a particular sample and are provided as an example; they *cannot* be used as a standard relationship for use with other samples.

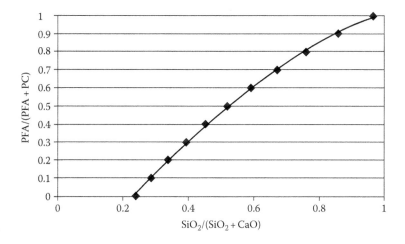

Figure 4.98 Relationship between silica and pfa contents of a concrete binder, based upon the example data given in Table 4.35. These data are derived from a particular sample and are provided as an example; they *cannot* be used as a standard relationship for use with other samples.

Table 4.36 Use of data in Table 4.35 and Figure 4.98 to determine the pfa content of the binder for the concrete example in question

	Compositions of the unknown concrete		
Calculated PFA/(PFA + PC) in the unknown sample	Measured SiO₂ content of binder	Measured CaO content of binder	Measured ratio SiO₂/ (SiO₂ + CaO) in binder
0.30	29.3	45.1	0.39

4.9.3 Applications of concrete composition by petrography

Determinations of concrete composition, either confirmation of constituents or quantification of mix proportions or both, are common requirements in concrete investigation. This chapter is now completed with some typical examples from real cases in which co-author Ian Sims acted as an expert witness: one is a civil dispute over the actual composition of a supplied ready-mixed concrete and the other one aspect of a criminal case, concerning the alleged source of constituents used for making a concrete that was used to encapsulate the remains of a murder victim.

4.9.3.1 Comparison of actual and design concrete mixes

A dispute arose between a contractor and his ready-mixed concrete supplier over concretes of strength grade RC40 for use in pile caps and ground beams and grade RC50 for superstructure columns and floor slabs. Following universal and non-marginal test cube failures and confirmatory tests on drilled cores, the developer instructed the contractor to remove all the RC40 (~80%) and RC50 (~20%) concretes placed to date, after which the contractor started proceedings against the ready-mixed concrete supplier. Basically, the contractor claimed that the concretes supplied had not complied with the project specification, whereas the supplier maintained that the contractor had deliberately added water to the mix on site, thereby raising the water/cement ratio and reducing the hardened concrete strength.

In joint work with the expert instructed by the other party, samples of all the cement, pfa, crushed limestone coarse aggregate, crushed limestone fines and natural sand constituents were obtained and trials carried out to investigate the effects on strength of the RC40 mix of varying the mix proportions including the water/cement ratio. During this joint process, a reference RC40 mix was prepared in the laboratory, using constituents from the same sources originally used by the supplier and also the mix design that had been declared by the supplier. The trial results for this reference mix suggested that it would probably achieve the specified characteristic strength requirements, albeit marginally.

Later, a representative ground beam, which had been cut out of the construction along with all the condemned concrete, but by chance retained on site, became available for investigation. Four cores were drilled, witnessed by the expert for the other party: three vertically through the full depth of the ground beam as placed and a fourth horizontally through the full width (Figure 4.99). All four cores were inspected in the laboratory, and one of the through-depth cores (C1, by chance) was selected as being visually consistent in terms of constituent distribution with depth and suitable to represent the full set of samples. Six relatively large-area thin sections (each 75 × 50 mm) were then prepared from positions evenly spaced along the 600 mm length of the core (depth of the beam) and each subjected to microscopical modal analysis using the point-counting technique (see Section 4.9.1). Similarly, large-area thin sections were made from one of the untested cubes of reference mix concrete made in the laboratory trials (see above in this section) and subjected to microscopic point-count analysis by the same operator during the same session.

The modal analysis results for the ex-site core and cube of reference concrete are summarised in Table 4.37 and Figure 4.100. Separate counts were made for the three aggregate types (coarse aggregate, fines and natural sand), the total binder (hydrated matrix, plus included residual visible grains of unhydrated cement *and* unreacted pfa), the unreacted grains of pfa and air voids. It is immediately apparent from the findings in Table 4.37 and Figure 4.100 that the ex-site and reference concrete mixes were substantially different.

Figure 4.99 100 mm diameter core drilling locations on the residual ground beam representing the condemned concrete. Note that the beam is shown on its side, having been turned over for drilling the 'horizontal' through-width sample. (Photograph courtesy of RSK Environment Ltd., U.K.)

Table 4.37 Case study investigation: summary of modal analysis results for the ex-site core and reference cube

	Relative proportions, % by volume[a]			
	Ground beam, core, C1		Trial mix, cube, TM1	
Constituents	Separate	Combination	Separate	Combination
Crushed limestone coarse aggregate	30		41	
Natural quartzitic sand	13		14	
Crushed limestone fines	44		23	
Cement matrix (cement and hydrates)	10		17	
Unreacted pfa	1		3	
Total binder (cement, pfa and hydrates)		11		20
Air voids	1		2	

Note: The data are from an actual case in the United Kingdom, but its identity and details must remain confidential.

[a] Overall data, based on six large-area thin sections from the ground beam core and two large-area sections from the trial mix cube.

The proportional contents of natural sand were similar, but otherwise, the ex-site concrete appeared to have a much higher content of limestone fines and a lower content of total binder.

In order to enable the volumetric modal analysis for the ex-site concrete to be converted into mix proportions, expressed as kg/m³ in the conventional manner and then calibrated against the reference concrete, corrections were devised by comparing the mix proportions calculated from the counting data for the reference mix with the known mix proportions used in the joint trial. This enabled the mix proportions calculated from the counting data for the ex-site concrete to be adjusted, to reflect the relationship established between the counting data and known mix proportions for the reference sample (Table 4.38). It was clear that the ex-site concrete, at least as represented by the sample analysed, had a substantially

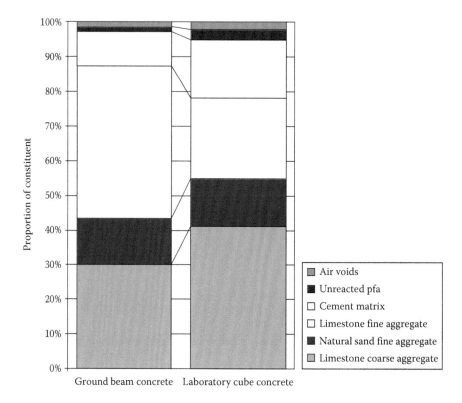

Figure 4.100 Chart summarising the modal analysis results in Table 4.37.

Table 4.38 Case study investigation: comparison of declared and actual mix proportions, using the modal analysis data given in Table 4.37

Constituents	Mix proportions, kg/m³, converted from % by volume			Declared mix proportions, kg/m³	Comparison of actual CI mix with declared mix
	Cube, TM I	Ground beam, CI			
	As found	As found	Adjusted[a]		
Limestone coarse aggregate	1107	810	751	1019	Lower
Natural sand	366	341	317	332	Similar
Limestone fines	621	1188	940	497	Much higher
Cement matrix	212	122	136	231	Lower
Unreacted pfa	—	—	40	99	Lower
Total binder	—	—	176	330	Much lower

Note: The data are from an actual case in the United Kingdom, but its identity and details must remain confidential.

[a] Adjusted using factors derived from comparing the declared mix proportions and the as-found data for the trial mix cube, which was made under controlled laboratory conditions using the declared mix proportions.

higher content of limestone fines than had been declared, while the total binder content (cement and pfa) was also significantly lower than declared. These findings were together considered sufficient to explain the lower-than-expected strengths, without recourse to the allegedly deliberate addition of mix water by the contractor.

Differentiation of the cement and pfa components of the binder was attempted, relating the separate counting data for residual unreacted pfa grains to the known content of pfa (from the same source) in the reference mix. Arguably, this should be fairly reliable in this case, given the availability of both pfa from the same source and a carefully controlled laboratory reference mix. If so, the adjusted data suggested that the pfa content of the binder in the ex-site concrete was rather lower (at 23%) than had been declared (30%).

4.9.3.2 Concrete in a murder case

A young man was accused and later convicted of murdering his father, then dismembering and incinerating the body, before interring the remains in a 'concrete' grave behind the garage at his father's property in central England. In this criminal case, Sims' involvement in 2010 was limited to the aspect of the three layers of mortar and a final layer of concrete that had allegedly formed the grave and particularly whether or not there was reliable evidence that the defendant had made that mortar and concrete.

During their excavation of the area behind the garage, the Scene of Crime Officers (SOCO) had noted the various mortar and concrete layers and taken samples for possible forensic investigation. Later, within the garage and in various parts of the main house, they discovered the residues of various building materials, including part-used bags of sand and various proprietary repair cements and mortars in plastic drums. Although the samples of mortar and concrete from the scene were not examined on behalf of the police, it was nevertheless suggested by the prosecution that the various materials found in the garage and house were remnants left over from mixing the mortar and concrete. Sims was asked by those representing the defendant to ascertain whether or not this was the case.

Obtaining contextually relevant samples was a challenge, because SOCO had completely excavated the 'grave' and then randomly backfilled the area. It was thus necessary to take various samples of mortar and concrete from the mixed backfill and then take them for visual comparison with the police samples at the forensic science laboratory in question. Once the close similarity between the defendant's and police samples had been established, it was possible to conduct detailed petrographic examination on the new samples. This established that the three mortar layers comprised closely similar materials, albeit placed at slightly different times.

The mortars were extremely well mixed and made using a relatively high binder content (~400 kg/m^3), comprising Portland cement and pfa (possibly a CEM II B-V cement in BS EN 197-1 [2000] parlance) and a sea-dredged sand (Figure 4.101). This quality of mixing and composition suggested a ready-mixed ('just add water') product, but there were key differences between the mortars and the residues found in the garage and house. Reference to datasheets for the residual products indicated that the sands were finer and more quartz rich than those found in the mortars, while the binders in the repair mortars contained significant amounts of high-alumina cement (HAC) and no pfa. It was clear that the mortars used to make the grave had no connection with the residues found in the garage and house.

Similarly, the concrete layer, which covered the mortar layers, also differed compositionally from the residues: obviously, the concrete contained a coarse aggregate (crushed granite), of which there was no residual material, again the sand was sea dredged and coarser than that found in part-used bags and once again the binder contained pfa and certainly no HAC.

Rogers (2011) has advised a slightly similar murder case in the Niagara area of Canada in 1991. The murderer encapsulated parts of his female victim's body in five concrete blocks inexpertly made using ready-mixed ('just add water') materials that he obtained from the local hardware store, then deposited the blocks in a nearby hydroelectric pump storage lake. The blocks were discovered when the water level in the lake dropped as a routine result of water being channelled into the power station the next morning, and the composition of the concrete was confirmed as being consistent with the bags of ready-mixed ingredients that the murderer had purchased using his credit card!

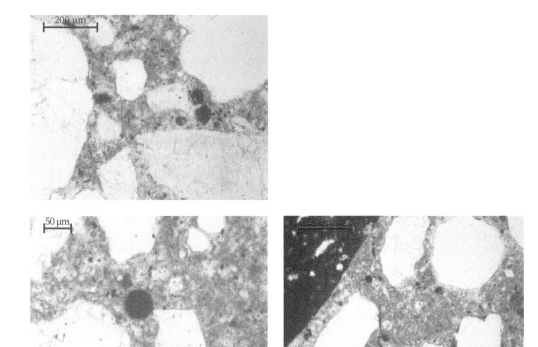

Figure 4.101 Thin-section views of the mortar forming the grave, showing the absence of any HAC, but the presence of some unreacted pfa.

REFERENCES

Aarre, T. 1995: Influence of measurement technique on the air void structure of hardened concrete. *ACI Materials Journal* 92, 6, 599–604.

AFNOR. 2010: *Métakaolin, addition pouzzolanique pour bétons – Définitions, spécifications, critères de conformité*, NF P18-513, Association Française de Normalisation, Saint-Denis Cedex, France (in French).

Agostinacchio, M. and Cuomo, G. 2004: Plastic shrinkage reduction in concrete pavements due to the addition of both steel and polymer-modified fibres, *Concrete* 38, 6, 44–46.

Alexander, M. and Mindess, S. 2005: *Aggregates in Concrete*, Modern Concrete Technology Series 13. Taylor & Francis, Oxford, U.K.

Aligizaki, K.K. 2005: *Pore Structure of Cement-Based Materials: Testing, Interpretation and Requirements*, Modern Concrete Technology. Taylor & Francis, Oxford, U.K., 432pp.

Ambroise, J., Martin-Calle, S. and Pera, J. 1982: Pozzolanic behaviour of thermally activated kaolin. In *Proceedings of the Fourth International Conference on Fly Ash, Silica Fume, Slag and Natural Pozzolans in Concrete*, Malhotra, V.M. (ed.). Istanbul, Turkey, 1, pp. 731–747.

Ambroise, J., Murat, M. and Pera, J. 1985a: Hydration reaction and hardening of calcined clays and related minerals. IV. Experimental conditions for strength improvement on metakaolinite minicylinders, *Cement and Concrete Research* 15, 83–88.

Ambroise, J., Murat, M. and Pera, J. 1985b: Hydration reaction and hardening of calcined clays and related minerals. V. Extension of the research and general conclusions, *Cement and Concrete Research* 15, 261–268.

American Concrete Institute. 1982a: *Guide to Durable Concrete*, ACI 201.2R-77 (reapproved 1982). ACI, Detroit, MI.

American Concrete Institute. 1982b: *Standard Practice for Concrete Highway Bridge Deck Construction*, ACI 345–82. ACI, Detroit, MI.

American Concrete Institute. 1983: *Building Code Requirements for Reinforced Concrete*, ACI 318–83. ACI, Detroit, MI.

American Concrete Institute. 1989: *Specifications for Structural Concrete for Buildings*, ACI 301-89 (revised version 1972 edition). ACI, Detroit, MI.

American Concrete Institute. 2002: *Standard Practice for Selecting Proportions for Normal, Heavyweight, and Mass Concrete*, ACI 211-1-91 (reapproved 2002). ACI, Detroit, MI, 38pp.

American Concrete Institute. 2006: *Guide for the Use of Silica Fume in Concrete*, ACI 234R-06. ACI, Farmington Mills, MI. (Replaced 1996, reapproved 2000, edition)

Andriolo, F.R. and Sgaraboza, B.C. 1987: The use of pozzolan from calcined clays in preventing excessive expansion due to the alkali-aggregate reaction in some Brazilian dams. In *Proceedings of the Seventh International Conference, Concrete Alkali-aggregate Reactions, 1986*, Grattan-Bellew, P.E. (ed.). Ottawa, Ontario, Canada, pp. 66–70, Noyes, NJ, 509pp.

Anon. 1976: Air entrainment and concrete. *Concrete Construction* March, 105–111.

Applied Petrography Group. 2010: *A Code of Practice for the Petrographic Examination of Concrete*, Special Report 2 (APG SR2). APG, Engineering Group, Geological Society, London, U.K., 20pp. (compiled by Eden, M.A., ppAPG).

Applied Petrography Group. 2010: *A Glossary of Terms and Definitions, Part 1. Aggregate Terminology*, Special Report 3, Part 1 (APG SR3 Part 1). APG, Engineering Group, Geological Society, London, U.K., 21pp. (edited by Poole, A.B., ppAPG).

Arnold, D.R. (Chairman) 1980: Pigments for integrally colored concrete (Task Group of ASTM Subcommittee Section C09.03.08.05 on methods of testing and specifications for admixtures). *Cement, Concrete, and Aggregates, CCAGDP* 2, 2, 74–77.

Aroni, S., de Groot, G.J., Robinson, M.J., Svanholm, G., Wittman, F.H. (eds.) 1993: *Autoclaved Aerated Concrete – Properties, Testing and Design – RILEM Recommended Practice – RILEM Technical Committees 78-MCA and 51-ALC*. E & F N Spon, London, U.K. (pp RILEM), 404pp.

Asbridge, A.H. 1994: *Metakaolin in Cement Mortars – The Effect on Chloride Ion Diffusion and Porosity*, BSc submission. University of Plymouth, Plymouth, U.K.

Asbridge, A.H., Chadbourn, G.A. and Page, C.L. 2001: Effects of metakaolin and the interfacial transition zone on the diffusion of chloride ions through cement mortars. *Cement and Concrete Research* 31, 1567–1572.

Asbridge, A.H., Jones, T.R. and Osborne, G.J. 1996: High performance metakaolin concrete: Results of large scale trials in aggressive environments. In *Proceedings of the International Conference on Concrete in the Service of Mankind, Dundee, U.K.*, Dhir, R.K. and Hewlett, P.C. (eds.). E & F N Spon (part of Taylor and Francis), Oxford, U.K., pp. 13–24.

Asbridge, A.H., Walters, G.V. and Jones T.R. 1994: Ternary blended concretes – OPC/ggbfs/Metakaolin, In *Proceedings of a Conference, Concrete across Borders*, Birkerød, Denmark (June 1994).

Asgeirsson, H. 1986: Silica fume in cement and silane for counteracting of alkali-silica reactions in Iceland. *Cement and Concrete Research* 16, 3, 423–428.

Ash, J.E. 1972: Bleeding in concrete – A microscopy study. *ACI Journal* 69, 4, 209–211. Title No. 69-19.

ASTM A820, 2006: *Standard Specification for Steel Fibers for Fiber Reinforced Concrete*. American Society for Testing and Materials, West Conshohocken, PA.

ASTM C33, 2011: *Standard Specification for Concrete Aggregates*. American Society for Testing and Materials, West Conshohocken, PA.

ASTM C88, 2005: *Test for Soundness of Aggregates by Use of Sodium Sulfate or Magnesium Sulfate*. American Society for Testing and Materials, West Conshohocken, PA.

ASTM C94, 2012: *Standard Specification for Ready-Mixed Concrete*. American Society for Testing and Materials, West Conshohocken, PA.

ASTM C125, 2007: *Standard Terminology Relating to Concrete and Concrete Aggregates*. American Society for Testing and Materials, West Conshohocken, PA.

ASTM C125, 2010: *Standard Terminology Relating to Concrete and Concrete Aggregates.* American Society for Testing and Materials, West Conshohocken, PA.

ASTM C150, 2012: *Specification for Portland Cement.* American Society for Testing and Materials, West Conshohocken, PA.

ASTM C151, 2009: *Test for Autoclave Expansion of Hydraulic Cement.* American Society for Testing and Materials, West Conshohocken, PA.

ASTM C231, 2010: *Standard Test Method for Air Content of Freshly Mixed Concrete by the Pressure Method.* American Society for Testing and Materials, West Conshohocken, PA.

ASTM C294, 2005: *Descriptive Nomenclature for Constituents of Concrete Aggregates.* American Society for Testing and Materials, West Conshohocken, PA.

ASTM C457, 2011: *Standard Test Method for Microscopical Determination of Parameters of the Air Void System in Hardened Concrete.* American Society for Testing and Materials, West Conshohocken, PA.

ASTM C595, 2012: *Standard Specification for Blended Hydraulic Cements.* American Society for Testing and Materials, West Conshohocken, PA.

ASTM C618, 2008: *Standard Specification for Coal Fly Ash and Raw or Calcined Natural Pozzolan for Use in Concrete.* American Society for Testing and Materials, West Conshohocken, PA.

ASTM C856, 2011: *Standard Practice for Petrographic Examination of Hardened Concrete.* American Society for Testing and Materials, West Conshohocken, PA.

ASTM C979, 2010: *Specification for Pigments for Integrally Colored Concrete.* American Society for Testing and Materials, West Conshohocken, PA.

ASTM C1723, 2010: *Standard Guide for Examination of Hardened Concrete Using Scanning Electron Microscopy.* American Society for Testing and Materials, West Conshohocken, PA.

Attiogbe, E.K. 1993: Mean spacing of air voids in hardened concrete. *ACI Materials Journal* 90, 2 174–181, Title No. 90-M19 (discussion in 91, 1, 1994, 121–123).

Axon, E.G. 1962: Method of estimating the original mix composition of hardened concrete using physical tests. *Proceedings of the American Society for Testing Materials* 62, 1068–1080.

Barnett, S.J., Lataste, J-F., Parry, T., Millard, S.G. and Soutsos, M.N. 2010: Assessment of fibre orientation in ultra high performance fibre reinforced concrete and its effect on flexural strength. *Materials and Structures* 43, 1009–1023.

Bassett, H. 1934: Notes on the system lime-works, and on the determination of calcium. *Journal of the Chemical Society, Part 11,* 276, 1270–1275.

Bate, S.C.C. 1974: *Report on the Failure of Roof Beams at Sir John Cass Foundation and Red Coat Church of England School, Stepney.* BRE Current Paper CP58/74, Building Research Establishment, Watford, U.K.

Beddar, M. 2004: Fibre-reinforced concrete – Past, present and future, *Concrete* 38, 4, 47–49.

Bensted, J. and Barnes, P. 2002: *Structure and Performance of Cements,* 2nd ed. Spon, London, U.K., 565pp.

Bentur, A. and Mindess, S. 1990: *Fibre Reinforced Cementitious Composites.* Elsevier Applied Science, London, U.K., pp. 22–32.

Berger, R.L. and McGregor, J.D. 1973: Effect of temperature and water-solid ratio on growth of $Ca(OH)_2$ crystals formed during the hydration of Ca_3SiO_5. *Journal of the American Ceramic Society* 56, 2, 73–79.

Berra, M. and Baronio, G. 1987: Thaumasite in deteriorated concretes in the presence of sulfates. In *Concrete Durability: Katharine and Bryant Mather International Conference,* Scanlon, J.M. (ed.). ACI SP-100, 2073–2089. Paper SP100-106. American Concrete Institute, Detroit, MI.

Berry, E.E. and Malhotra, V.M. 1987: Fly ash in concrete. In *Supplementary Cementing Materials for Concrete,* Malhotra, V.M. (ed.). Canada Centre for Mineral and Energy Technology (CANMET), Canadian Government Publishing Centre, Ottawa, Ontario, Canada, Chapter 2, pp. 35–163.

Bied, J. 1926: *Recherches industrielles sur les chaux, ciments en mortiers.* Dunod, Paris, France. (Completed by Chaix, M. after the death of Bied, L)

Blott, S.J. and Pye, K. 2008: Particle shape: A review and new methods of characterization and classification. *Sedimentology* 55, 31–63.

Bogue, R.H. 1955: *The Chemistry of Portland Cement,* 2nd ed. Reinhold, New York.

Bredy, P., Chabannet, M. and Pera, J. 1989: Microstructure and porosity of metakaolin blended cements. (1988 MRS Fall Meeting, Boston, USA, Pore Structure and Permeability of Cementitious Materials). *Materials Research Society Proceedings* 137, 431–436.

Brocklebank, I. 2012: *Building Limes in Conservation*, Donhead Publishing (Verlag). ISBN 978-1-873394-95-3.

Bromley, A.V. 2011: Personal communication.

Brouwers, H.J.H. 2004: The work of Powers and Brownyard revisited: Part 1. *Cement and Concrete Research* 34, 9, 1697–1716.

Brouwers, H.J.H. 2005: The work of Powers and Brownyard revisited: Part 2. *Cement and Concrete Research* 35, 10, 1922–1936.

Brown, B.V. 1982: Air entrainment (Part 1) (Concrete Society Current Practice Sheet no. 80). *Concrete* 16, 12, 59–60.

Brown, B.V. 1983: Air entrainment (Part 2) (Concrete Society Current Practice Sheet no. 81). *Concrete* 17, 1, 45–46.

Brunauer, S., Kantro, D.L. and Copeland, L.E. 1958: The stoichiometry of the hydration of β-dicalcium silicate and tricalcium silicate at room temperature. *Journal of the American Chemical Society* 80, 4, 761–767.

BS 12, 1991: *Specification for Portland Cement*. British Standards Institution, London, U.K.

BS 146, 1991: *Specification for Portland Blastfurnace Cement*. British Standards Institution, London, U.K.

BS 812, 1975: *Methods of Sampling and Testing of Mineral Aggregates, Sands and Fillers*. British Standards Institution, London, U.K. Superseded.

BS 812-102, 1989: *Testing Aggregates, Part 102, Methods for Sampling*. British Standards Institution, London, U.K.

BS 812-103, 1985: *Testing Aggregates, Part 103, Method of Determination of Particle Size Distribution, Section 103:1, Sieve Tests*. British Standards Institution, London, U.K.

BS 812-104, 1994: *Method for Qualitative and Quantitative Petrographic Examination of Aggregates, Part 104*. British Standards Institution, London, U.K.

BS 812-105:1, 1989: *Testing Aggregates, Part 105, Method for Determination of Particle Shape, Section 105:1, Flakiness Index for Coarse Aggregates*. British Standards Institution, London, U.K. See also BS EN 933-3.

BS 812-105:2, 1990: *Testing Aggregates, Part 105, Method for Determination of Particle Shape, Section 105:2, Elongation Index for Coarse Aggregates*. British Standards Institution, London, U.K.

BS 812-120, 1989/1995: *Method for Testing and Classifying Drying Shrinkage of Aggregates in Concrete, Part 120*. British Standards Institution, London, U.K. See also BS EN 1367-4.

BS 812-121, 1989: *Testing Aggregates, Part 121, Methods for Determination of Soundness*. British Standards Institution, London, U.K. See also BS EN 1367-2.

BS 882, 1992: *Specification for Aggregates from Natural Sources for Concrete*. British Standards Institution, London, U.K. See also BS EN 12620.

BS 882, 1201-2, 1973: *Specification for Aggregates from Natural Sources for Concrete (Including Granolithic)*. British Standards Institution, London, U.K.

BS 890, 1972 (1995): *Specification for Building Limes*. British Standards Institution, London, U.K.

BS 1014, 1975 (1986): *Specification for Pigments for Portland Cement and Portland Cement Products*. British Standards Institution, London, U.K. See also BS EN 12878.

BS 1198, 1199 and 1200, 1976: *Specification for Building Sands from Natural Sources*. British Standards Institution, London, U.K.

BS 1377, 1990: *Methods of Test for Soils for Civil Engineering Purposes, Part 3, Chemical and Electrochemical Tests*. British Standards Institution, London, U.K.

BS 1881-124, 1988: *Testing Concrete, Part 124: Methods for Analysis of Hardened Concrete*. British Standards Institution, London, U.K.

BS 3892-1, 1997: *Use of Pulverised Fuel Ash with Portland Cement*. British Standards Institution, London, U.K. Withdrawn in 2005. See also BS EN 450.

BS 4027, 1996: *Specification for Sulfate-Resisting Portland Cement*. British Standards Institution, London, U.K.

BS 4246, 1991: *Specification for High Slag Blastfurnace Cement*. British Standards Institution, London, U.K.

BS 4248, 1974: *Specification for Supersulphated Cement*. British Standards Institution, London, U.K.

BS 4550-3:7, 1978: *Methods of Testing Cement. Part 3, Physical Tests, Section 3:7, Soundness Test*. British Standards Institution, London, U.K. See also BS EN 196-3.

BS 5328-1, 1997: *Concrete, Part 1: Guide to Specifying Concrete*. British Standards Institution, London, U.K. See also BS EN 206-1.

BS 5497-1, 1987: *Precision of Test Methods. Guide for the Determination of Repeatability and Reproducibility for a Standard Test Method by Inter-Laboratory Tests* (ISO 5715-1986). British Standards Institution, London, U.K. Superseded.

BS 6089, 2010: *Assessment of In-Situ (SIC) Compressive Strength in Structures and Precast Concrete Components – Complementary Guidance to That Given in BS EN 13791*. British Standards Institution, London, U.K.

BS 6463, 1984–1987: *Quicklime, Hydrated Lime and Natural Calcium Carbonate, Part 1–4*. British Standards Institution, London, U.K.

BS 6588, 1996: *Portland Pulverized-Fuel Ash Cement*. British Standards Institution, London, U.K. Withdrawn in 2000.

BS 6610, 1996: *Specification for Pozzolanic Pulverized-Fuel Ash Cement*. British Standards Institution, London, U.K. Withdrawn in 2013.

BS 6699, 1986: *Specification for Ground Granulated Blastfurnace Slag for Use with Portland Cement*. British Standards Institution, London, U.K. Replaced by BS 6699: 1992.

BS 6699, 1992: *Specification for Ground Granulated Blastfurnace Slag for Use with Portland Cement*. British Standards Institution, London, U.K. Withdrawn in 2006. See also BS EN 15167.

BS 7583, 1992: *Specification for Portland Limestone Cement*. British Standards Institution, London, U.K.

BS 8110-1, 1997: *Structural Use of Concrete. Part 1, Code of Practice for Design and Construction*. British Standards Institution, London, U.K. See also BS EN 206-1 & BS 8500-1&2.

BS EN 196-3, 2005: *Methods of Testing Cement – Part 3: Determination of Setting Times and Soundness*. British Standards Institution, London, U.K.

BS EN 196-6, 1992: *Methods of Testing Cement. Part 6: Determination of Fineness*. British Standards Institution, London, U.K.

BS EN 196-7, 2007: *Methods of Testing Cement. Part 7: Methods of Taking and Preparing Samples of Cement*. British Standards Institution, London, U.K.

BS EN 197-1, 2000: *Cement. Composition, Specifications and Conformity Criteria for Common Cements*. British Standards Institution, London, U.K.

BS EN 197-1, 2011: *Cement. Composition, Specifications and Conformity Criteria for Common Cements*. British Standards Institution, London, U.K.

BS EN 206-1, 2000: *Concrete. Specification, Performance, Production and Conformity*. British Standards Institution, London, U.K.

BS EN 450, 1995: *Fly Ash for Concrete – Definitions, Requirements and Quality Control*. British Standards Institution, London, U.K.

BS EN 450-1, 2005(+A1, 2007): *Fly Ash for Concrete, Part 1, Definition, Specifications and Conformity Criteria*. British Standards Institution, London, U.K.

BS EN 450-2, 2005: *Fly Ash for Concrete, Part 2, Conformity Evaluation*. British Standards Institution, London, U.K.

BS EN 459-1, 2001: *Building Lime. Part 1: Definitions, Specifications and Conformity Criteria*. British Standards Institution, London, U.K.

BS EN 459-1, 2010: *Building Lime. Part 1: Definitions, Specifications and Conformity Criteria*. British Standards Institution, London, U.K.

BS EN 480-11, 2005: *Admixtures for Concrete, Mortar and Grout – Test Methods – Part 11: Determination of Air Void Characteristics in Hardened Concrete*. British Standards Institution, London, U.K.

BS EN 933-3, 1997: *Tests for Geometrical Properties of Aggregates – Determination of Particle Shape – Flakiness Index*. British Standards Institution, London, U.K.

BS EN 1367-2, 1998: *Tests for Thermal and Weathering Properties of Aggregates – Magnesium Sulfate Test*. British Standards Institution, London, U.K.

BS EN 1367-4, 1998: *Tests for Thermal and Weathering Properties of Aggregates – Determination of Drying Shrinkage*. British Standards Institution, London, U.K.

BS EN 8500-1, 2006: *Concrete. Complementary British Standard to BS EN 206-1. Part 1: Method of Specifying and Guidance for the Specifier*. British Standards Institution, London, U.K.

BS EN 8500-2, 2006: *Concrete. Complementary British Standard to BS EN 206-1. Part 2: Specification for Constituent Materials and Concrete*. British Standards Institution, London, U.K.

BS EN 12350-7, 2009: *Testing Fresh Concrete. Part 7: Air Content – Pressure Methods*. British Standards Institution, London, U.K.

BS EN 12504-1, 2009: *Testing Concrete in Structures, Part 1: Cored Specimens – Taking, Examining and Testing in Compression*. British Standards Institution, London, U.K. Includes National Annex NA, Guidance on the use of BS EN 12504-1.

BS EN 12620, 2002 (+A1, 2008): *Aggregates for Concrete*. British Standards Institution, London, U.K. See also BS PD 6682-1.

BS EN 12670, 2002: *Natural Stone Terminology*. British Standards Institution, London, U.K.

BS EN 12878, 2005: *Pigments for the Colouring of Building Materials Based on Cement and/or Lime – Specifications and Methods of Test*. British Standards Institution, London, U.K.

BS EN 13263-1, 2005 (+A1: 2009): *Silica Fume for Concrete. Definitions, Requirements and Conformity Criteria*. British Standards Institution, London, U.K.

BS EN 13791, 2007: *Assessment of In-Situ (SIC) Compressive Strength in Structures and Precast Concrete Components*. British Standards Institution, London, U.K. See also BS 6089.

BS EN 14650, 2005: *Precast Concrete Products – General Rules for Factory Production Control of Metallic Fibered (SIC) Concrete*. British Standards Institution, London, U.K.

BS EN 14721, 2005 (+A1: 2007): *Test Method for Metallic Fibre Concrete – Measuring the Fibre Content in Fresh and Hardened Concrete*. British Standards Institution, London, U.K.

BS EN 14889-1, 2006: *Fibres for Concrete – Part 1: Steel Fibres – Definitions, Specifications and Conformity*. British Standards Institution, London, U.K.

BS EN 14889-2, 2006: *Fibres for Concrete – Part 2: Polymer Fibres – Definitions, Specifications and Conformity*. British Standards Institution, London, U.K.

BS EN 15167-1, 2006: *Ground Granulated Blastfurnace Slag for Use in Concrete, Mortar and Grout, Part 1. Definitions, Specifications and Conformity Criteria*. British Standards Institution, London, U.K.

BS EN 15167-2, 2006: *Ground Granulated Blastfurnace Slag for Use in Concrete, Mortar and Grout, Part 2. Conformity Evaluation*. British Standards Institution, London, U.K.

BS PD 6682-1, 2009: *Aggregates. Part 1: Aggregates for Concrete – Guidance on the Use of BS EN 12620*. British Standards Institution, London, U.K.

Building Research Establishment. 1973: *Report on the Collapse of the Roof of the Assembly Hall of the Camden School for Girls*. Department of Education and Science, HMSO, London, U.K.

Building Research Establishment. 1981: *Assessment of Chemical Attack of High Alumina Cement Concrete*. BRE Information Paper IP22/81, Watford, U.K.

Building Research Establishment. 1994: *Assessment of Existing High Alumina Cement Concrete Construction in the UK*. BRE Digest 392, Watford, U.K.

Building Research Establishment. 2002: *Minimising the Risk of Alkali-Silica Reaction: Alternative Methods*. Information Paper IP1/02, BRE/CRC, Watford, U.K.

Building Research Establishment. 2004: *Alkali-Silica Reaction in Concrete*. Digest 330, Parts 1–4, IHS/BRE, Watford, U.K.

Bye, G.C. 1983: *Portland Cement – Composition, Production and Properties*. Institute of Ceramics/Pergamon Press, Oxford, U.K.

Bye, G.C. 1999: *Portland Cement – Composition, Production and Properties*, 2nd ed. Thomas Telford, London, U.K., 225pp.

Bye, G.C. 2011: *Portland Cement*, 3rd ed., Livesey, P. and Struble, L. (eds.). ICE Publishing (Thomas Telford), London, U.K., 217pp.

Cabrera, J.G. and Woolley, G.R 1985: A study of twenty-five year old pulverized fuel ash concrete used in foundation structures. *Proceeding – Institution of Civil Engineers*, Part 2, 79, 149–165, Paper 8885.

Cahill, J., Dolan, J.C. and Inward, P.W. 1994: The identification and measurement of entrained air in concrete using image analysis. *Petrography of Cementitious Materials*. ASTM STP 1215, 111-24, American Society for Testing and Materials, Philadelphia, PA.

Caldarone, M.A., Gruber, K.A. and Burg, R.G. 1994: High-reactivity meta-kaolin: A new generation mineral admixture. *Concrete International* 16, 11, 37–40.

Campbell, D.H. 1986: *Microscopical Examination and Interpretation of Portland Cement and Clinker*. Construction Technology Laboratories, Portland Cement Association, Skokie, IL.

Candlot, E. 1890: *Bulletin de la Societie d' Encouragement d'Industrie Nationale 5*, 685–716.

Carles-Gibergues, A., Saucier, F., Grandet, J. and Pigeon, M. 1993: New-to-old concrete bonding: Influence of sulfates type of new concrete on interface structure. *Cement and Concrete Research* 23, 2, 431–441.

Carlson, J., Sutter, L.L., Van Dam, T.J. and Peterson, K.W. 2006: Comparison of a flat-bed scanner and the RapidAir 457 system for determining air-void parameters of hardened concrete. In *Transportation Research Record: Journal of the Transportation Research Board*. No 1979, TRB, National Research Council, Washington, DC, pp. 54–59.

Carter, S. 1998: Pigmenting concrete. *Concrete* 32, 10, 33–34.

Cather, R (Chairman). 1997: Calcium aluminate cements in construction – A reassessment. Report of a Concrete Society Working Party, Technical Report No 46, The Concrete Society, Slough, U.K.

Chadbourn, G.A. 1997: Chloride resistance and durability of cement paste and concrete containing metakaolin. PhD Thesis, University of Aston, Birmingham, U.K.

Chancey, R.T., Stutzman, P., Juenger, M.C.G. and Fowler, D.W. 2010: Comprehensive phase characterization of crystalline and amorphous phases of a Class F fly ash. *Cement and Concrete Research* 40, 1, 146–156.

Chatterji, S. 1984: The spacing factor in entrained air-bubbles. Has it any significance? *Cement and Concrete Research* 14, 5, 757–758.

Chatterji, S. and Gudmundsson, H. 1977: Characterisation of entrained air bubble systems in concretes by means of an image analyzing microscope. *Cement and Concrete Research* 7, 423–428.

Chayes, F. 1956: *Petrographic Modal Analysis*. John Wiley & Sons, New York.

Chemical Society. 1975: *Recommendations for the Testing of High Alumina Cement Concrete Samples by Thermoanalytical Techniques*. Thermal Methods Group, Analytical Division of the Chemical Society, Lyme Regis, U.K.

Christensen, P., Gudmundsson, H., Thaulow, N., Damgard-Jensen, A.D. and Chatterji, S. 1979: Structural and ingredient analysis of concrete – Methods, results and experience. *Nordisk Betong* 3, 4–9. (C & CA translation from Swedish, T292, 1984).

CIRIA. 2001: *Freeze-Thaw Resisting Concrete – Its Achievement in the UK*. Report C559, Construction Industry Research and Information Association, London, U.K.

Coles, J.A. 1978: Colour and pigments. *Concrete* 12 3, 16–18.

Collins, R.J. and Gutt, W. 1988: Research on long-term properties of high alumina cement concrete. *Magazine of Concrete Research*, 40(145), 195–208 (and discussion, 41(149), 1989, 243–244).

Concrete Society. 1989: *Analysis of Hardened Concrete. A Guide to Tests, Procedures and Interpretation of Results*. Report of a joint working party of the Concrete Society and Society of Chemical Industry, Concrete Society Technical Report No 32, The Concrete Society, London, U.K. See also Figg 1989.

Concrete Society. 1991: *The Use of Ggbs and Pfa in Concrete*. Concrete Society Technical Report No 40, The Concrete Society, Slough, U.K.

Concrete Society. 1993: *Microsilica in Concrete*. Report of a Concrete Society Working Party. Technical Report No. 41, The Concrete Society, Slough, U.K., 54pp.

Concrete Society. 2005: *The Role of Water in Concrete and Its Influence on Properties* (Concrete Society discussion document, prepared by a working party of the Concrete Society), Ref. CS 156, The Concrete Society, Camberley, U.K.

Concrete Society. 2007: *Guidance for the Design of Steel-Fibre-Reinforced Concrete*. Report of a Concrete Society Working Group, Technical Report No. 63, The Concrete Society, Camberley, U.K.

Concrete Society. 2010: *Concrete Petrography – An Introductory Guide for the Non-Specialist*. Report of Concrete Society Working Party, Concrete Society Technical Report No. 71, The Concrete Society, Camberley, U.K.

Concrete Society. 2014: Analysis of hardened concrete – A guide to text procedures and interpretation of results, Technical Report No. 32, 2nd ed., The Concrete Society, Camberley, U.K. 76pp.

Concrete Technology Unit, University of Dundee. 2005a: Fly ash (Research Information Digest 3). *Concrete* 39, 7, 28–30.

Concrete Technology Unit, University of Dundee. 2005b: Conditioned fly ash (Research Information Digest 4). *Concrete.* 39, 9, 66–68.

Copeland, L.E. and Kantro, D.L. 1964: Chemistry of hydration of Portland cement at ordinary temperature. In *The Chemistry of Cements,* Vol. 1, Taylor, H.F.W.(ed.). Academic Press, London, U.K., Chapter 8, pp. 313–370.

Corish, A. 1989: European cement standards – A manufacturer's view. *Concrete* 23, 7, 19–20.

Cox, L. and van Dijk, S. 2002: Foam concrete: A different kind of mix. *Concrete* 36, 2, 54–55.

Crammond, N.J. 1984: Examination of mortar bars containing varying percentages of coarsely crystalline gypsum as aggregate. *Cement and Concrete Research* 14, 2, 225–230 (and personal communication 1990).

Crammond, N.J. 1985a: Quantitative x-ray diffraction analysis of ettringite, thaumasite and gypsum in concretes and mortars. *Cement and Concrete Research* 15, 3, 431–441.

Crammond, N.J. 1985b: Thaumasite in failed cement mortars and renders from exposed brickwork. *Cement and Concrete Research* 15, 6, 1039–1050.

Crammond, N.J. 1990: Personal communication. See Crammond (1984).

Crammond, N.J. and Currie, R.J. 1993: Survey of condition of precast high-alumina cement concrete components in internal locations in 14 existing buildings. *Magazine of Concrete Research* 45, 165, 275–279.

Crammond, N.J. and Halliwell, M.A. 1995: The thaumasite form of sulfate attack in concretes containing a source of carbonate ions: A microstructural overview. In *Advances in Concrete Technology, Second Symposium,* Malhotra, V.M. (ed.). ACI SP 154, Paper SP 154-19, 357–80. American Concrete Institute, Detroit, MI.

Cripwell, B. 1992: What is pfa? *Concrete* 26, 3, 11–13.

Crofts, D.B. 2006: Contemporary chemical techniques in concrete analysis. *Concrete* 40, 3, 40–42.

Crofts, D.B. 2011: Personal communication.

Crofts, D.B. 2014: *Testing and Analysis of Concrete: BS 1881-124 Revision,* Society of Chemical Industry (SCI) Construction Materials Group Meeting, held at IOM3, London, May 2014 (presentation available for free download from the RSK website: www.rsk.co.uk).

CROW. 2003: *Roads and Car Parks on Foam Concrete.* Record 22, CROW (The National Information and Technology Centre for Transport and Infrastructure: Acronym of title in Dutch), Ede, the Netherlands, 80pp.

Currie, R.J. and Crammond, N.J. 1994: Assessment of existing high alumina cement construction in the UK. *Proceedings of the Institution for Civil Engineers – Structures and Buildings* 104, 83–92.

Czernin, W. 1980: *Cement Chemistry and Physics for Civil Engineers,* translated by Amerongen, C. van (ed.), 2nd English ed. George Godwin Ltd, London, U.K. for Bauverlag Gmbh, Wiesbaden, Germany.

Davey, N. 1961: *A History of Building Materials.* Phoenix House, London, U.K.

Davey, N. 1965: *A History of Building Materials.* Chapter 12, Limes and cements. Phoenix House, London, U.K. 1961, reprinted 1965.

Deelman, J.C. 1984: Textural analysis of concrete by means of surface roughness measurements. *Materials and Structures* 17, 101, 359–367.

Demoulian, E., Gourdin, P., Hawthorn, F. and Vernet, C. 1980a: Influence of slags chemical composition and texture on their hydraulicity. In *Proceedings of the Seventh International Congress on the Chemistry of Cement,* Paris, France, Vol. III, pp. 89–94.

Demoulian, E., Vernet, C., Hawthorn, F. and Gourdin, P. 1980b: Slag content determination in cements by selective dissolutions. In *Proceedings of the Seventh International Congress Chemistry of Cements,* Paris, France, Vol. 2(111), pp. 151–156.

Deng, M., Hong, D., Lan, X. and Tang M. 1995: Mechanism of expansion in hardened cement pastes with hard-burnt free lime. *Cement and Concrete Research* 25, 2, 440–448.

Deng, M. and Tang M. 1994: Formation and expansion of ettringite crystals. *Cement and Concrete Research* 24, 1, 119–126.

Department of Transport. 1991: *Specification for Highway Works*. DoT, London, U.K.

Dewar, J.D. (Convenor). 1976 (1987): *Concrete Core Testing for Strength – Report of a Concrete Society Working Party*. Concrete Society Technical Report No 11, including Addendum (1987). Reprinted with addendum 1987. The Concrete Society. London, U.K.

Dewar, J.D. and Anderson, R. 1992: *Manual of Ready-Mixed Concrete*, 2nd ed. Blackie Academic & Professional (Chapman & Hall), Glasgow, U.K.

Dhir, R.K., Limbachiya, M.C., Henderson, N.A., Chaipanich, A. and Williamson, G. 1999: *Use of Unfamiliar Cements to ENV 197-1 in Concrete*. Report CTU/1098, Concrete Technology Unit, Dundee University, Dundee, U.K.

Dhir, R.K. and McCarthy, M.J. 1998: Use of BS EN 450 fly ash in structural concrete. *Concrete*. 32, 5, 26–29.

Dhir, R.K. and Paine, K.A. 2007: *Performance Related Approach to Use of Recycled Aggregates*. WRAP Project AGG0074, WRAP (Waste & Resources Action Programme), Banbury, U.K.

Diamond, S. 1976: Cement paste microstructure. In *Proceedings of a Conference Hydraulic Cement Pastes: Their Structure and Properties – An Overview at Several Levels*, Sheffield, U.K. Cement & Concrete Association, Slough, U.K., pp. 2–30.

Dolar-Mantuani, L. 1983: *Handbook of Concrete Aggregates – A Petrographic and Technological Evaluation*. Noyes Publications, Park Ridge, NJ.

Douglas, E., Malhotra, V.M. and Emery, J.J. 1985: Cementitious properties of non-ferrous slags from Canadian sources. *Cement, Concrete, and Aggregates CCAGDP* 7, 1, 3–14.

Dransfield, J. 2003: Admixtures for concrete, mortar and grout, In *Advanced Concrete Technology – Constituent Materials*, Newman, J. and Choo, B.S. (eds.). Elsevier, Oxford, U.K., Chapter 4, pp. 4/3–4/36.

Dransfield, J. 2006a: Admixture current practice – Part 1 (Current Practice Sheet No 149). *Concrete* 40, 8, 35–36.

Dransfield, J. 2006b: Admixture current practice – Part 2 (Current Practice Sheet No 149). *Concrete* 40, 10, 57–59.

Drissen, R. 1995: Determination of the glass content in granulated blastfurnace slag. *Zement-Kalk-Gips* 48, 1, 59–62.

Dunster, A. 2009: *Silica Fume in Concrete*. BRE Information Paper IP5/09, IHS/Building Research Establishment, Watford, U.K.

Dunster, A. and Weir, I. 1996: Appraisal of a building containing precast HAC concrete roof beams. *Construction Repair* May/June, 2–6.

Durning, T.A. and Hicks, M.C. 1991: Using microsilica to increase concrete's resistance to aggressive chemicals. *Concrete International* 13, 3, 42–48.

Eden, M.A. 2011: Personal communication.

Eglinton, M.S. 1987: *Concrete and Its Chemical Behaviour*. Thomas Telford, London, U.K.

Elsen, J., Lens, N., Aarre, T., Quenard, D. and Smolej, V. 1995: Determination of the w/c ratio of hardened cement paste and concrete samples on thin sections using automated image analysis techniques. *Cement and Concrete Research* 25, 4, 827–834.

Erlin, B. 1962: Air content of hardened concrete by a high pressure method. *Journal of Portland Cement Association* 5, 3, 240–249.

Erlin, B. and Campbell, R.A. 2000: Paste microhardness – Promising technique for estimating water-cement ratio. In *Water-Cement Ratio and Other Durability Parameters – Techniques for Determination*, Khan, M.S. (ed.). ACI SP-191, 43–55, Paper SP 191–4, American Concrete Institute, Farmington Hills, MI.

Erlin, B. and Stark, D.C. 1965/1966: Identification and occurrence of thaumasite in concrete. *Highway Research Record No.* 113, 108–113. Conference in 1965, published in 1966.

Fentiman, C.H., Mangabhai, R.J. and Scrivener, K.L. (eds.) 2008: *Calcium Aluminate Cements – Proceedings of the Centenary Conference 2008, Palais des Papes, Avignon, France, 30 June–2 July 2008*, Ref. EP94, IHS Building Research Establishment, Watford, U.K

Figg, J. 1983: Chloride and sulfate attack on concrete. *Chemistry and Industry* 17 October, 770–775.

Figg, J.W. (Chairman) 1989: *Analysis of Hardened Concrete – A Guide to Tests, Procedures and Interpretation of Results – Report of a Joint Working Party of the Concrete Society and Society of Chemical Industry*. Technical Report No. 32. The Concrete Society, London, U.K. See also Concrete Society.

Figg, J.W. and Bowden, S.R. 1971: *The Analysis of Concretes*. Building Research Establishment, HMSO, London, U.K.

Fookes, P.G. 1997: Aggregates: A review of prediction and performance, In *Prediction of concrete durability - Proceedings of STATS 21st Anniversary Conference, The Geological Society, London, UK*, Glanville, J. and Neville, A. (eds.), 16 November 1995, E & F N Spon, London, U.K., Chapter 6, pp. 91–170.

Fookes, P.G. and Collis, L. 1975a: Problems in the Middle East. *Concrete* 9, 7, 12–19.

Fookes, P.G. and Collis, L. 1975b: Aggregates and the Middle East. *Concrete* 9, 11, 14–19.

Fookes, P.G. and Walker, M.J. 2010: Concrete: A man-made rock? *Geology Today* 26, 2, 65–71.

Fookes, P.G. and Walker, M.J. 2011: Natural aggregates in the performance and durability of concrete: Physical characteristics. *Geology Today* 27, 4, 141–148.

Fookes, P.G. and Walker, M.J. 2012: Natural aggregates in the performance and durability of concrete: chemical characteristics. *Geology Today* 28, 1, 20–25.

Fraay, A.L.A., Bijen, J.M. and de Haan, Y.M. 1989: The reaction of fly ash in concrete – A critical examination. *Cement and Concrete Research* 19, 2, 235–246.

Frearson, J.P.H. and Sims, I. 1991: Sandberg on slag. *Concrete* 25, 6, 37–40.

French, W.J. 1991a: Concrete petrography: A review. *Quarterly Journal of Engineering Geology* 24, 1, 17–48.

French, W.J. 1991b: Comments on the determination of the ratio of ggbs to Portland cement in hardened concrete. *Concrete* 25, 6, 33–36.

French, W.J. 1992: Determination of the ratio of pfa to Portland cement in hardened concrete. *Concrete* 26, 3, 43–45.

French, W.J. 1993: Quantitative concrete petrography. *Science and Technology Conference*, The Institute of Materials Unpublished.

French, W.J. 1994: Quantitative petrography. Presented at *SCI Construction Materials Group Conference – Solving Construction Problems by Petrographic Analysis*, 24 February 1994.

Frias, M. and Cabrera, J. 2000: Pore size distribution and degree of hydration of metakaolin-cement pastes. *Cement and Concrete Research* 30, 4, 561–569.

Fundal, E. 1980: Microscopy of cement raw mix and clinker. *FLS-Review* 25. F.L. Smidth Laboratories, Copenhagen, Denmark.

George, C.M. 1983: Industrial aluminous cements. In *Structure and Performance of Cements*, Barnes, P. (ed.). Applied Science, London, U.K., pp. 415–470.

Germann Instruments A/S. 2011: RapidAir. Download from www.germann.org.

Gille, F., Dreizler, I., Grade, K., Kramer, H. and Woermann, E. 1965: M*icroscopy of Cement Clinker – Picture Atlas*. Benton-Verlag GmbH, English translation by P. Schmid 1980, Dusseldorf, Germany.

Goguel, R.L. and St. John, D.A. 1993a: Chemical identification of Portland cements in New Zealand concretes – I. Characteristic differences among New Zealand cements in minor and trace element chemistry. *Cement and Concrete Research* 23, 1, 59–68.

Goguel, R.L. and St. John, D.A. 1993b: Chemical identification of Portland cements in New Zealand concretes – II. The Ca-Sr-Mn plot in cement identification and the effect of aggregates. *Cement and Concrete Research* 23, 2, 283–293.

Goma, F. 1989: The chemical analysis of hardened concrete containing fly ashes, slags, natural pozzolans, etc. In Supplementary Papers, *Third CANMET/ACI International Conference: Fly Ash Silica Fume, Slag and Natural Pozzolans in Concrete*, Alasali, M. (ed.), Trondheim, Norway, pp. 828–845.

Gooding, P. and Halstead, P.E. 1952: The early history of cement in England. *Proceedings of the Third Symposium on the International Chemistry of Cement*, Cement and Concrete Association, London, U.K., pp. 1–29.

Goren, R., Baykara, T. and Marsoglu, M. 2001: Characteristic features and mineralogical structure of diatomite from Ankara region of Turkey. *British Ceramic Transactions* 100, 5, 237–239.

Grade, K. 1968: Determination of the fluorescence of granulated slags. In *Proceedings of the Fifth International Symposium on the Chemistry of Cement, Tokyo, Japan*, Vol. IV, pp. 168–172.

Granger, F. (ed. and translator) 1962: *Vitruvius on Architecture*, Vols I and II. Heinemann, London, U.K. 1st ed. 1931, reprinted 1962.

Grantham, M. 2005: External paving with polypropylene fibre concrete – A freeze/thaw problem? *Concrete* 39, 9, 47–49.

Grattan-Bellew, P.E. 1995: Laboratory evaluation of alkali-silica reaction in concrete from Saunders Generating Station. *ACI Materials Journal* 92, 2, 126–134.

Grattan-Bellew, P.E. and Mitchell, L.D. 2006: Quantitative petrographic analysis of concrete – The Damage Rating Index (DRI) method, a review. In *Marc-André Bérubé Symposium on Alkali-Aggregate Reactivity in Concrete*, Fournier, B (ed.), Montreal, Quebec, Canada, May 2006, pp. 321–334 (part of the *Proceedings of the Eighth CANMET International Conference on Recent Advances in Concrete Technology*, Montreal, Quebec, Canada, 31 May–3 June, 2006).

Grove, R.M. 1968: The identification of ordinary Portland cement and sulphate resisting cement in hardened concrete samples. *Silicates Industrials* 10, 317–320.

Gudmundsson, H. and Chatterji, S. 1979: The measurement of paste content in hardened concrete using automatic image analyzing technique. *Cement and Concrete Research* 9, 607–612.

Gutmann, P.F. 1988: Bubble characteristics as they pertain to compressive strength and freeze-thaw durability. *ACI Materials Journal* 85, 5, 361–366. (Title No. 85-M40)

Gutt, W. 1992: BS 6699: 1992 Specification for ground granulated blastfurnace slag for use with Portland cement. *Concrete* 26, 1, 37–38.

Gutt, W., Nixon, P.J., Smith, M.A., Harrison, W.H. and Russell, A.D. 1974: *A Survey of the Locations, Disposal and Prospective Uses of the Major Industrial By-Products and Waste Materials*. Current Paper CP 19/74, Building Research Establishment, Watford, U.K.

Hack, D.R. and Bryan, D.P. 2006: Aggregates, In *Industrial Minerals and Rocks*, 7th ed., Kogel, E.K., Trivedi, N.C., Barker, J.M. and Krukowski, S.T. (eds.). Society for Mining, Metallurgy and Exploration, Littleton, CO, pp. 1105–1119.

Halse, Y., Pratt, P.L., Dalziel, J.A. and Gutteridge, W.A. 1984: Development of microstructure and other properties in fly ash OPC systems. *Cement and Concrete Research* 14, 4, 491–498.

Hammersley, G.P. 1980: The identification of the primary constituents of hardened grouts, mortars and concretes. MSc Dissertation, Queen Mary College, University of London, London, U.K. (unpublished).

Hannant, D.J. 2003: Fibre-reinforced concrete. In *Advanced Concrete Technology – Processes*, Newman, J. and Choo B.S. (eds.). Elsevier, Oxford, U.K., Chapter 6, pp. 6/1–6/17.

Hansen, T.C. 1986: Physical structure of hardened cement paste, a classical approach. *Materials and Structures* 19, 114, 423–436.

Harrison, W.H. and Munday, R.S. 1975: *An Investigation into the Production of Sintered Pfa Aggregate*. BRE Current Paper CP2/75, HMSO/Building Research Establishment, Watford, U.K.

He, C., Osbaeck, B. and Makovicky, E. 1995: Pozzolanic reactions of six principal clay minerals: Activation, reactivity assessments and technological effects. *Cement and Concrete Research* 25, 8, 1691–1702.

Heinz, D. and Ludwig, U. 1987: Mechanisms of secondary ettringite formation in mortars and concretes subjected to heat treatment. In *Concrete Durability: Katharine and Bryant Mather International Conference*, Scanlon, J.M. (ed.), ACI SP-I00, 2059–2071, Paper No. SPlOO-I05, American Concrete Institute, Detroit, MI.

Hewlett, P.C. (ed.) 1998: *Lea's Chemistry of Cement and Concrete*, 4th ed. Arnold, London, U.K.

Highway Research Board. 1972: *Guide to Compounds of Interest in Cement and Concrete Research*. Highway Research Board Special Report 127, Washington, DC.

Hill, N., Holmes, S. and Mather, D. (eds.) 1992: *Lime and Other Alternative Cements*. Intermediate Technology Publications, London, U.K.

Hime, W.G. 1974: Multitechnique approach solves construction materials failure problems. *Analytical Chemistry* 46, 1230A.

Hime, W.G., Mivelaz, W.F. and Connolly, J.D. 1965: Use of infrared spectrophotometry for the detection and identification of organic additions in cement and admixtures in hardened concrete, In *Analytical Techniques for Hydraulic Cements and Concrete*, pp. 18–29. ASTM STP 395, American Society for Testing and Materials, Philadelphia, PA.

Hjorth, L. 1982: Microsilica in concrete. *Nordic Concrete Research* 1, 9.1–9.18.

Hodgkinson, L. and Rostam, O. 1991: Admixtures in air entrained concrete. *Concrete* 25, 2, 11–13.

Holland, T.C. 2005: *Silica Fume User's Manual*, 183pp. Silica Fume Association technical report FHWA-IF-05-016 for the Federal Highway Administration, Washington, DC.

Hooton, R.D. 1987: The reactivity and hydration products of blast-furnace slag. In *Supplementary Cementing Materials for Concrete*, Malhotra, V.M. (ed.). Canadian Centre for Mineral and Energy Technology (CANMET), SP 86-8E. Canadian Government Publishing Centre, Ottawa, Ontario, Canada, Chapter 4, pp. 245–288.

Hooton, R.D. and Emery, J.J. 1983: Glass content determination and strength development predictions for vitrified blast furnace slag. In *Proceedings of the CANMET/ACDI First International Conference on the Use of Fly Ash, Silica Fume, Slag and Other Mineral By-Products in Concrete*, Montebello, Quebec, Canada. ACI Special Publication SP-79, pp. 943–962.

Hover, K. 1993a: Why is there air in concrete? (Part 1 of 4). *Concrete Construction* 38, 1, 11–15.

Hover, K. 1993b: Air bubbles in fresh concrete. (Part 2 of 4). *Concrete Construction* 38, 2, 148–152.

Hover, K. 1993c: Measuring air in fresh and hardened concrete (Part 3 of 4). *Concrete Construction* 38, 4, 275–278.

Hover, K. 1993d: Specifying air-entrained concrete (Part 4 of 4). *Concrete Construction* 38, 5, 361–367.

Hubbard, F.H., Dhir, R.K. and Ellis, M.S. 1985: Pulverized-fuel ash for concrete: compositional characterisation of United Kingdom pfa. *Cement and Concrete Research* 15, 1, 185–198.

Hubbard, F.H., McGill, R.J., Dhir, R.K. and Ellis, M.S. 1984: Clay and pyrite transformations during ignition of pulverised coal. *Mineralogical Magazine* 48, 347, 251–256.

Idorn, G.M. 1967: *Durability of Concrete Structures in Denmark – A Study of Field Behavior and Microscopic Features*. Technical University of Denmark, Copenhagen, Denmark.

Idorn, G.M. 1969: *The Durability of Concrete*. Technical Paper PCS 46. The Concrete Society, London, U.K.

Ingham, J.P. 2003: Laboratory investigation of lime mortars, plasters and renders. *Journal of the Building Limes Forum* 10, 17–36.

Insley, H. 1936: Structural characteristics of some constituents of Portland cement clinker. *Journal of Research, National Bureau of Standards* 17, 353–361.

Insley, H. 1940: Nature of the glass in Portland cement clinker. *Journal of Research, National Bureau of Standards* 25, 295–300.

Insley, H. and Fréchette, van D. 1955: *Microscopy of Ceramics and Cements – Including Glasses, Slags, and Foundry Sands*. Academic Press, New York.

Ishai, I. and Tons, E. 1977: Concept and test method for a unified characterisation of geometric irregularity of aggregate particles. *Journal of Testing and Evaluation* 5, 1, 3–15.

Jakobsen, U.H. and Brown, D.R. 2006: Reproducibility of w/c ratio determination from fluorescent impregnated thin sections. *Cement and Concrete Research* 36, 1567–1573.

Jakobsen, U.H., Johansen, V. and Thaulow, N. 1995: Estimating the capillary porosity of cement paste by fluorescence microscopy and image analysis. *Materials Research Society Symposium Proceedings* 370, 227–236.

Jakobsen, U.H., Laugeson, P. and Thaulow, N. 2000: Determination of water-cement ratio in hardened concrete by optical fluorescence microscopy. In *Water-Cement Ratio and Other Durability – Techniques for Determination*, Khan, M.S. (ed.), ACI SP-191, 27–41, Paper SP 191–3. American Concrete Institute, Farmington Hills, MI.

Jakobsen, U.H., Pade, C., Thaulow, N., Brown, D., Sahu, S., Magnusson, O., De Buck, S. and De Schutter, G. 2006: Automated air void analysis of hardened concrete – A Round Robin study. *Cement and Concrete Research* 36, 8, 1444–1452.

Jana, D. 2007: A Round Robin test on measurements of air void parameters in hardened concrete by various automated image analyses and ASTM C457 methods. In *Proceedings of the 29th Annual International Conference on Cement Microscopy*, Québec City, Québec, Canada, 20–24 May 2007, International Cement Microscopy Association (ICMA), Farmington Hills, MI.

Janoo, V.C. 1998: *Quantification of Shape, Angularity and Surface Texture of Base Course Materials*. Special Report 98-1, U.S. Army Corps of Engineers, Cold Regions Research & Engineering Laboratory, Hanover, NH.

Jennings, H.M. and Pratt, P.L. 1980: Use of high voltage electron microscope and gas reaction cell for the microstructural investigation of wet Portland cement. *Journal of Materials* 15, 250–253.

Jensen, A.D., Eriksen, K., Chatterji, S., Thaulow, N. and Brandt, I. 1985: Petrographic analysis of concrete. *Danish Building Export Council* 12.

Jensen, O.-M. and Hansen, P.F. 2001: Water-entrained cement-based materials, I. Principles and theoretical background, *Cement and Concrete Research* 31, 4, 647–654.

Jensen, V. 1995: Personal Communication.

Jepsen, B.B. and Christensen, P. 1989: Petrographic examination of hardened concrete. *Bulletin of the International Association of Engineering Geology* 39, 99–103.

Jones, R., McCarthy, A., Kharidu, S. and Nicol, L. 2005: Foamed concrete – Developments and applications. *Concrete* 39, 8, 41–43.

Jones, T.R. 2002: Metakaolin as a pozzolanic addition to concrete. In *Structure and Performance of Cements*, Bensted, J. and Barnes, P. (eds.). Spon Press, London, U.K., pp. 372–398.

Jones, T.R., Walters, G.V. and Kostuch, J.A. 1992: Role of metakaolin in suppressing ASR in concrete containing active aggregate and exposed to saturated NaCl solution, In *The Ninth International Conference on Alkali-Aggregate Reaction in Concrete, London, Conference Papers*, Vol. 2, The Concrete Society (Ref CS.104), Slough, U.K., pp. 485–496.

Jowett, N. 2004: History of the use of concrete admixtures (Milestones in the History of Concrete Technology series). *Institute of Concrete Technology Yearbook* 2004–2005, 13–19.

Jun-yuan, H., Scheetz, R.E. and Roy, D.M. 1984: Hydration of fly ash – Portland cements. *Cement and Concrete Research* 14, 4, 505–512.

Kapralik, I. and Hanic, F. 1980: Studies of the system $CaO-Al_2O_3-MgO-SiO_2$. *Transactions and Journal of the British Ceramic Society* 79, 128–133.

Khatib, J.M., Sabir, B.B. and Wild, S. 1996: Some properties of metakaolin paste and mortar. In *Concrete for Environmental Enhancement and Protection*, Dhir, R.K. and Dyer, T.D. (eds.). E & F N Spon, London, U.K., pp. 637–643.

Khatib, J.M. and Wild, S. 1996a: Pore size distribution of metakaolin paste. *Cement and Concrete Research* 26, 10, 1545–1553.

Khatib, J.M. and Wild, S. 1996b: Sulphate resistance of metakaolin mortar. *Cement and Concrete Research* 28, 1, 83–92.

Klieger, P. 1980: Something for nothing – Almost (1979 Raymond E. Davis Lecture). *Concrete International Design and Construction* 2, 1, 15–23.

Kostuch, J.A., Walters, G.V. and Jones, T.R. 1993: High performance concretes incorporating metakaolin – A review. In *Concrete 2000, Dundee, Scotland*, Vol. 2, Dhir, R.K. and Jones, M.R. (eds.). E & F N Spon, London, U.K., pp. 1799–1811.

Koxholt, P.M. 1985: Iron oxide pigments – The largest colour. *Industrial Minerals (London)* Suppl., May 1985, 22–25.

Kozikowski, R.L., Vollmer, D.B., Taylor, P.C. and Gebler, S.H. 2005: *Factors Affecting the Origin of Air-Void Clustering*. PCA R&D Serial No 2789, Portland Cement Association, Skokie, IL.

Krumbein, W.C. 1941: Measurement and geological significance of shape and roundness of sedimentary particles. *Journal of Sedimentary Petrology* 11, 2, 64–72.

Krumbein, W.C. and Sloss, L.L. 1963: *Stratigraphy and Sedimentation*, 2nd ed. Freeman, San Francisco, CA.

Lambrechts, A.N. 2005: The technical performance of steel and polymer-based fibre concrete. *Institute of Concrete Technology Yearbook 2005–2006*. ICT, Camberley, U.K., pp. 31–35.

Lange, D.A., Jennings, H.M. and Shah, S.P. 1994: Image analysis techniques for characterization of pore structure of cement-based materials. *Cement and Concrete Research* 24, 5, 841–853.

Langer, W. 2009: Sustainability of aggregates in construction. In *Sustainability of Construction Materials*, Khatib, J.M. (ed.). Woodhead/CRC, Cambridge, U.K./Boca Raton, FL, Chapter 1, pp. 1–30.

Larbi, J.A. 1991: The cement paste – Aggregate interfacial zone in concrete. PhD Thesis, Technical University of Delft, Delft, the Netherlands.

Laurencot, J.L., Pleau, R. and Pigeon, M. 1992: The microscopical determination of air voids' characteristics in hardened concrete: Development of an automatic system using image analysis techniques applied to micro-computers. In *Proceedings of the 14th International Conference on Cement Microscopy*, Nashville, TN, pp. 259–273.

Lawrence, C.D. 1990: Sulfate attack on concrete. *Magazine of Concrete Research* 42, 153, 249–264.

Lawrence, C.D. 1995: Mortar expansions due to delayed ettringite formation – Effects of curing period and temperature. *Cement and Concrete Research* 25(4), 903–914.

Lawrence, C.D., Dalziel, J.A., and Hobbs, D.W. 1990: *Sulphate Attack Arising from Delayed Ettringite Formation*. BCA Ref. ITN 12, British Cement Association, Slough, U.K.

Le Chatelier, H. 1905: *Experimental Researches on Constitution of Hydraulic Mortars* (translated by J.L. Mack). McGraw-Hill, New York.

Lea, F.M. 1970: *The Chemistry of Cement and Concrete*, 3rd ed. Arnold, London, U.K. (see also Hewlett 1998 for 4th ed.)

Lee, A.R. 1974: *Blastfurnace and Steel Slag – Production, Properties and Uses*. Arnold, London, U.K.

Lees, G. 1964: The measurement of particle shape and its influence in engineering materials. *Journal of British Granite and Whinstone Federation* 4, 1–22.

Levitt, M. 1985: Pigments for concrete and mortar (Concrete Society Current Practice Sheet No. 99). *Concrete* 19, 3, 21–22.

Levitt, M. 1998: Pigments for concrete and mortar (Current Practice Sheet No 113). *Concrete* 32, 10, 31–32.

Lewis, R., Sear, L., Wainwright, P. and Ryle, R. 2003: Cementitious additions. In *Advanced Concrete Technology – Constituent Materials*, Newman, J. and Choo, B.S. (eds.). Butterworth-Heinemann (Elsevier), Oxford, U.K., Chapter 3, pp. 3/3–3/66.

Liu, J.J. and Khan, M.S. 2000: Comparison of known and determined water-cement ratios using petrography. In *Water-Cement Ratio and Other Durability Parameters – Techniques for Determination*, Khan, M.S. (ed.). ACI SP-191, 11–25, Paper SP 191–2, American Concrete Institute, Farmington Hills, MI.

Livesey, P. 2011: Building limes in the United Kingdom. *Proceedings of the Institution of Civil Engineers, Construction Materials* 164, CM1, 13–20.

Locher, F.W. 2006: *Cement: Principles of Production and Use*, 535pp. Verlag Bau + Technik GmbH, Düsseldorf, Germany.

Lynsdale, C.J. and Cabrera, J.G. 1989: Coloured concrete – A state of the art review. *Concrete* 23, 7, 29–34.

McCarthy, G.J., Swanson, K.D., Keller, L.P. and Blatter, W.C. 1984: Mineralogy of western fly ash. *Cement and Concrete Research* 14, 4, 471–478.

McCarthy, M.J., Halliday, J.E., Csetenyi, L.J. and Dhir, R.K. 2009: *Alkali-Silica Reaction Guidance for Recycled Aggregate in Concrete*. WRAP Project MRF 108-001, WRAP (Waste & Resources Action Programme), Banbury, U.K.

MacInnis, C. and Racic, D. 1986: The effect of superplasticizers on the entrained air void system in concrete. *Cement and Concrete Research* 16, 3, 345–352.

McIver, J.R. and Davis, D.E. 1985: A rapid method for the detection and semi-quantitative assessment of milled granulated blastfurnace slag in hardened concrete. *Cement and Concrete Research* 15, 3, 545–548.

Mailvaganam, N.P. 1984: Miscellaneous admixtures. In *Concrete Admixtures Handbook*, Ramachandran, V.S. (ed.). Noyes Publications, Park Ridge, NJ, pp. 480–557.

Majumdar, A.J. and Singh, B. 1992: Properties of some blended high-alumina cements. *Cement and Concrete Research* 22, 6, 1101–1114.

Majumdar, A.J., Singh, B. and Edmonds, R.N. 1990a: Hydration of mixtures of 'Ciment Fondu' aluminous cement and granulated blast furnace slag. *Cement and Concrete Research* 20, 2, 197–208.

Majumdar, A.J., Singh, B. and Edmonds, R.N. 1990b: Blended high-alumina cements. *Conferences on Advances in Cementitious Materials, Ceramic Transactions,* The American Ceramic Society, Westerville, OH, Vol. 16, pp. 661–678.

Male, P. 1989: Properties of microsilica concrete – An overview of microsilica concrete in the UK. *Concrete* 23, 8, 31–34.

Malhotra, V.M. (ed.) 1987: *Supplementary Cementing Materials for Concrete*. CANMET (Canada Centre for Mineral and Energy Technology), SP 86-8E, Canadian Government Publishing Centre, Ottawa, Ontario, Canada.

Malhotra, V.M. and Carette, G.G. 1983: Silica fume concrete – Properties, applications and limitations. *Concrete International Design and Construction* 5, 5, 40–46.

Malinowski, R. 1982: Durable ancient mortars and concretes. *Nordic Concrete Research, The Nordic Concrete Federation* 19, 1, 1–22.

Malinowski, R. and Garfinkel, Y. 1991: Prehistory of concrete. *Concrete International* 13, 3, 62–68.

Malisch, W.R. 1990: A contractors' guide to air-entraining and chemical admixtures – Their effect on concrete properties that affect job progress. *Concrete Construction* 35, 3, 279–286.

Massazza, F. 1993: Pozzolanic cements. *Cements and Concrete Composites* 15, 185–214.

Mayfield, B. 1990: The quantitative evaluation of the water/cement ratio using fluorescence microscopy. *Magazine of Concrete Research* 42, 150, 45–49.

Mehta, P.K. 1985: Influence of fly ash characteristics on the strength of Portland-fly ash mixtures. *Cement and Concrete Research* 15, 4, 669–674.

Midgley, H.G. 1957: A compilation of x-ray powder diffraction data of cement minerals. *Magazine of Concrete Research* 9, 25, 17–24.

Midgley, H.G. 1964: The formation and phase composition of Portland cement clinker. In *The Chemistry of Cements*, Vol. 1, Taylor, H.F.W. (ed.). Academic Press, London, U.K., Chapter 3, pp. 89–130.

Midgley, H.G. 1967: The mineralogy of set high-alumina cement. *Transactions of the British Ceramic Society* 66, 4, 161–187. (Also reproduced in Building Research Station, Current Paper 19/68, March 1968).

Midgley, H.G. 1979: The determination of calcium hydroxide in set Portland cements. *Cement and Concrete Research* 9, 1, 77–82.

Midgley, H.G. and Midgley, A. 1975: The conversion of high alumina cement. *Magazine of Concrete Research* 27, 91, 59–77.

Midgley, H.G. and Pettifer, K. 1971: The micro-structure of hydrated super sulfated cement. *Cement and Concrete Research* 1, 101–104.

Midgley, H.G. and Rosaman, D. 1960: The composition of ettringite in set Portland cement. In *Proceedings of the 4th International Symposium of Chemistry of Cement, Washington, DC*, National Bureau of Standards Monograph 43, U.S. Department of Commerce, Washington, DC, Vol. 205, pp. 259–262.

Mielenz, R.C. 1962: Petrography applied to Portland-cement concrete. In *Reviews in Engineering Geology*, Vol. 1, Flurh, T. and Legget, R.F. (eds.). The Geological Society of America, Boulder, CO, pp. 1–38.

Mielenz, R.C. 1983: Mineral admixtures – History and background. *Concrete International* 5, 8, 34–42.

Milestone, N.B. 1984: Identification of concrete admixtures by differential thermal analysis in oxygen. *Cement and Concrete Research* 14, 2, 207–214.

Minato, H. 1968: Mineral composition of blastfurnace slag. Supplementary Paper IV-llO. *Proceedings of the Fifth International Symposium Chemistry of Cement*, Tokyo, Japan, pp. 263–269.

Moir, G. 2003: Cements. In *Advanced Concrete Technology – Constituent Materials*, Newman, J. and Choo, B.S. (eds.). Butterworth-Heinemann (Elsevier), Oxford, U.K., Chapter 1, pp. 1/3–1/45.

Moir, G.K. and Kelham, S. 1989: *Performance of Limestone-Filled Cements: Report of Joint BRE/BCA/ Cement Industry Working Party*. Paper No. 7. Durability. Seminar held at the Building Research Establishment, 28 November.

Montgomery, D.G., Hughes, D.C. and Williams, R.I.T. 1981: Fly ash in concrete – A microstructural study. *Cement and Concrete Research* 11, 4, 591–603.

Moorhead, D.R. 1986: Cementation by the carbonation of hydrated lime. *Cement and Concrete Research* 16, 5, 700–708.

Moorhead, D.R. and Morand, P. 1975: Some observations of the carbonation of portlandite $(Ca(OH)_2)$ and its relationship to the carbonation hardening of lime products. In *Proceedings of the Third International Congress on Science and Research of Silicate Chemistry*.

Möser, B. 2009: Ultra high resolution scanning electron microscopy on building materials - Insulators and contaminating samples. In *Proceedings of the 17th International Conference on Building Materials (Ibausil)*, 23–26 September, Weimar, Germany.

Möser, B. and Stark, J. 2007: Nanoscale characterization of hydration processes by means of high resolution scanning electron microscopy imaging techniques, In *Proceedings of the 12th International Congress on the Chemistry of Cement*, 8–13 July, Montreal, Quebec, Canada.

Muller, D. and Morrison, I. 2000: Recent developments in glass fibre reinforced concrete, *Concrete* 34, 5, 47–49.

Murat, M. 1983a: Hydration reaction and hardening of calcined clays and related minerals. I. Preliminary investigation on metakaolinite. *Cement and Concrete Research* 13, 2, 259–266.

Murat, M. 1983b: Hydration reaction and hardening of calcined clays and related minerals. II. Influence of mineralogical properties of the raw-kaolinite on the reactivity of metakaolinite. *Cement and Concrete Research* 13, 4, 511–518.

Muthadhi, A. and Kothandaraman, S. 2010: Optimum production conditions for reactive rice husk ash. *Materials and Structures* 43, 9, 1303–1315.

Neville, A.M. 1975: *High Alumina Cement Concrete*. Lancaster, The Construction Press.

Neville, A.M. 2003: How closely can we determine the water-cement ratio of hardened concrete? *Materials and Structures* 36, 259, 311–318.

Neville, A.M. 2006: *Concrete: Neville's Insights and Issues*, 314pp. Thomas Telford, London, U.K.

Neville, A.M. and Brooks, J.J. 1987: *Concrete Technology*. Longman Group Limited, Harlow, U.K.

Neville, A.M. and Brooks, J.J. 2010: *Concrete Technology*, 2nd ed. Prentice Hall (Pearson), Harlow, U.K.

NIH (National Institutes of Health). 2010: ImageJ software. See http://rsb.info.nib.gov/ij/.

Nordtest. 1999: *Concrete, Hardened: Water-Cement Ratio*, NT Build 361, 2nd ed. Nordtest, Nordic Innovation, Oslo, Norway, pp. 1999–2011.

Oriol, M. and Pera, J. 1995: Pozzolanic activity of metakaolin under microwave treatment. *Cement and Concrete Research* 25, 2, 265–270.

Owens, P.L. 1980: Pulverised-fuel ash, Parts 1 & 2 (Concrete Society Current Practice Sheets 54 & 57). *Concrete* 14, 7 & 10, 33–34, 35–36.

Owens, P.L. 1982: Pulverised-fuel ash, Parts 3 & 4 (Concrete Society Current Practice Sheets 75 & 76). *Concrete* 16, 6 & 7, 39–41.

Owens, P.L. 1999: Pulverised-fuel ash, Part 1: Origin and properties (Current Practice Sheet No 116). *Concrete* 33, 4, 27–28.

Ozyildirim, C. and Halstead, W.J. 1994: Improved concrete quality with combinations of fly ash and silica fume. *ACI Materials Journal* 91, 6, 587–594. Title No. 91-M59.

Pade, C., Jakobsen, U.H. and Elsen, J. 2002: A new automatic analysis system for analyzing the air void system in hardened concrete. In *International Cement Microscopy Association Conference*, April 2002, San Diego, CA.

Parker, D.G. 1985: Microsilica concrete, Part 1: The material (Concrete Society Current Practice Sheet No. 104). *Concrete* 19, 10, 21–22.

Parker, D.G. 1986: Microsilica concrete, Part 2: In use (Concrete Society Current Practice Sheet No. 110). *Concrete* 20, 3, 19–21.

Parker, T.W. 1952: The constitution of aluminous cement. In *Proceedings of the Third International Symposium Chemistry of Cements*, London, U.K., pp. 485–529.

Parker, T.W. and Hirst, P. 1935: Preparation and examination of thin sections of set cement. *Cement and Cement Manufacture* 8, 10, 235–241.

Patel, H.H., Bland, C.H. and Poole, A.B. 1995: The microstructure of concrete cured at elevated temperatures. *Cement and Concrete Research* 25, 3, 485–490.

Payne, J.C. 2002: New admixtures technologies: An update. In *Institute of Concrete Technology Yearbook 2002–2003*. ICT, Crowthorne, U.K., pp. 71–77.

Peterson, K. 2008: Air-void analysis of hardened concrete: Helping computers to count air bubbles like people count air bubbles - Methods for flatbed scanner calibration. PhD dissertation, Michigan Technological University, Houghton, MI.

Peterson, K., Carlson, J., Sutter, L. and Van Dam, T. 2009a: Methods for threshold optimization for images collected from contrast enhanced concrete surfaces for air-void system characterization (Presented at *11th Euroseminar on Microscopy Applied to Building Materials*). *Materials Characterization* 60, 7, 710–715.

Peterson, K., Sutter, L. and Radlinski, M. 2009b: The practical application of a flatbed scanner for air-void characterization of hardened concrete. *Journal of ASTM International* 6, 9, 15.

Peterson, K.W., Swartz, R.A., Sutter, L.L. and Van Dam, T.J. 2001: Hardened concrete air void analysis with a flatbed scanner. *Transportation Research Record: Journal of the Transportation Research Board* 1775, 36–43.

Petrov, I. and Schlegel, E. 1994: Application of automatic image analysis for the investigation of auto-claved aerated concrete. *Cement and Concrete Research* 24, 5, 830–840.

Pettifer, K. 1991: Personal communication.

Philleo, R.E. 1983: A method for analyzing void distribution in air-entrained concrete. *Cement, Concrete, and Aggregates CCAGDP* 5, 2, 128–130.

Piasta, J., Sawicz, Z. and Piasta, W.G. 1989: Durability of high alumina cement pastes with mineral additions in water sulfate environment. *Cement and Concrete Research* 19, 1, 103–113.

Pigeon, M. and Pleau, R. 1995: *Durability of Concrete in Cold Climates*. Modern Concrete Technology 4, E & F N Spon (Chapman & Hall), London, U.K.

Pistilli, M.F. 1983: Air-void parameters developed by air-entraining admixtures, as influenced by soluble alkalies from fly ash and Portland cement. *ACI Journal* May–June, 217–222. (Title No. 80–22.)

Poole, A.B., Patel, H.H. and Shiekh, V. 1996: Alkali silica and ettringite expansions in 'steam-cured' concretes. In *Proceedings of the 10th International Conference on Alkali-Aggregate Reaction in Concrete*, Shayan, A. (ed.), Melbourne, Victoria, Australia, pp. 943–948.

Poole, A.B. and Sims, I. 2003: Geology, aggregates and classification. In *Advanced Concrete Technology – Constituent Materials*, Newman, J. and Choo, B.S. (eds.), Chapter 5, 5/3–5/36. Elsevier, Oxford, U.K.

Poon, C.-S., Lam, L., Kou, S.C., Wong, Y.-L. and Wong, R. 2001: Rate of pozzolanic reaction of metakaolin in high performance cement pastes. *Cement and Concrete Research* 31, 9, 1301–1306.

Powers, M.C. 1953: A new roundness scale for sedimentary particles. *Journal of Sedimentary Petrology* 23, 2, 117–119.

Powers, T.C. 1949a: The non-evaporable water content of hardened Portland cement paste; its significance for concrete research and its method of determination. *ASTM Bulletin* 158, 68–76.

Powers, T.C. 1949b: Air requirement of frost-resistant concrete. *Proceedings of the Highway Research Board* 29, 184–211.

Powers, T.C. and Brownyard, T.L. 1947/1948: *Studies of the Physical Properties of Hardened Portland Cement Paste*. Bulletin 22, Research Laboratories of the Portland Cement Association, Skokie, IL. (reprinted from *Journal of the American Concrete Institute*, 1947, 43, 101–132, 249–336, 469–504, 549–602, 669–712, 845–880 & 933–992).

Price, W. 2004: Portland-limestone cement – The UK situation (Current Practice Sheet No 136). *Concrete* 38, 3, 50–51.

Purnell, P. 2007: Degradation of fibre-reinforced cement composites. In *Durability of Concrete and Cement Composites*, Page, C.L. and Page, M.M. (eds.). Woodhead and Maney (Institute of Materials, Minerals & Mining), Cambridge, U.K., Chapter 9, pp. 316–363.

Raask, E. 1968: Cenospheres in pulverized-fuel ash. *Journal of the Institute of Fuel* 41, 339–344.

Ramachandran, V.S. (ed.) 1995: *Concrete Admixtures Handbook: Properties, Science and Technology*, 2nd ed. Noyes Publications, Park Ridge, NJ.

Ramlochan, T., Thomas, M.D.A., and Gruber K.A. 2000: The effect of metakaolin on alkali-silica reaction in concrete. *Cement and Concrete Research*, 30(3), 339–344.

Randolph, J.L. 1991: Thin section identification of various admixtures and their effects on the cement paste morphology in hardened concrete. In *Proceedings of the 13th International Conference on Cement Microscopy*, Tampa, FL, pp. 281–293.

Ravenscroft, P. 1982: Determining the degree of hydration of hardened concrete. *John Laing Forum* 4.

Rayment, D.L. and Majumdar, A.J. 1982: The composition of C–S–H phases in Portland cement pastes. *Cement and Concrete Research* 12, 6, 753–764.

Richardson, I.G. 2004: Tobermorite/jennite- and tobermorite/calcium hydroxide-based models for the structure of C–S–H: Applicability to hardened pastes of tricalcium silicate, β-dicalcium silicate, Portland cement, and blends of Portland cement with blast-furnace slag, metakaolin or silica fume. *Cement and Concrete Research* 34, 9, 1733–1777.

Richardson, I.G. 2008: The calcium silicate hydrates. *Cement and Concrete Research* 38, 2, 137–158.

Richardson, I.G. and Groves, G.W. 1990: The microstructure of blastfurnace slag/high alumina cement pastes. In *Calcium Aluminate Cements, Proceedings of a Symposium*, Queen Mary & Westfield College, University of London, QMW, London, U.K., pp. 282–293.

RILEM. 1988: Siliceous by-products for use in concrete – Final report (RILEM technical report – RILEM Committee 73-SBC). *Materials and Structures* 21, 121, 69–80.

Rittenhouse, G. 1943: A visual method of estimating two-dimensional sphericity. *Journal of Sedimentary Petrology* 13, 2, 79–81.

Rixom, M.R. (ed.) 1977: *Concrete Admixtures: Uses and Application*. Construction Press, Longman, New York.

Rixom, M.R. and Mailvaganam, N.P. 1999: *Chemical Admixtures for Concrete*, 3rd ed. E. & F.N. Spon, London, U.K.

Roberts, L.R. 1998: Developments in the use of special admixtures. In *Institute of Concrete Technology Yearbook 1998–1999*. ICT, Crowthorne (now Camberley), U.K., pp. 69–71.

Roberts, L.R. 2005: Recent advances in understanding of cement admixture interactions. In *Institute of Concrete Technology Yearbook 2005–2006*. ICT, Camberley, U.K., pp. 53–60.

Roberts, L.R. and Scali, M.J. 1984: Factors affecting image analysis for measurement of air content in hardened concrete. In *Proceedings of the Sixth International Conference on Cement Microscopy*, Albuquerque, NM, pp. 402–419.

Roberts, M.H. 1956: The constitution of hydraulic lime. *Cement and Lime Manufacture* 29, 3, 27–36.

Roberts, M.H. and Adderson, B.W. 1983: Tests on superplasticizing admixtures for concrete. *Magazine of Concrete Research* 35, 123, 86–98.

Roberts, M.H. and Jaffrey, S.A.M.T. 1974: *Rapid Chemical Test for the Detection of High-Alumina Cement Concrete*. BRE Information Sheet IS 15/74, Building Research Establishment, Watford, U.K.

Robson, T.D. 1962: *High-Alumina Cements and Concretes*. Contractors Record Limited, London, U.K.

Robson, T.D. 1964: Aluminous cement and refractory castables. In *The Chemistry of Cements*, Vol. 2, Taylor, H.F.W. (ed.). Academic Press, London, U.K., Chapter 12, pp. 3–35

Rogers, C.A. 2011: Personal communication.

Rooij, M.R.de. 2009: *Cementsteen – Basis Voor Beton*. Aeneas, Boxtel and Cement & Beton Centrum, Hertogenbosch, the Netherlands (in Dutch).

Rooij, M.R.de. 2011: Personal communication.

Rooij, M.R.de, Valcke, S.L.A., Suitela, W.L.D. 2011: An update on using image analysis to determine w/c ratio of concrete, In *Proceedings of the 13th Euroseminar on Microscopy Applied to Building Materials*, 14–18 June 2011, Ljubljana, Slovenia.

Royal Institution of Chartered Surveyors. 2015: *The 'Mundic' Problem – A Guidance Note – Recommended Sampling, Examination and Classification Procedure for Suspect Concrete Building Materials in Cornwall and Parts of Devon*, 3rd ed. RICS, London, U.K. See Stimson 1997 for 2nd ed.

Russ, J.C. and Dehoff, R.T. 2000: *Practical Stereology*, 2nd ed., 382pp. Springer-Verlag GmbH, Berlin, Germany.

Ryle, R. 1999: Metakaolin: A highly reactive pozzolan for concrete. *Quarry Management* December, 27–31.

Saad, M.N.A., DeAndrade, W.P. and Paulon, V.A. 1982: Properties of mass concrete containing an active pozzolan made from clay. *Concrete International* July, 59–65.

Sadegzadeh, M., Kettle, R. and Vassou, V. 2001: The influence of glass, polypropylene and steel fibres on the physical properties of concrete. *Concrete* 35, 4, 12–13.

Sahu, S., Badger, S., Thaulow, N. and Lee, R.J. 2004: Determination of water-cement ratio of hardened concrete by scanning electron microscopy. *Cement and Concrete Research* 26, 8, 987–992.

Salvador, S. 1995: Pozzolanic properties of flash-calcined kaolinite: A comparative study with soak-calcined products. *Cement and Concrete Research* 25, 1, 102–112.

Sandberg, A. and Collis, L. 1982: Toil and trouble on concrete bubbles. *Consulting Engineer* November, 32–35.

Sandberg, M. 1987: *The Estimation of Equivalent Water/Cement Ratio of Hardened Portland Cement Concrete by Fluorescence Microscopy*. Messrs Sandberg Test Procedure TP/G8/2, Messrs Sandberg, London, U.K.

Sandberg, M. 2010: *The Estimation of Equivalent Water/Cement Ratio of Hardened Portland Cement Concrete by Fluorescence Microscopy*. Messrs Sandberg Test Procedure TP/G8/2, Sandberg LLP, London, U.K. Check title and designation with Mike Eden.

Sanders, C.I., Sadeghi, S.S. and Nelson, N.I. 1995: Limitations of the carbonate extraction/UV spectrophotometric method for determining lignosulfonate-based admixtures in hardened concrete. *Cement, Concrete and Aggregates* 17, 1, 37–47.

Sarkar, S.L. and Aitcin, P.-C. 1990: The importance of petrological, petrographical and mineralogical characteristics of aggregates in very high strength concrete. In *Petrography Applied to Concrete and Concrete Aggregates,* Erlin, B. and Stark, D. (eds.). ASTM STP-1061, Philadelphia, PA, pp. 129–144.

Sato, T. and Beaudoin, J.J. 2007: The effect of nano-sized $CaCO_3$ addition on the hydration of cement paste containing high volumes of fly ash. In *Proceedings of the 12th International Congress on the Chemistry of Cement, Montreal, Quebec, Canada,* 8–13 July 2007; also published as NRCC-49700, National Research Council Canada, Ottawa, Ontario, Canada, 12pp.

Saucier, F., Pigeon, M. and Cameron, G. 1991: Air-void stability, Part V: Temperature, general analysis, and performance index. *ACI Materials Journal* 88, 1, 25–36 (Title No. 88-M4).

Scott, P.W., Critchley, S.R. and Wilkinson, F.C.F. 1986: The chemistry and mineralogy of some granulated and pelletised blastfurnace slags. *Mineralogical Magazine* 50, 141–147.

Scrivener, K.L. 2004: Backscattered electron imaging of cementitious microstructures: Understanding and quantification. *Cement and Concrete Composites* 26, 8, 935–945.

Scrivener, K.L. and Capmas, A. 1998: Calcium aluminate cements. In *Lea's the Chemistry of Cement and Concrete,* 4th ed., Hewlett, P.C. (ed.). Arnold, London, U.K., Chapter 13, pp. 709–778.

Scrivener, K.L. and Taylor, H.F.W. 1990: Microstructural development in pastes of a calcium aluminate cement. In *Proceedings of the International Symposium on Calcium Aluminate Cements* (dedicated to the late Dr H.G. Midgley), Mangabhai, R.J. (ed.), Queen Mary and Westfield College, University of London, E. & F.N. Spon (Chapman & Hall), London, U.K., pp. 41–51.

Sear, L. and Owens, P. 2006: Fly ash, Part 2: Standards (Current Practice Sheet No 146). *Concrete* 40, 2, 7–9.

Sear, L.K.A. (ed.) 2001: *(The) Properties and Use of Coal Fly Ash – A Valuable Industrial By-Product.* Thomas Telford, London, U.K.

Sellevold, E.J., Bager, D.H., Klitgaard Jensen, K. and Knudsen, T. 1982: *Silica Fume Cement Pastes: Hydration and Pore Structure,* Report BML 82.610, The Norwegian Institute of Technology, Trondheim, Norway, pp. 19–50.

Sellevold, E.J. and Nilsen, T. 1987: Condensed silica fume in concrete: A world review. In *Supplementary Cementing Materials for Concrete,* Malhotra, V.M. (ed.). The Canada Centre for Mineral and Energy Technology (CANMET), Canadian Government Publishing Centre, Ottawa, Ontario, Canada, Chapter 3, pp. 167–243.

Senior, S.A. and Franklin, J.A. 1987: *Thermal Characteristics of Rock Aggregate Materials.* Report RR 241, Ontario Ministry of Transportation and Communications, Downsview Toronto, Ontario, Canada.

Sibbick, R.G. 1993: The susceptibility of various UK aggregates to alkali silica reaction. PhD Thesis, Aston University, Birmingham, U.K.

Sibbick, R.G. 2010: *13th Annual QC/TS and Material Meeting,* 31 August–2 September 2010, Lehigh Hanson, Irving, TX, Need full details: See Sibbick CD: pp.

Sibbick, R.G. 2011: Personal communication.

Sibbick, R.G. 2012: Personal communication.

Sibbick, R.G., Brown, A.D., Dragovic, B., Knight, C., Garrity, S. and Comeau, R. 2007: Investigation into the determination of water to cementitious ratios by the use of the fluorescence microscopy technique in hardened concrete samples: Part 1. Presented at *11th Euroseminar on Microscopy Applied to Building Materials,* 5–9 June 2007, Porto, Portugal.

Sibbick, R.G. and Crammond, N.J. 2001: Microscopical investigations into recent field examples of the thaumasite form of sulfate attack (TSA). In *Proceedings of the 8th Euroseminar on Microscopy Applied to Building Materials,* Athens, Greece, pp. 261–269.

Sibbick, R.G., Crammond, N.J. and Metcalfe, D. 2003: The microscopical characterisation of thaumasite (presented at the *First International Conference on Thaumasite in Cementitious Materials,* June 2002, BRE, Watford, U.K.). *Cement and Concrete Composites* 25, 1059–1066.

Sibbick, R.G., Garrity, S., LaFleur, C. and Roberts, L.R. 2013: Determination of water to cementitious (w/cm) binder ratios by the use of the fluorescent microscopy technique in hardened concrete samples: Part III. In *Proceedings, 35th International Conference on Cement Microscopy, Chicago, IL,* International Cement Microscopy Association, Farmington Hills, MI.

Sibbick, R.G. (Ted), LaFleur, C. and Roberts, L.R. 2012: Determination of water to cementitious (w/cm) binder ratios by the use of the fluorescent microscopy technique in hardened concrete samples: Part II. In *Proceedings, 34th International Conference on Cement Microscopy, Halle/ Saale, Germany*, International Cement Microscopy Association, Farmington Hills, MI.

Sibbick, R.G. and Nixon, P.J. 2000: Investigation into the effects of metakaolin as a cement replacement material in ASR reactive concrete. In *Proceedings, 11th International Conference, Alkali- Aggregate Reaction in Concrete, Québec City, Quebec, Canada*, Bérubé, M.A., Fournier, B. and Durand, B. (eds.), June 2000, CRIB, Laval & Sherbrooke Universities, Sainte-Foy, Québec, Canada, pp. 763–772.

Sims, I. 1975: Analysis of concrete and mortar samples from Sparsholt Roman Villa, near Winchester, Hampshire. Report prepared for the Department of the Environment (not published).

Sims, I. 1977: High-alumina cement concrete: general discussion and testing procedures and case study examples. In PhD thesis, *Investigations into Some Chemical Instabilities of Concrete*, Queen Mary College, London University, London, U.K., Chapter 7, pp. 244–263. (unpublished)

Sims, I. 2001: Material evidence: The forensic investigation of construction and other geomaterials for civil and criminal cases. In *Forensic Engineering – The Investigation of Failures – Proceedings of the Second International Conference on Forensic Engineering, Institution of Civil Engineers, London*, Neale, B.S. (ed.), 12–13 November 2001, Thomas Telford, London, U.K.

Sims, I. 2006: Selection of materials for concrete in a hot and aggressive climate. Keynote Lecture in *Proceedings of the Eighth International Concrete Conference, Gulf Hotel Conference Centre, Bahrain*, 27–29 November 2006, Bahrain Society of Engineers, Manama, Bahrain, 62pp. Need Conference Title and check not incorrectly cited in text as 2008!

Sims, I. and Brown, B.V. 1998: Concrete aggregates. In *Lea's the Chemistry of Cement and Concrete*, 4th ed., Hewlett, P.C. (ed.). Arnold, London, U.K., Chapter 16, pp. 903–1011.

Singh, B. and Osborne, G.J. 1994: *Hydration and Durability of OPC/Metakaolin Blended Concrete*. BRE Client Report CR 291/94, Building Research Establishment, Watford, U.K.

Smith, M.R. and Collis, L. (eds.) 1993: *Aggregates: Sand, Gravel and Crushed Rock Aggregates for Construction Purposes*, 2nd ed. Geological Society Engineering Geology Special Publication No. 9, The Geological Society, London, U.K.

Smith, M.R. and Collis, L. (eds.) 2001: *Aggregates: Sand, Gravel and Crushed Rock Aggregates for Construction Purposes*, 3rd ed. (revised by Fookes, P.G., Lay, J., Sims, I., Smith, M.R. and West, G.) Geological Society Engineering Geology Special Publication No. 17, The Geological Society, London, U.K.

Smolczyk, H.G. 1980: Slag structure and identification of slags. In *Seventh International Congress on the Chemistry of Cement*, Paris, France, Volume I, Principal Reports, SubTheme Ill, pp. 1/3–1/17.

Sommer, H. 1979: The precision of the microscopical determination of the air-void system in hardened concrete. *Cement, Concrete, and Aggregates, CCAGDP* 1, 2, 49–55.

Sourie, A. and Glasser, F.P. 1991: Studies of the mineralogy of high alumina cement clinkers. *Transactions and Journal of the British Ceramic Society* 90, 3, 71–76.

Spence, F. 1988: Coloured concrete for architectural use. *Concrete Forum* 1, 15–18.

Stimson, C.C. (Chairman) 1997: *The 'Mundic' Problem – A Guidance Note – Recommended Sampling, Examination and Classification Procedure for Suspect Concrete Building Materials in Cornwall and Parts of Devon*, 2nd ed. The Royal Institution of Chartered Surveyors, London, U.K. See Royal Institution of Chartered Surveyors 2015 for 3rd ed.

St. John, D.A. 1990: The use of large-area thin sectioning in the petrographic examination of concrete. In *Petrography Applied to Concrete and Concrete Aggregates*, Erlin, B. and Stark, D. (eds.). ASTM STP-1061, American Society for Testing and Materials, Philadelphia, PA, pp. 55–70.

St. John, D.A. 1992: AAR investigations in New Zealand – Work carried out since 1989. In *Proceedings of the Ninth International Conference on Alkali-Aggregate Reaction in Concrete*, The Concrete Society, Slough, U.K., Vol. 2, pp. 885–893.

St. John, D.A. 1994a: *The Use of Fluorescent Dyes and Stains in the Petrographic Examination of Concrete*. Industrial Research Limited Report No. 243, The New Zealand Institute for Industrial Research and Development, Lower Hutt, New Zealand.

St. John, D.A. 1994b: *The Dispersion of Silica Fume*. Industrial Research Limited Report No. 244, The New Zealand Institute for Industrial Research and Development, Lower Hutt, New Zealand.

St. John, D.A., McLeod, L.C. and Milestone, N.B. 1996: The durability of DSP mortars exposed to conditions of wetting and drying. In *International Workshop, High Performance Concrete, Bangkok, 1994*, ACI SP-159, American Concrete Institute, Detroit, MI, pp. 45–56.

Stutzman, P. 2004: Scanning electron microscopy imaging of hydraulic cement microstructure, *Cement and Concrete Composites* 26, 8, 957–966.

Stutzman, P. 2007: Multi-spectral SEM imaging of cementitious materials. In *Proceedings of the 29th Annual International Conference on Cement Microscopy, Québec City, Quebec, Canada, 21–24 May 2007*, International Cement Microscopy Association (ICMA), Farmington Hills, MI.

Stutzman, P. 2012: Microscopy of clinker and hydraulic cements. In *Applied Mineralogy of Cement and Concrete*, Vol. 74, Broekmans, M.A.T.M. and Pöllmann, H. (eds.). Mineralogical Society of America/Geochemical Society, Chantilly, VA/St. Louis, MO, pp. 101–145.

Sutherland, A. 1974: *Air-Entrained Concrete*. C&CA Ref. 45.022, Cement and Concrete Association, Slough, U.K.

Swamy, R.N. 1988: *Natural Fibre Reinforced Cement and Concrete*, 288pp. Blackie, Glasgow, U.K.

Taylor, H.F.W. 1964: The calcium silicate hydrates. In *The Chemistry of Cements*, Vol. 1, Taylor, H.F.W. (ed.). Academic Press, London, U.K., Chapter 5, pp. 167–232.

Taylor, H.F.W. 1984: Studies on the chemistry and microstructure of cement pastes. *British. Ceramic Proceedings* 35, September, 65–82.

Taylor, H.F.W. 1990: *Cement Chemistry*. Academic Press, London, U.K.

Taylor, H.F.W. 1997: *Cement Chemistry*, 2nd ed. Thomas Telford, London, U.K.

Taylor, W.H. 1974: The production, properties and uses of foamed concrete. *Precast Concrete* February, 83–96.

Terry, R.D. and Chillingar, G.Y. 1955: Summary of 'concerning some additional aids in studying sedimentary formations' by M.S. Shvetsov. *Journal of Sedimentary Petrology* 25, 3, 229–234.

Thaulow, N. 1984: *Thin Section Investigation of Old Slag Cement Concrete*. Report 60-251-4420. Danish Technical Institute, Building Technology, Tastrup, Denmark.

Thaulow, N., Damgaard-Jensen, A., Chatterji, S., Christensen, P. and Gudmundsson, H. 1982: Estimation of the compressive strength of concrete samples by means of fluorescence microscopy. *Nordisk Betong* 2–4.

Thomas, M.D.A. and Matthews, J.D. 1991: *Durability Studies of Pfa Concrete Structures*. BRE Information Paper IP 11/91. Building Research Establishment, Watford, U.K.

Thomas, M.D.A. and Matthews, J.D. 1993: Performance of fly ash concrete in UK structures. *ACI Materials Journal* 90, 6, 586–593, Title No. 90-M59.

Törnebohm, A.E. 1897: The petrography of Portland cement. *Tonindustrie-Zeitung* 21, 1148–1150, 1157–1159. See also 1910–1911: *Baumaterielenkunde* 6, 142.

Troy, J. 2011: Air-entrained concrete – Long-term field exposure trials. *Concrete* 45, 5, 57–58.

True, G.F. 2011a: Personal communication.

True, G.F. 2011b: Development of image analysis techniques to assist evaluation of both air void structure and aggregate shape factors in concrete. PhD thesis, Wolverhampton University, Wolverhampton, U.K.

True, G.F. and Searle, D. 2012: Digital imaging and analysis – Cores, aggregate particles and flat surfaces. *Concrete* 46, 6, 16–20.

True, G.F., Searle, D., Sear, L. and Khatib, J. 2010: Voidage assessment of concrete using digital image processing. *Magazine of Concrete Research* 62, 12, 857–868.

Turriziani, R. 1964: The calcium aluminate hydrates and related compounds. In *The Chemistry of Cements*, Vol. 1, Taylor, H.F.W. (ed.). Academic Press, London, U.K., Chapter 6, pp. 233–286.

U.S. Bureau of Reclamation. 1955: p. 120. (referred to in Mielenz 1962).

Usher, P.M. (Convenor). 1980: *Guide to Chemical Admixtures for Concrete – Report of a Joint Working Party*. Concrete Society Technical Report No. 18, The Concrete Society, London, U.K.

Uthus, L., Hoff, I. and Horvli, I. 2005: Evaluation of grain shape characterization methods for unbound aggregate. In *Proceedings, Seventh International Conference on the Bearing Capacity of Roads, Railways and Airfields*, Trondheim, Norway, 12pp.

Van Aardt, J.H.P. and Visser, S. 1975: Thaumasite formation: A cause of deterioration of Portland cement and related substances in the presence of sulfates. *Cement and Concrete Research* 5, 3, 225–232.

Van der Plas, L. and Tobi, A. 1965: A chart for judging the reliability of point counting results. *American Journal of Science* 263, 1, 87–90.

Varma, S.P. and Bensted, J. 1973: Studies of thaumasite. *Silicates Industrials* 38, 29–32.

Verein Deutscher Zementwerke e.V. 1996: Activity report 1993–1996, Forschungsinstitut der Zementindustrie, Dusseldorf.

Vivian, R.E. 1987: The importance of Portland cement quality on the durability of concrete. In *Concrete Durability: Katherine and Bryant Mather International Conference*, Scanlon, L.M. (ed.), ACI SP-lOO, 1691–1701, Paper No. SP 100–86, American Concrete Institute, Detroit, MI.

Wadell, H. 1932: Volume, shape and roundness of rock particles. *Journal of Geology* 40, 443–451.

Wadell, H. 1933: Sphericity and roundness of rock particles. *Journal of Geology* 41, 310–331.

Wadell, H. 1935: Volume, shape and roundness of quartz particles. *Journal of Geology* 43, 250–280.

Walker, H.N. 1980: Formula for calculating spacing factor for entrained air voids. *Cement, Concrete, and Aggregates CCAGDP* 2, 2, 63–66.

Walker, H.N. 1981: Examination of Portland cement concrete by fluorescent light microscopy. In *Proceedings of the Third International Conference on Cement Microscopy*, Houston, TX, pp. 257–278.

Walters, G.V. and Jones, T.R. 1991: Effect of metakaolin on alkali-silica reactions (ASR) in concrete manufactured with reactive aggregates. In *Proceedings of the Second International Conference on the Durability of Concrete*, Montreal, Quebec, Canada, Vol. 2, pp. 941–947.

Warren, P.A. 1973: *The Effect of Voids on Measured Strength*. Technical Note 73, RMC Technical Services Ltd, Egham, Surrey, U.K.

Waste and Resources Action Programme (WRAP). 2005: *The Quality Protocol for the Production of Aggregates from Inert Waste*, 11pp. WRAP, Banbury, U.K. Produced by Quality Products Association, Highways Agency & WRAP.

Weigand, W.A. 1994: Progress toward a standard procedure for point-counting the phases in cement clinker with reflected light microscopy. In *Petrography of Cementitious Materials*, DeHayes, S.M. and Stark, D. (eds.). ASTM STP 1215, American Society for Testing and Materials, Philadelphia, PA.

Werner, O.R. (Chairman). 1987: Ground granulated blast-furnace slag as a cementitious constituent in concrete (ACl Committee 226 Report). *ACJ Materials Journal* 8, 4, 327–342, Title No. 84-M34.

Williamson, N. 2011: Steel-fibre-reinforced concrete floors. *Concrete* 45, 4, 20–22.

Winter, N.B. 2012a: *Scanning Electron Microscopy of Cement and Concrete*, 192pp. WHD Microanalysis Consultants Ltd, Woodbridge, U.K.

Winter, N.B. 2012b: Personal communication.

Winter, N.B. 2012c: *Understanding Cement – An Introduction to Cement Production, Cement Hydration and Deleterious Processes in Concrete*, 206pp. WHD Microanalysis Consultants Ltd, Woodbridge, U.K.

Wirgot, S. and Cauwelaert, F. Van. 1991: The measurement of impregnated cement fluorescence by means of image analysis. In *Proceedings of the Third Euroseminar on Microscopy Applied to Building Materials*, Barcelona, Spain.

Wirgot, S. and Cauwelaert, F. Van 1994: The influence of cement type and degree of hydration on the measurement of w/c ratio on concrete fluorescent thin sections. In *Petrography of Cementitious Materials*, DeHayes, S.M. and Stark, D. (eds.). ASTM STP 1215, American Society for Testing and Materials, Philadelphia, PA, pp. 91–110.

Wong, H.S. and Buenfeld, N.R. 2007: Estimating the water/cement (w/c) ratio from the phase composition of hardened cement paste. In *11th Euroseminar on Microscopy Applied to Building Materials*, 5–9 June 2007, Porto, Portugal, 11pp.

Wong, H.S. and Buenfeld, N.R. 2009: Determining the water-cement ratio, cement content, water content and degree of hydration of hardened cement paste: Method development and validation on paste samples. *Cement and Concrete Research* 39, 10, 957–965.

Wong, H.S., Matter, K.R. and Buenfeld, N.R. 2009: Estimating the water/cement (w/c) ratio of hardened mortar and concrete using backscattered electron microscopy. In *12th Euroseminar on Microscopy Applied to Building Materials*, 15–19 September 2009, Dortmund, Germany, 11pp.

Yingling, J., Mullings, G.M. and Gaynor, R.D. 1992: Loss of air content in pumped concrete. *Concrete International* 14, 10, 57–61.

Zhang, M.H. and Malhotra, V.M. 1995: Characteristics of a thermally activated aluminosilicate pozzolanic material and its use in concrete. *Cement and Concrete Research* 25, 8, 1713–1725.

Chapter 5

Appearance and textures of cementitious materials

5.1 INTRODUCTION

Building and construction with concrete has become a global prerequisite for social progress (Idorn 2004), and, consequently, the quality and long-term durability of such materials are of major importance in modern concreting practice (Mehta 1991). Yet the full value of the polarising microscope for examination of concrete in thin sections, together with the use of the scanning electron microscope, has still to be fully realised as essential tools in engineering practice for the investigation of concrete and related materials.

In the following sections, some of the more important microtextures in concrete and related materials are described, and the changes which may occur with time are illustrated to show how observation of these changes may have diagnostic value.

These observations using the polarising microscope and scanning electron microscope (SEM) require not only a knowledge of mineralogy to interpret aggregate textures but also an understanding of the chemistry and physics of Portland and other cements to interpret the texture of the hardened cement paste. Thus, the subject must be approached from the viewpoints of both chemical microscopy and classical mineralogy. Even where this type of systematic approach is used, many of the definitive observations of concrete microtextures observed in thin sections remain essentially qualitative in nature.

The petrographer is frequently required to examine the microtexture of concrete in order to explain why deterioration has occurred and relate this microtexture to the materials used or the methods of placement and the environmental history of the structure. Microscopic examination of concrete in thin sections can reveal a wealth of information on processes of deterioration, but certain limitations must be kept in mind. Two of these limitations are that thin sections are essentially two-dimensional so the third dimension must be inferred (see discussion by Elias 1971) and that the effective resolution of the polarising microscope is limited to about 1 μm. Since many of the reactions that occur in concrete result in products that are submicroscopic (i.e. less than 1 μm), their observation requires the use of electron microscopy. The scanning electron microscope can make use of both rough surface specimens and appropriately prepared thin sections (see Section 3.4). However, a major advantage of the SEM is that it is usually fitted with a microanalytical facility (EDS, see Section 2.5) allowing the small detailed changes in chemistry of micron-sized areas within the cement paste to be explored and mapped.

Petrographic examination of concrete using thin-section techniques can be a lengthy process requiring both careful observation and skill. For this reason, concrete petrographic investigation tends to be limited to structures and works where high standards of safety and durability are of concern. These works are usually constructed from concretes, made with specific cement types or cements containing particular mineral additions, and they will typically also contain chemical admixtures. The cement contents are often 300 kg/m³ or greater and the concrete will have been mixed and placed under supervision. In practice, petrographers do not get many opportunities to examine concretes with low cement contents or those that may have been both poorly placed and inadequately cured. It is for this reason that the concrete textures discussed and illustrated are largely confined to the better quality of structural concretes and the descriptions of textures are mostly limited to those observed using the polarising light microscope and backscattered electron images.

5.2 OPTICAL OBSERVATIONS OF THE HARDENED PORTLAND CEMENT PASTE MATRIX

Initially, the immediate reaction products of adding water to Portland cement are the formation of calcium hydroxide (see Section 5.2.5) and ettringite (Section 5.2.4). The latter then converts to monocalcium alumino-sulphate hydrate within the first 24 h. Concurrently, but more slowly, the anhydrous cement minerals hydrate forming the principal cementing phase, a group of essentially amorphous calcium silicate hydrates and calcium aluminate hydrates including those substituted with iron. The reaction forming calcium silicate hydrate is the principal source of calcium hydroxide. Apart from crystalline calcium hydroxide, these hydration products are normally beyond the resolution of the optical microscope and are best identified on fracture surfaces by SEM.

Calcium hydroxide has three distinct morphologies as a result of variations in hydration conditions: uniformly distributed anhedral small crystals (Figure 5.1), larger crystals that line voids and aggregate margins (Figure 5.2) and recrystallised relatively coarse crystals in cracks and cavities (Christensen et al. 1979) (Figure 5.3). In addition, electron microscope studies by Berger and McGregor (1973) suggest that small globular particles (~0.1 μm) scattered within the CSH are probably amorphous calcium hydroxide. This suggestion is supported by the work of Midgeley (1979) which appears to show that x-ray diffraction results underestimate the amount present compared to amounts determined by differential thermal analysis (DTA). As an example, one sample contained ~16% calcium hydroxide according to DTA analysis, but only ~13% determined by x-ray diffraction. The difference was ascribed

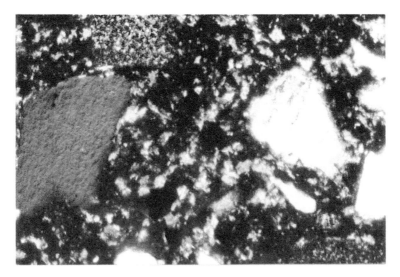

Figure 5.1 Anhedral crystals (yellow and white birefringences) of portlandite scattered through the cement matrix. Crossed polar illumination, width of field 1.3 mm. (Courtesy of M. Grove, private collection.)

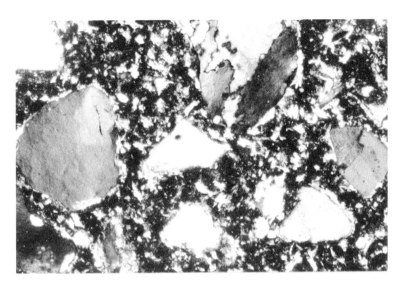

Figure 5.2 Lathlike crystals (yellow and white birefringences) of portlandite lining the margins of sand grains. Anhedral portlandite is also scattered through the cement matrix. Crossed polar illumination, width of field 1.3 mm. (Courtesy of M. Grove, private collection.)

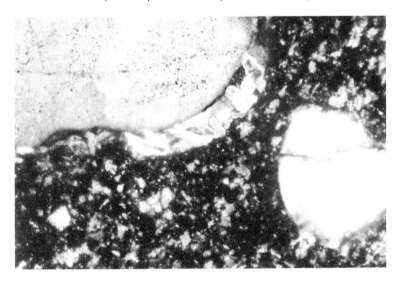

Figure 5.3 Large crystals (yellow, magenta and white birefringences) of recrystallised portlandite infilling a water-formed void. Crossed polar illumination, width of field 1.9 mm. (Courtesy of M. Grove, private collection.)

to the presence of amorphous calcium hydroxide. The development of calcium hydroxide (portlandite) during the hydration process is also affected by mix design, water/cement ratio and curing conditions (see Section 5.2.5).

The textural appearance and apparent density of hardened cement paste vary considerably as a result of differences in water/cement ratio and curing conditions. However, it is also affected by the thickness of the thin section. If the thin section is thicker than 30 μm, the paste will have a dark, milky appearance with many of the calcium hydroxide and remnant cement grains either partly obscured or hidden in the thickness of the section. Contrary to Section 11.1.1 of ASTM C856-95 (1995), Standard Practice for Petrographic

Examination of Concrete (now superseded by ASTM C856-04), there appears to be no justification for using section thicknesses greater than 30 μm as little useful detail can be observed in the hardened paste. When the thickness is reduced to 20–25 μm, the hardened cement paste becomes lighter in colour with an ill-defined background texture. Most calcium hydroxide crystals are reasonably clearly defined and the remnant and partially hydrated cement grains are easily seen. A thickness of 25 μm is preferred by many petrographers for the routine examination of thin sections of concrete and represents a compromise that yields the greatest information. Below 25 μm, the definition of remnant cement grains improves, but calcium hydroxide can appear skeletal unless considerable care is taken in grinding the section. A thin-section thickness of 20 μm is preferred by some petrographers especially when they are using smaller-sized polished thin sections as this thickness provides optimum resolution of the hardened cement paste and still allows satisfactory observation of the aggregate. Walker (1979) has shown that if a section is reduced by polishing to 8–14 pm in thickness, remnants of hydration shells become visible. These hydration shells are also known as 'Hadley shells' (Barnes et al. 1978) but are difficult to observe in 25–30 μm thick thin sections.

A routine optical examination of hardened cement paste in concrete is typically restricted to observations of porosity, entrapped and entrained air voids, cracks and fissures, calcium hydroxide, remnant and partially hydrated cement grains, the possible development of secondary crystals of ettringite and any alteration products. While this may appear limited, a surprising amount of diagnostic information is possible from this type of optical examination. There are claims by Randolph (1991) that the presence of chemical and finely divided mineral admixtures in concrete can be identified by changes in texture on the hardened cement paste including the distribution and morphology of the calcium hydroxide. The effect of chemical admixtures on the morphology of calcium hydroxide has been investigated by Berger and McGregor (1972) and Brandt (1993), but the effective use of these textural changes for the identification of admixtures is still the subject of investigation.

5.2.1 Typical composition of hardened Portland cement

The composition of a hardened cement paste at a water/cement ratio of 0.5 is shown in Table 5.1 (Taylor 1984). Volumetric values given in the last column of the table for calcium hydroxide, alite, belite and ferrite phases will not be completely visible because some small calcium hydroxide crystals and fragments of residual clinker may be obscured in the thickness of the thin section. Obviously the volumes of these phases observable in

Table 5.1 Calculated weight and volume percentages for a Portland cement paste hydrated for 91 days (water/cement = 0.5)

Phase	Saturated weight, %	Volume %	Dried at 110°C weight, %	Volume, %
Alite	6	4	8	4
Belite	4	3	6	3
Aluminate	0	0	0	0
Ferrite	2	1	3	1
Calcium hydroxide	14	12	19	12
Calcium carbonate	1	1	1	1
Calcium silicate hydrate	42	40	45	25
Hydrated aluminate and ferrite	18	16	19	13
Pores	12	24	0	42

Source: After Taylor, H.F.W., *British Ceramic Proceedings* 35, 65, 1984.

field concretes will also depend on the cement content, water/cement ratio and curing. However, the results given in Table 5.1 make a useful starting point for discussion of components in the cement matrix.

5.2.2 Remnant or oversize cement clinker grains

The remnant grains of cement clinker observed in the hardened paste of a mature concrete are usually those particles that are retained when cement is passed through a 90 µm sieve. Most of these oversized cement grains are surprisingly resistant to hydration and often will have barely observable hydration rims even where a concrete has been exposed to the weather for many years (Idorn and Thaulow 1983).

Both the quantities and particles sizes of remnant clinker grains in cements will vary according to their period of manufacture and the cement type. Lea (1970) quotes averaged results of residues recorded by Redgrave and Spackman as 10% on the 600 µm sieve and 45% on the 150 µm sieve as being typical for the year 1879. Lea then states that by 1910, the residue on the 90 µm sieve for many British and Continental cements had dropped below 10%, and by 1970, most ordinary Portland cements were averaging around 5% for 90 µm sieve residues and rapid hardening cements from 1% to 3%. Currently in those cement works with modern grinding circuits, it is the practice to exercise close control over the particle size distribution and to try and limit the residue on the 90 µm sieve to as little as 2%.

From the residues quoted earlier, we would expect that older concretes would have both a larger and a greater number of remnant grains of cement clinker embedded in their hardened pastes. This is illustrated in Figure 5.4. Further examples of remnant grains in more recent cements may be found in Figures 5.5 and 5.6 and in many of the other micrographs. A noticeable characteristic of remnant grains of cement clinker is the predominance of

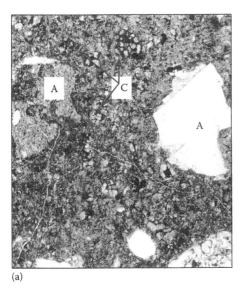
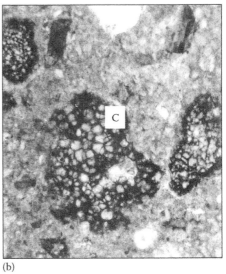

(a)   (b)

Figure 5.4 Examples of remnant grains of cement from two bridges built in the 1920s. Width of fields 1 mm. (a) The remnant cement grains (C) in this bridge concrete consist of clusters of belite together with some alite set in a dark interstitial matrix. Hydration rims are minimal. The microcracks are due to the reaction of the andesite aggregate (A) with cement alkalis. (b) The concrete from this hydro structure contains large remnant grains (C) consisting of nests of belite, some of which have dark hydration rims. Fragments of clinker as large as this are not common in modern concretes. Aggregates labelled A.

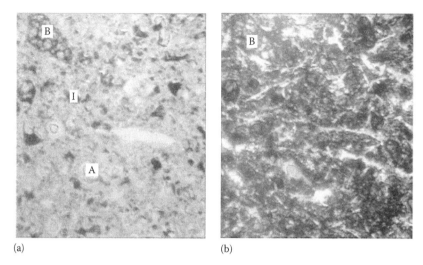

(a)   (b)

Figure 5.5 Partially hydrated grains of cement. A high cement content of 740 kg/m³ combined with a water/
cement ratio of 0.3 has resulted in self-desiccation and insufficient space for hydrate formation in
spite of moist curing this test concrete at 38°C for 7 years, This has further resulted in numerous
partially hydrated grains of cement still being present. Width of fields 0.44 mm. (a) The fragments
of alite (A) are surrounded by clear hydration rims while the fragments of dark interstitial phases
(I) arc irregular in shape and have no visible hydration rims. The belite cluster (B) at upper left is
a remnant cement grain. A sliver of chert is visible at middle right of photo. (b) Under crossed
polars the hardened cement matrix is well filled with calcium hydroxide which shows as irregular
white areas. The yellowish birefringence of the belite (B) in the remnant cement grain is clearly
visible and the biregringent texture of the sliver of chert is just discernible.

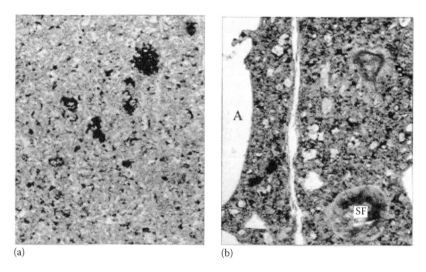

(a)   (b)

Figure 5.6 A comparison between the extremes of hydration. Width of fields 0.8 mm. (a) This geothermal
cement grout has been hydrated at 160°C for 3 months. The result is near complete hydration with
a few remnants of dark interstitial matrix of the cement and possibly some bentonite relics set in a
background of α-dicalcium silicate hydrate. Same area as shown in Figure 5.15b. (b) This super-high-
strength DSP mortar sample contains sufficient unhydrated cement to act as a fine filler. Grains as
small as 5–10 μm are visible showing hydration ceased early due to internal desiccation. The two
roughly circular areas in the texture are balls of undispersed silica fume (SF). Quartz aggregate is
also indicated (A). DSP sample mixed at water/cement = 0.23 with 20 silica fume and quartz sand as
aggregate. Moist cured for 28 days and then exposed to wetting and drying for 3 months.

rounded belite (C$_2$S) embedded in brown interstitial phase. In many cases, these grains have been derived from nests of belite due to oversized particles of quartz in the raw cement feed, and they may not represent the overall composition of the cement.

A convenient and arbitrary distinction can be drawn between those smaller residual grains of cement clinker which will eventually hydrate, provided they are in contact with sufficient water and there is space to accommodate their hydration products and those remnant grains which are resistant to hydration because of their large size. This distinction implies that the quantity and particle size range of partially hydrated cement grains observed will depend solely on the cement type, grain sizes, cement content and the degree of hydration of the concrete.

Summer et al. (1989) give typical particle size ranges for cements ground in modern conventional open-circuit mills as shown in Table 5.2. They claim that closed-circuit mills do not differ significantly in the particle size range produced so the figures quoted in the table form a useful general guide.

Lea (1970) quotes depths of hydration for Portland cement particles as being about 7 μm at 1 day and 20 μm at 7 days and other data to show that particles of cement less than 5–7 μm in diameter mainly contribute to 1–2 day strength, particles less than 20–25 μm to the 7-day strength and particles less than 25–30 μm to the 28-day strength. It is difficult to compare these data, but they suggest that in a concrete of moderate cement content, which has been adequately cured for 28 days, few partially hydrated grains of original size less than 30 μm should be present. This suggestion is in reasonable agreement with microscopic observation on adequately cured laboratory concretes.

However, where concrete is inadequately moist cured, more partially hydrated grains may be present in the outer layers compared to the inner core because the premature drying of the outer zones has halted hydration. Where a concrete contains a high cement content and is mixed at a low water/cement ratio, it may become almost impermeable to curing water and there may be insufficient water for complete hydration leading to a condition called self-desiccation. In addition, at very low water/cement ratios, pore space to accommodate only a limited volume of hydration products may be present. An illustration of the effect of high cement content and low water/cement ratio is shown in Figure 5.5.

The effect of extremes of hydration on the particles sizes and amounts of partially hydrated grains of cement is shown in Figure 5.6a. St John gives examples of a cement grout, mixed at a water/cement ratio of 0.62 that was exposed down a geothermal steam bore at 160°C for 3 months. Because of the prolonged exposure, only vestigial remnants of cement grains are present. This is an extreme that will be rarely met in normal concreting practice. By contrast, a super-high-strength densified system with small particles (DSP) mortar was mixed at a water/cement ratio of 0.23, moist cured for 28 days and then exposed for 3 months to daily cycles of wetting at 21°C and drying at 60°C. In spite of the area photographed in Figure 5.6b, being only

Table 5.2 Typical particle sizes and range of cement ground in open-circuit mills

μm	Typical, %	Range, %
<90	98	96–100
<63	95	90–98
<45	88	81–93
<30	75	69–81
<20	60	53–65
<15	48	44–52
<10	36	32–39
<5	21	18–23
<2	8	7–9

a few millimetres below the surface, small unhydrated grains showing sharp angular outlines are still present, and there is sufficient partially hydrated cement remaining to effectively act as a filler. This is the other extreme which only applies to special materials such as super-high-strength DSP mortar that are highly impermeable, and hydration is limited by self-desiccation.

Estimates for the amounts of residual grains of cement present in concrete cured at 20°C have been given by French (1991a). He claims that above a water/cement ratio of 0.61, hydration is virtually complete, with only 3%–4% of unhydrated material remaining for a water/cement ratio of 0.5; this then rises to 7%–8% for a water/cement ratio of 0.4. If the concrete is made at a temperature close to 0°C, the amount of unhydrated material increases, while in concrete cured above 40°C–50°C, hydration is generally complete. Addis and Alexander (1994) put forward a concept of cement saturation for concrete which concludes that hydration above water/cement ratios of 0.35–4.0 hydration should be complete because hydration is not then limited by lack of space or water. However, it should be noted that under normal conditions of manufacture and exposure, practical experience shows that concrete in most structures may only be 80%–90% hydrated irrespective of the water/cement ratio.

Lea (1970) quotes analyses of various size fractions of ground cement clinker which shows that the alite content tends to be higher in the fine fractions and the belite content higher in the coarse fractions. The rate of hydration of the phases is generally stated by Mindess and Young (1981) as being aluminate (C_3A) > alite (C_3S) > ferrite (C_4AF) > belite (C_2S). Combining these data and applying it to a well-cured concrete of moderate cement content suggests that the aluminate phase should be rare or absent and belite and ferrite phases the most common phases to be observed in remnant and partially hydrated grains. The amount of alite present should be less than the belite and ferrite, but this may be affected by some of the factors discussed earlier. This is in agreement with observation of well-cured concretes of moderate cement content where typically belite and ferrite are the most frequently observed residual clinker phases.

If a sample of concrete is polished and then etched either with HF vapour or with 1% nitric acid in ethanol, textural details of the clinker grains remaining in the hardened paste may be observed by examination in reflected light. Application of this technique to distinguish between sulphate-resisting and ordinary Portland cements was developed by Grove (1968) and is now incorporated in BS 1881-124 (1988). The practical considerations of using the technique are discussed in Section 4.4. It is claimed that the technique may also be used to determine the water/cement ratio by point counting residual grains of clinker as first described by Ravenscroft (1982).

This method for determining the apparent water/cement ratio of a concrete based on counting in thin sections the number of residual clinker particles per millimetre traverse of paste has also been proposed by French (1991a). He considers that although it may appear crude, in fact the method is very robust and gives satisfactory results, provided a wide range of standard materials are used for calibration. This method has been updated to allow an image analysis approach by Bromley (see Section 4.4).

In ASTM C856-04 (2004) under the heading 'Relict cement grains and hydration products', it states 'Cements from different sources have different colours of alumino ferrite and the calcium silicates have pale green or yellow or white shades. It should be possible to match cements from one source', presumably using these colours. There appears to be little published information on the validity or use of this technique.

5.2.3 Colour of hardened concrete

The normal, greenish, grey-black colour of cement clinker is mainly due to the ferric ion Fe^{3+} in the ferrite phase, which in thin sections is a deep brown colour as is illustrated by Campbell (1986). Lea (1970) also discusses the constituents, suggesting that it is the manganese and magnesium that darken the colour of the ferrite phase and so produce darker-coloured cements. More recently, Ichikawa and Komukai (1993) reported that substitution of Al^{3+} and Fe^{3+} by Mg and Zn changes the ferrite colour to grey, while an increase of the ratio of Si and Ti to the divalent ions prevents this colour change. The effect by various metallic ions on white cement is discussed by Kupper and Schmid-Meil (1988), and methods to deliberately colour cement clinker have been discussed by Laxmi et al. (1984).

Overall, the colour of fresh concrete is controlled by both the colour of the cement and the aggregate. Sand particles less than 0.5 mm have an important effect, while the colour of the cement itself is controlled by the raw materials used in its manufacture. Mineral admixtures may also affect colour. Silica fume and fly ash depending on their carbon content will darken the hardened cement paste, while interground limestone and slag will lighten the colour. The colour of the external surface of old concrete changes as it weathers, and it may become covered with efflorescence, moss and lichens.

Mielenz (1962), describing the appearance of fracture surfaces in a sound, well-cured concrete of low water/cement ratio, states: 'the fracture surfaces through the cement paste are grey, and the freshly fractured cement paste is somewhat vitreous and amorphous in appearance. Close inspection reveals black or dark-brown, vitreous, angular coarse particles of the cement embedded in the paste and still largely unhydrated... with an increase in water/cement ratio... the cement paste is weaker, softer, lighter in colour, and less vitreous in lustre and possibly somewhat granular in appearance, in contrast to equivalent concrete of high quality'. The differences in the texture of fracture surfaces as described earlier are visible under the low-power stereomicroscope. They are a useful rough guide to concrete quality and their evaluation should form part of the initial examination of a concrete.

The colour of the hardened cement paste in thin section is primarily a function of light absorption which is affected by both the thickness and the composition of the hardened paste. Typically in a 25 µm thick section, the paste is light brown in colour but may often have a greenish tint or, more rarely, a distinct yellow cast. As an example, two dark-coloured concretes, sampled from bridges by St John, when examined in thin section were yellow and contained distinctive small fragments of a reddish interstitial phase attached to the belite. The belite was generally of low birefringence. The dark colour suggested the cement used was high in manganese.

French (1991a) reports that where moisture passes through a concrete or where the water/cement ratio is high, the calcium silicate hydrates tend to be lighter in colour, more transparent and of a coarser texture than paste of a low hydration state.

There are difficulties in accurately reproducing the subtle colours observed in thin section as colour prints have a limited contrast ratio which is often exceeded in thin section. This results in 'green disease' and other colour distortions as discussed by McCrone and Delly (1973).

5.2.4 Ettringite in hardened concrete

Ettringite ($3CaSO_4 \cdot 3CaO \cdot Al_2O_3 \cdot 26{-}32H_2O$) is produced by a reaction that requires an excess of the sulphate ion SO_4^{2+} over the aluminate phase in the pore solution in the cement matrix. Its formation during the setting of concrete is expansive, but as the concrete is still in a plastic state, the volume increase is easily accommodated. As hardening proceeds, more of the aluminate phase moves into the pore solution converting the ettringite to monocalcium sulphoaluminate ($CaSO_4 \cdot 3CaO \cdot Al_2O_3 \cdot 12H_2O$). This conversion results in a small volume decrease and thus may occur in the hardened paste without damage. Both ettringite and monocalcium sulphoaluminate hydrate formed during setting and hardening are usually submicroscopic and require the use of the SEM or x-ray diffraction techniques for their identification. Only where ettringite is secondary and reformed from reaction of the SO_4^{2+} ion with monocalcium sulphoaluminate do the crystals become large enough to be observed in thin

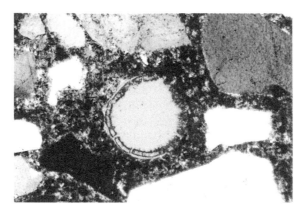

Figure 5.7 Typical appearance of secondary ettringite lining an air void. Crossed polar illumination, width of field 2 mm. (Courtesy of M. Grove, private collection.)

Figure 5.8 Secondary needle-like ettringite crystals forming at the margin of a large air void. Inclined polar illumination, width of field 0.3 mm. (Courtesy of M. Grove, private collection.)

section. A general review on the role of ettringite in concrete has been made by Negro (1985). Where ettringite is observed in cracks and voids, it appears as needles (Figures 5.7 through 5.9) which have a distinctive low birefringence and low refractive index and is often found in conjunction with calcium hydroxide. Under high magnification, well-crystallised needles of ettringite are revealed as rods (Kennerley 1965), as is shown in Figure 5.10, and as the SEM images, Figure 5.11. Because the ettringite is the stable phase with respect to monocalcium sulphoaluminate below 60°C, deposits of ettringite are more common in concrete than would be expected. Situations such as movement of water through a concrete or the formation of alkali–silica gel appear to be able to cause conditions that allow it to reform. In old concretes, these deposits can be massive as illustrated by the SEM images in Figure 5.11. However, as the ettringite is reformed in void space, its expansion can be accommodated.

This should be clearly distinguished from the formation of ettringite by sulphate attack, situations where a concrete has been steam cured or has been subjected to curing temperatures in excess of 65°C so that later reaction of carbon dioxide with monocalcium sulphoaluminate causes its formation (Kuzel 1990). This delayed ettringite formation (DEF) in the fabric of the hardened paste due to the ingress of the sulphate ion, as opposed to redeposition in pores, cracks and fissures as described earlier, is always potentially damaging (Min and Mingshu 1994; see also the discussion in Section 6.5.9). French (1991a) claims that significant

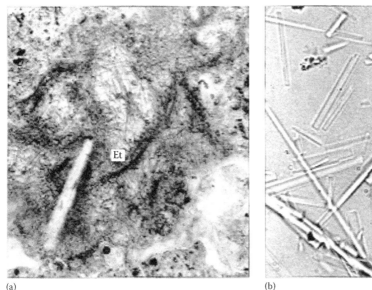
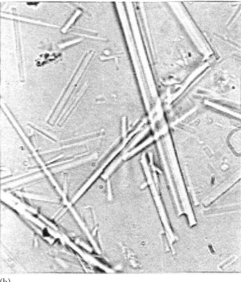

(a)   (b)

Figure 5.9 Ettringite. (a) The low refractive index plus low birefringence under crossed polars, identify the fine needles which fill the pore in this old concrete as ettringite (Et). There is no evidence of sulfate attack or alkali-aggregate reaction but some indication that moisture was moving slowly through the concrete in this seventy-year-old hydro structure. Width of field 0.75 mm. (b) At high magnification, ettringite needles are resolved as rods with a thickness of 1–3 µm. This sample was taken from a massive efflorescent deposit of ettringite which formed over a ten-year period in a submerged hydro dam gallery. The mechanism for the unusual formation of ettringite is discussed by Kennerley (1965). Width of field 0.04 mm.

formation of ettringite due to seepage of water through a concrete can, in some instances, lead to expansion and cracking. Ettringite is subject to carbonation, decomposing to calcite, gypsum and alumina gel (Grounds et al. 1988; Nishikawa et al. 1992) when in contact with water containing carbon dioxide. Under dry conditions, the fibrous habit of ettringite persists as its decomposition is retarded by these conditions. The special conditions which lead to the formation and damage associated with thaumasite $[Ca_3Si(OH)_6 \cdot 12H_2O]$ $(SO_4)(CO_3)$ which is sometimes associated with ettringite are discussed in Section 6.5.6.

5.2.5 Calcium hydroxide in hardened cement paste

Calcium hydroxide (or portlandite) is produced by the hydration of the silicate phases, tricalcium and dicalcium silicate, and free lime (CaO) as shown in the following approximate equations. No calcium hydroxide is produced from the hydration of the aluminate and ferrite phases:

$$2\left(3CaO{\cdot}SiO_2\right) + 6H_2O \rightarrow 3CaO{\cdot}2SiO_2{\cdot}3H_2O + 3Ca\left(OH\right)_2$$

$$2\left(2CaO{\cdot}SiO_2\right) + 4H_2O \rightarrow 3CaO{\cdot}2SiO_2{\cdot}3H_2O + Ca\left(OH\right)_2$$

$$CaO + H_2O \rightarrow Ca\left(OH\right)_2$$

From the first two equations, it can be seen that for weight, the hydration of tricalcium silicate $(3CaO \cdot SiO_2)$ produces three times the amount of calcium hydroxide as does the hydration of dicalcium silicate $(2CaO \cdot SiO_2)$. Thus, the ratio of these two phases will affect

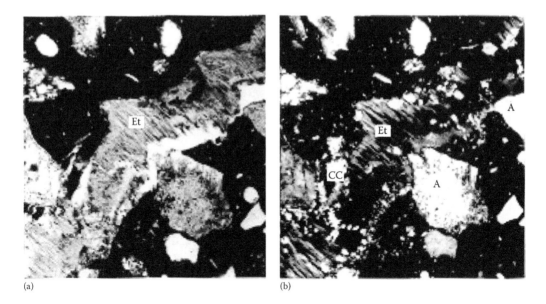

(a)   (b)

Figure 5.10 Massive enringite in the cracks of an eighty-year-old concrete from Lady Evelyn Lake Dam in Canada. Width of fields 0.8 rnm. (a) The ettringite (Et) deposited in the crack is massive and appears as curtain-like material. The concrete is undergoing an advanced alkali-aggregate reaction and the combination of long severe exposure and advanced reaction appears to have promoted this rather unusual formation. (b) Under crossed polars the typical low birefringence of the ettringite (Et) is clearly apparent together with unidentified dark material. There are also some isolated clusters of carbonation (CC) present in the texture. Aggregates labelled A.

the amount of calcium hydroxide formed. The range of the calcium silicates to be found in cements is given in Table 5.3 (Concrete Society 1987). The effects of these changes on the amounts of calcium hydroxide produced have been estimated for seven British cement plants as shown in Table 5.4 (Concrete Society 1988) and are less than would be expected mainly because the cements high in dicalcium silicate also have higher free lime contents.

The results quoted in Table 5.4 for cement with the maximum tricalcium silicate are typical of current ordinary Portland cements, while the cement composition with the lowest tricalcium silicate approximates to a low-heat cement. As these results were calculated for a fully hydrated cement paste with no free water present, they are somewhat artificial, and a more practical idea of the amount of calcium hydroxide likely to be present is given in Table 5.1. These results indicate that changes in cement composition over the years have not had a marked effect on the amount of calcium hydroxide present in concrete.

Some possible modifications to calcium hydroxide (portlandite) morphology in a Portland cement matrix due to hydration conditions have already been noted in the introduction of this section. Listed here are some other modifications resulting from mix design, admixture additions or curing conditions that have been reported:

- Addition of silica flour, some pozzolans (Lea 1970) or microsilica to the mix will reduce or suppress the development of portlandite and will alter its morphology to a lathlike form. Some reports suggest that microsilica increases the initial hydration of C_3S, and consequently, more calcium hydroxide forms though this reaction slows after two days (Andrija 1986; Sellevold et al. 1982). See Figure 5.12.

(a)

(b)

(c)

(d)

Figure 5.11 SEM back scattered images of secondary ettringite in a void at increasing magnifications (a), (b), and (c) and its x-ray spectrum (d). The spectrum is of the rectangular area Z3 as marked on the lower left of the SEM image (fracture surface specimen with gold coating). (Courtesy of I. Fernandes, University du Porto.)

Table 5.3 Percentages of tricalcium silicate and dicalcium silicate in cements for the period 1914–1984

Date	–1914	1914–22	1928–30	–1944	–1960	–1980	–1984
% tricalcium silicate	25	15–48	19–58	30–50	36–55	45–64	54–63
% dicalcium silicate	45	81–26	53–14	45–20	37–12	30–11	27–8
Ratio	0.6	0.2–1.8	0.4–4.1	0.7–2.5	1.0–4.6	1.5–5.8	2.0–7.9

Table 5.4 Effect of changes in cement composition on calcium hydroxide and calcium silicate hydrate gel contents of hydrated cement paste

Cement minerals (%)	Typical composition		Range of annual average compositions in 1988	
	1930	1980	Lowest tricalcium silicate cement	Highest tricalcium silicate cement
Free lime, CaO	2.6	1.7	2.2	1.0
Tricalcium silicate	39	53	45	64
Dicalcium silicate	32	18	21	11
Calcium hydroxide calculated	25	27	25	30
Calcium silicate hydrate gel calculated	52	49	47	50

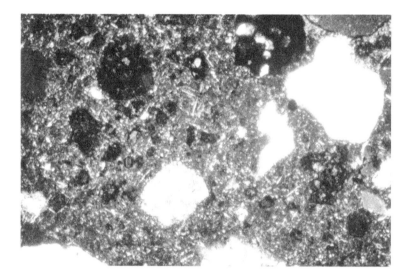

Figure 5.12 A concrete cement matrix containing microsilica. Note the reduction in portlandite and its change to a lathlike morphology. Crossed polar illumination, width of field 1 mm. (Courtesy of M. Grove, private collection.)

- At a given water/cement ratio, the development of portlandite crystals at aggregate boundaries tends to be thickest on dense impermeable aggregates like chert and thinner and impersistent on more porous aggregates, particularly carbonates (French 1991a). See Figures 5.2 and 5.13.
- Portlandite developed at aggregate surfaces tends to be more abundant in concretes where the aggregate grading and the presence of fines give the largest total aggregate surface area (French 1991a).
- High water/cement ratio mixes tend to allow large euhedral crystals to form, and low water/cement ratio tends to give small anhedral uniformly distributed portlandite crystals as in Figures 5.1 and 5.14.
- Elevated temperature curing tends to allow larger crystals of portlandite to form, and by contrast, smaller anhedral crystals are formed in low-temperature curing conditions.
- Portlandite may be dissolved and can be reprecipitated as aggregations of coarse platy crystals in open cracks and voids by percolating water and is an indicator of possible leaching. It is more soluble in water at low temperatures and may be associated with secondary ettringite and freeze–thaw damage. See Figure 5.3.

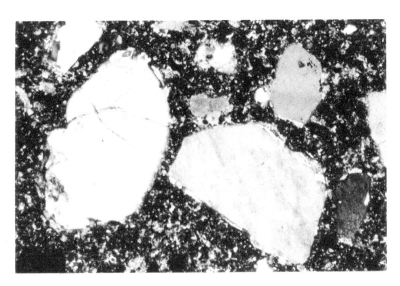

Figure 5.13 Large portlandite crystals along the margins of quartz grain aggregate particles. Crossed polar illumination, width of field 1.3 mm. (Courtesy of M. Grove, private collection.)

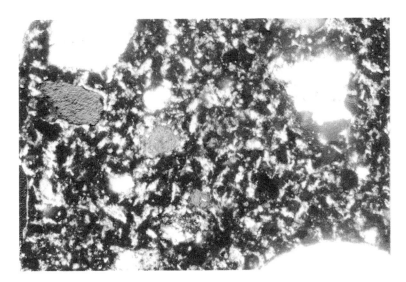

Figure 5.14 Randomly distributed portlandite crystals in a Portland cement matrix. Crossed polar illumination, width of field 3 mm. (Courtesy of M. Grove, private collection.)

When cement paste is autoclaved at 170°C–180°C, the principal phase formed is crystalline α-dicalcium silicate hydrate. Crystals of this hydrate may grow large enough to become visible under the microscope as shown in Figure 5.15.

Crystalline α-dicalcium silicate hydrate (α-$2CaO \cdot Si_2O_3 \cdot H_2O$) is not considered desirable in concrete products because of its low strength. Silica flour is added to the mix to prevent its formation in favour of a higher-strength tobermorite (calcium silicate hydrate)-type phase which is submicroscopic (St John and Abbott 1987).

Bache et al. (1966) have reviewed the occurrence of calcium hydroxide and its morphology in cement paste giving a series of micrographs showing its appearance in thin section. In pores where sufficient space and fluid are present to allow adequate crystallisation,

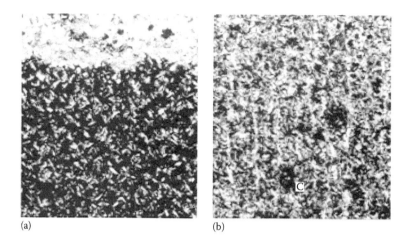

(a)   (b)

Figure 5.15 The appearance of α-dicalcium silicate hydrate in geothermal cement grout. Test grout. water: powder = 0.62, containing 3 per cent bentonite, exposed for three months to geothermal fluids at 160°C. Width of fields 1 mm. (a) The small white laths of α-dicalcium silicate hydrate can be clearly seen in the area below the zone of golden yellow carbonation at the top of the picture. The area photographed is completely leached of calcium hydroxide so that the laths are clearly visible against the dark background. Crossed polars. (b) An area from the centre of the same specimen. Here the α-dicalcium silicate hydrate crystals form a background in which are embedded aligned calcium hydroxide and the remnants of a few grains of cement (C) and bentonite relics which have managed to survive the extreme conditions. Circular polarisation.

large laths of calcium hydroxide occur or a crack may become infilled as shown in Figure 5.16. Where fluid has not filled the pore or fissure, calcium hydroxide crystals will only form on surfaces. However, the greater portion of the calcium hydroxide will be scattered through the hardened cement paste where it forms irregular clumps of variably sized crystals filling small void spaces. The maximum size of capillary void space varies with the water/cement ratio, and this will affect the sizes of calcium hydroxide crystals in the paste as shown in Figure 5.17.

Other examples of calcium hydroxide in the paste texture are illustrated in many of the figures in this chapter. Bache et al. (1966) point out that because the saturation concentration of calcium increases with decreasing calcium hydroxide crystal size, there will be a tendency towards the growth of large crystals at the expense of small ones dependent on the rate of diffusion of calcium hydroxide in the liquid-filled system. In the preceding discussion, it has been assumed that all the calcium hydroxide present in a hardened paste is crystalline and visible. Midgley (1979) estimated that about 18% of the calcium hydroxide present in a hardened paste is amorphous material by comparing the results of thermal analysis with those from x-ray diffraction. He considered this amorphous calcium hydroxide to be present as small-diameter globular particles of about 0.1 μm, but Taylor (1990) states that there does not appear to be any convincing evidence for this claim. Groves (1981) and Groves and Richardson (1994) claim to have observed the presence of microcrystalline calcium hydroxide, but this is disputed by Rayment and Majumdar (1982). However, it is wiser to assume that the upper limit for observable calcium hydroxide in thin section may only be 80% of the theoretical amount and the percentage observed in normal concrete thin sections may be considerably less than this.

The structure of hydrated Portland cement paste adjacent to aggregate surfaces differs from the bulk material. A number of investigators using a SEM report the presence of a thin continuous film of calcium hydroxide adhering to the surface which is in turn coated with

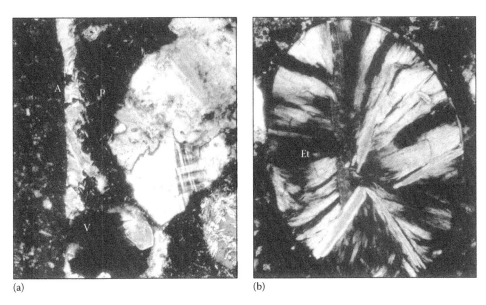

(a)   (b)

Figure 5.16 The appearance of well formed calcium hydroxide crystals in two sixty-year old concretes. Width of fields 0.8 mm. (a) The fissure present at the margin of the argillite aggregate (A) at the left is due to plastic settlement of the paste (P). It is filled with well crystallised calcium hydroxide. More commonly, fissures like this are only partly filled because they do not remain saturated. The interconnected pore (V) contains a tablet of calcium hydroxide and the paste at its margin is also filled with calcium hydroxide. Crossed polars. (b) Between crossed polars the large pore contains well formed laths of calcium hydroxide set in a fibrous mass of ettringite (Et). This pore will have been filled with fluid to allow crystals of this size to form. The range of interference colours of the calcium hydroxide are well illustrated.

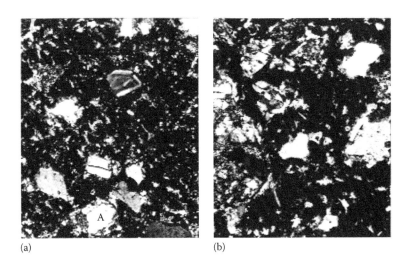

(a)   (b)

Figure 5.17 The effect of water/cement ratio on calcium hydroxide. Width of fields 0.8 mm. (a) Under crossed polars, this test concrete (water/cement ratio = 0.4), contains well dispersed calcium hydroxide throughout the paste texture which gives the appearance of only limited crystallisation at aggregate (A) margins. Remnant and partially hydrated cement grains are noticeable when the polars are uncrossed. (b) In contrast another test concrete (water/cement ratio = 0.8), contains larger aggregations of calcium hydroxide and crystallisation at aggregate margins is now obvious. There are far fewer remnant and partially hydrated grains in this sample compared to the one in Figure 5.17a.

a layer of rod-shaped C–S–H gel particles, the so-called duplex layer 1–1.5 μm in thickness (Barnes et al. 1978), though Scrivener and Gartner (1988) dispute the duplex nature of this film. A wider marginal zone of 40–50 μm is more porous than the bulk of the matrix with most cement grains in the zone being dissolved away or leaving empty hydrate shells (Hadley grains) (Barnes et al. 1978).

Mather (1952) suggests that the formation of calcium hydroxide crystals at aggregate interfaces is affected by the composition of the aggregate and that smaller crystals are observed on limestone aggregate interfaces than on siliceous aggregates. Bache et al. (1966) did not find this effect and considers that the surface texture of the aggregates may be of greater importance. French (1991a) considers that calcium hydroxide is more conspicuous on siliceous fine aggregates such as chert and quartz which may accumulate relatively thick layers, whereas on carbonate rocks, the layer is thin and irregular. In general, finer sands result in greater development of calcium hydroxide on surfaces. Serpentinites develop thick haloes and added materials such as the larger particles of fly ash and slag are often heavily coated with calcium hydroxide. Brandt (1993) observed the formation of an unusual spherulitic calcium hydroxide in voids and cracks where the concentration of superplasticiser exceeded 25 g/L of hardened paste. The effects discussed earlier are localised concentrations controlled by both the amount of water and the space at particle surfaces with crystallisation modified by the surface texture and porosity of the aggregate.

Calcium hydroxide is of considerable importance in concrete as its formation both in the fabric of the hardened cement paste and at interfaces affects strength. Bentur and Mindess (2006) have reviewed the effect that duplex layers of calcium hydroxide and calcium silicate hydrate, which form at aggregate and reinforcement faces, have on bond strength. The contribution of calcium hydroxide to the strength of concrete is controversial and has been reviewed by Barker (1984) who also discusses morphology. Yilmaz and Glasser (1991) have reported that the morphology of calcium hydroxide is strongly affected by superplasticisers. Many chemical admixtures modify calcium hydroxide formation, for example, bond strength is increased by the addition of 1.4 wt% PVA with a reduction of calcium hydroxide at cement/aggregate interfaces (Chu et al. 1994). Calcium hydroxide is also important in autogenous healing of cracks and porosity in concrete where it is undoubtedly the main agent involved (Lauer and Slate 1956) although Clear (1985) showed that carbonation of calcium hydroxide by bicarbonate in water percolating through cracks is also involved. The exact relationships of submicroscopic calcium hydroxide at interfaces and in the fabric of the paste to the calcium hydroxide that is observed under the microscope are typically poorly defined in examples taken from structural concretes.

5.2.6 Modification of calcium hydroxide in Portland cement concretes

Since calcium hydroxide is ubiquitous and one of the few easily visible components of hydration in cement paste, its quantitative estimation by microscopy would seem obvious. However, by its very nature, calcium hydroxide is best observed qualitatively. An attempt to determine the amount of calcium hydroxide in concrete by point counting has been reported by Larsen (1961). Using computer-aided image analysis techniques, the quantitative determination of crystalline calcium hydroxide in cement matrix should be more straightforward today.

Since calcium hydroxide tends to form in any space or on any surface where pore fluid is present, its absence or alteration can be indicative of the past history of the concrete.

The most obvious example is the carbonation of calcium hydroxide which is discussed in Section 5.4. Where it appears to be absent from a hardened cement paste, this may indicate that either high-alumina or supersulphated cement has been used because these cements do not produce calcium hydroxide during their hydration. Alternatively, the concrete may either contain a siliceous addition such as silica fume (Figure 5.6b) or be a high-pressure, steam-cured product containing a silica flour.

Although in these cases most of the calcium hydroxide will have reacted, it is not unusual for some residual calcium hydroxide to be present.

Porous aggregates may have a surrounding zone of paste (Figure 5.18) in which calcium hydroxide is deficient or absent and in rare cases calcium hydroxide will have crystallised in the porous rims of such aggregates. Where an aggregate has reacted to produce an alkali–silica gel which has bled into the surrounding paste (Figure 5.19), calcium hydroxide will be absent as the calcium is taken up by the alkali–silica gel.

The absence of calcium hydroxide from pores, cracks and fissures indicates that there has been insufficient fluid in these spaces for its formation. Along the margins of cracks through which water has percolated (Figure 5.18), calcium hydroxide will often be depleted by leaching. Careful observation of the way in which calcium hydroxide is present in localised areas of a hardened cement paste can provide useful information on the history of the concrete.

The solubility of calcium hydroxide is depressed by increasing temperature. The solubility of calcium hydroxide is given as 1.37 g/L at 0°C, 1.28 g/L at 15°C, 1.04 g/L at 40°C and 0.51 g/L at 100°C (Kaye and Laby 1995). This effect of decreasing solubility with temperature reduces the size of the individual calcium hydroxide crystals formed as noted

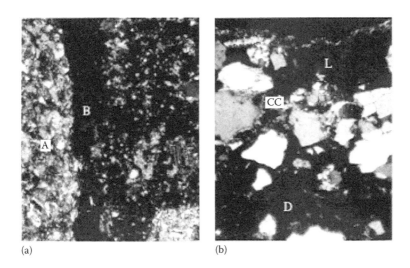

(a)   (b)

Figure 5.18 The absorption of calcium hydroxide into aggregate margins and the leaching of calcium hydroxide from crack margins. (a) Under crossed polars the dark band (B) of paste along the margin of the greywacke aggregate (A) in a thirty-five-year-old bridge is clearly deficient in calcium hydroxide. In cases such as this it is not clear whether the calcium hydroxide has been absorbed into the aggregate or the paste has been desiccated by the aggregate absorption. The lack of partially hydrated cement grains in the band suggests the former. Width of field 0.8 mm. (b) Under crossed polars the area (L) along the margins of this wide crack has been completely leached of calcium hydroxide and has undergone mild corrosion. There is a diffuse band of calcium carbonate (CC) backed by another area where calcium hydroxide is deficient (D). At the bottom of the picture normal paste texture is just visible. This is the effect of soft water flowing for sixty years through a large crack in an irrigation dam. Width of field 1.5 mm.

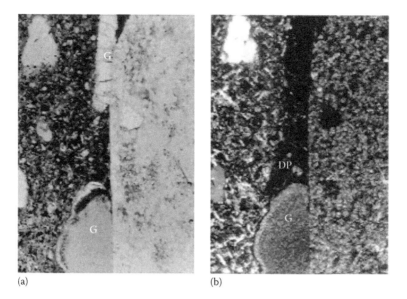

(a)   (b)

Figure 5.19 Absorption of calcium hydroxide by alkali-silica gel. Width of fields 0.8 mm. (a) The chert aggregate at the right has two patches of gel (G) along its margins. The lower patch is light brown with a darker margin while the upper is clear in colour with transverse cracks. The cement paste is full of remnant and partially hydrated cement grains with the brown interstitial phase being prominent. There are two grains of fine aggregate embedded in the paste. (b) Under crossed polars the texture of the chert aggregate shows clearly. The lower patch of gel (G) is crystallised while the upper patch has disappeared because it is a true gel and hence isotropic. Along parts of the aggregate margin the cement paste is dark (DP) due to gel having soaked into the paste and absorbed the calcium hydroxide. In the remaining paste a network of calcium hydroxide is clearly visible.

by Berger and McGregor (1973) in spite of the solubility being considerably reduced by the presence of alkali sulphates in the pore solution. Larger crystals of calcium hydroxide should form in concretes exposed to cooler environments.

Since calcium hydroxide forms in void spaces and fissures, it may be thought that it should be a good indicator for defining such space. However, these spaces must be filled or partially filled with sufficient fluid for crystallisation of the calcium hydroxide to occur, so its presence or absence in voids or fissures is as much an indication of moisture conditions and degree of hydration as the presence of the space itself. This can be of use if the environmental history of the concrete is known, but these observations by their very nature remain qualitative.

The depletion or absence of calcium hydroxide in the cement matrix usually indicates that either non-Portland cement or a siliceous admixture has been used. Apart from surface layers and along crack margins, calcium hydroxide leaching from the body of a concrete is usually restricted to porous water-retaining structures. If structural concretes are so porous that leaching is able to occur from normal exposure, carbonation occurs long before leaching of the body of the cement matrix takes place. The phrase 'the concrete is leached' should not be used and more correctly should be stated as 'the surface layers of the concrete are leached of their calcium hydroxide'.

Calcium hydroxide is highly susceptible to carbonation and the observation of the extent of this reaction is of great importance in the petrographic examination of concretes and as an indication of the permeability of the surface layers. When calcium hydroxide occurs abnormally, the cause should always be investigated. The ability to interpret such observations will depend on the experience of the petrographer.

5.3 CONCRETES CONTAINING MINERAL ADMIXTURES

The most common mineral additions and cement replacement materials have already been discussed in Section 4.6, including methods used in quantifying these materials in hardened concrete. Although there is some overlap, the following sections are complimentary to Section 4.6 in that they are largely concerned with the appearance of such materials under the optical and scanning electron microscope.

5.3.1 Portland limestone cement concrete

Between 10% and 35% of limestone may be used either as a filler (BS EN 206-1 2000), complying with BS 7979 (2001), or as an interground with Portland cement clinker (BS EN 197-1 2000). It primarily acts as a diluent allowing the production of cement requiring less energy input for its manufacture. As the limestone is softer than the cement clinker, intergrinding produces a finer particle size range which has a plasticising effect in concrete. Any loss in strength due to dilution of the cement with the limestone is reduced by lower water demand, less bleeding and increased workability. While some claim the durability of Portland limestone cements do not appear to differ significantly from Portland cements in equivalent concretes, problems of increased plastic shrinkage and cracking have been reported (Baron and Douvre 1987).

Finely ground powders can have a catalysing effect on the hydration of the silicate phases (Ramachandran 1988). The limestone may also react with the aluminate phases to form carboaluminates in a similar manner to gypsum, so the limestone acts as more than an inert filler in a number of ways. Surprisingly, many standards for ordinary Portland cement permit the addition of up to 5% fillers without restriction, but any limestone addition exceeding 5% classes the cement as a Portland/limestone cement. Typically, limestone addition to cement is in the range 10%–15% although European specifications permit additions of up to 35%. While standards may not require the limestone to contain more than 75%–80% calcium carbonate, in practice a minimum of 90% is recommended and limits on clay and dolomite contents may be imposed.

Data on the lowest level of limestone addition that can be observed in a thin section of concrete do not appear to be available. It is clearly visible in the texture of a concrete made from Portland/10% limestone cement given in Figure 5.20. Apart from the presence of the high-order birefringence of the fine particles of the limestone, the texture of the hardened paste appears a little different from that of an ordinary Portland cement (OPC). Where doubt exists, examination of a fracture surface of the concrete by SEM should reveal the presence of small rhombs of calcium carbonate.

5.3.2 Pozzolanic and Portland slag cements

Blended cements using ground granulated blast-furnace slag (ggbs) or pozzolanic materials mixed with Portland cement have become important and very widely used cementitious materials in the construction industry today. Blast-furnace slag is produced as a waste product from the smelting of iron ores. The compositions and properties of blast-furnace slags are discussed by Lea (1970) and more recently have been reviewed by Hooton (1987), Concrete Society (1991), Moranville-Regourd (1998) and Swamy (1993). Non-ferrous slags are also available but as yet have found little use (Douglas and Malhotra, 1987).

ASTM C595/595M-09 (2009) recognises and specifies three types blended with ggbs:

- Slag-modified Portland cement (<25% ggbs)
- Portland blast-furnace slag cement (25%–75% ggbs)
- Slag cement (>70% ggbs)

The BSI British Standards also recognise three main types but set different limits:

- Portland blast-furnace slag cement (65% ggbs); BS 146 (1996)
- Portland slag cement (6%–35% ggbs in nucleus); BS 146 (1996)
- High blast-furnace slag cement (50%–85% slag content); BS 4246 (2002)

Since pozzolanic materials react with the calcium hydroxide formed during cement hydration, there are practical limits to the relative proportions of pozzolan and cement that can be used; typically, 25%–30% of the mix is cement binder.

ASTM C595/595M-09 recognises three types:

- Pozzolana-modified Portland cement (<15% pozzolana)
- Portland pozzolan cement (<15%–40% pozzolana for general construction)
- Portland pozzolan cement (no limit if high early strength is not required)

The equivalent BSI British Standards only refer to pulverised-fuel ash (pfa) limits:

- Portland pfa (15%–35% pfa); BS EN 197-1 (2000)
- Pozzolanic pfa (35%–50% pfa); BS 6610 (1996)

A review of the chemical and physical requirements associated with all these specifications is given by Walker (2002) in Concrete Society Special Publication, CS 136.

The most common use of granulated slag is to replace between 25% and 70% of the Portland cement either by intergrinding or preferably as an addition of the separately ground material. The lime and alkalis in the Portland cement are sufficient to activate the granulated slag. Portland blast-furnace slag cements are covered in specifications such as BS 146 (1996), BS 4246 (2002), BS EN 15167-1 and 2 (2006). ASTM C595M-09 (2009), ASTM C989-09a (2009) and those from many other countries. A specific code of practice for its use in construction, Japan Society of Civil Engineers (1988), has been published in Japan.

Granulated slag activated by alkalis in the form of potassium and sodium hydroxide or carbonates or soluble glasses is finding increasing use in eastern Europe where chemical resistance and lower costs are required (Talling and Brandstetr 1989).

The product, known as F-cement, produces calcium silicate hydrates without any calcium hydroxide, and while it has good chemical resistance, it is more prone to carbonation than ordinary Portland cement. The carbonation products tend to form in the microcracking in the hardened paste (Byfors et al. 1989), and the final product of carbonation is a silica gel which can be reconverted to calcium silica hydrates by re-alkalisation (Bier et al. 1989).

If the slag is crystalline, a mineral of variable composition called melilite is formed together with a range of other minor minerals (Nurse and Midgley 1951; Minato 1968). Crystalline

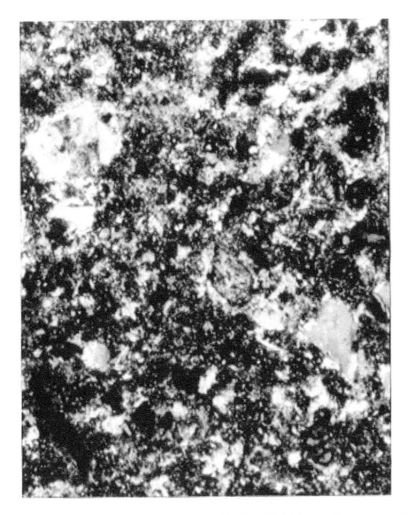

Figure 5.20 The texture of Portland limestone cement. Width of field 1 mm. The limestone interground with the Portland cement used in this test sample (317 kg/m³ of Portland/10 per cent limestone cement, water/cement ratio = 0.6) appears as small birefringent specks of light which on careful examination often have a rhombic shape. The texture of the limestone filler is similar to post-sampling carbonation and care will be required in sample preparation to prevent carbonation which can be confused with the limestone.

blast-furnace slag has little hydraulic activity and is principally used as an aggregate or for road fill. To give the slag hydraulic properties, it is granulated by rapid cooling of the molten slag with water to produce a glassy structure and then ground to a specific surface of approximately 4000 cm²/g. With the addition of activators, setting properties similar to those of Portland cement can be achieved.

Crystalline melilite forms a solid solution series with the end members akermanite $(2CaO \cdot MgO \cdot SiO_2)$ and gehlenite $(2CaO \cdot Al_2O_3 \cdot SiO_2)$, the percentages of which can be estimated by measuring the refractive indices of the melilite (Winchell and Winchell 1964). The melilite glass that forms during granulation of a slag approximates to a portion of the melilite solid solution series, and it is claimed that the composition can be estimated from the refractive index (Battigan 1986). However, the refractive indices of slag glasses are not the same as those of the crystalline melilite, and they should not be confused with each other. As many specifications set limits on the composition of slag cements, estimating their composition from the refractive indices of the glasses present would provide a rapid method of analysis.

Since ground granulated slag is whiter than Portland cement, it imparts a lighter colour to the concrete, and the presence of calcium sulphide interacting with the ferrite phase may impart a bluish colour (Powers-Couche 1992) which will fade with oxidation. Detection of the presence of granulated slag in concrete presents few problems in either thin or polished section because granulated slag hydrates more slowly compared with Portland cement and the larger grains tend to remain relatively intact and are often coated with calcium hydroxide. The effect on the texture of the hardened paste is to produce less calcium hydroxide than would be found in Portland cement alone, and this effect reaches a maximum at about 60% slag replacement (Taylor 1990). As the principal hydration products are similar to those of Portland cement, the appearance of the hardened paste texture will not be significantly different because of the presence of the slag, apart from the presence of residual grains. Supersulphated cement consists principally of granulated blast-furnace slag with activators. The minimum slag content allowed is 75%, but typically, it contains 10%–15% of anhydrite and about 5% of Portland cement as activators. Its principal use is for concrete which is required to withstand severe sulphate and chemical attack. The composition and other properties required are specified in BS 4248 (1974). The principal phases formed on hydration are laths of ettringite, up to 120 μm long and 0.5 μm thick, infilled by submicroscopic calcium silicate hydrate. No calcium hydroxide is formed (Midgley and Pettifer 1971). Thus, the texture of supersulphated cement in hardened concrete will be distinctive and consist of interlocked laths of ettringite set in a matrix of unresolvable calcium silicate hydrate with a complete absence of calcium hydroxide (see Figure 5.21). Supersulphated cement is more sensitive to carbonation during storage than ordinary Portland cements (Lea 1970). Concretes made from supersulphated cement lose strength due to carbonation (Manns and Wesche 1968), but as this is usually restricted to the outer surface of a structural member, it is not structurally significant.

Pozzolanic mineral additions are also widely used to modify cements used in construction. A wide range of materials, under the generic name of pozzolana, can be used to partially replace Portland cement. Common materials that have been used as pozzolana are the industrial waste products such as fly ash (Berry and Malhotra, 1987; Concrete Society 1991); silica fume (ACI Committee 226 1988; Concrete Society 1993); natural materials, for

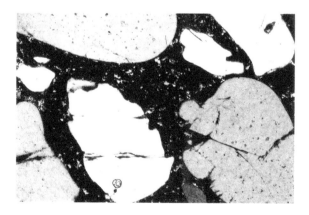

Figure 5.21 Supersulphated cement concrete. Note the hydrated paste is essentially isotropic, rare large ettringite crystals, but without portlandite or carbonation. Crossed polar illumination, width of field 2 mm. (Courtesy of M. Grove, private collection.)

example, diatomite, pumicite, tuff and trass (Mehta 1987); and calcined shale and calcined clays (Swamy 1992; Massazza 1993). The ACI Committee 232.1R, 232.2R (1995) and 234R (1995) and Lewis et al. (2003) have all reviewed the properties and uses of pozzolana cements, and German experience with blended cements is discussed by Schmidt et al. (1993).

The simplified reaction of pozzolanas such as silica fume or diatomite with the calcium hydroxide in Portland cement is as follows:

$$SiO_2 + Ca(OH)_2 = CaO \cdot SiO_2 \cdot nH_2O$$

The end product is a calcium silicate hydrate which may be indistinguishable from similar hydrates formed from the hydration of the Portland cement. This still remains the dominant reaction for those pozzolanas with an aluminosilicate-type composition, but calcium aluminosilicate and calcium aluminate hydrates are also formed which once again may be indistinguishable from similar cement hydrates. By contrast, the development of calcium hydroxide (portlandite) in the hardened cement matrix is reduced or suppressed.

While pozzolanas cover a considerable range of materials, the reactive constituents they contain are restricted to three mineral groups. These are species of amorphous silica, as found in silica fume and diatomite, glassy and amorphous aluminosilicates present in fly ash, pumicite, calcined shale and clay and altered aluminosilicates of a zeolitic nature which may be present in tuff and trass (silicified trachytic tuff). The difference between the last two groups is not well defined.

As discussed in Section 4.6.3, a pozzolana that has recently come into general use is silica fume produced as a by-product of the silicon and ferrosilicon industries (Aitcin 1983). It differs from other pozzolanas in consisting of spherical particles of reactive, amorphous silica which range from 20 nm to 1 µm in size and therefore below the resolution of the optical microscope. In theory this approximates to a mean particle size of about 0.1 µm which is about 100 times finer than any other pozzolana in use (ACI Committee 226 1988). Up to 10% of other material may be present mainly as coarser material which Bonen and Diamond (1992) report consists mainly of agglomerates of silica fume, quartz, cristobalite, silicon carbide, metallic silicon and carbon particles (see also Section 4.6.3).

Silica fume is collected as a loose powder with a bulk density of about 130 kg/m^3 which is 'densified' into a more manageable form by blowing air through the storage silo. After 1-day aeration, the bulk density increases to about 350 kg/m^3 which is marketed as 'undensified' silica fume, and after a week's aeration, the bulk density increases to about 700 kg/m^3 which is marketed as 'densified' silica fume (Figure 5.22). It is also densified by spraying water onto the fume in a drum pelletiser (Malhotra and Hemmings 1995). Silica fume is available for use in concrete both in powdered form and as a 50% stabilised slurry. The densification of silica fume is caused by agglomeration of particles which has a fundamental effect on its properties. It has been shown that the ability to disperse agglomerates of silica fume by ultrasonic dispersion varies with source and time of storage (Asakura et al. 1993; St John 1994) and that once some silica fumes are agglomerated, they cannot be redispersed. It has been suggested by St John (1994) that the tangling of the spherical particles fused into chains observed by transmission electron microscopy is primarily responsible for agglomerates that cannot be redispersed. He also proposed that the ultrasonic dispersivity of a silica fume as measured by laser diffractometry is a good indicator of whether a silica fume will disperse well in concrete.

Effectively, the mean particle size of densified silica fume varies between about 1 and 50 µm dependent on the source of the fume and time of storage. This applies both to powders and slurries. Sufficient individual particles remain and some agglomerates are so small that the special effects deriving from silica fume packing between cement grains is maintained, but this is highly

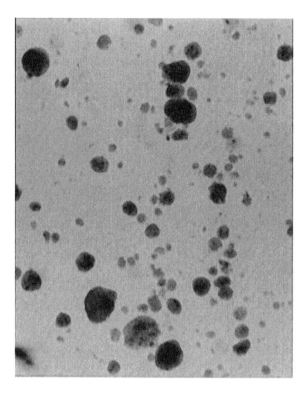

Figure 5.22 The appearance of silica fume. Width of field 1.2 mm. The actual particles are below the resolving power of the microscope but they aggregate into lumps which are visible at this magnification.

variable and is the probable explanation for the different efficiencies of silica fume in practice. When mixed into concrete and in super-high-strength DSP mortars, agglomerates larger than about 30 µm persist in both the concrete and mortar (Sveinsdottir and Gudmundsson 1993; Shayan et al. 1993). It may, under appropriate conditions of alkali and moisture content, lead to alkali–aggregate reaction (AAR) (St John et al. 1994).

The texture of concrete containing silica fume will vary according to the amount and dispersivity of the silica fume and the water/cement ratio. For ultra-high-strength DSP materials, which contain up to 25% of silica fume by weight of cement and are mixed at very low water/cement ratios, agglomerates will always be present irrespective of the method of curing and mixing (Figure 5.6). In addition, no calcium hydroxide will be observed. This can be confirmed by scanning electron microscope observation of fracture surfaces of DSP which show only a featureless hardened cement paste with no transition zone between the paste and the aggregates.

With high-strength concrete, which will normally contain only about 10% of silica fume by weight of cement, some calcium hydroxide may still be present dependent on age and curing. Agglomerates of silica fume will be present but may require a careful search of a large area of paste to be observed. These agglomerates are usually greater than about 25 µm in diameter, often have a light-brown interior and may show signs of scattered points of birefringence indicating partial reaction has occurred. Even in typical structural concretes, once the agglomerates have formed, they are persistent as shown by the report by Sveinsdottir and Gudmundsson (1993) on Icelandic concretes. It is also of interest to note that the silica fume used in these Icelandic concretes was interground with the cement used and was originally present as 7% by weight of cement.

Diatomite, which consists of the opaline silica shells of minute marine and freshwater algae, requires grinding to destroy the complex structures of the diatoms. Otherwise, its use as a pozzolana will cause excessive water demand in concrete. To achieve an acceptable water demand, the grinding must reduce the mean particle size to about 5 μm and residues must not exceed 5% on the 45 μm sieve. The residue may contain fragments of sponge spicules which are somewhat more resistant to grinding than the softer diatoms (Figure 5.23a and b). Diatomite is a less reactive material than silica fume, mainly because of its larger particle size. In concretes older than a few weeks, not many diatoms will be visible under the optical microscope, but careful searching by SEM may reveal fragments. The only fragments likely to be visible under the optical microscope are the remains of sponge spicules.

Pfa, commonly referred to as fly ash, is currently the pozzolana that finds greatest use. It varies considerably in composition dependent on the type of coal from which the ash is derived, and this is the basis of its specification in ASTM C618-08a (2008). Class F fly ash is produced from burning anthracite or bituminous coal and Class C from lignite or sub-bituminous coals. In broad terms, Class F is a low-lime ash and Class C a high-lime ash whose calcium oxide content varies from about 10% to 30% which gives it some weak cementitious properties. Class C ashes also contain a higher proportion of non-reactive material such as quartz and magnetite. Figure 5.24 shows optical images of pfa and Figure 5.25 is an SEM image which shows detail more clearly. The reactive materials in fly ash are glassy spheres consisting of an aluminosilicate glass in which incipient needles of mullite are present. The size range of the spheres varies from about 10 to 100 μm in diameter, but the actual particle size range present is dependent on the zones of the electrostatic precipitator from which the material is collected. Swamy (1993) has reviewed the requirements of European standards for the use of fly ash.

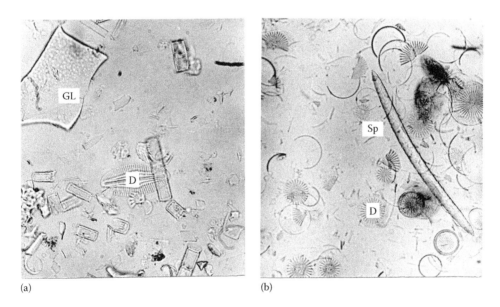

(a) (b)

Figure 5.23 The appearance of some diatoms and glass shards found in diatomite and pumicite. (a) This diatomite (Whirinakei, New Zealand) is a mixture of diatoms (D) rhyolitic glass (GL) and some clay. The glass shards have distinctive conchoidal fractures as shown in the large particle seen here. It is the unreacted ground-up remnants of the structures shown here that are used for their identification in concrete. Width of field 0.25 mm. (b) This is a cleaner diatomite (D), (Kaharoa, New Zealand), which also contains sponge spicules (Sp) which are less reactive than the diatoms and thus persist in concrete. They are often the only remnant that can be identified. A large, intact sponge spicule is clearly visible. Width of field 0.3 mm.

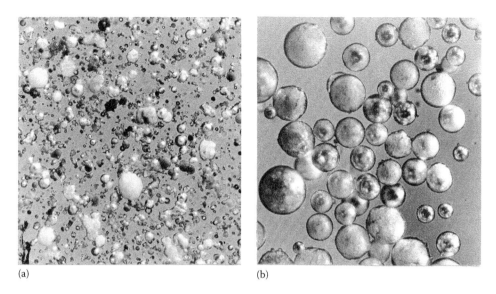

(a) (b)

Figure 5.24 The appearance of fly ash and its cenospheres. Width of fields 1 mm (a) This high lime fly ash (Meremere, New Zealand), contains typical glassy spheres, lumps of lime-rich material, opaque fragment s which are probably magnetite and other debris. It is remnants of the glassy spheres and particles of magnetite that are mainly used for identification in concrete. (b) If fly ash is separated in water many of the glass spheres float. These are known as cenospheres and consist of hollow spheres of glass material containing incipient crystals of mullite. Both these photomicrographs were taken with combined transmitted and oblique lighting.

Figure 5.25 Backscattered scanning electron image of pfa (fly ash). Note the broken shards in addition to the more obvious spheres. (Courtesy of L. Sear, UK Quality Ash Association.)

Even in old concrete, fly ash is easily recognisable under the microscope. Some of the larger spheres persist although they may be coated with calcium hydroxide and it may also be possible to observe particles of magnetite. Detection is easier because fly ash is used typically as a 20%–40% replacement, unlike silica fume and diatomite which are used in the 10%–15% range. Fly ash darkens the overall colour of concrete and this is an additional clue. The textural effects in the hardened cement paste are not as clear as with silica fume and diatomite. The amount of calcium hydroxide remaining at six months, for replacements varying from 0% to 60% fly ash, has been measured by Taylor et al. (1985). These results show that even at 60% replacement, 25% of the calcium hydroxide still remains, while at 25% replacement, some 80% remains. Thus, the effects of fly ash on concrete texture may not be apparent at lower replacement levels when attempting to observe the amount of calcium hydroxide and densification present in the texture.

Pozzolanas, based on volcanic aluminosilicate glasses such as pumicite, are slow-reacting materials, and residues can be observed in concrete under the microscope as isotropic fragments with conchoidal fracture (Figure 5.23a). Depletion of calcium hydroxide will be less than that occurring with the more reactive silica fume and diatomites. However, it is impossible to detect a pumicite where the concrete sand is derived from acid and intermediate volcanics as the glassy material present is similar in both materials. The effects on texture are similar to those found with fly ash. Pumicite and similar materials are typically used in concrete in the 15%–30% range, so there is a reasonable chance of their detection. The remaining types of pozzolanas require the same approach for their detection. In each case, detection will be dependent on being able to recognise some distinctive residue, and this requires knowledge of the microscopic appearance of the pozzolana both as raw material and in its ground state. The problem reduces itself to looking for 'needles in a haystack' and requires careful searching in the thin section.

In many cases, it is far more difficult to positively identify whether a pozzolana has been used than is commonly assumed. This is especially the case where lower cement content with mineral replacement has been used as the following case illustrates. A laboratory concrete, containing 190 kg/m^3 of cement and 27 kg/m^3 (12.5% replacement) of pumicite ground to optimum fineness, was examined after 25 years of exposure to the weather. The texture of the hardened cement paste was considerably depleted in calcium hydroxide, and the remaining calcium hydroxide crystals were small in size. No observable fragments of pumicite were visible in thin section, and after careful, prolonged searching of fracture surfaces by SEM, one fragment of volcanic glass was observed. It is most unlikely that this one fragment would have been located in a routine examination.

5.3.3 Petrographic examination of slags and pozzolanas

The quantitative estimation of slag content in concrete presents considerable problems which increase with the age of the concrete. Point counting grains of slag in thin or polished section is difficult even in young concretes, and the best methods available utilise a determination of the chemical composition of the hardened paste, but the interpretation of the analysis requires some knowledge of the compositions of the cement and slag used (Concrete Society 1988, French 1991a).

The microscopic examination of granulated blast-furnace slags can be used to determine both the composition and amount of glass present which are useful parameters for estimating the potential reactivity of the slag. Microscopic examination of concrete in thin section can identify the presence of slag, but quantitative estimation requires the use of chemical or electron probe microanalyses.

The percentage of glass in a granulated slag is considered to be an important indicator of its reactivity. Hooton and Emery (1983) have developed a microscopic method for estimating the glass content of granulated slags by crushing to about 50 µm and examining the grains in immersion oil, and Dengler and Montgomery (1984) also report some results using immersion oil techniques. The technique is described in the Canadian Standard (CSA 1983). The British Standards methods and the difficulties of quantifying the proportion of slag present have been discussed in Section 4.6.2. However, Drissen (1995) has found that the estimation of glass content using transmitted light has advantages over other methods and describes the techniques required.

In thin section, melilite crystals have a distinctive appearance (Figure 5.26a) with rectangular or square cross sections often zoned and show an anomalous blue interference colour under crossed polars (Insley and Fréchette 1955). Granulated slag is usually examined in reflected light where it is necessary to etch the polished section to reveal the details of structure. The truly glassy fragments do not etch but it will be found that many fragments are partially or completely crystalline dependent on the degree of vitrification of the slag. Many distinctive and rather beautiful patterns are formed by etching these areas of crystallinity (Figure 5.26b). It is difficult to identify these areas of crystallinity by optical means without the aid of x-ray analysis. The general appearance in transmitted light of ggbs in concrete is shown in Figure 5.27.

Another optical technique has been proposed for determining the hydraulic reactivity of granulated slags that is based on the fluorescence colours produced by UV light (Grade 1968). The technique appears to be suitable for ranking slags from the same works, but there is some question whether it is applicable to slags from different sources. It should be

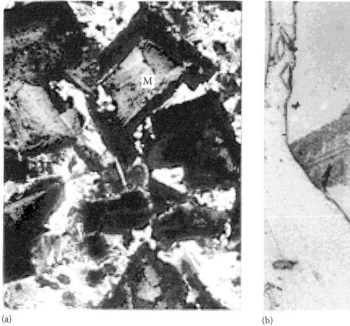
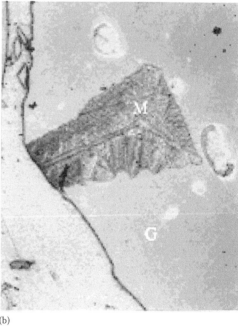

(a)   (b)

Figure 5.26 The appearance of melilite and granulated blastfurnace slag. Width of fields 0.8 mm. (a) Melilite crystals (M) are very distinctive in appearance. They have rectangular sections, zonation and a distinctive anomalous blue interference colour. The melilite is set in a matrix of crystals of dicalcium silicate, pyroxene and other minerals in this fragment of crystalline slag. Crossed polars. (b) When slag is granulated, i.e. rapidly cooling from the molten state, it forms a glass in which are interspersed incipient crystals of melilite and other material. In this polished section, which has been etched with HF vapour, an incipient crystallisation of melilite (M) is embedded in the glass (G). The reason for the different etching colours of the glass is not known.

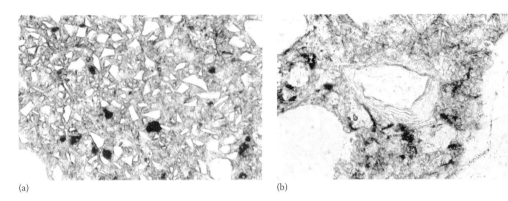

(a)   (b)

Figure 5.27 Ggbs cement replacement. The glassy shards are easily recognisable in thin section. Scattered diffuse green/blue patches of calcium sulphide interacting with the ferrite phase impart the bluish tinge to fresh surfaces. Plane-polarised light, width of fields (a) 1.0 mm and (b) 0.03 mm. (Courtesy of M. Grove, private collection.)

noted that the optical methods of estimating the reactivity of granulated slag are mainly applicable to the unground material. If the activity of the ground slag is required, the alkali activity test as specified in ASTM C1073-97a (2003) or by measurement of compressive strength as specified in BS 6699 (1992) may be used (this standard has now been replaced by BS EN 15167-1 and 2 2006).

Pozzolanas used in concrete have been described in Section 5.4.2, but unless the pozzolanic residues can be positively identified by microscopy or SEM, it is only possible to infer the presence of a pozzolana in a concrete from the change in texture caused by a reduction in the amount of calcium hydroxide present. Fortunately, in the case of fly ash and silica fume, which are the most commonly used pozzolanas, visible residues are usually present which allow unequivocal identification. In old concretes, many residues may be difficult to locate and identify without extracting them by dissolution of the hardened paste, and this difficulty increases with the age of the concrete. These difficulties are compounded when the original materials are no longer available for comparison purposes.

In examining the textural effects of incorporating pozzolana into concrete, the petrographer is faced with a complex situation. The two observable effects on texture are the removal of calcium hydroxide due to reaction of the pozzolana and the presence of pozzolanic residues. Observation of both of these largely depends on the reactivity of the pozzolana and the amount of residue remaining when a pozzolana is passed wet through a 45 μm sieve. As a rough indication, natural pozzolanas, which have been ground to give minimum water demand, should have a mean particle diameter not exceeding 10 μm and residues of not more than 12%.

Most pozzolanic residues can be identified in concretes which are less than one month old because the reaction has not had sufficient time to consume too much of the pozzolana. In these circumstances, quantitative estimates are possible either by chemical methods (Concrete Society 1989) or by electron microprobe analyses (Barker 1990 and French 1991a), but the reliability of these estimates is variable.

5.4 HIGH-ALUMINA OR CALCIUM ALUMINATE CEMENT CONCRETES

High-alumina cement, also known as calcium aluminate cement (CAC) or 'Ciment Fondu' (also discussed in Section 4.2.4), is usually manufactured from bauxite (Al_2O_3) and limestone melted or sintered together and ground to a fine powder. However, its composition is more variable than Portland cement because of the wider range of raw materials and melting/sintering processes used in its manufacture (Robson 1964; Scrivener and Capmas 1998). Initially, CAC contained 32%–45% Al_2O_3, but a range of cements, mainly used for refractory purposes, have since been developed containing 50%–90% alumina. CACs are also used in specialised applications combined with other minerals and admixtures including Portland cement, microsilica, ggbs, calcium sulphate and chemical additions such as latex.

The principal cementing compound, monocalcium aluminate ($CaO \cdot Al_2O_3$), typically forms between 40% and 70% of the composition with up to 10% of dicalcium silicate ($2CaO \cdot SiO_2$), 5%–20% of gehlenite ($2CaO \cdot Al_2O_3 \cdot SiO_2$) and 10%–25% of minor constituents. The last is a mixture of dodecacalcium hepta-aluminate ($12CaO \cdot 7Al_2O_3$), ferrous oxide (FeO), ferrite phase ($4CaO \cdot Al_2O_3 \cdot Fe_2O_3$) and pleochroite. The important properties of high-alumina cement concretes are their ability to produce strengths in 1 day which are equivalent to the 7-day strength of Portland cement, their sulphate resistance and their resistance to chemical attack. It can also be fired to produce a refractory material which is perhaps its most important application.

Concrete made with high-alumina cement has a serious disadvantage in that over time it can undergo a significant loss of strength when exposed to hot or warm moist conditions (Neville 1975) especially if it is mixed at a water/cement ratio exceeding 0.4. This loss of strength is primarily associated with a slow conversion of the two main hydrates $CaO \cdot Al_2O_3 \cdot 10H_2O$ and $2CaO \cdot Al_2O_3 \cdot 8H_2O$ to $3CaO \cdot Al_2O_3 \cdot 6H_2O$ which increases the porosity. This increase in porosity can affect both strength and durability. To detect and measure the extent of the conversion, it is necessary to analyse the concrete using x-ray diffraction and DTA techniques (Midgley and Midgley 1975; Redler 1991) (see also Chapter 4, Section 4.2.4). Recent work indicates that blending high-alumina cement with ggbs may avoid this problem of slow conversion of the phases (Majumdar et al. 1990a,b). An overview of the recent use of high-alumina cement has been given by Bensted (1993) and by Scrivener and Capmas (1998).

The hydration reactions of high-alumina cement are complex. The detailed reactions have been given by Midgley and Midgley (1975) and by Scrivener and Capmas (1998). An important point is that no calcium hydroxide is produced as any dicalcium silicate present reacts with alumina formed during the hydration. No calcium hydroxide is formed by hydrolysis. However, concretes made with high-alumina cements are still subject to carbonation as outlined in the reactions given in Section 5.5 (see also Figure 5.35).

The microscopic examination of high-alumina cement is more complex than Portland cement because of its greater variability in composition, solid solution between minerals and extensive formation of glasses. These complexities make it very difficult to calculate the potential phase composition from chemical analysis, though approximate methods have been proposed by Parker (1952), Calleja (1980) and Sorrentino and Glasser (1982). Examination of the mineral phases in high-alumina cement is best carried out on polished sections with selective etching (Parker 1952; Lea 1970). Scrivener and Capmas (1998) give some optical data on the mineral phases. Junior et al. (1990) and Venkateswaran et al. (1991) illustrate some textures in polished section. The appearance of a high-alumina cement mounted in immersion oil is illustrated in Figures 5.28 and 5.29, which show details of the texture of a grout made from high-alumina cement which after normal curing has been exposed to steam at 150°C. The texture consists of remnant and partially hydrated grains of the cement embedded in an undifferentiated matrix that appears isotropic at the magnification used. The thin-section photomicrographs in Figures 5.30 and 5.31 show the general features of the hydrated CAC and high-alumina cement.

There is considerable variation of crystallite size between grains and within single grains resulting from different cooling rates. In thin section, fine-grained samples give a general

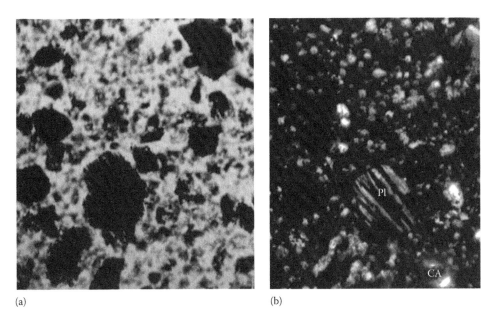

(a)   (b)

Figure 5.28 Anhydrous high-alumina cement mounted in oil with a refractive index of 1.54. Width of fields 0.44 mm. (a) The large particle in the centre has the fibrous character and pleochroism typical of pleochroite ($6CaO \cdot Al_2O_3 \cdot 4MgO \cdot SiO_2$). It is not possible to identify most of the other particles which are glass materials varying in colour from opaque to brown. (b) Under crossed polars the fibrous character of the pleochroite (Pl) stands out. The white birefringent particles are calcium aluminate (CA). Some of the brown-coloured birefringent particles may be a ferrite phase.

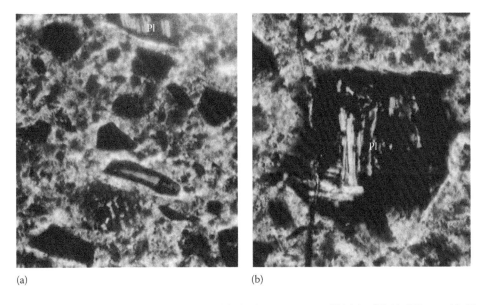

(a)   (b)

Figure 5.29 The texture of hydrothermally cured high-alumina cement. Width of fields 0.7 mm. (a) After 7 days at 150°C in this geothermal grout (water/cement ratio = 0.4) only the most resistant particles remain unhydrated. At the top, the fibrous particle of pleochroite (Pl) with its distinctive lilac-coloured pleochroism stands out. Other large particles are opaque or have well developed hydration rims. (b) The large particle of pleochroite (Pl) has fibres lying at right-angles to each other and shows the range of pleochroism that varies from white to lilac in colour. Most of the remaining particles visible are either isotropic or opaque.

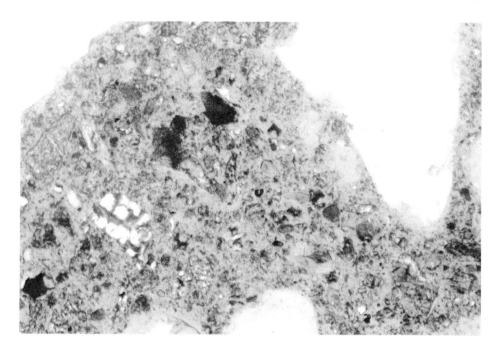

Figure 5.30 CAC concrete, the matrix contains brownish CAS crystalline material in an interstitial glassy phase and crystalline lamellae, probably pleochroite. Plane-polarised light, width of field 0.5 mm. (Courtesy of M. Grove, private collection.)

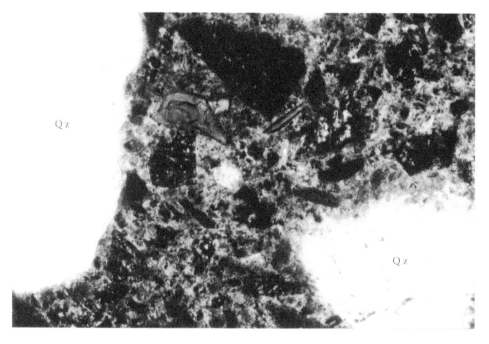

Figure 5.31 Thin section of high-alumina cement concrete. Note the angular residual clinker grains set in a dark cement matrix with rare transparent lathlike crystals, possibly of gehlenite. Plane-polarised light, width of field 0.5 mm. (Courtesy of M. Grove, private collection.)

appearance of transparent prismatic, lath and fibrous crystals set in a brown to opaque groundmass with angular residual clinker grains (see Figure 5.28). The larger transparent crystals are gehlenite and dodecacalcium hepta-aluminate ($12CaO \cdot 7Al_2O_3$) but are difficult to differentiate without recourse to x-ray diffraction analysis. Pleochroite is easily recognised by its fibrous and strong blue/violet pleochrism (Figures 5.28 and 5.29). Other crystallites are not easily identified with certainty in thin section, though ferrite is usually assumed to be present in the dark or opaque matrix.

Although the major features of hydrated CAC mortar and concretes can be studied using cut or polished samples and thin sections, the details of the phase distributions and fabric are best studied using a SEM with backscattered electron imaging, x-ray microanalysis and x-ray diffraction techniques.

5.5 CARBONATION OF PORTLAND CEMENT CONCRETES

The effects of carbonation are of importance in relation to the durability of reinforced concrete, and the subject continues to be widely investigated. An extensive literature review of the carbonation of reinforced concrete has been published by Parrott (1987) and a damage classification by Parrott (1990).

Carbonation of a hardened cement matrix most commonly takes place by the reaction of both calcium hydroxide and the cement hydrates with atmospheric carbon dioxide. The reactions are as follows:

$$Ca(OH)_2 + CO_2 \rightarrow CaCO_3 + H_2O$$

Calcium silicate hydrate $+ CO_2 \rightarrow$ various intermediates $\rightarrow CaCO_3 + SiO_2 \cdot nH_2O + H_2O$

Aluminate hydrates $+ CO_2 \rightarrow CaCO_3 +$ hydrated alumina

Ferrite hydrates $+ CO_2 \rightarrow CaCO_3 +$ hydrated alumina and iron oxides

These reactions show that not only Portland cement concretes but also high-alumina cements are subject to carbonation because of the reaction of the aluminate hydrates with carbon dioxide even though no calcium hydroxide is produced in the hydration of this type of cement. The detailed reactions that are possible between the components of concrete and carbon dioxide have been reviewed by Calleja (1980). Pozzolanic materials may decrease the permeability of concrete (Massazza 1998), unless it is inadequately cured, leading to a reduced rate of carbonation. Carbonation acts through solution and some pore fluid, such as a film on internal surfaces, as a prerequisite for the carbonation reaction to proceed. However, the rate of carbonation is minimal for a saturated concrete, because the permeation of carbon dioxide is hindered by saturation of pores. The rate increases to a maximum where the concrete is allowed to dry to the range 70%–50% RH. Below this humidity range, the rate reduces rapidly (Verbeck 1958) because there is no longer sufficient water in the pores and capillaries of the paste.

For the practical purpose of explaining the depth of carbonation observed, the major controlling factors are the water/cement ratio, the porosity of the aggregates, curing regime, ambient temperature and the environment to which the structure is exposed. However, Brown (1991) found in an extensive survey of 400 structures that carbonation depths correlated well with concrete quality but there was no significant relationship with reported exposure conditions. In general, any factor which increases the permeability of a concrete to carbon dioxide will also increase the rate of carbonation provided that the necessary internal moisture conditions are present. Thus, if a concrete is kept continuously moist, the effective permeability to carbon dioxide may be low and the rate of carbonation will be minimal. The same concrete protected from the weather and exposed to relative humidities of 50%–70% and higher ambient temperature will carbonate more rapidly, while the normal outdoor exposure of wetting and drying will result in intermediate rates of carbonation.

For structural concretes, the water/cement ratio has the greatest influence on carbonation. The effect of water/cement ratio on the porosity and permeability of a cement paste and its relationship to durability has been discussed by Hansen (1989) who used the classical work of Powers and co-workers to illustrate the effect of permeability on durability:

$$p = \frac{w/c - 0.36m}{w/c + 0.32m}$$

where

p is the porosity
w/c is the water/cement ratio
m is the degree of cement hydration, $0 < m < 1$

If we assume $m = 0.75$ for a typically cured concrete, $m = 0.5$ for a poorly cured concrete and $m = 1.0$ for the theoretical limit of hydration, porosities can be calculated from the preceding equation as shown in Table 5.5.

Powers et al. (1958) postulated that a paste with a porosity of less than 30% would be watertight, because the capillary pores in the paste become discontinuous. They also determined the time required to moist cure a concrete so that $p = 30\%$ for a range of water/cement ratios as shown in Table 5.6. The data in Table 5.6 also show that at low water/cement ratios, carbonation will be largely controlled by a degree of curing readily achievable in practice. As the water/cement ratio exceeds 0.5, the time required to reduce the paste

Table 5.5 Porosity of cement paste calculated from water/cement ratio and degree of cement hydration (m)

WIC	Percentage porosity		
	m = 0.5	m = 0.75	m = 1.0
0.4	30	18	6
0.45	35	23	12
0.5	39	28	17
0.55	42	32	22
0.6	46	36	26
0.65	48	39	30
0.7	51	42	33

Table 5.6 Times required to cure cement paste to obtain a porosity of 30%

WIC	Time	Degree of hydration
0.4	3 days	0.5
0.45	7 days	0.6
0.5	14 days	0.7
0.6	6 months	0.9
0.7	1 year	1.0

Source: After Powers, T.C. et al., *J. Portland Cem. Assoc. Res. Dev. Lab.*, 1(2), 38, 1958.

porosity to 30% or less exceeds the practical time of curing, and the depth of carbonation will be controlled by the water/cement ratio.

Parrott (1987) quotes the results of numerous investigations which obtained good correlations between carbonation depth and the strength of the concrete. In most respects, this is merely a restatement of the principles stated earlier. Higher-strength concretes use more cement and require lower water/cement ratios than lower-strength concretes. For practical engineering purposes, this method is useful as the strength of the concrete is often known. However, it can be in error if the quality of the cover concrete differs markedly from the concrete in the core of the structure because inadequate compaction can lead to an increased water/cement ratio in the outer layers. This situation is quite common in field concretes, and in these cases, it will be found that there may be poor correlation between depth of carbonation and overall water/cement ratio (French 1991a). However, Parrott (1992) found a near-linear relationship between carbonation depths and water absorption in cover concrete in specimens stored for up to one and a half years in indoor and outdoor exposure. Further work by Parrott (1994) shows that the air permeability and water absorption rates are very sensitive to the moisture content of a concrete.

The chemical reactions of carbonation result in volume changes. Calcium carbonate has a molecular volume which is 11% greater than the calcium hydroxide which it replaces, but approximately 2.5% less than the cement hydrates. The newly formed calcium carbonate with its increased volume now fills previously empty pores and capillaries reducing permeability. In spite of these volume changes, carbonation results in an overall shrinkage of a concrete indicating the processes are more complicated. The theory of carbonation shrinkage is beyond the scope of this book, and reference should be made to a discussion of the various theories by Ramachandran et al. (1981) and Lawrence (1998).

While it is generally agreed that the overall rate of carbonation is controlled by the permeability of the concrete, little information on the individual rates applying to calcium hydroxide and the cement hydrates appears to be available. French (1991a,b) states that in practice it is usual to find the calcium silicates alter to carbonate as readily as the hydroxide and even that residual unhydrated clinker particles become pseudomorphed to amorphous silica.

5.5.1 Effects of carbonation on porosity and strength

Carbonation of an ordinary Portland cement hardened paste increases the strength of the paste and decreases its porosity as shown by a number of studies quoted by Parrott (1987). In these respects, it improves the properties of the paste in concrete with the disadvantages that it lowers the pH of the pore solution and may lead to shrinkage. The decrease in porosity will also decrease the permeability of the paste, but where the outer layer of a concrete is highly porous, this reduction in permeability is usually insufficient to prevent oxygen and chloride ions penetrating to the reinforcement. Dense well-compacted concrete and increase in cement contents will usually reduce the depth of carbonation. The effect of carbonation on high-alumina cement is claimed to be similar to that on Portland cement (George 1983) although the carbonation products formed are sometimes carboaluminates and not calcium carbonate. It has been claimed (Cussino and Negro 1980) that the inclusion of fine limestone enhances the formation of the carboaluminates and overcomes the strength losses accompanying conversion. In the case of blast-furnace slag cements, carbonation may have a deleterious effect on the hardened cement paste as it appears to decrease strength and increase porosity (Meyer 1968; De Ceukelaire and Van Nieuwenberg 1993). Carbonation causes a marked increase in porosity and reduction in strength of supersulphated cement probably due to the carbonation of the ettringite (Manns and Wesche 1968). Further references to the effects of carbonation on hardened pastes are given by Parrott (1987).

5.5.2 Typical textures of carbonation

The calcium carbonate formed in hardened cement paste is almost invariably present as clumps of crystals. In its initial stages, carbonation appears in thin section as diffuse birefringent material. This texture occurs along the margins of cracks and at surfaces which have been coated with a sealant soon after hardening (Figure 5.32). It is also observed as an artefact in thin sections where the specimen has not been adequately protected against carbonation during storage and preparation. As carbonation becomes more intense and proceeds inwards from the surface, the crystals form a dense texture which darkens the paste. Under crossed polars, the carbonation now appears as a birefringent mass of light which eliminates microscopic detail in the hardened matrix. This is the typical carbonation which will be found in the outer layer of concrete (Figures 5.32a and 5.33). In its final stage, the carbonated paste takes on a fretted appearance and may appear lighter in colour or show banding (Figures 5.33b and 5.34). At this stage of carbonation, it is probable that the cement hydrates have been largely destroyed. In many cases, remnant cement grains and islands of uncarbonated or partially carbonated cement paste may be observed, but overall where the carbonation is intense, the most observable detail of the hardened cement paste is either obscured or destroyed. A mechanism to explain why islands of non-carbonated material occur in carbonation zones has been proposed by Houst and

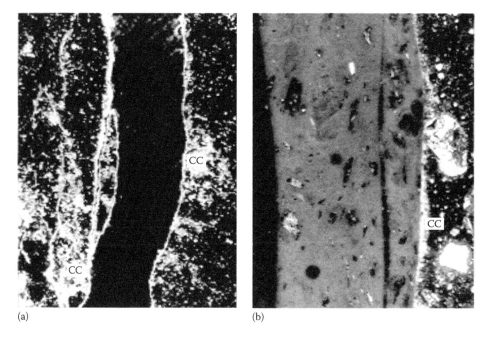

(a) (b)

Figure 5.32 Calcium carbonate along a crack margin and at the outer surface of a very young concrete. Width of fields 0.9 mm. (a) Under crossed polars this 250 μm wide crack located 50 mm deep in this fifty-year-old concrete illustrates the typical appearance of the initial stages of carbonation (CC) which has not proceeded. The crack margins can be regarded as surfaces available to the atmosphere but protected from the weather. The depth and intensity of carbonation in such cases is largely dependent on the permeability of the concrete. (b) Under crossed polars an early stage of carbonation (CC) is visible. In this case the young test specimen had to be briefly dried to allow an epoxy coating to be applied. It can be clearly seen that the epoxy has been applied in two layers and contains a filler. Adhesion between the epoxy and concrete is excellent which is typical for these types of resin systems.

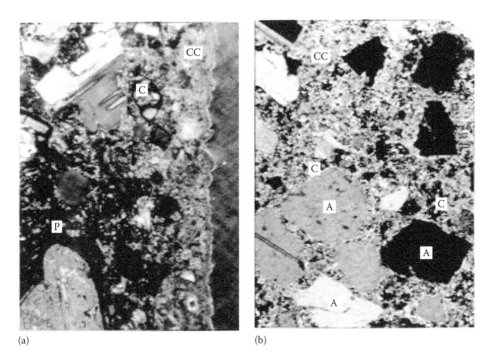

(a)   (b)

Figure 5.33 Typical carbonation at the surface of concrete and an example of intense carbonation deep inside a concrete. Width of fields 0.8 mm. (a) Under crossed polars. Classical carbonation of a ten-year-old structural concrete. The surface is weathered and carbonation (CC) forms a band that varies irregularly between 250 and 500 μm in width grading into the sound paste (P). The surface has dried out so that the rate of carbonation has exceeded the rate of hydration resulting in the presence of prominent partially hydrated cement grains (C). (b) 50 mm deep inside this sixty-year-old concrete the paste is intensely carbonated (CC). The calcium carbonate crystals appear discrete and the texture has a fretted appearance due to remnant grains (C). It is probable that most of the paste texture has been destroyed by intense carbonation which is due to the high porosity of this concrete. Aggregates labelled A.

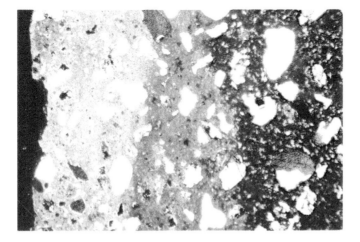

Figure 5.34 Depth of carbonation is shown by the birefringent zone on the left. Since carbonation depends on the permeability of the concrete, it will sometimes appear as banding in the carbonation layer (as illustrated here). Crossed polar illumination, width of field 2.3 mm. (Courtesy of M. Grove, private collection.)

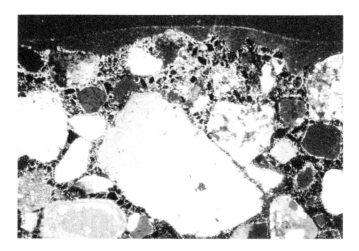

Figure 5.35 Carbonated HAC concrete, the normally largely isotropic matrix becomes highly birefringent due to the development of calcium carbonate. Crossed polar illumination, width of field 3 mm. (Courtesy of M. Grove, private collection.)

Wittmann (1994). Figure 5.35 is of a carbonated high-alumina concrete; the zone of carbonation extends inwards from the external surface as atmospheric carbon dioxide alters the calcium aluminium and silicate hydrates.

5.5.3 Carbonation of the outer layers of concrete

Almost as soon as the cement matrix starts to harden, carbonation commences if superficial drying of the surface is allowed to occur. This is graphically illustrated in Figure 5.36b, where a thin layer of diffuse carbonation has developed when the surface was briefly dried prior to sealing with a polymeric coating. Even in mortar bars mixed to comply with ASTM C227-90 (1990) and stored under continuously moist conditions, carbonation penetrates about 15 μm into cast surfaces and about 0.5 mm into the top surface. In a well-compacted, dense concrete, the average depth of carbonation, even after years of exposure, will rarely exceed 2 mm and is often as little as 0.5 mm. However, at the margins of aggregates located near the surface, carbonation may extend along crack caused by plastic settlement and bleeding. In lesser-quality concretes, carbonation often penetrates more than 2 mm, while a carbonation depth greater than 5 mm indicates porous concrete in the outer layer.

Some of these effects have been illustrated by Strunge and Chatterji (1989) who concluded that thin-section analysis gives more detailed information on carbonation than that detected by phenolphthalein.

When carbonation depths exceed 5 mm, the nature of the carbonation zone may change. Islands of uncarbonated cement paste are often present in the carbonation zone, and isolated areas of carbonation may be seen beyond the carbonation zone in the paste and around some aggregates. These isolated patches of carbonation indicate that a porous pathway was present out of the plane of the section. The margins of the carbonation zone are irregular and carbonation will penetrate deeply into cracks if these are present. This behaviour is an expression of the localised variability which occurs in concrete especially in the porous surface areas of a concrete.

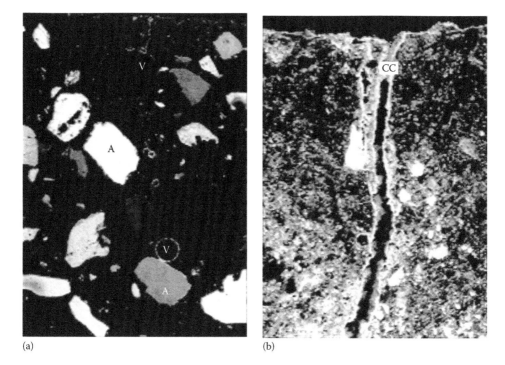

(a) (b)

Figure 5.36 Carbonation associated with surface cracking. Width of fields 0.75 mm. (a) Under the crossed polars the surface of this three-year-old bridge concrete is only lightly weathered and carbonated although some of this is hidden as the section is a little thick. The drying shrinkage crack, about 2 mm deep and <10 μm wide, intersects some pores (V), and is filled with carbonation for its entire length. Aggregates labelled A. (b) Under crossed polars this tension crack, which is about 30 μm wide is plugged with carbonation (CC). The surface is slightly weathered and the depth of carbonation indicates that locally the thirty-year-old bridge concrete is of only moderate quality. This is a very typical surface texture.

5.5.4 Carbonation associated with crack systems

A surface crack provides a pathway for external elements to penetrate the body of the concrete. Fine cracks of less than 10 μm in width may carbonate to depths of 1–2 mm even where the average surface depth of carbonation is less than 0.5 mm. These fine cracks are primarily due to shrinkage (Figure 5.36a) and are invariably carbonated at the margins. As the crack width increases, the depth of carbonation also increases and the crack will be open unless fluid has been present to enable crystallisation to occur (Figure 5.36a). A crack 60 μm wide may be carbonated along its margins up to a depth of 50–100 mm, allowing carbonation of aggregate margins and areas of porosity close to the crack. Cracks wider than 100–150 μm clearly allow the atmosphere to penetrate to considerable depths. The extent of reaction between the atmosphere and the hardened paste in these wide cracks will depend on the crack width, the severity of that environment and the quality of the concrete (Figure 5.37).

The width of a crack at the surface of concrete may appear to be less than in the interior, because calcium carbonate can partially or completely block the entrance. In thin section, cracks which appear to be closed at the surface will quite often prove to be open for the remainder of their length (Figure 5.37b). Carbonation of the paste along crack margins may vary from a light dusting of carbonate to irregular bands of carbonated paste extending the full depth of the crack. Crystals of calcite will often line crack

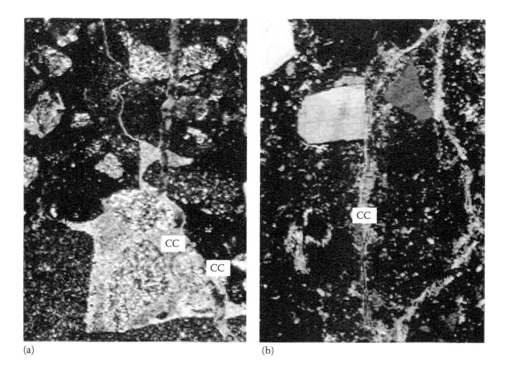

(a)   (b)

Figure 5.37 Carbonated paste deep inside concrete associated with cracks. (a) At a depth of 50 mm in this fifty-year-old bridge concrete, a 250 μm wide crack is not carbonated along its margins but intersects areas of paste that are clearly carbonated (CC). This indicates that the large crack is a later occurrence but other evidence strongly suggests it is not an artefact. The reasons for the uncarbonated margins are not easy to explain but may be related to environmental conditions. This complex behaviour of carbonation is not uncommon. Crossed polars. Width of field 2 mm. (b) At a depth of 30 mm in this forty-year-old bridge concrete, a splintered tension crack is almost wholly filled with carbonation (CC). Although it does not exceed 25 μm in width for its whole length it is carbonated to a depth of nearly 40 mm. Areas of paste adjacent to the crack are also spotted with carbonation. This is in contrast to the adjacent photomicrograph where carbonation is largely confined to discrete areas of paste along the margins of the crack. Crossed polars. Width of field I mm.

margins, but a crack wider than 20 μm is rarely completely filled with carbonation products unless the concrete has been saturated for a considerable period of its history. These observations on carbonation are in general agreement with the recommendations of the discontinued BS 5337 (1976) (now replaced by BS 8007 1987), which limited crack widths to 100 μm in locations alternating between wet and dry. In BS 8007 (1987) which in turn has been replaced by BS EN 1992-3 (2006), these recommendations have been omitted. However, ACI Committee 224 (1985) also recommends a similar limit for water-retaining structures.

5.5.5 Carbonation associated with aggregate types in concrete

For normal-density concretes, the porosity of the aggregates does not significantly contribute to carbonation. However, there is one form of carbonation that is intrinsic to concrete which is not dependent on atmospheric CO_2. This is the formation of basic calcium carbonate rims around some carbonate rock aggregates. This is discussed more fully in connection with alkali–carbonate reaction in Section 6.9.

Table 5.7 Mix details, length changes and carbonation areas of 50 × 50 × 300 mm test beams of concrete containing lightweight aggregate from outdoor exposure for 5 years

Specimen	Cement content, kg/m³	Water/ cement ratio	Unit weight, kg/m³	Length change, %	Carbonated area, %	Alkali content, kg/m³
F8/1/1	290	1.1	1120	+0.014	54	0.92
F8/1/3	350	0.83	1215	+0.044	28	1.12
F8/1/4	435	0.67	1275	+0.088	12	1.39
F8/1/6	450	0.60	1340	+0.071	10	1.44
F8/1/5	530	0.60	1445	+0.060	5	1.69
F8/1/7	560	0.56	1415	+0.126	8	1.79

If vesicular lightweight aggregates such as scoria, pumice or manufactured lightweight aggregates (St John and Smith 1976) are used, the effect on carbonation can be significant. This effect is illustrated in Table 5.7 where the depth of carbonation is clearly related to the bulk density of the concrete. Work by Swamy and Jiang (1992) correlated carbonation depth to the pore structure of lightweight concrete. However, in structural lightweight concrete, it is common practice to use coarse lightweight aggregates mixed with normal-density fines. It is claimed that provided cement contents of 350 kg/m³ or more and low water/cement ratios are used, the carbonation of structural lightweight concretes is little different from that of normal-weight concretes (Clarke 1993). Mays and Barnes (1991) inspected forty structures containing structural lightweight concrete constructed prior to 1977 and concluded it was not less durable than normal-weight concrete although its durability may be more sensitive to poor workmanship.

5.5.6 Surface exudations and efflorescence

A very common feature observed on the surface of concrete is the presence of a white or coloured coating or stalactite-like growths extending from soffits and overhangs where they result from leaching of salts from within the concrete (Figures 5.38 and 5.39). Alternatively, efflorescences may be observed on surfaces within the first metre or so of ground level. These may be due to salts present in groundwaters rising up through the concrete by capillary action and/or leaching by of salts from within the concrete.

Figure 5.38 The white efflorescence on the corner of this structure is due to movement of water under gravity through cracks and joints dissolving calcium hydroxide and depositing it at the surface as calcium carbonate as the moisture evaporates.

Figure 5.39 A concrete rail was viaduct constructed in 1903; Glenfinnan, Scotland, UK. Cracks and joints have opened up because of cold joints, load stresses and possibly thermal stresses. The margins of cracks are corroded from the soft acid rainwater. Carbonation of calcium hydroxide leads to the white streaks of calcium carbonate efflorescence.

This efflorescence is typically a white incrustation on the surface of concrete consisting of salts precipitated from the pore solution. Obviously, for visible efflorescence to occur, pore solution must be able to move to the surface to allow evaporation and crystallisation of the salts which are formed from the dissolved calcium, sodium, potassium and sulphate ions from the interior of the concrete. However, the salts containing sodium, potassium and sulphate ions are readily soluble in water and will be rapidly rain washed from concrete surfaces unless they are protected. The quantities of these salts are also limited by their concentration in the concrete. Colour effects are almost always due to dissolution of iron from corroding steel reinforcement, from pyrite impurity in the aggregate or more rarely from other metals used on the structure. Further details of efflorescences, their causes and significance are given in Chapter 6, and for concrete blocks, see Figures 7.3 and 7.4 in Section 7.2.

Where massive exudation occurs, it is usually due to the leaching action of water moving through joints of a structure and then depositing the leachates as they evaporate. Although water other than the pore solution is also involved, the main source of the efflorescence is still from pore solution. Good examples of massive efflorescence will be found on many old dams, reservoirs, mass concrete structures and wet tunnels, which may be streaked with spectacular deposits composed of calcium carbonate (Figure 5.39). The white deposits are often discoloured by dirt, algae and rust.

The soluble efflorescent salts, such as sodium sulphate, potassium sulphate, sodium carbonate, potassium carbonate, trona (Figg et al. 1976), syngenite and calcium langbeinite, are found only in protected situations such as under soffits and in tunnels. In wet tunnels with considerable air movement through them, the drying effect of the air can draw groundwater up the walls of porous concrete, and bands of efflorescence may occur due to evaporation, particularly around joints (Figure 5.40).

Figure 5.40 The complex halo visible surrounding the vertical joint in the sixty-year-old tunnel wall has been caused by water wicking up the joint from the saturated rail ballast and then evaporating leaving white crystalline material at the margins of the halo. This was identified as thenardite and is derived from ground water in which sulphate levels were quite low. The concrete was found to be reasonably dense and thus has been able to resist the build up of sulfate.

This is a case where efflorescence is not wholly derived from the pore solution and strictly speaking should be regarded as a deposit. Salts such as thenardite and mirabilite (sodium sulphate hydrates) and possibly gypsum are likely to be found in these cases as well as calcium carbonate (St John 1982).

Alkali–silica gel is sometimes found on the surface of concrete undergoing AAR although much less commonly than supposed. This is not efflorescence but an exudation which will often be intermixed with calcium carbonate. Most of the gel observed in these cases will be carbonated. This is discussed in Section 6.8.

5.6 INTERFACES WITHIN CONCRETE

It is convenient to differentiate the intrinsic interfaces that are formed between paste, aggregate and reinforcing steel, from other interfaces such as those formed by joints, mortars and coatings. The nature of the bonds between paste and aggregates, reinforcing and other materials has been the subject of considerable investigation (Massazza and Costa 1986), primarily because in some cases they may have fundamental effects on the strength and durability of concrete.

The observation of interfaces in concrete and mortars in thin section can provide information on the degree of contact, absorption into the substrate, calcium hydroxide, porosity, void spaces, dirt and debris. Optical observations are limited to dimensions exceeding about two microns. Thus, the broader details of interfaces may be observed by optical microscopy but these may need to be correlated with those made by SEM and chemical or infrared analysis, and are particularly important where chemical admixtures or surface coatings have been used.

5.6.1 Aggregate/cement paste interface

The nature of the interface is controlled by the properties of both the aggregate and the cement paste. Aggregates used in concrete have varying degrees of porosity, shape and surface roughness as well as the possibility of adherent dust and dirt. Even within one batch of aggregate, a considerable variation will be found. For instance, porosity may vary unless the lithology of the parent rock is uniform and many aggregates are polymictic or a mixture of natural rounded pebbles and crushed materials.

The cement paste in contact with the aggregate (or reinforcing steel) forms a distinct surface layer (see Section 5.2.5). Also, the finer portions of the paste will tend to concentrate in this layer. Physical contact and/or degree of separation of the paste from the aggregate will be largely dependent on the amount of water concentrated at the interface and the amount of plastic settlement that takes place before setting. Both these factors, which are closely interrelated, are creating space at the interface in which crystallisation from the pore solution can occur. In Figure 5.41, the duplex film, which is about 1 μm thick, consists of calcium hydroxide and calcium silicate hydrate, which coats much of the surface of the steel. On this duplex film lies another semi-continuous layer about 3 μm thick of stacked calcium hydroxide crystals with their c-axes semi-perpendicular to the steel interface. The zone of calcium hydroxide lies against a weak, porous layer of cement paste enriched in ettringite which changes gradually into normal paste at a distance that appears to vary from about 20 to 40 μm in width. This description is probably applicable to most non-porous surfaces that are reasonably smooth and unreactive. In the case of aggregates that are both porous and rough surfaced, much greater variation in the details of the aggregate–paste interface would be expected. The only details observable optically are the calcium hydroxide layer and any voids or fissures that may be present.

The interfacial regions of paste surrounding aggregates have been shown to be different from those of the bulk paste and have been called an 'aureole of transition'. Details of this transition aureole are clearly shown against steel reinforcement in the diagram by Massazza and Costa (1986) (Figure 5.41), and it should be noted that some of the details are still controversial (Diamond 1986); see also Section 5.2.5. Most of the finer detail is beyond the resolving power of the light microscope, but the larger crystals of calcium hydroxide including their orientation, and possibly increases in paste porosity, are details that should be observable.

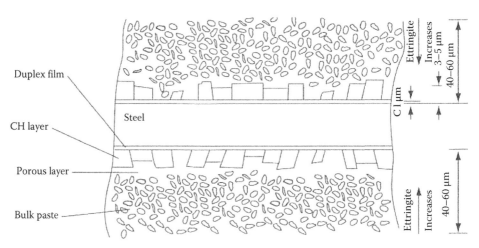

Figure 5.41 Schematic illustration of steel/paste interface. (From Massazza, F. and Costa, U., Bond, paste-aggregate, paste-reinforcement and paste fibres, In *Eighth International Congress Chemistry of Cements*, Rio de Janeiro, Brazil, Vol. 1, pp. 159–180, 1986.)

Carles-Gibergues et al. (1993) found that differences between the types of sulphate present in the cement affect the bonding of new-to-old concrete causing variation in both the nature of the calcium hydroxide crystallisation and the amount of ettringite present.

In thin section, the aggregate–paste interface tends to be more variable than suggested by many of the models proposed. An important variable is the absence or presence of meniscus-like voids underneath coarse aggregates due to plastic settlement and bleeding. This is why it is preferable to cut a thin section from the vertical plane in the core sample to reveal the presence of this settlement and to use thin sections of large surface area to show the true extent of this type of variation. Other voids due to air trapped at the aggregate surface and ragged fissures caused by bleeding may also be present. Whether any of these types of void spaces are filled or lined with calcium hydroxide will be dependent on the amount and time that pore fluid has been present in the void spaces. If a void space has been filled with pore fluid for a considerable period during the life of the concrete, it may be infilled by well-crystallised calcium hydroxide. In other cases, voids may only be lined with calcium hydroxide, and it is not uncommon to find plastic settlement fissures and air voids with only the merest hint, or even apparently devoid, of any visible calcium hydroxide. The dimensions of these voids and fissures usually exceed 20 µm, and it could be argued that they are faults and not part of the aggregate–paste interface. In concretes with a high water/cement ratio, these faults are common and must affect strength. There appears to be no information on systematic investigation by microscopy to relate the incidence of settlement and bleeding faults around aggregates to factors such as water/cement ratio and sand properties except for a report by Hoshino (1988).

Where the paste is in contact with the aggregate surface and not disrupted by bleeding and/or settlement, most aggregate surfaces will be covered with a semi-continuous layer of calcium hydroxide provided that the aggregate is not porous (see Figures 5.13 and 5.42). This layer can vary from a few microns up to about 20 µm thick even around the same aggregate and is usually in intimate contact, forming a bridge between the aggregate surface and the paste. There appears to be little information on systematic investigation by light microscopy of the orientation of the calcium hydroxide layers. A study of the orientation of calcium hydroxide layers surrounding aggregates in thin section should be possible, but the small size of calcium hydroxide crystals involved could make confirmation by the use of interference figures difficult or impossible. Many aggregates have sufficient water absorption to require this factor to be taken into account in the concrete mix design. In some concretes, the aggregate appears to be surrounded by an aureole of darkened paste and no calcium hydroxide is observable at the aggregate interface. The darkening of the paste is probably illusory and due to a lack of visible calcium hydroxide, the formation of which appears to have been prevented by suction from the porous aggregate. The level of aggregate water absorption at which this effect occurs is not known, but the phenomenon is not uncommon in concrete. Sarkar et al. (1992) have reported that paste–aggregate bonding is dependent on the nature of the external shell of a lightweight aggregate and that the absence of oriented calcium hydroxide crystals at the interface is related to the water absorption of the aggregate.

Some aggregates, such as limestone (Grandet and Ollivier 1980), interact with the pore solution of the concrete, and it is claimed that the sizes of calcium hydroxide crystallising at aggregate interfaces are also affected by the composition of the aggregate (French 1991a). At a practical level, the main determinants of the incidence of faults and the amount of calcium hydroxide crystallising at aggregate/cement interfaces are undoubtedly the water/cement ratio and the bleeding capacity of the concrete. In typical structural concretes, localised variations of mixing and compaction can result in a range of aggregate/paste interfaces being present in a single large-area thin section. This variability decreases with lower water/cement ratios and adequate compaction of the concrete.

Figure 5.42 Paste interfaces with kraft pulp fibre and a low porosity aggregate. Width of fields 0.8 mm. Around this pyroxene aggregate (P) is a semi-continuous layer of calcium hydroxide which is typical of low porosity aggregates in mortars of high cement content. The pyroxene aggregate has been positioned at extinction to highlight the calcium hydroxide at its interface. All the particles in this sample are derived from a crushed pyroxene olivine basalt. Cross polarised light.

5.6.2 Steel reinforcement/cement paste interfaces

The interface between steel and paste is similar in many respects to that of aggregates except that steel has practically zero porosity, apart from any rust layer on its surfaces. Steel surfaces are more uniform than aggregate surfaces so that the details of any interfacial layers are more likely to be observed. The paste–steel interface will still be subject to the effects of bleeding and plastic settlement. However, it is rarely possible to observe paste–steel interfaces by transmitted light microscopy because of the difficulties involved in the preparation of thin sections containing steel. One technique has been proposed by Garrett and Beaman (1985) for retaining diametral cross sections, but ideally adequate examination of the steel/paste interface requires longitudinal sectioning of the steel reinforcing. This is one of the neglected areas of concrete petrography and will remain so until adequate sample preparation methods are available. Figure 5.43 shows development of zoned rust products adjacent to corroding steel reinforcement in a concrete. Note the thin development of carbonation at the margin between the corrosion products and hardened cement paste.

Figure 5.43 Details of the boundary between the hydrated iron oxides and the concrete showing the Liesegang type of banding zonation of the corrosion products and the peripheral bright carbonation band in the adjacent cement paste matrix. Width of field 0.8 mm plane polarised light.

5.6.3 Fibre reinforcement/cement paste interfaces

Steel, carbon, glass, asbestos, kraft pulp (cellulose wood fibres) and polymer fibres are commonly used to reinforce cement products. The examination of their interfaces in thin section can provide useful information, but on the whole, these interfaces are better observed by SEM. Fibre reinforcement is discussed in Section 7.7, but reviews covering these materials will be found in references, for example, Bentur and Mindess (2006), Massazza and Costa (1986), Diamond (1986), Walker (2002) and Brandt (1995).

Some observations with the optical microscope are possible in the cases of glass, asbestos and kraft pulp or other vegetable fibres (see Section 7.7), but as reported from scanning electron microscope studies (Barnes et al. 1978; Diamond 1986), little if any paste or calcium hydroxide can be observed penetrating between the fibres, as shown in Figures 7.11 and 7.13, unless the individual fibres of a glass strand debond. Along the length of the strands or the debonded fibres, copious calcium hydroxide crystallises. In the case of asbestos fibre, the high wetting capacity of this fibre combined with the small cross section of individual fibres appears to prevent the crystallisation of any but large crystals of calcium hydroxide at the fibre interface. It is not possible to make statements about the interface of kraft pulp fibre as the products in which the fibre is incorporated are usually steam cured at high pressure.

Many products containing fibrous reinforcement are steam cured at temperatures of around 180°C and also contain silica flour as a mineral addition. Effectively, this removes much of the calcium hydroxide from the cement by reaction with the silica flour and results in an interface deficient in calcium hydroxide. As indicated in Figure 7.11 shows that the cement paste appears to be in direct contact with the asbestos fibres with little calcium hydroxide present at the interface. At ambient temperatures, siliceous mineral additions can also have similar effects. The effect will be greatest for a highly active addition such as silica fume and will decrease with the coarser or less reactive additions.

5.6.4 Render interfaces

A traditional external rendering consists of a number of coats of cement plaster laid over substrates of concrete or masonry. A general description of current UK techniques will be found in Robinson et al. (1992) and American techniques in ACI Committee 524 (1993). Renders which are applied as a single coat are now common in Europe (Mehlmann 1989). Ignatiev and Chatterji (1992) have discussed the compatibility of mortar and concrete. These interfaces are always characterised by carbonation of the substrate and some depletion of calcium hydroxide in the plaster layer itself due to suction from the substrate. Contact along the interface is never continuous and voids and intermittent fine fissures are evident. The deposition of calcium hydroxide, so often found associated with aggregate interfaces, is not as common at render interfaces probably because of suction from the substrate.

Rendering failures often occur because of the loss of adhesion between the render and the concrete substrate. Modern concreting frequently uses fair-faced formwork coated with a release agent to give a smooth surface to the concrete. These surfaces do not make a good substrate for rendering unless all vestiges of the release agent have been removed by either surface preparation or weathering. Residues of the release agent causing the problem are not observable by microscopy and require solvent extraction and chemical or infrared analysis for their detection and identification. Thus, petrographic examination is of limited use in the diagnosis of adhesion failures and made more difficult in that almost invariably only the render and not the concrete is made available for examination. Meaningful observation of an interface requires both surfaces to be available for examination, best achieved by coring through both the render and the substrate.

The special type of sprayed render, known as shotcrete or gunite, is a pneumatically applied mortar or concrete used as a rendering, rock face stabilisation and also for the repair of damaged concrete surfaces (Schrader and Kaden 1987). See Section 8.6, for further details. If a thick layer is required, the mortar is applied over a preceding layer soon after initial set has occurred, and these interfaces can be identified in thin section by a diffuse line of calcium hydroxide (Figure 8.6a and b) and on occasion by some darkening due to absorption by the substrate. It is now common to include silica fume and a range of fibres in shotcrete to improve its rebound and physical properties, and these additions will considerably affect its appearance in thin section as discussed in Sections 5.6.3 and 7.7.

5.6.5 Other cementitious interfaces

Tiles have traditionally been fixed to surfaces with a bedding mortar, which now often includes a latex polymer to improve adhesion (Robinson et al. 1992) (see Figure 5.44). To level the tiles and ensure it is in good contact with the mortar, the tiler drives each tile against the mortar by beating. Thin-section examination shows there is an intimate interface between the tile and hardened paste only separated by a semi-continuous layer of calcium hydroxide which is generally less than 5 μm in width (Figure 5.45). Along the interface, void areas are present due to entrapped air. Where the tile has been insufficiently beaten during laying, only the tile nibs will bite into the mortar leaving recesses with large bladed crystals of calcium hydroxide attempting to bridge the gap.

Portland cement-based renders are usually mixed with lime, air-entraining agents and/or other plasticisers. With internal and external plasters, there is a need to distinguish between cement-based and gypsum-based mixes. Since gypsum is slightly soluble in water, gypsum-based mixes are usually restricted to internal applications. Details of the interfaces between substrates and gypsum plasters are discussed in Section 9.2.

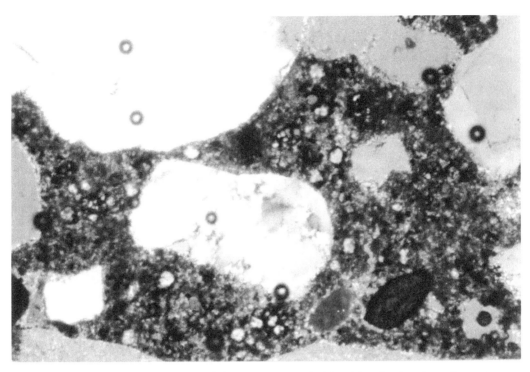

Figure 5.44 Latex polymer admixture added to improve workability. Identification depends on the concentration used. Here the latex appears as rounded/spherical particles yellowish to orange in colour in the cement mortar matrix. (Note: Several included air bubble artefacts, top left and centre). Crossed polar illumination, width of field 1.3 mm. (Courtesy of M. Grove, private collection.)

Joints between pours of concrete separated by a time interval are often characterised by a line of carbonation in the concrete substrate and depletion of calcium hydroxide in the contact zone of the overlying concrete due to absorption by the substrate.

These joints are usually irregular because the underlying surface is sometimes mechanically roughened prior to concreting. Voids, cracks, fissures, dirt and debris are often also observed at such interfaces. Polymers such as epoxy resin sometimes used as a coating or topping are characterised by some penetration of the polymer into the substrate with excellent adhesion resulting (see Figure 5.32b). The interface in this case is to the existing carbonated surface and does not involve further development of calcium hydroxide.

The petrographer is sometimes required to ascertain whether a visible pour line on an exposed concrete surface is a non-critical surface feature or a cold joint. This is readily determined from a core cut at right angles to the visible line so as to contain the line. Visual and microscopic examination of the core will determine how far the line penetrates into the concrete and whether there is debris, carbonation and changes in composition or microstructure at the interface which would be indicative of a cold joint.

5.7 VOIDS IN CONCRETE

The range of voids and pore types found in concrete are listed in Table 5.8 together with some of their effects on the properties of concrete. The gel capillaries and pores may be observed using electron microscopy. The porosity involving capillaries exceeding 10 nm will

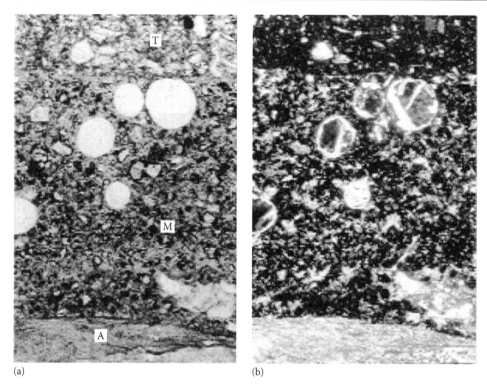

(a)   (b)

Figure 5.45 An interface between a ceramic tile and bedding mortar. Width of fields 0.65 mm. (a) The area of the rich bedding mortar (M) between the bottom surface of the tile (T) and a sand particle (A) contains some entrapped air and numerous partially hydrated and remnant cement grains. The paste is in close contact with the aggregate and the tile but in many other areas the interface contained voids and had parted from the tile. (b) At the tile interface a thin line of calcium hydroxide is just visible and to a lesser extent at the aggregate margin. The entrapped air voids contain laths of calcium hydroxide. It is possible the styrene-butadiene latex used in the mortar has had some effect on the way the calcium hydroxide has crystallised at the interface. Crossed polars.

Table 5.8 Classification of pore sizes in cement pastes

Designation	Diameter	Description	Role of water in void	Paste properties affected
Voids	2–0.25 mm	Foamed systems	Behaves as bulk water	Strength, density
Voids	Large–0.010 mm	Entrapped air voids	Behaves as bulk water	Strength
Voids	1.0–0.005 mm	Entrained air voids	Behaves as bulk water	Strength
Capillary pores	10–0.05 μm	Large capillaries	Behaves as bulk water	Strength, permeability
Capillary pores	50–10 nm	Medium capillaries	Moderate surface tension forces generated	Strength, permeability, shrinkage at high humidities
Gel pores	10–2.5 nm	Small (gel) capillaries	Strong surface tension forces generated	Shrinkage at 50% RH
Gel pores	2.5–0.5 nm	Micropores	Strongly absorbed water: no menisci formed	Shrinkage; creep
Gel pores	<0.5 nm	Micropores 'interlayer'	Structural water involved in bonding	Shrinkage; creep

Source: Adapted from Mindess, S. and Young, F.J., *Concrete*, Prentice-Hall, Englewood Cliffs, NJ, 83p, 1981.

affect durability and is important. The conditions under which porosity in concrete crosses over from connected to discrete capillaries have already been discussed in Section 5.5. Where the capillary pore structure is not interconnected, its principal effect is on strength (Mindess and Young 1981). However, where the capillary pore structure is interconnected, it not only affects the strength but also the durability of the concrete because of increased permeability. This especially applies to the outer layer of a concrete, where durability will be reduced because the interconnected capillaries may allow air and vapour to penetrate to a sufficient depth to corrode the reinforcement.

As the optical microscope tends to be restricted to a maximum resolution of about 1–2 μm with thin sections of concrete, effective observation is restricted to dimensions greater than 1 μm. Thus, only those categories in Table 5.8 whose dimensions exceed 5 μm can be easily examined. In practice, the capillary and coarser gel pores require the use of SEM, usually of fracture surfaces, while the finer gel pores may be measured by mercury porosimetry.

The basic difference between a capillary pore and an air void in plastic concrete is that capillary pores are elongated spaces filled with water, while air voids are roughly spherical pores formed by the entrapment or entrainment of air. Hence, the difference in the minimum dimensions quoted. The minimum size of an air void is limited by its internal gas pressure which in turn is controlled by its radius and the surface tension of the liquid. Small bubbles have high gas pressures and in practice bubbles of less than 4 μm tend to disappear (Dolch 1984). An arbitrary distinction is made between entrapped air voids, that is, those voids which are trapped during mixing, and entrained air voids produced by admixtures which have a foaming or air-entraining action in the plastic cement paste.

5.7.1 Entrapped air voids

Entrapped air voids are invariably produced in the mixing and placing of concrete as has been described in Section 4.5. In an adequately graded and compacted concrete mix, the volume present is determined by the maximum size of the coarse aggregate. In examples where 65 mm aggregate has been used, entrapped air will be about 0.3%–0.5% in volume. This increases to about 3% for 10 mm aggregate. The dimensions of entrapped air voids range from about 10 mm downwards; they will have relatively smooth margins and be spherical to irregular in shape. An increasing number of larger and irregular voids typically correlates with increasing water/cement ratios in the concrete.

In cement mortars, depending on the sand grading used, entrapped air voids occupy a volume of about 3%–8%. Because of the absence of coarse aggregate and a resultant higher cement-to-aggregate ratio, the entrapped air voids in mortar tend to have a more regular distribution and in many cases will appear similar to an entrained air system. However, it is not as stable as an entrained system and is more easily removed during placement of the mortar. Measurement of bubble size distributions will also show that entrapped air is deficient in the smaller bubble sizes.

From examinations of the air void system in a thin section, it is possible to characterise the parameters of the system and observe whether voids contain calcium hydroxide and other products. The general appearance in a laboratory mortar of different percentages of entrained air voids (measured by an air metre on the plastic mix) is shown in Figure 5.46. These laboratory mortars were mixed with a cement/aggregate ratio of 1:2.75 and a water/cement ratio of 0.5 using an OPC, Leighton Buzzard graded quartz sand and Vinsol resin as the air-entraining agent. The cubes were compacted by rodding in three layers (Figure 5.47).

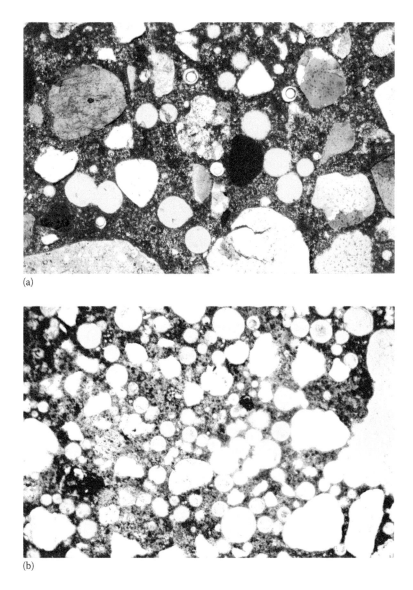

(a)

(b)

Figure 5.46 Examples of (a) normal and (b) excessive air entrainment in the cement matrix impregnated with yellow-dyed resin as seen in transmitted light. Inclined polar illumination, width of fields 1.8 mm. (Courtesy of M. Grove, private collection.)

5.7.2 Entrained air void systems

The entrainment of a stable, well-spaced air void system in plastic concrete and mortar mixes is of considerable importance in those countries subject to severe frost conditions. In addition, entrained air has a lubricating effect on plastic cement paste and produces more workable concretes and mortars. This is especially important for plaster renderings. There is a large literature on the properties of air entrainment in concrete, and reference should be made to codes of practice such as ACI Committee 212.3R (1989) and for a general overall discussion of the chemistry and other properties to Dolch (1984).

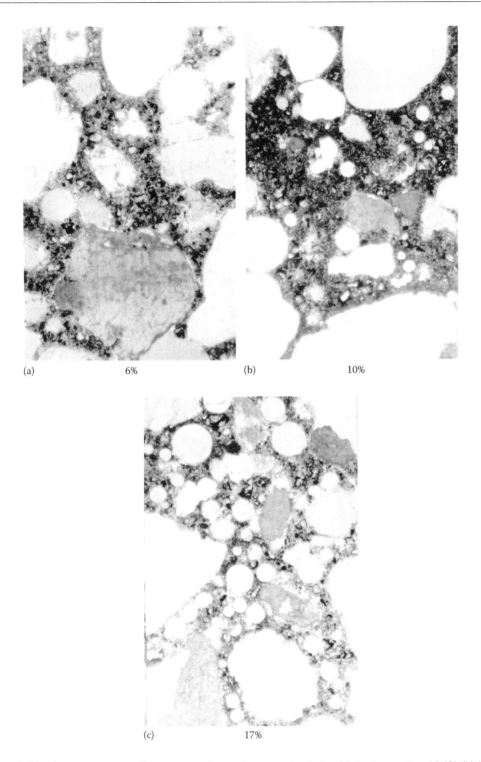

(a)   6%   (b)   10%

(c)   17%

Figure 5.47 Laboratory-prepared mortar samples to show entrained air voids in thin section: (a) 6%, (b) 10% and (c) 17% air voids, respectively. Plane polar illumination, width of field 1.4 mm (Courtesy of St John).

As described in Sections 2.4 and in detail in Section 4.5.3, measurement of the air void system is traditionally carried out by observing finely ground sections of concrete with the aid of a low-power stereo-binocular microscope (ASTM C457-90 1990). To measure voids down to the smaller diameters so that the system is adequately characterised requires the use of a magnification at least as high as 120x (Bruere 1960), and the image of the ground surface tends to be of poor quality at these magnifications. However, finishing the surface with a finer grit size (see Section 3.4.2) will produce a significant improvement, and a binocular microscope of a higher quality will also improve observation of the detail of small voids. A method of measurement of bubble sizes from measured chord lengths using the Rosiwal linear traverse has been proposed by Lord and Willis (1951).

The problem of resolution does not exist in transmitted light microscopy where sharp images are readily available at high magnifications. It is not widely recognised that large-area thin sections will allow the detailed examination of air void system in concrete (see Figure 5.46a and b). Used with a suitable point-counting stage or image analysis system (see Section 2.6) and appropriate softwear, accurate quantification is also possible using thin sections. From examination in thin section, it is possible to measure the air void system and observe its relationship in the texture and whether voids contain calcium hydroxide and other products. In addition, variations of the air void system in the outer and inner layers of the concrete are easily observed.

The magnifications of the photomicrographs in Figure 5.48 are in the range of magnification specified by ASTM C457-90 (1990) and show the type of detail that can be observed in thin section.

5.7.3 Capillary and gel pores

Capillary voids are not easy to observe in thin section as they are well dispersed and their dimensions are at or beyond the working limit of observation. In practice, this is overcome by incorporating a fluorescent dye in the epoxy resin used for impregnation and visually estimating the amount of fluorescence (Dubrolubov and Romer 1985) and relating this to the water/cement ratio (see also Section 4.4). Strictly speaking, this is not observation of capillary voids but an assumption that the coarser capillaries can be satisfactorily impregnated with epoxy resin (see Section 3.3.4).

As indicated in Table 5.8, the capillary and gel pores present in the cement paste matrix range from 10 μm to 10 nm for capillary and from 10 nm downwards for gel pores, so direct study of these pores is only possible using electron microscopy. A direct method used by Parrott et al. (1984) involved impregnation of an alite paste with resin, acid etching a polished surface to remove the paste and viewing the etched surface using SEM. Most information about pore sizes and volumes reported in the literature has been obtained indirectly by methods such as mercury intrusion porosimetry and water or nitrogen sorption isotherms. A review of various methods is given in Taylor (1990). The relevance of study of pores and pore size distributions relates to cement hydration, strength and diffusion of ions and gases.

5.7.4 Other air voids in concrete

Other voidages, such as cracks (further discussion of voids is given in Section 4.5), bleeding channels, voids and fissures due to plastic settlement, excessive air entrainment and inadequate compaction, are all faults associated in varying degrees with unsatisfactory mix design, dispensing of constituents, mixing and placement practices.

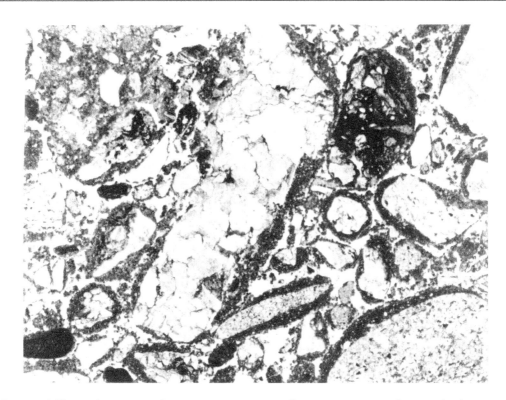

Figure 5.48 This is the texture of concrete containing insufficient mixing water. As a result, the cement was not dispersed during mixing and adhered to the aggregates as sticky rims. Compaction was insufficient to fully compact the concrete leaving minimal contacts between cement rims. The set concrete was porous, weak and crumbly in texture and superficially appeared not to have set. As piles for a major structure, the concrete was condemned and replaced (Courtesy of St John). Plane-polarised illumination, width of field 12 mm.

Bleeding channels and plastic settlement are present in most concretes but become more prominent in concretes with high water/cement ratios and will tend to increase in frequency in the upper portions of a lift or slab. They tend to be faults and do not need highlighting as they are of sufficient dimension to be clearly visible in thin section. To observe the full extent of plastic settlement, the plane of section must be vertical. On the other hand, bleeding channels tend to the vertical and so are more likely to be intersected by a horizontal section. In practice, the most information appears to be gained from a vertical section.

Fissures due to plastic settlement tend to form at the bottom margins of aggregates, may have well-defined margins and are often meniscus-like in shape. If pore fluid has been present in them for any length of time, they will be lined or filled with calcium hydroxide. Bleeding channels are irregular and tortuous, and because a section represents only one plane, they will appear to be discontinuous. The channel margins are diffuse and slowly grade into the cement paste suggesting that some of the cement paste at the margins has been partially dissolved during the bleeding process. It is unusual to observe much calcium hydroxide associated with bleeding channels probably because of leaching by the bleed water. There is a problem in deciding at what point plastic settlement and bleeding are significant from observation made in thin section. Useful interpretation requires knowledge of the incidence of these faults for the typical materials used in any one area and the examination of surfaces of large sample cross sections from a range of sampling points in a structure.

Severe overdosage with an admixture, for example, producing excessive entrained air, has a spectacular effect on a concrete. Visually, the concrete appears finely honeycombed with air bubbles (Figure 5.47). By the time the concrete has reached air contents of around 25%, it may have little strength and can be crumbled easily and may superficially appear not to have set. Since each 1% of air reduces strength by roughly 5%, the reason for this effect is obvious. In thin section, such concretes appear as a mass of small bubbles, sometimes coalesced together, with only thin septums of paste separating bubbles. Because of the close spacing of the air voids, these septums do not have sufficient dimensions to provide any strength. Excess voidage can also be found in concrete due to incomplete compaction but is largely restricted to dryer concrete mixes that do not flow easily. The extreme condition is where insufficient water is used so that the cement and water are not fully dispersed during mixing giving rise to rich layers of sticky cement surrounding aggregates which are almost impossible to compact. This effect is illustrated in Figure 5.48 in a sample of concrete from an in situ rammed pile which had the appearance of having not set. In thin section, it was observed that the concrete was only held together by narrow porous bridges between the rich layers of undispersed cement paste surrounding the aggregates (St John 1983). All the margins of the cement paste facing into voids were ragged and porous, and in places very large crystals of calcium hydroxide were present. Less spectacular cases of incomplete compaction consist of ragged void space which may contain completely unconsolidated material. The bridging spaces between the ragged void spaces may be porous so that in effect they act as weakened bridges. Kaplan (1960) investigated the effects of incomplete compaction on concrete and showed that an unconsolidated concrete with 25% void space loses about 90% of its compressive strength. Segregation of the fine from the coarse aggregate can occur with harsh mixes, mixes that are too wet, or too dry, and those that are sand deficient. Where the segregation is on a wide-scale microscopy is inappropriate, and the fault (honeycombing) should be observed by comparison of large cut faces. Small localised patches of segregation are not uncommon in concrete and are usually ignored by the petrographer, unless their frequency suggests problems in the mix design and/or placement of the concrete.

5.7.5 Aerated and no-fines concretes

Aerated concretes, also known as gas, foamed or cellular concretes and mortars are used as semi-structural lightweight and insulating materials (Short and Kinniburgh 1978; Regan and Arasteh 1990; Namdiar and Ramamurthy 2006). The aeration, which consists of discrete bubbles which are one to two magnitudes larger in size than air-entrained bubbles, is emplaced by mechanical or chemical foaming. Mechanical foaming may be carried out by adding a preformed foam or by air foaming in situ. Chemical foaming, usually based on the addition of aluminium powder to the mix, generates the foam by the evolution of gas bubbles in situ. Many of the aerated products are not concretes but mortars and include additions of silica or blast-furnace slag. Precast products of this type are often autoclaved. The texture illustrated in Figure 5.49 is of a commercial sample of autoclaved lime silica block rather than Portland cement, but its aerated structure consists predominantly of pores with diameters of 1.5–2 mm with some interspersion of smaller pores. The smallest pore observed was 0.25 mm in diameter.

No-fines concrete is made with 10–20 mm aggregate and a cement/aggregate ratio by volume of about 1:8 (Malhotra 1976) (see also Section 7.2). Its texture, illustrated in Figures 5.50 and 8.7, consists of a packing of aggregate coated with cement paste which acts as a bridge to cement the aggregates together. Between the individual aggregates, there are large irregular voids with smooth margins.

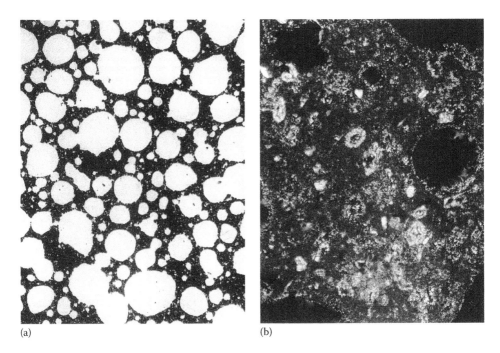

Figure 5.49 The texture of an aerated concrete. (a) The void system in this commercial lime silica brick (possibly Siporex) has probably been produced by the use of aluminium powder. The pore structure is predominantly in the 1.5–2 mm range but pores as small as 0.25 mm are present. Width of field 16 mm. (b) Detail of the lime silica texture of the aerated brick. The hardened paste has carbonated due to many years of storage in the laboratory. Relics of the silica flour and lime used are still visible. Width of field 0.75 mm.

5.8 CRACKING IN CONCRETE

Cracking in concrete is often an important indicator of some form of deterioration. Consequently, cracking has been the subject of numerous reviews, for example, Fookes and Collis (1977), Hollis et al. (2004) and Concrete Society Technical Report No. 22 (1992, 2010). This last draws a distinction between structural cracking in concrete resulting from stresses arising from externally imposed loadings and intrinsic cracking arising by the nature of the constituent materials and their interactions. Structural cracking can be limited to acceptable levels in terms of structural integrity and aesthetics by attention to design and detailing. Intrinsic cracking can also be minimised partly by attention to design and detailing, but in addition, the appropriate selection of materials, mix design and placing and curing methods must also be considered.

Unexplained cracking is a common reason for engineering inspection and core sampling of concrete structures. The causes of cracking can be complex and their examination forms an important part of a petrographic examination. Because of this complexity of causes, it is vital that the petrographer either inspects the structure or is given an accurate and detailed description of all cracking observed. Too often the inspecting engineer makes sampling decisions without reference to the petrographer. This may result in either the need for further sampling or an unreliable outcome from the petrographic examinations because of insufficient information.

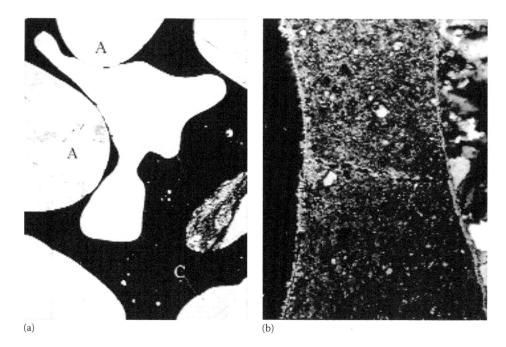

(a)   (b)

Figure 5.50 The texture of no-fines concrete. (a) In no-fines concrete large irregular voids are formed in the cement paste (C) which in this laboratory sample has cracked due to drying and/or carbonation shrinkage. The rounded quartz aggregate used (A) is clearly visible. Width of field 16 mm. (b) Detail of (a). The cement paste has carbonated and cracked due to long storage in the laboratory. Shrinkage has also caused cracking at the interface of the aggregate. Cement clinker remnants are clearly present in the hardened paste. Width of field 0.9 mm.

5.8.1 Cracks resulting from tensile strain

Hardened concrete cracks when its tensile strain capacity is exceeded by stresses applied from any source. The tensile strain capacity of hardened concrete varies from about 0.008% to 0.015%, but whether the concrete actually cracks for any given set of stresses is complex and depends on rate of strain, restraint by aggregates, the age of the concrete, creep and of course reinforcing and other factors. The relationship of many of these factors to the strain capacity of a concrete is beyond the scope of this chapter, and reference should be made to Concrete Society Technical Report No. 22 (1992, 2010) for further information. Goltermann (1995) has proposed a prediction method for the classification of crack patterns related to strain in the hardened cement paste and aggregates.

Because the surface of concrete is effectively an anisotropic interface for strain, it is subject to superficial cracking. This is especially the case for localised drying shrinkage, which may penetrate the concrete to a depth of only a few millimetres. It is an important part of petrographic examination to differentiate superficial surface cracking from other types of cracking which may have more serious and long-term consequences.

5.8.2 Differentiating between structural and non-structural cracks

To distinguish between structural and non-structural cracks requires a systematic approach to diagnosis. Where the cracking is obviously structural/mechanical in nature (see Figure 5.51), its diagnosis lies within the realm of structural engineering. Further information

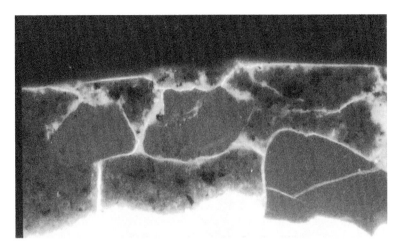

Figure 5.51 Mechanically induced cracking. A petrographic thin section in transmitted light showing micro-cracking due to explosive damage, undertaken to determine the extent of damage prior to repair. UV illumination, width of field 2 mm. (Courtesy of M. Grove, private collection.)

concerning some of the common causes of structural cracking is discussed in a simple manner by Kelly (1981) and Gustaferro and Scott (1990), and more extensive reviews have been given by ACI Committee 224.1R (1985), RILEM Committee 104-DDC (1994) and Campbell-Allen (1979). The cracking of concrete associated with corrosion of reinforcement has been discussed by Pullar-Strecker (1987) and is also described in Section 6.3.

Where the cause is in doubt and where the properties of the materials may be involved, it is necessary for the engineer to follow the examination and analysis procedures as detailed in the flow chart (Figure 5.52) to enable structural cracking to be clearly distinguished from

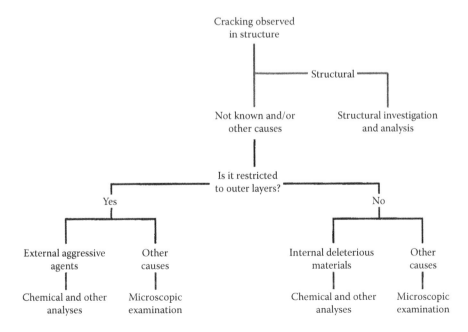

Figure 5.52 Outline of method required for investigating cracking in structures.

non-structural cracking. It is rarely possible to determine the cause of non-structural cracking without carrying out some type of examination of cut surfaces.

Where the relationship between cracking and deterioration of a concrete fabric has to be established, thin-section examination is often required. The exceptions are cases of expansion due to hard burnt lime and excess magnesia in the cement and some cases of attack by external aggressive agents where chemical analysis is obligatory and examination of sections may not be justified.

5.8.3 Non-structural cracks in concrete

There is no entirely satisfactory scheme for classifying the causes of non-structural cracking in concrete. Figure 5.53 is a modification of the classification given in Concrete Society Technical Report No. 22 (1992, 2010). It has been modified to emphasise the relationship between the internal and surface cracking which needs to be established in petrographic examination. The outer layer of concrete for the purposes of this table is the thickness of concrete penetrated by the environment. In practice, this is a layer that can vary in depth from 1 to 2 mm in high-grade concretes to as much as 50 mm or greater in lower-quality concretes.

The cracking as commonly observed in thin section or on cut surfaces may be due to a number of causes. These can only be evaluated by considering both the visible effects on the structure and the observations made in thin section and other available data concerning the structure and its environment. It cannot be overemphasised that adequate diagnosis of cracking requires examination of both the external surfaces and internal fabric of a concrete. The following explanations provide a summary of the most common causes, but do not cover all circumstances.

5.8.3.1 Cracking subparallel to, and restricted to, the near surface

1. *Scaling and exfoliation:* Sulphate solutions, freeze/thaw attack or any expansive agent that penetrates the surface layer can cause scaling and exfoliation. Scaling which is primarily associated with mild freeze/thaw attack and fretting with salt recrystallisation are merely the initial stages of exfoliation. A surface attacked by frost is rough and irregular, and the concrete remaining will be cracked, often with a subparallel mesh of cracks, to a depth which is dependent on the degree of attack and the permeability of the concrete (Figure 5.54a,b). The subparallel cracking usually grades relatively abruptly into sound concrete. Where severe exfoliation occurs due to sulphate attack, sheets of concrete can be detached from the surface. In thin section, the exfoliated area consists of a mass of subparallel cracks which run through both the hardened paste and aggregates. Apart from being able to estimate the depth to sound concrete, there is little else to be observed. Sampling exfoliated concrete for thin-section examination presents difficulties in that with core samples, much of the exfoliated texture is lost unless special procedures are used to consolidate the surfaces before coring (see Section 3.4.3).

   In cases of sulphate attack, crystals of gypsum and/or ettringite may be present filling or lining cracks, pores and fissures. It is rarely possible to observe gypsum or ettringite intimately interspersed in the hardened paste, so diagnosis should always be confirmed by x-ray diffraction analysis and where necessary supplemented by SEM and chemical analyses for the sulphate content of the affected areas.

   A very common cause of cracking, scaling and spalling is the result of rusting of steel reinforcement close to the concrete surface (see Section 6.3). The rust occupies several

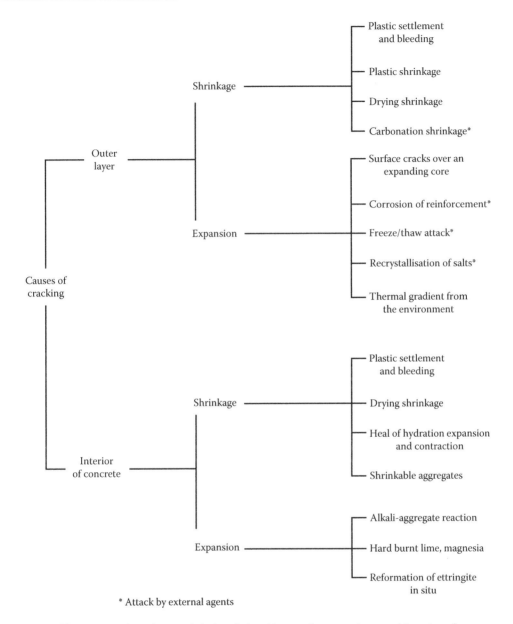

Figure 5.53 The causes of cracking and their relationships to the outer layer and interior of a concrete. Modified from Concrete Society 1992/2010.

times the volume of the original steel with the consequent staining cracking and possible spalling of the surface (see Figure 5.56).

2. Pop-outs: A pop-out is distinct from scaling and spalling and results from cracks subparallel to the surface caused by the expansion of an aggregate, a contaminant, corroding reinforcement or may be due to expansive alkali-silica gel forming on an aggregate particle just below the concrete surface or by the freezing of water in a porous chert or similar particle lying under the surface of the concrete. In the initial stages, a pop-out appears as a semicircular crack with vertical displacement of the semicircular margin. As the pop-out develops, the crack becomes roughly circular

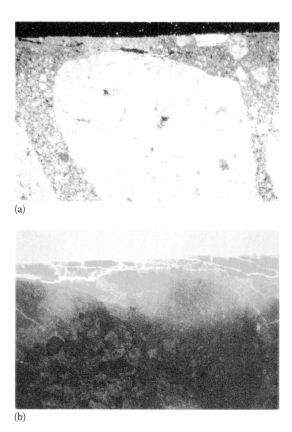

(a)

(b)

Figure 5.54 (a) Frost damage as seen in thin section with cracking subparallel to the surface. Inclined polar illumination. (b) A similar (a) but in UV illumination, width of fields 3 mm. (Courtesy of M. Grove, private collection.)

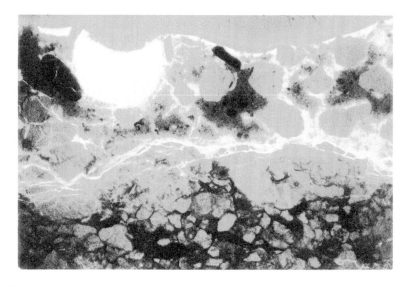

Figure 5.55 Microcracking of a paved surface, cracking is parallel to the surface and further cracking has occurred within the detached layer. UV illumination, width of field 2 mm. (Courtesy of M. Grove, private collection.)

Figure 5.56 Spalling as a result of steel reinforcement corrosion, due to the steel being too close to the concrete surface. Note a spall is developing on the left, but the spalled surface has completely detached from the rusted steel to the right.

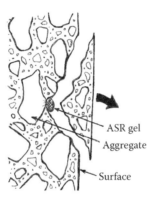

ASR gel

Aggregate

Surface

Figure 5.57 A diagrammatic cross section of a concrete spall due to development of an expansive ASR gel.

and an inverted cone of material gradually protrudes above the surface and may be removed leaving a crater with remnants of an expansive aggregate at its apex (Figures 5.57 and 5.58). The dimensions of pop-outs range from a few millimetres to 100 mm in diameter and up to 40 mm in depth. Expansive materials that cause pop-outs are alkali-reactive particles, iron sulphide minerals and porous aggregates that undergo freeze/thaw attack. Examination of the area of a pop-out in thin section is rarely justified as visual examination of the apex of the pop-out crater will often reveal the cause. In thin section, the expansive aggregate (which is usually partially disintegrated) can be identified. If alkali–silica gel is present, it is also readily identifiable, but the iron sulphide minerals may be difficult to observe and may require identification by x-ray diffraction or SEM microanalysis techniques.

5.8.3.2 Cracking approximately perpendicular to, and restricted to, the near surface

1. *Fine, filled cracks:* Fine pattern cracking, often widely distributed and attributable to drying and/or carbonation shrinkage, is common on plastered surfaces and may also be present where the finished surface of the concrete has been too wet or too rich in cement (Figure 5.59). Its external appearance is of white lines, often raised because of

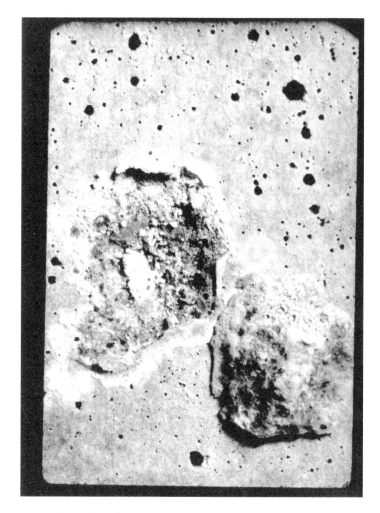

Figure 5.58 Pop-out on the surface of a test prism. The detached 'cone' has been placed face down on the adjacent concrete surface. Width of photograph 25 mm (Courtesy of St John).

Figure 5.59 Drying shrinkage cracks in a paving slab, with a coloured mortar worked into the upper surface before initial set. The cracks are highlighted by raised white lines of carbonation (Courtesy of St John).

infilling by carbonation, or, where moisture is absent, dark lines because of infilling dirt or simply contrast with a lighter surrounding surface.

Some very fine cracking may not be apparent to the naked eye on rough surfaces but can be seen with the aid of a hand lens or stereomicroscope. When observed in thin section, this cracking rarely exceeds 10 mm in depth and 10 µm in width and is invariably filled or contains carbonate if the surface has been exposed to the weather for any period. It is not uncommon for the cracks to end at aggregate margins. Where doubt exists about the cause of this type of cracking, thin-section examination is required to show its relationship to the remainder of the concrete.

A more severe form of the aforementioned cracking is typical of carbonation shrinkage referred to in Section 5.5.4. There are few concretes in which some superficial examples of this type of surface cracking cannot be observed in thin section as shown in Figure 5.60.

2. Fine open cracks: These often tend to form patterns in localised areas. They may be simple surface drying shrinkage cracks due to overworking of the surface or too rapid drying leading to a cement-rich layer on the surface which shrinks as the cement sets.

Such cracks usually only involve the top 1–3 mm of the surface (see Figure 5.60), and the cement-rich layer is also often the cause of 'dusting' of floor surfaces in use. If the cracking is open at the surface, careful examination will often reveal some vertical displacement at the crack margins. Where the crack is closed because it is filled with carbonation deposit, this displacement may not be observable. In thin section, these cracks usually range from about 15 to 40 µm in width and can penetrate the concrete to depths of about 50 mm. Dependent on exposure conditions, they may have carbonated margins or be filled with carbonation products for much of their depth. The cracks tend to skirt aggregate margins and take the line of least resistance as shown in Figure 5.37.

This type of cracking can be typical of the first stages of AAR in which the core of the concrete has expanded just sufficiently to exceed the tensile strain of the surface in localised areas. In the outer layers of the concrete, expansive reaction may be absent due to leaching and/or movement of alkalis. If examination of the concrete reveals that potentially reactive aggregates are present, it can be assumed that the first stages of AAR may be present in the concrete even if no alkali–silica gel or aggregate reaction is observed. Examples of cracking in the various stages of AAR development are discussed and illustrated in Section 6.8.

3. Cracking associated with steel reinforcement corrosion: This type of cracking is reflective of the underlying reinforcement that is usually open and can be up to several millimetres wide. The edges of the cracks are often displaced vertically in relation to each other, and rust staining may be present but this is not invariable and less common than would be expected. At the corners and edges of members, a levering action is evident which tends to spall the concrete (see Figure 5.56). In its early stages, it may be difficult to diagnose, but when more advanced is usually obvious. Although cracks due to reinforcement corrosion can easily be examined (see Figure 5.61), it is rarely possible to successfully make thin sections of samples containing corroded steel reinforcement. Concrete slice specimens can be prepared cutting the reinforcement, but additional cracking at the steel–concrete interface is likely to be induced by the cutting. A rapid method of field investigation is to undertake either half-cell potential mapping or resistivity testing. Alternatively, the affected cover concrete can be physically removed to examine the reinforcement for signs of corrosion, as described by Pullar-Strecker (1987).

Figure 5.60 Cross section showing shrinkage crack crazing due to the over-trowelling of a floor surface. These cracks only extend downwards for 1–2 mm from the surface. Plane-polarised illumination, width of field 6 mm. (Courtesy of M. Grove, private collection.)

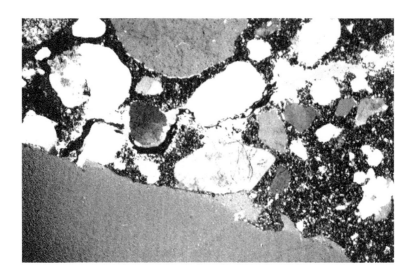

Figure 5.61 A diagonal crack in concrete resulting from the expansion caused by corroding steel. Note the marginal carbonation and iron oxide/hydroxide infilling the lower part of the crack. Crossed polar illumination, width of field 1.3 mm. (Courtesy of M. Grove, private collection.)

5.8.3.3 Cracking reflecting interior expansion of the concrete

Surface cracking due to expansive forces in the interior concrete may initially be identified by field inspection. Core surfaces and cut slabs provide clearer evidence, but careful examination of thin sections usually allows positive evidence of the causes to be established. The details of individual cracks and the several types of pattern and the orientations of the cracks in three dimensions will provide key evidence as to the stresses involved and the causes of the cracking (Figures 5.62 and 5.63):

1. Rectangular or triangular patterns of cracks: This type of crack is usually open, with widths varying from 60 μm to a few millimetres. The pattern of cracking may either be restricted to localised areas or be quite widely spread but is never uniform over the structure. A typical example is shown in Figure 5.62, and many other examples of cracking due to AAR will be found in the literature, for instance, in Hobbs (1988) and Swamy (1992). However, there is a tendency for authors to illustrate the more spectacular cases of cracking. It should be realised that external cracking even in cases of severe AAR is often not spectacular. Vertically displaced margins to the cracks are common and margins may be darkened especially after rain. Where larger cracks are present and water has percolated through the crack system, efflorescence may be present which may be confused with carbonated alkali–silica gel. Clearly visible, external, alkali–silica gel is not common with AAR unless the reaction produces large amounts of gel. In thin section, it will be found that in many cases, the surface cracks transect a layer of concrete up to 75 mm in depth in which AAR is visually absent.
2. Cracking restricted to a single principal direction: This type of cracking is usually open, with widths varying from about 60 μm to 2 mm. This type of cracking may be the result of steel reinforcement corrosion below the surface, and hence the cracking will reflect the pattern of the reinforcement. In the cases of reactive expansive aggregate particles, the most common being AAR if the expansion is partially influenced by the steel reinforcement or by an imposed loading, cracks may be restricted to one direction (see Figures 5.63 and 5.64).

Figure 5.62 Highway viaduct supports, Quebec. Note the rectangular pattern of cracking of the concrete pile cap contrasts with cracking subparallel to maximum stress direction in the inclined deck support columns.

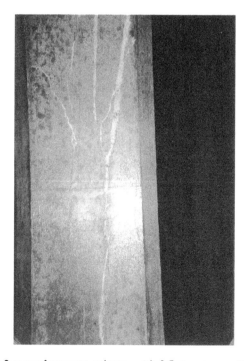

Figure 5.63 A section of an 8 m roof support column, with 0.5 m cross section that is supporting a concrete roof of a covered water reservoir. The vertical reinforcing rods were tied together by stirrups at 0.5 m centres. The roof loading has caused the ASR cracking (now filled with calcium carbonate) to orientate in the direction of principal stress.

(a)

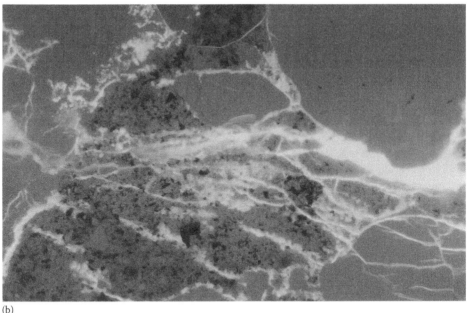

(b)

Figure 5.64 A reactive aggregate particle on the left produces a series of microcracks in the adjacent gel–soaked cement matrix. A large horizontal partially gel-filled crack cuts the cement paste between aggregate particles. The crack margins may be leached of calcium hydroxide. It is not always possible to determine the cause of such cracks by thin section alone without additional evidence. (a) In plane polar illumination. (b) The same view in UV illumination showing more clearly the extent of the cracking. Width of field 1.3 mm. (Courtesy of M. Grove, private collection.)

Because the expansion is constrained, the cracks may be narrow. However, in the plane perpendicular to the direction of cracking, the typical complex pattern of internal cracking may be observed.

3. Patterns of fine cracks within the concrete: These are often open, but are sometimes carbonated, and may have vertically displaced margins. In thin section, a relatively closely spaced pattern of discontinuous cracking may extend to the concrete surface, but the pattern remains approximately uniform at particular depths within the concrete. The width of these cracks is limited and typically less than 0.5 mm at the surface; wider tension cracks tend to be absent. These patterns are typical of expansive reactions (including AAR) under conditions which have led to reaction throughout the depth of the concrete (see Figure 5.65). In the case of AAR, cracking requires fairly continuous moist conditions.

Swamy (1992) illustrates this type of cracking for various expansive strains. Alkali–silica gel and evidence of reacting particles should be present before making a diagnosis of AAR.

4. Simple single or subparallel sets of cracks: These are relatively wide open cracks or are subparallel sets with variable spacings with widths usually not less than 100 μm. In thin section, these cracks usually run completely through a thin section and in the structure will penetrate either through the member or to considerable depths. The cracks are usually continuous with limited subparallel splintering and carbonation and will have penetrated to depths controlled by the crack width. A typical example is shown in Figure 5.65. Long-term drying shrinkage cracks are a common feature of hardened concrete panels and beams where expansion/contraction joints and reinforcement have been inadequate to restrain the cracking. If water has flowed through the crack, the margins may be leached of calcium hydroxide. In many cases, it is not possible to determine the cause of these cracks by thin-section examination as they do not embody much information, and due to their wide spacing, it is unlikely that more than one crack will be included per thin section.

Typical causes are long-term drying shrinkage of unreinforced slabs, excessive thermal gradients, temperature cycling, settlement of foundations and other structural problems that are best examined on-site. The purpose of any examination will often be directed to first eliminating other possible causes associated with the materials and evaluating the effect of the environment on the concrete surrounding the crack system before making a diagnosis of the cause.

5.8.3.4 Cracking limited to the interior of the concrete

1. *Fine cracks in the interior of the concrete*: These typically exhibit few visible effects on the surface and are best examined using appropriate thin sections, since they may be missed on surfaces of ground slices. Because the cracks are largely restricted to the interior of the concrete, they are limited to widths of less than about 15 μm and tend to occur only in localised areas. Wider and more general cracking must result in visible surface cracking. Cracking due to internal drying shrinkage tends to be confined to localised areas of the hardened cement paste and often forms a rectilinear pattern. Where cracks run from aggregate to aggregate, or aggregate to void, they are typical of the earliest stage of AAR, but the diagnosis must be supported by the presence of other evidence of ASR in the concrete (Palmer 1992). Fine cracks restricted to the body of the concrete can also fall into the category of 'cement matrix quality' faults (Figure 5.66). There are a number of causes for this type of cracking which have been

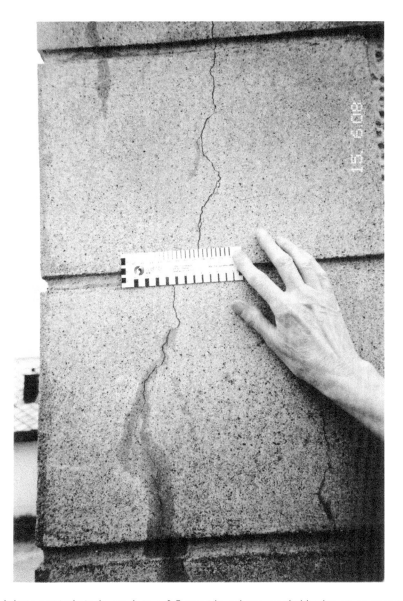

Figure 5.65 A long vertical single crack in a 0.5 m wide column, probably due to unrestrained drying shrinkage in this example. Note staining due to rain leaching. The crack width gauge reads 0.75 mm at the point measured. Swamy (1992) illustrates this type of cracking for various expansive strains. Various expansive mechanisms are possible, but additional evidence (e.g. that of alkali–silica gel for ASR) would be necessary before a diagnosis of the particular expansive mechanism can be made.

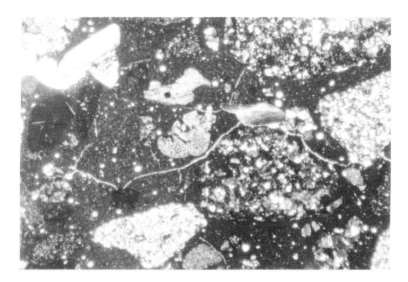

Figure 5.66 Internal cracking due to localised drying shrinkage in a 50-year-old concrete. Cracks do not exceed 20 μm in width. Plane-polarised illumination, width of field 1.4 mm (Courtesy of St John).

reviewed by Chatterji (1982). They may also be artefacts resulting from overheating during thin-section manufacture (see also Section 5.8.5, and Section 3.4.5).

2. *Cracking peripheral to aggregate particles*: This form of cracking is best examined in thin section. Radial cracking is often associated with the peripheral crack round the particle. The possible causes of this form of cracking can be summarised as follows:
   • Small movements of the aggregate during setting in immature concrete
   • Excess water or air voids at aggregate surfaces
   • Dust or other coatings on aggregate surfaces
   • Calcium hydroxide (portlandite) leached from cement/aggregate interfaces by percolating water
   • Significant mismatch of thermal expansions between aggregate and cement matrix
   • Moisture-sensitive shrinking aggregates
   • Aggregate particles with soft or friable surfaces

Shrinking and/or absorptive aggregates can cause localised shrinkage effects in the surrounding cement paste as described in Section 6.2.4. Many commercial aggregates contain isolated particles that fall into this category. If a substantial portion of the aggregate present in a concrete has shrunk, extensive surface cracking may occur as reported by Stutterheim et al. (1967). The external appearance of this map cracking is shown in Lea (1970) and the internal textures by Roper et al. (1964).

5.8.4 Petrographic examination and the interpretation of crack systems

Cracks and crack systems are common and readily observed features of concrete structures. If interpreted correctly, it will provide information as to their initiation, development and possible causes. A detailed petrographic examination of crack patterns and crack morphology in three dimensions carried out on-site, on crack surfaces, on cut samples and ground slices and in petrographic thin section will often give definitive information as to initiation, development and causes of cracks and crack systems.

At the microscopic level, there are a number of attributes of cracks, channels and fissures that can provide information on the time when they formed and also give clues as to their cause. Hardened concrete will crack as a brittle material (like broken pottery), but in the plastic state, cracks only form as channels, fissures and voids before setting. This provides a simple method of determining whether crack formation took place before or after setting of the concrete.

Under the microscope, plastic/early shrinkage cracking which has occurred during the setting of the concrete is typical of malleable materials. Cracks will go round rather than across aggregate particles. They will be typified by smeared or rounded margins so that they no longer fit together exactly. There may be partial bridging of the gaps in the way in which clay will 'tear' when pulled apart. In hardened concrete, cracking is clearly identified by well-defined matching margins. In higher-strength concretes, cracking may run through aggregates, but in lower-strength concrete usually skirts aggregate margins. Voids formed by pockets of air are easily identified by their shape and generally have smooth margins, but a margin can be diffuse if its orientation is such that it forms a thin wedge of hardened paste during sectioning. These thin wedges of hardened paste can be mistaken for alkali–silica gel by an inexperienced observer. Channels are irregular tunnels with diffuse margins that carry bleeding water upwards in a meandering fashion. In thin section, they are rarely continuous because they meander in and out of the plane of the section (see Figures 5.67 and 5.68) like material which requires care to distinguish it from the cement paste. Fissures are cracks in the plastic state that have not been filled with water. Often channels are difficult to observe because they are filled with diffuse laitance highly irregular margins, which may vary from sharp to diffuse and do not run through aggregates. Unlike channels, they should be continuous in thin section.

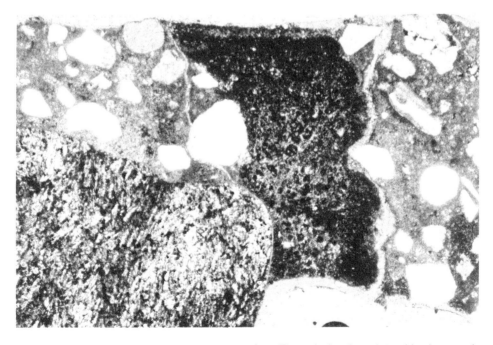

Figure 5.67 Microcracks running up to a concrete surface. The right-hand crack is a bleed water channel running from a water-formed void at the bottom of the crack. The crack on the left is due to drying shrinkage. Note the width is constant along its length contrasting with the variability in width of the bleed channel. Plane polar illumination, width of field 2.3 mm. (Courtesy of M. Grove, private collection.)

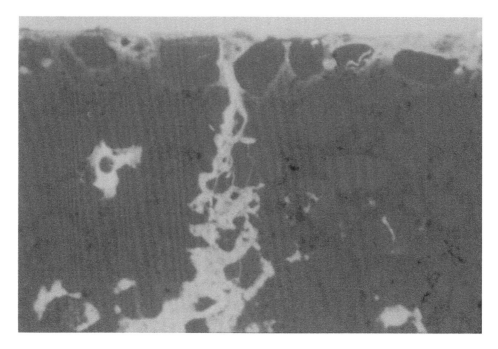

Figure 5.68 A thin section in UV illumination illustrates how bleed channels tend to be narrower at the concrete surface, but widen near to the source of free water. Width of field 2.3 mm. (Courtesy of M. Grove, private collection.)

Fissures are primarily due to excessive drying of concrete while still plastic and, because of this, may be concentrated in surface areas. The subject of plastic shrinkage, settlement and the associated cracking is discussed by Rivard et al. (2002) and is well illustrated in the Concrete Society Technical Report No. 22 (1992). A staining method for their detection of such cracks in ground surfaces has been given by Ash (1972).

DEFINITIONS OF CRACK TYPES AND WIDTHS IN CONCRETE AND MORTAR FOR USE IN THIN-SECTION EXAMINATION:

Fine microcracks	<1 μm		
Microcracks	1–10 μm	Microcracks[a]	<100 μm
Fine cracks	10–100 μm	Fine cracks[a]	10–100 μm
Cracks	100 μm–1 mm	Wide cracks[a]	>100 μm
Large cracks	>1 mm		

Source: After RSK Environment Ltd., Hemel Hempstead, U.K., 2011.

[a] Jensen et al. (1985).

The width of cracks should be recorded during petrographic examinations and during site inspections where a crack gauge, such as illustrated in Figure 5.65, and a vernier calliper gauge are useful. They should also be measured during the petrographic examination of thin sections. A suitable fivefold series of practical definitions for crack widths is in current use by RSK Ltd., U.K. while Jensen et al. (1985) categorised cracking into three groups and used colour photographs to illustrate the categories.

Table 5.9 Relationship between crack width and depth of penetration of carbonation in field structures

Crack width, μm	Carbon dioxide penetration, mm	Surface cracks filled	Cracks in body filled
≤20	<10	Nearly always	Variable
20–60	<100	Often plugged	Rare
≥100	>100	Rarely plugged	Very rare

The minimum crack width in hardened cement paste readily observable by optical microscopy is about 2 μm. Cracks narrower than this tend to merge into the background of the hardened cement paste unless delineated, although they can sometimes be distinguished by darkening of the crack margins. Narrow cracks can be delineated by dyes incorporated into the epoxy resin used to impregnate the concrete. However, in a concrete that is saturated internally, it is doubtful whether cracks narrower than 5 μm persist for any length of time as they will be filled by deposition of calcium hydroxide and merge into the fabric of the paste. In practice, most non-structural cracks observed in the body of a concrete will be found to be not less than about 10 μm in width with the majority falling into the 20–60 μm range.

Where expansive patterns of cracks have formed those 100 μm in width or wider tend to form a grid of large dimension over the pattern of the finer cracks. Isolated cracks, especially those that run deep into the concrete, will have minimum widths of about 100 μm and maximum widths up to a few millimetres. At the surface of concrete, drying and/or carbonation shrinkage cracking is usually limited to a range of 5–15 μm in width. Typical ranges for carbon dioxide penetration, plugging and filling of cracks with calcium hydroxide or carbonate are given in Table 5.9. This requires some qualification as it only represents data for structures such as bridges and buildings exposed to a temperate climate. The deposition of calcium hydroxide in water-retaining structures will be much greater and is more difficult to categorise.

Dimensions of channels and fissures are more difficult to specify. Since channels have no clearly defined margins and often appear discontinuous in thin section, dimensions are variable. In practice, their widths are observed to be up to a few hundred microns in size. The minimum width of a fissure is not likely to be less than 100 μm because of the way it is formed in the plastic state. This also implies that in practice, widths observed will be highly variable (see Figures 5.64 and 5.68).

5.8.5 Cracking as artefacts of sampling and preparation

Cracks may also be introduced as artefacts (see also Section 3.4.5) because hardened cement paste is brittle and easily cracks during sampling and specimen preparation unless reasonable care is exercised. In those concretes which have undergone normal exposure to climatic conditions, most cracks will contain some calcium hydroxide or carbonation. There are usually observable signs to show whether a crack is new or old. If these signs cannot be observed, it is then possible that the cracks may have been caused by either sampling or sectioning. A typical example of sampling cracking occurs in the sides of cores due to chattering of the diamond coring bit. These appear as short clean cracks at the edges of the thin section.

Drying cores at too high a temperature or overheating the thin section during preparation and some preparation techniques used for SEM examination can cause shrinkage microcracking of the cement paste. As a general rule, concrete used for sectioning should never be heated above 60°C. When this occurs, it is usually obvious as the paste has a distinctive 'shrunk' look about it. Another form of cracking that appears to be related to sectioning is often observed at the outer surface of the core. The corroded and carbonated layer undergoes a patchy delamination which in most cases is clearly an artefact.

A question that may be raised with the petrographer concerns whether the cracking observed in a thin section has formed when strains in the concrete were released by core sampling. In making a decision about this, the same principle is applied as discussed earlier. However, experience suggests that cracking due to release of strain is uncommon, although the width of existing cracks may be increased. This is to be expected as releasing strain into existing cracks is more likely to occur than the initiation of new cracks. A more probable situation is that the sampling will miss the cracking present in the concrete altogether. Even apparently badly cracked test samples will often intersect very few cracks when thin sectioned in the laboratory. The probability of intersecting a crack by the plane of the thin section is somewhat less than would be intuitively expected, and some skill is required in selecting the best plane. This is an even more acute problem when sampling a large structure.

PETROGRAPHIC EXAMINATION CHECKLIST FOR CRACKS AND CRACK SYSTEMS

- Examine the intersection of both sides of the crack with the concrete surface.
- Investigate the changes in width of the crack along its length and with depth.
- Compare both crack walls; do the sides fit together? Is there displacement?
- Examine the detailed morphology of the crack surfaces.
- Investigate and identify the infilling if present and its relation to the walls.
- Examine the nature of the crack adjacent to aggregate particles.
- Examine the way in which cracks intersect and bifurcate.
- Identify the orientations of crack patterns in relation to gross stress directions.

GENERAL PETROGRAPHIC EXAMINATION CHECKLIST OF KEY FEATURES IN CONCRETE THAT SHOULD BE RECORDED BY THE PETROGRAPHER

- Aggregates (coarse and fine), types, mineralogy, shapes, sizes proportions and distribution
- Features of aggregate/cement interface zones
- Anhydrous and hydrating cement clinker particles, composition, mineralogy, sizes and distribution
- Portlandite (if present) morphology, sizes, shape, distribution and quantity
- Mineral admixtures (if present), type and distribution
- Air voids, sizes, distribution, shapes, infillings (if present) and proportions
- Cracking, types, sizes, relationships, distribution, displacement and infillings (if present)
- Fabric of the cementitious matrix, texture, colour and mineral phase interrelationships
- Secondary minerals (if present), types and relationships
- Carbonation, extent, variability, texture and relationships
- Any features indicative of degradation, alteration or damage

REFERENCES

ACI Committee 212. Report 212R 1989: *Chemical Admixtures for Concrete*. American Concrete Institute, Detroit, MI. (*ACI Materials Journal* 86, 3, 297–327.)

ACI Committee 224.1R. 1985: *Guide to the Use of Waterproofing, Damp-Proofing*. American Concrete Institute, Detroit, MI. (*ACI Materials Journal* 81, 3, 211–230.)

ACI Committee 226. 1988: *Measurement of Properties of Fibre Reinforced Concrete*. American Concrete Institute, Detroit, MI. (*ACI Materials Journal* 84, 2, 158–166.)

ACI Committee 232. 1995: *Guide for the Use of Silica Fume in Concrete*. American Concrete Institute, Detroit, MI. (*ACI Materials Journal* 91, 4, 410–426.)

ACI Committee 234. Report 234R. 1995: *Guide for the Use of Silica Fume in Concrete*. American Concrete Institute, Detroit, MI. (Abstract in *ACI Materials Journal* 92, 4, 437–440.)

ACI Committee 524. 1993: *Guide to Portland Cement Plastering*. American Concrete Institute, Detroit, MI. (*ACI Materials Journal* 90, 1, 69–93.)

Addis, B.J. and Alexander, M.G. 1994: Cement-saturation and its effects on the compressive strength and stiffness of concrete. *Cement and Concrete Research* 24, 5, 975–986.

Aitcin, P. (ed.) 1983: *Silica Fume 52*. Les Editions de L'universite de Sherbrooke, Sherbrooke, Quebec, Canada.

Andrja, D. 1986: Hydration of alite and C_3A and changes of some structural characteristics of cement paste by the addition of silica fume. In *Proceedings of the Eighth International Congress on Chemistry of Cement*, Rio de Janeiro, Brazil, Vol. 4, pp. 279–285.

Asakura, K, Yoshida, H., Nakato, T. and Namamura, T. 1993: Effect of characters of silica fume on physical properties of cement paste. *JCA Proceedings of Cement and Concrete* 47, 178–183 (in Japanese, English abstract).

Ash, J.E. 1972: Bleeding in concrete – A microscopy study. *ACI Journal* 69, 4, Title No. 69-19, 209-211, American Concrete Institute, Detroit, MI.

ASTM C227-90, 1990: *Potential Test Method for Alkali Reactivity of Cement-Aggregate Combinations (Mortar Bar Method)*. ASTM, Philadelphia, PA.

ASTM C457-90, 1990: *Standard Test Method for Microscopical Determination of Parameters of the Air Void System in Hardened Concrete*. ASTM, Philadelphia, PA.

ASTM C595/595M-09, 2009: *Standard Specification for Blended Hydraulic Cements*. ASTM, Philadelphia, PA.

ASTM C618-08a, 2008: *Standard Specification for Coal Fly Ash and Raw Calcined Natural Pozzolan for Use in Concrete*. ASTM, Philadelphia, PA.

ASTM C856-95, 1995: *Standard Practice for Petrographic Examination of Hardened Concrete*. ASTM, Philadelphia, PA.

ASTM C856-04, 2004: *Standard Practice for Petrographic Examination of Hardened Concrete*. ASTM, Philadelphia, PA.

ASTM C989-09a, 2009: *Standard Specification for Slag Cement for Use in Concretes and Mortars*. ASTM, Philadelphia, PA.

ASTM C1073-97a, 2003: *Standard Test Method for Hydraulic Activity of Ground Slag by Reaction with Alkali*. ASTM, Philadelphia, PA.

Bache, H.H., Idorn, G.M., Nepper-Christensen, P. and Nielsen, J. 1966: Morphology of calcium hydroxide in cement paste. *Proceedings of the Symposium on Structure of Portland Cement Paste and Concrete*. Highway Research Board Special Report No. 90, National Research Council, Washington, DC, 154–174.

Barker, A. 1984: Structural and mechanical characterisation of calcium hydroxide in set cement and the influence of additives. *World Cement* 15, 1, 25–31, Palladian Publications, Farnham, U.K.

Barker, A. 1990: The use of electron-optical techniques to study the hydration characteristics of blended cements. In *Proceedings of the 12th International Conference on Cement Microscopy*, International Cement Microscopy Association, Vancouver, British of Columbia, Canada, pp. 107–120.

Barnes, R.D., Diamond, S. and Dolch, W. 1978: The contact zone between Portland cement paste and glass aggregate surfaces. *Cement and Concrete Research* 8, 2, 233–243.

Baron, J. and Douvre, C. April 1987: Technical and economical aspects of the use of limestone filler additions in cement. *World Cement* 18, 3, 100–114, Palladian Publications, Farnham, U.K.

Battigan, A.F. 1986: The use of the microscope for estimating the basicity of slags in slag-cements. In *Eighth International Congress Chemistry of Cement*, Rio de Janeiro, Brazil, Theme 3/4, pp. 17–21.

Bensted, J. 1993: High alumina cement – Present state of knowledge. *Zement-Kalk-Gips* (9), 560–566.

Bentur, A. and Mindess, S. 2006: *Fibre Reinforced Cementitious Composites,* 2nd ed. Taylor and Francis Group/E & F. N. Spon, London, U.K., pp. 1–601

Berger, R.L. and McGregor, J.D. 1972: Influence of admixtures on the morphology of calcium hydroxide formed during tricalcium silicate hydration. *Cement and Concrete Research* 2, 43–55.

Berger, R.L. and McGregor, J.D. 1973: Effect of temperature and water-solid ratio on growth of $Ca(OH)_2$ crystals formed during the hydration of Ca_3SiO_5 *Journal of the American Ceramic Society* 56, 2, 73–79.

Berry, E.E. and Malhotra, V.M. 1987: Fly ash in concrete. In *Supplementary Cementing Materials for Concrete,* Malhotra, V.M. (ed). Canada Centre for Mineral and Energy Technology (CANMET), Canadian Government Publishing Centre, Ottawa, Ontario, Canada, Chapter 2, pp. 35–163.

Bier, T.A., Kropp, J. and Hilsdof, H.K. 1989: The formation of silica gel during carbonation of cementitious systems containing slag cements. In *Proceedings of Third International Conference on Fly Ash, Silica Fume, Slag and Natural Pozzolans in Concrete,* Norway, ACI SP-114, Vol. 2, pp. 1413–1428.

Bonen, D. and Diamond, S. 1992: Investigation on the coarse fraction of a commercial silica fume. In *Proceedings of the 14th International Conference on Cement Microscopy,* International Cement Microscopy Association, Orange country, CA, pp. 103–113.

Brandt, A.M. 1995: *Cement-Based Composites-Materials, Mechanical Properties and Performance.* E. & F.N. Spon, London, U.K.

Brandt, I. 1993: Connection between chemical admixtures and the structure of $Ca(OH)_2$ in thin sections of hardened concrete. In *Proceedings of the Fourth Euroseminar on Microscopy Applied to Building Materials,* Lindqvist, J.E. and Nitz, B. (eds.), Swedish National Testing and Research Institute, Building Technology SP Report 1993, Borås, Sweden, p. 15.

Brown, J.H. 1991: The effect of exposure and concrete quality: Field survey results from some 400 structures. In *Proceedings of the Fifth International Conference on the Durability of Materials and Components, Brighton, U.K., 1990,* E. & F.N. Spon, London, U.K., pp. 249–259.

Bruere, G.M. 1960: The effect of type of surface-active agent on the spacing factors and surface areas of entrained bubbles in cement pastes. *Australian Journal Applied Chemistry* 11, 289–294.

BS 146, 1996: *Specification for Portland Blastfurnace Cement.* BSI, London, U.K.

BS 1881, 1983: *Testing Concrete Part 106, Methods for Determination of Air Content of Fresh Concrete.* BSI, London, U.K.

BS 1881-124, 1988: *Testing Concrete Part 124. Methods of Analysis of Hardened Concrete.* BSI, London, U.K.

BS 4246, 2002: *Specification for High Slag Blastfurnace Cement.* BSI, London, U.K.

BS 4248, 1974: *Specification for Supersulphated Cement.* BSI, London, U.K.

BS 5337, 1976: *Code of Practice for the Structural Use of Concrete for Retaining Aqueous Liquids.* (Note: replaced by BS 8007: 1987). BSI, London, UK.

BS 6610, 1996: *Specification for Pozzolanic Pulverised-Fuel Ash Cement.* BSI, London, U.K.

BS 6699, 1992: *Specification for Ground Granulated Blastfurnace Slag for Use with Portland Cement.* BSI, London, U.K.

BS 7979, 2001: *Specification for Limestone Fines for Use with Portland Cement.* BSI, London, U.K.

BS 8007, 1987: *Code of Practice for Design of Concrete Structures for Retaining Aqueous Liquids.* BSI, London, U.K.

BS EN 197-1, 2000: *Concrete – Part 1: Specification, Performance, Production and Conformity Criteria for Common Cements*. BSI, London, U.K.

BS EN 206-1, 2000: *Concrete – Part 1: Specification, Performance, Production and Conformity*. BSI, London, U.K.

BS EN 1992-3, 2006: *Code of practice for design of concrete structures for retaining aqueous liquids*. BSI, London, U.K.

BS EN 15167-1 and 2, 2006: *Ground Granulated Blast Furnace Slag for Use in Concrete, Mortar and Grout. Definitions, Specifications and Conformity Criteria*. BSI, London, U.K.

Byfors, K., Klingstedt, G., Lehtonen, V., Pyy, H. and Romben, L. 1989: Durability of concrete made with alkali activated slags. In *Proceedings Third International Conference on Fly Ash, Silica Fume, Slag and Natural Pozzolans in Concrete*, Norway, ACI SP-114, Vol. 2, pp. 1429–1466.

Calleja, J. 1980: Durability. Sub-theme VII-2. In *Seventh International Congress Chemistry of Cement*, Paris, France, 1, Sub-theme VII-2, pp. 2/1–2/48.

Campbell, D.H. 1986: *Microscopical Examination of Interpretation of Portland Cement and Clinker*. Construction Technology Laboratories, Portland Cement Association, Skokie, IL.

Campbell-Allen, D. 1979: *The Reduction of Cracking in Concrete*. University of Sydney and Cement and Concrete Association of Australia, Sydney, New South Wales, Australia, 165pp.

Carles-Gibergues, A., Saucier, F., Grandet, J. and Pigeon, M. 1993: New-to-old concrete bonding: Influence of sulfates type of new concrete on interface structure. *Cement and Concrete Research* 23, 2, 431–441.

Chatterji, S. 1982: Probable mechanisms of crack formation at early ages of concretes: A literature survey. *Cement and Concrete Research* 12, 271–376.

Christensen, P., Gudmundsson, H., Thaulow, N., Damgard-Jensen, A.D. and Chatterji, S. 1979: Structural and ingredient analysis of concrete – Methods, results and experience. *Nordisk Betong* 3, 4–9. (C & CA translation from Swedish, T292, 1984).

Chu, J.K., Kim, J.H., Park, C. and Robertson, R.E. 1994: Changes induced by PVA in cement microstructure surrounding aggregate. *Proceedings of Materials Research Society (MRS)*, 370, 10, 347–355.

Clarke, J.L. 1993: *Structural Lightweight Aggregate Concrete*. Blackie Academic and Professional, London, U.K.

Clear, C.A. 1985: The effects of autogenous healing upon the leakage of water through cracks in concrete. Cement and Concrete Association Technical Report No. 559, Slough, U.K, pp. 1–31.

Concrete Society. 1987: Concrete core testing for strength. Concrete Society Technical Report No. 11.

Concrete Society. 1988: Analysis of hardened concrete. A guide to tests, procedures and interpretation of results. Report of joint working party of the Concrete Society and Society of the Chemical Industry. Concrete Society Technical Report No. 32.

Concrete Society. 1989: Analysis of hardened concrete. Report of a Joint Working Party of the Concrete Society and Society of Chemical Industry. Concrete Society Technical Report No. 32, 117pp.

Concrete Society. 1991: The use of ggbs and pfa in concrete. Concrete Society Technical Report No. 40.

Concrete Society. 1992, 2010: Non-structural cracks in concrete. Report of a Concrete Society Working Party. Concrete Society Technical Report No. 22.

Concrete Society. 1993: Microsilica in concrete. Report of a Concrete Society Working Party. Concrete Society Technical Report No. 41.

CSA. 1983: *Cementitious Hydraulic Slag*. A363-M1983, Canadian Standards Association, Toronto, Ontario, Canada, pp. 1–11.

Cussino, L. and Negro, A. 1980: Hydration of aluminous cement in the presence of silicic and calcareous aggregate. In *Seventh International Congress on the Chemistry of Cements*, III, Paris, France, Editions Septima, Vol. 62, pp. 62–67.

De Ceukelaire, L. and Van Niewenberg, D. 1993: Accelerated carbonation of a blast-furnace cement concrete. *Cement and Concrete Research* 23, 2, 442–452.

Dengler, T.R. and Montgomery, J.R. 1984: The estimation of activity and glass content of blast furnace slag by microscopic observation. *Proceedings of the Sixth International Conference on Cement Microscopy*, Albuquerque, NM, pp. 265–275.

Diamond, S. 1986: The microstructure of cement paste in concrete. In *Eighth International Congress on the Chemistry of Cement*, Rio de Janeiro, Brazil, Vol. 1, pp. 123–147.

Dolch, W.L. 1984: Air entraining admixtures. In *Concrete Admixtures Handbook* Ramachandran, V.S. (ed.). Noyes Publications, Park Ridge, NJ, pp. 269–302.

Douglas, E. and Malhotra, V.M. 1987: A review of the properties and strengths of non-ferrous slags-Portland cements binders. In *Supplementary Cementing Materials for Concrete*, Malhotra, Y.M. (ed.). CANMET, Ottawa, Ontario, Canada.

Drissen, R. 1995: Determination of the glass content in granulated blastfurnace slag. *Zement-Kalk-Gips* 48, 1, 59–62.

Dubrolubov, G. and Romer, B. 1985: Guidelines for determining and testing the frost as well as frost-salt resistance of cement-concrete. *Bulletin of Betonstrassen AG. Research and Consulting in Concrete Road Construction (Concrete Roads Ltd.)* Special Number, June. Wildegg/Switzerland.

Elias, H. 1971: Three-dimensional structure identified from single sections. *Science* 174, 993–1000.

Figg, J.W., Moore, A.E. and Gutteridge, W.A. 1976: On the occurrence of the mineral trona $(Na_2CO_3 \cdot NaHCO_3 \cdot H_2O)$ in concrete deterioration products. *Cement and Concrete Research* 6, 5, 691–696.

Fookes, P.G. and Collis, L. 1977: *Problems in the Middle East*. Viewpoint Publication, Cement and Concrete Association, Slough, U.K., pp. 1–7.

French, W.J. 1991a: Concrete petrography: A review. *Quarterly Journal of Engineering Geology* 24, 1, 17–48.

French, W.J. 1991b: Comments on the determination of the ratio of ggbs to Portland cement in hardened concrete. *Concrete (Journal of the Concrete Society)* 25, 6, 33–36.

Garrett, H.L. and Beauman, D.R. 1985: A method for preparing steel reinforcement mortar or concrete for examination by transmitted light microscopy. *Cement and Concrete Research* 15, 917–920.

George, C.M. 1983: Industrial aluminous cements. In *Structure and Performance of Cements*, Barnes, P. (ed.). Applied Science, London, U.K., pp. 415–470.

Goltermann, P. 1995: Mechanical predictions of concrete deterioration-Part 2: Classification of crack patterns. *ACI Materials Journal* 92, 1, 58–63, American Concrete Institute, Detroit, MI.

Grade, K. 1968: Determination of the fluorescence of granulated slags. In *Proceedings of the Fifth International Symposium on the Chemistry of Cement* IV, Tokyo, Japan, pp. 168–172.

Grandet, J. and Ollivier, J.P. 1980: Etude de la formation du monocarboaluminate de calcium hydrate au contact d 'un granulat calcair dans une pate de ciment Portland. *Cement and Concrete Research* 10, 6, 759–770.

Grounds, T., Midgeley, H.G. and Nowell, D.V. 1988: The carbonation of ettringite by atmospheric carbon dioxide. *Thermochimica Acta* 135, 347–352.

Grove, R.M. 1968: The identification of ordinary Portland cement and sulphate resisting cement in hardened concrete samples. *Silicates Industriels* 10, 317–320.

Groves, G.W. 1981: Microcrystalline calcium hydroxide in Portland cement pastes at low water/cement ratio. *Cement and Concrete Research* 11, 713–718.

Groves, G.W. and Richardson, I.G. 1994: Microcrystalline calcium hydroxide in pozzolanic cement pastes. *Cement and Concrete Research* 24, 6, 1191–1196.

Gustaferro, A.H. and Scott, N.L. 1990: Reading structural concrete cracks. *Concrete Construction* December, 994–1003.

Hansen, T.C. 1989: Marine concrete in hot climates – Designed to fail. *Materials and Structures* 22, 344–346.

Hobbs, D.W. 1988: *Alkali-Silica Reaction in Concrete*. Thomas Telford, London, U.K.

Hollis, N., Walker, D., Lane, D.S. and Stutzman, P.E. 2004: *Petrographic Methods of Examining Hardened Concrete: A Petrographic Manual*. Publication No. FHWA-HRT-04-150, U.S. Department of Transportation, Federal Highway Administration, Charlottesville, VA.

Hooton, R.D. 1987: The reactivity and hydration products of blast-furnace slag. In *Supplementary Cementing Materials for Concrete*, Malhotra, V.M. (ed.). Canadian Centre for Mineral and Energy Technology (CANMET), SP 86-8E, Canadian Government Publishing Centre, Ottawa, Ontario, Canada, Chapter 4, pp. 245–288.

Hooton, R.D. and Emery, J.J. 1983: Glass content determination and strength development predictions for vitrified blast furnace slag. In *Proceedings of the CANMET ACDI First International Conference on the Use of Fly Ash, Silica Fume, Slag and Other Mineral By-Products in Concrete,* Montebello, Quebec, Canada. ACI Special Publication SP-79, American Concrete Institute, Detroit, MI, pp. 943–962.

Hoshino, M. 1988: Difference of the w/c ratio, porosity and microscopical aspect between the upper boundary paste and the lower boundary paste of the aggregate in concrete. *Materials and Structures* 21, 336–340.

Houst, Y.F. and Wittmann, F.H. 1994: Influence of porosity and water content on the diffusivity of CO_2 and O_2 through hydrated cement paste. *Cement and Concrete Research* 24, 6, 1165–1176.

Ichikawa, M. and Komukai, Y. 1993: Effect of burning conditions and minor components on the colour of Portland cement. *Cement and Concrete Research* 23, 933–938.

Idorn, G.M. 2004: Comments regarding the AAR research 2004 and onwards. In *Proceedings of the 12th International Conference on Alkali-Aggregate Reaction in Concrete*, International Academic Publishers, World Publishing Co., Beijing, Japan, pp. 1–7.

Idorn, G.M. and Thaulow, N. 1983: Examination of 136-years-old Portland cement concrete. *Cement and Concrete Research* 13, 739–743.

Ignatiev, N. and Chatterji, S. 1992: On the mutual compatibility of mortar and concrete in composite members. *Cement and Concrete Composites* 14, 179–183.

Insley, H. and van Fréchett, D. 1955: *Microscopy of Ceramics and Cements – Including Glasses, Slags and Foundry Sands.* Academic Press, New York.

Japan Society of Civil Engineers. 1988: *Recommendation for Design and Construction of Concrete Containing Ground Granulated Blast-Furnace Slag as an Admixture.* Translation from the Concrete Library No. 63, Japan Society of Civil Engineers, Tokyo, Japan.

Jensen, A.D., Eriksen, K., Chatterji, S., Thaulow, N. and Brandt, I. 1985: Petrographic analysis of concrete. Danish Building Export Council 12.

Junior, K.M. Munhoz, F.A.C., Junior, J.S. and Placido, W.F. 1990: Characterization and quantitative determination of calcium aluminate clinker phases through reflected light microscopy. *Proceedings of the 12th International Conference on Cement Microscopy,* Vancouver, British of Columbia, Canada, 1–15 April, 1990.

Kaplan, M.F. 1960: Effects of incomplete consolidation on compressive and flexural strength, ultrasonic pulse velocity, and dynamic modulus of elasticity of concrete. *Journal American Concrete Institute* 31, 853–867.

Kaye, G.W.C. and Laby, T.H. 1995: *Tables of Physical and Chemical Constants,* 16th ed. Longman, Essex, England.

Kelly, J.W. September 1981: Cracks in concrete. *Concrete Construction* 26, 9, 725–734.

Kennerley, R.A. 1965: Ettringite formation in dam gallery. *Journal American Concrete Institute* December, 559–576.

Kupper, D. and Schmid-Meil, W. 1988: Process engineering and raw materials factors affecting the lightness and hue of 'white' clinker and cement. *Zement-Kalk-Gips* 2, 30–32.

Kuzel, H.J. April 1990: Reactions of CO_2 with calcium aluminate hydrates in heat treated concrete. In *Proceedings of the 12th Annual International Conference on Cement Microscopy,* Vancouver, British of Columbia, Canada, pp. 219–227.

Larsen, G. 1961: Microscopic point measuring: A quantitative petrographic method of determining the $Ca(OH)_2$ content of the cement paste in concrete. *Magazine of Concrete Research* 13, 38, 71–76.

Lauer, K.R. and Slate, F. 1956: Autogenous healing of cement paste. *Journal American Concrete Institute* June, 1083–1098.

Lawrence, C.D. 1998: Physicochemical and mechanical properties of Portland cements. In *Lea's Chemistry of Cement and Concrete*, 4th ed., Hewlett, P.C. (ed.). Arnold, London, U.K., Chapter 8, pp. 343–420.

Laxmi, S., Ahluwalia, S.C. and Chopra, S.K. 1984: Microstructure of coloured clinkers. In *Proceedings of the Sixth International Conference on Cement Microscopy,* Albuquerque, NM, pp. 61–77.

Lea, F.M. 1970: *The Chemistry of Cement and Concrete,* 3rd ed. Edward Arnold Ltd., London, U.K.

Lewis, R., Sear, L., Wainwright, P. and Ryle, R. 2003: Cementitious additions. In *Advanced Concrete Technology, Constituent Materials*, Newman, J. and Choo, B.S. (eds.). Elsevier, Oxford, U.K.

Lord, G.W. and Willis, T.F. 1951: Calculation of air bubble size distribution from results of a Rosiwal traverse of aerated concrete. *ASTM Bulletin* October, 56–61.

Majumdar, A.J., Singh, B. and Edmonds, R.N. 1990a: Hydration of mixtures of 'Ciment Fondu' aluminous cement and granulated blast furnace slag. *Cement and Concrete Research* 20, 2, 197–208.

Majumdar, A.J., Singh, B. and Edmonds, R.N. 1990b: Blended high-alumina cements. *Conferences on Advances in Cementitious Materials, Ceramic Transactions* 16, 661–678.

Malhotra, V.M. 1976: No-fines concrete – Its properties and applications. *ACI Journal* November, 628–644.

Malhotra, V.M. and Hemmings, R.T. 1995: Blended cements in North America – A review. *Cement and Concrete Composites* 17, 25–35.

Manns, W. and Wesche, K. 1968: Variation in strength of mortars made of different cements due to carbonation. In *Proceedings of the Fifth International Symposium on the Chemistry of Cements* III, Cement Association of Japan, Tokyo, Japan, pp. 385–393.

Mather, P.K. 1952: Applications of light microscopy in concrete research. *Symposium on Light Microscopy*, ASTM STP 143, West Conshohocken, PA, pp. 51–70.

Massazza, F. 1993: Pozzolanic cements. *Cements and Concrete Composites* 15, 185–214.

Massazza, F. 1998: Pozzolana and pozzolanic cements slag. In *Lea's Chemistry of Cement and Concrete*, 4th ed., Hewlett, P.C. (ed.). Arnold, London, U.K., Chapter 10, pp. 471–631.

Massazza, F. and Costa, U. 1986: Bond, paste-aggregate, paste-reinforcement and paste fibres. In *Eighth International Congress Chemistry of Cements*, Rio de Janeiro, Brazil, Vol. 1, pp. 159–180.

Mays, G.C. and Barnes, R.A. 1991: The performance of lightweight aggregate concrete structures in service. *The Structural Engineer* 69, 20, 351–361.

McCrone, W.C. and Delly, J.G. 1973: *The Particle Atlas*, 1 &2. Ann Arbor Science Publishers, Ann Arbor, MI.

Mehlmann, M. 1989: Current state of research on renderings and mortars. *Zement Kalk Gips* 3, translation (1/89), 70–72.

Mehta, P.K. 1987: Natural pozzolans. In *Supplementary Cementing Materials for Concrete*, Malhotra, V.M. (ed.). CANMET (Canada Centre for Mineral and Energy Technology), Canadian Government Publishing Centre, Ottawa, Ontario, Canada, Chapter 1, pp. 1–33.

Mehta, P.K. 1991: Durability of concrete – Fifty years' progress? *Durability of Concrete, Second International Conference*, ACI SP-126, Montreal, Quebec, Canada, pp. 1–31.

Meyer, A. 1968: Investigations on the carbonation of concrete. *Proceedings of the Fifth International Symposium on the Chemistry of Cement*, Tokyo, Japan, Vol. 3, pp. 394–401.

Midgley, H.G. 1979: The determination of calcium hydroxide in set Portland cements. *Cement and Concrete Research* 9, 1, 77–82.

Midgley, H.G. and Midgley, A. 1975: The conversion of high alumina cement. *Magazine of Concrete Research* 27, 91, 59–77.

Midgley, H.G. and Pettifer, K. 1971: The micro-structure of hydrated super sulfated cement. *Cement and Concrete Research* 1, 101–104.

Mielenz, R.C. 1962: Petrography applied to Portland-cement concrete. In *Reviews in Engineering Geology*, Flurh, T. and Legget, R.F. (eds.), Vol. 1. The Geological Society of America, Boulder, CO, pp. 1–38.

Min, D. and Mingshu, T. 1994: Formation and expansion of ettringite crystals. *Cement and Concrete Research* 24, 1, 119–126.

Mindess, S. and Young, F.J. 1981: *Concrete*. Prentice-Hall, Englewood Cliffs, NJ, 83p.

Minato, H. 1968: Mineral composition of blastfurnace slag. Supplementary Paper IV-ll0. *Proceedings of the Fifth International Symposium Chemistry of Cement*, Tokyo, Japan, pp. 263–269.

Moranville-Regourd, M. 1998: Cements made from blastfurnace slag. In *Lea's Chemistry of Cement and Concrete*, 4th ed., Hewlett, P.C. (ed.). Arnold, London, U.K., Chapter 11, pp. 633–669.

Nambiar, E.K.K. and Ramamurthy, K. 2006: Influence of filler type on the properties of foam concrete. *Cement and Concrete Composites* 28, 5, 475–480.

Negro, A. 1985: Modele mathematique de l'expansion de l'ettringite. *Ciments, Betons, Platres, Chaux* 756, 319–327.

Neville, A.M. 1975: *High Alumina Cement Concrete*. The Construction Press, Lancaster, U.K.

Nishikawa, T., Suzuki, K. and Ito, S. 1992: Decomposition of synthesized ettringite by carbonation. *Cement and Concrete Research* 22, 1, 6–14.

Nurse, R.W. and Midgley, H.G. 1951: The mineralogy of blastfurnace slag. *Silicates Industries* 16, 7, 211–217.

Palmer, D. 1992: *The Diagnosis of Alkali-Silica Reaction*. Report of a Working Party, British Cement Association, Wexham Springs, Slough, U.K.

Parker, T.W. 1952: The constitution of aluminous cement. In *Proceedings of the Third International Symposium on the Chemistry of Cements*, BRE/DSIR/C&CA, London, U.K., pp. 485–529.

Parrott, L.J. 1987: *A Review of Carbonation in Reinforced Concrete*. Cement and Concrete Association, Slough, U.K., 42 and appendix.

Parrott, L.J. 1990: Damage caused by carbonation of reinforced concrete. *Materials and Structures* 23, 135, 230–234.

Parrott, L.J. 1992: Water absorption in cover concrete. *Materials and Structures* 25, 149, 284–292.

Parrott, L.J. 1994: Moisture conditioning and transport properties of concrete test specimens. *Materials and Structures* 27, 127, 460–468.

Parrot, L.J., Patel, R.C., Killoh, D.C. and Jennings H.M. 1984: *Journal of the American Ceramic Society* 67, 4, 233–237.

Powers, T.C., Copeland, L.E. and Mann, H.M. 1958: Capillary continuity and discontinuity in cement pastes. *Journal of the Portland Cement Association Research and Development Laboratories* 1, 2, 38–48.

Powers-Couche, L.J. 1992: Microscopical examination of a slag cement concrete. *Proceedings of the 14th International Conference on Cement Microscopy*, Orange Country, CA, pp. 256–258.

Pullar-Strecker, P. 1987: *Corrosion Damaged Concrete: Assessment and Repair*, 97pp. Butterworths, London, U.K.

Rasmachandran, V.S. 1988: Thermal analyses of cement compounds hydrated in the presence of calcium carbonate. *Thermochimica Acta* 127, 385–394.

Ramachandran, V.S., Feldman, R.F. and Beaudoin, J.J. 1981: *Concrete Science. Treatise on Current Research*. Heyden, London, U.K., pp. 380–387.

Randolph, J.L. 1991: Thin section identification of various admixtures and their effects on the cement paste morphology in hardened concrete. *Proceedings of the 13th International Conference on Cement Microscopy*, Tampa, Florida, pp. 281–293.

Ravenscroft, P. 1982: Determining the degree of hydration of hardened concrete. *John Laing Forum*, 4.

Rayment, D.L. and Majumdar, A.J. 1982: The composition of C–S–H phases in Portland cement pastes. *Cement and Concrete Research* 12, 6, 753–764.

Redler, L. 1991: Quantitative x-ray diffraction analysis of high alumina cements. *Cement and Concrete Research* 21, 5, 873–884.

Regan, P.E. and Arasteh, A.R. 1990: Lightweight aggregate foamed concrete. *The Structural Engineer* 68, 9, 167–173.

RILEM Committee 104-DC. 1994: Draft recommendations for damage classification on concrete structures. *Materials and Structures* 27, 170, 362–369.

Rivard, P., Fournier, B. and Ballivy, G. 2002: The damage rating index method for ASR affected concrete – A critical review of petrographic features of deterioration and evaluation criteria. *ASTM, Cement, Concrete and Aggregate* 24, 2, 81, Paper ID CCA11228-242.

Robinson, G., Vaughan, F. and Hordell, G. 1992: Tiling. In *Specification*. MBC Architectural Press and Building Publications, London, U.K., 92pp.

Robson, T.D. 1964: Aluminous cement and refractory castables. In *The Chemistry of Cements*, Taylor, H.F.W. (ed.), Vol. 2. Academic Press, London, U.K., Chapter 12, pp. 3–35.

Roper, H., Cox, J.E. and Erlin, B. 1964: Petrographic studies on concrete containing shrinking aggregates. *Journal of the PCA Research and Development Laboratories* 6, 3, 2–18.

Sarkar, S.L., Chandra, S. and Berntsson, L. 1992: Interdependence of microstructure and strength of structural lightweight aggregate concrete. *Cement and Concrete Composites* 14, 239–248.

Schmidt, M., Harr, K. and Boeing, R. 1993: Blended cement according to ENV 197 and experiences in Germany. *Cement, Concrete and Aggregates. CCAGDP* 15, 2, 156–164.

Schrader, E. and Kaden, R. 1987: Durability of shotcrete. In *Concrete Durability, Katherine and Bryant Mather International Conference*, Scanlon, J.M. (ed.). ACI SP-100 2, American Concrete Institute, Detroit, MI, pp. 1071–1101.

Scrivener, K.L. and Capmas, A. 1998: Calcium aluminate cements. In *Lea's Chemistry of Cement and Concrete*, 4th ed., Hewlett, P.C. (ed.). Arnold, London, U.K., Chapter 13, pp. 709–778.

Scrivener, K.L. and Gartner, E.M. 1988: Microstructural gradients in cement paste around aggregate particles. *Materials Research Society Symposium Proceedings* 114, 11–20.

Sellevold, E.J., Bager, D.H., Klitgaard Jensen, K. and Knudsen, T. 1982: *Silica Fume Cement Pastes: Hydration and Pore Structure*. Report BML 82.610, The Norwegian Institute of Technology, Trondheim, Norway, pp. 19–50.

Shayan, A., Quick, G.W. and Lancucki, C.J. 1993: Morphological, mineralogical and chemical features of steam-cured concrete containing densified silica fume and various alkali levels *Advances in Cement Research* 5, 20, 151–162.

Short, A. and Kinniburgh, W. 1978: *Lightweight Concrete*, 3rd ed. Applied Science Publishers, London, U.K.

Sorrentino, F.P. and Glasser, F.P. 1982: The phase composition of high alumina cement clinkers. In *Proceedings of the International Seminar on Calcium Aluminates*, Turin, Italy, pp. 14–43.

Strunge, H. and Chatterji, S. 1989: Microscopic observations on the mechanisms of concrete neutralization leds. Berntsson et al., Durability of concrete: Asbestos of admixtures and industrial byproducts. *Second International Seminar*, pp. 229–235. Swedish Council for Building Research, Stockholm, Sweden.

Stutterheim, N., Laurie, J.A.P. and Shand, N. 1967: Deterioration of a multiple arch dam as a result of excessive shrinkage of aggregate. In *Proceedings of the Ninth International Conference on Large Dams*. Istanbul, Turkey, pp. 227–224.

St. John, D.A. 1994: *The Dispersion of Silica Fume*. Industrial Research Limited Report No. 244, The New Zealand Institute for Industrial Research and Development, Lower Hutt, New Zealand.

St. John, D.A. 1982: An unusual case of ground water sulfate attack on concrete. *Cement and Concrete Research* 12, 5, 633–639.

St. John, D.A. 1983: The petrographic examination of an unusual texture in some failed concrete piles. In *Proceedings of the Fifth International Conference Cement Microscopy*, Nashville, TN, pp. 204–15.

St. John, D.A. and Abbott, J.H. 1987: The microscopic textures of geothermal cement grouts exposed to steam temperatures of 150°C and 260°C. In *Proceedings of the Ninth International Conference on Cement Microscopy*, Reno, NV, pp. 280–300.

St. John, D.A. and Smith, L.M. 1976: Expansion of concrete containing New Zealand argillite aggregate. *The Effect of Alkalis on the Properties of Concrete*. Proceedings of Symposium, London, U.K., C&CA, Slough, U.K., pp. 319–354.

Summer, M.S., Hepher, N.M. and Moir, G.K. 1989: The influence of a narrow particle size distribution on cement paste and concrete water demand. *Ciments, Betons, Platres, Chaux* 778, 164–168.

Sveinsdottir, E.L. and Gudmundsson, G. 1993: Condition of hardened Icelandic concrete. A microscopic investigation. In *Proceedings of the Fourth Euroseminar on Microscopy Applied to Building Materials*, Lindqvist, J.E. and Nitz, B. (eds.), Building Technology SP Report 1993: 15, Swedish National Testing and Research Institute, Borås, Sweden.

Swamy, R.N. (ed.) 1992: *The Alkali-Silica Reaction in Concrete*. Blackie and Sons/Van Nostrand Reinhold, Glasgow, U.K./New York.

Swamy, R.N. 1993: Fly ash and slag: standards and specifications – Help or hindrance. *Materials and Structures* 26, 164, 600–613.

Swamy, R.N. and Jiang, E.D. 1992: Pore structure and carbonation of lightweight concrete after 10 years exposure. In *Structural Lightweight Aggregate Concrete Performance*. ACI SP-/36, American Concrete Institute, Detroit, MI, pp. 377–395.

Talling, B. and Brandstetr, J. 1989: Present state and future of alkali-activated slag concretes. In *Proceedings of the Third International Conference on Fly Ash, Silica Fume, Slag and Natural Pozzolans in Concrete*, ACI SP-114, 2, Trondheim, Norway, pp. 1519–1545.

Taylor, H.F.W. September 1984: Studies on the chemistry and microstructure of cement pastes. *British Ceramic Proceedings* 35, 65–82.

Taylor, H.F.W. 1990: *Cement Chemistry*. Academic Press, London, U.K.

Taylor, H.F.W., Mohan, K. and Mori, G.K. 1985: Analytical study of pure and extended Portland cement paste: Fly ash and slag-cement pastes. *Journal American Ceramic Society* 68, 12, 685–690.

Venkateswaran, D., Narayan, S., Chakraborty, I.N. and Biswas, S.K. 1991: Microstructure of high purity alumina cement clinkers and its effect on cement properties. In *Proceedings of the 13th International Conference on Cement Microscopy*, Tampa, FL, pp. 71–85.

Verbeck, G.J. 1958: Carbonation of hydrated Portland cement. In *Cement and Concrete*. ASTM STP 205, ASTM International, Philadelphia, PA, pp. 17–36.

Walker, H.N. 1979: *Evaluation and Adaptation of the Method of Microscope Examination of Hardened Concrete*. Virginia Highway and Transportation Research Council VHTRC 79-R42, Charlottesville, VA (Reprinted in 1992, as FHWA-HRT-04-150).

Walker, M. (ed.) 2002: *Guide to the Construction of Reinforced Concrete in the Arabian Peninsula*. CIRIA C577, Concrete Society Special Publication CS136, London, U.K.

Winchell, A.N. and Winchell, H. 1964: *The Microscopical Characters of Artificial Inorganic Solid Substances: Optical Properties of Artificial Minerals*. Academic Press, New York.

Yilmaz, V.T. and Glasser, F.P. 1991: Refraction of alkali-resistant glass fibres with cement. Part 1. Review, assessment, and microscopy. Part 2. Durability in cement matrices conditioned with silica fume. *Glass Technology* 32, 4, 91–98, 138–147.

Chapter 6

Examination of deteriorated and damaged concrete

6.1 INTRODUCTION

6.1.1 Background to concrete durability

The durability of concrete is a broad, extensively researched subject and there is much literature available. Codes of practice to assist engineers in designing durable concrete have been published by the British Standards Institution (previously BS 8110 1985 BS 5328 1990 and now BS EN 206-1 2000 and BS 8500-1, 2 2006), the American Concrete Institute (annual publication), the Comité Euro-International du Beton (1992) and the Commission Internationale des Grands Barrages (1989). Good general reviews of concrete durability include Kropp and Hilsdorf (1995), Glanville and Neville (1997), Hobbs (1998), Walker (2000), Neville (2001), Rendell et al. (2002), Newman and Choo (2003), Sims 2003, Page and Page (2007) and Neville (2006).

A discussion of the approach used in the USSR to deal with aggressive fluids has been given by Ivanov (1981) and in South Africa by Alexander et al. (1994). Concrete Society Technical Report 22 provides a valuable review of the types and causes of non-structural cracks in concrete (1992 and 2010). Monographs by Eglinton (1987) and Mays (1992), intended for the engineer, are useful texts that deal with many aspects of the durability of concrete. Robery (2009a) has provided guidance for ensuring concrete durability in extreme exposure conditions.

The durability of concrete involves both intrinsic and extrinsic reactions. Intrinsic attack occurs from deleterious substances or properties incorporated into the concrete mix, although an intrinsic component such as alkali may be later augmented from external sources. Extrinsic attack takes place from agents outside the concrete and, with the exception of physical agencies, usually involves deleterious solutions. This latter type of attack produces a zone of deterioration that moves in from the outer surface of the concrete. From a textural aspect, this leads to two important differences for the petrographer. Intrinsic attack will affect a considerable portion of the concrete texture, while extrinsic attack is restricted to zones adjacent to outer surfaces. This helps to explain why laboratory testing has on the whole been more successful in predicting durability for an intrinsic reaction such as alkali–aggregate reaction (AAR) than for an extrinsic problem such as sulphate attack (Cohen and Mather 1991; Mehta 1993), although also some forms of deterioration leave clearer diagnostic signatures in the concrete for later identification by the petrographer.

The concept of an external layer of deterioration is important when investigating the effects of external agents on concrete. In many cases, this outer layer consists of three zones: a an outer corrosion zone where the attack has gone to completion; an intermediate zone where attack on calcium hydroxide, the hydrates and clinker occurs and salts are deposited; and an inner zone where leaching of calcium hydroxide and possibly other materials from the hardened paste takes place.

When examined petrographically, the outer zone usually consists of amorphous forms of hydrated silica, alumina and iron in which are embedded aggregates. In most cases, only hydrated silica and aggregate remain as both the iron and alumina compounds are more readily removed. It is rarely intact as it often suffers damage during sampling, and in some cases, such as attack by citric acid, the zone is of minimal width because silica is dissolved by this acid. The outer zone merges into the intermediate zone. It is the most complex of the zones as recrystallisation of salts, removal of calcium hydroxide and decomposition of the cement hydrates all occur together in this zone. It is highly variable in depth depending on the reaction, often contains carbonated material and may be stained by amorphous, hydrated iron compounds. Observable calcium hydroxide is rare or absent, indicating that the pH will be below that required for the stability of the cement hydrates. Where salt recrystallisation has occurred, the intermediate zone may be extensively exfoliated. The intermediate zone grades into apparently sound hardened cement paste, but careful examination will often detect an inner zone where the paste appears to be darkened due to a diminution of calcium hydroxide. This is the chemical gradient which invariably occurs where hardened paste is undergoing dissolution; it has been described in detail by Pavlik (1994).

In many cases, lack of durability in concrete is a chemical problem that arises during manufacture or later, although to the engineer it is the changes in volume and loss of mechanical integrity that are of importance. The hardened cement paste and many aggregates are stable only under certain conditions, of which pH is by far the most important. Mehta et al. (1992), in reviewing the 'performance and durability of concrete systems', summarised some relevant data from Reardon (1990) and Gabrisova et al. (1991) on the stability of the cement hydrates as follows:

> In the pH range 12.5 to 12.0, calcium hydroxide, calcium monosulphate hydrate and aluminate hydrates, dissolve in sulphate solution from which ettringite precipitates out. Next, gypsum precipitates out in the pH range 11.6 to 10.6; below pH 10.6, ettringite is no longer stable and will start decomposing. Note that the lowering of the pH below 12.5 will also cause the C-S-H phase to be subjected to cycles of dissolution and reprecipitation (Ca/Si ratio is 2.12 at pH 12.5, and 0.5 at pH 8.8) which continue until it is no longer stable at or below pH 8.8. All the foregoing chemical changes in a hardened Portland cement product must be taken into consideration when attempting to understand their physical manifestations, such as expansion, cracking, and loss of strength or mass.

Low permeability in the surface zone is of crucial importance to reducing the rate of attack by corrosive agents. Products formed by the attack may be beneficial or deleterious depending on circumstances. For instance, carbonation may make a concrete more impermeable while reducing the pH, but if the concrete still remains sufficiently permeable after carbonation, oxygen and moisture can penetrate and corrode the reinforcing steel. On the other hand, the formation of a product such as ettringite can be expansive and disrupt the concrete. When considering whether an increase or decrease in permeability is likely to occur, comparison of molar volumes of reactants and reaction products provides useful information.

Intrinsic reactions sometimes occur between aggregates and the pore solution of a concrete because of internal sulphate or salt attack. Unsoundness due to hard burnt lime (CaO) and/or magnesia (MgO, periclase) is now rare and, as cement manufacturers now control production to avoid these problems, they will not be discussed in detail in this chapter. A concrete or mortar made with a cement containing excess hard burnt lime or magnesia can expand for months or years after setting, leading to the development of cracks. The expansion is due to the slow hydration of the hard burnt lime or magnesia/periclase leading to volumetric expansion. Deng Min et al. (1995) have proposed a mechanism for the expansion that occurs when hard burnt lime hydrates in hardened cement paste, and McKenzie

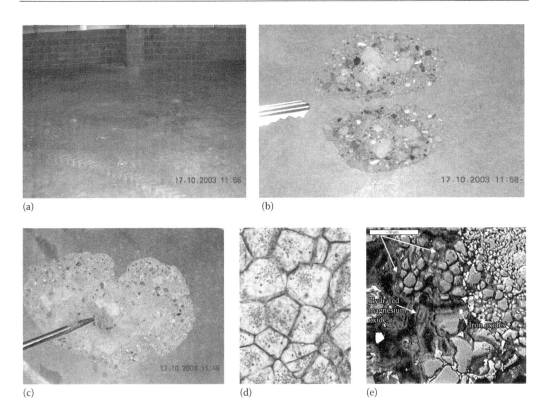

Figure 6.1 Concrete damaged by the gradual hydration of periclase (MgO): (a) surface pop-outs on a concrete floor surface; (b) close-up of pop-out pit and its 'lid'; (c) close-up of pop-out pit, highlighting the expanding particle at its base, and in a separate case; (d) periclase in a steel slag aggregate, in thin sections; (e) periclase altering to brucite (Mg(OH)$_2$), in SEM.

(1994) has described the microscopic methods for studying lime burning and quality. The textures of hard burnt lime illustrated will be applicable to lime in cement.

Similar problems can be caused by impurities of lime or periclase in aggregates, especially some waste/by-product materials. The authors have investigated several cases of concrete damage caused by periclase, variously as contamination of otherwise inert aggregate and as a constituent of steel slag material (Figure 6.1).

6.1.2 Durability investigation and classification

While the concept of extrinsic and intrinsic reaction (see Section 6.1.1) is useful to the petrographer, some form of classification of concrete deterioration, for example, that proposed by Popovics (1987), is of more practical use for engineering purposes. Popovics' classification is a useful summary of the types of attack likely to be encountered and is reproduced in Table 6.1.

The petrographic investigation of durability requires an assessment of the quality of the in situ concrete. If the problem involves the materials, it is necessary to identify cement type, mineral additions and mineralogy of the aggregates; to estimate the cement content, water/cement ratio and air content; and to make a qualitative assessment of aggregate quality and concrete compaction. The data from such an assessment will show that, in a surprisingly large number of cases, the concrete was not mixed and placed according to specification and/or accepted practice, although the deviations may not be large.

Table 6.1 Classification of concrete deterioration

Class I	Leaching by soft water
Class II	Non-acidic reactions
Class IIA	Base exchange, that is, magnesium salts
Class IIB	Saponification reactions from fats and oils
Class IIC	Reactions with sugars
Class III	Reactions involving excessive expansion
Class IIIA	Sulphate attack
Class IIIB	AAR
Class IV	Reactions with acidic water
Class V	Physical processes other than mechanical
Class VA	Salt solutions causing efflorescence and/or spalling or cracking
Class VB	Freezing and thawing
Class VI	Mechanical deterioration
Class VIA	Abrasion and wear
Class VIB	Excessive shrinkage, uneven thermal expansion, overload, repeated loading, etc.

Source: Popovics, S., A classification of the deterioration of concrete based on mechanism, *Concrete Durability – Katherine and Bryant Mather International Conference*, ACI SP-100, American Concrete Institute, Detroit, MI, 1987, Vol. 1, pp. 131–42.

Where severe or unusual service conditions apply, non-compliance with accepted practice can lead to deterioration of the concrete.

The second part of the petrographic investigation involves identification of the conditions that are causing the materials to deteriorate. In most cases, identification of the cause of the aggression and the extent of deterioration is possible *provided that the investigative methods used are adequate*. However, an estimate of the remaining useful life of the materials in a structure requires rates of deterioration to be determined. In many cases, it is difficult, if not impossible, to estimate the rate of attack, and any prediction of the remaining useful life of the concrete will need to be based on previous experience.

Given the potential complexity of investigating problems of durability, it is pertinent to define the scope of concrete petrography. Where the in-service conditions are unusual, petrographic examination combined with chemical investigation of field concretes can provide information that may lead to the better design of concretes. In cases where the mechanism of deterioration is still not fully understood, such as with expansive processes in large structures such as dams, engineering and petrographic field data are vital and often might take precedence over laboratory testing. The applicability of these field data is largely dependent on being able to characterise the condition of the concrete, and an important part of this characterisation will be provided by examination of microtextures in thin section.

Often, the most obvious manifestations of concrete distress and/or deterioration will be cracks, some of which will have structural causes, but others of which will have non-structural origins that relate variously to the nature, quality and behaviour of the concrete material. An excellent explanatory and diagnostic guide to non-structural cracking is provided in Concrete Society Technical Report 22 (1992, 2010), including a schematic diagram illustrating typical crack patterns and locations (see Figure 6.2) and a chart demonstrating that the time to the appearance ('age') of cracks can be a helpful diagnostic element (see Table 6.2). Practising petrographers will find that exceptions to these 'rules of thumb' will sometimes be encountered, but the guidance is nonetheless sound and helpful. Fookes (1977) prepared a similar earlier guide to concrete cracking in hot arid climatic conditions, and Fookes and Walker (2011) provided a useful simplified chart (Figure 6.3).

Sibbick (2011) has correctly commented that, in practice, the petrographer will also and typically be confronted with combinations of mechanisms, variously occurring concurrently

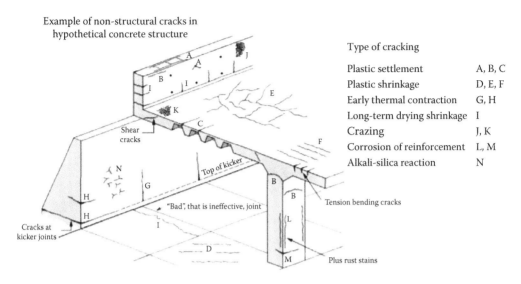

Example of non-structural cracks in
hypothetical concrete structure

Type of cracking

Plastic settlement	A, B, C
Plastic shrinkage	D, E, F
Early thermal contraction	G, H
Long-term drying shrinkage	I
Crazing	J, K
Corrosion of reinforcement	L, M
Alkali-silica reaction	N

Figure 6.2 Diagram illustrating non-structural crack types and typical locations. (From Concrete Society, Non-structural cracks in concrete, Report of a Concrete Society Working Party, Concrete Society Technical Report No 22, 4th & current edition, The Concrete Society, Camberley, U.K., 2010.)

Table 6.2 Time to appearance of various types and causes of non-structural cracks (simplified and developed from Concrete Society Technical Report No 22 [2010])

Type	Crack pattern (see Figure 6.2)	Time to appearance (UK conditions)	Primary cause (simplified)	Secondary cause (simplified)
Plastic settlement	A, B, C	10 min to 3 h	Excess bleeding	Rapid early drying
Plastic shrinkage	D, E	30 min to 6 h	Rapid early drying	Low bleeding rate
	F		Rapid early drying and near-surface steel	
Early thermal contraction	G	1 day to 2 to 3 weeks	Excess heat generation	Rapid cooling
Early shrinkage by self-desiccation		Hours to days	Very low water/cement ratios	
Long-term drying shrinkage	I	Weeks or months, up to several years	Inefficient stress relief	High shrinkage concrete
Surface crazing		Days to months	Impermeable formwork	Poor quality concrete
Reinforcement corrosion	L M	More than 2 years	Inadequate cover, carbonation, salts Calcium chloride	Poor quality concrete
Delayed ettringite formation (DEF)	Can be similar to ASR	Months to years	Excess heat generation	Other factors and subsequent wetting
Alkali-carbonate reaction (ACR)	Can be similar to ASR	Months to years	Reactive carbonate aggregate	Mechanism not fully understood
Alkali-silica reaction (ASR)	N	More than 5 years	Reactive aggregates and high alkalis	Other factors and subsequent wetting

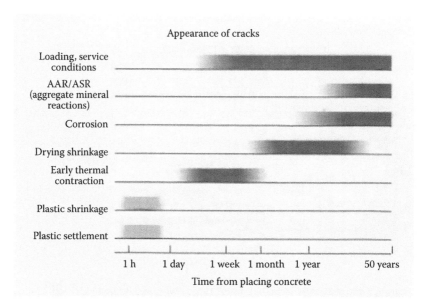

Figure 6.3 Chart showing the relationship between time and the appearance of cracking from various causes. (From Fookes, P.G. and Walker, M.J., *Geol. Today*, 27(4), 141, 2011.)

or sometimes as synergistic interactions between mechanisms. Common examples would include freeze–thaw cycling and salt scaling, freeze–thaw cycling and surface abrasion, AAR and calcium leaching processes, the thaumasite form of sulphate attack (TSA) and aggressive carbonation, AAR followed by seawater and/or sulphate attack (Sibbick 2009), alkali–silica reaction (ASR) and delayed ettringite formation (DEF) (Martin 2010; Martin et al. 2012a,b) and AAR followed by de-icing salt scaling and reinforcement corrosion. These various mechanisms and others will be separately discussed in the succeeding subsections of this chapter, but the petrographer must remain alert to combinations and will sometimes need to ascertain which mechanism has been the main causal and/or initiating factor. In the case of the sometimes almost 'symbiotic' relationship between ASR and DEF, for example, Thomas et al. (2007, 2008a) have devised a testing procedure to help determine the primary mechanism.

6.1.3 Quantification of deterioration and damage

Petrography is sometimes employed to diagnose a particular issue, such as the cause of surface degradation or the cause of cracking and/or manifestations of expansion, but on other occasions, the examination will form part of a routine, pre-purchase or pre-change-of-use condition survey. Whichever applies, some degree of quantification of the findings will be helpful to the structure managing engineer, variously for comparative purposes, and/or to gain insight into the degree to which a concrete is affected by the mechanism(s) concerned and/or to act as a benchmark for future surveys, but no standard procedures have been developed. Sims et al. (1992) suggested a method for quantifying evidence of ASR in concrete, but the quantitative system that has gained the most international recognition was developed by Grattan-Bellew (1995) and this 'damage rating index' (DRI) is described in the immediately following text, with kind assistance from Paddy Grattan-Bellew.

The ASTM C856-11 (2011) standard method for the petrographic evaluation of concrete provides a sound basis for determining the condition and mineralogical composition of concrete cores taken from a structure. However, this method does not provide a procedure for quantifying the amount of deterioration that has occurred in the concrete. The DRI method was developed in the 1990s to address this shortcoming in ASTM C856 (Grattan-Bellew 1995). The DRI method was developed initially and primarily to evaluate concrete affected by ASR, but can also be used to evaluate concrete affected by other mechanisms of deterioration, such as alkali–carbonate reaction (ACR), DEF, damage caused by cycles of freezing and thawing and sulphate attack (Grattan-Bellew and Mitchell 2006). The DRI method consists essentially of counting the numbers of features indicative of concrete deterioration on the polished surface of a concrete core.

6.1.3.1 Methodology

A series of cores should be taken from the structure under investigation so that an estimate of the mean damage to the structure can be obtained. The minimum core diameter should be 10 cm (100 mm). The surface area of the sawn surface of the core should be 200 cm², so that a reasonably representative area of the concrete from each core is examined. The core should be sawn in two axially and the sawn surface ground using a series of grits down to about 40 μm, or its diamond equivalent, and/or polished. In many concretes, polishing has been found to be unnecessary. The finished surface of the core is then coated with uranyl acetate, let stand for two to three minutes, and rinsed off with water (Natesaiyer and Hover 1988). Uranyl acetate is adsorbed preferentially by any alkali–silica gel (and some other compounds; see also Section 6.8), which then fluoresces when viewed in ultraviolet (UV) light. This greatly assists in observing narrow cracks filled with gel that might otherwise be missed.

The polished surface of the core is mounted under a stereo-binocular microscope on a mechanised stage, if available, and the entire surface is scanned, one field of view at a time at a magnification of ~16×. The numbers of features indicative of damage to the concrete are counted in each field of view and then modified by the factors shown in Table 6.3. The purpose of applying factors to the raw numbers is to attempt to relate, on the basis of experience, the measured features to the actual damaging effect that they have on the concrete. For example, reaction rims are indicative that ASR has occurred, but usually do not actually contribute to the damage to the concrete; for this reason, a relatively low factor of 0.5 is applied. In the case of coarse aggregate particles, cracking almost always occurs even in freshly cast concrete, and for this reason, the number of cracks is discounted by a factor of 0.25. By contrast, open cracks, peripheral spaces at aggregate/cement interfaces (indicative of debonding) and gel deposits associated with cracking in the cement paste matrix are all considered especially important and allocated higher weighting factors of 3.0 or 4.0.

Upon completion of the factoring of the numbers of damage features, the results are normalised for an area of 100 cm². The DRI is the normalised sum of the factored numbers of damage features. If a mechanical stage is not available, a one centimetre grid, that is

Table 6.3 DRI damage features and their weighting factors

Damage feature measured	Factor
Cracks in coarse aggregate	0.25
Cracks with gel in coarse aggregate	2.0
Open cracks in coarse aggregate	3.0
Debonding around coarse aggregate	3.0
Reaction rims around aggregates	0.5
Cracks in the cement paste	2.0
Cracks with gel in the cement paste	4.0
Gel in air voids	0.5

approximately the area observed in one field of view at a magnification of 16×, can be drawn on the surface and the numbers of damage features recorded in each square. Following the practice adopted by Grattan-Bellew, the results of the DRI measurements are shown on a bar chart. Depending on the type of aggregate in the core, additional damage features can be added. For example, corrosion occurs in some particles of sandstone from Norway. It was assumed that the corrosion of the particles would contribute to the deterioration of the concrete.

An actual example of the application of DRI to cores from a Canadian dam structure has been provided by Paddy Grattan-Bellew and is illustrated in Figures 6.4 and 6.5. Some images of the key damage features observed under the microscope are shown in Figure 6.4, and the resultant DRI bar charts, for cores from various different parts of the dam, are shown in Figure 6.5.

In this case, as in most, there is a wide variation in the DRIs of cores taken from different parts of the structure. Assuming that ASR is the main mechanism responsible for the observed deterioration, the variation in the DRIs can be related to a number of possible causes, including variations in the cement content of the concrete in different components of

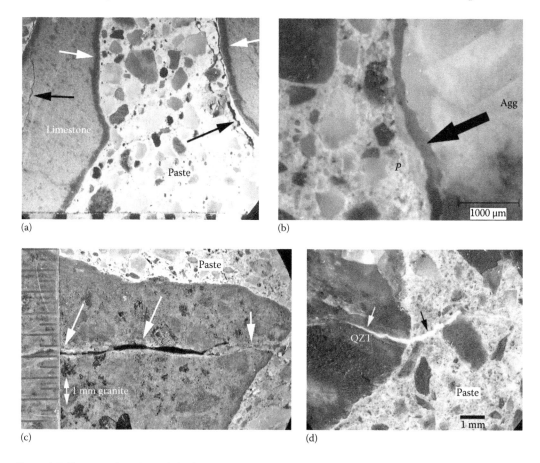

Figure 6.4 Photographs of a polished, uranyl acetate-treated surface of a core viewed under a stereobinocular microscope showing (a) reaction rims around coarse aggregate particles indicated by white arrows, while the black arrows point to cracks in the aggregate and in the cement paste, (b) an open crack at the aggregate/paste interface caused by debonding of the aggregate, (c) an open crack in a coarse aggregate particle indicated by white arrows and (d) a crack filled with alkali–silica gel in a quartzite coarse aggregate particle and in the cement paste (in UV light).

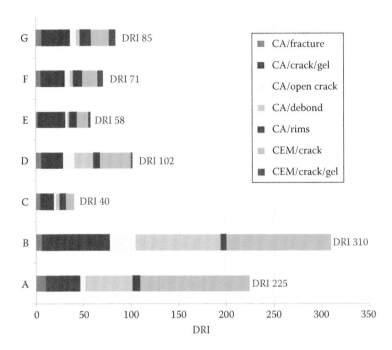

Figure 6.5 DRI charts for the cores taken from different parts of a dam, showing the determined variation in damage ratings.

the structure, variations in the alkali content of the cement and variations in the moisture content of the concrete.

There is gel in cracks in the coarse aggregate particles in all the cores, but in only a few of the cracks in the cement paste. It should be noted that in cores A and B with the highest DRIs, 225 and 310, respectively, gel was not observed in the many cracks in the cement paste. In cold climates, such extensive cracking in the cement paste can usually be attributed to cycles of freezing and thawing of concrete above the waterline in the dam. Further evidence that the cracking in the cement paste is mainly due to cycles of freezing and thawing is the presence of multiple cracks subparallel to the surface of the core (see Figure 6.6).

6.1.3.2 Interpretation and reproducibility

Shrimer (2006) found a reasonable correlation between a visual damage rating of structures in the field and the DRI values obtained separately for cores taken from them (see Figure 6.7). However, it is not known how much expansion has occurred in the structures.

The AAR expansion occurring in structures from which cores have been taken for DRI measurements is generally not known. Therefore, instead, DRIs were determined for concrete prisms stored at 38°C and ~100% relative humidity (RH), to assess AAR expansion (see also Section 6.8), and the findings were compared. Similar assessments were made using DRI and expansion data from a number of authors (Rivard and Ballivy 2005; Dunbar et al. 1996 and an unpublished report by Grattan-Bellew 2011). The resultant correlations are shown in Figure 6.8. The following expansions were calculated from the trend lines fitted to graphs of expansion versus DRI for a DRI of 100: Rivard and Ballivy (2005), 0.10%; Dunbar et al. (1996), 0.09%; and Grattan-Bellew, 0.18%. The results from the first two authors were essentially the same, while those from Grattan-Bellew were significantly higher. The reason

Figure 6.6 Cracks subparallel to the surface of the core caused by frost action. The core diameter is ~10 cm (100 mm).

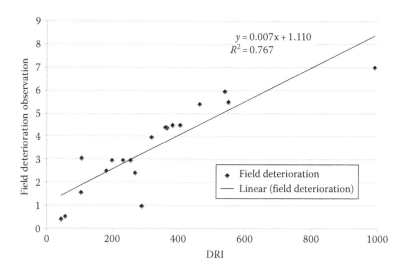

Figure 6.7 Chart showing the correlation between the measured DRI values and the visually observed deterioration ratings of assorted concrete structures. (Drawn using data from Shrimer, F.H., Development of the damage rating index method as a tool in the assessment of alkali-aggregate reaction in concrete: A critical review, Fournier, B. (ed), *Marc-André Bérubé Symposium on Alkali-Aggregate Reactivity in Concrete*, Montreal, Canada, May 2006, pp. 391–401.)

for the differences between the calculated expansions has not been established, but may be related to the types of aggregates tested.

The single operator reproducibility for the DRI method is very good. A coefficient of variation of 6% was obtained for 6 repeat measurements on a polished core. However, when multiple operators (again 6) made measurements on the same core using the same equipment and after discussion on what should be counted, the coefficient of variation rose to 22%

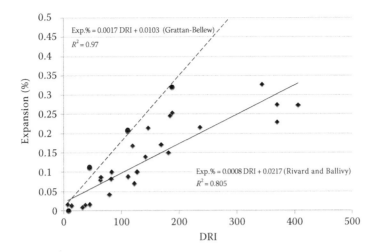

Figure 6.8 Correlations between the expansions of concrete prisms and their measured DRIs.

(Grattan-Bellew and Mitchell 2006). When DRI measurements were made by several operators in different laboratories on the same polished cores, the coefficient of variation became unacceptably high. This result was not really unexpected, given the subjective nature of counting the various damage features. However, for this reason, the DRI method has not so far been adopted as a standard test. There are several factors that affect the measured DRIs in fairly unpredictable ways, including the light sources (natural light alone or in combination with UV illumination), the quality of the polished specimen surface and the magnification of the microscope (Grattan-Bellew and Mitchell 2006).

To date, attempts to reduce the inter-laboratory coefficient of variation of the DRI method to an acceptable level have not been successful, but it is certainly desirable that a suitably reproducible standardised procedure is devised. Recent practical research at Laval University in Québec, Canada, has been endeavouring to find ways of reducing the multi-operator variation with the DRI method: Bérubé et al. (2012) found very good relationships between DRIs and expansions caused by ASR and freeze–thaw, but more difficulty with DEF, while Villeneuve et al. (2012) have demonstrated some improvements using a modified method.

6.1.3.3 Alternative methods

A method for quantitative assessment of the damage to concrete caused by ASR using image analysis was developed by Rivard et al. (2000) and involves a two-step process. In step 1, the core is sawn to provide a flat surface that is then treated with uranyl acetate, and measurements of the amount of gel-filled cracks are made in UV light using an image analyser. In step 2, the surface is then cleaned ultrasonically and impregnated with a fluorescing epoxy resin. When the epoxy has hardened, the surface is ground using a series of diamond wheels followed by a second treatment with uranyl acetate. The polished surface is then placed in UV light and image analysis used to quantify the crack density in the concrete. Good correlation was found between the crack densities measured in cores from a dam and the expansions determined for those cores. The main drawback to this method is the complexity of the two-stage sample preparation and separate scanning, which is time consuming and, in cases where a large number of cores must be evaluated, relatively expensive. Probably for this reason, although the method removes the subjective nature of the DRI method, it has not so far been widely applied, even by its author.

6.2 PLASTIC AND DRYING SHRINKAGE

6.2.1 Overview

The loss of moisture from a concrete or mortar either during setting or after hardening leads to a volume shrinkage which, if of sufficient magnitude, may lead to the development of cracks. The basic mechanism first involves the loss of free water which causes little or no shrinkage, but as drying continues, absorbed water is removed leading to progressive shrinkage. Although the shrinkage is directly proportional to the loss of water for simple laboratory cement paste specimens, cement type, aggregate type, grain sizes, mix design and curing conditions all influence the amount and pattern of drying shrinkage. In some cases, cement properties can have particular influence on the propensity for cracking caused by shrinkage and thermal contraction (see Section 6.10.1); a controversy over 'crack-prone' cement has existed for many years in the United States (Burrows 2007; Eastwood 2007). Petrographic examination of almost all concrete and cementitious materials will encounter shrinkage cracks or microcracks, variously forming in the plastic or hardening phases (see Chapter 5).

6.2.2 Macroscopic effects of plastic and drying shrinkage

Loss of water due to bleeding and excessive drying during setting may lead to plastic shrinkage cracks, which appear within a few hours of the placement of the concrete. Dependent on exposure conditions, drying shrinkage develops more slowly over a period from weeks to months once the concrete has set (Figure 6.9). Commonly, drying shrinkage occurs within a few days of casting as a network of fine cracks or crazing restricted to the cement-rich surface layers (Figure 6.10). Reviews of and detailed guidance on the effects of shrinkage and the development of cracking in concrete are given by Fookes (1977), Concrete Society (1992, 2010), Highways Agency (1987: thermal), Neville and Brooks (1987, 2010), Alexander (2002a, b), Day and Clarke (2003: plastic and thermal), Brooks (2003: elasticity, shrinkage, creep and thermal), CIRIA (2007: thermal) and Concrete Society (2008a).

Figure 6.9 Example of drying shrinkage crack at the centre of a wall panel.

Figure 6.10 Crazing on the end of a cast in situ wall panel.

6.2.3 Petrographic examination of shrinkage cracking

Visual examination on-site is often sufficient to make a satisfactory diagnosis of the type of cracking and its cause (see guidance in CSTR22: Concrete Society 1992, 2010). Petrographic examination of thin sections cut perpendicular to the surface and including examples of the cracks will allow confirmation of the field diagnosis. Fluorescent microscopy (see also Chapters 2 and 4) is particularly useful for identifying bleeding channels, highly microporous localised areas of bleed water entrapment within concrete and laitance development (see the diagram in Figure 6.11 and some examples in Figure 6.12).

Surface crazing does not normally extend much more than a millimetre into the surface and is confined to the cement-rich surface layer, which is usually carbonated. Plastic shrinkage cracks form thin wedges which narrow inwards from the surface and may extend from a few millimetres to several centimetres into the concrete. Although they form before the final setting of the concrete, the cracking may develop calcium hydroxide crystals on margins. Samples taken from hardened concrete are typically carbonated both on the outer surface and along the margins of the cracks. The depth and distribution of the carbonation will depend on the age of the concrete and its permeability.

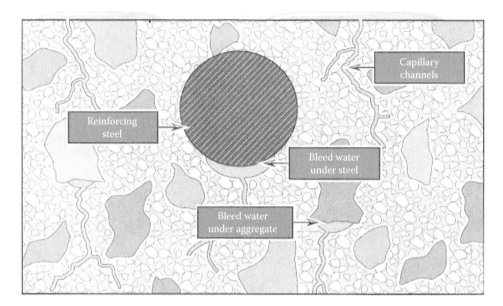

Figure 6.11 Schematic diagram of bleeding channels and bleed water entrapment in concrete. (After Holland, T.C., Silica fume user's manual, Report No FHWA-IF-05-016, Silica Fume Association, Lovettsville, VA, for the Federal Highway Administration [FHWA], U.S. Department of Transportation, Washington, DC, p. 193.)

Cracks that initiate and continue to develop during the plastic phase and/or during the early part of the hardening phase will typically (but not invariably) follow a path between the coarse and fine aggregate particles, whereas cracks that initiate or propagate after hardening are more likely to follow a path that transects some of the aggregate particles. Cracks that are induced in a dense, strong concrete well after the main hardening and strength gain process has been completed might well follow a path that cuts across the well-bonded aggregate/paste texture, frequently transecting aggregate particles. These comparatively simple concepts are illustrated in Figure 6.13, but many factors will influence the path of a crack through concrete and its interpretation will not always be clear.

The correct diagnosis of the longer-term drying shrinkage cracks will be based on the direct observation made of the structure and on evaluation of the mix design and concrete quality. Petrographic examination of specimens containing the cracking will only confirm that such cracks formed after hardening of the concrete. Calcium hydroxide crystals will not normally be present in cracks unless a crack forms a pathway for moisture which may also allow carbonation of crack margins.

6.2.4 Shrinkage of aggregates in concrete

Some aggregate types, especially those containing clay and clay-like minerals, are capable of substantial moisture movements themselves, thus contributing to the initial drying shrinkage of concrete and sometimes causing excessive shrinkage cracking. A sandstone aggregate from South Africa was reported to exhibit excessive shrinkage in concrete (Stutterheim 1954) and found by Roper et al. (1964) to contain the swelling clay, montmorillonite (smectite). High concrete shrinkage associated with glacial gravel aggregates in Scotland was found to be associated with greywacke, mudstone and sometimes altered basalts and

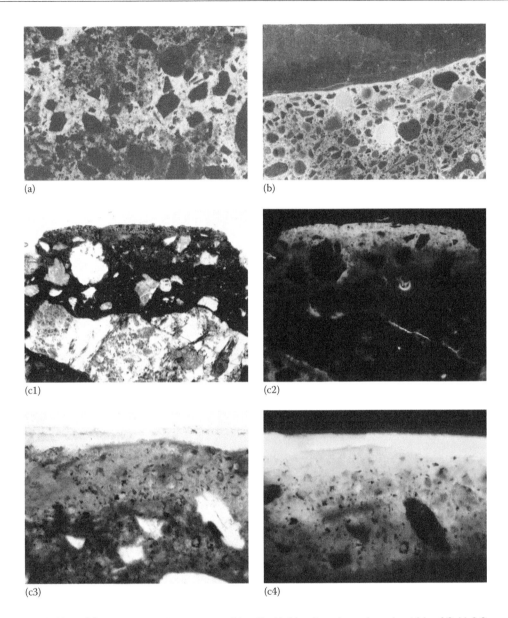

Figure 6.12 Use of fluorescence microscopy to identify (a) bleeding channels, uvl, width of field 0.8 mm; (b) bleed water entrapment, uvl, width of field 4 mm; and (c) development of highly porous laitance, in this case, on a finished concrete patio surface, leading to severe surface scaling in service (images c1 and c2, width of field 2.2 mm, ppl and uvl, respectively; images c3 and c4, width of field 0.6 mm, ppl and uvl, respectively). (Photomicrographs courtesy of RSK Environment Ltd [a and b] and Dr Ted Sibbick of W R Grace and Co [c1 to c4].)

dolerites (Edwards 1970). Cole and Beresford (1980) have described weathered basalt aggregates from Australia that sometimes exhibit considerable moisture instability associated with smectites and chlorite–smectites.

Although potentially shrinkable rock constituents can be identified by routine aggregate petrography, it is common in the United Kingdom for a specialised shrinkage test to be carried out. This test, which is a concrete drying shrinkage procedure using standardised

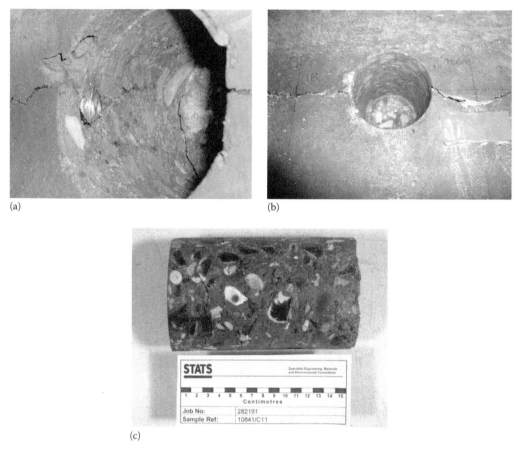

(a)

(b)

(c)

Figure 6.13 Examples of crack paths in concrete: (a) cracks initiated during the plastic and/or early hardening phase, (b) cracks that initiate or propagate after hardening, (c) cracks induced in strong mature hardened concrete, in this case caused by ASR. (Photographs courtesy of Paul Bennett-Hughes, RSK Environment Ltd., Hemel Hempstead, U.K.)

mix proportions, was developed by the Building Research Station (BRS), following the performance problems encountered in Scotland, and published in their Digest 35 (Building Research Establishment 1963, 1968, 1971). This guidance was superseded by Digest 357 (Building Research Establishment 1991) and a modified version of the test was also published as BS 812-120 (1989), which unusually included an appendix on the classification of the aggregate shrinkage findings: category 'A' for values up to 0.075%, considered generally suitable for all concrete uses, and category 'B' for values >0.075%, for which usage is limited. A European standard test method is also now available as BS EN 1367-4 (1998).

The pattern of surface cracking of unrestrained concrete caused by aggregate shrinkage is similar to that of conventional drying shrinkage: map cracking. However, the presence of reinforcement will be likely to modify the crack pattern, and secondary damage, for example, by frost action in Scotland, can greatly worsen and obscure the original shrinkage cracking (Figure 6.14).

Concrete shrinkage induced or exacerbated by aggregate shrinkage can be diagnosed by petrographic examination of the interior, when aggregate shrinkage creates a fairly distinctive pattern of microcracking (Figure 6.15). Building Research Establishment (BRE) Digest

Figure 6.14 Cracking and deterioration of concrete caused by shrinkable aggregates. Further deterioration would occur as the result of freeze–thaw action facilitated by the shrinkage cracking. In this view, the shrinkage cracks are highlighted by white deposits of calcium carbonate, which have resulted from leaching of the cement paste by water percolating through the crack system. (After Sims, I. and Brown, B.V., Concrete aggregates, Hewlett, P.C. (ed), *Lea's the Chemistry of Cement and Concrete*, 4th edn., Edward Arnold, London, U.K., Chapter 16, 1998, pp. 903–1011; Photograph courtesy of Dennis Palmer, retired, British Cement Association.)

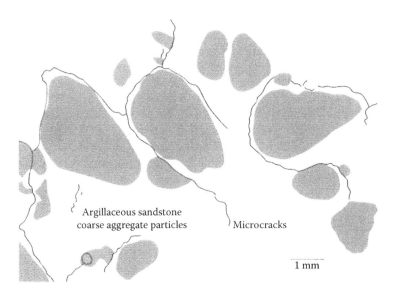

Argillaceous sandstone
coarse aggregate particles

Microcracks

1 mm

Figure 6.15 Pattern of microcracking characteristic of aggregate shrinkage in concrete: a drawing taken from a ground section of concrete containing shrinkable coarse aggregate, 14x magnification. (From Roper, H. et al., *J. PCA Res. Dev. Lab.*, 6(3), 2, 1964.)

357 (1991) suggests that if examination identifies shrinkable rock types as aggregate particles and 'clear evidence' of peripheral cracking (along aggregate particle boundaries) or internal cracking along bedding planes in sedimentary rocks, or both, then *it is probable that significant shrinkage has occurred.* Care should then be taken to discount other possible causes, for example, DEF, which can also induce peripheral cracking, although caused

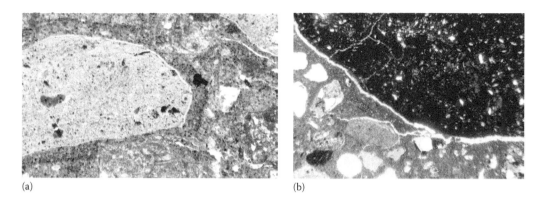

(a) (b)

Figure 6.16 Examples of aggregate shrinkage in concrete in thin sections (peripheral cracks infilled with yellow-dyed impregnating resin): (a) concrete from Puerto Rico and (b) concrete from Northern Ireland.

by cement paste expansion rather than aggregate shrinkage, but when secondary ettringite might also be present and the aggregate types might not be shrinkable in character (see Section 6.5.9).

Several examples of cracking induced or exacerbated by aggregate shrinkage have been investigated by the authors in recent years. In some bridge piers in Puerto Rico, cracking that had been alleged to be caused by the contractor adding excess water to ready-mixed concrete was found actually to be caused by the shrinkage of aggregate including proportions of geologically altered, clay-rich rock (Figure 6.16). In another case, involving a concrete floor dispute in Northern Ireland, cracking and joint opening in the floor was much greater than predictable from the mix design and floor layout. Petrography revealed an internal crack pattern that was characteristic of aggregate shrinkage (also see Figure 6.16), and moisture movement tests on the hardened concrete confirmed an unusual capacity for reversible wetting expansion and drying shrinkage. The gravel aggregates were found to be obtained from deposits that were geologically related to the shrinkable gravels in Scotland that had given rise to the BRS research in the 1960s and 1970s (Edwards 1970).

6.3 CORROSION OF STEEL REINFORCEMENT

Corrosion of steel reinforcement is one of the most common cause of the deterioration of concrete structures, which in certain circumstances may lead to costly repairs or premature failure of the structure. The expansive forces resulting from rusting of reinforcement cause the surrounding concrete to crack and spall, the bond between the steel and the concrete is lost and the steel reinforcement itself is weakened. The first visual indications of rusting of the reinforcement are fine cracks followed by unsightly brown rust stains reflecting the pattern of the underlying reinforcement on the concrete surfaces. However, visible rust staining is not always present at the surface, and this can make initial diagnosis difficult in such cases. As the damage progresses, the staining becomes more severe, the cracks widen and propagate and segments of concrete overlying the rusting steel are spalled off. The macroscopic appearance of concrete damaged by corrosion of steel reinforcement is illustrated in Figure 6.17.

Figure 6.17 Spalling of concrete cover caused by corrosion of reinforcement: (a) fence post with no visible rust stains, which is not uncommon; (b) spalling of soffit, probably promoted by chlorides; and (c) coastal jetty in the Caribbean.

Shrier et al. (1994) have defined corrosion as the *'undesirable deterioration of a metal or alloy, that is, by an interaction with the environment which adversely affects those properties of the structure which are to be preserved'*. In the case of reinforced concrete structures, these are the integrity of the reinforcement and, associated with that, the integrity of the surrounding concrete matrix. Reinforcement corrosion and its prevention have been the subject of many dedicated texts, including Glass (2003) and Broomfield and Atkins (2007).

The macroscopic features of reinforcement corrosion are dependent on the nature of the corrosion process that is taking place and, in particular, the availability of oxygen at the corrosion site, to fully oxidise and produce crystalline corrosion products.

Figure 6.18 Example of surface damage caused by oxidation of 'black rust' corrosion in an industrial water-retaining facility as the result of drying during an outage period.

Where oxygen (and moisture) is readily available, steel will corrode to produce orange/red corrosion products (rust), which is accompanied by a large increase in volume of up to ten times that of the original metal (Leeming 1990). This expansion, initially impeded by the surrounding cement paste, results in the development of internal stresses, which disrupt the matrix of the concrete causing it to crack and spall. Examples of this common form of reinforcement corrosion are manifold and include those shown in Figure 6.17.

Where oxygen is not readily available, for example, in fully water-saturated concrete, the corrosion products that are produced in such environments remain liquid (there being no oxygen to generate expansive crystalline compounds) and, therefore, cracking and spalling will not occur. The corrosion products formed in such environment are frequently termed 'black rust' and, while they will not disrupt the surrounding concrete cover, their production can be accompanied by significant loss of reinforcement cross section. In severe cases, this can result in the reduction or loss of structural integrity with no outwardly visible signs of the corrosion process occurring. When cases of saturated concrete structures containing such 'black rust' are subsequently dried, such as in the case of periodically emptied water-retaining facilities, oxidation and resultant expansive stress can rapidly lead to surface damage, often localised at construction joints where water ingress was maximised (see Figure 6.18).

6.3.1 Role, depth and quality of concrete cover

In the majority of reinforced concrete structures, under most conditions of atmospheric exposure, burial in the ground or immersion in water (fresh or saline), provided there is a good depth of cover by good-quality concrete, the extent of corrosion of the reinforcing steel is negligible. This is commonly attributed to the 'passivation' of a surface film, usually referred to as the 'passive film', on the steel in the alkaline environment of the enveloping concrete (see Section 6.3.2 for further discussion on the composition and structure of the passive film on steel).

The purpose of the cover concrete is therefore twofold: (1) to provide an environment suitable for the preservation of the passive film and (2) to inhibit, for a period of time, the penetration of aggressive agents from the external environment (in which the reinforced concrete structure is to operate) and thus prevent them from reaching the surface of the steel; that is, the cover provides barrier protection.

It is generally accepted that passivity is not a static state but is a dynamic balance between breakdown and repair of the passive film. Numerous studies have observed that a high critical pH, in excess of approximately 11.5 (experimental figures differ slightly), is required to ensure passivity is maintained (Gouda and Monrad 1973, Pourbaix 1974, Alvarez and Galvele 1984, Babushkin et al. 1985). Experience has shown that the pH of the aqueous solutions within the pores of good-quality concrete lies within the range 12.5–13.6, which is directly attributable to the hydration of the clinker minerals and small quantities of alkaline salts in the cement. In hardened concrete, this high pH of the pore solutions is buffered by the presence of solid calcium hydroxide, which will readily dissolve to release further hydroxyl ions if the pH falls. Consequently, good-quality concrete provides an ideal environment for passivity to be established and maintained.

Although now considered somewhat simplistic, the most widely used model for the corrosion of steel in concrete structures is that published by Tuutti (1982), who proposed that corrosion of reinforcement can be thought of as a two-stage process (see Figure 6.19).

The first stage, called the initiation period (t_i), recognises the changes to the characteristics of the cover concrete so that corrosion is possible. These changes are caused by loss of alkalinity, or chloride penetration, or a combination of both. The initiation period is followed by the propagation period (t_p) in which the reinforcement corrodes to a defined level deemed to be the end of the design life.

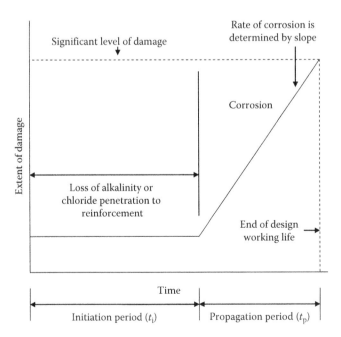

Figure 6.19 Two-phase model of reinforcement corrosion. (From Tuutti, K., *Corrosion of Steel in Concrete*, No 4–82, Swedish Cement and Concrete Institute [CBI], Stockholm, Sweden.)

The duration of the initiation period, that is, the ability of the concrete to resist the changes to its characteristics (ingress of aggressive agents), such that corrosion can be initiated, is directly related to two factors. The concrete 'quality', that is, its chemical composition and physical structure, particularly porosity and permeability, and the depth of concrete (or thickness of concrete cover) through which the aggressive agents have to penetrate to reach the surface of the steel.

It has long been recognised that producing concrete with good chemical resistance to acidic agents, which would reduce the pH, and low permeability that would prevent the penetration of aggressive salts, is largely dependent on a combination of compositional factors: (1) the cementitious materials used (i.e. cement type and any mineral additions), (2) the cement/binder content and (3) the achievement of a low water/cement (or water/binder) ratio (which might be facilitated by the inclusion of admixtures). However, even with an appropriate mix design, the importance of good workmanship and correct curing of the concrete should not be overlooked, as poor practice in these respects can negate all the beneficial effects of designing a corrosion-resistant concrete mix.

As stated earlier, the concrete cover also provides simple barrier protection to the reinforcement, although concrete is generally considered to be an imperfect barrier as some penetration of aggressive agents will always occur, at a greater or lesser rate, depending on the quality of the concrete achieved. Consequently, within limits, the thicker the barrier (i.e. the greater the depth of cover), the longer the time period required before any changes to the protective environment, provided by the concrete, reach the reinforcement and corrosion can be initiated. However, thicker concrete cover is also more liable to plastic and/or drying shrinkage cracking, which could reduce the effectiveness of, or even breach, its protective or barrier function. Therefore, guidance is available on the minimum cover depths to be specified for particular circumstances (see Table 6.4) (e.g. see Hobbs 1998 and Marsh 2003).

6.3.2 Steel reinforcement corrosion mechanisms and factors

Corrosion of reinforcement in concrete is an electrochemical process with anodic and cathodic areas developing on the steel as a result of inhomogeneity in the steel itself, for example, at locations where the oxide layer is damaged by abrasion with aggregate particles, or as the result of local variations within the adjacent concrete such as permeability or sulphate/chloride concentrations. The electrochemical cells with the concrete pore fluid as electrolyte may develop on scales ranging from microns to metres in size, depending on the particular circumstances.

The electrochemical corrosion process is dependent on three conditions that must be simultaneously satisfied: (1) the uninhibited dissolution (oxidation) of iron at the 'anode' (accompanied by the production of electrons), (2) the presence of an electrolyte of high ionic conductivity (the concrete pore fluid) and (3) the occurrence of a suitable reaction at the 'cathode' (i.e. a chemical reaction that will accept electrons produced at the anode in order for the overall reaction to proceed without the accumulation of electrical charge, which is most commonly the reduction of oxygen, but in certain circumstances can be the liberation of hydrogen).

Where two dissimilar metals are connected, for example, between reinforcement and the tie wires, at welds or where galvanised structural members are in contact with the reinforcement, a corrosion cell can develop again with the concrete pore fluid acting as the electrolyte. Corrosion cells on macro- and microscales developing on reinforcing steel need first to break down the protective oxide layer (passive film).

Table 6.4 Some guidance on minimum depths of concrete cover

Use	Exposure class[a]	Exposure description[a]	Nominal cover (mm)	Minimum designated concrete[b]
Reinforced and prestressed concrete inside enclosed buildings	XC1	Dry or permanently wet	15[c]	RC20/25
External reinforced and prestressed vertical elements of buildings	XC3/XC4 and XF1	Moderate humidity or cyclic wet and dry	20[c]	RC40/50
		Moderate water saturation, no deicing	25[c]	RC32/40
			30[c]	RC28/35
Horizontal elements with high saturation without deicing agent and subject to freezing while wet	XC4 and XF3	High water saturation, no deicing	20[c]	RC40/50XF
			30[c]	PAV2
			35[c]	PAV1
Reinforced or prestressed buried foundation in AC-1, where hydraulic gradient is not >5	XC2/ XA:AC-1[d]	Wet, rarely dry/low sulphate ground	50 (against blinding)	RC25/30
			75 (directly against soil)	
Reinforced or prestressed buried foundation in AC-2 or more aggressive	XA:AC-2 to AC-5m[d]	Chemical attack: moderate to high sulphates	50 (against blinding)	C25/30, or use a 'designed' concrete
			75 (directly against soil)	

Source: Based upon BS 8500-1, *Concrete, Complementary British Standard to BS EN 206-1, Part 1: Method of Specifying and Guidance for the Specifier*, British Standards Institution (BSI), London, U.K., 2006.

Note: Mainly from Table A.3 therein, referring to a working life of at least 50 years and 'designated' concretes; more detailed guidance for 'designed' concretes is given in Tables A.4 and A.5 therein.

[a] Exposure classes defined in Table A.1 of BS 8500-1:2006.
[b] See Table A.14 of BS 8500-1:2006.
[c] Cover depth might need to be increased for prestressing steel: refer to appropriate design code.
[d] See BRE Special Digest 1 (2005).

Bloom and Goldberg (1965) considered this to be a two-layer film with a conduction film of magnetite (Fe_3O_4), in direct contact with the steel, covered, in turn, by an insulating layer of ferric oxide (Fe_2O_3). There remains general agreement that the passive film comprises a thin film of hydrated iron oxides; however, there is not full agreement on its exact composition and structure. A variety of mono-, bi- and tri-layer structures, as well as colloidal compositions, have been proposed over many years of research (Cahan and Chen 1982). Irrespective of the exact composition and structure of the passive film, its breakdown can result from a loss of alkalinity or the presence of chloride ions in excess of a critical concentration. Damaged areas of the film may repair themselves, first by the formation of ferrous hydroxide ($Fe(OH)_2$), which will oxidise through intermediate forms, to ferric oxide which is impermeable to ions and so passivates the steel.

The corrosion cell is illustrated diagrammatically in Figure 6.20. The movement of electrons from the anode to the cathode requires the sites to be electrically connected forming a complete electronic (through the metal) and ionic (through the concrete) circuit, generally termed a 'corrosion cell'. These arise due to heterogeneities in either the steel (e.g. inclusions of different chemical compositions) or the concrete (in particular variations in the rate or

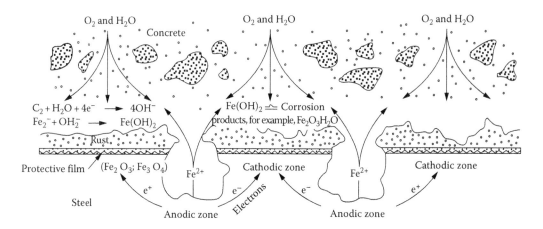

Figure 6.20 Diagrammatic representation of pitting corrosion of steel in concrete. Oxidation of iron to ferric ions is the principal initial anodic process. At the cathode, relatively soluble ferrous hydroxide develops which is then modified and oxidised into the familiar brown rust minerals as hydrated ferric oxides such as goethite and lepidocrocite. (Diagram courtesy of Dr Darrell Leek, Mott Macdonald, Croydon, U.K.)

extent of the penetration of reactants for the anodic and cathodic reactions through more or less permeable sections of cover concrete).

If the concrete has carbonated or chloride ions are present, the steel reinforcement will corrode if the oxide layer is damaged and moisture and oxygen are available. The anodic reaction may be expressed by

$$3Fe + 8OH^- \rightarrow Fe_3O_4 + 4H_2O + 8e^-$$

and the complementary reaction at the cathode by

$$8e^- + 4H_2O + 2O_2 \rightarrow 8OH^-$$

A typical corrosion cell set out by Wilkins and Lawrence (1980) comprised the elements shown in Table 6.5.

In many cases, the oxidation and reduction reactions occur sporadically over the same surface resulting in general corrosion. In other cases, they can be separated on scales ranging from microns (microcell) to many metres (macrocell) on the same reinforcement bar, or on two separate bars (Figure 6.21) provided they are electrically connected, and can result in intense localised, or pitting, corrosion, depending on circumstances.

Table 6.5 Elements of corrosion cell

Steel (anode— oxidation of iron)	Less permeable concrete High Cl^- Low O_2	More permeable concrete High Cl^- High O_2	Steel (cathode— reduction of oxygen)

Source: Described by Wilkins, N.J.M. and Lawrence, P.F., Fundamental mechanisms of corrosion of steel reinforcement in concrete immersed in seawater, Technical Report No. 6, *Concrete in the Oceans*, Cement and Concrete Association, Crowthorne, U.K., 1980.

Figure 6.21 Reinforcement corrosion in the Arabian Gulf, showing the concrete broken out to reveal parallel steel bars, the cathodic bars being clean and non-corroded, while the anodic bars are corroding. (Photograph courtesy of Professor Peter Fookes, private consultant, Winchester, U.K.)

6.3.3 Carbonation, loss of alkalinity and corrosion of reinforcement

A common cause of corrosion of reinforcement is where carbon dioxide from the atmosphere slowly penetrates the concrete and converts the calcium hydroxide and cementing phases to calcium carbonate ('carbonation' is discussed in Section 5.5). On carbonation reaching the steel reinforcement, the calcium hydroxide layer adjacent to the steel (as shown in Figure 5.41) is replaced by calcium carbonate with a pH of 9.5, which is insufficient to maintain the protective layer on the steel. Provided moisture and oxygen are available, corrosion is able to occur (Figure 6.22). As explained in Sections 5.5 and 6.3.1, the extent to which carbonation proceeds inwards from the exposed surface depends on the permeability and depth of the concrete cover over the reinforcement, cracking and the RH of the exposure conditions. Sibbick (2011) recalls that carbonation can sometimes induce additional cracking of the cover (Parrott 1987), which can further compromise its effectiveness.

Concrete is a naturally alkaline material, with a pH in excess of 12.5, which protects and preserves the passive film on steel reinforcement (see Section 6.3.1). Alkalinity can, however, be progressively lost in concrete structures that are in contact with acidic environments, either by nature of the location in which they are constructed (e.g. naturally occurring acidic environments and contaminated industrial areas, either atmospheric or within the ground and groundwater) or in view of the materials that they may contain (e.g. acidic liquids, including sewage).

The most common, naturally occurring, compounds that can result in acid attack of concrete include organic acids in groundwater in, or close to, areas where peat deposits are found and from carbon dioxide either as a gas in the atmosphere or dissolved in high concentration in groundwater. There is a wide range of industrially produced compounds or waste products of industrial processes that can attack concrete, including gaseous oxides of sulphur and nitrogen produced by the combustion of carbon-based fuels and emitted from chimneys, mine spoil and waste water. Of all of the aforementioned, as mentioned at the outset of this subsection, perhaps the most common cause of the loss of alkalinity in concrete structures, resulting in corrosion of reinforcement, is 'carbonation' (see also Chapter 5). Carbonated concrete has a pH of 9.5, well below the critical pH for the maintenance of the

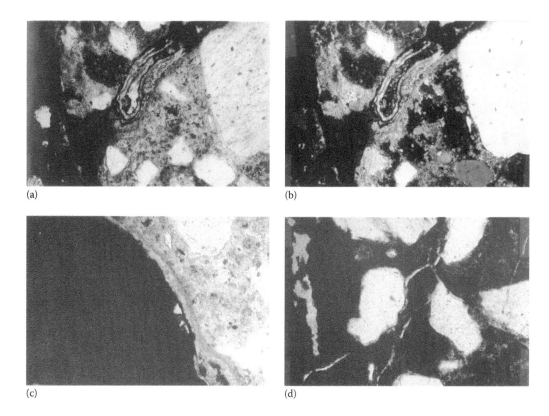

Figure 6.22 An example of the effects of corrosion of steel reinforcement on the adjacent concrete, widths of field ~0.8 mm: (a) the reinforcing steel is to the left and is separated from the hardened cement paste by a dark brown band of hydrated iron oxides which extrude into the concrete towards an elongated void; (b) under crossed polars, the relationship of the corrosion products and associated carbonation are visible; (c) detail of the boundary between the hydrated iron oxides and the concrete showing zonation in the corrosion products and an adjacent dark band of carbonation in the hardened cement paste; (d) hydrated iron oxides have penetrated the hardened cement paste, which is cracked because of the volume increases caused by formation of the corrosion products.

passive film (pH 11.5). Consequently, in the presence of sufficient oxygen and moisture, corrosion can occur of reinforcement that is covered or enveloped by carbonated concrete.

Carbonation will not progress from the surface into the cover concrete in conditions where the pore structure of the concrete is either saturated with moisture or is very dry (i.e. where there is less than 65% RH, at which point capillary pores and air voids can dry out [Moll 1964]). The rate of carbonation is a function of the permeability of the concrete, and in particular the binder, and the supply of carbon dioxide. The depth of carbonation for a particular concrete is traditionally taken as increasing at a progressively slower rate equal to the square route of time.

Petrography is particularly useful in the assessment of the extent to which concrete cover has been carbonated (Sims 1994) and the methods were addressed in Chapter 5. Even a rudimentary examination can include an indicative measurement of 'depth of carbonation' using phenolphthalein indicator solution (see Figures 6.23 and 6.24), which can also be carried out on-site, testing the outermost zones of core holes when they are freshly exposed and prior to any backfilling. However, while simple and rapid, the phenolphthalein method is indirect (actually detecting alkalinity above about pH = 9 [strictly, pH = 8.3–10], with

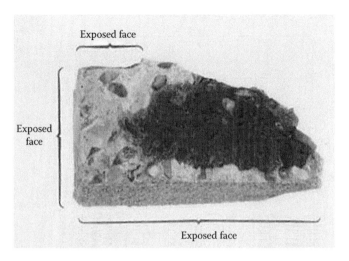

Figure 6.23 Depth of carbonation using phenolphthalein. (From Building Research Establishment, *Carbonation of Concrete Made with Dense Natural Aggregates*, Information Sheet, IS 14/78, Building Research Station, Watford, U.K., p. 3, 1978.)

'carbonated' material [strictly, pH = <8.3] remaining unstained), and the apparently clear boundary shown between 'carbonated' and 'uncarbonated' concrete can sometimes be misleading. A quantitative carbonation profiling technique is potentially available, using drilling dust samples and X-ray diffraction (XRD) for the contents of portlandite and calcite (Parrott 1990), and Sims (1994) gives an example, but it is a relatively expensive approach and there are some technical limitations.

Sims (1994) has shown that assessments of the degree, depth and profile of carbonation can be more sophisticated when carried out in thin section under the petrological microscope. In some cases, there is an intermediate zone of 'partial carbonation', which could be interpreted as 'uncarbonated' by the staining method (see Figure 6.25). The potential importance of recognising zones of partial carbonation is shown in Figure 6.26. Also, during thin section examination, the pattern of localised deeper penetration of carbonation along the borders of cracks, joints and other features can be recorded (see Figure 6.27). As well as helping to assess the residual protective capability of the concrete cover to reinforcement, this pattern of carbonation ingress, which is typically very slow along cracks and similar discontinuities owing to their limited exposure to carbon dioxide (compared with outer surfaces that are exposed to a continuously replenished atmospheric supply), has been used to assess the age at which cracking took place.

6.3.4 Chlorides and corrosion of steel reinforcement

Chloride ions, in sufficient concentration (usually termed the 'threshold concentration'), at the surface of the reinforcement can cause localised breakdown of the passive film even where the pH values of the concrete are well above the critical value for the maintenance of the passive film. The generally small areas where breakdown of the passive film occurs then become anodes, while the larger unaffected areas become cathodes. The large cathode to anode area ratio can result in intense localised corrosion with the formation of deep corrosion 'pits' in the surface of the steel. Specialised accounts of the influence of chloride and chloride-rich environments on reinforcement corrosion or its prevention may be found in Hobbs and Matthews (1998), Costa and Appleton (1999) and Cather (2000).

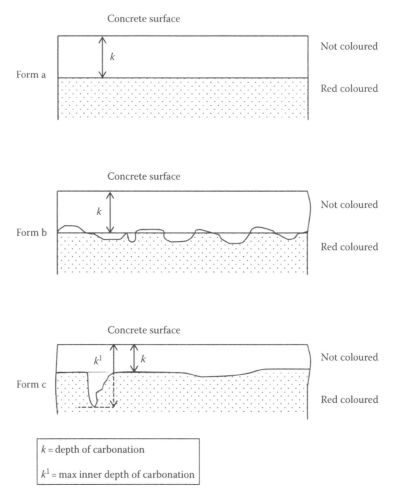

Figure 6.24 Definitions used in the assessment of depth of carbonation using phenolphthalein indicator solution. (After Meyer, A., *Proc. fifth Int. Symp. Chem. Chem.*, 3, 394, 1968.)

A number of theories have been proposed to explain the mechanism of chloride breakdown of the passive film. These have included dissolution of the film at pores formed by the entry of chloride ions (Venu et al. 1965), chloride forming a high energy complex around a cation in the surface of the film (Hoar and Jacob 1967) and chloride migration through the passive film causing breakdown on reaching the passive film–metal interface (McBee and Kruger 1974). In a detailed electron microscope and microanalysis study, Leek and Poole (1990) showed that chloride ions were progressively incorporated into the passive film and that this change of composition resulted in it debonding from the surface of the metal (Figure 6.28).

Electrochemical studies of the phase equilibria relationships between the various components of the metal/electrolyte system in terms of electrical potential and hydrogen ion concentration have been presented as potential/pH diagrams (Pourbaix 1974). These shown that the presence of chloride modifies the system, such that a zone of corrosion exists across a wide range of pH environments (pH 1–14), so that a high (critical) pH alone will not result in passivity of the steel. A detailed scanning electron microscope study coupled with microanalysis by Leek (1997) determined elemental concentrations in the regions around the anodic and cathodic areas as illustrated in Figure 6.29.

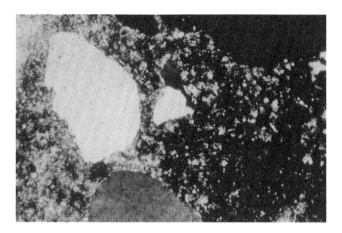

Figure 6.25 Thin-section (xpl) view of the interface between uncarbonated (right) and carbonated (left) paste in concrete, showing the occasional presence of an intermediate zone of 'partial carbonation'.

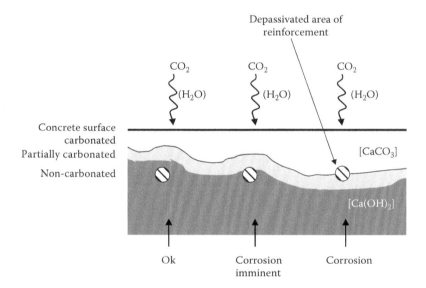

Figure 6.26 Diagram illustrating the potential significance of a partially carbonated zone to an assessment of residual degree of protective cover: the bar shown as at risk of 'imminent' corrosion appears to be covered by non-carbonated cover, but in reality it is only protected by partially carbonated concrete.

The presence of chloride in concrete can arise from a number of sources, either from the original constituents of the concrete mix (e.g. some inadequately drained or washed marine aggregates, or previously from the use of chloride-based admixtures) or from an external saline environment by ingressing into the concrete from its surface. Salt contamination of the aggregates or from airborne salt on unprotected reinforcing steel gave rise to numerous corrosion problems in the early developments in the Arabian Gulf region (Kay et al. 1982; Sims 2006; Walker 2002, 2012).

External sources of chloride include marine or near-marine environments (which can be divided into three zones of exposure: splash, tidal and submerged), saline soil deposits

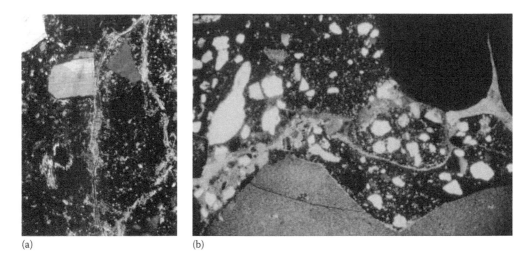

(a)   (b)

Figure 6.27 Thin-section views showing the deeper penetration of carbonation along cracks and other discontinuities in two concrete examples: (a) shows buff-coloured carbonation, whilst (b) shows a crack filled with yellow-dyed resin, bordered by orange-brown carbonation.

Figure 6.28 SEM micrograph showing the incorporation of chlorides into the passive film. (From Leek, D.S. and Poole, A.B., The breakdown of the passive film on high yield mild steel by chloride ions, Page, C.L., Treadaway, K.W.J. and Bamforth, P.B. (eds), *Corrosion of Reinforcement in Concrete*, Elsevier Applied Science. (pp Construction Materials Group, Society of Chemical Industry), New York, 1990, pp. 65–73.)

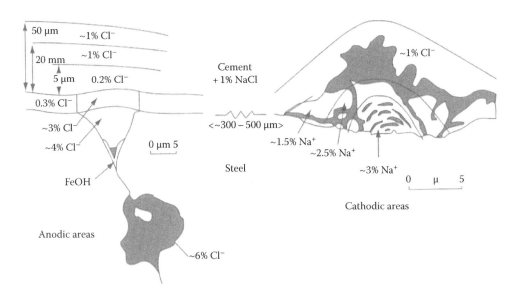

Figure 6.29 Diagrammatic illustration of chloride and sodium concentrations close to anodic and cathodic areas on a laboratory-prepared specimen treated with a 1 per cent salt solution. The shaded areas represent corrosion products. (Diagram courtesy of Dr Darrell Leek, Mott Macdonald, Croydon, U.K.)

and de-icing salts used to treat roads and other pavements in some regions. The penetration of chloride ions in solution occurs through the pore structure of the concrete (see Section 6.3.1) usually by means of a combination of permeability under pressure, diffusion and/or capillary rise. Internal microcracks within the concrete can form preferred pathways for chloride movement. Summaries of the complex processes whereby chlorides can penetrate concrete structures, especially concrete cover, and methods for their assessment, may be found in Kropp and Hilsdorf (1995), Kropp (1995), Bamforth et al. (1997) and Glass (2003).

On aerially exposed surfaces, chlorides can arrive through capillary rise (where in contact with the ground), in water droplets or as wind-blown dust which can be wetted by dew and thus penetrate the concrete by absorption. In environments subject to humid/dry cycles, repeated wetting/drying of surfaces may lead to a build-up of salts at surfaces. Whether this results in precipitation of salts at concrete surfaces or gradual penetration of ions into the concrete (or both) depends, among other factors, on the relative duration of the humid/dry periods. But this is a mechanism whereby chloride concentration can gradually increase and exceed the initiation threshold at the reinforcement surface.

Not all of the chloride ions present in, or penetrating, the concrete will be available to initiate corrosion of the reinforcement. A proportion can become chemically bound into the cement paste in the form of calcium chloroaluminate hydrate ($3CaO \cdot Al_2O_3 \cdot CaCl_2 \cdot 10H_2O$) dependent on the amount of aluminate phase present in the cement. It should be noted, however, that subsequent carbonation of the cement paste can release this bound chloride, increasing the concentration of chloride in the pore fluid ahead of the advancing carbonation front.

Historically, the use of calcium chloride as a setting accelerator in concretes has given rise to corrosion problems and is rarely used in modern concrete for that reason. It also attacks the calcium hydroxide in the concrete and forms calcium oxychloride which can cause disruption of the concrete (Collepardi and Monosi 1992).

6.3.5 Prevention of reinforcement corrosion

The principal means of ensuring durability of reinforced concrete structures is for them to be designed in accordance with prevailing national and international codes of practice. However, in particularly aggressive conditions, or where corrosion has already been initiated in existing structures, there is a variety of techniques by which corrosion may be either prevented or mitigated (see Table 6.6).

These techniques include the use of applied voltage for 'cathodic protection' and the use of sacrificial anodes and corrosion inhibitors, which appear either to stabilise the protective film on the steel or to absorb oxygen from the system. Various impermeable organic coatings have been used in attempts to minimise the diffusion of oxygen and moisture into the concrete. Electrolytic methods have been used to re-alkalise or remove chlorides from concrete to prevent or arrest steel reinforcement corrosion (Vennesland 1994). These methods appear to be successful but may have the disadvantage of activating localised concentrations of sodium ions which may be sufficiently high to activate dormant AAR in the concrete (Al-Kadhimi and Banfill 1996).

6.4 FROST AND FREEZE–THAW ACTION

Hardened concrete may be damaged by frost particularly if it is subjected to freeze–thaw cycles typical of the northern temperate latitudes. In particular, the use of de-icer salts can lead to scaling (Marchand et al. 1994). The damage usually starts with a superficial flaking of the surface layers resulting from the development of irregular cracks subparallel to the exposed surface, and as the damage becomes more severe, significant spalling of exposed surfaces and edges occurs (Figure 6.30). Such primary or secondary (worsening effects initiated by an earlier mechanism) damage from freeze–thaw cycles, or occasionally from freezing of a saturated material ('freezing when wet'), is a major threat to concrete durability in some climatic regions of the world, including some areas with extensive concrete-based development (such as parts of northern Europe, much of Canada, parts of the United States, parts of Russia and parts of China). Specialised accounts of the freeze–thaw performance of concrete will be found in Pigeon and Pleau (1995), Marchand et al. (1997), Hobbs et al. (1998), Harrison et al. (2001), Pigeon et al. (2003), Richardson (2007) and Neville and Brooks (2010).

Table 6.6 Some techniques for reducing or mitigating the risks of reinforcement corrosion

Admixtures	Reinforcement	Surface	Electrochemical
Waterproofing admixtures Corrosion inhibitors incorporated into the original concrete mix	Unreinforced design Corrosion-resistant reinforcement (stainless steel, galvanised steel, and epoxy-coated steel)	Controlled permeability formwork Sealers (silane, etc.) Coatings or membranes Migrating corrosion inhibitors	Corrosion/chloride monitoring Cathodic protection, including provision for future installation Cathodic protection at the outset (cathodic prevention) Realkalisation of the concrete

Source: Courtesy of Darrell Leek.

(a)

(b)

(c)

Figure 6.30 Examples of freeze–thaw damage: (a) affecting the exposed surfaces on a column of a railway viaduct (note the cracking parallel to the exposed corner and associated spalling [pen scale is 130 mm in length]), (b) exposed lower surfaces of a concrete storage tank, (c) exposed surfaces of a concrete coastal wall, Trondheim, Norway.

6.4.1 Freeze–thaw mechanisms and factors

Moisture must be present in the pore structure of the concrete for damage caused by freezing to occur. Thus, with well-made dense concretes with low water/cement ratios, damage is minimal. Air-entrained concrete (see Chapter 4) in well-made low water/cement ratio mixes, with carefully controlled spacing between air voids, is a widely used method of minimising the effects of freeze–thaw damage.

A variety of explanations for the mechanism of freeze–thaw damage have been proposed. Early theories suggested that the volumetric expansion of pore water on freezing in a confined zone beneath an already frozen surface produced the disruption. This idea was modified by Collins (1944), suggesting that the disruption of the surface layers could derive from the growth of ice crystals resulting from the migration of water towards them because of vapour pressure differences. Hydraulic pressure mechanisms caused by water migration through the capillary system, resulting in cracking and disruption, have been proposed by several authors. Powers (1956) proposed that the expansion due to freezing in the pores within the surface layers forced water inwards through small pores with sufficient hydraulic pressure to disrupt the surrounding material. The effectiveness of air entrainment in overcoming this lends support to the validity of the hypothesis.

The theories of frost action have been reviewed by Pigeon and Pleau (1995) who also discuss testing techniques and methods for producing concrete that is durable in frost conditions. More recent studies by Bager and Sellevold (1986, 1987) using low-temperature calorimetry indicate that the inward migration of the ice front from the surface is much affected by the previous drying history of the concrete. Mitchell (2000) has claimed that the use of polypropylene fibres in concrete can assist with the resistance of freeze–thaw cycles, and certainly this would be consistent with the potential effectiveness of fibres in controlling the formation of cracks and microcracks in concrete (see Section 4.8).

It is probable that the observed damage to hardened concrete is the result of the interaction of several of these mechanisms. Petrographically, no changes to the cement or aggregate mineralogy can be observed, and the damage is usually restricted to a series of irregular cracks which lie subparallel to the exposed surfaces and/or corners. Such crack systems typically propagate through the cement paste and round the margins of aggregate particles. The distance between the individual subparallel cracks depends on the aggregate sizes and the freeze–thaw rates but they are usually sensibly equispaced. The cracks furthest from the surface are impersistent, but as the exposed surface is approached, the cracking becomes more persistent and more open and bridging cracks approximately perpendicular to the surface become common.

6.4.2 Macroscopic and microscopic evidence of freeze–thaw action

The crack pattern, the details of the local environment, the concrete sample and case history, together with the petrographic information relating to the concrete, usually provide sufficient evidence for the correct diagnosis of frost or freeze–thaw damage. Observations on-site, together with a laboratory study of the crack pattern developed on resin-impregnated slabs cut normal to, and including, the surface, are the best aids to diagnosis, though large-size petrographic thin sections can be used in place of the cut slabs. The DRI example in Section 6.1.3 is from a Canadian dam and includes freeze–thaw cracking.

Freeze–thaw cycles lead to three main visible physical effects on exposed concrete surfaces:

1. Surface scaling and progressive delamination, with sets of cracks beneath and parallel or subparallel to the surface;
2. Suites of cracks adjacent to and following joints, edges and other discontinuities, often term 'D' or 'D-line' cracking;
3. Surface pop-outs, associated with porous or microporous, frost-susceptible aggregate particles just beneath the exposed concrete surface.

Delamination, effect (1), appears to have a complex causation (Pigeon et al. 2003; Richardson 2007), including the differential behaviour of layers above and below a frozen zone and/or the formation of ice lenses, with expansion of frozen layers being analogous to the 'frost heave' mechanism in soils (Henry 2000). D-line cracking, effect (2), is rather distinctive (Figure 6.31) and appears to be related to the freezing of susceptible coarse aggregates at depth, with larger particle sizes typically creating the worst damage (Schwartz 1987; Van Dam et al. 2002; Thomas and Folliard 2007). Pop-outs, effect (3), are superficial features associated with discrete susceptible aggregate particles and can sometimes be induced by freezing of saturated surfaces, without a need for prolonged freeze–thaw cycling (see further discussion below in this section).

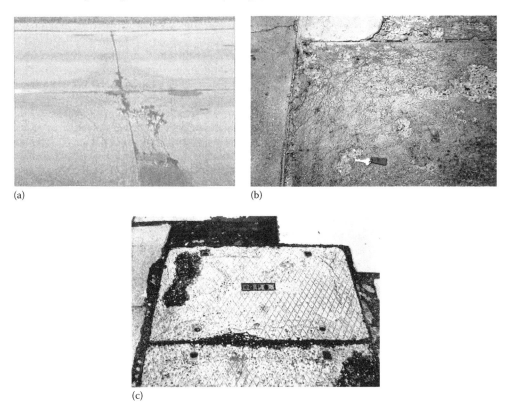

(a) (b)

(c)

Figure 6.31 Examples of D-line cracking on concrete surfaces: (a) pavement. (From Thomas, M.D.A. and Folliard, K.J., Concrete aggregates and the durability of concrete, Page, C.L. and Page, M.M. (eds), *Durability of Concrete and Cement Composites*, Woodhead & Maney (pp Institute of Materials, Minerals & Mining), Cambridge, U.K., Chapter 7, pp. 247–281, 2007.), (b) detail of a pavement and (c) precast concrete manhole covers in London.

In practice, of course, affected concrete structures could exhibit all of these effects in potentially confusing combination, not to mention the additional possibility of other deteriorative mechanisms also being present, and experience also shows that the use of de-icing salts multiplies freeze–thaw cycles and thus exacerbates the resultant damage (Dubberke and Marks 1985).

The effects of cold weather on the setting of concrete are reflected in the microfabric of the cement paste. In petrographic thin section, the cement paste in concrete that has been placed and cured in cold conditions is typically lighter in colour than might normally be expected for a given water/cement ratio. The effect is due to the more open porous nature of the hydrates formed and may be a consequence of slower than normal hydration. Unless care is taken, this effect can introduce errors into the method for estimating water/cement ratio in hardened concrete by calibrating the intensity of fluorescence from a dyed resin impregnating the pore spaces of the cement paste against a series of standards (see Sections 2.5 and 4.4.4). Concrete that has set in cold conditions might also display exceptionally large portlandite crystals (see Chapter 4), but this can also be caused by alternative factors. Sibbick (2011) adds that there is often a concentration of secondary minerals (ettringite, calcite and portlandite) associated with freeze–thaw-induced cracking.

In freezing conditions, mixing of concrete becomes difficult; the mix water freezes so that the hydration of the cement minerals is partial and uneven with the ice crystals in the mix occupying a larger volume than the water they replace. Thus, the concrete does not set properly and remains a friable mass.

Jensen (2011) reports that, in Scandinavia, even when the overall concrete quality is sufficient to resist the development of cracking considered characteristic of freeze–thaw cycles, certain non-frost-resistant aggregate particles located near the exposed concrete surface can expand during the freezing process and form 'pop-outs' superficially similar to those sometimes caused by ASR. The authors are familiar with similar occurrences in the United Kingdom, especially when particles of microporous flint cortex, some oolitic limestone and some clay-containing aggregate particles, located within the concrete at or just beneath a weathering surface, can absorb water during wet weather and then expand on freezing shortly afterwards (see Figures 6.32 and 6.33).

In one case, following problems with and concerns over 'foreign object debris' (FOD) on some concrete airfield runways and taxiways in Southern England, assessment methods and specifications had to be developed to limit the content of microporous flint cortex within suitable aggregates (in much of Southern and Eastern England, the abundant aggregates are fluvioglacial and marine sands and gravels, dominated by flint, with varying contents of the

(a)

(b)

Figure 6.32 Site views of concrete with pop-outs caused by freezing when wet: (a) flint aggregate concrete steps in London and (b) detail of steps, showing an abundance of incipient pop-outs that will eventually combine to cause general surface spalling.

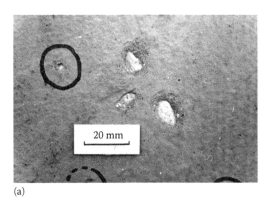

(a)   (b)

Figure 6.33 Pop-outs caused by microporous flint cortex just beneath the exposed concrete surface (a) causing loss of the covering concrete matrix, in precast concrete paving slabs, and (b) pop-out caused by a flint cortex particle just beneath a pavement surface; the frost susceptible particle has split.

microporous variety known as 'cortex'). In the end, control of cortex content was monitored using quantitative aggregate petrography (using the method now published as BS 812-104 1994), but research also applied the RILEM test method (RILEM 1995, 1997) and attempts at beneficiation using flotation, based on the fact that flint cortex has a lower density than typical 'dense' flint (Dolar-Mantuani 1983).

In another case, precast panels on a central reservation in the English Midlands exhibited a severe pop-out incidence, gradually leading to overall surface scaling, as the result of wetting and de-icing salts from the adjacent road carriageway and the presence within the concrete of a frost-susceptible oolitic limestone aggregate (see Figure 6.34).

Figure 6.34 Surface degradation as a result of freezing while wet, exacerbated by de-icing salts. The concrete surface is the central reservation between two road carriageways in the English West Midlands; scaling is caused by freeze–thaw action on a concrete made with a weak and porous oolitic limestone aggregate, which was greatly worsened in those areas most subjected to the wheel-splashed accumulation of water and de-icing salt. (From Sims, I. and Brown, B.V., *Concrete aggregates*, Hewlett, P.C. (ed), *Lea's the Chemistry of Cement and Concrete*, 4th edn., Edward Arnold, London, U.K., Chapter 16, 1998, pp. 903–1011.)

6.5 SULPHATE ACTIONS

6.5.1 Overview

The literature on sulphate attack is complex and confusing. There does not appear to be a consensus on some details of the mechanisms involved in spite of an investigation that now extends over 60 years. Recent general texts on sulphate actions include Hobbs and Matthews (1998), BRE (2005), Neville (2006) and Bensted et al. (2007).

The mechanisms by which the various external sulphates attack concrete are still a matter of some controversy (Odler and Gasser 1988; Lawrence 1990; Cohen and Mather 1991; Mehta 1993). However, for practical purposes, the extent of attack on any one type of hardened cement paste largely depends on the amount of sulphate in solution, which in turn is related to the cations and other anions in solution. Thus, the aggressiveness of the sulphates in descending order approximately related to their solubility is ammonium, magnesium, sodium, potassium and calcium. A common form of sulphate attack is from groundwater in gypsum-bearing soils, which attacks concrete to form ettringite in the hardened cement paste as follows:

$$\text{Calcium aluminate hydrates} + \underset{\text{(gypsum)}}{CaSO_4 \cdot 2H_2O} \rightarrow \underset{\text{(ettringite)}}{3CaO \cdot Al_2O_3 \cdot 3CaSO_4 \cdot 32H_2O}$$

The ferrite phase can undergo a similar reaction to form analogous compounds to ettringite. This reaction is the basis for the formulation of sulphate-resisting Portland cement (SRPC) in which the content of the aluminate phase is limited to a maximum of 3% by mass, thus restricting the amount of damaging ettringite that can be formed. Other common sulphates are derived from magnesium, sodium and potassium sulphates with the limited possibility of ammonium sulphate either as an effluent or from over-fertilised soil. These sulphates in solution will form gypsum by interaction with the calcium hydroxide in the concrete which may then interact, in turn, with the hydrated aluminate phases to form ettringite as given earlier:

$$\left(NH_4\right)_2 SO_4 + Ca\left(OH\right)_2 + 2H_2O \rightarrow CaSO_4 + 2NH_4OH$$

$$MgSO_4 + Ca\left(OH\right)_2 + 2H_2O \rightarrow CaSO_4 \cdot 2H_2O + Mg\left(OH\right)_2$$

$$Na_2SO_4 + Ca\left(OH\right)_2 + 2H_2O \rightarrow CaSO_4 \cdot 2H_2O + 2NaOH$$

$$K_2SO_4 + Ca\left(OH\right)_2 + 2H_2O \rightarrow CaSO_4 \cdot 2H_2O + 2KOH$$

Magnesium sulphate also attacks the hydrated silicate phases to form gypsum, brucite and hydrated silica and, below a pH of 10.6, it also attacks ettringite:

$$\text{Calcium silicate hydrates} + MgSO_4 \rightarrow CaSO_4 \cdot 2H_2O + Mg\left(OH\right)_2 + \text{hydrated silica}$$

$$\text{Ettringite} + MgSO_4 \rightarrow CaSO_4 \cdot 2H_2O + Mg\left(OH\right)_2 + \text{hydrated alumina}$$

The reaction of magnesium sulphate with ettringite tends to be restricted to the outer layer where the removal of calcium hydroxide occurs and the pH is lower. It is also claimed that brucite can very slowly interact with hydrated silica gel to form a magnesium silicate as the final stage in sulphate attack. Lea (1970) states that in practice this step may be reached only after long periods of time.

At typically low temperatures and under conditions of high humidity, sulphate solutions can also attack the silicate hydrates in concrete and mortar to produce thaumasite, which can be a rapid and serious type of sulphate action on concrete. This 'thaumasite form of sulphate attack' (TSA) is discussed in detail in Section 6.5.6.

The physical mechanisms by which deterioration occurs from sulphate attack are two-fold. These are (1) the expansive formation of either ettringite and/or gypsum in the hardened fabric of the concrete, causing cracking and exfoliation, and (2) the softening and dissolution of hydrated cementing compounds to form a mush, variously owing to direct attack on these compounds by magnesium sulphate, or their decomposition by removal of calcium hydroxide reacting with the sulphates, or through TSA. Either or both mechanisms can apply, depending on the temperature and other environmental conditions and types and concentrations of sulphate in solution and the composition of the concrete. Also, it has been assumed the pH of the groundwater is near neutral. Sulphate attack can be greatly increased by the presence of carbonic, sulphuric, hydrochloric and other acids in water.

Sulphate attack is not restricted to solutions and, with increasing air pollution attack from airborne sulphur compounds, is becoming an increasing problem and can be a particular problem for cementitious masonry jointing mortars. The mechanism of this type of attack has been discussed by Zappia et al. (1994).

6.5.2 Ground and groundwater sulphates

The sulphates of greatest concern for the durability of concrete are found in soils and groundwaters. In the United Kingdom until 2001, the assessment of ground for its aggressiveness towards buried concrete was covered by BRE Digest 363 (1991, 1996), but urgent research following the discovery of serious cases of TSA on the foundations of several motorway bridges (see Section 6.5.6) eventually led to a completely revised replacement, known as BRE Special Digest 1 (2001, 2003, 2005). BRE SD1 has now become established in the United Kingdom and beyond as a valued, though perhaps quite conservative, source of guidance for designing concrete to minimise the risk of damage from sulphate actions. BRE SD1 includes a good review of the various sulphate actions and then detailed field and laboratory procedures for assessing the aggressiveness of the ground at a particular site, leading to classifications of both the 'aggressive chemical environment for concrete' (ACEC) in respect of the ground conditions and the 'design chemical' (DC) class (plus some 'additional protective measures' [APMs]) for specifying concrete accordingly (Table 6.7a–d).

Of the common sulphates found in soils, only calcium sulphate dihydrate (gypsum) is effectively limited by its solubility. This is equivalent to about 1200 ppm of SO_3 in groundwater. If greater concentrations of SO_3 are present, other cations such as Mg, Na and K must be present to balance the sulphate in solution. However, where the SO_3 content is less than 1200 ppm, it must not be assumed that the sulphate has been solely derived from gypsum. Any or all the appropriate cations can also be present in the solution. It should be noted that magnesium and sodium sulphates have solubilities that are 150–200 times greater than gypsum. A major discovery of the research that led to BRE SD1 was that the sulphate content (or class) of the reworked ground had increased since its original assessment, apparently owing to the oxidation of pyrite (iron disulphide) in that ground (Figure 6.35). Accordingly, the cautious approach of BRE SD1 includes calculating the 'total potential sulphate' content, by assuming that all the pyrite or 'oxidisable sulphide' could eventually be converted to sulphate.

Biczoc (1967) states that the sulphate content (SO_4) of soil is usually in the range of 0.01%–0.05% but there are soils in which up to 5% sulphate is present. There are many regions around the world in which soils contain several percent of sulphates. This sulphate is derived from the oxidation of sulphide minerals such as pyrite, the biological decomposition

Table 6.7 Selected guidance from BRE SD1 (2005) on classification of aggressiveness of ground conditions and design measures for suitably resistant concrete: (a) ACEC classes in natural ground conditions (Table C1 in BRE SD1), (b) ACEC classes in brownfield locations (Table C2 in BRE SD1), (c) DC classes and APMs for general in situ use of concrete (Table D1 in BRE SD1) and (d) qualities of general in situ concrete for resisting chemical attack (Table D2 in BRE SD1) (consult BRE SD1 [2005] for detailed guidance)

Table 6.7a ACEC classification for natural ground locations[a]

| Design sulphate class for location | Sulphate | | | Groundwater | | ACEC class for location |
| | 2:1 water/soil extract[b] | Groundwater | Total potential sulphate[c] | Static water | Mobile water | |
1	*2* (SO$_4$ mg/L)	*3* (SO$_4$ mg/L)	*4* (SO$_4$ %)	*5* (pH)	*6* (pH)	*7*
DS-1	<500	<400	<0.24	≥2.5		AC-1s
					>5.5[d]	AC-1[d]
					2.5–5.5	AC-2z
DS-2	500–1500	400–1400	0.24–0.6	>3.5		AC-1s
					>5.5	AC-2
				2.5–3.5		AC-2s
					2.5–5.5	AC-3z
DS-3	1600–3000	1500–3000	0.7–1.2	>3.5		AC-2s
					>5.5	AC-3
				2.5–3.5		AC-3s
					2.5–5.5	AC-4
DS-4	3100–6000	3100–6000	1.3–2.4	>3.5		AC-3s
					>5.5	AC-4
				2.5–3.5		AC-4s
					2.5–5.5	AC-5
DS-5	>6000	>6000	>2.4	>3.5		AC-4s
				2.5–3.5	≥2.5	AC-5

[a] Applies to locations on sites that comprise either undisturbed ground that is in its natural state (i.e. is not brownfield – Table 6.7b) or clean fill derived from such ground.

[b] The limits of design sulphate classes based on 2:1 water/soil extracts have been lowered relative to previous Digests (Box C7 of BRE SD1 [2005]).

[c] Applies only to locations where concrete will be exposed to sulphate ions (SO$_4$) which may result from the oxidation of sulphides (e.g. pyrite) following ground disturbance (Appendix A1 and Box C8 of BRE SD1 [2005]).

[d] For flowing water that is potentially aggressive to concrete owing to high purity or an aggressive carbon dioxide level greater than 15mg/L (Section C2.2.3 of BRE SD1 [2005]), increase the ACEC Class to AC-2z.

Explanation of suffix symbols to ACEC class:

• Suffix 's' indicates that the water has been classified as static.

• Concrete placed in ACEC classes that include the suffix 'z' primarily have to resist acid conditions and may be made with any of the cements or combinations listed in Table 6.7d.

Table 6.7b ACEC classification for Brownfield locations[a]

Design sulphate class for location	Sulphate and magnesium					Groundwater		ACEC class for location
	2:1 water/soil extract[b]		Groundwater		Total potential sulphate[c]	Static water	Mobile water	
1	2 (SO$_4$ mg/L)	3 (Mg mg/L)	4 (SO$_4$ mg/L)	5 (Mg mg/L)	6 (SO$_4$%)	7 (pH)[d]	8 (pH)[d]	9
DS-1	<500		<400		<0.24	≥2.5		AC-1s
							>6.5[d]	AC-1
							5.5–6.5	AC-2z
							4.5–5.5	AC-3z
							2.5–4.5	AC-4z
DS-2	500–1500		400–1400		0.24–0.6	>5.5		AC-1s
							>6.5	AC-2
						2.5–5.5		AC-2s
							5.5–6.5	AC-3z
							4.5–5.5	AC-4z
							2.5–5.5	AC-5z
DS-3	1600–3000		1500–3000		0.7–1.2	>5.5		AC-2s
							>6.5	AC-3
						2.5–5.5		AC-3s
							5.5–6.5	AC-4
							2.5–5.5	AC-5
DS-4	3100–6000	≤1200	3100–6000	≤1000	1.3–2.4	>5.5		AC-3s
							>6.5	AC-4
						2.5–5.5		AC-4s
							2.5–6.5	AC-5
DS-4m	3100–6000	>1200[e]	3100–6000	>1000[e]	1.3–2.4	>5.5		AC-3s
							>6.5	AC-4m
						2.5–5.5		AC-4ms
							2.5–6.5	AC-5m
DS-5	>6000	≤1200	>6000	≤1000	>2.4	>5.5		AC-4s
						2.5–5.5	≥2.5	AC-5
DS-5m	>6000	>1200[e]	>6000	>1000[e]	>2.4	>5.5		AC-4ms
						2.5–5.5	≥2.5	AC-5m

[a] Brownfield locations are those sites, or parts of sites, that might contain chemical residues produced by or associated with industrial production – Section C5.1.3 of BRE SD1 (2005).

[b] The limits of Design Sulphate Classes based on 2:1 water/soil extracts have been lowered relative to previous Digests – see Box C7 of BRE SD1 (2005).

[c] Applies only to locations where concrete will be exposed to sulphate ions (SO$_4$), which may result from the oxidation of sulphides such as pyrite, following ground disturbance (Appendix A1 and Box C8 of BRE SD1 [2005]).

[d] An additional account is taken of hydrochloric and nitric acids by adjustment to sulphate content – see Section C5.1.3 of BRE SD1 (2005).

[e] The limit on water-soluble magnesium does not apply to brackish groundwater (chloride content between 12,000 mg/L and 17,000 mg/L). This allows 'm' to be omitted from the relevant ACEC classification. Seawater (chloride content about 18,000 mg/L) and stronger brines are not covered by this table.

Explanation of suffix symbols to ACEC class

• Suffix 's' indicates that the water has been classified as static.

• Concretes placed in ACEC classes that include the suffix 'z' have primarily to resist acid conditions and may be made with any of the cements in Table D2 of BRE SD1 (2005).

• Suffix 'm' relates to the higher levels of magnesium in Design Sulphate Classes 4 and 5.

Table 6.7c Selection of the DC Class and the number of APMs for concrete elements where the hydraulic gradient due to groundwater is 5 or less: for general in situ use of concrete[a,b,c]

ACEC class (from Tables 6.7a and 6.7b)	Intended working life	
	At least 50 years[d,e]	At least 100 years
AC-1s, AC-1	DC-1	DC-1
AC-2s, AC-2	DC-2	DC-2
AC-2z	DC-2z	DC-2z
AC-3s	DC-3	DC-3
AC-3z	DC-3z	DC-3z
AC-3	DC-3	DC-3 + one APM of choice
AC-4s	DC-4	DC-4
AC-4z	DC-4z	DC-4z
AC-4	DC-4	DC-4 + one APM of choice
AC-4ms	DC-4m	DC-4m
AC-4m	DC-4m	DC-4m + one APM of choice
AC-5z	DC-4z + APM3[f]	DC-4z + APM3[f]
AC-5	DC-4 + APM3[f]	DC-4 + APM3[f]
AC-5m	DC-4m + APM3[f]	DC-4m + APM3[f]

Note: For specification of DC Class, see Table 6.7d. For choice of APMs, see Table D4 in BRE SD1 (2005).

[a] Where the hydrostatic head of groundwater is greater than five times the section thickness, one step in DC class or one APM over and above the number indicated in the table should be applied except where the original provisions included APM3. Where APM3 is already required, an additional APM is not necessary.

[b] A section thickness of 140 mm or less should be avoided in in situ construction, but where this is not practical, apply a one step higher DC class or an additional APM except where the original provisions included APM3. Where APM3 is already required, an additional APM is not necessary.

[c] Where a section thickness greater than 450 mm is used and some surface chemical attack is acceptable, a relaxation of one step in DC class may be applied, provided that for reinforced concrete, APM4 (sacrificial layer) is applied – see Section D6.5 in BRE SD 1 (2005).

[d] The concrete quality given in column 'at least 50 years' is also adequate for foundations to low-rise domestic housing with an intended working life of 'at least 100 years'.

[e] Structures with an intended working life of 'at least 50 years' but for which the consequences of failure would be relatively serious, should be classed as having an intended working life of at least 100 years for the selection of the DC Class and APM (Section D7 in BRE SD1).

[f] Where APM3 is not practical, see Section D6.1 for guidance.

Explanation of suffix symbols to DC Class:

• Concrete placed in ACEC Classes that include the suffix 'z' primarily must resist acid conditions and may be made with any of the cements listed in Table D2 of BRE SD1 (2005).

• Suffix 'm' relates to the higher levels of magnesium in DS Classes 4 and 5.

Table 6.7d Concrete qualities to resist chemical attack for the general use of in situ concrete: limiting values for composition

DC class	Maximum free-water/cement or combination ratio	Minimum cement or combination content (kg/m³) for maximum aggregate size				Recommended cement and combination group
		≥40 mm	20 mm	14 mm	10 mm	
DC-1	—	—	—	—	—	A–G inclusive
DC-2	0.55	300	320	340	360	D, E, F
	0.50	320	340	360	380	A, G
	0.45	340	360	380	380	B
	0.40	360	380	380	380	C
DC-2z	0.55	300	320	340	360	A–G inclusive
DC-3	0.50	320	340	360	380	F
	0.45	340	360	380	380	E
	0.40	360	380	380	380	D, C
DC-3z	0.50	320	340	360	380	A–G inclusive
DC-4	0.45	340	360	380	380	F
	0.40	360	380	380	380	E
	0.35	380	380	380	380	D, G
DC-4z	0.45	340	360	380	380	A–G inclusive
DC-4m	0.45	340	360	380	380	F

Grouped cements and combinations

	Cements	Combinations
A	CEM I, CEM II/A-D, CEM II/A-Q, CEM II/A-S, CEM II/B-S, CEM II/A-V, CEM II/B-V, CEM III/A, CEM III/B	CIIA-V, CIIB-V, CII-S, CIIIA, CIIIB, CIIA-D, CIIA-Q
B	CEM II/A-L[a], CEM II/A-LL[a]	CIIA-L[a], CIIA-LL[a]
C	CEM II/A-L[a], CEM II/A-LL[a]	CIIA-L[a], CIIA-LL[a]
D	CEM II/B-V+SR, CEM III/A+SR	CIIB-V+SR, CIIIA+SR
E	CEM IV/B(V), VLH IV/B(V)	CIVB-V
F	CEM III/B+SR	CIIIB+SR
G	SRPC	—

Note: For cement and combination types, compositional restrictions and relevant standards, see Table D3 of BRE SD1 (2005).

[a]   The classification is B if the cement/combination strength class is 42.5 or higher and C if it is 32.5.

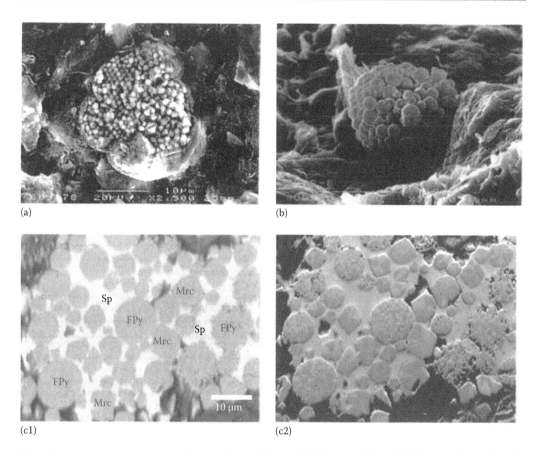

Figure 6.35 Appearance of susceptible pyrite and its oxidation, creating solid iron oxides/hydroxide residues and leading to the release of sulphates: (a) SEM image of a pyrite poly-framboid with well-formed octahedral crystallites and partial recrystallisation (Lower Lias clay), (b) SEM image of a weathered pyrite framboid resting on gypsum consisting of partly dissolved spherical crystallites (weathered Oxford Clay) ((a) and (b) are from Czerewko, M.A. et al. *Q. J. Eng. Geol. Hydrogeol.*, 36, 331, 2003.), (c) SEM images ([c1] backscatter Z contrast, [c2] forescatter OC) of pyrite framboids (FPy) occurring in a pyrite–marcasite (Mrc) clump, enveloped by massive sphalerite (Sp). (From Ohfuji, H. et al., *Am Mineral.*, 90, 1693, 2005.)

of organic material containing proteins and from agricultural and industrial pollution interacting with minerals in the soil. The oxidation of sulphide minerals and biological decomposition of organic matter require oxygen so that usually much higher sulphate concentrations are present in the groundwater of the upper soil layers than in the deeper artesian waters.

Sampling and analysis of soil and groundwaters for sulphate content and the evaluation of results need to be carried out with some care, as detailed in BRE SD1 (2005). This is because water movements may be vertical or horizontal depending on seasonal variations in rainfall and on the geology of the site and its environs. Also a low sulphate content in the soil does not eliminate the possibility of sulphate attack since groundwater may flow from adjacent areas, particularly if the soil is disturbed by construction; the 'mobility' of any groundwater is an important consideration in SD1. There are also problems in obtaining samples that are not diluted with surface water.

Differing terminologies are used in referring to sulphates. For instance, the term sulphate can refer to either SO_3 or SO_4. Results reported as SO_4 may be converted to SO_3 by multiplying by the factor 0.833. Concentrations may be reported as parts per million, parts per hundred thousand and g/L (or g/L). The unit g/L (or g/L) may be converted to these other units by multiplying by 1000 and 100, respectively.

Because of the widespread nature of the problem of sulphate attack, many countries have specifications and codes of practices for its prevention. Lea (1970), Lawrence (1990), Harrison (1992a,b), Hobbs and Matthews (1998), Marchand and Skalny (1999), Skalny et al. (2002), Stark (2002) and especially BRE SD1 (2001, 2003, 2005) are some convenient sources for appropriate recommendations.

6.5.3 Conventional sulphate attack

This subsection addresses the actions of sulphates on the calcium hydroxide and calcium aluminate hydrate elements of Portland cement concrete, typically leading to the secondary formation of mainly ettringite and sometimes gypsum within hardened concrete.

6.5.3.1 Properties of concrete relating to sulphate attack

There is overwhelming evidence to show that concretes of low permeability have good resistance to sulphate attack. This can be achieved by using a low water/cement ratio and high cement content, and resistance can be further increased by the use of sulphate-resisting cement or cements blended with blastfurnace slag and pozzolanas. However, while sulphate-resisting cements provide a good protection against attack, much of this protection can be negated if the concrete is too permeable or, historically, admixtures such as calcium chloride are used. Where pozzolanas are used, highly reactive, low-calcium pozzolanas are required, and the concrete must be adequately cured to prevent increased permeability (Harboe 1982; Stark 1982; Dunstan 1982; Mather 1982; Harrison 1992). Al-Amoudi et al. (1995) found that concretes made with blended cements, particularly those made with blastfurnace slag, gave an inferior performance against magnesium sulphate attack when compared to plain cement mortar.

In general, aggregates do not have a decisive effect on sulphate resistance. Porous aggregates which allow hydration and dissolution of salt solutions in their pores are subject to cracking and the development of pop-outs in the surfaces of concrete (Stark 1982). Piasta et al. (1987) found that porous limestone aggregates increased sulphate resistance. However, Fiskaa et al. (1971) found that replacement of fine aggregates with limestone was unfavourable. Aggregates containing kaolinised feldspars can form ettringite (Coutinho 1979).

Curing affects sulphate resistance principally because well-cured concretes tend to be more impermeable. High-pressure steam curing is also effective, but where pozzolanas and blastfurnace slag are used, autoclaving may reduce durability (Stark 1982). Osborne (1991) found that carbonation is beneficial in reducing the effect of sulphate attack, because of the removal of calcium hydroxide from the surface layers of the concrete and ettringite is unstable below pH 10.6. This is why a number of investigations have found an increased durability from concretes allowed to dry for some weeks after moist curing. Such drying enhances surficial carbonation.

As would be expected, the most common point for sulphate attack is on concrete just above or below ground. Concrete cast in contact with the soil, such as foundation footings and piles, is particularly prone to sulphate attack because the permeability of the surfaces cannot be controlled (Harrison and Teychenné 1981). It has been emphasised that where ordinary (as opposed to sulphate-resisting) Portland cement is used in below-ground concrete, it is necessary to provide adequate resistance to moisture entry (Hamilton and Handegord 1968; Novokshchenov 1987). Results of long-term tests reported by Harrison (1992) confirm the importance of using low-permeability concretes especially in cases where the hydrostatic head is applied to the concrete or internal drying of basement floors induces a moisture gradient.

6.5.3.2 External appearance of concrete attacked by sulphate

The external appearance of sulphate attack appears to be quite variable as would be expected from such a complex reaction. Lea (1970) states that the action of soluble sulphates on cements results in the formation of calcium sulfoaluminate and gypsum accompanied by a marked expansion. In general, concrete attacked by sodium and calcium sulphate eventually becomes reduced to a soft mush, but when magnesium sulphate is the main destructive agent, the concrete remains hard though it becomes much expanded (Figure 6.36). This differentiation of attack by magnesium sulphate is often noted and is particularly applicable to seawater attack on concrete (see Section 6.5.4).

The visible effects of sulphate attack are exfoliation of the outer layers of the concrete, corrosion and reduction of the hardened paste to what is commonly described as a 'soft pug' or a 'mush' and the deposition of salts on surfaces and in exfoliation cracks. These effects suggest that essentially two physical processes are occurring in sulphate attack: the expansive recrystallisation of salts in the surface of the concrete and direct corrosive attack on the hardened cement paste. However, which process predominates under any given set of conditions is still an open question.

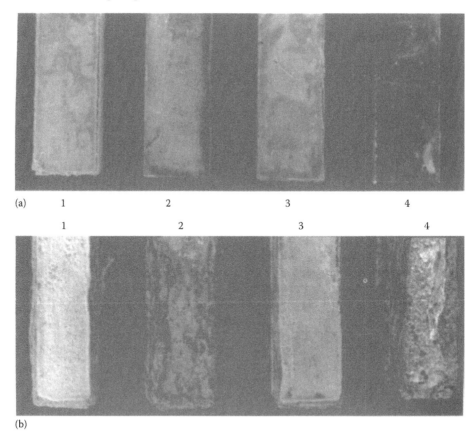

(a)   1   2   3   4

(b)

Figure 6.36 Some differences between sodium sulphate and magnesium sulphate attack. Laboratory specimens exposed for 140 days to (a) sodium sulphate solution and (b) magnesium sulphate solution. 1 = OPC, 2 = OPC + 15% silica fume, 3 = sulphate-resisting cement and 4 = sulphate-resisting cement +15% silica fume all mixed at a water/cement ratio of 0.3. The specimens attacked by magnesium sulphate have exfoliated and expanded much more than those exposed to sodium sulphate. The results are a good example of the variability that can occur in sulphate attack. (From Cohen, M.D. and Bentur, A., *ACI Mater. J.*, 85(3), 148, 1988.)

Figure 6.37 Fence posts in sabkha with a saline water table less than 0.5 m beneath the ground surface. Note the worst damage occurs just above ground level.

Descriptions of sulphate attack (and/or salt weathering: see Section 6.7.3) on field concretes above ground vary across a spectrum of examples where exfoliation of surfaces predominates (Figure 6.37) to cases where the hardened paste is corroded and softened and exfoliative processes appear absent. However, in many cases, illustrations of sulphate attack depict spectacular deterioration and disintegration of concretes that are the terminal results of attack where much of the evidence for the mode of attack has been destroyed. The examination of below-ground concretes presents special problems (Figure 6.38). Examination of concrete surfaces below ground is difficult because removal of soil will invariably remove much of the more fragile areas of attack at the concrete–soil interface. Thus, it is not possible to gain an accurate picture of the sulphate attack unless special care is taken in sampling below-ground concrete, leaving intact soil in place on the sample surface and later carrying out careful examination in the laboratory using a combination of systematic sub-sampling and then thin sectioning, scanning electron microscopy (SEM) and chemical analysis.

Since much of the evidence for relating the nature of visible attack to types of sulphate and other conditions appears to have been based on experience from accelerated laboratory testing, with thin section examination being rarely carried out, it is not surprising that controversy exists over the nature of sulphate attack.

In many cases of sulphate attack, salts are observed on surfaces. For instance, Novak and Colville (1989) found a range of efflorescent sulphates on floors and basements built over sulphate-rich soils. Analysis of these external deposits should always be made, but interpretations based on these analyses should be made with caution as they do not necessarily represent the processes of deterioration taking place inside the concrete.

Figure 6.38 Destruction of evidence caused by removal of adherent material. This pipe collapsed because of acid sulphate attack. When the pipe was removed from the ground, the staff sampling the pipe carefully removed the clay adhering to the outside of the pipe and destroyed vital evidence.

There is another, somewhat indirect, variant of sulphate attack which can cause heaving and cracking in foundations and floors. This is where the hardcore, fill or soil contains iron sulphides (typically pyrite), and sulphates are formed expansively in situ, which cause the subsoil to heave. In many cases, in the absence of a waterproof separating membrane, direct sulphate attack on the concrete underside also occurs. Lea (1968) and Nixon (1978) discussed the effect on ground-floor slabs laid over fill materials containing sulphates, and Stark (1987) has reported on a case of heaving of a pavement laid over a Portland cement–stabilised slag sub-base. In recent years, apparent problems of this nature, variously involving pyritic rock foundations and fill materials, have been reported in the Canadian province of Québec (Ballivy et al. 2002) (Figure 6.39); in Japan (Yamanaka et al. 2002), where bacteria were found to be involved in the pyrite decay process, and in the United States (Anderson 2008).

Similar cases have been sporadically reported in the British Isles, usually in association with notably pyritous rock fill materials. Some recent disputes in the Dublin area have centred on an allegation of damage to concrete floors from the pyrite-induced heave of an underlying fill or hardcore material, but there was no evidence found of any sulphate attack on the buried concrete elements, despite these being made without any deliberate intention to create sulphate-resisting concrete. Following the new principles in SD1 and the experiences in Québec and concerns in Ireland, the BRE has recently greatly modified its advice on hardcore for supporting ground floors of buildings (Longworth 2011). Also, in Ireland, particular guidance has now been issued by the Association of Consulting Engineers of Ireland (2012), and a report has been prepared for the Irish government by the Pyrite Panel (Tuohy et al. 2012).

6.5.3.3 Internal textures of concrete attacked by sulphate

Other than TSA (see Section 6.5.6) and DEF (see Section 6.5.9), there appear to be only a few published reports of petrographic examinations of sulphate-attacked concrete specimens using thin sections. Nielsen (1966) has presented brief descriptions, together with

Figure 6.39 Examples of damage to concrete foundations and floor slabs in Québec from heave caused by oxidation of pyrite in the underlying shales or rock fill, plus sulphate attack on the underside of the slab: (a) heaving of the concrete slab, (b) localised heaving leading to 'star cracking', (c) sulphate deposits along cracks and joints indicative of sulphate attack on the concrete, (d) displacement of foundation walls. (From Ballivy, G. et al., *Can. J. Civ. Eng.*, 29, 246, 2002 © Canadian Science Publishing or its licensors.)

photomicrographs of thin sections, showing the effects of eight months of attack on cement pastes (water/cement ratios 0.30, 0.34 and 0.38) by 0.07 molar solutions of sodium, magnesium and ferrous sulphates. Cracking was generally parallel to the surface suggesting the onset of exfoliation. In the case of sodium sulphate attack, the one crack present was filled with gypsum and only limited deterioration had taken place. Some ettringite was observed and appeared to have formed after the occurrence of the cracks. In the case of magnesium sulphate attack, gypsum was always precipitated on the surface of the specimen, and the magnesium hydroxide (brucite) was precipitated on the gypsum. An air void near the surface was filled with gypsum and ettringite was not observed. In the case of ferrous sulphate attack, gypsum was still the dominant reaction product together with various unidentified iron hydroxides. Ettringite was not observed. The outer layer of the specimen was undergoing exfoliative cracking, and the general impression was that deterioration was much greater than in the other solutions. A detailed representation of the microstructure of the zone of deterioration caused by magnesium sulphate attack has been given by Bonen and Cohen (1992). There appears to be a general agreement between these two reports.

Knofel and Rottger (1985) give a range of micrographs of textures of laboratory concretes that have been attacked by atmospheres enriched with carbon and sulphur dioxides. They found that carbonation penetrated more deeply into the hardened paste than sulphate attack, which was no more than a few millimetres deep. Blended blastfurnace slag cements performed slightly less favourably than OPC in their tests.

In most reports, examination is limited to visual assessment supplemented by XRD identification of the sulphates precipitated in the concrete. For instance, in the long-term investigation of sulphate attack at Northwick Park, Harrison and Teychenné (1981) of the BRE used XRD and a staining technique devised by Poole and Thomas (1975) for the identification of sulphate minerals. O'Neill (1992) has illustrated the appearance of some of the common sulphate minerals and concluded that their examination as powder mounts in refractive index oils is a useful method of identification.

Examination of thin sections cut from a concrete of which one side had been exposed to a mist containing high concentrations of sodium sulphate together with lesser amounts of sodium chloride gave the following textures. The surface zone was undergoing an extensive exfoliation (Figure 6.40) with many of the cracks filled with gypsum, which graded abruptly into apparently sound concrete. Some of the aggregates in this exfoliation zone were exfoliated and also filled with gypsum. No ettringite was observed in cracks or the texture generally. An extensive AAR was also occurring in the exfoliation zone involving some of the volcanic aggregates present (Figure 6.41). However, in spite of the extent of the AAR, the cracking was dominated by the sulphate attack, and the typical pattern cracking associated with expanding aggregates was absent. Since low-alkali cement had been used, the alkali

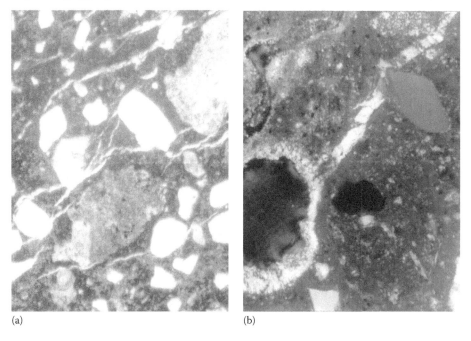

(a)   (b)

Figure 6.40 The texture of attack by sodium sulphate: (a) the zone of sulphate attack contains an extensive system of fine cracks that form an exfoliation texture; many of the cracks are filled with crystals of gypsum, width of field ~2.2 mm; (b) at a higher magnification and under crossed polars, it is seen that the gypsum tends to align itself across the cracks, width of field ~0.8 mm.

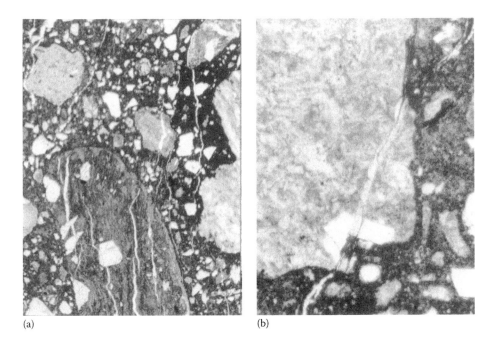

(a)   (b)

Figure 6.41 AAR associated with attack by sodium sulphate: (a) in the aggregates undergoing AAR, the cracking was often aligned with the exfoliation caused by the sulphate attack, width of field ~6.5 mm; (b) detail of another aggregate undergoing AAR; in this region of the sample, sulphate attack was not as apparent, width of field ~0.2 mm.

required for AAR was derived from the release of sodium from the sodium sulphate, and possibly the sodium chloride, reacting with the calcium hydroxide.

A similar texture was found by St John (1982) where concentration of sodium sulphate occurred by wicking and evaporation from porous concrete surfaces in some tunnels, of groundwater containing less than 50 ppm SO_4. In this case, non-reactive aggregates were present in the concrete so AAR was absent. The same pattern of exfoliation with some cracks filled with gypsum was present, and once again, the deteriorated exfoliated zone graded rapidly into sound concrete. Serious attack was restricted to porous concrete, as some denser concretes in adjacent rail tunnels subject to the same conditions were largely unaffected.

St John (1991) examined the acid sulphate attack on some storm water pipes (Figure 6.42), which has caused localised failures in portions of the business district of Rotorua, New Zealand. In this case, geothermal gases containing hydrogen sulphide and carbon dioxide dissolved in the groundwater at temperatures of about 50°C to form sulphuric and carbonic acids, although sulphuric acid probably predominates at the lower pH levels. It should be noted that the attack in this case is the reverse of that normally encountered in sewers where the crown is corroded owing to oxidised hydrogen sulphides (Section 6.5.5) and the concrete covered by *effluent* is often relatively sound. The case to be described indicates the complexity of sulphate attack and the type of information that can be gained by the examination of the concrete texture.

Two centrifugally spun pipes, corroded to the point of failure, were examined. One pipe (A) contained coarse olivine basalt and sand consisting of basalt and acid and intermediate volcanic particles. The other pipe (B) contained a weathered pyroxene andesite coarse aggregate and a sand consisting of acid and intermediate volcanic particles. The texture of both pipes suffered from faults typical of fabrication by centrifugal spinning (see Chapter 7), and both were extensively corroded and had collapsed as a result. The outsides of the pipes were extensively corroded,

(a)

(b)

Figure 6.42 Acid sulphate attack on storm water drainage pipes: (a) Because of acid sulphate attack, portions of the storm water drainage system collapsed. The inside surface of the crown of the pipe appears sound but the invert has undergone severe attack. On the piece that has broken away, the inside surface is 'puggy', and it can be seen from the fracture surfaces in the main portion of the pipe that both the inside and outside surfaces have been attacked to considerable depths. (b) Some idea of both the internal and external depths of attack can be gauged from the sawn surface of the sample.

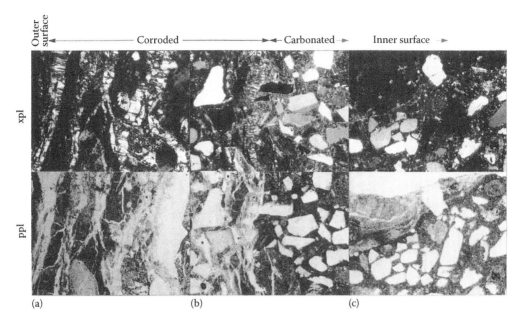

Outer surface

←———————— Corroded ————————→ ←— Carbonated —→   Inner surface →

xpl

ppl

(a)   (b)   (c)

Figure 6.43 Micrographs comparing the textures of the attack on the inside and outside surfaces of pipe A: (a) the hardened paste in the outer zone of the pipe has been completely corroded to an exfoliated amorphous silica with embedded aggregates. The cracks contain aligned gypsum crystals which are clearly visible under crossed polars, widths of field ~0.8 mm. (b) The texture shown in the these micrographs extends about 5 mm into the wall where it grades abruptly into a 5–6 mm wide zone of carbonation. The interface with this carbonation zone is shown here and stands out clearly under crossed polars, widths of field ~2 mm. (c) On the inside surface of the invert, the texture is more typical of acid attack. The texture shown is at the edge of a corrosion pit and consists of a corroded area separated by a thin zone of carbonation from the hardened paste. A leached zone is not apparent, widths of field ~2.8 mm.

even though this was not fully apparent from visual examination. Internally, severe corrosion of the inverts had occurred and the texture was soft and mushy. The surfaces of the crowns appeared intact apart from corrosion pits, which increased in frequency towards the inverts.

Examination of thin sections revealed a complex texture with some differences between the two pipes, which may have occurred because of differences in corrosive conditions and quality of the pipes (Figure 6.43). For these reasons, the attack on the two pipes may not be directly comparable. Externally, pipe A was completely corroded to a depth of more than 5 mm. The corroded layer contained dispersed sulphides and amorphous silica, but the aggregates were still intact. Some minor and limited exfoliation and cracking were present. There was a 5–7 mm carbonated zone behind the corroded layer, backed by a zone leached of calcium hydroxide. Internally, a 10 mm layer of the invert was also corroded in a similar manner to the outside of the pipe with some signs of vestigial gypsum. The transition zone to the hardened paste consisted of a narrow irregular and diffuse zone of carbonation, with a distinct yellowing of the texture backed by a narrow zone leached of calcium hydroxide. The remaining intact portion, about one-third of the wall thickness, contained veins of ettringite mainly deposited in fissures at aggregate margins.

In contrast to pipe A, the corrosion of the external surfaces of pipe B exhibited an exfoliation texture with aligned gypsum that was present in both paste and aggregates (Figure 6.44). The hardened cement paste had been completely corroded to amorphous silica and ettringite was not visible. A highly irregular 2–7 mm zone of carbonation was present behind

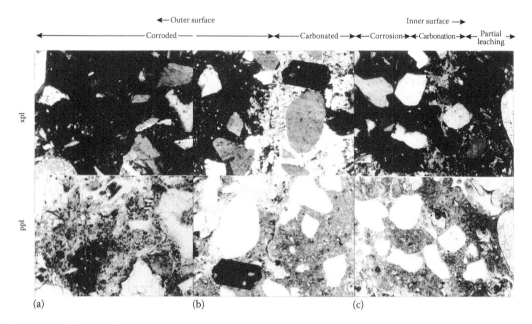

Figure 6.44 Micrographs comparing the textures of the attack on the inside and outside surfaces of pipe B, widths of field ~0.8 mm: (a) The texture shown is typical of the corroded zone. The hardened paste has been corroded to amorphous silica as shown under crossed polars. There is some darkening and yellowing due to the attack but gypsum is absent. (b) About 5 mm into the wall, a well-defined band of carbonation was present. The texture is typical of acid attack. (c) On the inside of the pipe, the corrosion zone consisted of amorphous silica bounded by a narrow irregular zone of yellowing. Under crossed polars (xpl), it is seen that this yellowing consists of a mixture of carbonation and an amorphous material. The zone of leaching is not well shown.

the corrosion layer. Irregular veins of carbonation were also present in the body of the wall. Internally, corrosion varied from pits in an otherwise intact surface, which was carbonated to a depth of about 0.5 mm, to complete corrosion of the surface at the invert. The remains of the corrosion layer were relatively narrow and consisted of a texture of amorphous silica containing exfoliation cracking and some aligned gypsum. The transition to sound paste was discontinuous and consisted of a yellow zone and a thin layer of carbonation. Although the crown was not sectioned, there were indications that the body of the concrete would also contain veins of carbonation. Given the moist conditions, high concentration of carbon dioxide in the geothermal gases and the fact that pipe B was fissured along many aggregate margins, this type of deep carbonation was to be expected.

The cases described illustrate the complexity of sulphate attack on concrete. Attack varied from simple exfoliation of concrete to complex corrosion. Where simple exfoliation occurred, ettringite was not detected and it is of interest in this respect to note that Ludwig (1980) claims sulphate attack is always initiated by ettringite forming just below the surface causing cracks. The ettringite is then decomposed and the cracks are filled with gypsum. The gypsum formation is a secondary reaction and may possibly contribute to the total expansion.

The scarcity of other published reports on the examination of thin sections of sulphate attack is surprising as this type of examination is essential for providing useful information on the textural aspects of deterioration. Where it is necessary to determine the depth of deterioration, large-area thin-section examination will be required. Considerable care is required to see that the delicate and fragile surface areas are not damaged during sampling and are retained in the section for examination. Because of the possibility that some secondary ettringite may be present at the submicroscopic level, optical microscopical examination must usually be supplemented by detailed examination of selected areas using SEM, supported by EDX-type microanalysis, XRD and general chemical analyses, as necessary. However, using SEM, care will be required to distinguish any *secondary* ettringite from *primary* ettringite formed during the cement hydration process.

It will be seen in Section 6.5.6 that many cases previously thought to be 'unusual' cases of conventional sulphate attack were actually fairly classic examples of the TSA.

6.5.4 Seawater attack

Seawater attack on concrete has been the subject of investigation since before the turn of the last century. The state of many of these investigations was reported in the international conference on 'Performance of Concrete in Marine Environment' (American Concrete Institute 1980). More recently, the potential problem has become of considerable significance for the construction and durability of offshore oil installations.

Open ocean water contains approximately 3.5% of soluble salts by weight consisting of approximately 1.1% sodium, 0.04% potassium, 0.1% magnesium, 0.04% calcium, 2% chloride, 0.3% of sulphate (SO_4) and traces of other materials. The pH of seawater varies between 7.8 and 8.3 with about 10 ppm of carbonate, 80 ppm of bicarbonate and small amounts of free carbon dioxide present. In areas of partially enclosed water, for example, the Arabian Gulf, salt concentrations can rise to 5% or even 10% in localised areas (Fookes et al. 1986; Sims 2006).

Deterioration caused by saltwater attack on concrete is most severe in the intertidal and splash zones, because wetting and drying can drive salts into the concrete surface and cause salt crystallisation. Wave impact and the physical abrasion from suspended particles and, in some cases, ice can destroy the protective surface layer of brucite and aragonite which forms in contact with seawater. In tropical climates, rates of attack are considerably increased and, in cold climatic regions, the areas subject to wetting and drying are also especially prone to cracking due to freeze–thaw attack. However, in practice, the principal cause of deterioration of concrete in seawater is due to the ingress of chloride ions causing corrosion of steel reinforcement. This is fully discussed in Section 6.3.

The principal components in seawater responsible for the chemical deterioration of hardened cement paste are magnesium and sulphate. These can attack the calcium hydroxide and aluminate phases according to the equations given in Section 6.5.3, but in practice, seawater attack does differ in some respects from sulphate attack. Lea (1970) considered the action of seawater as one of several reactions proceeding concurrently:

> Leaching actions remove lime and calcium sulphate while reaction with magnesium sulphate leads to the formation of calcium sulphoaluminate which may cause expansion, rendering the concrete more open to further attack and leaching. The deposition of magnesium hydroxide blocking the pores of the concrete probably tends to slow up the action of dense concretes though on more permeable concretes it may be without much effect.

He also states that unless the pH falls below 7, as can occur in sheltered estuaries where organic matter may cover the seabed, attack by free carbon dioxide is minimal.

A study of mortar samples immersed in artificial seawater for periods of up to 3 years (Conjeaud 1980) indicated that SO_4 and Cl^- diffuse rapidly into the mortar. The rate of diffusion is rapidly slowed by the development of an almost impermeable layer of aragonite forming over brucite in the surface of the mortar, owing to attack by bicarbonate and magnesium ions. This mechanism and the reduction in permeability have been confirmed by Buenfeld and Newman (1984, 1986) who also illustrate some of the textures (Figure 6.45).

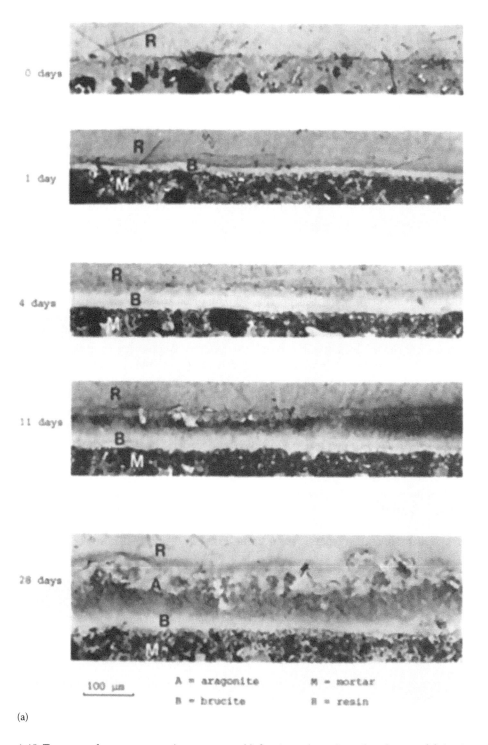

(a)

Figure 6.45 Textures of seawater attack on mortar. (a) Sections through surface layers of 0.4 w/c ratio OPC mortar after various periods of immersion in sea-water. (After Buenfeld, N.R. and Newman, J.B., *Mag. Concrete Res.*, 1984; Buenfeld, N.R. and Newman, J.B., *Cement Concrete Comp.*, 16(5), 721, 1986.)   (*Continued*)

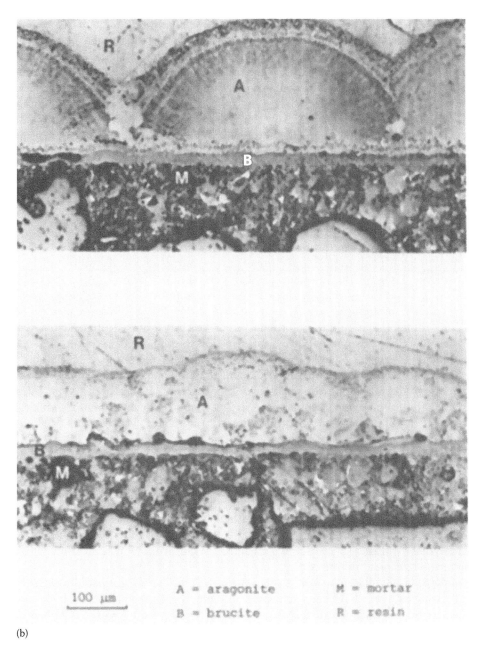

(b)

Figure 6.45 (Continued) Textures of seawater attack on mortar. (b) Sections through surface layers of 0.4 w/c ratio OPC concrete after 33 weeks of immersion in sea-water. (After Buenfeld, N.R. and Newman, J.B., *Mag. Concrete Res.*, 1984; Buenfeld, N.R. and Newman, J.B., *Cement Concrete Comp.*, 16(5), 721, 1986.) (Note: Unfortunately the original photographs are no longer available for reproduction.)

Although these layers protect submerged concrete from further deterioration, concrete and mortar surfaces in the intertidal zone are subject to greater damage due to salt crystallisation, impact and abrasion.

Clearly, the permeability of the concrete is the most important factor in determining the resistance of concrete to chemical attack by seawater. Dense impermeable concrete is most resistant to damage, and Massazza (1985) concludes that pozzolanic or slag cement (>60% slag) makes a more durable marine concrete than OPC and noted that lime–pozzolana mortars used by the Romans in sea works are still in good condition. Concrete made with high-alumina cement (HAC) under conditions which prevent strength regression is highly resistant to seawater and sulphate attack. Although calcium aluminate hydrates are slowly attacked by sulphate solutions with the formation of ettringite, there is no calcium hydroxide in the cement paste fabric to react with sulphates and consequently the concrete remains relatively impermeable.

The chemical reactions within the hardened cement paste can produce gypsum, calcium chloride and calcium bicarbonate, all of which are soluble in seawater and may be leached. Ettringite and calcium carbonate are also formed. The ettringite may be associated with cracking and expansion of the concrete and calcium carbonate develops as aragonite rather than calcite in the presence of magnesium. The magnesium hydroxide (brucite) formed is almost insoluble in seawater and remains in situ.

Brucite can be difficult to identify positively in petrographic thin sections of marine concretes because it is sometimes present as small crystallites. However, by contrast, Sibbick (2009) found large sheets of secondary brucite in a number of investigations of marine attack. Figures 6.46 and 6.47 are taken from two of the case studies presented by Sibbick (2009). He thought that this thick brucite layer, which was found often associated with aragonite and calcite precipitated layers, may well have even acted in a protective manner to the underlying cement paste, possibly reducing further calcium leaching. Sibbick validated the brucite identity using SEM–EDX chemical mapping (see Figure 6.48). Sibbick also confirmed that a combination of factors, working together, exacerbated the marine-based degradation (he found evidence of some ASR, poor mixing, cement balling and variable water/cement ratio); if one or more of these aggravating factors was reduced or absent, the resulting marine degradation was mitigated or seriously reduced.

The birefringence of brucite is moderate ($\eta_\xi - \eta_\omega = 0.019$) and it may exhibit anomalous reddish-brown colours, but because of the fine-grained scale-like nature of the crystallites in the cement paste, these features are often masked. Because the refractive indices and birefringence of calcite and aragonite are similar, it is only possible to report the presence of a carbonate. Usually, the size of the crystallites in the cement paste prevents other optical properties from being determined so that XRD, electron microscopy with EDX-type microanalysis or infrared spectroscopy is required to confirm the identifications of brucite, aragonite and calcite in marine concretes.

The alteration of concrete in zones adjacent to the outer surface is usually clearly developed in concretes which have been subject to seawater attack, and these zones develop and widen with time provided the surface is not removed by abrasion. Figure 6.49 illustrates an example of these zones, which extend up to 10 mm into the mass concrete of a boat slipway which had been exposed to seawater for 60 years. The concrete is of poor quality in spite of a high cement content. The cement paste is completely altered to fine scale-like brucite crystals in a zone extending almost 2 mm inwards from the surface. There is then a zone 5–6 mm wide in which large crystallites of aragonite are developed as irregular patches in a microporous paste, evidenced by a darkening of the paste because the particle size of the remaining materials is small enough to cause destructive interference of light. In the final innermost zone, the cement paste becomes pale coloured because of the extensive development of very small ettringite crystals in the paste itself and in air voids.

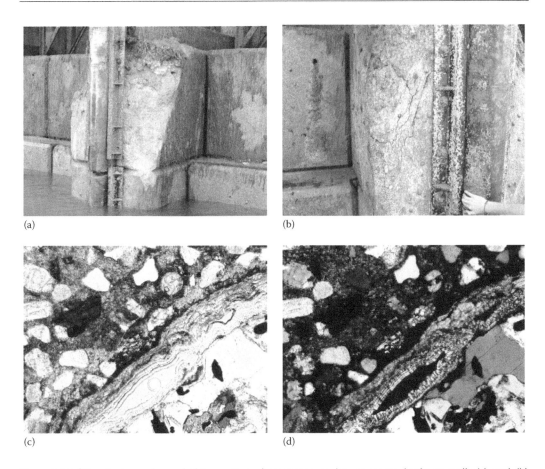

(a)   (b)

(c)   (d)

Figure 6.46 Site photographs and thin-section photomicrographs, concrete harbour wall: (a) and (b) surface deterioration, (c) and (d) photomicrograph of a large microcrack surrounding a coarse aggregate particle, filled with secondary calcite and brucite. The adjacent cement paste is also heavily depleted, ppl (c) and xpl (d), width of field ~1.8 mm. (From Sibbick, R.G., Microscopic examination of degradation processes in cement based materials associated with the marine environment, Presented at the *31st International Cement Microscopy Association (ICMA) Conference*, St Petersburg, FL, 2009.)

Figure 6.50 illustrates a similar pattern of zonation, but because this concrete is dense and impermeable, and has been exposed to seawater for less than 10 years, the zones only extend 3 mm into the concrete from its outer surface.

6.5.5 Sulphate attack in sewerage systems

Normal sewage effluents are alkaline and do not attack concrete in contact with them directly. However, anaerobic bacteria present in sewerage pipes can decompose the inorganic and organic sulphur compounds present, releasing hydrogen sulphide to the atmosphere above the effluent. This gas is absorbed by the water film coating the pipe walls, which contain aerobic bacteria. The aerobic bacteria oxidise the hydrogen sulphide to sulphurous and sulphuric acids, which attack and dissolve the hardened cement paste (Thistlethwayte 1972; McClaren 1984). The result is the development of a concrete surface on the crowns of sewer pipes consisting of aggregates standing proud of severely corroded cement paste.

Figure 6.47 Site photographs and thin-section photomicrographs, mortar bedding joints in stone masonry of sea and estuary walls: (a) Sampling masonry mortar from the intertidal zone. (b) Photomicrograph of a mortar top surface, exhibiting extensive surface-parallel microcracking and abundant calcite precipitation, ppl, width of field ~3.5 mm. (c) Photomicrograph of the seaward-exposed mortar top surface, showing white/colourless brucite lining voids, ppl, width of field ~3.5 mm. (d) and (e) Photomicrograph of a large void extensively filled with brucite (crystalline and gel type) and scattered 'popcorn' calcite crystals, ppl (d) and xpl (e), width of field ~3.5 mm. (From Sibbick, R.G., Microscopic examination of degradation processes in cement based materials associated with the marine environment, Presented at the *31st International Cement Microscopy Association (ICMA) Conference*, St Petersburg, FL, 2009.)

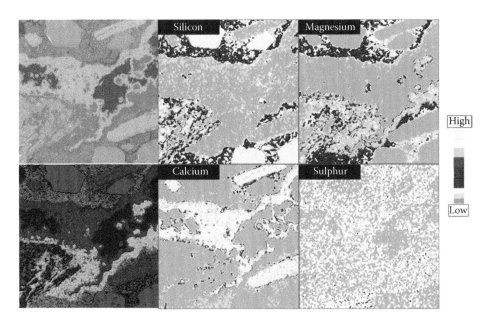

Figure 6.48 Validation of brucite composition using SEM–EDX chemical mapping on polished surfaces: top left shows the backscattered electron image, while the chemical maps for the same area are shown (as labelled) for silicon, magnesium, calcium and sulphur, with the colour scale relating to relative intensities for each element; a combined composition map is included in the bottom left, with silicon highlighted in blue, calcium in red and magnesium (indicating the presence of brucite) shown in green. (Material courtesy of Dr Ted Sibbick.)

This acid corrosion of the crown of a sewerage pipe will be most severe if the sulphur content of the effluent is high and the flow rates slow or stagnant, so that the sewer becomes septic as shown in Figure 6.51. Temperature and oxidation–reduction potential will also influence the growth of bacteria and hence the rate of production of sulphuric acid. A detailed study of corrosion of sewerage pipes was undertaken by CSIR South Africa (1967), and these studies were followed up by Barnard (1967), Thistlethwayte (1972) and Pomeroy (1977).

One case examined involved an acid attack on the crowns of some 1300 mm diameter main sewerage pipes made with 50 MPa concrete and greywacke aggregate. In this case, the acid attack on the concrete was stopped before the pipes were too severely damaged, by running the pipes full of sewage and thus preventing any further oxidation of the hydrogen sulphide.

The example shown (Figure 6.52) has been taken from the crown of a large-diameter concrete sewerage pipe, which has undergone acid attack from the oxidisation of hydrogen sulphide by aerobic bacteria. The acid dissolves the hardened cement paste leaving the larger aggregates standing proud of the surface. In severe cases, the aggregates fall out as the depth of attack increases.

The surfaces examined had been in contact with sewage for some period, and it is likely that some of the outer corroded material may have been removed. The external appearance of the corroded surfaces (Figure 6.52) was brownish yellow, and aggregates were standing proud of the corroded cement paste. In thin section (Figure 6.53), the texture of the corroded zone consisted of an approximately 250 μm wide outer layer of amorphous brown material intermixed with carbonation products, backed by a 2 mm zone of hardened paste heavily leached of calcium hydroxide. Behind the leached layer, there was a zone of intense carbonation varying from 250 μm to 2 mm in width, which graded into sound paste.

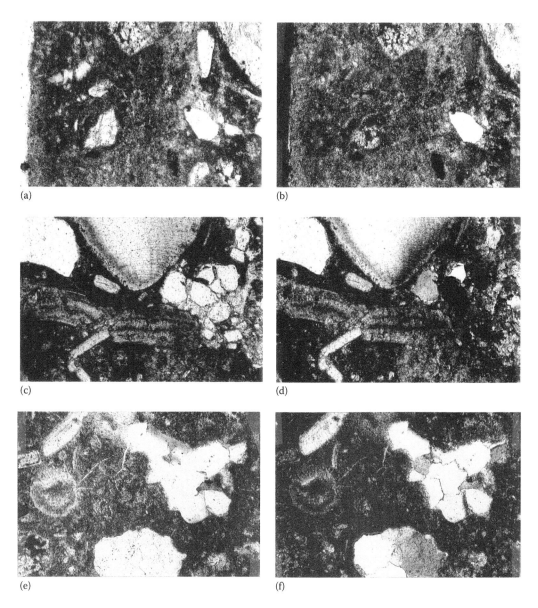

Figure 6.49 The zones of alteration in a concrete exposed to seawater attack. The width of field of each photomicrograph is ~1.4 mm: (a) The hardened cement paste in the outer zone is largely altered to fine-grained brucite and carbonation; the exterior surface is at the left-hand edge. (b) Under crossed polars, the carbonation is clearly visible but the brucite is too fine grained to be visible. (c) Part of the zone between 3 and 7 mm from the surface showing patchy darkening of the paste due to leaching of calcium hydroxide and the development of aragonite. (d) Under crossed polars, the band of aragonite is visible and calcium hydroxide is largely absent from the hardened cement paste; shell fragments form part of the aggregate. (e) Innermost zone showing development of acicular ettringite crystals in the matrix and small air voids; the small shrinkage cracks are due to drying of the specimen. (f) Under crossed polars, low birefringence distinguishes the ettringite in the pore.

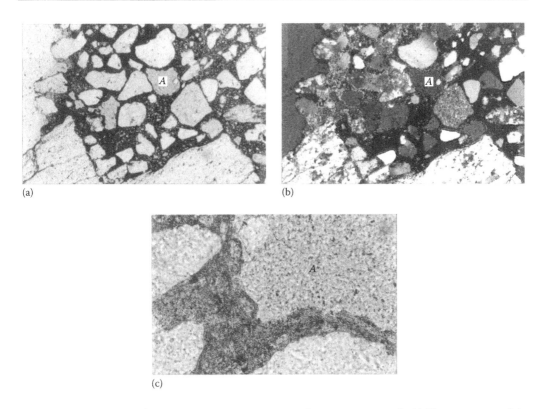

(a)

(b)

(c)

Figure 6.50 The texture of a precast concrete pipe exposed to seawater attack: (a) The outer zone (left-hand side) is altered to brucite and calcite, which merges into a darkened porous zone leached of calcium hydroxide with development of ettringite in the cement matrix and air voids, width of field ~3.8 mm. (b) Under crossed polars, the outer zone is clearly shown by the carbonation, width of field ~3.8 mm. (c) Enlarged view of the area around the air void (*A*) showing the ettringite development both in the void and in the surrounding matrix, width of field ~0.2 mm.

The texture of the corroded surface of the sewerage pipes is similar to that observed by Chandra (1988) and De Ceukelaire (1992b) and is probably typical of most forms of acid attack, where the hardened cement paste is solubilised. A fragile outer zone is formed where the acid attack has gone to completion, and only hydrated amorphous silica and alumina remain, which may be lightly stained by iron compounds. This outer corroded zone can vary from as little as 20 μm to several millimetres in depth. Inside the corroded zone is a reaction layer consisting of carbonated paste intermixed with brown amorphous material derived from the iron-bearing phases in the hardened paste. Although the corrosion zone may appear as separate layers of brown amorphous material and carbonation, careful examination will usually show that this separation can rarely be justified. Inside the reaction layer is a zone of hardened paste, which is depleted in calcium hydroxide and merges into sound paste. This last zone is often missed in examination, but must always be present as acid attack inevitably sets up chemical gradients. The most obvious evidence of the chemical gradient is some depletion of calcium hydroxide.

The bacterial activity that is described in this section is probably the best known form of 'biodeterioration' that affects concrete materials and products, but the deleterious involvement of such microorganisms is probably much more widely implicated in a range of degradation processes, which similarly affect building stone. A useful, well-referenced and indeed award-winning overview on this subject was recently published

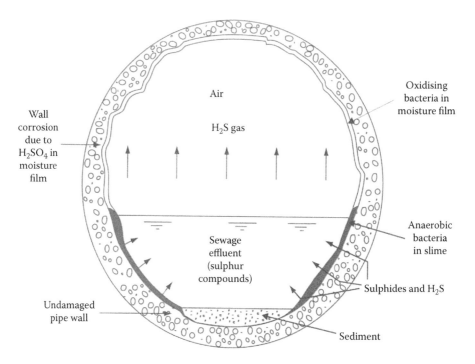

Air

H₂S gas

Oxidising bacteria in moisture film

Wall corrosion due to H₂SO₄ in moisture film

Anaerobic bacteria in slime

Sewage effluent (sulphur compounds)

Sulphides and H₂S

Undamaged pipe wall

Sediment

Figure 6.51 Diagrammatic representation of the mechanisms leading to corrosion in a sewerage pipe.

Figure 6.52 The effect of bacterial acid attack on the crown of a sewerage pipe.

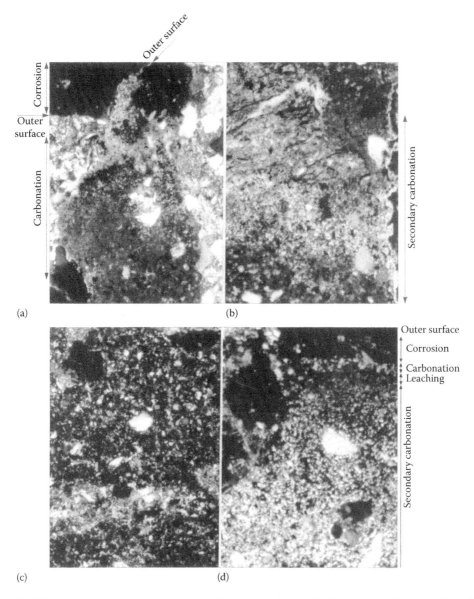

Figure 6.53 The corrosion texture of attack by the bacterial oxidation of hydrogen sulphide, xpl, widths of field ~0.9 mm: (a) At the outer surface between two greywacke aggregates, which are standing proud, the hardened cement paste is altered and consists of a complex mixture of carbonation and staining from iron compounds. Behind this is a layer of paste depleted in calcium hydroxide with red specks of iron compounds. Overlaying parts of these zones is an irregular layer of amorphous corroded paste (not shown here) which is often stained with yellow and red material. (b) Varying in depth of 0–2 mm behind the zonation caused by the acid attack is a separate zone of carbonation which merges into sound cement paste. This interior zone of carbonation has probably occurred subsequent to the acid attack and shows that the carbon dioxide present in the sewer has been able easily to penetrate the acid corrosion layer. Note how carbonation has concentrated along a fissure at the aggregate margin at the right side of the micrograph. (c) The sound hardened paste in the interior of the pipe wall consists of a rich dense texture with numerous remnant and partially hydrated cement grains in a close packing of aggregates. (d) In some areas, the subsequent carbonation has merged with the initial acid attack. Unless large sample areas are examined, it is these variations in texture that can lead to erroneous conclusions. The textures shown in (a) and (b) dominated the surfaces examined.

by De Belie (2010), also including explanations of ways in which some bacteria can be used beneficially for cleaning or consolidation purposes.

6.5.6 Thaumasite sulphate attack (TSA)

The importance of the TSA, as a separate and often more rapid and more destructive mechanism of sulphate action on concrete, only became recognised shortly after the original publication of this book. After serious concrete degradation of the buried elements of some M5 motorway bridges in western England was identified in 1998 as being associated with the occurrence of thaumasite, the responsible U.K. government department, then known as the Department of Environment, Transport and the Regions (DETR), appointed an 'expert group' of engineers and scientists, under the leadership of Professor Leslie Clark, in order to establish the causes of this unusual deterioration and how it might be prevented in future concrete construction.

Clark's first report (DETR 1999) remains an essential reading on the topic of TSA, though some of the initial recommendations of his Thaumasite Expert Group (TEG) have since been modified (DETR 2000, 2002; Nixon and Longworth 2003; Building Research Establishment 2005). We have already seen (Section 6.5.2) how the research arising from the TEG work led to the publication of BRE Special Digest 1 (2001, 2003, 2005), with new procedures for assessing the aggressiveness of ground towards buried concrete. Much associated research on thaumasite was published in association with the First (and so far the only) International Conference on Thaumasite in Concrete, which was held at the BRE (2002) (from which some of the papers were subsequently published in Cement and Concrete Composites in 2003 and 2004). Introductory guides on thaumasite can be found in Hartshorn and Sims (1998), Sims and Hartshorn (1998), Bensted (1999) and Crammond (2002). The implications for concrete specification were considered by Hobbs (2003).

6.5.6.1 Nature of TSA

TSA is different from the conventional form of sulphate attack (see Section 6.5.3) in that

- It occurs preferentially at low temperatures (typically below 15°C);
- The chief reaction product is thaumasite, although ettringite or gypsum may also be present;
- It degrades the hardened cement paste, primarily the main calcium silicate phases (rather than the calcium aluminate phases), resulting in a soft paste-like matrix;
- It requires availability of carbonate ions as well as sulphate ions.

In tunnels and other underground works where high humidity and low temperatures prevail, the possible presence of thaumasite should always be considered. Hagelia et al. (2001) have reported the presence of extensive TSA degradation of concrete and shotcrete tunnel linings in the presence of sulphide-bearing alum shales. Sibbick (2011) advises that the most severe cases of TSA occur when there is an abundant source of carbonate. Such carbonates have been shown to have been sourced from limestone aggregates (DETR 1999), limestone fillers in the cement (Hartshorn et al. 1999) and bicarbonate in groundwater (Collett et al. 2004). French and Crammond (2000) and French (2002) have noted the common concurrence of thaumasite and 'subaqueous carbonation' in concrete in wet ground. An account of the occurrence of TSA in the United Kingdom is given by Crammond (2002).

In conventional sulphate attack, the nucleation and deposition of excess ettringite cause expansion and cracking while maintaining the bulk of the concrete's strength, whereas TSA can result in severe loss in strength and loss of section of affected members. Softening and loss of surface material can result in reduced cover to reinforcement and, ultimately, reinforcement corrosion. Loss of surface material can also affect the frictional restraint of buried concrete piles, for example, potentially leading to structural issues.

TSA, therefore, appears potentially to have more serious consequences than the formation of ettringite. Given conducive conditions, the hardened cement paste can become completely disintegrated by softening. SRPC may not provide full protection against this form of sulphate attack, as it is not the aluminate component of the cement hydrates under attack, but the all-important calcium silicate constituents. Concretes containing ground granulated blast furnace slag (ggbs) have been shown to have good resistance to TSA, as well as the conventional form of sulphate attack (DETR 1999; Higgins and Crammond 2002; Higgins 2002).

Recently, a number of historic cases of 'unusual' conventional sulphate attack have been reassessed in the light of current understanding and found to be classic TSA and/or TF degradation (Hagelia and Sibbick 2009). In fact, according to Sibbick (2011), many more cases examined in recent times have appeared to be TSA rather than conventional sulphate attack.

As explained earlier, TSA was acknowledged widely as a potentially serious durability issue in 1998 following its identification in concrete bridge foundations on the M5 motorway in Gloucestershire, United Kingdom. The DETR urgently appointed a group of experts (the TEG) to investigate and report on the safety, commercial and scientific issues which were raised by this apparently 'new' problem. The TEG published their report in January 1999 (DETR 1999), giving guidance on the identification of structures at risk, the diagnosis of TSA and its assessment, the prognosis for affected structures and guidance for the specification of concrete which would mitigate the effects of TSA. Further research was initiated, and reviews of the TEG report were commissioned at 1-year (DETR 2000) and 3-year (DETR 2003, Nixon and Longworth 2003) periods, to update the scientific knowledge based on new research findings and to issue further guidance based on the initial report.

The TEG report (DETR 1999) identifies two stages in the deterioration of concrete as a result of thaumasite: thaumasite *formation* (TF, see Figure 6.54) and the *TSA* (see Figures 6.55 and 6.56). TF identifies a condition in which thaumasite is present in the voids of structural concretes but is not actively contributing to the degradation of the matrix. TF is widespread in many foundation concretes associated with sulphate-bearing ground or groundwater conditions. While the occurrence of TF may be a precursor to TSA, the

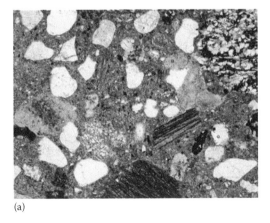
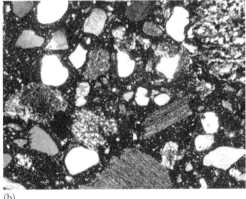

(a)   (b)

Figure 6.54 TF in thin section; note thaumasite occurs as void infillings but does not affect the integrity of the cementitious matrix: (a), ppl; (b), xpl, widths of field 4 mm. (Photomicrographs courtesy of Dr Ted Sibbick.)

Figure 6.55 TSA in hand specimen. (From DETR, The thaumasite form of sulphate attack: Risks, diagnosis, remedial works and guidance on new construction, Report of the Thaumasite Expert Group (TEG), Department for the Environment, Transport and the Regions (DETR), London, U.K., 1999.)

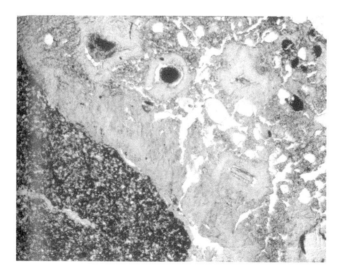

Figure 6.56 TSA in thin section, ppl, x25 magnification: a dolomite coarse aggregate (left) is surrounded by a thick deposit of thaumasite (yellow in ppl); fine aggregate particles of carbonate and shell fragments are also surrounded by thaumasite haloes. (From DETR, The thaumasite form of sulphate attack: Risks, diagnosis, remedial works and guidance on new construction, Report of the Thaumasite Expert Group (TEG), Department for the Environment, Transport and the Regions (DETR), London, U.K., 1999.)

development of the more serious TSA condition is not inevitable. Ground conditions and concrete chemistry must be favourable (Crammond 1985) for the nucleation and growth of thaumasite and the consequential conversion of sound cement paste to mush. Unlike ettringite, thaumasite contains a carbonate component which can be sourced from the aggregate, from the cement (Hartshorn et al. 1999) or from the atmosphere (Yang et al. 1999) or from bicarbonate-rich groundwaters (Collett et al. 2004).

The M5 bridges in the United Kingdom were not the first structures to have been affected by TSA, and there are numerous published articles which identify thaumasite as the main sulphate-bearing phase in the affected concrete. The first reported occurrence in concrete was identified by Erlin and Stark (1965/1966) in mortar and concrete that had undergone sulphate attack. The occurrence was associated with ettringite either as overgrowths or bands of thaumasite in the ettringite crystals. Van Aardt and Visser (1975) identified rapid thaumasite deterioration of autoclaved mortar bars containing dolomitic aggregate exposed to sulphate solution at 5°C. Lukas (1975) and Berra and Baronio (1987) identified thaumasite in deteriorated tunnel concretes, Gouda et al. (1975) in soil cements and Crammond (1985) and Moriconi et al. (1994) in mortars and renderings from exposed brickwork. Thaumasite has also been identified in hard wall gypsum plasters in contact with Portland cement (Leifeld et al. 1970; Ludwig and Mehr 1986). Deterioration of concrete due to the formation of thaumasite from a sulphide-bearing slate aggregate has been described by Oberholster et al. (1984). The occurrence of thaumasite in North America is discussed by Bickley et al. (1994). TSA is most commonly encountered in underground locations (such as foundations, tunnels and pile caps) where increased moisture and lower temperatures are more common.

Crammond (1985) states that thaumasites can form rapidly if the temperature is around 4°C, there is constant high humidity, the initial reactive alumina (i.e. the alumina in the calcium aluminate phase) is between 0.4% and 1.0% and there is an adequate supply of available sulphate and carbonate anions. It is not clear whether or not thaumasite forms by conversion from ettringite, but Taylor (1990) considers that ettringite is only required as a nucleating agent. The microscopic and XRD identification of thaumasite is difficult and it can be easily confused with ettringite as it has a similar fibrous character, but in favourable circumstances, as shown in Figure 6.57, its first-order yellow interference colour in 30 μm thick sections under crossed polars aids identification.

Care should be taken in interpreting sulphate attack in areas where the preferred conditions are not met, for example, in dry, high-temperature regions. TSA has been identified in regions where it is not apparent that the preferred conditions prevail (Sims and Huntley 2002, 2004; Torres et al. 2002; Vieira et al. 2010). While most cases of TSA occur in relatively low-temperature conditions, such as less than 15°C, with perhaps around 4°C seeming to be an optimum level for TSA development, severe degradation has still been found at up to 20°C. In some cases, this has been linked to another factor, such as an excessive source of sulphate (Sibbick and Crammond 2001).

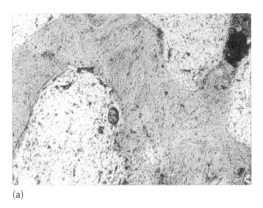

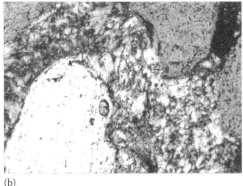

(a)   (b)

Figure 6.57 The appearance of thaumasite, widths of field ~0.75 mm: (a) The matted felt of fibrous looking crystals is quite similar to massive deposits of ettringite. (b) Under crossed polars, it is immediately apparent that the fibrous material has a higher birefringence than ettringite and is showing traces of first-order yellow birefringence which helps to identify it as thaumasite.

6.5.6.2 Petrography of TSA

Owing to the included water within the thaumasite and ettringite structures, it is critical that samples are dried carefully to avoid desiccation and denaturing of the sulphate minerals. Samples should be dried at 40°C, followed by vacuum impregnation with a dyed epoxy resin. Fluorescent dyes, while useful for other microscopical examinations, are not recommended for TSA diagnosis as these may mask the subtle colours of thaumasite under the optical microscope. Once impregnated, microscopical subsamples should be subject to the same careful handling as other concrete materials.

Thaumasite is a complex calcium silicate carbonate sulphate hydrate mineral with the chemical formula $CaSiO_3 \cdot CaCO_3 \cdot CaSO_4 \cdot 15H_2O$. It is a naturally occurring, but rare, mineral in metamorphic rocks and has also been identified as a reaction product in cementitious materials which have undergone sulphate attack (see Section 6.5.6.1).

Figure 6.55 shows an advanced case of TSA where, in hand specimen, the thaumasite is abundant as distinctive white haloes of crystalline thaumasite surrounding limestone coarse aggregate particles. Thaumasite also appears in voids and within areas of the cement matrix itself. In buried structures, typically, thaumasite can be observed in subparallel bands or lenses within the concrete, parallel to the outer surface of the affected concrete unit in contact with the ground (Figures 6.56 to 6.58). Thin-section microscopical examination of TSA-affected concrete also typically shows carbonate aggregate particles surrounded by haloes of thaumasite, with thaumasite intergrowths in cracks and voids (Figure 6.59). Although it is now recognised that TSA can occur in concretes without carbonate aggregates (Sims and Huntley 2004), there is sometimes clear evidence for the consumption of limestone aggregates in the formation of thaumasite (Figure 6.60).

The TEG report (DETR 1999) identifies three distinct types of thaumasite in affected concrete:

- **Type I** thaumasite is colourless to pale yellow with a disordered, porous nature and very low birefringence (0.000–0.005), being almost isotropic in places (Figure 6.61). Ettringite, by comparison, has a birefringence of 0.006, so it is easy to understand how the two phases can be confused;

Figure 6.58 Thin-section appearance of banded TSA: width of field ~4 mm, ppl. (Photomicrograph courtesy of Dr Ted Sibbick.)

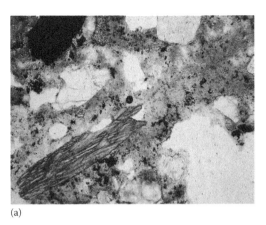
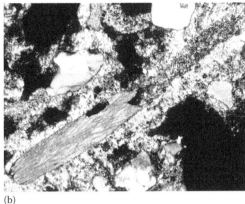

(a)   (b)

Figure 6.59 Thin-section view of thaumasite halo around a shell fine aggregate particle, width of field ~2 mm: (a), ppl; (b), xpl. (Photomicrographs courtesy of Dr Ted Sibbick.)

- **Type II** thaumasite is colourless to straw yellow, with higher birefringence (0.005–0.007, white to light yellow) (Figure 6.62). The crystals grow in more ordered rosette-like bunches, displaying a much less porous nature than the Type I variety;
- **Type III** thaumasite crystals have a well-structured wavy appearance (Figure 6.63). They are colourless to yellow in plane polarised light, with a birefringence (under crossed polars) of 0.007–0.027 (first-order yellow to second-order yellow). Typically, TSA in affected concretes shows a combination of the three types, and sometimes ettringite is also present.

Chemical analysis of concrete for sulphate content is useful to confirm that sulphate attack is the primary cause of concrete deterioration, but cannot distinguish TSA from other forms of sulphate attack. Other instrumental methods must be employed.

Conventional petrographic (optical microscopy) techniques have proven useful in differentiating thaumasite and ettringite; however, care should be exercised by the petrographer as thaumasite and ettringite have similar crystal structures and optical properties. Both thaumasite and ettringite have a trigonal crystal system, forming acicular (needle-like) crystals. In natural samples, the birefringence of thaumasite (0.036) is much higher than that of ettringite (0.006), but in deteriorated concrete, the difference in birefringence is sometimes difficult to discern, particularly in samples where both phases are present, or where mixed crystals (ettringite/thaumasite solid solution) have formed. In this respect, it can sometimes be very difficult to distinguish one from the other using optical microscopy alone.

Based upon their extensive practical investigations of samples from structures, Sibbick et al. (2002) demonstrated that TSA develops within hardened concrete through four progressive stages of degradation, which they characterised as Zones 1 to 4, for the least to the most affected, respectively (see Table 6.8 and Figure 6.64). As TSA comprises deterioration of concrete from the exposed outer surface inwards, at any one time, the most affected material will be in the surface zone, with the least affected area being at the maximum penetration into the concrete, adjacent to the unaffected concrete at greater depth. Figure 6.64 shows an idealised sectional representation of a severely affected but previously good-quality concrete covering reinforcement, which is still protected by some residual depth of unaffected and hardly affected (Zone 1) concrete material. Sibbick et al. found that the zones are usually still apparent, but less well defined, within lower-quality concrete or other cementitious materials (such as mortars and renders).

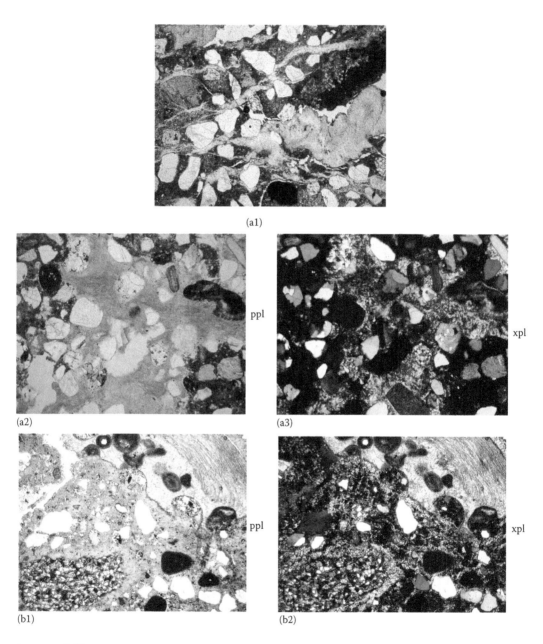

Figure 6.60 Thin-section photomicrographs showing advanced TSA, in which the whole binder is converted to thaumasite: (a) width of field ~3.5 mm: (a1) banded ppl, (a2) ppl, (a3) xpl. (b) Another example width of field ~3.5 mm, b1: ppl, b2: xpl. (Photomicrographs courtesy of Dr Ted Sibbick.)   (*Continued*)

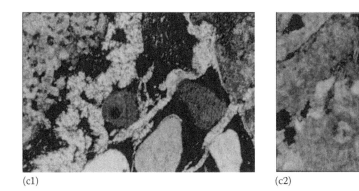

(c1)   (c2)

Figure 6.60 (Continued) Thin-section photomicrographs showing advanced TSA: (c) another example, width of field ~2 mm and xpl: (cl) complete replacement of cement matrix by thaumasite, (c2) consumption of limestone aggregate in TSA, note quartz inclusions in the original limestone, some of which are now isolated within the thaumasite product and some of which are partly left in the limestone and partly within the thaumasite-replaced matrix. (Photomicrographs courtesy of RSK Environment Ltd.)

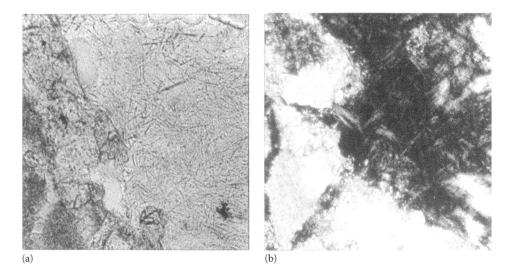

(a)   (b)

Figure 6.61 Example of Type I thaumasite, 320x magnification: (a) ppl, (b) xpl. (From DETR, The thaumasite form of sulphate attack: Risks, diagnosis, remedial works and guidance on new construction, Report of the Thaumasite Expert Group (TEG), Department for the Environment, Transport and the Regions (DETR), London, U.K., 1999.)

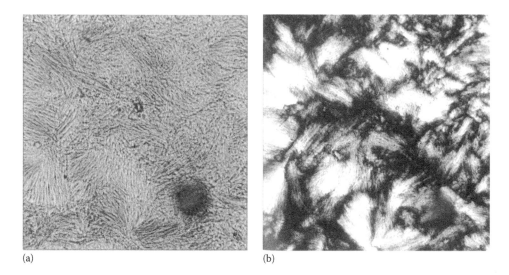

(a)   (b)

Figure 6.62 Example of Type II thaumasite, 320x magnification: (a) ppl, (b) xpl. (From DETR, The thaumasite form of sulphate attack: Risks, diagnosis, remedial works and guidance on new construction, Report of the Thaumasite Expert Group (TEG), Department for the Environment, Transport and the Regions (DETR), London, U.K., 1999.)

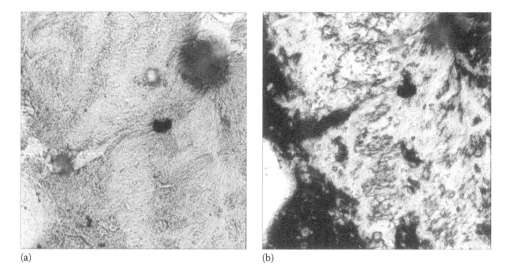

(a)   (b)

Figure 6.63 Example of Type III thaumasite, 320x magnification: (a) ppl, (b) xpl. (From DETR, The thaumasite form of sulphate attack: Risks, diagnosis, remedial works and guidance on new construction, Report of the Thaumasite Expert Group (TEG), Department for the Environment, Transport and the Regions (DETR), London, U.K., 1999.)

Table 6.8 Four-stage degradation sequence for TSA development (after Sibbick et al. 2002)

Zone 1: No visual evidence of attack; petrographic examination can reveal occasional voids and adhesion cracks around aggregate particles lined with thaumasite or ettringite.

Zone 2: Thin cracks lined with white thaumasite begin to appear running subparallel to the concrete surface. Calcium carbonate is sometimes precipitated into these cracks. Little portlandite is observed within the cement paste matrix. There is no evidence of other sulphate-bearing minerals.

Zone 3: An abundance of subparallel cracks filled with thaumasite becomes wider, and the amount of unattacked cement paste matrix is greatly reduced. Haloes of white thaumasite can be seen around coarse and fine aggregate particles. Calcium carbonite is sometimes precipitated into the cracks. Little portlandite is observed in the still unattacked cement paste. There is no evidence of other sulphate-bearing minerals.

Zone 4: Complete transformation of the cement paste matrix thaumasite. All that remains are occasional aggregate particles embedded in extremely soft white mush (thaumasite) and a few residual 'islands' of heavily depleted cement paste.

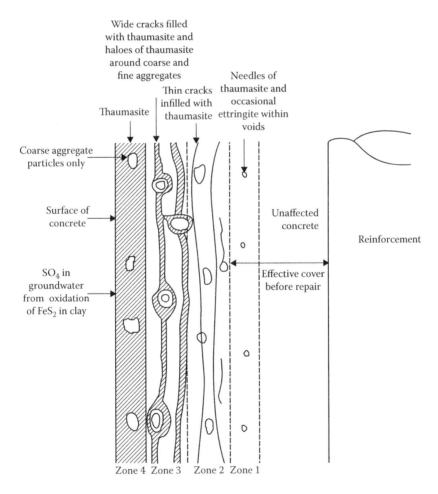

Figure 6.64 Schematic diagram showing the idealised progression of TSA degradation with high-quality structural concrete. (After Sibbick, R.G. et al., The microscopical characterisation of thaumasite, Holton, I. (ed), *Proceedings of the First International Conference on Thaumasite in Cementitious Materials*, BRE, Garston, U.K., June 2002, pp. 19–21.)

When considering cases of suspected or probable TSA, petrographers are recommended to follow this excellent four-stage Sibbick et al. scheme. It should be recognised that, in severely attacked cases, the outermost parts of the concrete may be fragile and will only have survived into the laboratory if the sampling was carried out with care and informed expertise; ideally, the petrographer will have been involved in that sampling, both to ensure recovery of the full sample, including the fragile outermost components, and also to observe the context from which the samples each derive (Hartshorn and Sims 1998; Sims and Hartshorn 1998).

As described in Table 6.8 and depicted in Figure 6.64, the least affected (**Zone 1**) material will only display thaumasite (and/or ettringite) lining some voids and perhaps pre-existing spaces at cement–aggregate interfaces (this might be described as TF). **Zone 2** is characterised by additional thin cracks, subparallel to the outer surface and each other, which are lined with thaumasite and possibly calcite; there is little portlandite within the cement matrix. In **Zone 3**, these subparallel cracks become abundant and appear to be wider and filled with thaumasite and possibly calcite, as the amount of unattacked but portlandite-depleted matrix is much reduced; haloes of white thaumasite develop around coarse and fine aggregate particles. Finally, in any **Zone 4**, which might have been lost in the case of poor sampling, more or less all of the cement matrix will have been transformed into soft white thaumasite, with embedded particles of unreacted aggregate and occasional remnants of depleted matrix. Sibbick et al. also reported that this four-zone development can sometimes occur beneath an outer crust of relatively sound concrete, which easily debonds from the degraded concrete when it is disturbed.

According to Sibbick et al., the three types of thaumasite (I, II and III, see earlier and Figures 6.61 to 6.63) were found in close association with each other within the zonal system, but the types tended (but not invariably) to develop in association with particular occurrences. For example, Type I and the most common Type II primarily formed the main masses of thaumasite arising from alteration of the cement matrix, whereas the haloes around aggregate particles and deposits within voids and pre-existing spaces were usually Type III. They also found examples of the coexistence of TSA degradation with other mechanisms, especially including combinations with conventional sulphate attack (see Section 6.5.3) or seawater attack (see Section 6.5.4) or 'popcorn calcite carbonation' (PCD: see Section 6.6.1) or AAR (see Sections 6.8 and 6.9). The association of TSA with PCD, which had also previously been reported by French and Crammond (2000), appears to be particularly interesting, as it is possible that Zone 4 secondary constituents could ultimately themselves be replaced by the friable clusters of stable calcite that characterise the PCD texture.

XRD has been used very successfully to distinguish thaumasite and ettringite (Sims and Hartshorn 1998), including mixtures of the two phases (Barnett 1998; Barnett et al. 2002). Ettringite/thaumasite mixed crystals have also been identified, indicating an apparent solid solution between the two. The XRD traces are similar but there are distinct differences, with a shift in 2-theta values dependent upon composition (Figure 6.65).

SEM and EDX-type microanalysis can be used positively to identify thaumasite and to distinguish thaumasite from ettringite in deteriorated concretes (Figure 6.66). SEM analysis can be carried out in conjunction with petrography on polished thin-section offcuts or uncovered thin sections. Analysis with the SEM has the advantage over many other techniques, such as XRD or IRS, in that the sample does not have to be crushed, avoiding the dilution of the thaumasite with the other concrete constituents. For this reason, the effective limit of detection for thaumasite is lower than that for other techniques.

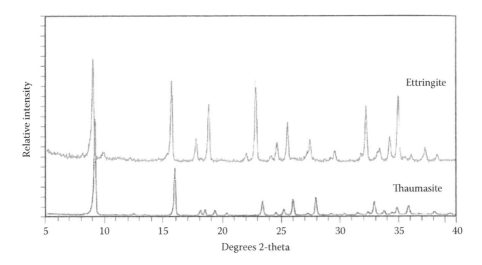

Figure 6.65 X-ray diffractograms for thaumasite and ettringite. (From Hartshorn , S.A. and Sims, I., Concrete, 32(8), 24, 1998.)

Infrared spectrophotometry (IRS) can also be used readily to distinguish thaumasite from ettringite, owing to the presence of octahedrally coordinated silicon atoms in thaumasite.

6.5.7 Internal sulphate attack

Internal sulphate attack on concrete does not require an external source of sulphate and is instead caused by the inadvertent inclusion in the concrete mix of materials containing sulphate, usually in the form of aggregate constituents or contaminants, or materials (such as pyrite or pyrrhotite) that can oxidise over time to produce sulphate. A particular case of such internal sulphate attack experienced in the United Kingdom, which has become known as 'the mundic problem' ('mundic' is the Cornish word for pyrite), is dealt with separately in Section 6.5.8. Also, the important type of internal sulphate action known as 'DEF' has been the subject of considerable research and investigation since the original edition of this book and is now considered in detail in Section 6.5.9. Scrivener and Skalny (2005) provide a useful review of internal sulphate attack and DEF.

The inclusion of excess sulphate in concrete occurs most commonly from contaminated aggregates. Where possible, concrete aggregates contaminated with sulphates are avoided, and their inadvertent inclusion usually indicates inadequate investigation and/or control of aggregate supplies. However, in some areas where good-quality aggregates are in short supply, contaminated aggregates may have to be used. For instance, in parts of the Arabian Peninsula, crushed sabkha contaminated with calcium and magnesium sulphates is sometimes used for aggregate (Fookes and Collis 1975; Sims 2006). Samarai (1976) found that where the gypsum content of an aggregate exceeds 0.6% for OPC and 0.7% for SRPC, the expansive formation of ettringite can lead to deterioration of concrete. He suggested that the total SO_3 content of concrete, expressed by weight of an OPC, should not exceed 5.0% (here rounded from his cited 5.2%). Crammond (1984), from experimental work related to this problem, recommends a maximum limit of 4% SO_3 by weight of ordinary Portland cement and a 2.5% limit for gypsum by weight of aggregates.

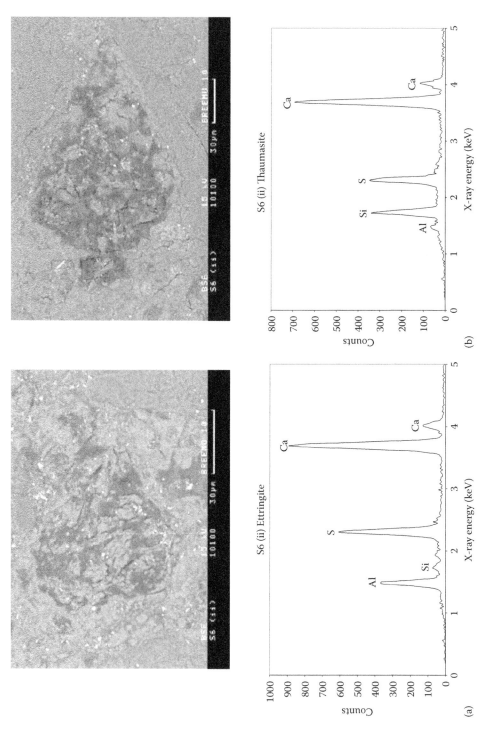

Figure 6.66 Example of use of SEM/EDX to distinguish thaumasite from ettringite in a single sample of deteriorated concrete: (a) Ettringite (note the main peaks for calcium (Ca), sulphur (S) and aluminium (Al) in the spectrum). (b) Thaumasite (note the silicon (Si) peak, in addition to calcium and sulphur).

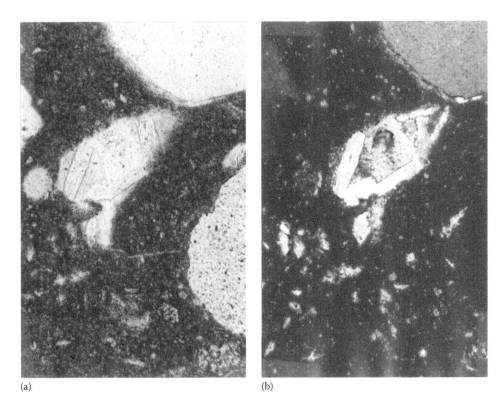

(a)   (b)

Figure 6.67 The texture of internal sulphate attack, widths of field 0.8 mm: (a) Laths of calcium hydroxide and gypsum have crystallised in a pore and microcracking is present in the hardened cement paste. (b) Under crossed polars, the gypsum is distinguished by its low birefringence and is seen to be scattered throughout the paste.

The texture of concrete undergoing sulphate attack from contaminated aggregates is shown in Figure 6.67. Crammond (1984) reports that petrographic evidence shows that coarse crystalline gypsum present as aggregate reacts with the cement alkalis to form calcium hydroxide on the gypsum crystals. The calcium hydroxide crystal then grows inwards towards the centre of the gypsum crystal. This conversion of gypsum to calcium hydroxide releases sulphate to form ettringite with the aluminate hydrates.

A rarer cause of excess sulphate being mixed into concrete, mortars and plasters is when workmen add gypsum during mixing. This is due either to adding the wrong 'grey powder' to the mix or mistakenly adding gypsum plaster in the form of the hemihydrate to accelerate setting and hardening. The resulting expansion occurs within weeks to months because finely ground gypsum reacts quickly. The reaction is spectacular and requires complete removal of the contaminated concrete or mortar. Efflorescence, scaling, spalling and cracking are widespread. Analysis, usually by XRD, will show that large amounts of gypsum and ettringite are present and this is usually sufficient to identify the cause. Examination of the texture of this type of failure in thin section will not provide any additional information to that found from XRD and chemical analyses.

Some sulphide minerals such as pyrite (FeS_2), marcasite (FeS_2), pyrrhotite ($Fe_{1-x}S$) and chalcopyrite ($CuFeS_2$) in the presence of oxygen and the alkaline pore solution of concrete can oxidise to produce ferric hydroxide ($Fe(OH)_3$) and a range of sulphates (Figure 6.68). BS EN 12620 (2002) states that pyrrhotite is an unstable form of FeS and, when present in aggregates, special precautions are necessary: the upper limit on total sulphur is reduced

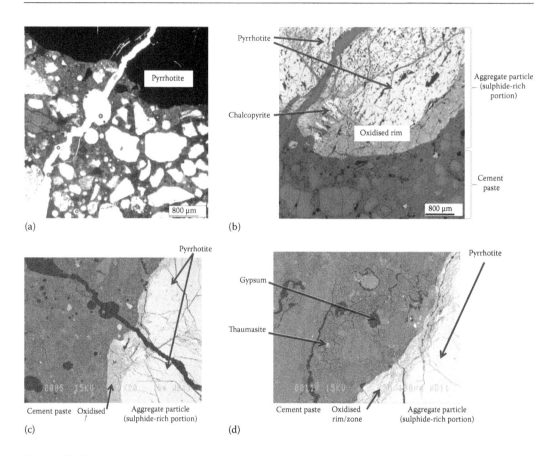

Figure 6.68 Examples of pyrrhotite decay within concrete: (a) Polished thin-section photomicrograph, showing a crack running into the cement paste from a pyrrhotite grain (black portion in the upper part of the view) in the gabbro aggregate particle, ppl. (b) Polished thin-section photomicrograph, showing a sulphide-rich portion (mainly pyrrhotite with some chalcopyrite) of a gabbro aggregate particle (upper portion of the view) with an oxidised rim zone and large crack running from the particle into the cement paste, reflected light. (c) Polished thin-section micrograph under the SEM, showing a sulphide-rich portion (mainly highly microcracked pyrrhotite) of a gabbro aggregate particle (right-hand portion of the view) with an oxidised rim zone and large crack running from the particle into the cement paste, backscattered electron image. (d) Polished thin-section micrograph under the SEM, showing a sulphide-rich portion of a gabbro aggregate particle (right-hand portion of the view) with an oxidised rim zone and highly microcracked adjacent area of cement paste, with secondary deposits of gypsum and thaumasite (all showing desiccation cracking). (Images courtesy of Andreia de Almeida Rodrigues, Josée Duchesne and Benoit Fournier at Laval University, Québec, Canada.)

from the usual 1% by mass to only 0.1% by mass. As the volume of the products is greater than the original mineral, pop-outs occur above aggregates near the surface, and in severe cases, expansion and pattern cracking may also occur as illustrated by Fulton (1977: see also Addis and Owens 2001). The ferric hydroxide, often present as goethite, produces a characteristic rust stain associated with the pop-outs. The following equation shows the pyrite oxidation reaction:

$$2FeS_2 + 7O_2 + 2H_2O \rightarrow 2FeSO_4 + 2H_2SO_4$$

The sulphuric acid and ferrous sulphate then react with the alkaline constituents present in the pore solution of the concrete to give a range of alkaline and calcium sulphates and ferrous hydroxide ($Fe(OH)_2$) which then oxidises to ferric hydroxide, $Fe(OH)_3$. In some cases, ettringite may be formed. These reactions are more complex than given earlier as among the products that may be found associated with the pop-outs are jarosite ($KFe_3(SO_4)_2(OH)_6$) and other complex sulphates as reported by Shayan (1988) and De Ceukelaire (1991).

Usually, damage caused by the presence of sulphide minerals in the aggregates is restricted to pop-outs and iron staining (Figure 6.69). Fulton (1977: also see Addis and Owens 2001) recommends that the total sulphide in aggregates should be limited to 2% for coarse aggregates and less for fine aggregate especially when mine tailings are used as aggregate. However, Midgley (1958) investigated the presence of pyrite in Thames river gravels and found that only some of the pyrite was reactive and that immersion of grains in saturated lime water distinguished reactive pyrite by the formation of a brown precipitate. Cases of severe deterioration of low-quality 'mundic' concrete blocks have been ascribed in part to decay of sulphides in mine waste aggregate (Stimson 1997, Royal Institution of Chartered Surveyors 2015) (see Section 6.5.8). Cases of more serious concrete deterioration attributed

Figure 6.69 The typical appearance of pop-outs and surface rust staining caused by aggregates containing pyrite.

to internal sulphate attack, variously involving pyrite and especially pyrrhotite, with some minor chalcopyrite and pentlandite, finely disseminated in a meta-gabbro aggregate, have recently been investigated by Rodrigues et al. (2012) in Québec (see Figure 6.68), who are now working on a performance test for aggregates. In their example, Rodrigues et al. found that only the pyrrhotite was oxidised, not the pyrite, and thought that thin layers of siderite (ferrous carbonate) coating the sulphides might be associated with the formation of thaumasite (see Section 6.5.6 for a full discussion of the TSA).

The petrographic investigation of pop-outs involves examination of the reaction products located around the apex of the pop-out and its socket. Identification is best carried out by microscopic examination of grains in oil immersion or preferably by XRD analysis and, where necessary, examination and analysis by SEM. Where jarosite is present, its yellow colour is distinctive and a good indication that sulphide minerals are involved. It is necessary to ensure that adequate analysis is carried out to determine the cause of pop-outs, as some shale and/or microporous aggregates undergo large volume increases on wetting and/or freezing while wet and can also produce pop-outs (Dolar-Mantuani 1983) (see also Section 6.4).

6.5.8 Mundic problem

The so-called 'mundic problem' affects concrete buildings in a small region of the United Kingdom (Cornwall and parts of Devon) and seems primarily to relate to an example of internal sulphate attack (see Section 6.5.7) arising mainly from the use of mining waste as aggregate, although some other mechanisms might also be involved. Similar cases of concrete deterioration have been witnessed by the authors in other regions and countries, but the exceptional concentration in Cornwall and parts of Devon is a consequence of the former metallic (especially tin) mining industry in that region, giving rise to large quantities of waste material that will have seemed ideal for use in low-cost concrete and concrete blocks. The resultant 'mundic problem' ('mundic' is an old Cornish word for pyrite, FeS_2) is of great interest to concrete petrographers, because the system of assessment and classification of concrete in buildings, adopted in the region by the Royal Institution of Chartered Surveyors (RICS), is based on a petrographic examination scheme (Stimson 1997, RICS 2002/2005, 2015).

6.5.8.1 Background to the mundic problem

Sulphate attack resulting from decomposition of mineralised aggregates is especially important in Southwest England (Cornwall and South Devon). It is widely called the 'mundic block problem' (Bromley and Pettifer 1997). Degradation is associated with the in situ oxidation of largely pyrite (and other sulphide minerals) in mine waste aggregates and sulphuric acid attack on the cement (see Section 6.5.7 for reported problems with pyrrhotite in Canada). It is most commonly seen in domestic and small commercial properties. Deterioration is occasionally so severe that concrete becomes structurally unsafe. Lending institutions (building societies and banks) require petrographic screening of all properties built with concrete blocks before the mid-1950s. Houses built with concrete containing potentially deleterious aggregates may not be considered mortgageable, even if the concrete shows no obvious signs of degradation.

Since the decline in use of natural stone at the beginning of the twentieth century, most domestic and small commercial properties in the region were built with concrete blocks or mass concrete. Before the Second World War (1939–1945), aggregates were rarely transported more than 20 km, and until the mid-1950s, concrete blocks were often locally made, sometimes by individual builders. Shuttered concrete was mixed on-site.

In many parts of Cornwall and South Devon, ample supplies of cheap and often suitably graded aggregate were available as waste materials from the region's former metalliferous mining industry. The use of mining, and more particularly ore processing, wastes as aggregates is central to the problem of accelerated concrete degradation in the region.

Mine wastes in Southwest England are geologically complex materials, made up from mineralised veinstones (from lodes), altered wall rock and unaltered country rock in varying proportions (Figure 6.70). Mining wastes *sensu stricto* from shaft sinking, cross cutting and development are composed largely of unaltered country rocks. They are usually very coarse. The maximum fragment size is the largest piece of rock that could be manhandled into a wagon. Such wastes, unless they were re-crushed, would have been quite unsuitable for use as concrete aggregates.

Most waste materials used as concrete aggregates were tailings products from various ore concentration processes. Ore concentration takes place in two stages. First, the mineralogically complex ore is crushed to achieve liberation. Then it is subjected to beneficiation, when the valuable material is separated from the gangue to produce a concentrate. Mineral concentration processes work most efficiently with closely graded feeds, so careful screening and regrinding led to the province-wide generation of millions of tonnes of wastes that were often of a convenient size for use as concrete aggregates. Unlike coarse mining wastes, which are usually just ordinary rock, mineral processing wastes are composed mostly of veinstone or lode minerals and of fragments of altered wall rocks that border the mineral lodes.

Many kinds of tin, copper and arsenic processing wastes have been identified in concrete aggregates. Some are benign. The wastes from granite-hosted tin mineralisation are safe if used as aggregates because the ores have low sulphide mineral content; the gangue and altered wall rocks are made from quartz and stable silicate minerals. Exogranitic tin and copper mining wastes were abundant in many parts of Southwest England. The ores were rich in sulphide minerals, in veins and altered wall rock. Fine-grained and potentially very reactive sulphide minerals were not recovered before the introduction of flotation separation techniques in the early part of the twentieth century. In the eighteenth and nineteenth centuries, rich copper ores, with coarse chalcopyrite, were often broken by hand and sorted before grinding and further beneficiation. The waste, typically with 50–100 mm maximum particle size, contained some residual chalcopyrite and other sulphide minerals including pyrite and arsenopyrite. Hand-cobbed copper wastes were used as aggregates in low-fines mass concrete in the important Camborne–Redruth mining district. Poor construction and high sulphide mineral content have sometimes led to very serious degradation.

Southwest England was never an important lead-producing region, but two former lead mines, near Newquay and in East Cornwall, respectively, are of enormous consequence in terms of deleterious aggregates. At both mines, the lead ore, galena (PbS), occurred in narrow veins with quartz, calcite and fluorite. The wall rocks were mudstones and these were impregnated with fine-grained, disseminated pyrite during mineralisation. Dense galena was separated from the lighter gangue and pyritic wall rocks in jigs. Before jigging, the ore was crushed to give a product evenly graded between about 1 and 10 or 15 mm. The processing techniques did not change through the lifetime of the mines, so enormous quantities of similar waste accumulated at both sites. The material was ideally graded for concrete block aggregate, and manufacturing plants were established at both mine sites in the 1920s. Concrete degradation results from oxidation of disseminated wall rock pyrite. Soluble sulphur species react with calcium in pore fluids in the concrete, resulting in expansive growth of gypsum in and around mudstone aggregate fragments.

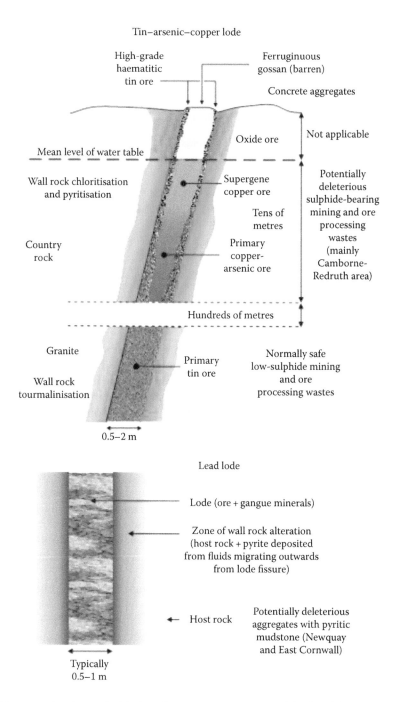

Figure 6.70 Diagrams showing typical lodes of tin–arsenic–copper and lead and locations of pyrite and other sulphides.

6.5.8.2 Petrographic examination of mundic concrete

Simple petrographic screening, to identify potentially deleterious aggregates, can often be carried out using a low-power stereomicroscope. The study of degradation, and especially the assessment of any potential for future deterioration, requires conventional transmitted light polarising microscopy. Reflected light microscopy is essential to study the behaviour of sulphides and other opaque minerals. Recourse to the scanning electron microscope may be necessary for investigation of fine-grained secondary products.

It is very important, when making petrographic assessments of the potential stability of sulphide-bearing aggregates, to appreciate that the walls and foundations of a simple house are complex physiochemical systems. Identical concrete with unstable aggregates may show contrasting behaviour in different parts of the same structure. An idealised section through the foundation and wall of a typical domestic property is shown in Figure 6.71. There are many variations, particularly in older houses, such as absence of a damp-proof membrane or solid walls built with mass concrete or concrete blocks.

Physiochemical conditions are strongly contrasted in different parts of the structure. The pore fluids in concrete with uncarbonated cement are very alkaline. The pH is greater than 12.5. In time, the cement becomes carbonated by reaction with carbon dioxide in the air. When the cement is fully carbonated, its pH falls to about 8.5.

Mass concrete used in foundations and even in some walls often has cement/aggregate ratios and macroscopic pore volumes comparable with those of modern structural concrete. Carbonation depth is often a few millimetres, even after 70–80 years. Foundation concrete is often saturated with groundwater from which oxygen has been removed by reaction with

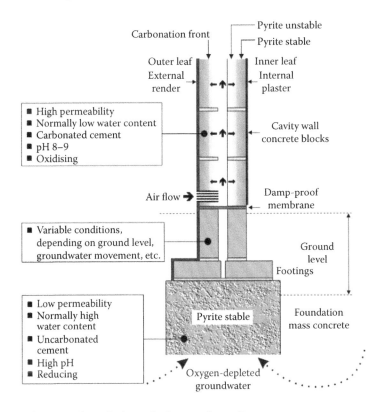

Figure 6.71 Idealised section through the wall of a typical mundic property.

Fe(II) minerals and organic matter in soil. Pyrite in wet mass concrete foundations usually shows only superficial oxidation. The stability of pyrite in contact with uncarbonated cement paste may be a result of oxygen depletion or build-up of a layer of oxidation products that are insoluble under alkaline conditions, effectively passivating the pyrite. Additionally, it is possible that pyrite oxidation is facilitated by the action of sulphur bacteria that are dormant under alkaline conditions.

Low-strength concrete blocks and much poorly constructed mass concrete in the region have cement: aggregate proportions between 1:6 and 1:15. Void volume is typically 5%–15% and can be as high as 25%. The concrete typically has a 'honeycomb' structure. In such materials, carbonation might be expected to proceed very rapidly because air can circulate freely through the strongly interconnected voids, but nevertheless, it is not always complete in blockwork walls, even after 70–80 years. In cavity wall construction, where the outer leaf is usually protected by sound render and the inner leaf is covered by mortar or plaster, carbonation usually begins at the cavity. Moisture-laden air flowing through the cavity diffuses into the blockwork. Carbonation fronts migrate from the cavity towards the protected surfaces of the blocks. In single block walls protected externally by sound render, the progress of carbonation depends mainly on the nature of the internal covering. If the inner surface of the blockwork is exposed, carbonation proceeds from the inner face towards the external render. If both surfaces are protected by mortar render or if the inner surface is covered in gypsum plaster, carbonation normally gradually proceeds inwards leaving a 'core' of uncarbonated cement.

Rapid pyrite oxidation follows carbonation of the cement paste. Degradation begins at the cavity. Debris spalls from the inner faces of the blockwork and accumulates in the bottom of the cavity. It is only when pyrite oxidation has already penetrated deep into the blockwork that the characteristic 'mundic' cracking appears in the external render (Figure 6.72).

Two main degradation mechanisms are recognised, each associated with specific types of pyritic aggregate. The first is associated mainly with tin–copper–arsenic wastes like those which characterise the former Camborne–Redruth mining district. Degradation is caused by oxidation of liberated and easily accessed pyrite and other sulphide minerals in tailings products and re-crushed mining wastes and direct sulphate attack on carbonated cement paste. In low-strength, high-voidage concrete (such as mass concrete with low cement content or blockwork), degradation results mainly from sulphide oxidation and direct sulphate attack on the cement. Secondary gypsum grows at aggregate–cement interfaces and in the carbonated cement matrix. Some degradation is caused by expansion resulting from replacement of calcite by gypsum. Rupture of cement–aggregate bonds, disintegration of cement 'bridges' between aggregate fragments and pore volume collapse are also important. Initial pyrite concentrations of at least about 1% are generally required for serious degradation to occur.

The second mechanism is associated with the use of mesothermal, mudstone-hosted lead ore processing wastes like those used near Newquay and in East Cornwall. These aggregates carry little liberated sulphide but have abundant graphitic mudstone wall rock with fine-grained, disseminated, sometimes framboidal pyrite (Figure 6.73). Degradation results from oxidation of fine pyrite and bulk expansion of the mudstone aggregate. Expansion is caused by the growth of secondary minerals in lensoid veinlets parallel with bedding and/or cleavage in the mudstone (Figures 6.74 to 6.76) and at the interface between the aggregate and the cement. Gypsum is the most important secondary mineral. Water-soluble phases including iron sulphates and potash alum, formed by direct sulphate attack on clays and micas in the mudstone, are also found.

Figure 6.72 Typical examples of 'mundic' cracking evident on the rendered exterior of dwellings and buildings in Cornwall, UK.

6.5.8.3 RICS classification scheme

Most mortgage lending organisations in Southwest England now require a special 'mundic' assessment of concrete properties built before about the 1950s, and this is included in the building survey in accordance with RICS guidance (Stimson 1997, RICS 2002/2005, 2015). A chartered surveyor takes concrete samples as an additional part of his survey, and these are examined and classified by a petrographer, with the surveyor being responsible for the final overall advice on mundic condition of the property in question. The classification scheme is based upon characterisation of the aggregate composition and the condition of the concrete, yielding four classes: A (sound concrete without any mundic-related aggregate), A/B (sound concrete with a minority of mundic-related aggregate), B (sound concrete with a majority of, or wholly comprising, mundic-related aggregate), or C (unsound concrete, which may be unsound for susceptibility mundic or other reasons). The RICS guidance leading to this classification is summarised in Tables 6.9 to 6.11.

A flow chart for the RICS petrographic classification procedure is shown in Figure 6.77. It is a staged approach in order to minimise expenditure by house vendors or mortgagors.

Figure 6.73 Photomicrograph showing fine-grained, disseminated pyrite (including some framboidal pyrite) in a fine siltstone aggregate fragment, reflected light. The pyritised siltstone is from wall rock adjacent to a lead–zinc lode (East Cornwall, United Kingdom). (Image courtesy of Alan Bromley.)

Figure 6.74 Photomicrographs showing mudstone aggregate particles with cleavage-parallel and peripheral secondary gypsum veins. Fine-grained fibrous gypsum has grown following reaction between soluble sulphur species and calcium ions in pore fluids. The growth of gypsum is expansive and has caused dilation of the mudstone fragment, with propagation of microcracks in the surrounding binder. Concrete block from a house in Torpoint, East Cornwall, United Kingdom: (a) ppl, (b) xpl. (Images courtesy of Alan Bromley.)

Stage One involves visual examination of the surveyor's core samples by an experienced petrographer, using only the unaided eye and low-power microscopy, when it is frequently possible to establish either Class A or Class C material with confidence (the original classification scheme has been modified in the latest edition of the RICS scheme: see Table 6.11 at the final paragraph of this subsection). Stage Two, when necessary, involves the additional

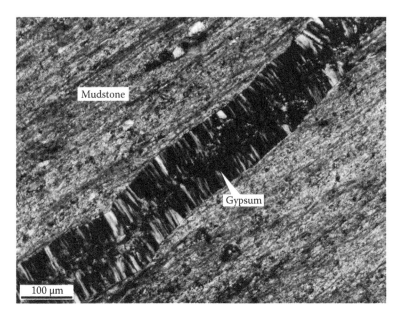

Figure 6.75 Detail showing the fibrous structure of gypsum in a cleavage-parallel vein (see also Figure 6.74). Concrete block from a house in Torpoint, East Cornwall, United Kingdom: xpl. (Image courtesy of Alan Bromley.)

Figure 6.76 Scanning electron microscope image (secondary electron image), showing spalling and expansion of a small mudstone aggregate fragment following growth of secondary gypsum in lenses along cleavage planes. Concrete block from a house in Liskeard, East Cornwall, United Kingdom. (Image courtesy of Alan Bromley.)

examination of thin sections under a high-power petrological microscope (and sometimes but rarely supported by SEM and other analyses), when samples can be classified in accordance with the RICS scheme. Classes A/B and B can only be determined after Stage Two examination. Bromley (2002) prepared an excellent compendium, based on his unparalleled experience and knowledge of mine waste and other aggregates in Southwest England, which

Table 6.9 Table showing aggregate type classification in RICS mundic guidance (after Table 1 in RICS [2015]; consult RICS [2015] for detailed guidance)

Group 1	1-1	China clay waste
	1-2	Crushed granite and related igneous rocks (e.g. elvan)
	1-3	Crushed basic and metabasic igneous rocks (e.g. epidiorite, serpentinite)[1,2]
	1-4	Furnace clinker or coking breeze[3]
	1-5	Beach or river sands and gravels
	1-6	Other (e.g. Group 2, reclassified as a result of current knowledge and/or further investigation)[5]
Group 2	2-1	Crushed sedimentary or metasedimentary rocks (killas)[2,4,5]
Considered potentially deleterious	2-2	Most metalliferous mining and/or proceeding wastes[6,7]
	2-3	Slags (largely non-ferrous) and incinerator wastes[8]

1 Meta-basic igneous rocks from the Newlyn–Penzance area have a long history of use as aggregates, and there is little evidence that they have been associated with concrete degradation, despite an often significant content of well-crystallised pyrite. It is for the petrographer to judge in specific cases as to whether or not chemical analysis should be carried out.

2 See guidance on chemical analysis in Section 3.10 of RICS (2015).

3 If a particular furnace clinker is found to contain unstable or potentially unstable constituents, it should be classified as Group 2–3.

4 Certain quarried sedimentary and meta-sedimentary rocks in North Cornwall (Newquay, north-eastwards) have been used widely as aggregates, and apparently there is little evidence that they are associated with accelerated concrete degradation. It is for the petrographer to judge whether to include the subject material within Group 1–6, usually following a Stage 2 examination.

5 Crushed limestone or sandstone (except greywacke) should usually be classified within Group 1–6.

6 Granite mining and/or processing waste with a low sulphide content would be an example of an exception (see Section 3.10 of RICS (2015) for a method for checking sulphide content); provided the waste is made essentially from quartz and stable silicate and oxide minerals, the sulphide minerals are present mainly as coarse discrete grains, and the concrete is judged to be sound. It is for the petrographer to judge whether to include the subject material within Group 1–6, usually following a Stage 2 examination and/or chemical analysis.

7 Most metalliferous mining/processing wastes are readily apparent to the petrographer because of their mineralogical constituents. However, in some cases, such indicative minerals are present in only low concentration and/or are very finely disseminated (e.g. the silver–lead mine wastes of East Cornwall), so that Stage 2 examination may be required to establish a reliable identification.

8 Incinerator wastes may derive from the burning of industrial and/or domestic refuse, and are characterised by the diversity of their constituents and the variability of their composition; they may also contain amounts of furnace clinker and various slag-like components, as well as a variety of other debris.

Table 6.10 Table showing concrete condition classification in RICS mundic guidance (after Table 4 in RICS [2015]; consult RICS [2015] for detailed guidance: tick the various boxes as appropriate for the sample in question, then trace the column bearing the ticks appearing nearest to the right of the table to the bottom of the table, where this assessed concrete condition is given)

	Occurrence			
Observed feature	*None*	*Rare*	*Common*	*Abundant*
1. Sulphide decay and associated staining of surrounding matrix				
2. Concrete matrix alteration, including recrystallisation				
3. Secondary sulphate mineral development				
4. Evidence of moisture susceptibility in fine-grained metasediment				
5. Physical incoherence				
6. Cracking (other than externally induced)				
Condition assessment	*Sound*		*Unsound*	

Table 6.11 Table showing concrete mundic classification in RICS mundic guidance (after Table 3 in RICS [2015]; consult RICS [2015] for detailed guidance)

Aggregate(s)	Concrete condition	Former	New
Group 1 only	Sound	A	A_1
Group 1 plus up to 30% group 2	Sound	A/B	A_2
Greater than 30% group 2	Sound (passed stages)	A/B	A_3
Greater than 30% group 2	Sound	B	B
Mainly group 2	Unsound	C1	C_1
Mainly group 1	Unsound	C2	C_2

comprises described scans from sawn and ground surfaces prepared along the lengths of a typical range of these RICS core samples.

Petrographically, there was no reason for supposing that a Class B material, still sound after so many years in service despite its content of mundic-related aggregates, was any less acceptable for mortgage purposes than Class A material (many other types of potentially unstable building materials are accepted as mortgageable around the country), but the mortgage lending organisations were more comfortable with concrete material that only contained a minority (i.e. 30% or less) of mundic-related aggregates, so that the intermediate Class A/B was introduced after negotiation. Today, most lending organisations operating in Southwest England will agree to mortgage any concrete properties comprising only Class A or Class A/B materials.

Originally, the RICS scheme in Cornwall and parts of Devon had been conceived to enable 'bad apples' to be separated from 'good apples', in the belief that 'mundic' concrete was mainly unacceptable in the long-term, but needed to be differentiated for mortgaging purposes from all the concrete in the region that was no different from concrete anywhere else in the country. Over 15 years of working in Cornwall and Devon and examining about 12000 concrete samples using this RICS method, one petrography practice has found that more than 80% represented Class A material (Figure 6.78). Nevertheless, surveyors and petrographers were surprised at the frequency with which they encountered Class B concrete (about 15% of the 12,000 samples in the case of the practice mentioned earlier), that is, containing mundic-related aggregates, but nevertheless still sound after many decades of service. In an endeavour further to evaluate these Class B materials, after research by Sibbick, then at the BRE (Bromley and Sibbick 1999; Lane et al. 1999), a moisture-sensitivity test was introduced, as a Stage Three of the RICS procedure (RICS 2002/2005). Some examples of Stage Three results are shown in Figure 6.79.

Material that gives a satisfactory performance in this test can be reclassified from Class B to Class A/B and thus becomes mortgageable. An adjustment to the classification scheme in the latest version of the RICS guidance (RICS 2015) will see Classes A and A/B changed to A_1 (the original A), A_2 (the original A/B) and A_3 (reclassified after successful Stage Three testing of Class B, also some other special reasons for reclassification), in order to emphasise the intended equivalence of the various routes to a Class A outcome (see Table 6.11).

6.5.9 Delayed ettringite formation (DEF)

In the original version of this book, a short account of DEF was included in the subsection on 'internal sulphate attack', when it was thought that the mechanism was mainly an occasional problem with precast concrete subjected to accelerated curing. However, since then, it has become apparent that DEF is a more frequently encountered, though sometimes controversial,

Note: The flow chart is simplified and reference should be made to the RICS (2015) text for full details

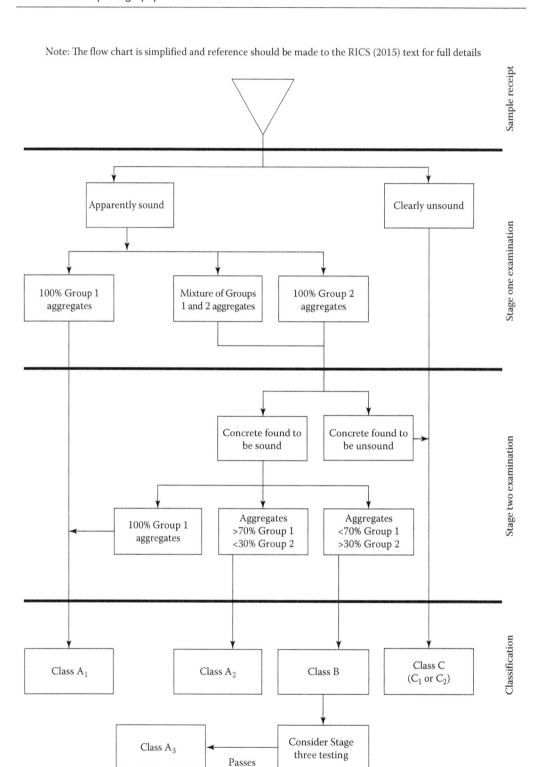

Figure 6.77 **Flow chart from the RICS (2015) mundic guidance.**

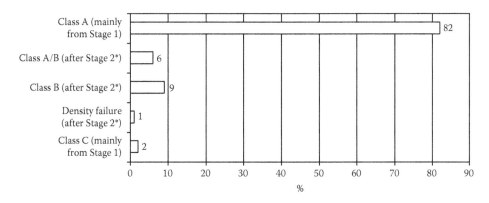

* Based on proportions returned by Stage 2 testing on about 44% of the 18.5% of the samples
subjected to Stage 1 testing and recommended for Stage 2 testing (Stage 2 testing of the other
56% was not commissioned, so that the samples remained unclassified).

Figure 6.78 Statistical chart, showing frequency of Classes A, A/B, B and C, based upon data for about 12,000 samples over 15 years. (Courtesy of Petrolab Limited.)

problem, which can also occur with large sections of concrete and which additionally might have been misdiagnosed in some previous structure investigations (including some cases of alleged ASR). Moreover, research has found that the concurrence of AAR and DEF is quite common (Martin 2010, Martin et al. 2012a,b), with their respective contributions to causation of damage being a challenge to determine. General accounts of DEF will be found in Day (1992), Thaulow et al. (1997), Quillin (2001), Thomas (2001) and Scrivener and Skalny (2005).

6.5.9.1 Background and mechanism of DEF

The sulphate present in Portland cement under normal conditions will give rise to the formation of hydrated sulphoaluminate compounds, and these form while the concrete is in a plastic state, or just after the concrete has begun its initial set, and do not cause any significant localised disruptive action (Collepardi 2003).

Ettringite is a hydrous sulphoaluminate mineral ($3CaO \cdot Al_2O_3 \cdot 3CaSO_4 \cdot 26–32H_2O$) that is very commonly found as a secondary ettringite within voids in almost any concrete subjected to moisture ingress. DEF is a specific form of ettringite formation that occurs in concrete *after* it has hardened and is distinct from a normal secondary ettringite formation. Since moisture is required for ettringite formation, DEF is generally triggered in affected concrete by moisture ingress. DEF is also sometimes referred to as 'heat-induced internal sulphate attack' and can give rise to a significant damaging expansion in concrete structures (Figure 6.80).

It is generally acknowledged that DEF occurs as a result of high temperatures during curing that interfere with the normal hydration reactions of C_3S (alite) and calcium sulphates (Taylor 1997). The critical temperature above which there is a risk of DEF is around 65°C. It follows from this that DEF may occur either in concrete deliberately cured at elevated temperatures, such as in steam-cured precast concrete, or in large-volume concrete pours where the heat generated during cement hydration cannot dissipate.

Steam curing is often used in precast concrete plants to obtain high early strengths and thus increase production rates. DEF in concrete can occur when the curing temperature exceeds about 65°C or 75°C. The conditions and mechanism of DEF depend on a number of factors, as shown by Ghorab et al. (1980), Heinz and Ludwig (1986), Brown and Bothe (1993) and Glasser et al. (1995). Not all cements are affected. It is claimed that the molar

Property 12-Expansion values for 75 mm cores at 38°C excluding first seven days

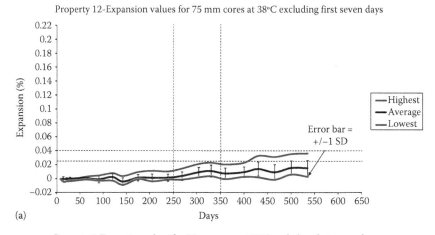

(a)

Property 6-Expansion values for 75 mm cores at 38°C excluding first seven days

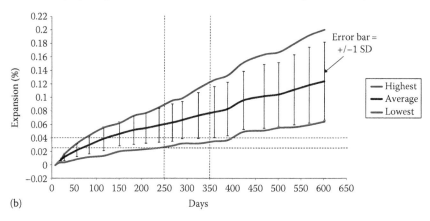

(b)

Figure 6.79 Example curves from the RICS (2015) Stage three moisture-sensitivity test, showing maximum, minimum and average curves for the set of cores in each case from different properties: (a) limited expansion in the test, enabling the concrete to be reclassified from B to A/B (or now A₃), (b) expansion in the test, such that the concrete remains Class B. (Courtesy of Dr Ted Sibbick.)

ratio of SO_3/Al_2O_3 in the cement has to be greater than 0.55 for a significant DEF to occur. Thus, cements low in alumina, such as SRPCs or under-sulphated cements, are less likely to be affected. However, Odler and Chen (1995) found that expansion depended on the sulphate and aluminate contents but not on their ratio.

The extent of expansion caused by DEF increases as the temperature during curing is raised to 90°C, which is about the maximum attainable for atmospheric steam curing of products. For higher temperatures, such as those used for autoclaving products, the sulphate and aluminate phases are different and as a result DEF has not been reported. Lawrence (1993) concluded from laboratory studies that limiting concrete temperatures to a maximum of 70°C was a secure way of avoiding the problem. Later work by Lawrence (1995) found a complex relationship between composition and expansion, but concluded that the involvement of ASR (some evidence of which often occurs alongside DEF) in the expansion was not an essential feature.

A number of cases of damaging expansion resulting from DEF have been reported, including widespread damage to railway sleepers and other precast products (Tepponen and Ericksson 1987, Neck 1988, Sylla 1988, Johansen et al. 1994). Oberholster et al. (1992)

Figure 6.80 Examples of DEF damage to concrete structures: (a) and (b) Columns of a concrete viaduct in southern United States. (From Thomas, M.D.A. et al., *Cement Concrete Res.*, 38, 841, 2008.) (c) and (d) Pile caps from highway bridges in the Indian subcontinent region. (e) Anchor block for a television transmitter mast in the United Kingdom, originally interpreted as thermal cracking, but petrographic evidence consistent with DEF.

investigated cracking in low-pressure steam-cured railway sleepers in South Africa and concluded that, in this case, AAR was the primary cause and not DEF. Similarly, Shayan and Quick (1992) found that railway sleepers in Australia had also cracked due to AAR. Stark et al. (1992) found that in concretes with a low water/cement ratio, SO_3 contents as low as 3.3% and curing temperatures of 60°C, DEF may occur. DEF has been reported in precast concrete bridges (Knights and Baldwin 2010) as well as in some cases of cast in situ concrete (Grantham et al. 1999, Eden et al. 2007, Thomas et al. 2007). There are also some rare examples of DEF in concrete heated as a result of fire (Eden 2011).

Recently, the authors have helped to investigate severe cracking to bridge pile caps and large piers on a major new transportation project in a country with a hot–wet tropical climate. This cracking, which developed within a year of casting the concrete, had the superficial appearance of ASR, but was found by petrography and SEM–EDX to be caused by DEF (Figure 6.81). However, although all thin sections from deep within the concrete units exhibited abundant evidence of DEF, some of the thin sections also displayed small developments of ASR gel.

This relationship between ettringite and ASR has been recognised for some years (Blackwell and Pettifer 1992). In order for ettringite to form, sulphate ions need to be carried by moisture gradients and concentrated at nucleation sites. The hydrophilic properties of ASR gels provide

Figure 6.81 Cracking caused by DEF that looks superficially like ASR; example from the Indian subcontinent region.

a suitable mechanism for generating such moisture gradients within the concrete (Poole et al. 1996). We shall see that it is sometimes necessary to augment petrographical examination with specialised testing in order to decide whether DEF or ASR has been the primary cause of damage to concretes showing evidence of both mechanisms (Thomas et al. 1997).

Temperatures in the range 75°C–90°C are typical of low-pressure steam curing and also occur in large volumes of freshly placed concrete when temperatures generated by the heat of hydration are not controlled. It is suggested that DEF occurs because the sulphate and aluminate ions become bound in the calcium silicate hydrate structures which under moist conditions are later slowly released expansively to form ettringite (Heinz and Ludwig 1986; Fu et al. 1995). A different mechanism was proposed by Kuzel (1990), who suggested that carbonation of the low calcium sulfoaluminate hydrate to a monocarbonate, $4CaO \cdot Al_2O_3 \cdot CO_2 \cdot 11H_2O$, and gypsum occurs. The gypsum formed by this reaction is then available to react later with the remaining low calcium sulfoaluminate hydrate to form ettringite. However, this proposal does not seem to have been confirmed by later work.

There is disagreement in the literature regarding the mechanism of expansion resulting from DEF (see the excellent review by Thomas 2001). There are three common theories on the mechanism of expansion in concrete suffering from DEF as follows:

- The expansion relates to the growth of ettringite on aggregate surfaces and in veins in the paste (Diamond 1996a).
- The expansion relates to general expansion of the paste leading to the development of voids on the aggregate particle surfaces where ettringite can form (Scrivener and Taylor 1993; Glasser et al. 1995; Taylor 1997).
- Collepardi (2003) has proposed that pre-existing microcracks, formed, for example, by thermally induced stresses, allow ettringite formation to begin in microcracks and subsequently to contribute to expansion.

Observations by the present authors (see the following section) indicate support for the expansion mechanism proposed by Scrivener and Taylor (1993) and Taylor (1997), whereby DEF initially causes general expansion in the paste, opening partings at the aggregate–paste interfaces, which subsequently become filled with relatively coarse ettringite, creating the characteristic haloes. Eden et al. (2007) found that the extent of ettringite formation in cracks in the paste and partings around aggregates increased as DEF-induced expansion of concrete proceeds.

The principal risk factors that contribute to the potential for the occurrence of DEF are:

1. High temperatures during curing, which might occur in precast concrete cured at elevated temperature or in cast in situ concrete placed in large volumes;
2. Moisture availability, where ettringite formation requires a source of moisture.

The following additional factors may also contribute to the development of DEF:

3. High cement content. Cement hydration is an exothermic reaction and the amount of heat generated by the hydration of the cement in concrete increases with increasing cement content (cements also vary in their heat of hydration property);
4. AAR. AAR commonly leads to expansive cracking and the development of such cracks promotes moisture ingress and the mobilisation of sulphates present in the cement paste (some ASR is often found associated with DEF);
5. High sulphate content of the cement. The amount of sulphate in the concrete available for ettringite formation is critical to determining the amount of expansion that might occur due to DEF.

It follows from these factors that prevention of DEF will mainly be achieved by controlling temperature development during curing (minimising peak temperature and both its time of occurrence and endurance), plus consideration of the mix constituents, especially cement composition and the avoidance of AAR (see Section 6.8). However, some work has been carried out into the prohibitive effect of using pozzolanas and/or slags in the concrete mix (Ramlochan et al. 2003, 2004).

6.5.9.2 Petrographic investigation of DEF

In spite of some continuing controversy over mechanisms and controlling factors, there is little doubt that cracking of concrete caused by DEF does occur. As it is an intrinsic reaction in which the whole paste expands (Taylor 1997), cracking is typically accompanied by ring fissuring around aggregates in which redeposition of ettringite occurs. Given the appropriate conditions, it is possible for alkali–aggregate reactivity to occur concurrently, and the findings of Brown and Bothe (1993) and Poole et al. (1996) suggest that AAR may increase the likelihood of DEF. It is not uncommon in cases of AAR to find that considerable redeposition of ettringite has occurred in some cracks but ring fissures around aggregates are absent. The formation of ring cracks around aggregates is an important distinction, and DEF should not be diagnosed with confidence unless this textural aspect of the concrete is present.

A sample was examined from a cracked structure where it was believed that internal temperatures in the concrete may have exceeded 90°C during early curing. Over a storage period of a year at 20°C and 100% RH, a core sample from the structure expanded over 0.5%. In thin section, the coarse limestone aggregate particles and many of the quartz sand particles were surrounded by a parting or fissure, infilled with ettringite, up to 100 μm in width. This effect was uniformly spread across the whole of the large-area thin section. The ettringite needles were usually orientated at right angles to the smooth margins of the cement paste surrounding the fissures. The needles of ettringite always seemed to have grown from the surface of the paste margins, rather than from the aggregate surfaces, and gave the appearance of having grown into an existing space (Figure 6.82a), though there is always the possibility that some dehydration shrinkage during thin-section preparation

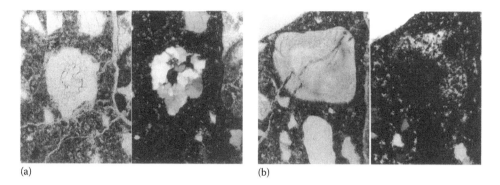

(a)   (b)

Figure 6.82 The texture of DEF, widths of photomicrographs 0.9 mm, left ppl, right xpl: (a) The cracking and fissures have been delineated by the use of a yellow dye. A fissure, filled with aligned ettringite, runs along the margin of the large limestone aggregate to the right of the micrograph and less well-defined fissures, also filled with ettringite, surround most of the other aggregates. The cracking in the hardened paste is also filled with ettringite. Most of the ettringite is barely visible here but can be clearly seen when viewed under the microscope. (b) The chert aggregate in the centre of the micrograph is clearly undergoing AAR. There is a fissure along the margin of limestone aggregate on the right-hand side of the micrograph filled with clearly visible aligned ettringite. Some of the other aggregates present are also ringed with filled fissures.

can cause some space to form within such ettringite-filled fissures. The hardened cement paste contained fine pattern microcracking which was also filled with ettringite and had the appearance of a typical expansion texture for in situ formation of ettringite.

A minor AAR was occurring in some chert and quartzite grains consistent with the high alkali content of the concrete, but was not considered sufficient to cause the amount of cracking and expansion observed (Figure 6.82b). The conclusion from this petrographic examination was that DEF had clearly occurred in the hardened paste causing cracking and expansion. In addition, some mechanism was operating to cause the formation of ring fissures around most of the aggregates which provided space for the subsequent growth of ettringite.

Johansen et al. (1994) have reported similar observations in railway sleepers (Figure 6.83) and put forward a hypothesis that the fissuring around the aggregates is due to the overall expansion of the hardened cement paste. In effect, if each aggregate is considered equivalent to a void and an essentially isotropic expansion of the paste occurs, each void must grow in volume proportional to the expansion. It follows that the width of the ring fissure formed around each aggregate will be proportional to the size of the aggregate. Thus, the paste gives the appearance of having shrunk away from the aggregates, forming ring fissures which then provide space for the growth of the ettringite needles. The texture they illustrate is similar to that described earlier for the sample from a cracked structure. Scrivener and Taylor (1993) have also concluded DEF is driven by processes occurring within the paste and not by the formation of ettringite at aggregate interfaces.

DEF can generally be recognised in petrographic thin sections from the characteristic development of ettringite in peripheral cracks around the surfaces of aggregate particles as well as within cracks in the paste (Figures 6.84 and 6.85). Eden et al. (2007) used measurements of the percentages of coarse and fine aggregate particles with ettringite growth around their surfaces to rate the severity of DEF in concrete from a bridge in Malaysia.

Although the acicular ettringite crystals are commonly observed in thin sections to have orientated perpendicular to the walls of a crack or aggregate as described earlier, scanning electron micrographs show that in three dimensions, the orientations are more variable, for example, the crystals sometimes form as felted masses aligned parallel to the crack surfaces (Figure 6.86).

Figure 6.83 External cracking of railway sleepers caused by DEF. (Photograph courtesy of Niels Thaulow.)

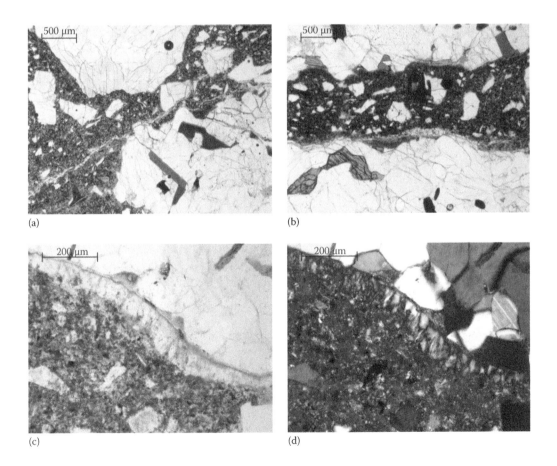

Figure 6.84 Examples of ettringite in peripheral cracks around the surface of gneiss particles in concrete undergoing DEF. In ppl (a–c), the ettringite is shown white. In (d) xpl, the ettringite is shown as a fibrous mottled grey crack filling; images (c) and (d) are the same view, one in ppl and one in xpl. (Photographs courtesy of Paul Bennett-Hughes, RSK Environment Ltd.)

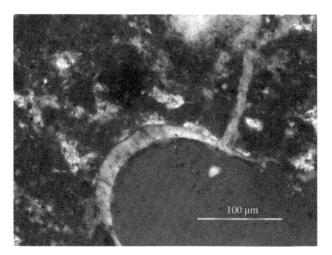

Figure 6.85 An example of ettringite in a peripheral crack around the surface of a quartz sand particle in concrete undergoing DEF. The ettringite-filled crack continues into the surrounding cement paste. The ettringite occurs as a fibrous mottled grey crack filling. Thin section, transmitted light, crossed polars view with the scale bar representing 0.1 mm. (Photograph courtesy of Mike Eden, Sandberg LLP.)

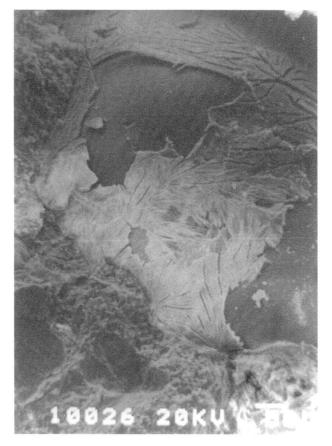

Figure 6.86 Scanning electron micrograph of a crack surface coated with ettringite crystals. The crack has formed as a 'ring crack' at the surface of a rounded aggregate particle.

SEM is now commonly used to augment optical microscopy in the investigation of concrete suspected to be exhibiting DEF, when the identity of secondary ettringite can be confirmed using EDX-type microanalysis, as well as studying its form and modes of occurrence (Figures 6.87 and 6.88). In one important case involving relatively large concrete units in which early high temperatures had developed, microanalytical comparisons were made of the paste for samples from the (hotter) centre and from the (cooler) surface regions. It was confirmed that the paste became relatively depleted of sulphate in the internal parts of the concrete that exhibited a higher frequency of coarser ettringite variously infilling aggregate–paste partings, microcracks and voids; in other words, in those 'hotter' internal regions of the concrete, sulphate had effectively migrated from the paste into these spaces to form ettringite (Table 6.12).

Using the SEM, it is also possible to detect the formation of ettringite in the paste as well as within the peripheral partings, cracks and voids (Figure 6.89). X-ray mapping for sulphur can then be used to assess the distribution of ettringite in the cement paste (Figure 6.90).

It has been suggested that the potential for future expansion in concrete affected by DEF can be assessed from core expansion testing at 38°C (Eden et al. 2007). However, in their approach to differentiating the roles of DEF and ASR when they occur together in concrete,

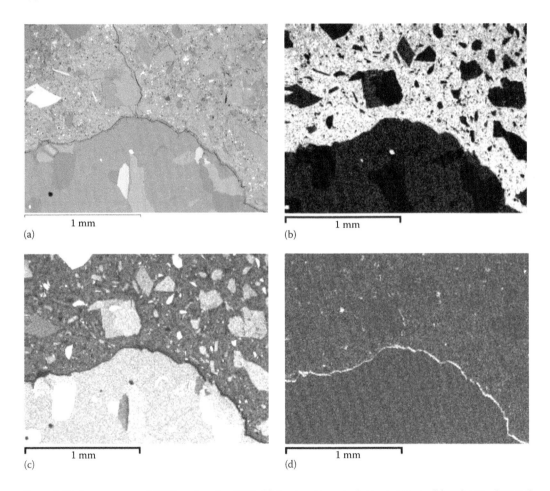

Figure 6.87 Appearance of DEF under the SEM (a) backscattered electron image, (b) calcium chemical map, (c) silica chemical map, (d) sulphur chemical map; a sulphur-rich crack is shown near the interface between the coarse aggregate and the cement matrix. (Images courtesy of Paul Bennett-Hughes, RSK Environment Ltd.)

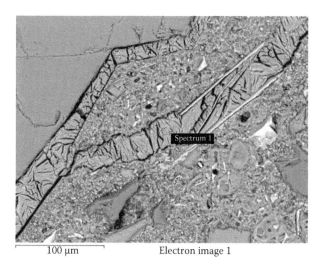

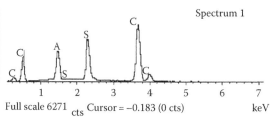

Figure 6.88 SEM image and EDX spectrum from the sample shown in Figure 6.87, confirming ettringite. (Image and spectrum courtesy of Paul Bennett-Hughes, RSK Environment Ltd.)

Table 6.12 EDX microanalysis count data, comparing the paste composition between relatively 'hotter' internal regions (with greater evidence of DEF) and relatively 'cooler' near-surface regions – note the depletion of the sulphate (SO_3) in the paste where DEF is more abundant

Laboratory sample ref.		9832/C1	9832/C3	9885/C1	9885/C3
Client sample ref.		B16-A1	B16-A2	B16-A1	B16-A2
Condition		Cracked pile cap	Uncracked pile cap	Cracked pile cap	Uncracked pile cap
Overall abundance of DEF features[a]		Sporadic	Sporadic	Abundant	Sporadic
Elemental composition, %	SO_3	1.4	1.6	0.6	1.0
	Na_2O	0.7	0.6	0.5	0.4
	MgO	2.1	0.7	2.1	1.2
	Al_2O_3	6.2	6.4	6.0	6.6
	SiO_2	28.2	28.3	27.6	27.8
	K_2O	1.7	1.0	0.9	0.7
	CaO	55.5	57.2	58.2	58.0
	TiO_2	0.4	0.4	0.4	0.5
	Fe_2O_3	3.8	3.8	3.8	3.8

[a] For example, air voids, partings, and/or microcracks partially infilled or lined by secondary ettringite deposits.
Glossary of terms used in the table:

Frequency	Rare – Only found by thorough searching
	Sporadic – Only occasionally observed during normal examination
	Common – Easily observed during normal examination
	Frequent – Easily observed with minimal examination
	Abundant – Immediately apparent to initial examination

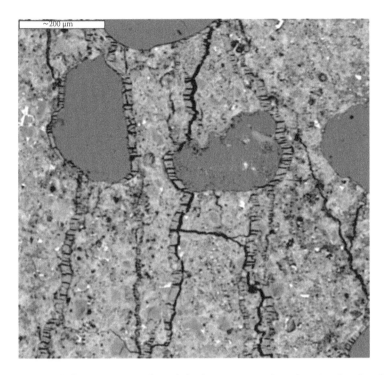

Figure 6.89 Backscattered electron image of a polished concrete surface showing the abundant formation of ettringite in microcracks in the paste in concrete affected by DEF. The scale bar measures 0.2 mm. (Image courtesy of Mike Eden, Sandberg LLP.)

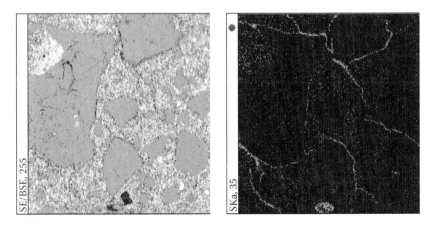

Figure 6.90 X-ray map for sulphur in a pile cap sample showing high expansion after expansion testing. A backscattered electron image illustrating the area mapped for sulphur is shown on the left and illustrates fine aggregate particles surrounded by cement paste. The X-ray map for sulphur is shown in the image on the right, where the areas of high concentrations of sulphur appear pink and indicate the abundance of ettringite on aggregate surfaces and in microcracks in the paste. Each image measures about 1 mm × 1 mm. (Images courtesy of Mike Eden, Sandberg LLP.)

Thomas et al. (2008b) chose to encourage any potential for DEF expansion by storing cores in lime water at 20°C, conditions that would not sustain ASR. In carrying out any such expansion testing, it is always essential to verify the cause of the expansion measured by conducting petrographic examinations before and after the expansion testing.

It is common for there to be an association of DEF with AAR (Thomas et al. 2007), and in some cases, there can be ambiguity in the interpretation of the primary cause of expansion (Figures 6.82b, 6.91 and 6.92). While it is likely that in some cases DEF has been incorrectly diagnosed as AAR in the past, equally, since the recognition of DEF, it is probable that some genuine cases of AAR have been misdiagnosed as DEF, because the observed ettringite distribution reflects the formation of secondary ettringite within cracks opened by ASR. Knowledge of the curing history of the concrete and temperatures reached during the curing of the concrete are critical to the petrographic diagnosis of DEF.

Thomas et al. (2008) have considered procedures for diagnosing cases in which evidence of both DEF and AAR is present in the concrete. Based upon some actual structures, they provide examples of the SEM appearance (Figure 6.93) and the chemical mapping also carried out (Figure 6.94).

Thomas et al. (2007, 2008) also described the use of core testing to help evaluate the respective contributions of DEF and ASR to the expansion. This is based on the principle that cores stored at room temperature in lime water solution will tend to expand if there is a potential for DEF within the concrete, whereas comparable cores stored at 80°C in an alkaline (sodium hydroxide) solution will tend to expand if there is a potential for ASR (Figure 6.95). The present authors have applied this approach, including petrographic examination before and after testing, when there was certainly a relationship between any expansion in lime water and the increased amount of microscopic evidence of DEF. In the particular case under investigation, the authors did not find a similar relationship between expansion in sodium hydroxide solution and any signs of ASR: some cores exhibited expansion, but no apparent evidence of any ASR.

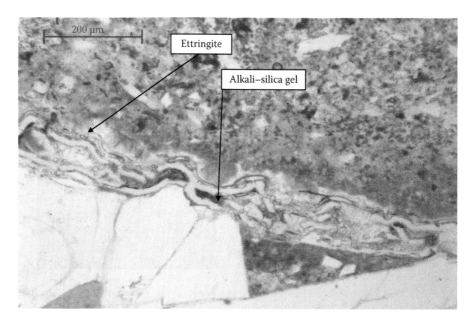

Figure 6.91 Example in thin section (ppl) of the concurrence of dominant DEF and minor ASR in thin sections; from highway structures in the Indian subcontinent region. (Photomicrograph courtesy of Paul Bennett-Hughes, RSK Environment Ltd.)

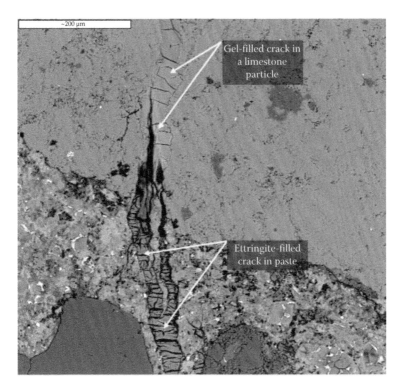

Figure 6.92 Backscattered electron image of a polished concrete surface showing an example of concrete showing symptoms of both AAR and DEF. A gel-filled crack in a bituminous limestone particle at the top of the field of view continues as an ettringite-filled crack in the paste. The scale bar measures 0.2 mm. (Image courtesy of Mike Eden, Sandberg LLP.)

6.6 ACID AND ALKALINE ATTACKS

Hardened cement and calcareous types of aggregate are rapidly attacked by acidic solutions with a pH of less than 4.0. Where the hardened matrix contains HAC, it is more resistant above pH 4.0, but below this pH, attack is greater than that found with Portland cement (Lea 1970). The composition of ordinary Portland cement, including the use of replacement materials such as fly ash, appears to be less important than the water/binder ratio in determining performance in acid conditions (Matthews 1992). Similarly, Ballim and Alexander (1991) found that while ground blastfurnace slag and fly ash reduced leaching from carbonic acid attack, the resulting porosity was a little different between ordinary Portland and blended cements. Some abstracts on the subject of chemical attack were collated in Appendix 1 of the original version of this book (St John et al. 1998). A more recent account of the resistance of concrete to acids and alkalis may be found in Kelham (2003).

The most important factors in corrosive attack are the amount of fluid flowing over the exposed surface of the concrete and the pH. If a significant flow rate is occurring, the attack on the concrete may be considerable, even for mildly acidic conditions. This is especially the case where soft acidic waters are permeating through porous concrete and leaking joints in hydraulic structures (Commission Internationale des Grands Barrages 1989). However, if one of the corrosion products formed is insoluble in the attacking fluid, it can provide a protective layer on the surface of the concrete. This occurs in marsh water (Lea 1970) where, if humic acid is present, it forms an insoluble coating of calcium humate. This may also apply

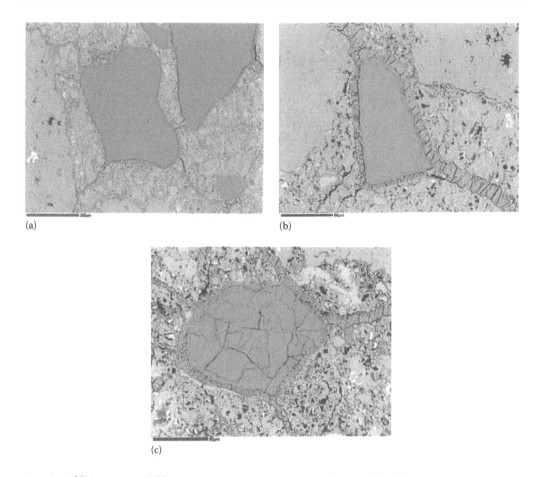

(a)   (b)

(c)

Figure 6.93 SEM images of DEF in cast in situ concrete bridge columns: (a) backscattered electron image of damage attributed to DEF alone; (b) backscattered electron image, showing ettringite-filled spaces around an aggregate particle, from a concrete column with no sign of AAR; (c) backscattered electron image, showing ettringite-filled spaces surrounding a reacting aggregate particle. (From Thomas, M.D.A. et al., *Cement Concrete Res.*, 38, 841, 2008.)

to some extent to many acids which form an insoluble layer of gelatinous silica on the surface of concrete (Grube and Rechenberg 1989).

Between pH 7.0 and 8.5, attack is essentially limited to superficial leaching of calcium hydroxide from the concrete surface. Even at a pH of 6.0, attack is minimal unless the flow rate is significant and is only potentially serious for thin-walled products such as concrete pipe and highly permeable mass concretes. At a pH of 5.0, attack is serious for concrete pipes and, depending on the flow rate of the water, will limit the life of such products. The former BS 8110 (1997) recommended that Portland cement should not be used without protection in conditions where the pH is consistently less than 5.5. In practice, it has been found that most calcareous cementing systems will not withstand a pH of less than 4.0 without the provision of an adequate protective coating. However, for massive concrete that is reasonably impermeable, while attack in the pH range 4.0–7.0 may corrode concrete surfaces, the depth affected is usually insufficient to be serious unless it leads to corrosion of reinforcement. BS 8110 (1997) has been superseded by BS EN 1992-1-1 (2004).

Although Portland cement concretes are themselves alkaline, they are attacked by strong alkaline solutions where the concentration exceeds 20% but are largely unaffected if the

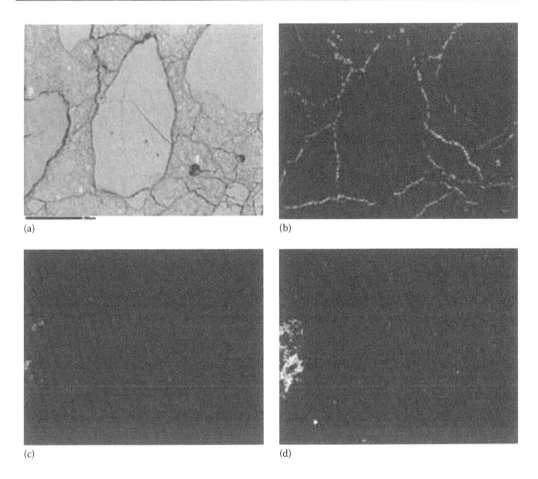

Figure 6.94 Backscattered electron image (a) and its X-ray maps: (b) sulphur, (c) sodium, (d) potassium. (From Thomas, M.D.A. et al., *Cement Concrete Res.*, 38, 841, 2008.)

concentration is less than 10% (Biczoc 1967). Concretes containing HAC are less resistant to alkaline solutions than those made with Portland cements. Alkaline solutions slowly act on calcium aluminate cements (CACs) by dissolving the alumina gel and attacking the hydrated calcium aluminates (see **Chapter 4**).

There are three factors that need to be determined where concrete structures are undergoing corrosion. These are the composition of the attacking fluid determined by chemical analysis, the extent of the corrosive attack on the concrete texture and estimation of the rate of attack. Petrographic examination is able to delineate the nature and depth of the damage to the concrete and in favourable circumstances may allow an approximate rate of attack to be determined.

6.6.1 Natural acid waters and 'aggressive carbonation'

Most natural waters have acidities that fall into the pH range 4–8.5. The pH will rarely fall below pH 5 owing to dissolved carbon dioxide, unless it is under pressure. Below pH 5, the acidity is usually due to the presence of humic acid, which is limited in aggression because calcium humate is almost insoluble in water and forms a protective layer on the concrete (Lea 1970).

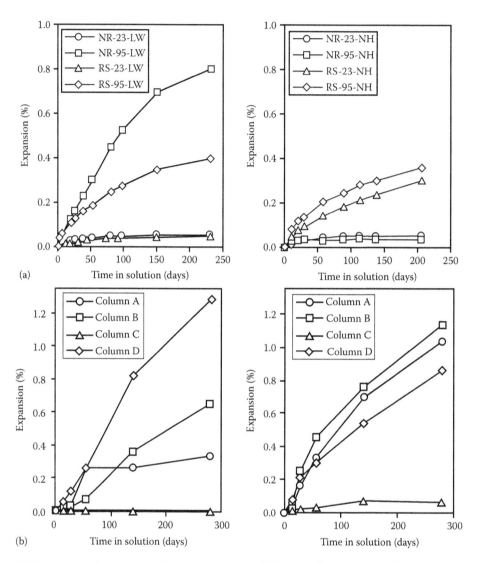

Figure 6.95 Examples of core expansion tests to assess DEF and ASR expansion: (a) Reference laboratory concretes prepared with non-alkali-reactive or alkali-reactive aggregate combinations and
cured at 23°C or 95°C, variously to induce DEF or ASR. (b) Site concrete from cracked bridge
columns; in each case, the left-hand curves are for storage in lime-saturated water at room
temperature, while the right-hand curves are for storage in 1 M NaOH solution at 80°C. (From
Thomas, M.D.A. et al., *Cement Concrete Res.*, 38, 841, 2008.)

Where groundwater percolates through decaying organic matter, the carbon dioxide produced is absorbed by the water. The effect of the absorbed carbon dioxide on the pH of the
water depends on the amount of bicarbonate already present, because only the free carbon
dioxide, that is, the carbon dioxide not required to stabilise the bicarbonate in solution,
affects the pH. The less the water is buffered by bicarbonate, the less the amount of carbon
dioxide required to lower the pH. A good example is distilled water which has close to zero
buffering. Absorption of a few ppm of carbon dioxide from the atmosphere is sufficient to
give an equilibrium pH of about 5.7. Distilled water is highly aggressive to concrete as it has
both a high solvent capacity because of its low buffering and is also acidic. This situation

occurs in the water of some mountain and high moorland areas, which are relatively free of dissolved salts (Commission Internationale des Grands Barrages 1989). However, buffering of waters is also affected by other calcium salts such as gypsum and the salts present in seawater, as discussed by Terzaghi (1949), so in practice, the situation may be more complicated than that described earlier.

Confusions as to both nomenclature and methods of measurement have occurred in the past when groundwaters were analysed for relative aggressiveness to concrete. Reference should be made to Terzaghi (1949) who first identified some of the problems and Rogers (1973) and Cowie and Glasser (1991/92) who have reviewed the subject. Large errors are possible in the determination of free carbon dioxide and pH, and it is essential that all measurements be checked by equilibrium calculations, as described by Stumm and Morgan (1970) and the American Public Health Association (1992). Similarly, groundwaters may be turbid when sampled, and measurement of pH can be affected by suspension effects at the liquid junction to the reference electrode of the pH metre (Bates 1964). Many groundwaters, especially those in aquifers, are under pressure so that measurement of carbon dioxide in surface samples may not fully represent the carbon dioxide actually present in the subsurface waters that are in contact with the buried concrete. Special techniques for sampling subsurface waters to avoid the loss of CO_2 have been developed by John et al. (1977). Because of the problems involved, it is essential that sampling and analysis of groundwaters for dissolved carbon dioxide, alkalinity and pH be carried out by experienced water analysts.

When natural waters containing carbon dioxide attack concrete, both the calcium hydroxide and the hardened cement paste are first carbonated and then dissolved. The two attacking species are carbonic acid H_2CO_3 and calcium bicarbonate $Ca(HCO_3)_2$. As calcium bicarbonate is soluble, it can be removed by the attacking fluids providing sufficient carbonic acid is present to stabilise the bicarbonate formed. This process, which can be very destructive to concrete (Figure 6.96), also appeared to be greatly enhanced by the sheer high volumes of water flow and the raised water pressure levels involved (Thaulow and Jacobsen 1997; Sibbick and Crammond 2003). The reactions can be simplified as:

$$Ca(OH)_2 + H_2CO_3 \rightarrow CaCO_3 + 2H_2O$$

$$Ca(OH)_2 + Ca(HCO_3)_2 \rightarrow 2CaCO_3 + 2H_2O$$

However, calcium carbonate, whether present due to the carbonation of calcium hydroxide or as limestone aggregate, and the hydrated cement phases are also attacked by carbonic acid:

$$CaCO_3 + H_2CO_3 \rightarrow Ca(HCO_3)_2$$

and thus overall,

$$\text{Hydrated cement} + H_2CO_3 \rightarrow \text{intermediates} \rightarrow \text{Si hydrates} + \text{Al hydrates} + \text{Fe hydrates}$$

6.6.1.1 Petrographic textures

The end products of the attack are gelatinous forms of hydrated silica and alumina and iron hydroxides. In practice, most of the hydrated alumina and iron hydroxides are removed by the attacking fluid, so that the only remaining product is gelatinous silica. It is the formation of soluble calcium bicarbonate and its removal by the fluid that are responsible for the actual corrosion of concrete surfaces or, in one particularly severe case, the complete consumption of concrete blocks (but strangely not the mortar joints) from selected areas of the foundations of a property within 18 months of construction (Sibbick and Crammond 2003).

This calcium bicarbonate material can be completely lost to the material by being transported out with the water or moisture passing through it, potentially in very large amounts. It can also be redeposited as a secondary calcite edging to voids, joints, outer surfaces, microcracks or other locations where it can precipitate.

However, as seen in Figure 6.96, it can also form part of an unusual feature known as the 'popcorn calcite deposition' (PCD) texture, which forms within the depleted residual cement paste matrix. This residual cement paste from these decalcification processes is usually a silica gel with no significant cementing capacity, hence the weak and friable nature of the end product and its potential to corrode and abrade easily. The PCD texture is a residual pattern of tight clusters of calcite crystals (rosettes), which develop relatively evenly within

(a)

(b)

(c)

(d)

(e)

(f)

Figure 6.96 Attack on concrete by carbonic acid and calcium bicarbonate: (a) and (b) Bedding mortar in stone masonry within the intertidal zone of a sea wall, with development of 'PCD' with depletion of the cement paste matrix. (c) and (d) Concrete blocks within the intertidal zone of a sea wall, with secondary PCD apparently taking advantage of cracking induced by AAR. (e) and (f) Poor-quality (porous) concrete coping units exposed to weathering and leading to extreme PCD, together with freeze–thaw damage. *(Continued)*

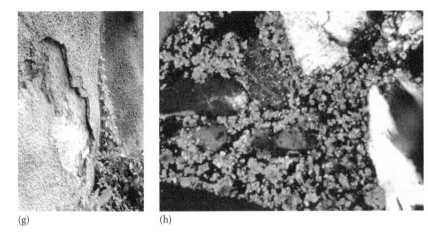

(g)   (h)

Figure 6.96 (Continued) Attack on concrete by carbonic acid and calcium bicarbonate: (g) and (h) Norwegian sprayed concrete tunnel linings, exhibiting a combination of PCD and TSA (see Section 6.5.6). (From Sibbick, R.G. and Crammond, N.J., The petrographical examination of popcorn calcite deposition (PCD) within concrete mortar, and its association with other forms of degradation, *Ninth Euroseminar on Microscopy Applied to Building Materials*, Trondheim, Norway, 2003.)

the residual decalcified cement paste matrix (now largely a silica gel). The sizes and concentration of these calcite clusters exhibit a modest degree of variability.

This potentially very destructive form of 'aggressive carbonation', with the characteristic PCD texture of the incompetent reaction product, has been well described by Bromley and Pettifer (1997), Thaulow and Jacobsen (1997), French and Crammond (2000) using the term 'subaqueous carbonation', Thaulow et al. (2001) using the term 'bicarbonation' and Hagelia et al. (2001). Sibbick and Crammond (2003) discuss the circumstances in which PCD may occur, using chemical mapping under the SEM to demonstrate the decalcification of the depleted cement paste and the secondary concentration of calcium into the clusters or 'rosettes' that provide the distinctive 'popcorn'-like texture (Figure 6.97).

This distinctive PCD texture is generally associated with a number of processes in which the pH of the cementitious system is reduced. A close association has been found with TSA, seawater attack and ASR, in addition to environments in which high volumes of water are flowing through a system. It has been reported by Gerdes and Wittmann (1992) that deterioration of concrete through reactions similar to those detailed earlier can be induced by long-lasting electric fields in mortar coatings lining water reservoirs.

When a thin section is examined from more homogeneous, dense concrete corroded by soft acidic water, it will be observed that the non-calcareous aggregates are left standing proud of the corroded surface of the hardened cement paste. However, many limestone aggregates will be dissolved more readily than the surrounding cement paste. As the attack proceeds, smaller aggregates near the surface may become dislodged as the typical rough corrosion surface develops. The texture of the outer layers of the hardened paste consists of an irregular outermost zone consisting primarily of gelatinous silica lying over a zone of carbonation, which merges into another zone of hardened paste depleted in calcium hydroxide as illustrated in Figure 6.98.

The zones are not very wide and the gelatinous outermost zone may be partially absent, as it is easily damaged and lost in sampling and may be difficult to observe because it is isotropic. In addition, it is often colourless, although it can contain patches of yellow to dark brown occluded material. Occasionally, a thin line of reddish-brown material may separate the outer zone from the carbonation zone. While the carbonated layer is obvious in thin section, the zone leached of calcium hydroxide is usually not well defined and is often difficult to observe. A method has been proposed by Koelliker (1985) to overcome the problem of retaining the outer, gelatinous layer during sampling.

This type of zonation is *always present* on the undisturbed surfaces of dense concretes corroded by soft acidic waters. The outermost zone of gelatinous silica represents the completion of the attack. The inner zone depleted of calcium hydroxide is where there is a concentration gradient occurring as a result of the attack, and the zone of carbonation is where the actual dissolution of material occurs.

6.6.1.2 Estimation of rates of attack

Classification of the aggressiveness of natural waters containing carbon dioxide was made by Werner and Giertz-Hedstrom (1934) based on experience in Scandinavia. Subsequent classifications, such as those given by Lea (1970), are refinements of this basic work and will indicate the degree of protection required by use of concrete in new works. However, where attack has occurred on concrete, the situation may arise where the extent of attack needs to be determined so that some estimate of an expected lifetime may be made (see Figures 6.99 to 6.103).

A number of methods, both theoretical and practical, have been proposed for estimating the rate of attack. All the practical methods depend on estimation of the amount of material removed by the corrosion. The simplest test is to measure weight loss of laboratory specimens (Tremper 1932; Werner and Giertz-Hedstrom 1934) and relate these losses to

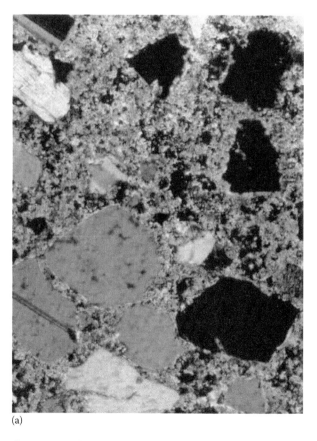

(a)

Figure 6.97 Examples of popcorn calcite deposition (PCD): (a) Typical appearance in thin sections, xpl.

(*Continued*)

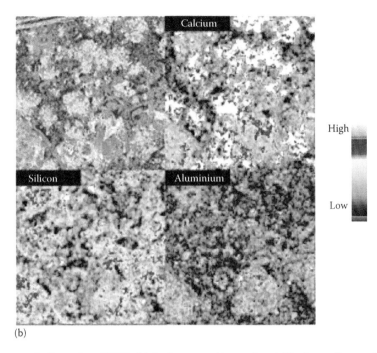

(b)

Figure 6.97 (Continued)   Examples of PCD: (b) Backscattered image (top left) of a small area of PCD textured matrix, with chemical maps for calcium, silicon and aluminium for the same area, showing the concentration of calcium into the 'popcorn' clusters and the depletion of calcium from the residual cement paste regions. (From Sibbick, R.G. and Crammond, N.J., The petrographical examination of popcorn calcite deposition (PCD) within concrete mortar, and its association with other forms of degradation, *Ninth Euroseminar on Microscopy Applied to Building Materials*, Trondheim, Norway, 2003.)

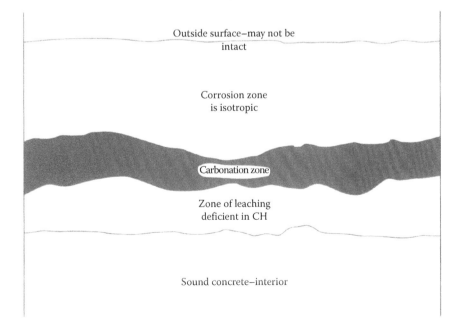

Figure 6.98   Diagram illustrating the zones of soft water attack.

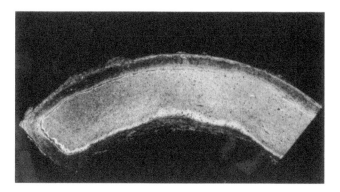

Figure 6.99 The texture of the soft water corrosion of asbestos cement pipes from St John and Penhale (1980), width of view ~170 mm: photograph of a negative print of a large-area thin-section cut from an asbestos cement pipe. The pipe had been buried for 20 years on a site with soft acidic water flows (pH ~6). The outer dark zone is where the hardened cement paste has corroded to amorphous silica and the brightish white zone is carbonation. The zone of leaching is not apparent but is clearly visible under the microscope (see Figures 6.100 and 6.101). See Figure 6.102 for a thin-section of a pipe exposed to similar conditions.

products and structures. Al-Adeeb and Matti (1984) measured the free lime content and permeability of asbestos–cement pipes to estimate rate of attack. Where water is percolating through a structure, the increase in calcium in the leachate has been used to estimate weight loss from the concrete (Commission Internationale des Grands Barrages 1989), but this is restricted to those cases where a leachate can be collected.

St John and Penhale (1980) have used another approach. The depth of external corrosion on a 3-year-old concrete sewerage system was estimated by systematic core sampling and

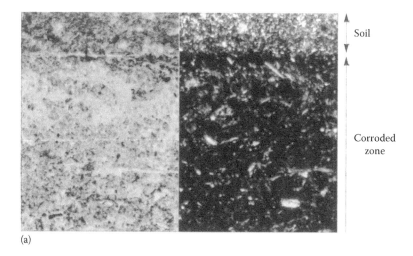

Soil

Corroded zone

(a)

Figure 6.100 The zonation of soft water attack on an asbestos cement pipe. The fine texture of these pipes, which were not autoclaved, clearly illustrates the zones associated with the attack, widths of fields 0.9 mm, left ppl, right xpl: (a) The interface between the soil and the corroded zone often contains deposits of brownish-yellow iron compounds, which are visible at the soil interface on this sample. The corroded zone of the pipe consists of amorphous silica, asbestos fibres and some indistinct relics of miscellaneous materials. Under crossed polars, the birefringence of the soil minerals and the interface stained with iron compounds stand out clearly. The corroded zone is almost entirely isotropic and only remnants of asbestos fibre are visible.   (*Continued*)

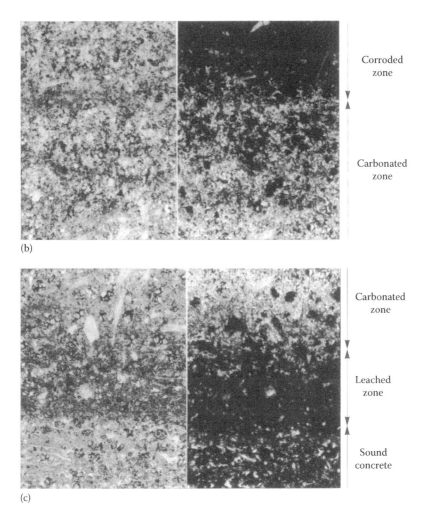

Corroded
zone

Carbonated
zone

(b)

Carbonated
zone

Leached
zone

Sound
concrete

(c)

Figure 6.100 (Continued) The zonation of soft water attack on an asbestos cement pipe. The fine texture of these pipes, which were not autoclaved, clearly illustrates the zones associated with the attack, widths of fields 0.9 mm, left ppl, right xpl: (b) The interface between the corroded and carbonated zones is well defined by a layer of iron staining and the transition to the carbonated zone. Under crossed polars, the isotropic nature of the corroded zone, the iron staining and the carbonation are clearly illustrated. The dark spots in the carbonated zone are relics of cement grains. Asbestos fibres are present but is not easy to observe as it tends to merge into the texture at this magnification. (c) The leached zone appears as a darkened layer of cement paste which under crossed polars is deficient in calcium hydroxide. The transitions from carbonated to leached zones and from leached to sound hardened paste are clearly visible under crossed polars. The sound hardened paste is full of partially hydrated and remnant cement grains as well as calcium hydroxide. Asbestos fibres are just visible in all the zones.

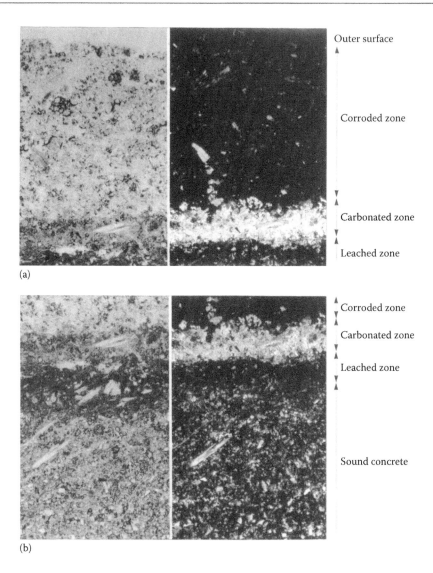

Figure 6.101 Variations in the texture of corrosion of a non-autoclaved asbestos cement pipe, widths of field 0.9 mm, left ppl, right xpl: (a) Outer corroded zone, including relicts of cement clinker grains, which is unusual. Since this pipe contains no silica flour, it is rich in remnant grains which may be part of the reason for their presence in the corroded zone. Normally, the corroded zone contains little if any part of the cementitious system apart from amorphous silica. Note the absence of iron staining and the way the corroded zone terminates in a ragged manner at the outer edge because adherent earth was cleaned off during sampling. Under crossed polars, the corroded, carbonated and leached zones are clearly defined. (b) A view across the three corrosion zones, grading into sound material. Remnant cement grains and asbestos fibres are visible. The zones are clearly defined under crossed polars.

measuring corrosion depths in thin section. The corrosion depth was defined as the sum of the corroded, carbonated and leached layers, on the assumption that when the leached layer penetrated to the steel reinforcement, corrosion would occur. Groundwater analyses were also performed which showed that the water flowing around the pipes had a range of corrosivity. The results were compared, using the same petrographic techniques, with test samples of pipes that had been buried under similar conditions for many years under known conditions.

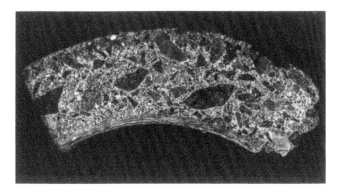

Figure 6.102 The texture of the soft water corrosion of concrete pipes from St John and Penhale (1980), width of view ~170 mm: a large-area thin section from a concrete pipe buried for 20 years on a site with soft acidic water flows (pH ~6). The corrosion and carbonated zones are barely apparent, but still clearly visible under the microscope (see Figure 6.103). The two photographs in Figures 6.99 and this 6.102 illustrate that zones of deterioration are more apparent in fine-textured fabrics, hence the need for adequate microscopical examination of concrete textures, where alteration may not be apparent in sawn or fracture surfaces because of the presence of aggregate.

The rate equations for attack by soft waters are all similar and the simple equation proposed by Valenta and Modry (1969) is adequate for most purposes. It is $t = ax^2$ where t is the time in years, a is the rate constant and x is the depth of corrosion. As the corrosive conditions were variable, a range of rate constants was derived that ranged from 0.7 to 1.18. This translates to a minimum life of thirty years for the most aggressive condition found to a maximum life of 500 years for the least aggressive condition. The results obtained were in a reasonable agreement with those of Tremper (1932) and Werner and GiertzHedstrom (1934).

6.6.2 Carbonation and corrosion by geothermal fluids

Carbonation and corrosion of concrete by geothermal fluids containing carbon dioxide are much more rapid than at ambient temperatures. If the fluids are insufficiently buffered, the carbon dioxide will give rise to acidic conditions and severe attack occurs within months. The cement grouting systems used to anchor geothermal bore casings are one of the products most likely to be severely damaged, as the temperatures of fluids surrounding the boreholes can range up to 300°C at pressures of up to 50 bars (Milestone et al. 1986). Geothermal cement grouts usually consist of Portland cement mixed at water/cement ratios of about 0.6 or greater and may contain added silica flour in the form of ground quartz to suppress the formation of ex-dicalcium silicate hydrate and bentonite and chemical additions to improve pumpability. Because of their formulation, these cement grouts are fairly permeable.

6.6.2.1 Textures of grouts exposed to geothermal fluids

The textures of grouts exposed to geothermal fluids containing carbon dioxide graphically illustrate many aspects of the corrosion mechanisms at elevated temperatures. In addition, the effect of temperatures ranging from 160°C to 260°C on the hydration products formed, with and without the addition of silica flour to the grouts, is of importance for both carbonation and corrosion.

St John and Abbott (1987) examined sections of both autoclaved grout specimens in the laboratory and field samples exposed to downhole geothermal fluids at New Zealand's Broadlands Geothermal Development. The textures are similar to those produced by corrosion from soft

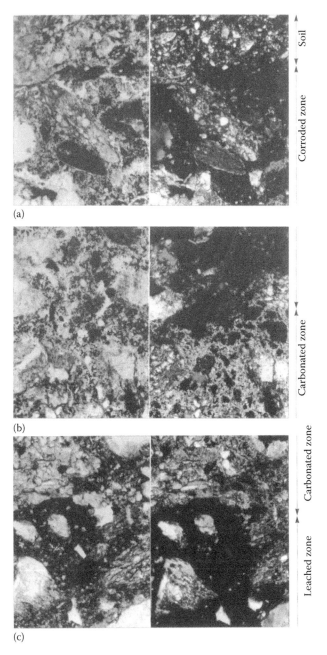

Figure 6.103 The zonation of soft water attack on centrifugally spun concrete pipe. The close-packed aggregate
texture of these pipes makes the zones of attack difficult both to see and to illustrate, widths of
field 0.12 mm, left ppl, right xpl: (a) The soil–corrosion interface. A layer of iron staining covers the
surface of the pipe at the greywacke-derived soil interface. The texture of the corroded cement
paste is stained brown but otherwise is isotropic under crossed polars. The aggregates present
are greywacke. (b) The corrosion/carbonation interface. The change in colour from corroded to
the carbonated zones is subtle. The portion of the corroded zone is darkened by possible relicts
of cement grains. Even under crossed polars, the two zones are not clearly defined and appear to
merge into one another. (c) The carbonation-leached zone interface. Once again, these zones are
not well differentiated. Even under crossed polars, the carbonation is muted. In concrete specimens, it is easy to overlook the presence of the leached zone because of the crowded texture.

acidic water at ambient temperatures and are shown for the autoclaved specimens in Figure 6.13 of St John et al. (1998). The mode of carbonation at these high temperatures is largely controlled by the amount of calcium hydroxide present, which in turn is controlled by the amount of silica flour or other material that can react with calcium hydroxide.

When calcium hydroxide and the hydrated cement in the outer margins are carbonated, a chemical gradient forms between the carbonated and inner zones to give a clearly defined zone leached of calcium hydroxide as shown, for example, in Figure 6.101b. Where silica flour has been added to the grout, most of the calcium hydroxide will have reacted, and direct carbonation of the cement hydrates is the dominating process. At 260°C, the carbonation zone may sometimes consist of a series of complex carbonation bands rather than a single band, irrespective of the composition of the grout.

Where silica flour has been added, the direct carbonation of cement hydrates does not set up the same visible chemical gradient and as a result the leached zone is absent. At 160°C, there is an ill-defined zone of alteration, which is the interface between the carbonation zone and the uncarbonated grout. At 260°C, this ill-defined alteration zone is obscured by the presence of 'clumped carbonation', but the carbonation zone is still observable because of the deposition of dark material at its inner margin. Apart from this, the carbonation grades imperceptibly into a texture where clumps of calcium carbonate crystals become less frequent towards the centre of the specimen.

Irrespective of temperature, considerable amounts of silica flour remain unreacted in the outer zones of the grout samples because carbonation removes calcium hydroxide before it can react with the silica. The quantities remaining suggest that initial rates of carbonation are rapid. It is surprising that, even after three months of exposure at 160°C, modest amounts of quartz as small as 10 μm in size can be observed at the centre of samples. Only at 260°C has the temperature been sufficient to ensure an almost complete reaction of the silica flour. Similarly, the presence of remnant cement grains is surprising. Some cement grains appear to be highly resistant to hydration and only at 260°C has essentially complete hydration been observed.

Where the attacking fluid is acidic, even if only mildly acidic, a corroded outer zone of gelatinous silica is formed which appears to be porous. It is very fragile and rarely remains intact at thicknesses of greater than 3–4 mm and appears to offer a little barrier to the access of fluids at these temperatures. More rapid corrosion occurs when silica flour is added to the grouts because carbonation of the cement hydrates formed under these conditions leads to increased permeability (Milestone et al. 1986). Later laboratory work indicates that any addition that reacts with calcium hydroxide or increases permeability has similar effects. It has been found that data obtained in the laboratory give similar textures and rates of attack to those found with downhole specimens.

The hydrated cement phases present in the grouts can only be identified by XRD analysis as, apart from α-dicalcium silicate hydrate, all the phases are submicroscopic. The types of high-temperature phases found have been discussed by Milestone et al. (1986).

6.6.3 Acid-type attack from sulphates, sulphides, brine and microbial action

Cases of acid attack on concrete are not restricted to naturally occurring carbonic acid. Thornton (1978) has reported on a case of a sulphate-rich lake water flowing through concrete outlet tunnels and being converted to sulphuric acid by bacteria. Attack had decomposed surfaces to depths of 30 mm. This is similar to the acid attack that occurs in sewers (see Section 6.5.5). Eglinton (1987) briefly discusses the conditions under which pyrite in rocks, soils and wastes can be oxidised to form sulphuric acid, rather than forming sulphates, and mentions cases where the resulting low pH values have caused corrosion.

Concentrated brines containing magnesium chloride, associated with some geologic sequences, may be encountered when drilling for gas and oil. Oberste-Padtberg (1985) has illustrated the textures of oil well grouts made from either high sulphate-resistant or ordinary Portland cement blended with blastfurnace slag and fly ash. These textures show that not only does the typical oriented crystallisation of brucite occur at the surface, but magnesium ions can penetrate the interior probably through cleavages and microfissures. This not only allows the expansive formation of brucite internally but also results in acidification of the pore solution by the decrease of hydroxyl activity to a level where the calcium silicate hydrate phases may be no longer stable.

The attack by acid slimes on building stone, which will also be applicable to concrete, is reported by Blaschke (1987) and Bock and Sand (1986). The slimes are produced by fungi and algae that collect polluted air, leading to sulphate attack, and by nitrifying bacteria producing nitric acid. Unexpected complexity can be encountered with new materials; for example, Hughes (2012) has explained how microbial (algal) assemblages can foul and degrade some polymer fibres used for reinforcing marine concrete, ironically leading to deterioration of surfaces that had been fibre reinforced to improve durability.

6.6.4 Industrial chemical attack

Industrial chemical attack is here defined as the attack by chemicals that either are not found in nature or whose concentration has been increased above natural levels by industrial activity. It covers a wide range of substances and no attempt is made to cover the whole field. For detailed information on the attack by individual chemicals, reference should be made to Biczoc (1967). Recommendations for protection against a range of chemicals are made in the report by the ACI Committee 515.1R (American Concrete Institute 1994), and a listing of the effect of many chemicals on concrete has been made by the Portland Cement Association (2007). Concrete and mortars are often used in conjunction with masonry to provide protection against chemicals and information on this topic is given by Sheppard (1982).

A fundamental problem in considering the chemical resistance of calcareous hydraulic materials is to know where to set the limits to useful discussion. Because concrete is a relatively cheap building material, there is a tendency for the construction industry to extend its use into areas where it is no longer an appropriate material for the exposure conditions. In these cases, the problem is not whether the concrete will be attacked but how to determine the rate of attack so that the time to repair or replacement can be estimated.

As discussed in Section 6.6.1, concrete should not be exposed to fluids or vapours of less than pH 4, without either suitable protection or the use of more chemically resistant materials such as supersulphated cement, because unacceptably high rates of attack may occur. However, this generalisation covers a complex and variable situation. Kuenning (1966) found that the chemical composition of the aggressive liquid was at least as important as pH and also found no correlation between pH and weight loss from mortar bars. Resistance of mortar was increased by a longer curing time and by a decrease in water/cement ratio.

Kuenning's second observation confirms the point that has been repeatedly observed by investigators. That is, the more impermeable the concrete, the lower the rate of attack for any given exposure condition. Over recent decades, super-high-strength materials, such as the 'densified system with small particles' (DSP), have been developed which have increased resistance to aggressive fluids (St John et al. 1993, Fattuhi and Hughes 1986) and show promise as protective linings with increased durability. DSP is so impermeable that attack is limited to the surface, and the aggressive fluids cannot penetrate any distance into the fabric of the mortar, thus limiting the rate of attack as shown in Figure 6.104.

Acetic acid   Citric acid   Lactic acid

Distilled water

Hydrochloric acid   Nitric acid   Sulphuric acid

Ammonia   Sodium hydroxide   Phosphoric acid

Tap water

Sodium chloride   Sodium tripolyphosphate   Sodium sulphate

Figure 6.104 Chemical attack on ultra-high-strength DSP mortar. The small DSP prisms have been exposed to 5% concentrations of the chemicals shown (except for sodium chloride solution which was saturated), for 16 weeks at 21°C. The chemical solutions were renewed every 4 weeks. Significant volume losses due to citric (75%) and sulphuric acids (33%) occurred. Weight losses for the strong acids varied from 77% to 15% indicating that significant leaching occurred even where volume losses of less than 3% were measured.

Some other aspects of the chemical resistance of concrete have gradually become important. The ability of concrete structures safely to contain fluids and toxic solids and the solidification of toxic wastes using cementitious materials have become important, as a result of national and international legislations for environmental protection (Morrey 1991; Commission of the European Communities Council 1993).

Macro- and micro-encapsulations of hazardous and radioactive wastes have led to the development of a wide range of specialist cement-based materials designed to overcome problems of permeability and mechanical and chemical long-term stability. In many cases, cement replacements are used, such as ground granulated blastfurnace slag and fly ash, together with hydration-modifying admixtures, in order to meet the special requirements of such materials. Roy (1988), Fairhall (1989, 1994) and Davidovits (1994) have reviewed the containment of nuclear wastes. Petrographic study of these materials presents no special difficulties, and the methods described elsewhere in this book may be applied without modification.

The use of cement to solidify and stabilise hazardous wastes is a well-established practice, though some wastes have been shown significantly to retard or poison the normal hydraulic and pozzolanic reactions (Hills 1993; Jones 1988) and doubt has arisen concerning the long-term stability of some of these materials (ENDS 1992, Webb 1993). Reviews of the methods in current use for the fixation and solidification of hazardous wastes are given by Connor (1990), Spence (1993) and the Portland Cement Association (2001). The petrographic techniques applicable to cement solidified hazardous wastes have been reviewed by Wakely et al. (1992), but petrographic observations are hampered by the common occurrence of iron-rich or carbonaceous particles in the wastes, and particularly in incinerated wastes, these tend to mask the details of the cementitious matrix microstructure. In many examples, the solidified wastes are more closely allied to cement-stabilised soils than to normal concretes or mortars. Since chemical stabilisation and fixation are as important as mechanical stability, a chemical approach is more often the most useful method of study (Cocke 1990).

6.6.4.1 Petrographic investigation

When the petrographer is faced with the evaluation of the chemical attack on concrete, mortars and calcareous materials, caution is necessary. Where new construction is involved, the experience from the literature may be used. However, where deterioration is occurring in an existing structure owing to chemical attack, it is necessary not only to evaluate the state of the concrete but also to investigate the chemistry of the aggressive fluids involved. It may be advisable to consult a specialist for advice as the way in which some chemical species attack concrete can be unexpected. The factors requiring evaluation are the composition of the aggregates, the cement content and type, the water/cement ratio, the air void system and the identification of any mineral additions present, plus any textural faults in the hardened cement paste. Textural faults, such as porous surfaces as shown by excessive depths of carbonation, excessive drying shrinkage cracking of surfaces and plastic settlement around aggregate margins will be likely to increase rates of attack.

Identification and concentration of the major ions in solution and the pH must be determined and any temperature fluctuations of the attacking fluid estimated. Where the fluid is in the form of condensation, it may be difficult to sample. Condensates may be absorbed into filter paper which is then immediately placed in distilled water to provide a diluted solution for later analysis. Condensates do not necessarily adequately represent the attacking fluids, as they will usually be modified by interaction with the surface of the concrete.

There is only a limited number of published cases of attack by industrial chemicals where the texture of the concrete has been examined in thin section.

Chandra (1988) and De Ceukelaire (1992) have described the textures resulting from attack by hydrochloric acid. Chandra used mortars and exposed them to 15% hydrochloric acid for 5 days, and De Ceukelaire studied the attack by 1% hydrochloric acid over a period of 2 years on ordinary Portland cement, sulphate-resisting cement and blastfurnace slag cement.

Both observed two main zones of attacked concrete on the surfaces of the specimens, with some differences in detail. The outer corroded zone consisted of mainly amorphous silica and alumina with varying degrees of colouration. Chandra found a yellow colour that he identified as iron chloride, which was absent from De Ceukelaire's samples. The second zone was brown in colour owing to the presence of amorphous material, which was considered to be ferric hydroxide. De Ceukelaire also observed a thin layer of carbonation between the brown zone and the sound concrete.

A detailed chemical description of zonation caused by acid attack has been given by Pavlik (1994), who found that attack by nitric and acetic acids gave similar patterns. In the case of nitric acid attack, the outer corroded zone largely consisted of silica and graded into a brown area where hydrated iron oxide was present. Below this, there was a rapid transition to a layer depleted in calcium hydroxide and enriched in sulphate. Attack by acetic acid differs in that, in the outer corroded zone, the silica content is lower and the zone is preferentially enriched in hydroxides of alumina and iron. No brown layer is reported, but the zone depleted in calcium hydroxide and enriched in sulphate is still present. Unfortunately, no microscopical examination was made of the corrosion zones, but the results support the generalised scenario of an outer corroded zone, an intermediate reaction zone where iron is concentrated and an inner zone leached of calcium hydroxide. Strunge and Chatterji (1989) have described and illustrated the texture of surfaces attacked by acid rain. They identified four zones consisting of an innermost zone of unaltered concrete, bordered by a zone of carbonation, followed by a zone free of crystalline material. The outermost zone consisted of a porous material.

6.7 WEATHERING AND LEACHING

6.7.1 General aspects of weathering and deterioration

Deterioration of concrete with time is dependent on factors ranging from faults at the design and construction stage through choice of materials to environmental conditions. Typically, several mechanisms operate together contributing towards the overall deterioration. Concrete structures are rarely built under ideal conditions, and defects in the concrete material, arising from imperfections in mix design and casting, caused by plastic shrinkage, plastic settlement cracking, blow holes, honeycombing and scouring of surfaces, can provide pathways for ingress of water or other fluids and areas of weakness that can be preferentially exploited by aggressive environmental conditions.

The choice of inappropriate aggregates or cement/binder may also lead to problems of concrete deterioration which develop in the longer term. One example, AAR, will be discussed in Sections 6.8 and 6.9, but other examples include the use of aggregates containing soft or unsound particles or 'shrinkable' aggregates that exhibit excessive moisture movement (see Section 6.2). Another example is the incorporation of small percentages of 'reactive' pyrite in concrete aggregates, which can lead to unsightly, if usually superficial, brown stains on concrete surfaces (see Section 6.5).

Use of HAC (or now 'CAC'), in conditions where the conversion of the metastable hydrate $CaO \cdot Al_2O_3 \cdot 10H_2O$ to the more stable $3CaO \cdot Al_2O_3 \cdot 6H_2O$ is likely to lead to an unacceptable loss of strength and possible failure, is an example of unsuitable materials that have been used in the past (see Chapters 4 and 5).

Climatic conditions, which involve wetting and drying, freezing and thawing and thermal movement, may have a major effect on rates of deterioration of concrete structures. Consequently, the effects of deterioration are often most severe on the exposed weather sides of structures. In condition surveys, plotting the exposure conditions of the structure on a plan using a scale of five levels from most severe to most sheltered can be a useful aid to the prediction of the future pattern of deterioration of the structure.

Particularly aggressive environments such as sulphate-bearing groundwaters, attack by acids, damage from de-icing salts and other types of chemical attack have been dealt with in various preceding sections. In most examples, the deterioration progresses inwards from the outer surface of the concrete and along existing cracks, joints between lifts and other pre-existing defects. Usually, the deterioration affects the cement matrix, sometimes leading to a complete breakdown of the structure of the surface layers, allowing the aggregate particles to fall away.

Cracking, whether induced by overloading, cyclic stressing leading to fatigue, impact or other causes such as freeze–thaw action or ASR is the most commonly observed sign of deterioration. An analysis of the causes of various types of crack pattern has already been covered in Section 6.1 and is also described and discussed in a number of publications, for example, in Concrete Society Report No 22 (1992, 2010), Fookes (1977) and Fookes and Walker (2011). The correct interpretation of crack patterns during the field survey of structures is perhaps the most important aid to the diagnosis of deterioration. Other observations that may assist include vertical displacement of crack margins, discolouration or degradation of surfaces and the occurrence of pop-outs.

The careful correlation of visible features of deterioration with respect to position within the structure, including their relative severities, and the correlation of deterioration with the original constructional sequence and the severity of environmental exposure will provide clear guides towards diagnosis. However, a field examination of structures, even when constructional details are available, is rarely sufficient to allow an unequivocal and clear diagnosis of the cause of a deterioration to be made without a more detailed laboratory investigation of cored samples taken from carefully selected locations on the structure. In many cases, this will require petrographic examination of those samples, when there are clear advantages for the petrographer, whenever sensibly feasible, to have had an opportunity to observe the sample locations in context on the structure in question.

6.7.2 Water leaching

Water alone, with neutral acidity (pH = 7), no excessive content of dissolved salts (say, potable or tap water) and no suspension load (such as clays, silts or sands, as are transported by rivers), is no great threat to the durability of concrete. However, as was shown in Section 6.6, waters that are especially pure (such as distilled water), or acidic, or even alkaline can be aggressive towards concrete, mainly attacking the binder (but also some types of aggregate) and causing surface erosion by dissolution and/or more complex stages of degradation and eventual loss. Evidence of such water leaching is commonly observed in concrete that has been exposed to water migration in service, and the various particular mechanisms were described in Section 6.6. Some of the typical textures displayed by

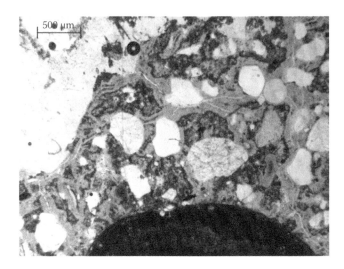

Figure 6.105 Victorian concrete from foundations subject to extensive leaching and water percolation. The areas within the binder that are leached are shown light brown. Unaffected areas are shown dark brown.

concrete subjected to such leaching, including wholly or partially dissolved areas of paste associated with recrystallised and redeposited materials, are illustrated in Figure 6.105.

6.7.3 Salt weathering

We have already seen that 'salts' can have important influences on the durability of concrete. Sodium chloride (common salt) can have a profound influence on the risks of corrosion of steel reinforcement (see Section 6.3.4) and obviously also plays a part in seawater attack (see Section 6.5.4). Sulphates can have a variety of actions on concrete, typically entering into complicated and usually deleterious reactions with cement phases (see Section 6.5). However, in addition to these particular 'salt' issues, the general geomorphological process of 'salt weathering' can affect concrete in certain regions and/or circumstances. This salt weathering process is one of the natural agencies whereby rock is gradually reduced to soil, and it is especially an engineering problem in hot arid climates, where evaporation exceeds precipitation and, as a result, salts are caused to migrate from the ground into materials (such as aggregates or stone or concrete) that are variously placed on to, or buried beneath and within, that ground. A full text on salt weathering will be found in Goudie and Viles (1997).

A particularly severe variety of salt damage arises where concrete structures have foundations immersed in saline groundwaters, which commonly occur at the coastal margins of desert regions such as the Arabian Gulf region (Fookes and Collis 1977; Walker 2002; Sims 2006; Walker 2012). Saline water is drawn up the surface layers of the concrete by capillary action, and evaporation leads to the crystallisation and build-up of salts on the concrete surface and within the surface layers. This 'salt weathering' process causes surface spalling and cracking, exposing reinforcement to corrosion exacerbated by the presence of chloride. Typically, the worst damage occurs a few hundred millimetres from the ground surface (see earlier Figure 6.2). A full treatment of the diagnosis and effects of this type of salt damage is given by Kay et al. (1982). Sodium chloride salt crystals forming

Figure 6.106 Scanning electron micrograph of a cubic sodium chloride crystal formed within the hardened cement paste of a concrete and apparently generating a microfracture.

within the concrete can be shown to cause propagation of microcracks within the cement paste as is illustrated in Figure 6.106.

6.8 ALKALI-AGGREGATE REACTION (AAR)

6.8.1 Overview

At the time that this book was originally published, it was appreciated that the most common form of AAR is 'ASR' and accepted that another earlier-described type of AAR, termed 'alkali–silic*ate* reaction', was better seen as a variant of ASR, with 'ACR' being recognised as a different and much less well-understood mechanism. There has since been considerable research and much debate about ACR, including the suggestion that it too might be just a variant of ASR. Accordingly, in this revised edition of the book, 'ACR' is now covered in Section 6.9 ('AAR Involving Carbonate Aggregates'). In this present subsection, AAR is frequently used as a general term to cover all the reactions between cement/binder alkalis and aggregates, but the more specific term ASR is also used.

There is a very large literature on AAR that has developed over more than 70 years since it was first identified and which has previously been referenced by bibliographies published by Figg (1977), British Cement Association (1992) and Diamond (1992). Books dedicated to AAR include Hobbs (1988); Swamy (1992), which is currently being updated and expanded; West (1996); and Blight and Alexander (2011). Recent summary overviews have been published by Fournier and Bérubé (2000), Oberholster (2001), Sims and Poole (2003), Thomas and Folliard (2007) and Stark et al. (2010). Undoubtedly, one of the most valuable resources for current researchers is the collection of proceedings from the series of thirteen international conferences on AAR (see Table 6.13), which started in 1974, with the 14th ICAAR taking place in Austin, TX, in 2012.

Several national or regional reports or guides on AAR have also been published, and these are generally useful internationally, especially as aggregates are a key component in the AAR mechanism and political boundaries have no relevance to these geomaterials. That said, despite a strong thread of AAR technical consistency, climatic conditions, non-aggregate concrete constituents and construction practices do vary between regions and countries, so that actually the detailed experience of AAR has proved to be different for

Table 6.13 List of ICAAR proceedings (1974–2016)

Number	Year	Place	Chairman	Nations	Delegates	Papers
1	1974	Koge, Denmark	Gunnar Idorn	4	23	13
2	1975	Reykjavik, Iceland	Haraldur Àsgeirsson	7	25	20
3	1976	Wexham Springs, UK	Alan Poole	13	48	27
4	1978	Purdue, Indiana, USA	Sidney Diamond	8	54	26
5	1981	Cape Town, South Africa	Bertie Oberholster	12	52	38
6	1983	Copenhagen, Denmark	Gunnar Idorn	21	187	56
7	1986	Ottawa, Canada	Paddy Grattan-Bellew	23	~300	101
8	1989	Kyoto, Japan	Kioshi Okada	16	327	136
9	1992	London, UK	Alan Poole	29	320	150
10	1996	Melbourne, Australia	Ahmad Shayan	26	220	130
11	2000	Québec City, Canada	Marc-André Bérubé	24	230	142
12	2004	Beijing, China	Tang Ming-shu	22	178	168
13[a]	2008	Trondheim, Norway	Børge Wigum	27	~132	133
14[b]	2012	Austin, Texas, USA	Kevin Folliard	28	194	128
15[c]	2016	Sao Paolo, Brazil	Haroldo de Mayo Bernardes	tba	tba	tba

[a] Proceedings available on a CD (which included a print-on-demand book option).
[b] Proceedings available on a memory stick.
[c] Scheduled for 3–7 Jul 2016.

individual countries. Some of the more helpful national or regional publications available in English include Cement and Concrete Association of New Zealand (1991) and Munn (2003); Concrete Society Technical Report 30 (1999), BRE Digest 330 (2004), BRE IP1/02 (2002), Institution of Structural Engineers (ISE, 1992/2010), BS 8500-1&2 (2006) and BS 7943 (1999); Helmuth et al. (1993), Stark et al. (1993) and Fournier et al. (2010); Laboratoire Central des Ponts et Chaussées (LCPC 1994); Nijland and Siemes (2002); Richardson (2003); and Norwegian Concrete Association (2008).

Petrographers will find some published AAR picture atlases particularly useful: Le Roux et al. (1999), Lorenzi et al. (2006) and Norwegian Concrete Association (2008). A RILEM technical committee (TC 219-ACS) has recently completed an expansion of the Lorenzi et al. (2006) atlas, which currently includes reactive aggregate types from around Europe, to provide a more global coverage (Broekmans et al. 2009; Fernandes et al. 2012, 2015).

AAR is the expansive reaction between the alkalis sodium and potassium in the pore solution of a concrete and minerals in the aggregates. The principal source of alkalis is derived from the cement itself, but any source of sodium or potassium (including those from mineral additions or 'supplementary cementitious materials' [SCMs], admixtures and even some aggregates) can contribute to the reaction provided that the alkali can move into the pore solution of the concrete and create the necessary hydroxyl ion concentrations required. French (2005) has suggested that alkalis involved in ASR can in due course be recycled within the concrete, to perpetuate the process, and we shall see that this might be a factor in some reactions affecting large mass structures, such as dams, over an extended period.

Three types of reaction are identified in AAR:

1. The classic 'ASR'. The reaction between alkalis and varieties of silica minerals such as opal, chalcedony, cristobalite and tridymite and volcanic glasses to produce expansive alkali–silica gels.
2. The 'slow reactive/expansive' form of ASR. Formerly termed 'alkali–silicate reaction', the reaction of alkalis with disordered forms of silica and quartz, often submicroscopic, in sedimentary and metamorphic rocks to produce expansive alkali–silica gels.

There is no clear dividing line between Types 1 and 2, the behaviour of which differs in several ways. Jensen (1993) suggested that Type 2 should include those aggregates in which the expansive force is caused by internal crystallisation pressure rather than by water uptake by gel. French (2005) has described three types of ASR, depending on the reactive aggregate: cherts and volcanic glasses (Type 1), rocks such as recrystallised sandstones with 'complexly intersutured grain boundaries or sometimes a fine-grained porous recrystallised matrix' (Type 2) and rocks such as greywackes and argillites, also some cataclasites and metaquartzites (Type 3). It thus seems that French would subdivide Type 1 (into his Types 1 and 2), while his Type 3 would equate to Jensen's Type 2.

3. The 'ACR'. Apparent reaction rims at the margins of limestone aggregate particles are not uncommon in concretes, but the most common deleterious, expansive reactions have involved argillaceous dolomitic limestones. The mechanisms involved remain uncertain and are the subject of some controversies, including suggestions for replacing the term ACR with 'so-called ACR' (Katayama and Sommer 2008) or other names. ACR is discussed fully and separately in Section 6.9.

Concrete damage resulting from the ASR reactions is caused by expansion from the formation of alkali–silica gel in the aggregate or at its margins and the subsequent absorption of pore fluid, which causes the gel to swell and exert pressure. This internal pressure may be large enough to crack the concrete and just occasionally sufficient to fracture reinforcing steel (Miyagawa et al. 2006). The reaction can only proceed if *all* the following three factors are present in the concrete:

1. Sufficient moisture, here defined as not less than an 85% RH, in the pore structure of the concrete;
2. A sufficiently high alkalinity in the pore fluid surrounding the reacting particle;
3. Reactive mineral(s) in the aggregate, which often may not react unless it is present in a critical proportional range.

If *any one* of these three factors is absent, then the ASR will not proceed. If either factor 1 or factor 2 ceases to apply during the course of an existing ASR, the reaction will stop, but may recommence if the necessary conditions in factors 1 and/or 2 later become satisfied again. French (2005) suggested that the reaction was not limited by the amount of alkali in the concrete, because the initially formed ASR gel gradually reacts in turn with the cement paste to form calcium silicates, thus releasing or 'recycling' the alkalis to react with further susceptible aggregates. Gradual carbonation of ASR gel might have a similar effect. Charlwood and Scrivener (2011) have speculated that such alkali recycling could possibly be one explanation for the persistent expansion observed for many large concrete dams.

Many types of reactive aggregates will only react if they are present in a critical volume proportion range. The proportion of reactive aggregate which produces the maximum expansion for a given alkali content in the concrete is known as the 'pessimum' proportion, and some typical examples are illustrated in Figure 6.107 (see also Figure 16.24 in Sims and Brown [1998] and one practical investigation in Merriaux et al. [2003]).

6.8.2 Chemistry of the alkali–silica reaction

The larger portion of the alkali metal ions in Portland cement is usually present as sodium sulphate (Na_2SO_4), potassium sulphate (K_2SO_4) and as the mixed salts $(Na,K)_2SO_4$ and calcium langbeinite ($K_2Ca_2(SO_4)_3$) depending on the amounts and ratio of sodium

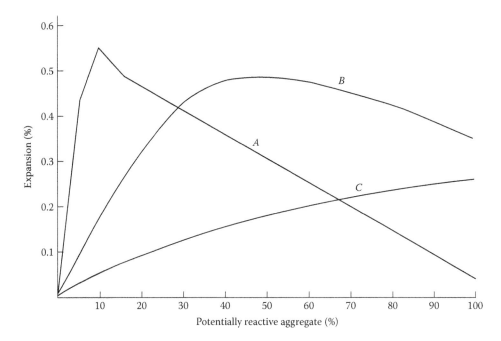

Figure 6.107 ASR pessimum proportion curves are shown for three volcanic aggregates: *A* is a rhyolite that exhibits a pronounced pessimum at around 10%; *B* is an andesite exhibiting a broad pessimum between about 40% and 60%; *C* is a fresh glassy andesite with no pessimum proportion. Some aggregate types exhibit marked pessimum proportions over a narrow range of concentration, but in general most aggregates have a wider pessimum proportion range and andesite *C* is unusual. The curves are the plots of the 12-month expansions obtained from testing some New Zealand volcanic materials by ASTM C227 at 1.5% Na_2O equivalent. (From Cement and Concrete Association of New Zealand, Alkali-aggregate reaction. Minimising the risk of damage to concrete. Guidance notes and model specification clauses, Cement and Concrete Association Report No TR3. See also Munn 2003 for 2nd edn., 1991.)

and potassium present. In this form, they are readily soluble. Smaller amounts occur in solid solution in the cement minerals and may be substantially released within 28 days as the cement hydrates. Alkalis may also be introduced as sodium chloride, for example, owing to inadequate draining or washing of marine-derived aggregates or less significantly from de-icing salts or deposited salt spray. Additions (SCMs) and admixtures are other potential sources of alkalis. Some aggregates themselves may also be potential sources of alkalis over a period within concrete.

When water is added to Portland cement, the alkali sulphates react with the hydrating tricalcium aluminate and calcium hydroxide to precipitate ettringite, releasing sodium and potassium ions into the pore solution. This pore solution contains considerable concentrations of these ions and very low concentrations of other ions such as calcium, sulphate and chloride. The considerable concentration of alkalis in the pore solution results in a range of hydroxyl concentrations of about pH 13–14. It is these hydroxyl concentrations that provide the chemical driving force for ASR.

Sodium chloride is readily soluble in the pore solution and can react with the aluminate phases to form chloroaluminates, so that the sodium ions are released into the pore solution, further to contribute to the overall alkali concentration. The extent to which the chloroaluminates form is dependent on both the concentrations of chloride and the aluminate phases present (Chatterji et al. 1987a,b; Kawamura et al. 1996; Sibbick and Page 1998).

At the maximum levels of chloride permitted in reinforced concrete by many standards (say 0.1%–0.15% for moist environments), complete conversion to chloroaluminates with release of sodium ions can be assumed. This reaction of sodium chloride can take place in both the plastic and hardened states of concrete. Nixon et al. (1987a,b) confirmed that sodium chloride added to cement paste rapidly converts to alkali hydroxide in the pore solution and thus exacerbates ASR.

ASR is essentially an attack by sodium and potassium ions, with concomitant hydroxyl ions, on varieties of silica to produce alkali–silica gel. The rate of this attack will depend on both the types of silica present and the concentration of the alkalis in the pore solution. Only at the upper end of the pH range does significant attack develop. The gels that form consist of a calcium–sodium–potassium silicate of variable compositions. The formation of the swelling form of these gels appears to require a minimum amount of calcium hydroxide to be present in the concrete. Pozzolanas effectively compete for the available calcium hydroxide, and this is seemingly one of the reasons why they can suppress ASR expansion, as without sufficient calcium hydroxide only non-swelling gels are formed. However, Duchesne and Bérubé (1994) found no correlation between reduction in concrete expansion and calcium hydroxide depletion when testing SCMs.

The swelling types of gel are capable of absorbing water into their structure and expanding. It is this expansive pressure which creates a tensile strain within the concrete and causes cracking when the local, unrestrained, tensile strength of the concrete is exceeded. The expansive forces are believed rarely to exceed 6–7 MPa (Diamond 1989), but the force varies both with the composition of the gel, in a way which is not fully understood, and with the total amount of gel present in the concrete. Spatially, the reaction between aggregate and alkalis to form gel is rarely uniform and nearly always appears as separated point sources of reaction. The point sources are often restricted to a few local areas but in more severe cases will be widespread. It is this tendency to occur as point sources of expansive force that gives the typical, somewhat random to semi-regular pattern of internal cracking seen in concretes affected by ASR, as shown in Figure 6.108.

The amount of gel also depends on the amount of available reactive silica, and, therefore, up to a point, an increase in the amount of reactive silica produces an increase in expansion. Above a certain proportion of reactive silica to alkali, so much of the alkali is absorbed that the concomitant concentration of hydroxide in solution is insufficient to maintain the same severity of attack and the expansion decreases again. This is the reason that is given to account for the critical, or 'pessimum', proportion that occurs with so many aggregates (Concrete Society 1999; Cement and Concrete Association of New Zealand 1991). Other discussions on the possible explanation of pessimum behaviour are given by Hobbs (1978 and 1981) and French (1980).

6.8.3 External appearance of concrete affected by ASR

Cracking caused by ASR is never distributed uniformly over the surface of a concrete even in test specimens stored under closely controlled laboratory conditions. In the field, its appearance can vary from just one or two cracks to isolated areas of pattern cracking, and in severe cases, extensive pattern cracking is evident. It is common to find that only a limited portion of the surfaces appears to be affected, while the remainder of the structure shows little sign of the reaction in spite of the same mix design and concreting materials ostensibly being used throughout. It is also common to find the reaction is more severe on the weather side of the structure, possibly owing to the greater extent of wetting and drying (Wood et al. 1996). However, measurements of RH for several Norwegian road bridges, dams and office buildings (Jensen 2004a,b, 2005) have shown a high variation in humidity, which can explain

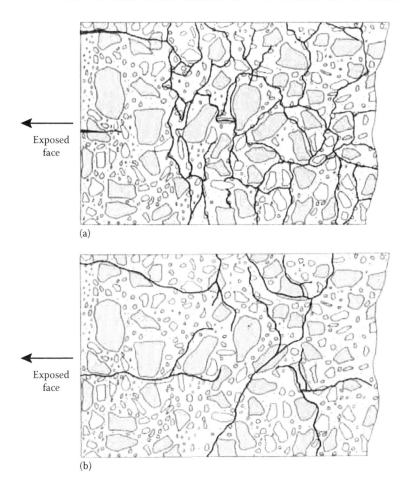

Exposed
face

(a)

Exposed
face

(b)

Figure 6.108 Internal crack patterns caused by ASR. In the earliest stage of reaction, the internal pattern cracking will be less developed, perhaps only isolated cracks will be present, perpendicular to the external surface, extending from that surface and terminating at depths of 50–75 mm. As the reaction develops, the internal crack patterns develop: (a) Crack pattern caused primarily by reactive fine aggregate (sand) particles. (b) Crack pattern caused primarily by reactive minerals in the coarse aggregate. (From British Cement Association, *Alkali-Aggregate Reaction 1977–1992, a Definitive Bibliography*, Ref: CIS/B8, BCA, Slough (later Crowthorne), U.K., 1992a; British Cement Association, *The Diagnosis of Alkali-Silica Reaction, Report of a Working Party*, 2nd edn., Ref: 45.042, BCA, Wexham Springs, U.K., 1992b.)

the non-uniformly distributed cracking seen in many concrete structures damaged by ASR. Moreover, from about 50 mm depth from the surface, internal RH was found to be constant over a period of years and no wetting and drying cycles were observed. At Elgeseter bridge in Trondheim, Norway, RH varied from 100% on west-facing columns exposed to rain to less than 80% in east-facing surfaces of the same column. On the west-facing surfaces, several vertical wide cracks caused by AAR could be followed from the ground level to the lower bridge deck (10 m), whereas on the east-facing surfaces, visible cracking was absent.

In severe cases, the differential expansion of the concrete is sufficient to cause visible displacements of structural elements (Figure 6.109). Where concrete is heavily reinforced, cracking may be wholly restrained or only present in the direction of the reinforcement. Examples of various types of external cracking caused by ASR are illustrated in British

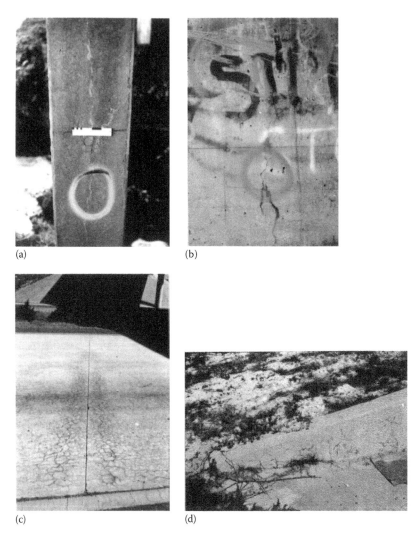

(a)

(b)

(c)

(d)

Figure 6.109 The appearance of external cracking due to moderate AAR in a temperate climate: (a) The slender pier of this 20-year-old bridge has cracking restricted to a few areas parallel to its length, probably due to pre-stressing. The ASR is in its earliest stage and no gel was visible in thin sections. Because the margins of the crack are carbonated, they appear white. (b) More typical signs of ASR are the localised somewhat irregular cracks with darkened crack margins. The pier foot of this 30-year-old bridge within the tidal zone of brackish water was more severely cracked. Internally, clear signs of a minor to moderate ASR were present. (c) This pavement of a South African motorway is extensively pattern microcracked, especially around the transverse joints. It is difficult to assign a prime cause to this cracking from visual examination, as the cracking could have a number of causes exacerbated by vehicular loading. (d) However, examination of an adjacent drainage culvert, containing similar concrete, reveals the typical, somewhat random pattern cracking typical of AAR. Note that the crack margins are darkened. Thin-section examination of core samples from the pavement revealed that extensive and severe AAR was occurring.   *(Continued)*

(e)

(f)

Figure 6.109 (Continued) The appearance of external cracking due to moderate AAR in a temperate climate: (e) Looking across the 25-year-old hardstanding area of Whenuapai airbase in New Zealand, displacement of one-half of the hardstand relative to the other is clearly apparent along the construction joints. This differential expansion between reacting and non-reacting slabs continues unabated. (f) However, signs of visible cracking are limited and unspectacular. This is in spite of the slab being unreinforced apart from transfer pegs at joints. Thin-section examination of core samples from the hardstand identified the non-reactive and reactive areas of the pavement and showed that a severe ASR was occurring.

Cement Association (1992), and the assessment of expansion using crack widths has been discussed by Jones and Clark (1996).

Survey investigations in Norway and Switzerland using measurements of crack widths on numerous concrete structures with similar aggregate types have given important information on the extent of AAR and the degree of damage (Jensen and Merz 2008). The structures investigated by Jensen and Merz were dominated by construction ages around 20–60 years. Because the graphical distribution of the maximum crack width is influenced by an increased number of structures in some periods relative to periods with fewer structures (false peak height), the average maximum crack widths in 323 Norwegian structures have been calculated for each decade as shown in Figure 6.110, where the number of structures in each decade is representative for statistical analyses (Jensen 2011). Structures without cracks and without measurable cracks are not included in the calculations. Note that the average crack width increases significantly from 0.25 mm after 10 years to 2.25 mm until 60 years (about 2 mm at 50 years) and shows an exponential development. Gel is water soluble and in many or most cases will be washed away from concrete surfaces by rainwater. Therefore, gel exudations as well as cracks are often only observed in water-protected locations, such as the soffits of bridge decks.

Where the ASR is still active, cracks will often be open and margins may be vertically displaced. On fine cracks, this displacement can be felt by rubbing the fingertips across the crack. Older cracks may be filled with carbonated calcium hydroxide (calcite), which can be difficult to distinguish in the field from carbonated gel (see Figure 6.111). Exudations of alkali–silica gel are not common and, where these are present, indicate that there has been sufficient moisture to carry the gel through to the surface. This exuded gel is usually white in colour because the originally transparent gel carbonates to white crystalline material on contact with air. Removal of exudation followed by

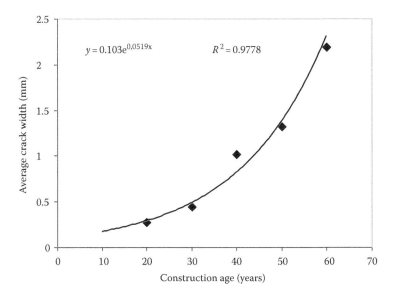

$y = 0.103e^{0.0519x}$ $R^2 = 0.9778$

Figure 6.110 Average maximum crack width versus construction age (in decades) for 323 Norwegian structures, showing an exponential curve and a best fit R^2 of 0.98. (Figure courtesy of Dr Viggo Jensen.)

Figure 6.111 ASR cracking of a concrete structure, with associated white exudations of mainly calcite and some carbonated alkali–silica gel. A typical section of the old Maentwrog Dam in Wales, United Kingdom, since replaced by a larger new dam downstream.

microscopical examination in a suitable refractive index oil will sometimes show a clear gel present beneath the carbonated exterior. The texture of carbonated gel in thin section is shown in Figure 6.112.

Although gels may not be visible, a sufficient amount may have been absorbed at crack margins to darken them, especially when the concrete surface is wetted. Thus, enhanced delineation of cracks may be noted after rain which persists for some period after the

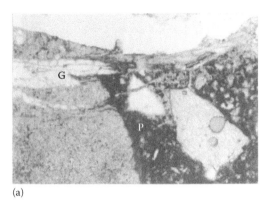

(a) (b)

Figure 6.112 Carbonated gel, widths of field 1.3 mm: (a) Photomicrograph in ppl of a fracture surface at the bottom of a core sample. Gel (G) runs along a crack in the rhyolite aggregate across to the fracture surface, where it has carbonated during storage (paste: P). (b) In xpl, the gel near the surface can be seen to be carbonated. The carbonation extends into the porous hardened paste. Gel extruded onto an outer surface will tend to have the same texture of carbonation as shown here.

remainder of the concrete surface has dried. Also, there may be a tendency for a pink/brown colouration to form along the margins of cracks, possibly due to alkali bleaching the lichens which may be present on the concrete surfaces (see Figure 6.113).

A number of other surface features can occur with ASR. Where sufficient moisture is present and the alkalis are not depleted from the outer layer of the concrete, pop-outs can occur as a result of the formation and swelling of gel. Pop-outs sometimes occur in laboratory specimens which have been continuously moist cured, but are uncommon in structures. In some cases where a structural member is heavily restrained by reinforcing, the only evidence of AAR will be internal evidence of gel and rare surface pop-outs. An unusual occurrence of staining with pop-outs associated with granite aggregate was reported as an AAR case by Mindess and Gilley (1973).

Where unexplained cracking is observed in a structure, especially if some of the features as described earlier are visible, core sampling and further investigation should be undertaken. The sites for this sampling must be selected with care, to obtain a reliable diagnosis of the cause(s) of cracking, as described in Chapter 3 and Sections 6.8.4 to 6.8.6. In a very short paper referring to concrete bridges in the United Kingdom, Pearson-Kirk et al. (2004) outlined a 'total assessment rating' scheme, including various parameters in addition to specific ASR features, for use in developing structure maintenance strategies.

6.8.4 Microscopic textures and features of concrete affected by ASR

The microscopic textures of concretes affected by ASR are variable and depend on the extent of the reaction, the size and distribution of the reacting aggregates and the environment of the concrete structure. The three types of feature observed in thin sections by the petrographer are cracking, alkali–silica gel and signs of reaction in aggregate particles. These should *all be present* for unequivocal identification of ASR in thin section. In practice, meaningful observation is possible only if adequate sampling has been performed and sufficient surface areas of the axial planes of the core samples are examined (Sims 1992; Palmer 1992).

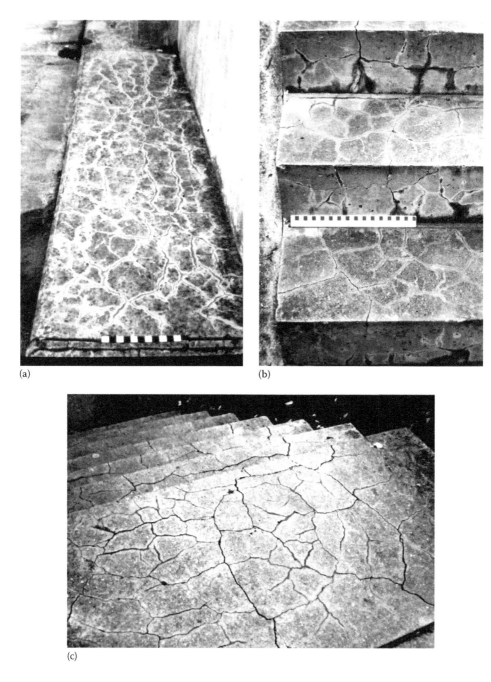

(a)

(b)

(c)

Figure 6.113 ASR map cracking of concrete structures, showing characteristic bordering: (a) unreinforced pavement in Plymouth, United Kingdom, (b) and (c) unreinforced steps in Exeter, United Kingdom.

In the early stages, cracking caused by ASR will be limited to a few isolated cracks which may penetrate up to 50–75 mm into the concrete. Examination in thin section may fail to observe gel, any cracking apart from the surface cracks and reacting aggregates. This type of cracking occurs because the expansive effects of the ASR have not generally exceeded the tensile strength of the concrete. However, the outer surface of a concrete is depleted in alkalis early in its life as a result of leaching and wetting and drying effects. Thus, ASR is absent from the outer 50–75 mm of concrete, and this outer shell is in tension with respect to the inner portion of the concrete where the expansive reaction is occurring. This situation typically leads to the localised formation of a few isolated cracks in the outer shell in the early stages of ASR (see Chapter 5). Cases of this type of cracking have been reported by St John (1990) where all the conditions for ASR are present and no other cause for the cracking can be found.

With an increasing development of reaction and consequent expansion, internal cracking develops and the frequency of surface cracking also increases, but this surface cracking has a more widely spaced pattern than that found internally. A detailed description of this type of cracking is given in Chapter 5. Finally, in cases of severe expansion, major cracks several millimetres in width may develop, which penetrate deeply into or through the concrete. The location of a major crack is largely controlled by the configuration, design and loading of the structure and the way in which it interacts to accommodate the large internal expansion caused by the ASR.

A complex pattern of cracking will be found on the surface where environmental conditions have not caused leaching of alkalis from the concrete shell. In these cases, fairly closely spaced, fine pattern cracking will be present on the surface, and there will be little gradation in the spacing and frequency of the cracking from surface to the interior. In addition, both gel and reacting aggregates will be found in the outer shell and pop-outs may occur.

The preceding textural description of the cracking due to ASR is for unrestrained concrete exposed to temperate weather conditions. Where frost attack occurs, cracking initiated by ASR is enhanced and may even lead to severe disintegration of the concrete. If restraint is present, most commonly owing to the presence of steel reinforcement, cracking of the concrete will be partially or completely suppressed in the direction of the restraint. Internal microcracking is also affected by the size and shape of the reacting aggregate as shown in Figure 6.108. Where a portion of the fine aggregate has reacted, the spacing and widths of the internal microcracking are smaller than where coarse aggregate has reacted (Palmer 1992).

6.8.4.1 Observations on cores and polished slices

Before proceeding with the preparation and detailed examination of thin sections, much useful information can be obtained by the careful inspection of core samples and polished slice specimens impregnated with fluorescent resin as follows (Jensen 2011):

Some possible observations on **Cores** (Figure 6.114):

- White reaction rims on reacted aggregates (reacted aggregates parted along the crack with reaction products, seen on broken faces);
- Gel exudations in air voids, cracks and broken faces and on the curved outer faces of cores, showing as damp or 'sweaty' patches;
- Cracking from surface and into the concrete;
- Internal cracking in reacted aggregates (easily seen in slow-reacting aggregates in late stages);
- Cracking in paste.

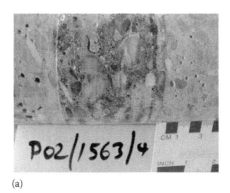

(a)

(b)

(c)

Figure 6.114 Concrete cores taken from a severely ASR-effected structure which demonstrate (a) sweaty and heavily degraded porous chert aggregate particles, (b) severe ASR-induced expansive cracking developing subparallel to the outer concrete surface at right, and (c) severely ASR-cracked aggregate particles and associated now dried ASR gel deposits on the outer surface of the core.

Some possible observations on **polished slices** (Figure 6.115):
(Many of the same features can be observed as in cores, but with a better clarity.)

- Reaction rims on reacted aggregates seen as darker thick rims; can be mistaken as weathering rim, but not when crushed aggregates have been used. Reaction rims are later visible in fluorescence impregnated thin section;
- By the use of fluorescence impregnation, cracking becomes very visible, for example, cracks down to 1 μm can be observed by the unaided eye using UV illumination;
- Location of thin sections should be selected in areas with signs of reaction. Dependent on the size of thin sections, often more than one thin section should be taken. It is important that gel is not removed by the preparation of thin sections and as little water as possible should be used during preparation.

Figure 6.115 View of a sawn slice of concrete (100 mm width) showing cracks through several coarse and fine aggregate particles. 'Sweaty' patches within aggregate patches are shown dark grey. (Photo courtesy of Paul Bennett-Hughes, RSK Environment Ltd.)

6.8.4.2 Observations on thin sections

The alkali–silica gel found in concrete is a material of variable composition, consisting of SiO_2, Ca, Na and K together with other minor elements (Poole 1992). In thin section, typical alkali–silica gel is colourless, usually conchoidally cracked due to drying of the gel in sample preparation, and has a mean refractive index of 1.48–1.49 with an upper limit of 1.51 and a lower limit of 1.45 (Idorn 1961). Sibbick (2011) advises that, while the actual ASR gel maybe colourless, gel deposits in various concretes can exhibit multiple colours, probably derived from minor contaminates picked up from the aggregate and/or cement paste. When the ASR gel is deposited in cracks and pores, it is distinctive in texture and immediately recognisable in thin sections (Figure 6.116). However, in older concretes, the appearance of matured gel is often more variable. West (1996) includes some very helpful labelled diagrams to explain ASR features in thin sections (see one example in Figure 6.116).

Perhaps the most distinctive variation in the texture of the gel is that present within reacting aggregates. In the interior of reacted aggregates, cracks are usually empty apart from crack entrances which are often plugged by layers of granular, birefringent material, clear gel and dark gel as shown in Figure 6.117. The three layers have been shown to be similar in chemical composition and probably represent stages in the formation of the gel and its interaction with the cement paste as described by Davies and Oberholster (1986), Andersen and Thaulow (1990) and Katayama and Bragg (1996). However, the granular, birefringent layer is not always present and may be difficult to observe in some types of reactive aggregate and where the aggregate particles are small.

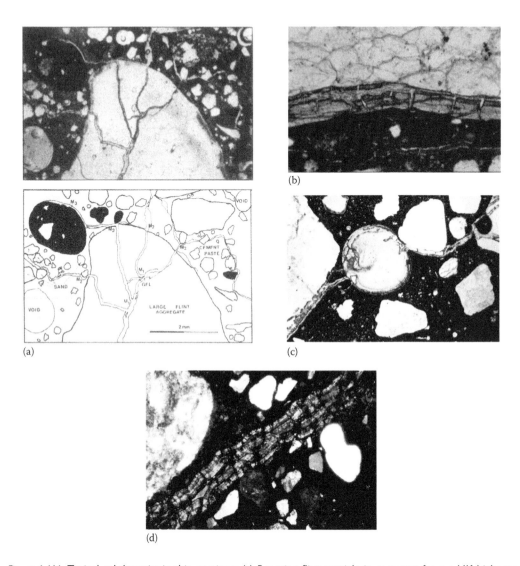

Figure 6.116 Typical gel deposits in thin sections: (a) Reacting flint particle in concrete from a UK highway, showing a well-developed system of gel-filled cracks within the aggregate and propagating into the surrounding matrix. (From West, G., *Alkali-Aggregate Reaction in Concrete Roads and Bridges*, Thomas Telford, London, U.K., 1996, p. 163.) (b) Gel deposits with characteristic desiccation cracks filling a crack peripheral to a quartzite aggregate particle (upper part of view), within concrete from a motorway bridge in the English Midlands; the bridge was later replaced. (c) Gel deposits filling cracks and lining air voids within concrete from a motorway bridge in northern France; the bridge was later replaced. (d) Gel deposits filling a crack within the same concrete structure as (c), but exhibiting a distinct layering; the cracking was initiated by deflection of an under-reinforced bridge deck and the ingress of moisture with the generation and movement of gel was thought possibly to be seasonal.

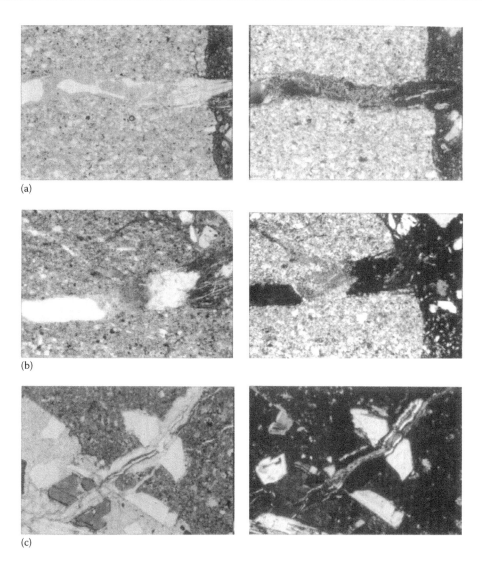

Figure 6.117 Some textures of typical gel plugs associated with reacting aggregates, widths of field ~1.4 mm, left ppl, right xpl: (a) The hornfelsed shale coarse aggregate in this South African hydro structure has a crack which is mainly empty apart from a plug of gel which protrudes into the cement paste. Inside the aggregate, there is an area of material that has a granular appearance. Where the gel protrudes into the paste, it is clear and then becomes darkened. Under crossed polars, the granular material inside the coarse aggregate is crystalline and birefringent and hence is not a true gel. The plug of gel which protrudes is isotropic irrespective of whether it is clear or darkened. There is the appearance of a typical gel plug. (b) The blue-green argillite coarse aggregate in this 8-year-old Canadian dam has a crack which is empty apart from an area of granular material and a plug of gel which protrudes into the cement paste. In this case, darkened gel protrudes into the aggregate as well as where it contacts the paste. Under crossed polars, the granular material inside the coarse aggregate is crystalline and birefringent and grades into isotropic gel. (c) The andesite aggregate in this 60-year-old bridge has a crack filled with clear gel, which is difficult to distinguish from the glassy matrix of the rock. Displacement of the crack margins is clearly shown by the pyroxene and feldspar crystals. There is only minimal darkening of the gel in contact with the cement paste. Under crossed polars, where the gel runs through the paste, it is partially crystallised and birefringent granular material is not present within the crack. This absence of birefringent gel is fairly typical of reacting volcanic materials.

In the cracks and voids of the hardened paste, alkali–silica gel can also be variable in texture and colour and is prone to crystallisation. This crystallisation has a much lower birefringence than the granular, birefringent material discussed earlier. The differences are illustrated in Figures 6.117 and 6.118. The darkening of gel in contact with cement paste has already been mentioned. This darkening can range from a light tan to a deep brown-black colour. Gel is sensitive to carbonation, so that cracks in the outer shell of concrete associated with ASR are often carbonated to depths considerably greater than would be expected. In general, older gels tend to be more crystallised, but the circumstances in which gel crystallises are not fully understood and, as shown in Figure 6.118, clear gel and crystallised gel can be in intimate contact. A possible explanation for gel crystallisation has been proposed by Dron and Brivot (1996). It should not be assumed that all gel in contact with cement paste is darkened as clear gel may be observed in cracks running through the hardened paste. Gel can soak into the hardened paste at the margins of aggregates and cracks, causing an apparent darkening probably due to the absorption of calcium hydroxide.

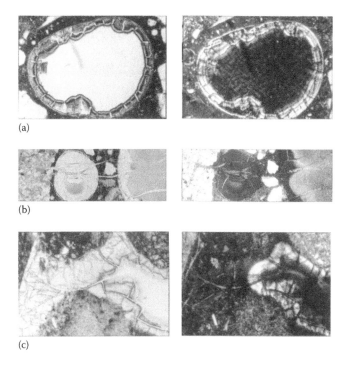

(a)

(b)

(c)

Figure 6.118 Some of the microscopic textures of alkali–silica gels, left ppl, right xpl: (a) Gel has flowed into the entrapped air void and coated its inner surface. It has curved (conchoidal) cracks due to drying and is layered and stained brown. Gel will often be found rimming pores and can be present as a thin layer which can be difficult to distinguish from hardened cement paste when it is thinned to a wedge during sectioning. Under crossed polars, the gel is crystallised in spite of the fact that it has been continuously moist cured and the sample is only 2 years old. This photomicrograph was taken from a section of an ASTM C227 (2010) mortar bar specimen containing basalt coarse aggregate and rhyolitic sand. Width of field 1.44 mm. (b) The horn-felsed shale aggregate contains a crack filled with a typical gel plug, which has extruded gel into two connected pores. The pores contain clear gel with conchoidal cracks together with some gel which appears to be stained brown. Under crossed polars, the brown gel is seen to be crystallised. Inside the crack, the granular infilling material is also birefringent. Width of field 2.8 mm. (c) Clear gel with conchoidal cracks fills part of the large entrapped void space in this 40-year-old bridge and shows no sign of darkening in contact with the hardened paste. Under crossed polars, its outer margin is shown to be crystallised. Width of field 1.4 mm.

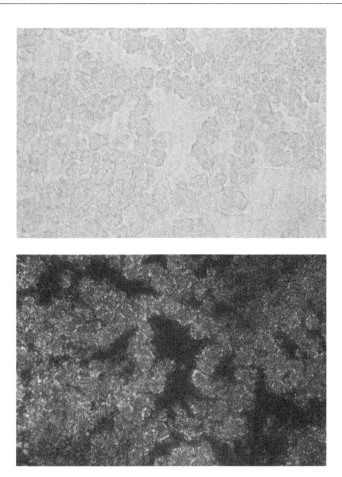

Figure 6.119 Cryptocrystalline reaction products within an air void in reacted concrete, upper view ppl, lower view xpl, width of each view 0.6 mm. Note the intergrown reaction products, globular agglomerates and 'salt and pepper' texture under polarised light. (Photomicrographs courtesy of Dr Viggo Jensen.)

In Norwegian 'slow-expanding' aggregates, a brownish-coloured reaction product occurs in cracks and air voids in reacted concretes, both from field structures and from laboratory concrete prism test samples (Jensen 1993). In polarised light, the reaction products appear granular with a 'salt and pepper' texture (similar to the appearance of chert and flint) and with cryptocrystalline grain sizes (and therefore called 'cryptocrystalline reaction product'). Sometimes, and by use of high magnification (400×), individual crystals and agglomerates can be observed under the petrographic microscope (Figures 6.119 and 6.120).

Using SEM, the reaction product consists mostly of platy crystals and globular agglomerates (Jensen 2011). EDX-type microanalysis gives a composition of silicon, potassium, sodium and calcium, with minor aluminium, iron, magnesium and sulphur. Calcium content is relatively low compared with gel located in the cement paste. In the cement paste, gel occurs as clear amorphous gel in cracks and air voids. These gels are sometimes re-crystallised and laminated suggesting development by several 'gel pulses'.

Cryptocrystalline reaction products only occur in reacted aggregates and air voids (it has probably flowed into the air voids from reacted aggregates located nearby). In cracks in the aggregate–cement paste contact zone, cryptocrystalline reaction product is

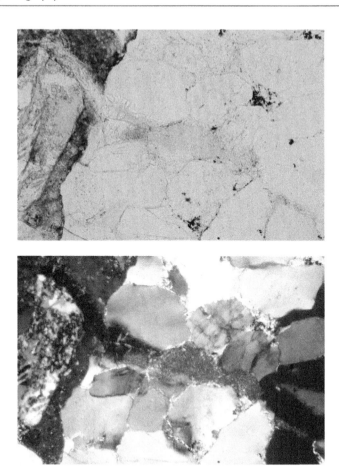

Figure 6.120 Reacted sandstone, upper view ppl, lower view xpl, width of each view 2.4 mm. Right part of view shows an empty crack (yellow in ppl) where constituents have been removed by the reaction. The central view shows light brown cryptocrystalline reaction product within a crack, exhibiting 'salt and pepper' texture in polarised light. Note that the cryptocrystalline reaction products turn into clear amorphous gel with shrinkage cracks near the contact zone with the cement paste and within the cement paste. (Photomicrographs courtesy of Dr Viggo Jensen.)

succeeded by clear amorphous gel, which also occurs in the cracks and air voids in the cement paste.

Based on the observation of slow-expanding aggregates, it can be hypothesised that cryptocrystalline reaction products originated from a precursor gel arising from a reaction internal to the aggregates. In contact with the more basic cement paste, the precursor gel (or cryptocrystalline reaction products) transforms into a clear amorphous gel, as seen in cracks in the contact zone to the cement paste (see Figure 6.120). Air voids filled with cryptocrystalline reaction products also show a 'rim', with clear amorphous gels in contact with the cement paste (Jensen 1993).

In the early stages of ASR, the fact that alkali–silica gel is not observed does not necessarily mean that it is absent. Unless gel is present in cracks or voids, it can be difficult to observe in thin sections with the optical microscope. This effect can be demonstrated by visual observation of fracture surfaces of core samples where in some cases reaction rims may be clearly visible around some aggregates (Figure 6.121). However, when the core sample is thin sectioned, these reaction rims often effectively disappear and cannot be observed.

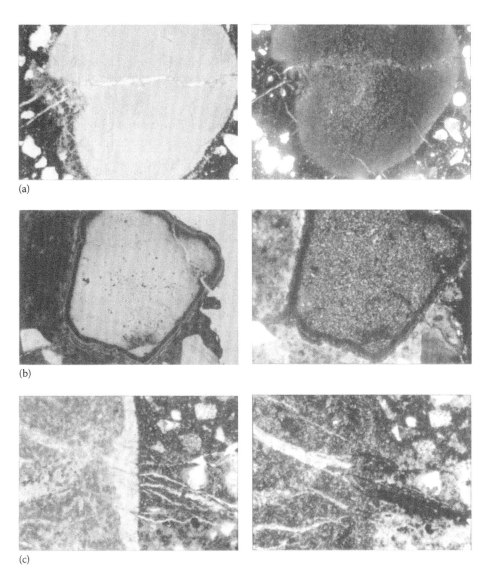

Figure 6.121 The appearance of reaction and weathering rims in thin section, left ppl, right xpl: (a) This chert aggregate from a 30-year-old bridge in the Midlands of the United Kingdom has a somewhat diffuse brown margin intersected by numerous fine cracks filled with gel. The dark rim could still be due to weathering, but the balance of evidence suggested that it is caused by reaction as it possibly contains gel. Under crossed polars, the centre of the chert is showing some signs of disintegration and the cracking is highlighted. Width of field 3.2 mm. (b) By contrast, the dark rim on the chert aggregate from a structure in the southwest of the United Kingdom is definitely caused by weathering, as was apparent when it was compared with other unreacted aggregates present. However, the aggregate is reacting, as the weathering rim is penetrated by fine cracks with gel. Under crossed polars, note how post-sampling carbonation has obscured the texture of the hardened paste. Width of field 1.4 mm. (c) The rhyolite coarse aggregate in this 60-year-old concrete dam in the United States appears to have a light-coloured reaction rim. Inside the aggregate, granular material and gel fill stranded cracks and extrude as clear gel into the hardened paste. Under crossed polars, the paste texture is partly obscured by post-sampling carbonation. The granular material is birefringent, the gel is isotropic and the assumed reaction rim is obviously altered. Width of field 3.2 mm.

For this reason, the amount of gel observed in thin sections frequently cannot be correlated with the severity of the ASR, and the work by Johansen et al. (1994) indicates that damage may occur when limited amounts of gel form. Martineau et al. (1996) also found that it is not possible systematically to relate expansions to the quantity of gel found by fluorescence. Sibbick (2011) has confirmed that some of the concrete containing generally slower-reacting greywacke and siltstone aggregates in the United Kingdom often contained little residual ASR gel in both the aggregate and the surrounding cement paste. This seemed to have little effect on the resulting total expansions developed. Such problems will obviously present challenges to attempts to quantify damage by 'counting' features including gel deposits (e.g. the DRI scheme in Section 6.1.3).

Sibbick (2011) has found that, if thin sections are not fully impregnated, especially deep within the cracking developed within the aggregate, the high-alkali 'lower-viscosity' gels in that area will often be more easily lost during thin-section preparation than the surrounding more rigid calcium-rich ASR gels found in the cracks in the surrounding cement paste. Generally, he finds that, in otherwise identical concrete samples, more ASR gel is retained in thin sections achieving optimum impregnation.

In terms of interpretation, it will be seen later that the matter is even more complicated, because just an absence or apparent absence of alkali–silica gel cannot be taken as proof of the absence of ASR from a given concrete, nor does the identification of gel deposits prove that ASR expansion and damage have occurred (Palmer 1992). It is necessary to demonstrate a causal link between the reaction and any cracking.

It is difficult to make meaningful observations on the mobility of gel. There is no doubt that under appropriate conditions, gel can move considerable distances, but the relationship of gel viscosity varies with aggregate type, the point in the life of the structure and the local environment within the concrete. When gel is observed in cracks and voids with no obvious aggregate source, it must be remembered that the thin section is a plane and the actual source may be close by, just below or above the plane of the section. In a well-developed ASR, innocuous aggregates may also be cracked because they cannot withstand the expansive forces generated. Mobile gel can flow into these aggregate cracks but should not be taken as evidence that the aggregate has reacted.

Ettringite is commonly found intimately associated with alkali–silica gel, but much of this ettringite is observable only with the aid of the scanning electron microscope and cannot be seen in thin sections. There is a range of microproducts associated with gel that have been observed with the SEM and, while these are of importance in elucidating the mechanism of reaction and gel formation, they are generally not observable by the light microscope (Regourd-Moranville 1989). However, bands and accretions of ettringite are also quite commonly found in the cracks and voids of concretes undergoing ASR, not necessarily intimately associated with gel and these are observable in thin sections. There is a report by Marfil and Maiza (1993) of the identification of a zeolite of the heulandite–clinoptolite group in microfissures and pores of a concrete highly fissured by AAR. No details are given of the aggregates. The micrographs given illustrate materials that could easily be confused with ettringite. The concurrence of ASR and ettringite is also discussed in Section 6.5.9.

The presence of alkali–silica gel is one of the more important observations required for the identification of ASR. Lightweight and cellular concretes are claimed to have sufficient void space to accommodate the gel without expansion, but this claim must be treated with caution as there is still no clear understanding of whether the expansive pressures generated are primarily due to the formation of the gel or to its subsequent absorption of water. However, there is experimental evidence to show that expansion decreases with increasing porosity of aggregates and that air entrainment has a similar effect, at least initially (Jensen et al. 1984; Collins and Bareham 1987; Building Research Establishment 2002).

6.8.4.3 Reactive aggregates

ASR occurs between the alkaline pore fluid of a concrete and one or more reactive minerals in the aggregate. In many cases, the reactive mineral forms a small part of the aggregate, or, alternatively, only a small portion of the reactive mineral present in the aggregate is available for reaction. Since the reactive mineral will be dispersed in the rock texture, it would be expected that aggregate porosity would be important in that it may effectively control the rate of access of the pore fluid and in addition may affect the magnitude of the expansive forces generated. For example, Rayment (1992) found that the porous 'cortex' was typically the most reactive component of the flint fragments within the United Kingdom sand and gravel aggregates, with the expansions obtained for particular gravel deposits (Rayment and Haynes 1996) usually being correlated with their porous cortex contents. However, Rayment and Haynes (1996) also found a small number of flint gravels with low contents of porous cortex that produced anomalous expansion in their tests. This finding led to all flint aggregates in the United Kingdom being regarded as exhibiting 'normal' reactivity (Building Research Establishment Digest 330 1997/1999), with a previous empirical assumption that aggregate combinations containing >60% flint (Sims 1992) could be regarded as 'non-reactive', being abandoned.

In ASR, there are really only two generalised classes of minerals that are known to be expansively reactive with the alkalis in concrete. These are metastable types of silica including some disordered forms of quartz and aluminosilicate glasses with a silica content usually in excess of 60%–65% and not less than 50% (Katayama et al. 1989). The natural forms of metastable silica are opal, chalcedony, tridymite and cristobalite (Figure 6.122) and of the aluminosilicate glasses, the matrix of fresh intermediate to acid volcanic rocks (Figure 6.123). Most commercial glasses, including some glass fibres and mineral wools (Keriene and Kaminskas 1993), are also potentially reactive (see Section 4.8). For instance, the commercial aluminosilicate glass known as 'Pyrex' is so reactive that it is used as the standard reactive aggregate for comparison purposes in ASTM C 441-05 (2005). Synthetic cristobalite materials have also been used for the same purpose (Lumley 1989; Berra et al. 1995).

The identification of opal, tridymite and cristobalite, and of the glass matrix of fresh acid/intermediate volcanic rocks, theoretically presents few problems to the petrographer. The practical problem is that small quantities of some of these reactive minerals can be important. It has been shown that as little as 1%–2% of opaline material (Stanton et al. 1942) and only a few percent of tridymite and cristobalite can cause ASR (Kennerley and St John 1988), which tested as potentially deleterious in the ASTM C289 (2007) chemical test, suggestive of a marked pessimum proportion of the aggregate (Katayama et al. 1989; Katayama 2010b). Thus, care is required when carrying out petrographic examination of aggregates likely to contain these minerals to see that they are not overlooked.

Where doubt exists about the presence of opal, cristobalite and tridymite, they can be quantitatively determined as phosphoric acid residues (Talvitie 1964; Cement Association of Japan 1971, 1996). The amounts of cristobalite, tridymite and quartz in volcanic aggregates, as determined by a combination of the phosphoric acid method and XRD analysis, have been reported from Japan, New Zealand (Katayama et al. 1989; Katayama 2010b) and Iceland (Katayama et al. 1996a). In ASR-affected field concretes, reaction sites of cristobalite and tridymite in andesite aggregates have only been identified in Japan, by means of direct polarising microscopy and SEM observation on polished thin sections (Katayama 2010b). In other countries, thin sections prepared by concrete petrographers or geologists were too thick (at the standard 30 µm) to allow reliable identification of these fine-grained silica minerals contained within the groundmass of volcanic rocks (Katayama [2011]: see **Chapter 3** for discussion on the use of thin sections ground thinner than usual).

Also, the 'gel pat test' described by Jones and Tarleton (1958) has now been adapted within BS 7943 as a method for the detection of opal (or similarly highly reactive silica) in aggregates (BS 812-123 1999) (see Figure 6.124).

Determination of the potential reactivity of chalcedony and disordered forms of quartz in flints and cherts, and alteration of the glassy matrix of volcanics, presents petrographic problems for which there are no ready solutions. The petrographer is rarely able to relate the presence of these minerals to an assessment of potential reactivity. In the case of chalcedony, which is often submicroscopic, its presence does not necessarily indicate potential reactivity. As mentioned earlier, BS 7943 includes an adaptation of the 'gel pat test' for identifying highly reactive silica, but this provides no assistance with evaluating the likelihood of reaction with flints, cherts, glassy volcanics and other more moderately reactive rocks.

The work by Rayment (1992, 1996) has shown that other forms of disordered silica present in English flints and cherts are the more likely cause of reactivity, and Groves et al. (1987) identified reactive, disordered structures in opal by transmission electron microscopy (TEM). Zhang et al. (1990) also used TEM to study the relationship between silica microstructure and reactivity, when they found that the density of dislocations in quartz

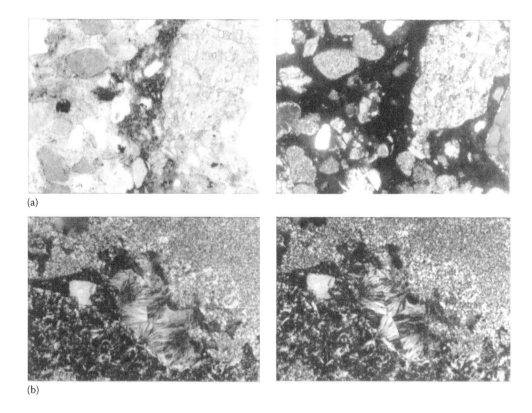

(a)

(b)

Figure 6.122 The appearance of the silica minerals, opal, chalcedony, tridymite and cristobalite, left ppl, right xpl. (a) Opal. The opal forms the brown matrix to this glauconitic gritstone. A more common occurrence in AAR is where the opal forms rims around particles. Under crossed polars, the opal is seen to be isotropic. Opal is a rather nondescript material and is mainly distinguished by its low refractive index of 1.4–1.46 and lack of other distinguishing characteristics. Width of field 1.4 mm. (b) Chalcedony. The fibrous texture and low refractive index identify the chalcedony present at the edge of this English chert aggregate. Under crossed polars, the low birefringence of the chalcedony is revealed, and the typical cryptocrystalline nature of the chert is also illustrated. Width of field 1.4 mm.   *(Continued)*

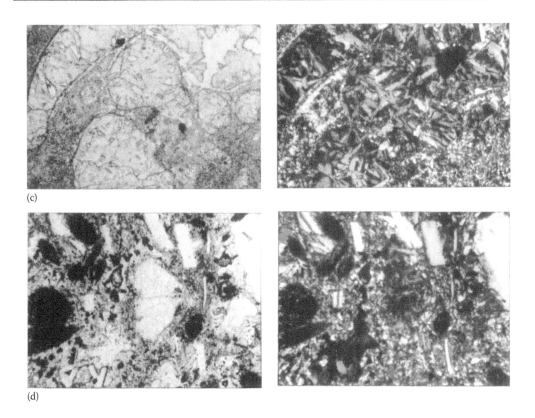

Figure 6.122 (Continued) The appearance of the silica minerals, opal, chalcedony, tridymite and cristobalite, left ppl, right xpl. (c) Tridymite. The low refractive index suggests the presence of a pool of tridymite in this Whakamaru rhyolite from New Zealand. Under crossed polars, the wedge-shaped crystals with simple twinning and low birefringence clearly identify the material to be tridymite. The orange-yellow crystals are pyroxene and the white material probably quartz. There is also some fine matrix materials present. Width of field 1.4 mm. (d) Cristobalite. Cristobalite tends to form small pools of low refractive index material with a texture that appears like overlapping roof tiles. The pools can be quite small and it needs a keen eye to distinguish cristobalite in the matrices of volcanic rocks. This also tends to be the case for the other silica minerals. Under crossed polars, the pools of cristobalite have a low birefringence which, combined with the low refractive index and tile-like texture, identifies it in this Daisen andesite from Japan. Width of field 3.1 mm.

crystals was an indicator of reactivity potential. Similarly, the *degree* of strain in quartz has previously been claimed to correlate with reactivity (Dolar-Mantuani 1983), but this is now believed to be incorrect, and it is generally accepted that submicroscopic silica in veins and at crystal boundaries is responsible for reactivity in rocks that might also display strained quartz (Grattan-Bellew 1992; Kerrick and Hooton 1992) (see Figure 6.125).

Thomson et al. (1994) investigated the microscopical and XRD analysis of potentially reactive rocks and concluded that there is a good relationship between expansivity and crystallinity index (CI) of quartz and that silica dissolution could be related to grain size and crystal defects and deformation. They also claimed that petrographic examination is effective in predicting reactivity. Similarly, Wigum (1995, 1996) has correlated microstructural features of cataclastic rocks with their alkali reactivity. The techniques for measuring strain in quartz have been reviewed by Smith et al. (1992), and a new method was proposed

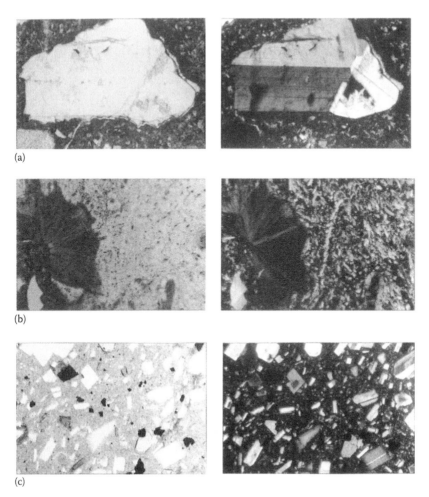

(a)

(b)

(c)

Figure 6.123 Typical matrices of intermediate and acid volcanic rocks, widths of field 1.4 mm, left ppl, right xpl: (a) The lower margin of the feldspar crystal is attached to an irregular rim of clear, acid volcanic glass, which has reacted and formed a crack, and some associated gel which is not visible. The feldspar crystal is set in a poorly differentiated matrix of this particle, which is probably derived from rhyolite. Under crossed polars, the glass rim is completely isotropic, and the matrix consists of a mixture of isotropic materials in which are embedded finely crystalline materials. The fissure on the upper right of the feldspar crystal is filled with calcium hydroxide. Whether either the glass rim or the matrix of the rock has reacted, or perhaps both, is impossible to distinguish. (b) The matrices of many fresh rhyolites exhibit wide ranges of texture, the most spectacular being spherulitic texture which consists of radiating areas of feather-like material. The remaining matrix is also often partially devitrified as shown here and filled with minute crystals of feldspar, quartz and ferromagnesian phases. It is claimed that one of the devitrification products of spherulitic texture is cristobalite, but in practice, it is the brown undifferentiated matrices of rhyolite that are most commonly associated with AAR as shown in Figure 6.127a. Under crossed polars, details of the crystallisation of the matrix are revealed. (c) The matrix of this fresh Egmont andesite from New Zealand is a relatively clear brown glass in which are set crystals of feldspar, magnetite and pyroxenes. Both cristobalite and tridymite are usually absent from this rock, which is the main reason why the aggregate has no pessimum proportion. Under crossed polars, the glass matrix has partially devitrified as minute crystals of material scattered throughout in spite of the freshness of the rock. The glassy matrices of many dacites, andesites and basalts are often filled with devitrification products, often to such an extent that it is difficult to distinguish the glass. This is one of the reasons why in these cases petrography is unable to do more than indicate the need for further testing in respect of assessing ASR potential.

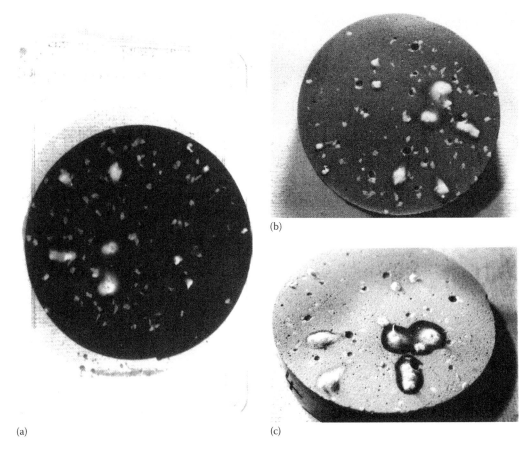

(a)   (b)   (c)

Figure 6.124 Examples of gel pats with varying amounts of highly reactive silica, betrayed by the generation of white spots of gel during the test. © British Standards Institution (BSI–www.bsigroup.com). Extract reproduced with permission. Source: BS 7943: 1999 Guide to the interpretation of petrographical examinations for alkali silica reactivity.

that eliminates some of the uncertainty in the measurements. The effect on reactivity of the degree of alteration and devitrification of the glassy matrix of volcanic rocks is not certain; however, extensively altered and devitrified volcanic rocks are often non-reactive.

Jensen (2011) advises that strain in quartz is the first and, in many cases, only sign of defor-mation of rocks. The more advanced deformation is cataclasis under high pressure and 'low' temperature. Quartz turns into sub-grains or crushes to smaller particles, re-crystallises or in some cases turns into glassy phases. Feldspar often survives as lenticular particles 'flowing' in an orientated matrix of deformed, crushed or re-crystallised quartz. The cataclastic rock types are classified as 'cataclasite' when non-preferred orientation occurs and as 'mylonite' with preferred orientation. Cataclasite and mylonite are highly alkali-reactive when quartz is present. The microtexture of cataclasite and mylonite is easily identified under the petro-graphic microscope. However, some recrystallised cataclasite/mylonite rocks can be difficult to distinguish from other recrystallised quartz-bearing rocks (see Figure 6.126).

In slow reactive/expansive aggregates, cryptocrystalline and/or microcrystalline quartz is the reactive constituent. This has been verified by reaction in 'pure' quartz-bearing rocks and by SEM observation of dissolution groups in quartz from reacted meta-rhyolite. Image analyses of grain sizes carried out in thin sections from 96 different reacted aggregates have

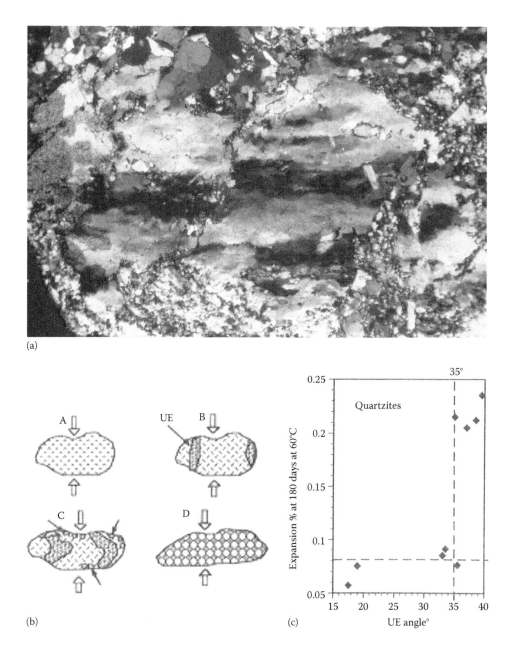

(a)

(b)

(c)

Figure 6.125 Relationship between strained quartz and reactive quartz: (a) strained quartz in thin section, xpl, (b) development of undulatory extinction and sub-grain boundaries in response to stress, (c) variation of mortar bar expansion and undulatory extinction angle. (From Grattan-Bellew, P.E., Microcrystalline quartz, undulatory extinction and the alkali-silica reaction, *Proceedings of the 9th International Conference on Alkali-Aggregate Reaction in Concrete*, London, U.K., 1992, pp. 383–94.)

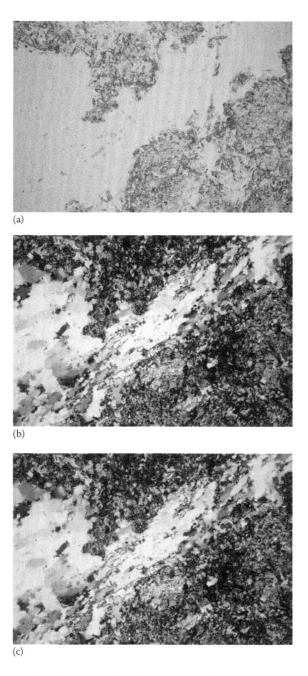

(a)

(b)

(c)

Figure 6.126 Appearance of mylonite under the petrological microscope, width of view 2 mm: (a) Photomicrograph of a thin section in ppl, showing mylonite rock originating from quartz diorite. Plagioclase occurs as irregular areas with numerous secondary zoisite, epidote and sericite inclusions separated with 'elongated bands' or veins of 'clean' quartz. (b) Photomicrograph of the same view in xpl. Quartz is highly strained, with development of sub-grain and recrystallised quartz due to cataclasis. Plagioclase is only slightly affected by the deformation process, but altered. (c) Photomicrograph of the same view using a gypsum plate. By use of the gypsum plate and an increase to second-order interference colours, the lattice preferred orientations of quartz crystals display similar interference colours (yellow-reddish), showing that quartz crystals obtain the same optical orientation due to the deformation.

shown geometrical averages of D_{min} and D_{max} from 12 to 60 μm (Jensen 1993). However, larger particles are often observed reacted in sandstones and some quartzites. Jensen (2011) adds that feldspar has sometimes been implemented in ASR associated with some reacted non-deformed granites, but this remains uncertain and has not so far been demonstrated in a publication; however, feldspar (and especially altered or weathered feldspar) is commonly identified as a possible source of additional 'releasable' alkalis within concrete.

This inability to correlate petrographically chalcedony, disordered silica, altered glasses and other aggregate constituents with potential reactivity has resulted in a range of chemical, mortar and concrete tests being developed to determine the reactivity of aggregates. These have previously been reviewed by Grattan-Bellew (1989, 1996), Nixon and Sims (1992, 2000), Sims and Nixon (2006) and Sims et al. (2004). ASTM C289 (2007) has often proved unreliable for rocks containing chalcedony and disordered forms of silica, but modification of this test by Sorrentino et al. (1992) showed promise in overcoming some of the problems. In Denmark, a 'chemical shrinkage' test has been used to identify reactive flints and cherts (Knudsen 1992), and in Germany, a sodium hydroxide dissolution test is used for the detection of opaline materials in particular materials. Because the mortar bar test, ASTM C227 (2010), can give variable results when testing some of these rocks, an alternative mortar test carried out at 80°C (Oberholster and Davies 1986) has been developed as ASTM C1260 (2007). An up-to-date outline of international testing procedures for AAR is given in Section 6.8.6.

While these tests are of varying adequacy, when examining concretes undergoing ASR, the petrographer is faced with the problem of often being unable precisely to identify the reactive mineral(s) in the aggregate. However, while the reactive mineral constituent(s) may not be identifiable, the examination of concrete in thin sections does indicate whether rocks as opposed to minerals have reacted. Such information from structures can be of vital importance in identifying reactive aggregates and indicating the need for laboratory testing. Where satisfactory results from testing aggregates for potential reactivity are not available, identification from field concretes may be the only source of reliable information.

Because of the problems associated with the identification of reactive minerals, the *texture* of reacting aggregates has assumed some importance in petrographic examination. Idorn (1961, 1967) was one of the first to use thin sections to investigate systematically the texture of reacting aggregates and published many excellent micrographs of disintegrating flints and cherts together with some data on phyllites, basalts and shale. Since then, there have been numerous petrographic investigations of concrete, but illustrations of textures of reacting aggregates tend to be limited. Only a few systematic studies similar to that undertaken by Idorn appear to be readily available, including Jensen (1993) and Broekmans (2002).

Attack in its simplest form consists of a simple major crack system transecting the aggregate with the crack normally being empty of gel apart from crack entrances, which may be plugged by gel layers as described earlier (see Figure 6.117). Reaction rims and disintegration of rock texture are not observed. It might be thought that this simple cracking is typical of the early stages of reaction, but this is not the case. For a surprising number of aggregates such as andesites, dacites, greywackes, sandstones, quartzites, granites and coarser-grained rocks in general, the visible attack on the rock texture is limited, no matter how advanced the ASR (see Figure 6.127).

At the other extreme are those aggregates that are severely disintegrated as a result of ASR. Much of the internal texture may have been consumed by the reaction, which deposits gelatinous material inside a microcracked shell. The finer-grained rocks, such as flints, cherts, argillites, shales, phyllites and rhyolites, often exhibit this texture that is typical of those rocks, where the reactive minerals form a substantial portion of the rock texture (Figure 6.127). Good examples of this type of disintegration texture in flints are given by Idorn (1967). However, in the earlier stages of ASR, these aggregates may exhibit simple

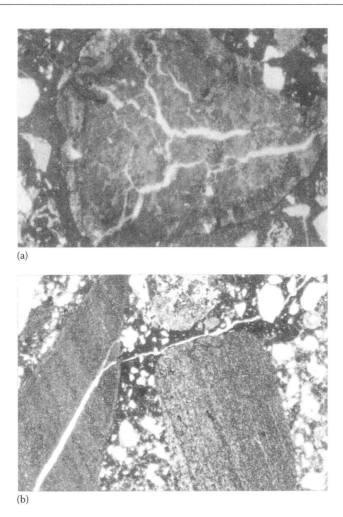

(a)

(b)

Figure 6.127 Two types of cracking in reacting aggregates: (a) The texture of the rhyolite particle is disintegrating into a complex of internal cracks and fissures with minute tongues of gel protruding into the paste, width of field 2.3 mm. (b) The blue-green argillite coarse aggregate has one main crack with gel plugs at the crack entrances. While the cracking tended to run along bedding planes of the aggregate, in this case, it is clearly crossing the bedding, width of field 2.4 mm.

cracking with little disintegration, so that a range of reaction textures may be found even in the same thin section.

Jensen (2011) explains that a 'blacklist' of alkali-reactive aggregates and minerals was published by Mielenz et al. (1947). These alkali-reactive aggregates contained one or more of the following components: opal, chalcedony, volcanic glass, devitrified glass, tridymite and possibly hydromica. This group of aggregates and minerals should be designated as 'fast-reacting/expansive' ASR (i.e. Type 1 as shown in Section 6.8.1).

In the late 1950s and 1960s, some reports on unidentified cement–aggregate reactions considered to be different from traditional ASR were published. The aggregate types were siliceous rocks, such as argillite, phyllite, greywacke, granite, quartzite, schist and sandstone, and the alkali-reactive constituents were suggested to be cryptocrystalline/microcrystalline quartz, strained/granulated quartz and/or possibly certain feldspars. These aggregates should be designated as 'slow-reacting/expansive' ASR (i.e. Type 2 as shown in subsection 6.8.1). A list of potentially alkali-reactive aggregates and minerals has been published by RILEM (Sims 2003), but will be excluded from the revised version of AAR-1 (see Section 6.8.6) (RILEM 2015), with RILEM stressing that the experience with alkali-reactive aggregates varies from country to country, so that wherever possible these should be assessed using national or regional criteria.

ASR rims are not necessarily observable in thin sections and should not be confused with rims due to weathering (Figure 6.121). The presence of weathering is easily checked by observing unreacted aggregates present in the section. To observe the presence of reaction rims, it is necessary to examine either, or both, a polished slice impregnated with fluorescent resin (examined under UV light; see Figure 6.128) or a fracture surface of the concrete preferably supplemented by testing for the fluorescence of gel with uranium salts as described by Natesaiyer and Hover (1989), or under the SEM to analyse the rims directly by EDX-type microanalysis.

In the groundmass of glassy volcanic rocks or the fine matrix of sedimentary rocks, the texture often gives the appearance of having been consumed by the ASR, with an abrupt

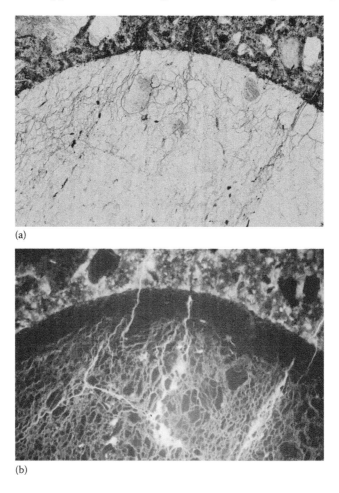

(a)

(b)

Figure 6.128 Reacted mylonite, widths of view 5.8 mm: (a) upper view in ppl, (b) lower view, same area in fluorescence light. Note the intense interfacial cracking between particles/mineral phases and the light brown cryptocrystalline reaction products in some of the cracks. Yellow cracks/areas in plane polarised light become more visible under fluorescence light. Note the dark green dense reaction peripheral zone in the aggregate and the intense internal cracking within the aggregate. (Photomicrographs courtesy of Dr Viggo Jensen.)

transition to unaffected matrix. In some cases, delineation with fluorescent dyes shows the presence of fine cracks through the unaffected matrix leading to the attacked areas. Where veins of alteration feeding into cracks are observed, which could possibly be caused by ASR, it is often the case that these same veins can also be found in unreacted aggregates (Figure 6.129), and so the relationship of this alteration to the ASR cannot be established. A more reliable identification of ASR gel within cracks in aggregates should be carried out by SEM observation on polished thin sections at high magnifications, coupled with EDX-type analysis (if possible aided by the chemical mapping of elements), and not only by optical microscopy using covered thin sections of normal thickness (Katayama 2010b).

In the United Kingdom during the 1990s, various cases of slow reactive/expansion (Type 2) ASR were found to be associated with greywacke aggregates, including the Maentwrog Dam (see Figure 6.130), which was later replaced by a larger dam downstream (Blackwell and Pettifer 1992). However, it was found to be impossible definitively to differentiate reactive and non-reactive varieties of greywacke by petrography alone, although EDX-type microanalysis of the clay-grade matrix can help to indicate the potentially reactive more silicified rocks.

Owing to the problem in identifying the alkali reactivity of greywacke aggregates by conventional petrography, a protocol was agreed in the United Kingdom (British Cement Association 1999) for testing particular sources using a scheme based on the BS 812-123 (1999) concrete prism test. It was also discovered that greywacke in the form of natural gravel/sand aggregate was inherently less reactive and less expansive than similar materials used as a crushed rock aggregate. Accordingly, Concrete Society Technical Report 30 (1999) and BRE Digest 330 (Building Research Establishment 1999, 2004) placed a 10% limit on the content of crushed greywacke in 'normal' reactivity aggregate; greywacke aggregates containing >10% crushed greywacke were to be classified as 'high' reactivity unless shown

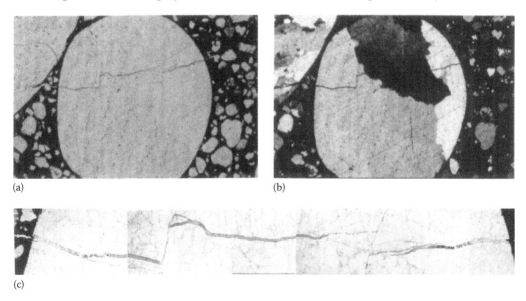

(a)

(b)

(c)

Figure 6.129 Veins of alteration and reaction in a quartzite aggregate: (a) The round quartzite aggregate is cracked. The entrances to the cracks are plugged with gel as illustrated in Figure 6.117. The remainder of the crack is filled with a brown material which is vaguely birefringent. Width of field 8.5 mm. (b) Under crossed polars, the metamorphic character of the quartzite is apparent. (c) A montage detail of the crack shows that it is difficult to distinguish reaction from alteration. Similar veins of brown material could be observed in other aggregates where reaction was absent. However, interpretation in these cases is limited by observation being restricted to the plane of the section; length 4.2 mm.

Figure 6.130 Old Maentwrog Dam in Wales, United Kingdom, an example of long-term alkali–greywacke reaction, exacerbated by leaching and freeze–thaw action. The original structure was superseded by a new larger dam downstream and is now under the enlarged reservoir.

otherwise using the BCA protocol. Current BS 8500-2 (2006) refers to BRE Digest 330 in respect of ASR.

Recently, Katayama (2012) has demonstrated that SEM examination of concrete, using polished thin sections rather than fracture surfaces, is particularly suitable for distinguishing minute flakes of clay minerals/phyllosilicates from reactive crypto- to microcrystalline quartz that has converted to ASR gel, and later crystallised into rosettes, within the greywacke aggregate.

RILEM TC 219-ACS has prepared a picture atlas of alkali-reactive mineral, rock and aggregate types from around the world (Fernandes et al. 2012, also Broekmans et al. 2009; Fernandes et al. 2012, 2015). This Atlas (designed AAR-1.2 by RILEM) will build on the very useful atlas previously produced and published for European aggregates by an EU-funded research project, 'PARTNER' (Lorenzi et al. 2006). A localised picture atlas has been published by Le Roux et al. (1999) for French materials. BS 7943 (1999) provides an experience-based guidance on the reactivity of UK aggregates, in support of the BS 812-104 (1994) method for petrographic examination of aggregates, based upon the principle that many or most UK-sourced aggregates (such as the extensively used flint- and quartzite-based gravel/sands) could be taken as exhibiting 'normal' reactivity, whereas some aggregates might be recognised by petrography as likely to exhibit relatively either 'high' or 'low' reactivity (for which expansion testing would be required for clarification).

6.8.4.4 Lightweight aggregates

Concretes containing significant proportions of lightweight aggregates such as pumice, scoria, expanded shales and sintered fly ash, and also concretes with substantial air void contents, frequently do not undergo expansion even if alkali–silica gel is present. This is based on an early work by Vivian (1947), as well as field experience over the years. It is possible that the substantial void space in the aggregates or air void system can accommodate the gel and any resulting expansion. However, there are a few cases reported where lightweight concretes have expanded and caused problems.

De Ceukelaire (1992) has reported on a bridge in Belgium where chert, present in the sand used in a lightweight concrete bridge, reacted and expanded to such an extent that the bridge had to be replaced. The expanded shale coarse aggregate was not able to accommodate the gel sufficiently to prevent expansion. In Kansas, United States, Crumpton (1988) reported that several bridge decks made with expanded shale lightweight aggregate had to have as much as a third of a metre or more of their length sawn off because of concrete expansion. Initially, expansion was slow and AAR was not evident, but as expansion continued, abundant alkali–silica gel was found. It is not stated whether normal weight materials were also present in these concretes. St John and Smith (1976) found that test specimens made with an expanded argillite aggregate slowly expanded in spite of the low-alkali cement having been used. No alkali–silica gel was observed. Thus, expansion in lightweight concretes can occur under appropriate conditions. The problem is that these conditions have not, as yet, been defined, and it is not always clear whether other aggregates may be involved.

The petrographic examination of lightweight concrete is more difficult than normal weight concrete because of the glassy texture of the aggregates used. In the case of expanded shales, the aggregate margins are defined by a denser oxidised rim of material which is reasonably smooth and presents a definable interface with the cement paste, providing it has not been crushed. Natural pumice and scoria are usually crushed materials and the interface with the hardened paste is highly irregular. The glassy texture in thin sections has a poorly defined texture, and it can be difficult to trace cracks and differentiate alkali–silica gel, as shown in Figure 6.131.

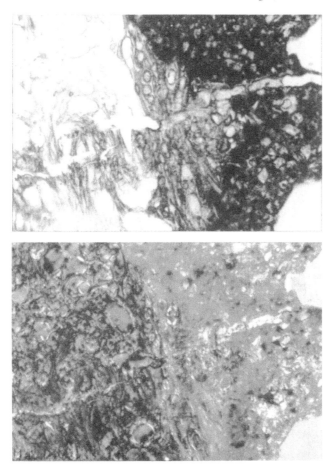

Figure 6.131 The texture of gel associated with a pumice aggregate within lightweight concrete, widths of field 1.4 mm, upper ppl, lower xpl: The glassy vesicular structure of the pumice aggregate particle has an irregular crack which emerges at its apex. The gel plug is difficult to distinguish until it protrudes into the hardened cement paste. Under crossed polars, the gel is even more difficult to distinguish and is seen to be crystallised in contact with the hardened paste.

In addition, both surface carbonation and carbonation down cracks can be more extensive than in normal weight concretes, as described in Chapter 5. In the more severe cases, the carbonation can obscure the concrete texture and make interpretation difficult.

The potentially beneficial influence of air entrainment on ASR expansion was reviewed by a working group arranged by the BRE and its findings published as Information Paper IP1/02 (BRE 2002). It was found that the effects were limited, with only a small proportion of the entrained voids becoming filled with gel, which might mean that ultimately reliance cannot be placed on the accommodation of alkali–silica gel within lightweight and other voided concretes.

6.8.5 Examination of concretes suspected of ASR

6.8.5.1 Interpretation of ASR textures

If *adequate* sampling and petrographic examination of concrete are carried out, the only cases where unequivocal identification of ASR is not possible are those where either the reaction is in its earliest stage or the concrete contains considerable void space. It is the experience of the authors, gained from an examination of a wide range of concretes from many countries, that the identification of ASR is not a problem, provided the procedures in this subsection are followed (St John 1985). Difficulties in making unequivocal identification usually indicate faulty petrographic practice, although there may be problems in determining the extent to which any damage can be attributed to the ASR identified (Sims 1996).

Identification of any ASR is only one of the aims of petrographic examination, and, as discussed in Section 6.1, determination of the quality of the concrete is necessary for interpretation. This is especially the case where the concrete is exposed to severe climatic conditions, aggressive environments and other causes of deterioration. In cases where damage is the result of more than one cause, the problem of deciding which is the primary cause can be difficult, but is considerably aided by determining the quality of the concrete. The problem of differentiating ASR and DEF as causes of damage in some concretes has already been discussed (see Section 6.5.9). If no other cause of damage is identified, the extent and pattern of cracking, linked to the evidence of reaction sites, are a reasonable indicator of the *severity* of the ASR in the concrete. The minimum extent of localised strain in two dimensions can be crudely estimated for an expanding concrete from summing the crack widths on selected orthogonal axes, but this requires a planar area of at least 75 × 75 mm and is only practicable in thin sections. Such estimates should be treated with caution as many other factors may influence the results.

Since it is only possible to observe gel in open spaces such as voids, cracks and fissures and the conditions under which alkali–silica gel crystallises were not well understood in the classical literature on ASR, statements such as 'the gel is old' or 'the reaction has undergone several episodes because of gel layering' should be made and regarded with caution, when chemical compositions of the ASR gel have not been established by SEM–EDX analysis.

Petrographic examination cannot provide information about whether ASR has ceased or whether residual expansion is still possible. This requires the expansive movement in the structure and residual expansion on core samples to be determined (Thaulow and Geiker 1992). Nielson (1994) has proposed a method based on dimensions of the structural member for determining if and when repair will be required in cases of ASR, and Chrisp et al. (1994) proposed a method of non-destructive testing to quantify the damage resulting from AAR. The British Cement Association (BCA) guidance on diagnosis (1992b) includes an unrestrained core expansion test, but this is only intended for identifying any *potential* for further expansion and cannot help with forecasting the performance of concrete in a given structure, for which procedures have been recommended by the U.K.'s Institution of Structural Engineers (1992/2010).

From a thin-section examination, it can be concluded that some aggregate particles in the concrete have reacted, but any statement about aggregates being 'innocuous' needs to be made and regarded with caution and requires supplementary data to define the conditions. One of the important merits of petrographic examination of concrete thin sections, both with polarising microscopy and SEM on polished thin sections, is that this can roughly establish the extent and stage/sequence of ASR in the concrete: (1) rim formation of aggregate (cryptic stage), (2) internal cracking of aggregate, (3) extension of cracks from reacted aggregate into the surrounding cement paste (here, concrete starts cracking) and (4) deposition of ASR gel within air voids and along cracks remote from the reacting aggregate (Katayama et al. 2004b, 2008; Katayama 2010a). In this regard, particles of coarse aggregate should not be separated from concrete by crushing, for petrographic examination, but instead, a representative portion of the concrete sample, containing both coarse aggregate and cement paste, should be used to prepare a thin section for examination.

Supplementary data such as chemical analysis of the concrete to determine alkali content, information on possible pessimum proportion and data from condition surveys are required to validate statements made about aggregates being innocuous (e.g. Katayama et al. 2004b). In this respect, the use of fluorescent resin-impregnated polished surfaces and/or uranium salt solutions on fracture surfaces and/or examination by SEM may show gel and attack inside aggregates that are not apparent from microscopical examination, and these techniques should always be considered in cases of doubt. However, it should be noted that fluorescent-dyed resin only penetrates into open spaces (e.g. open cracks, open air voids and capillary pores), so that closed cracks filled with ASR gel cannot be impregnated by the fluorescent resin. In Section 6.1.3, it was mentioned that staining of any ASR gel by uranyl acetate is routinely used in the DRI method.

Based on numerous thin-section analyses of concretes with 'slow reactive/expansive' aggregates, Jensen (1993, 2011) has been able to classify the harmfulness of ASR as reaction types: 'minor and/or preliminary' ASR is when reaction products are observed in air voids in the cement paste and internal in cracks in the aggregates, while 'deleterious' ASR is when cracks caused by ASR are propagating into the surrounding cement paste and, in more advanced stages, connect one or several reacted aggregates. Identification of reacted aggregates is possible when reaction products appear in internal cracks in the aggregates and cracks running out in the cement paste, but in cases where reaction products are exuded into isolated air voids in the cement paste, ASR can be documented but not the reactive aggregates.

As explained in Section 6.1.3, methods have been developed for quantifying the evidence for AAR. A method using large-area thin sections was suggested by Sims et al. (1992), but the DRI scheme devised by Grattan-Bellew (1995) for polished specimens has been used most routinely in North America and sometimes in other parts of the world. The 'Norwegian crack counting method' uses fluorescent resin-impregnated polished cores examined under UV light (Jensen 1993, 2004a,b, 2005). This system involves counting aggregates containing cracks and cracks in the cement paste (Figure 6.132). It is important that only cracks assessed to be developed by ASR are counted. In processing the results, the number of ASR-cracked coarse (i.e. >4 mm) aggregates is expressed as a percentage of the total number of coarse aggregates and then plotted against the number of cracks running out into the cement paste, expressed per square centimetre (cracks/cm^2) of the investigation area. An example outcome is shown in Figure 6.133, showing 93 results that are variously classified as ASR not observed, 'minor and/or preliminary' ASR or 'deleterious' ASR (as defined earlier). As might be expected, concretes

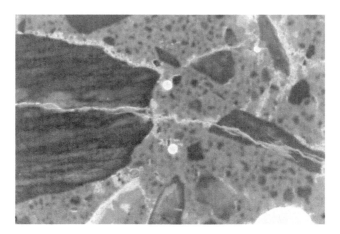

Figure 6.132 An example of a surface prepared for the Norwegian crack counting method for quantifying the evidence of AAR, showing cracks on a polished surface under UV light. (Photomicrograph courtesy of Dr Viggo Jensen.)

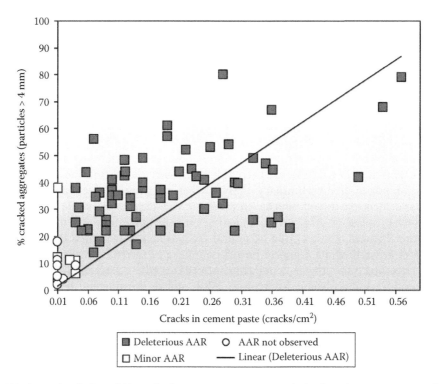

Figure 6.133 A graphical plot of % cracked aggregates against cracks/cm² in the cement paste from 93 fluorescent resin-impregnated slabs. (Figure courtesy of Dr Viggo Jensen.)

without and with only minor ASR cluster close to the zeros, while concretes with deleterious ASR are distributed with higher values; in Figure 6.133, the best fit linear correlation is shown for % cracked aggregates and paste cracks/cm².

The interpretation of thin-section examination of concretes undergoing AAR should not be extended to making statements about structural integrity. While a concrete may be severely microcracked internally, aggregate interlock, localisation of cracking, restraint by reinforcement and other factors can reduce losses in strength. It is now generally agreed that the structural effects of AAR, apart from movement, have been overestimated, but this does not obviate the need for an adequate structural assessment (Institution of Structural Engineers 1992/2010; Walton 1993).

However, there is a growing recognition that large concrete dams, typically with design and service lives that are much longer than most other building and civil engineering structures, might represent a special case (Charlwood 2007; Charlwood and Scrivener 2011; Charlwood et al. 2012). These often seem to involve a relatively low level of reactivity (sometimes with concrete mixtures that might normally be regarded as non-deleterious) causing expansive movement that is relentlessly prolonged unabated over many decades. It seems that, in such long-service structures, a slow reaction might gradually cause an ongoing damage, with possibly other factors, such as alkali release from aggregates and alkali recycling helping to perpetuate the mechanism. In the future construction of concrete dams, measures to minimise the risk of AAR might have to be more stringent than previously thought necessary (Sims et al. 2011, 2012a,b).

Structures undergoing AAR do not generally collapse, even though they may require repair and in severe cases replacement. There is only one case of collapse reported and that involved both AAR and sulphate attack (Haavik and Mielenz 1991). However, Sibbick (2011) correctly states that severe ASR can 'open up' an otherwise sound concrete to other externally derived degradation processes, such as seawater attack, calcium leaching, external sulphate attack and freezing and thawing cycles, including exacerbation from de-icing salts, and can also potentially facilitate corrosion of reinforcements.

6.8.5.2 Practical examination

Failure to follow the necessary procedures when carrying out a petrographic examination of a concrete suspected of undergoing ASR may lead to erroneous conclusions. If a number of cores, all taken from an area of concrete with a moderate cracking caused by ASR, are examined, a marked variability will be found. In some of the cores, little if any cracking and gel will be found, while in others, the reaction is clearly present and commensurate with the external cracking observed. A careful selection of sampling sites is imperative and the petrographer should ideally be involved with inspecting the structure and choosing the sample sites. If this is not possible, the selection must be carried out by personnel who have the skill and experience to make the necessary judgements.

Experience indicates that the outer 150–200 mm of a core sample frequently provides the most information on quality, weathering and external attack. If the external portions of a core sample are missing, ideally, the core should be rejected as unsuitable for reliable petrographic examination, as the relationship of ASR to external factors cannot be ascertained. If facilities for large-area thin sectioning (minimum size 75 × 75 mm) of concrete are not available, it may be better only to use thin sections for aggregate identification and to rely on examination of ground surfaces, which can be impregnated with fluorescent-dyed epoxy resins. While this may give less information than can be obtained from examination of large-area thin sections, it will not lead to erroneous conclusions because too small an area has been examined. Sims et al. (1992) have discussed the sampling and methods required for quantifying the microscopic examination of AAR. A useful practical guidance on the investigation of concrete for ASR diagnosis was published by an expert working party assembled by the British Cement Association (1992b), and a thorough guidance for application to transportation structures in North America was prepared by Fournier et al. (2010). Detailed new international guidance, designated AAR-6.1, has been published by RILEM TC 219-ACS (Godart et al. 2013).

6.8.5.3 Other considerations

One of the questions that inevitably arises in the investigation of ASR is the amount of alkali contributed by the cement to the concrete. Most aggregates require around 3.0 kg/m³ of Na_2O equivalent or greater for significant reaction to occur, although a number of aggregates have been reported that require as little as 2.0 kg/m³ (Woolf 1952; Nixon and Sims 1992). In tests of various UK aggregates, Sibbick and Page (1992) identified alkali 'thresholds' (as Na_2Oeq) ranging from 3.5 kg/m³ for silicified siltstone and limestone materials, through 4.5 kg/m³ for greywacke and strained granite samples, to 6 or 7 kg/m³ for flint (Thames Valley) and quartzite/chert (Trent Valley) gravels.

Consider a hypothetical example of a typical bridge concrete supposedly containing 360 kg/m³ of cement, which contains an alkali content of 0.55% as Na_2O equivalent. This concrete, if correctly mixed, would have initially contained 2.0 kg/m³ of alkali. However, parts of the bridge concrete are now undergoing ASR in spite of laboratory data indicating that a significant reaction of the aggregate used requires at least 3.0 kg/m³ of alkali. There are at least four possibilities to consider in this case.

The first and most important possibility is that a high-alkali cement was used instead of the low-alkali cement specified. This occurs more commonly than would be expected. The second is the possibility of the laboratory test data being wrong, and in fact, the aggregate is significantly reactive at the lower alkali content. Many aggregates are difficult to test in the laboratory because the results are sensitive to pessimum proportion of the reactive component and to leaching of alkali from the test specimen during storage (Rogers and Hooton 1989; Lindgard et al. 2012). The third possibility is that alkali has been preferentially concentrated in one portion of the concrete, although petrography should resolve this question because the signs of reaction will be restricted to those layers. The fourth possibility is that there is another source of alkali apart from the cement. This alkali can be derived externally from salts or be internally derived from aggregates, mineral additions (SCMs) or salt contamination of aggregates. These are all possibilities which must be eliminated for the effective analysis of AAR in concrete.

In a series of investigations involving the authors in the United Kingdom, the first of these possibilities (unintentional use of a higher-alkali cement than planned) had been caused by erroneously low alkali levels being declared by the cement manufacturer for a particular works. This occurrence was immediately notified to its customers by the manufacturer as soon it was recognised, but this was only after the use of the affected cement in numerous projects, including some in which the BS 8500 (BSI 2006) measures (based upon the guidance given in BRE Digest 330 [2004] and Concrete Society Technical Report 30 [1999]) for minimising the risk of ASR had been applied. These measures determine maximum binder contents and other concrete mix parameters, depending on the quality-assured manufacturer's declared alkali content for the cement. Consequently, there were concerns that, at least in some cases, the measures taken to avoid ASR problems might prove to be inadequate in the longer term, owing to the inadvertently higher concrete alkali content than the generalised BS criteria permitted for 'normal' reactivity aggregates. Reassurance was provided to several owners and/or contractors, using petrographic assessment to demonstrate that the aggregate combination could be regarded as having 'low' reactivity (Sims et al. 2008).

Some of the methods of determining the alkali content of a concrete are to identify the origin of the cement, determine the cement content by petrography and use alkali results from historical cement analyses (see Chapter 2). Where this technique can be applied, it may unequivocally resolve the question of the alkali content derived from the cement irrespective of whether construction records are available.

The only other alternative is to try to estimate the alkali content by analysing the concrete. Suitable methods of analysis are discussed in Chapter 2. Great care is required in sampling to avoid areas of concrete which are affected by leaching of alkalis. Surface areas which are invariably leached by wetting and drying and also localised areas of concrete undergoing significant ASR should be avoided. On most structures, it is possible to find less reacted areas immediately adjacent to severe areas of cracking which are likely to contain levels of alkalis more representative of the concrete as mixed. The guidance is given by the British Cement Association (1992b) on the analysis of concrete for alkali content, and Wood et al. (1996) have reported that an analysis of an adequately numerous set of adequate samples is required for the overall value to be a reliable indicator of the original concrete alkali content.

Other internal sources of alkali in concrete are salt contamination from marine-derived aggregate (mainly when either inadequately dewatered or unwashed) and the alkali content present in some admixtures, especially superplasticisers. The determination of the chloride content from deep inside a concrete will indicate the amount of salt contamination from aggregate, because salt spray and de-icing salts are normally limited to penetration depths of about 100 mm in better-quality concretes. The identification of a superplasticiser in an old concrete is very difficult, and there is currently no standard technique available for quantification (but also see Chapters 2 and 4). The amounts of alkali that may be contributed from chemical additives have been estimated by the Cement and Concrete Association of New Zealand (1991).

It is also now fairly well established that the aggregates themselves can contribute alkalis to the concrete, partly because of AAR attacking aggregates. The possibility that volcanic aggregates will release alkali into the pore solution without themselves undergoing expansive AAR has been reported by St John and Goguel (1992). In particular, some New Zealand basalts have released sufficient alkali to initiate expansive reaction of other aggregates present in concrete structures. Goguel (1996) and Bérubé et al. (1996a) found that a range of aggregates may release alkalis. Gillott and Rogers (1994) reported that dawsonite ($NaAlCO_3(OH)_2$) releases alkali by reaction with calcium hydroxide in the pore solution of concrete, although expansion may have also resulted from increases in the solid volume of the dawsonite reaction itself. A keynote paper by Bérubé and Fournier (2004) thoroughly reviewed the background to these alkalis 'releasable' from aggregates and considered various procedures for their assessment, finding that the methods based on lime-saturated solutions produced underestimates, while those using alkali solutions (NaOH for K-release and KOH for Na-release) appeared more realistic. RILEM TC 219-ACS is currently preparing a test method, designated AAR-8, for assessing the releasable alkali content of an aggregate sample.

There is a number of conditions that may impact on the severity of AAR, which must be considered in the petrographic examination. These are frost attack and/or cyclic freezing and thawing of the concrete and frequent condensations on surfaces (e.g. under desert conditions), salt-laden winds with minimal rainfall and pavements, especially those which are effectively sealed on their underside restricting moisture movement to the top surface.

The relationship of damage caused by combined freezing and thawing of concrete and AAR is not well documented. Where the surface of a concrete is disintegrating, the primary cause must be frost or freeze–thaw attack, even if AAR is present. AAR does not by itself usually cause disintegration of surfaces apart from pop-outs. However, more severe cracking seems to occur in countries with cold winter climates, and this suggests that freezing and thawing are opening up the cracks initiated by AAR to depths which may be controlled by moisture penetration (Bérubé et al. 1996b). The contribution to the overall damage in the structure from freezing and thawing can be evaluated only by characterising the air void system and measuring the porosity and permeability of the concrete and the frost susceptibility of the aggregates (Hudec 1991). Sibbick (2011) finds that ASR adventitiously combined with other problems, such as freeze–thaw cracking, total alkali levels enhanced by de-icing salts and construction failures such as poor joint development, can all lead to the enhanced development, or even initiation, of deleterious ASR in an otherwise less potentially deleterious concrete. Katayama et al. (2004b) presented an example of estimating alkali contributions in ASR-affected concretes under the influence of de-icing salt.

Desert conditions can provide a very severe environment for concrete. Hot days followed by cold nights lead to considerable condensation on concrete surfaces. This daily condensation constitutes a pumping effect on the moisture in the surface of the concrete. If salt deposition occurs, as in parts of the Arabian peninsula, these salts are transported into the concrete and both augmentation of AAR and sulphate attack may occur (Fookes and Collis 1975; Fookes 1993). Salt-laden winds are common in areas adjacent to the sea and may penetrate considerable distances inland. The airborne salt gets deposited on surfaces and these salts are available to move into the concrete when dew forms, also salts can migrate into concrete by capillary rise from the ground, in both cases facilitating corrosion of reinforcement and augmenting the alkalis in the concrete. However, despite the high temperatures and the availability of additional alkalis derived from salts, Sims (2006) has noted the apparent rarity of cases of AAR in such regions. In wet temperate climates, such as New Zealand, this does not appear to be a potential problem, as chloride levels measured in exposed concrete indicate that frequent rain is removing the salt before it can penetrate. Unless the salt deposition is severe, its effect on ASR is probably not significant.

Pavements, because they are relatively thin and have large horizontal surfaces, are more sensitive to climatic conditions and deposition of salts. If the pavement is effectively sealed on its bottom surface, moisture and salts can only move in and out of the concrete via the exposed upper surfaces. There is little information about what effect this configuration has on the distribution of alkalis and its relationship to AAR. Concentration of alkalis causing ASR in a pavement has been reported by Hadley (1968) and in laboratory specimens by Nixon et al. (1979). Torii et al. (1996) found that cathodic protection significantly increased expansion in concrete containing reactive aggregate, and similarly, Al-Kadhimi and Banfill (1996) reported that an electrochemical treatment to 're-alkalise' concrete protecting steel reinforcement had the effect of worsening ASR. Hobbs (1988) argues that increases in alkali concentration from these causes are likely to be only transitory and will not be significant in ASR. Given the difficulties in making reliable alkali analyses of concrete, published data should be treated with caution.

The effect of temperature on ASR has been investigated by Poole (1992). Temperatures of 38°C led to faster rates of reaction, while temperatures of 10°C–15°C, which are more typical of ground temperatures, give slower reaction as would be expected. However, the total amount of expansion is sometimes reported to be greater at the lower temperature. This suggests that concrete below ground should eventually be more severely cracked than a similar concrete above ground.

Expansion essentially ceases in a concrete that has dried to an internal RH of about 80% or less. Stark (1985) measured relative humidities in some dams in the southwestern United States and found that only the concrete within a few inches of the exposed surfaces was sufficiently dry to preclude expansion, even under desert conditions. It seems to be accepted that the larger structural concrete members do not dry out sufficiently to prevent expansion. This may have some bearing on the long-term behaviour of AAR in large dam structures, which was discussed earlier. Some data on the factors influencing the RH in concrete have been reported by Parrott (1991).

6.8.6 International schemes for assessment, diagnosis and specification

In the years following the discovery of AAR in the United States in 1940, a series of aggregate tests for assessing AAR potential was devised by the ASTM, and these are still continuously reviewed, updated and widely used. However, as countries around the world started to have their own experience of AAR, involving a range of aggregate types each behaving slightly differently, a plethora of proposed new tests and test variants appeared in the literature and in national standards. In 1988, RILEM established a Technical Committee (TC 106), charged with endeavouring to establish aggregate tests for AAR that could be accepted and used internationally. In accordance with RILEM rules, which require a periodic renewal of TCs, this initial TC was first superseded by TC 191-ARP and then, from 2007, by TC 219-ACS. This TC was superseded by TCAAA in 2014.

In these later TCs (191-ARP and 219-ACS), the objectives were broadened to include the development of international guidance for AAR diagnosis, structural appraisal (including modelling) and specification, while the work has continued to provide an internationally applicable assessment scheme for aggregates and, in particular, a practicable project-specific AAR performance test for actual concrete mixes. The recommendations of these RILEM TCs have been given designations with the prefix 'AAR-' and are listed in Table 6.14.

Table 6.14 Recommendations of the RILEM Technical Committees on AAR

RILEM designation	Date published	Date revised	Topic
AAR-0	2003[a]	2015[b]	Outline guide to the use of RILEM methods in assessing the alkali-reactivity potential of aggregates
AAR-1.1	2003[a]	2015[b]	Petrographic method
AAR-1.2	2015[c]	—	Petrographic atlas
AAR-2	2000 (TC 106-2)[a]	2015[b]	Accelerated mortar-bar test (AAR-2.1 and 2.2 for short–fat and long–thin mortar-bar options)
AAR-3	2000 (TC 106-3)[a]	2015[b]	38°C Method for aggregate combinations using concrete prisms (AAR-3.1 and 3.2 for aggregate test and alkali threshold options)
AAR-4.1	2015[b]	—	60°C Method for aggregate combinations using concrete prisms
AAR-5	2005[a]	2015[b]	Rapid preliminary screening test for carbonate aggregates
AAR-6.1	2013[d]	—	Guide to diagnosis and appraisal of AAR damage to concrete in structures: Part 1 – Diagnosis
AAR-6.2	Expected 2016	—	Guide to diagnosis and appraisal of AAR damage to concrete in structures: Part 2 – Appraisal and Repair
AAR-7.1	2015[b]	—	International specification to minimize damage from alkali reactions in concrete: Part 1 – Alkali–Silica Reaction
AAR-7.2	2015[b]	—	International specification to minimize damage from alkali reactions in concrete: Part 2 – Carbonate Aggregates
AAR-7.3	2015[b]	—	Preliminary international specification to minimize damage from alkali reactions in concrete: Part 3 – Concrete dams and other hydro structures
AAR-8	tba	—	Determination of alkalis releasable by aggregates in concrete
AAR-9	tba[e]	—	Modelling of structures affected by AAR
AAR-10	tba	—	Reformance testing of concrete regarding AAR potentiale[e]

[a] Published in Materials and Structures unless indicated otherwise.
[b] To be published in a separate RILEM book in 2015 (the methods and guidance will probably also be available for download from the RILEM website).
[c] To be published as a separate RILEM book in 2015 (Fernandes et al. 2015).
[d] Published as a separate RILEM report (Godart et al. 2013).
[e] To be published as part of AAR-6.2.
tba, to be advised/announced.

In terms of the assessment of aggregates, AAR-0 is the key, being a guide to using the test methods AAR-1 to AAR-5 and providing tentative criteria for their interpretation. A flow chart for the assessment scheme is shown in Figure 6.134. It is a three-stage programme, which starts with a petrographical examination (AAR1.1: RILEM 2003, 2015), in order to allocate the aggregate to one of three broad reactivity classes:

Class I: Very unlikely to be alkali reactive
Class II: Potentially alkali reactive or alkali reactivity uncertain
Class III: Very likely to be alkali reactive

The petrographer is also required to ascertain whether any Class II or III aggregates primarily comprise siliceous constituents (class suffixed '-S'), or carbonate constituents (class suffixed '-C'), or a mixture of both (class suffixed '-SC'). Class II and III materials may then be subjected to testing, with Stage Two being an optional rapid screening procedure,

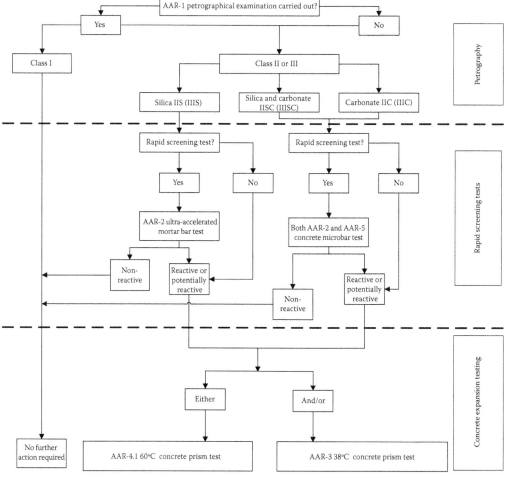

*If no petrographical examination has been carried out, assume class II (or III)

Figure 6.134 Flow chart for the RILEM aggregate assessment scheme. (From RILEM AAR-0, Refer to the RILEM book (due in 2015), including the new AAR-0, new AAR-1.1, new AAR-2, new AAR-3, AAR-4.1, new AAR-5, AAR-7.1, AAR-7.2, and AAR-7.3 [RILEM 2015].)

employing different methods for -S, -C and -SC aggregates (see Table 6.14). Stage three involves expansion tests using concrete prism specimens, either the well-established but often too lengthy (12 months or more) method involving storage at 38°C or the promising newer method involving storage at 60°C, which can provide interpretable results at 15–20 weeks (see Table 6.14). RILEM is also developing a test (AAR-8) for assessing the releasable alkali content of aggregates.

Petrographers should also refer to RILEM AAR-6.1 (Godart et al. 2013), which provides an up-to-date catalogue of methods for use in the diagnosis and prognosis of AAR in concrete. Petrography is not used directly in the recommended international specifications, AAR-7.1 for ASR and AAR-7.2 for carbonate aggregates (RILEM 2015), where aggregate reactivity is established as 'low', 'medium' or 'high' on the basis of the determined alkali threshold using a programme of AAR-3 concrete prism testing (RILEM 2015). A broadly similar approach has been described in North America by Thomas et al. (2008). Preliminary guidance has also been drafted (as AAR-7.3) for minimising the risk of AAR in concrete dams and other hydro structures (RILEM 2015).

6.9 AAR INVOLVING CARBONATE AGGREGATES

Deleteriously expansive 'ACR' has been documented for certain dolomitic coarse aggregates since the late 1950s (Swenson 1957). This reaction has recently been termed 'so-called ACR', because investigations of concretes from the type localities of ACR in Ontario have indicated that no ACR exists that is free from ASR. The real cause of the expansion for the phenomenon called ACR seems to be ASR of fine-grained quartz concealed within the dolomitic aggregate (Katayama 2004, 2006, 2010a, 2011a; Grattan-Bellew et al. 2008, 2010; Katayama and Grattan-Bellew 2012). This means that a redefinition of the conventional term ACR is necessary.

Earlier researchers simply confused the deleteriously expansive ACR with an elementary reaction called 'de-dolomitisation', so that care should be taken in reviewing classical papers on this subject. The traditional petrographic techniques, which basically relied on polarising microscopy of covered thin sections with normal thickness in transmitted light, are not really capable of identifying the very fine reaction products and expansion cracks that extend from the reacted dolomitic aggregate. Until recently, numerous experiments with rock powders and rock cylinders immersed in alkaline solution have been carried out, but none of these methods can elucidate the site and the sequence of the reaction (i.e. within or outside the aggregate, elementary reaction or not, earlier or later, etc.).

Hence, to review and investigate the expansion mechanism of dolomitic aggregates, it is recommended to examine concrete samples, most preferably using advanced analytical techniques suitable for characterising both the loci of the reaction and the compositions of the reaction products. One of these techniques is a combination of observation and chemical analysis using the same polished thin section by means of polarising microscope, SEM with an energy-dispersive spectrometer (EDS) and electron probe microanalyser (EPMA) (Poole and Sotiropoulos 1980; Katayama 2004, 2010a) (see Chapter 2 for further explanation of these EDX-type analytical options). This technique is also applicable to the 'slow reactive/expansive' ASR characterised by fine cracks with a few visible reaction products (Katayama 2012b). In the typical examples of ACR in Ontario, expansion cracks in field concretes are now found to be associated with ASR gel derived from cryptocrystalline quartz, and not with the de-dolomitisation of carbonates (Katayama 2004, 2006, 2010a).

6.9.1 Background and current position on 'ACR'

ACR is principally the interaction of cement alkalis with argillaceous dolomitic limestones. It should be distinguished from the reaction of siliceous limestones, where there is a readily observable alkali–silica gel in thin sections and cracking formed by attacks on the siliceous component of the rock. The dividing line between the ASR and ACR reactions is considered by some workers to be arbitrary (Katayama 1992), but most workers in this field in China still consider it to be a separate reaction (Tang and Deng 2004).

Reactive argillaceous dolomitic rocks in a quarry may be in beds of limited depth and represent only a small proportion of the output of the quarry. Thus, the proportion of reactive dolomitic aggregate present can be highly variable, which is a complicating factor in petrographic examination. The incidence of ACR is not as widespread as ASR, but can be suspected wherever argillaceous dolomitic rocks are interbedded with other types of dolomite or limestone (Dolar-Mantuani 1983).

There are at least three reactions associated with carbonate rocks in concrete, plus one similar reaction known in geological literature, as follows:

6.9.1.1 De-dolomitisation

This is the expansive reaction by cement alkalis with impure dolomitic limestones. Dolomite contents range from as little as 5% to over 80%. The dolomite crystals, which average about 50 µm in size, are set in a fine-grained matrix of calcite, usually less than 5 µm in size, together with disseminated clay and silica. The clays are usually illite and mixed-layer chlorite, and the silica consists of various forms of quartz, some of which may be disordered and reactive. These clays and silica comprise the acid-insoluble residue of the rock, which often tests as reactive by ASTM C289 (2007). There is strong evidence to suggest that a minimum of 5%–10% insoluble residue is required for expansive ACR to occur in concrete (Hadley 1961; Katayama 1992).

The alkaline pore solution in the concrete reacts with the dolomite $(CaMg(CO_3)_2)$ to de-dolomitise the rock, form brucite $(Mg(OH)_2)$ and calcite and regenerate the hydroxyl ion (OH^-). The de-dolomitisation reaction consists of two elementary reactions: one within the aggregate that forms a reaction rim and the other within the cement paste producing a carbonate halo (Katayama 2004, 2010a):

$$\text{Aggregate (reaction rim): } CaMg(CO_3)_2 + 2OH^- \rightarrow Mg(OH)_2 + CaCO_3 + CO_2^{2-}$$

$$\text{Cement paste (carbonate halo): } CO_2^{2-} + Ca(OH)_2 \rightarrow CaCO_3 + 2OH^-$$

It is claimed that this formation of brucite and calcite is not expansive, and it is believed that de-dolomitisation opens up the texture and facilitates reaction of the alkalis with the matrix. It has been proposed that the actual expansion is caused by the re-wetting of the clays (mainly illite) in the matrix of the dolomite (Swenson and Gillott 1967). However, this is disputed by Tang (1992), who considers that the formation of the brucite and calcite is expansive because of a topochemical in situ reaction. Katayama (1992, 2011a) considers neither of these theories adequate and has suggested that the reaction of fine-grained (not necessarily disordered) silica in the matrix of the dolomite to produce alkali–silica gels is the cause of the expansion.

In later papers, Deng and Tang (1993) and Tang et al. (1994) claim that expansion occurs due to de-dolomitisation and that the rate and amount of expansion increase with increasing pH. Prince et al. (1994) showed that the products formed from de-dolomitisation depend on equilibria which are controlled by the calcium/sodium ratio in the pore solution: higher ratios lead to the formation of calcite and brucite and lower values to the formation of brucite and pirssonite, $CaNa_2(CO_3)_2 \cdot 2H_2O$. Products of reaction between dolomite powders and alkaline solutions have been identified by XRD analysis, including brucite, calcite, gaylussite, $CaNa_2(CO_3)_2 \cdot 5H_2O$ and buetschliite, $Ca_2K_6(CO_3)_5 \cdot 6H_2O$ (although current usage of this mineral name is for $CaK_2(CO_3)_2$) (Sherwood and Newlon 1964).

Alkali–carbonate rocks do not exhibit 'pessimum' proportions and may be reactive with cement alkali contents as low as 0.4% Na_2O equivalent. Marked rimming of aggregates may be seen, but the typically expansive aggregate from Kingston, Canada, rarely shows a reaction rim (Katayama 2004). There is one major difference from ASR. Alkali–silica gel has rarely been observed in expansive concretes containing these impure dolomitic limestones. It has been suggested (Katayama 1992) that the gel is fixed by a brucite component to form the aggregate rims commonly observed which are enriched in silica (Poole and Sotiropoulos 1980). However, brucite is extremely insoluble and any interaction between hydrous silicate anions and magnesium cations will be very slow. Thus, it is surprising that alkali–silica gel is not observed more often in the early papers on ACR. A combined formula of these two elementary reactions, as shown in the following, might be used to assess what compounds would form after the overall reactions between the powder sample and alkaline solution:

Inside and outside aggregate: $CaMg(CO_3)_2 + Ca(OH)_2 \rightarrow 2CaCO_3 + Mg(OH)_2$.

6.9.1.2 Alkali–silica reaction

This is the expansive reaction by cement alkalis with siliceous limestones to give a classical ASR. The siliceous portion of the limestone produces alkali–silica gels. It should be possible for both alkali–silica and ACRs to be present in the same concrete, if both types of aggregate are present. Such a concurrence of ASR and ACR has been reported for some Chinese aggregates (Tang et al. 1996; Tong and Tang 1996). In a strict sense, this should be called concurrence of ASR and de-dolomitisation, instead of ACR, so as not to cause confusion to the readers of papers. It is essential to distinguish de-dolomitisation from ACR (Katayama 2004, 2010a).

Gali et al. (2001) simulated results of their experiments using powders of dolomite and silica and suggested that antigorite-like minerals crystallise in the reaction between silica and brucite, although they do not give an analysis. Quantitative SEM–EDS analysis of ACR-affected concretes identified the presence of Mg–silicate gel as a reaction product between a brucite component and alkali–silica gel, having a composition of sepiolite ($4MgO \cdot 6SiO_2 \cdot 7H_2O$) (Katayama 2004, 2006, 2010a, 2011a) and antigorite ($3MgO \cdot 2SiO_2 \cdot 2H_2O$) (Katayama 2011b). If the composition of virgin alkali–silica gel before contact with cement calcium is represented by one of the kanemite alkali–silicate hydrates ($NaHSi_2O_4(OH)_2 \cdot 2H_2O$ or $Na_2O \cdot 4SiO_2 \cdot 7H_2O$) for convenience, the formation of the magnesium–silicate gel could be written as follows (Katayama 2010a):

Mg-Si gel: $4Mg(OH)_2 + 1.5(Na_2O \cdot 4SiO_2 \cdot 7H_2O \rightarrow 4MgO \cdot 6SiO_2 \cdot 7H_2O + 3NaOH + 6H_2O$.

This reaction does not seem to produce a volume increase in the solid phases, when the molar volume of sodium hydroxide is omitted because of its water solubility. Magnesium–silicate gel has also been identified in the 'dolostone' (dolomite rock) and dolomitic limestone aggregates in ASR-affected concrete structures in the Quebec province, along with a distinctive alkali–silica gel and de-dolomitisation (Katayama 2006). In the past, the reaction due to carbonate aggregates in this region was regarded as ASR of siliceous limestone (Bérubé and Fournier 1986; Durand and Berard 1987). ASR gel occurs as crystalline phases of rosette aggregation filling cracks of the aggregates and presents XRD peaks similar to those of okenite, $CaO \cdot 2SiO_2 \cdot 2H_2O$ (Bérubé and Fournier 1986). Their chemical compositions obtained by EDS analysis (Durand and Berard 1987) are close to a modified composition of this mineral, with a substitution $(2Na, 2K, Fe)O = CaO$ (Katayama 2010a).

It would be useful to understand the general trends of alkali–silica gel that evolve in concrete undergoing typical ASR. In general, alkali–silica gel migrates along cracks from the reacted aggregate into cement paste, leaching alkalis but gaining calcium from the cement paste, and finally of CSH gel. This process can be illustrated on the Ca/Si–Ca/(Na+K) diagram, where ASR gel approaches compositions in the field concretes presents a compositional trend line passing three points: *low-Ca ASR gel* (Ca/Si = 1/3–1/6, Ca/(Na+K) = 1), *high-Ca ASR gel* (Ca/Si = 0.8–1.0, Ca/(Na+K) = 10) and a *convergent point* (Ca/Si = 1.3–1.6, Ca/(Na+K) = 100) at which an apparent chemical equilibrium is attained with CSH gel (Katayama et al. 2004, 2008; Katayama 2008, 2010c). The inclination of the trend line and the position of the convergent point on the diagram vary according to the degree of weathering of concrete, such as leaching, carbonation and freezing and thawing (Katayama 2008, 2010a, 2012b).

The alkali–silica gel having been crystallised into rosette-like aggregations resembles several naturally occurring alkali–calcium silicate hydrate minerals, among which mountainite, $KNa_2Ca_2Si_8O_{19}(OH) \cdot 6H_2O$ or $Na_2O \cdot 0.5K_2O \cdot 2CaO \cdot 8SiO_2 \cdot 6.5H_2O$ according to its classic formula (Gard and Taylor 1957), is most popular, since the earlier description in concrete (Mullick and Samuel 1987). On the compositional trend line at the intercept Ca/(Na+K) = 1.0, alkali–silica gel with Ca/Si = 1/3 corresponds to mountainite, with $2(Ca_{4/6}, 2Na_{1/6}, 2K_{1/6})O \cdot 4SiO_2 \cdot 3H_2O$, while alkali–silica gel with Ca/Si = 1/6 corresponding to $(Na,K)_2O \cdot 2CaO \cdot 12SiO_2 \cdot nH_2O$) could be makatite ($Na_2O \cdot 4SiO_2 \cdot 5H_2O$) or kanemite ($Na_2O \cdot 4SiO_2 \cdot 7H_2O$), if a replacement of $2CaO = 2(Na, K)_2O$ is possible (Katayama 2010a,b). At the intercept Ca/(Na+K) = 1.0, there have been missing links in the natural counterparts with Ca/Si = 1/4 and 1/5.

Recently, mountainite family minerals with Ca/Si = 1/4 have been discovered from an alkaline complex in Russia (Zubkova et al. 2010): shlykovite ($KCaSi_4O_9(OH) \cdot 3H_2O$ or $K_2O \cdot 2CaO \cdot 8SiO_2 \cdot 7H_2O$) with Ca/(Na+K) = 1 and cryptophyllite ($K_2O \cdot CaO \cdot 4SiO_2 \cdot 5H_2O$) with Ca/(Na+K) = 1/2. Rosette-like crystals with these compositions are formed within cracks of reacted aggregate (Katayama 2011a, 2012b). The formation of alkali–silica gel with a classic mountainite-family composition from silica produces about 100% volume increase of solid phases, enough to explain the larger expansion by ASR (Katayama 2010a):

Low-Ca ASR gel: $12SiO_2 + 2NaOH + 2KOH + 4Ca(OH)_2 + 3H_2O \rightarrow$

$3[2(Ca_{4/6}, 2Na_{1/6}, 2K_{1/6})O \cdot 4SiO_2 \cdot 3H_2O]$.

Besides, when alkali-silicate gel (e.g. $Na_2O \cdot 4SiO_2 \cdot nH_2O$) resembling a water glass absorbs water, hydrostatic osmotic pressure is generated up to 10 MPa, which is enough to crack concrete (Katayama 2012b).

6.9.1.3 De-calcitisation

This is the alkaline dissolution of the surface of a crystalline limestone (or marble), forming a rim of birefringent material that has the appearance of calcium hydroxide. This reaction is generally believed to be non-expansive. Chen and Wang (1987), based on SEM observations, revealed that portlandite $Ca(OH)_2$ forms on the rim within the limestone by a dissolution/precipitation process, *not* by epitaxial growth. The following reaction, which may be called here 'de-calcitisation', has been suggested:

Aggregate (reaction rim): $CaCO_3 + 2OH^- \rightarrow Ca(OH)_2 + CO_2^{2-}$

This de-calcitisation reaction was formerly believed to be unlikely, as calcium hydroxide is more soluble than calcium carbonate. Monteiro and Mehta (1986) identified an $11.2°$ 2θ ($CuK\alpha$) peak by XRD analysis of the transition zone between the limestone aggregate and cement paste and attributed it to the peak ($11.4°$ 2θ) of one of the dehydrated phases of the basic calcium carbonate $2CaCO_3$ $Ca(OH)_2$ $1.5H_2O$. Schimmel (1970a,b), who originally identified this basic carbonate, does not give optical or solubility data for this compound, but St John et al. (1998) presumed that this carbonate has similar optical properties to the calcium hydroxide for which it is likely to be mistaken. However, an XRD analysis by Milanesi et al. (1996) indicates that $Ca(OH)_2$ actually forms from $CaCO_3$ that appears during de-dolomitisation in highly alkaline conditions (2 M and 10 M NaOH solutions).

In the system limestone–Portland cement, monocarboaluminate, $3CaO \cdot Al_2O_3 \cdot CaCO_3 \cdot 11H_2O$ ($11.6°$ 2θ peak), is the commonest reaction product that forms between calcite and aluminate in the cement paste. This hydrate occurs in the carbonate front within the cement paste, in which carbonate ions released from the reacting carbonate aggregate diffuse into the surrounding cement paste (Hadley 1961). Other carbonate varieties of hydrocalumite ($3CaO \cdot Al_2O_3 \cdot Ca(OH)_2 \cdot 12H_2O$ ($11.0°$ 2θ peak)) may also form there as an intermediate phase, which include hemicarboaluminate ($3CaO \cdot Al_2O_3 \cdot Ca[(CO_3)_{0.5}, (2OH)_{0.5}] \cdot 11.5H_2O$ ($10.8°$ 2θ peak) and $3CaO \cdot Al_2O_3 \cdot Ca[(CO_3)_{0.67}, (SO_4)_{0.33}] \cdot 11H_2O$ ($11.3°$ 2θ peak). Some of these aluminate hydrates have been directly identified by quantitative SEM–EDS analysis from ACR-affected concretes in Ontario (Katayama 2010a, 2011):

Cement paste (carbonate front):

$CaCO_3 + 3CaO \cdot Al_2O_3 + 11H_2O \rightarrow 3CaO \cdot Al_2O_3 \cdot CaCO_3 \cdot 11H_2O$

or,

$3CaO \cdot Al_2O_3 \cdot CaSO_4 \cdot 12H_2O + CO_3^{2-} \rightarrow 3CaO \cdot Al_2O_3 \cdot CaCO_3 \cdot 11H_2O + H_2O + SO_4^{2-}.$

6.9.1.4 Natural de-dolomitisation

Another type of de-dolomitisation, which may be termed 'calcitisation of dolomite', is known to occur in dolomitic formations in geological time. López-Buendia et al. (2008) claim that de-dolomitisation itself can produce a volume increase, because calcite has larger d-spacings than those of dolomite: that is, $d_{104} = 3.03Å$ with calcite, while $d_{104} = 2.89Å$ with dolomite. They believe that de-dolomitisation in concrete is likewise a volume increase reaction. In order to compare the unit cell volumes of dolomite and calcite, mass balance should be considered, where one molar dolomite is equivalent to two molars of calcite. Then the de-dolomitisation meeting this condition could be expressed as follows (Katayama 2010b):

Hypothetical de-dolomitisation: $CaMg(CO_3)_2 + Ca^{2+} \rightarrow 2CaCO_3 + Mg^{2+}$

In this case, magnesium ions liberated from dolomite are carried away into solution without forming brucite within the dolomite crystals, and more than 10% of volume increase may result. However, such an entire replacement of dolomite by calcite does not occur in the dolomitic aggregate in concrete; instead, nearly equal amounts of brucite and calcite form (Katayama 2004, 2010a). Actually, the concrete specimens of López-Buendía et al. (2006), using a Spanish dolomitic limestone aggregate, did not expand but showed contraction.

Natural de-dolomitisation develops in the near-surface weathering of dolomitic formations by actions involving karst-meteoric water, hydrothermal water and seawater, for some dolomites. At the outcrops, numerous solution pits and vugs form by the leaching, and purplish pigments owing to oxidation of ferrous iron are visible (Matsumoto et al. 1988). Under the microscope, later calcites replacing dolomite rhombohedra, called de-dolomites, have a pseudomorphic texture of the original rhombohedral outlines of the dolomite. This suggests that these later calcites are precipitated within dissolution voids after the dolomite rhombs (Evamy 1967), without accompanying volume increase. Such a type of geological de-dolomitisation is a dissolution/precipitation process in the open system, fairly different from the de-dolomitisation in concrete. Instead of the earlier formula, the reaction without a volume increase could be written as follows (Katayama 2010b, 2011a):

Natural de-dolomitisation (no volume increase):

$$CaMg(CO_3)_2 + 0.75Ca^{2+} \rightarrow 1.75CaCO_3 + Mg^{2+} + 0.25CO_3^{2-}$$

If the dolomite is of a ferroan variety (typically, ankerite), after de-dolomitisation, one molar dolomite will produce one molar calcite with a quarter mole of limonite or goethite $(FeO \cdot OH)$ that produces a rusty stain (Al-Hashimi and Hemingway 1973):

Natural de-dolomitisation (ferroan species):

$$4[Ca(Mg,Fe)(CO_3)_2] + 2H_2O + (½)O_2 \rightarrow 4CaCO_3 + Fe_2O_3 + 2Mg^{2+} + 2CO_3^{2-} + 2H_2CO_3$$

$$Fe_2O_3 + H_2O \rightarrow 2(FeO \cdot OH)$$

In this case, about 30% of the solid volumes decrease after the reaction, suggesting that a porous texture is produced within the dolomite rhombohedra. In the weathered portions of the dolomitic rocks, rhombohedral cavities have often been found (Evamy 1967; Al-Hashimi and Hemingway 1973), which are the products of the selective leaching of the dedolomitic calcites that had formed after the dolomite rhombohedra.

6.9.2 Practical examination of concretes suspected of ACR

The external appearance of concretes undergoing an expansive ACR is of a somewhat finely spaced map cracking as shown in Figure 6.135. It resembles a concrete undergoing ASR that has been exposed to continuously moist conditions, although frost may open up some of the cracks. External deposits of alkali–silica gel are extremely rare and, if they are present, suggest the possibility that an ASR must be considered. Recent investigations of an experimental concrete pavement (sidewalk) placed in Kingston, Ontario, have clearly shown that ASR is responsible for the expansion of what has been called ACR (Grattan-Bellew et al. 2008, 2010; Katayama and Grattan-Bellew 2012).

The internal texture of concrete undergoing ACR has not been well described in the literature especially with respect to the relationship of the outer shell of the concrete to the interior. It would be expected that the outer shell would be less reacted in view of the leaching of alkalis by weathering, as for other types of AAR, but the fine mesh of cracks visible externally suggests that this may not be an operative factor. Katayama (2010a) identifies products of de-dolomitisation and alkali–silica gel from the outer shell (depth <50 mm) of typical ACR field concretes in Ontario, in which leaching of alkalis and ingress of chloride ions from de-icing salts are evident (Katayama and Sommer 2008). This means that these processes do not affect identification of ACR or even ASR.

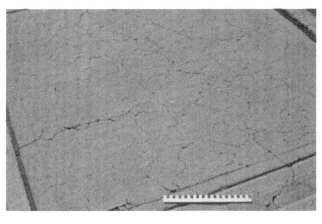

Figure 6.135 External pattern cracking due to the 'alkali–carbonate' reaction usually appears as a mesh of closely spaced polygonal cracks somewhat similar to map cracking in ASR where the concrete has been continuously moist cured. This Canadian concrete footpath and displaced kerb contain Kingston limestone, one of the classical aggregates which give rise to 'alkali–carbonate' reaction. However, research has now demonstrated this to be a form of carbonate-hosted ASR.

6.9.2.1 General features

In ACR, the coarse aggregate produces most of the expansion and cracking, which is in contrast to many ASR reactive aggregates, where greater reaction is usually found in the fine aggregate. Cracks originate in the aggregates and extend into the hardened paste and, in severe cases, form the typical network of expansive cracking. Early papers on ACR generally maintained that alkali–silica gel was absent under the optical microscopes, although there are reports of gel being present in some cases (Katayama 1992). Gel plugs at crack entrances were formerly thought to be absent, but under the SEM on polished thin sections, they can be seen, although less frequently than in typical ASR (Katayama 2010a; Katayama and Grattan-Bellew 2012). Ettringite is occasionally observed, but is not common. From field concretes, Friedel's salt and monocarboaluminate are identified by SEM–EDS and XRD analyses (Katayama 2010a, 2011). In some cases, cracks contain a poorly defined reaction product under the polarising microscope, which suggests that movement of materials has occurred.

The most distinctive features of ACR are the presence of cracking associated with argillaceous dolomitic limestone in the coarse aggregate. The expansive dolomitic limestones are

Figure 6.136 Photomicrograph of the 'characteristic' texture of expansive dolomitic limestone: rhombohedra of dolomite in a matrix of calcite, quartz and clay.

claimed to have a distinctive texture (Dolar-Mantuani 1983). This texture consists of well-developed rhombohedra of dolomite, in the size range 100–25 μm or less, enveloped by a fine-grained matrix of calcite, quartz and clay (Figure 6.136). The most typical rock type is characterised by the dolomite rhombohedra of silt size (Hadley 1961, CSA A23.1 1994, 2009). In practice, reliable recognition of this texture may require previous petrographic experience with these rocks. In the past, destructive analyses were commonly used in the study of ACR, which include the etching of reaction rims by a concentrated hydrochloric acid solution (ASTM C856 2011), staining to identify the presence of dolomite with copper nitrate solution (Fairbanks 1925) and calcite with alizarin red S solution (Friedman 1959) and removal of the aggregate from concrete for XRD analysis of dolomite and chemical analysis. Generally, none of these methods is recommended, since the concrete texture is broken and the reaction products undergo alteration. Instead, a combined application of polarising microscopy and SEM–EDS analysis using the same polished thin section (with a thickness of 15 μm, not 30 μm) is suitable for characterising ACR in concrete.

6.9.2.2 Reaction rims in aggregates

Reaction rims are the sites at which de-dolomitisation (the first elementary reaction: $CaMg(CO_3)_2 + 2OH^- \rightarrow Mg(OH)_2 + CaCO_3 + CO_2^{2-}$) occurs within the inner periphery of dolomitic aggregate particles (Figure 6.137). SEM investigation of polished thin sections reveals that spots of brucite and calcite form in nearly equal amounts following the preceding reaction and that they produce a 'pseudomorphic texture', keeping the original outlines of dolomite rhombohedra within the aggregate (Katayama 2004, 2010a, 2012a) (Figure 6.138). What is important here is that none of the expansion cracks appears through the de-dolomitisation itself (see Figure 6.137). This is evident because the total molar volume of the solid phases involved decreases about 5% by volume after the reaction, so that the formation of brucite within the dolomitic aggregate is not the cause of expansion of the aggregate.

At the interface of dolomite and cement paste, hydrotalcite, $6MgO \cdot Al_2O_3 \cdot CO_2 \cdot 12H_2O$, forms a narrow rim. This can be demonstrated by the mapping of elements Al and Mg using an EPMA or EDS on the polished thin section (Katayama 2010a, 2011a). If finely divided silica (cryptocrystalline quartz) is present in the dolomitic aggregate, Mg–silicate gel forms through the reaction between brucite (or its gel) and alkali–silica gel. This Mg–silicate gel can be seen within the entire section of the fine aggregate but is largely confined to the

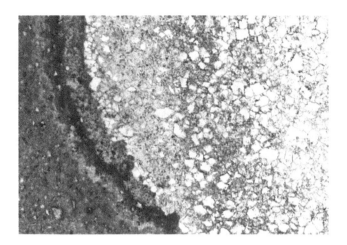

Figure 6.137 De-dolomitisation producing distinctive reaction rims on an Austrian dolomite coarse aggregate and carbonate haloes in the surrounding cement paste. No expansion cracks are visible in the cement paste surrounding the reacted aggregate. Photomicrograph of the polished thin section of a RILEM AAR-5 concrete 'microbar'. (From Sommer, H. et al., *Mater. Struct.*, 38(282), 787, 2005.), taken under the polarising microscope. Width of field 1 mm. (From Katayama, T., Diagnosis of alkali-aggregate reaction – Polarizing microscopy and SEM-EDS analysis, Castro-Borges, P., Moreno, E.I., Sakai, K., Gjørv, O.E. and Banthia, N. (eds), *Proceedings of the Sixth International Conference on Concrete under Severe Conditions (CONSEC'10)*, Merida, Mexico, 2010b, pp. 19–34; Specimen courtesy of Hermann Sommer.)

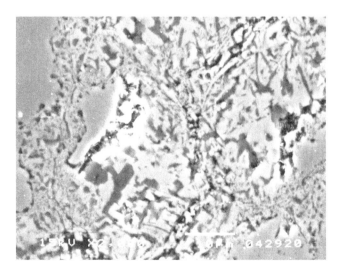

Figure 6.138 De-dolomitisation in the reaction rim of an Austrian dolomite coarse aggregate. A pseudomorphic texture after the dolomite rhombohedra, composed of dark spots of brucite and bright spots of calcite. No expansion cracks are formed in the cement paste. SEM micrograph of the polished thin section of a RILEM AAR-5 concrete 'microbar'. (From Sommer, H. et al., *Mater. Struct.*, 38(282), 787, 2005.) Scale bar 10 μm. (From Katayama, T., Diagnosis of alkali-aggregate reaction – Polarizing microscopy and SEM-EDS analysis, Castro-Borges, P., Moreno, E.I., Sakai, K., Gjørv, O.E. and Banthia, N. (eds), *Proceedings of the Sixth International Conference on Concrete under Severe Conditions (CONSEC'10)*, Merida, Mexico, 2010b, pp. 19–34; Specimen courtesy of Hermann Sommer.)

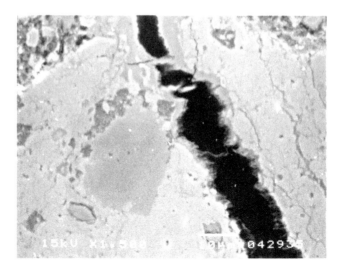

Figure 6.139 Mg–silicate gel (dark patches) rimming or replacing the dolomite rhombohedra in the typi-cal 'alkali–carbonate' reactive Kingston dolomitic limestone coarse aggregate. The expansion crack is lined with alkali–silica gel broadening into the cement paste, while small cracks are filled with alkali–silica gel derived from the reaction with the particles of cryptocrystalline quartz in the argillaceous matrix. SEM micrograph of the polished thin section of a RILEM AAR-5 concrete 'microbar'. (From Sommer, H. et al., *Mater. Struct.*, 38(282), 787, 2005.) Scale bar 10 µm. (From Katayama, T., Modern petrography of carbonate aggregates in concrete – Diagnosis of so-called alkali-carbonate reaction and alkali-silica reaction, Fournier, B. (ed), *Marc-Andre Berube Symposium on Alkali-Aggregate Reactivity in Concrete*, part of *Eighth CANMET/ ACI International Conference on Recent Advances in Concrete Technology*, Montreal, 31 May–3 June 2006, Supplementary paper (corresponding to 423–444, 2006; Specimen courtesy of Hermann Sommer.)

reaction rim of the coarse aggregate (Figure 6.139). Because these reaction products are finely grained, generally a few micrometres across and less than the thickness of thin sec-tions, traditional optical microscopy is not suitable for identifying them.

6.9.2.3 Carbonate haloes in cement paste

Carbonate haloes form within the cement paste in contact with the reacting dolomitic aggre-gate (see Figure 6.137). It is the carbonation of cement paste that takes place in the open system (the second elementary reaction: $CO_2^{2-} + Ca(OH)_2 \rightarrow CaCO_3 + 2OH^-$), in which carbonate ion (CO_2^{2-}) released by de-dolomitisation is a mobile component migrating into the cement paste. In this process, portlandite ($Ca(OH)_2$) dissolves and its calcium precipi-tates as calcite ($CaCO_3$), *not* replacing the portlandite in situ, but filling pore spaces within adjoining cement paste and further carbonating the CSH gel locally, which results in the formation of a porous crust of calcite (Figure 6.140). As a result, no volume increase occurs within the cement paste (Katayama 2010a, 2011a, 2012a). In this carbonate halo, concen-trations of alkalis can be confirmed by the mapping of elements by EPMA on the laboratory concrete specimen (Katayama 2004). It should be noted that conventional experiments that immerse rock powders in alkali solution inevitably miss this important elementary reaction that forms a carbonate halo outside the aggregate.

The formation of the carbonate halo around the margins of aggregates, which some workers originally called 'carbonate rims', has been noted as another feature of ACR

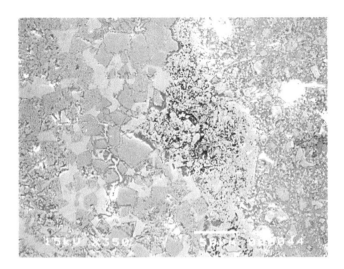

Figure 6.140 Carbonate halo in the cement paste adjoining a de-dolomitised Kingston dolomitic limestone aggregate. Note the porous texture of the calcite crust replacing the cement paste. Dolomite rhombohedra contacting the carbonate halo (originally cement paste) have a narrow dark rim of hydrotalcite. SEM micrograph of the polished thin section of a RILEM AAR-2 mortar bar (RILEM 2000a). Scale bar 50 µm. (From Katayama, T., *Cement Concrete Res.*, 40, 643, 2010a; Specimen courtesy of Hermann Sommer.)

(Mather et al. 1964). In weathered concretes, this comprises a soft, friable, buff-coloured zone in the surrounding concrete paste, consisting mostly of fine-grained calcite (Sims and Poole 1980). Carbonate haloes do not surround all aggregates or necessarily completely ring an aggregate. Examination of a test specimen in thin sections indicates clearly that these carbonate haloes are not due to carbonation by carbon dioxide from the atmosphere. Many haloes are isolated by uncarbonated hardened paste, suggesting they are an in situ formation of calcite from the reaction of the alkalis with the dolomitic aggregate.

6.9.2.4 Alkali–silica gel

Alkali–silica gels in concretes containing dolomitic aggregates tend to be carbonated to a fragile material, so that these are often lost during the process of sample preparation (cutting, grinding and polishing). Even then, however, a detailed petrographic examination, including polarising microscopy (Figure 6.141), SEM investigation and quantitative EDS analysis using the same polished thin sections, enables it to identify the presence of alkali–silica gel. Examples are provided for both typical ACR field concretes in Ontario, Canada (Figure 6.142), and in concrete 'microbar' test specimens made with ACR-reactive Kingston aggregate (see Figure 6.139) (Katayama 2004, 2006, 2008, 2010a, 2011, 2012a). Under the SEM, alkali–silica gel is visible lining the expansion cracks that extend from the reacted aggregate into the cement paste or coming from isolated particles of cryptocrystalline quartz in the argillaceous matrix of the aggregate (see Figure 6.139). It is now accepted that the cause of expansion of dolomitic aggregates, traditionally known as ACR, is in reality an ASR and has nothing to do with de-dolomitisation. This means that the 'ACR' is actually a combined phenomenon of harmless de-dolomitisation and deleteriously expansive ASR of cryptocrystalline quartz concealed within the argillaceous matrix of the dolomitic aggregates (Katayama 2004, 2006, 2010a, 2011a). Previous research into ACR confused the chemical reaction (de-dolomitisation) with the expansion mechanism (ASR).

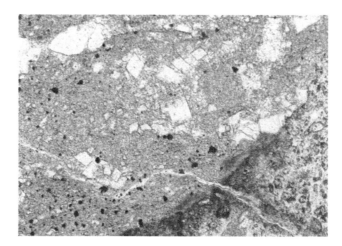

Figure 6.141 Alkali–silica gel exuding along the expansion crack that extends from the typical 'alkali–carbonate'-reactive Kingston dolomitic limestone coarse aggregate into the cement paste. Photomicrograph of the polished thin section of a RILEM AAR-5 concrete 'microbar'. (From Sommer, H. et al., *Mater. Struct.*, 38(282), 787, 2005.) Width of field 0.5 mm. (From Katayama, T., *Mater. Character.*, 53(2–4), 85, 2004; Specimen courtesy of Hermann Sommer.)

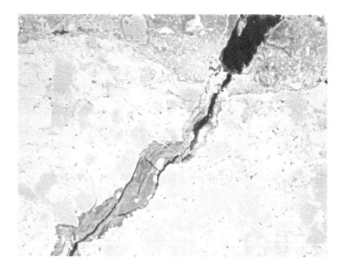

Figure 6.142 'ACR' field concrete in Ontario, Canada, showing alkali–silica gel filling the expansion crack of a dolomitic limestone coarse aggregate, which is carbonated near the cement paste owing to intense weathering of the concrete. This alkali–silica gel comes from the reaction involving isolated particles of cryptocrystalline quartz in the argillaceous matrix of the aggregate. SEM micrograph of the polished thin section of field concrete. Width of field 0.18 mm. (From Katayama, T., *Cement Concrete Res.*, 40, 643, 2010a.)

Alkali–silica gel found in the ACR concretes, both field and laboratory, presents a compositional trend line on the Ca/Si–Ca/(Na+K) diagram, spanning from low-Ca alkali–silica gel to CSH gel. The atomic ratios of Ca/Si at which compositional lines intersect Ca/(Na+K) = 1.0 range between 1/2 and 1/6 (Katayama 2010a), which are similar to those of the typical ASR in other rock types of aggregate previously mentioned (Katayama and Bragg 1996; Katayama et al. 2004a, 2008; Katayama 2008). What is interesting to note is that the alkali–silica gel in the dolomitic aggregate remains intact when it is within cracks of the coarse aggregate, while in the fine aggregate, it reacts with the brucite component on the dolomite rhombohedra, converting to the Mg–silicate gel, which is probably less expansive than the pure alkali–silica gel. This explains why, in ACR, the fine aggregate produces less expansion than does the coarse aggregate. The alleged scantiness of the reaction products in ACR is caused partly by the fact that alkali–silica gel is generally trapped around the fine, silt-sized de-dolomitised rhombohedra, unless excessive amounts of alkali–silica gel exude into the cement paste along the expansion cracks, so that observations using conventional 'thick' thin sections (30 μm) under the polarising microscope are likely to miss these alkali–silica gel products (Katayama 2011a).

6.9.2.5 Cryptocrystalline quartz

In thin sections under transmitted light, cryptocrystalline quartz is concealed within a highly birefringent matrix composed of clay minerals and carbonates, which is beyond the resolution of even a top-quality optical microscope. Cryptocrystalline quartz in the typical ACR-reactive aggregate from Kingston can be extracted either by phosphoric acid (Katayama 2004) (Figure 6.143) or hydrochloric acid (Grattan-Bellew et al. 2008). The CI of the extracted quartz, as determined by XRD analysis after the method of Murata and Norman (1976), is generally high (9.9) for the bulk reactive aggregate with the presence of highly crystalline detrital quartz (Katayama et al. 1996), whereas it is low (4.5) for the most reactive layer dominated by the cryptocrystalline quartz in the aggregate from the same quarry (Grattan-Bellew et al. 2008, 2010). Grattan-Bellew et al. performed a petrographic examination of the concrete sidewalk (pavement) in Kingston, applying the uranyl acetate fluorescence method (e.g. ASTM C856 2011), and found a widespread occurrence of alkali–silica gel within cracks in and around the reacted dolomitic limestone aggregate. They confirmed the conclusion by Katayama (2004, 2006) that 'ACR' is actually ASR of the

Figure 6.143 Cryptocrystalline quartz extracted by the phosphoric acid treatment from typical 'alkali–carbonate'-reactive Kingston dolomitic limestone coarse aggregate. This quartz concealed within the argillaceous and carbonate matrix of the aggregate produces ASR, which is the real cause of the deleterious expansion of 'ACR' with this aggregate. Scanning electron micrograph of the polished thin section. Scale bar 10 μm. Image courtesy of Tetsuya Katayama.

cryptocrystalline quartz. Within the same concrete cores that Grattan-Bellew et al. studied, crack-filling alkali–silica gel and its carbonation process were seen by SEM–EDS analysis of polished thin sections (Katayama and Grattan-Bellew 2012).

6.9.3 Ancillary techniques for assessing ACR potential

Katayama (2010a) has explained that traditional petrographic techniques are unsuitable for the reliable diagnosis of ACR in concrete. Problems arise inter alia from (1) poor sample preparations that may lose reaction products; (2) stereomicroscopic observations under only low magnification; (3) polarising microscopy based on covered and relatively 'thick' thin sections (i.e. standard 30 μm) that obscure the details of carbonate aggregates with highly birefringent constituents; (4) impregnation of thin sections with fluorescent dye, which overlaps cracks and reaction products that were present in the oblique direction within the thickness of the thin section and obscures the details by its strong colour; and (5) SEM observations on the fracture surface providing non-random section and poor information on the rock texture and reaction products. Therefore, as explained earlier, the best method for concrete petrography of suspected ACR will be the combination of polarising microscopy to examine the rock texture and SEM–EDS investigation of reaction products, continuously using the same 'thin' (15 μm) and polished thin sections of concrete. In the first three of the following subsections, some procedures are described that can augment the petrographic examination process.

Also, in addition to petrography, numerous engineering test methods have been proposed to evaluate the expansion potential of carbonate aggregates, but little has been done on the scientific aspects to interpret the nature of the elementary reactions involved in ACR, based on observations and analyses of the reaction products. Because the mechanism of ACR has not been well understood (Jensen 2009), often by researchers citing old references (Katayama 2010a, 2011), concrete engineers and petrographers have relied heavily on supplementary tests for the identification of reactive dolomitic rocks. The final three of the following subsections address some of these additional methods.

6.9.3.1 Acid-insoluble residues

To assist the petrographic examination of carbonate aggregates, it is recommended to extract acid-insoluble residues for SEM investigation. The phosphoric acid method (JCAS I-31 1996) liberates quartz from rocks, dissolving feldspars, clay minerals and sulphides. A variety of quartz forms (e.g. chalcedony, cryptocrystalline quartz, microcrystalline quartz, detrital quartz) are separated from the dolomitic aggregates, including ACR-reactive dolomitic limestone from Kingston (Katayama 2004, 2010a) (see Figure 6.142). This technique is also applicable to the volcanic rocks, to extract the alkali-reactive silica minerals cristobalite and tridymite, but dissolves 'opal CT'.

Hydrochloric acid is most frequently used in extracting the acid-insoluble residue from carbonate aggregates. A popular method to describe compositions of the carbonate rocks is to use a triangular diagram to plot the contents of calcite, dolomite and acid-insoluble residue. The acid-insoluble residue is subject to XRD analysis to identify the presence and types of any clay minerals contained therein (e.g. illite, chlorite, smectite) and further determine the CI of quartz (see next subsection). These processes are helpful to the petrographer, because both the clay minerals and the cryptocrystalline quartz are very fine grained and cannot be easily identified in thin sections under the polarising microscope (Grattan-Bellew et al. 2008, 2010).

6.9.3.2 Crystallinity index (CI) of quartz

The CI of quartz can be estimated using XRD, from the relative intensities of the quintuplet peaks at 67°–69° 2θ (CuKα) (Murata and Norman 1976). This method does not measure a fundamental physical parameter of the quartz, but possibly still provides a practical value in assessing alkali reactivity of such rock types as chert and flint, in which crystallinity of quartz continuously increases from poorly ordered, fine-grained chalcedonic to crypto-crystalline quartz (originally deposited as opaline fossils), to highly ordered coarse-grained quartz during *burial diagenesis* and metamorphism. For this reason, this method is not applicable to those rocks in which quartz crystals of different origin are intermixed, such as sandstones with reactive interstitial authigenic quartz and non-reactive detrital quartz, and granites, gneisses and mylonites with reactive microcrystalline quartz interposed between the non-reactive coarse-grained quartz. Serious concerns over the reliability of this method have been raised by Guggenheim et al. (2002) and Marinoni and Broekmans (2013), the latter recommending discontinuation of the Murata & Norman approach.

In the circum-Pacific regions, Cainozoic cherts have been characterised by having lower CI than the Mesozoic cherts (Murata and Norman 1976), thus presenting higher alkali reactivity (Katayama and Futagawa 1989). Without such a geological background of knowledge on the diagenesis, it would be difficult to understand the reason why the alkali reactivity of cherts and flints varies considerably both between countries and between areas within the same country, even if numerous engineering expansion tests are performed. Mesozoic cherts with a regional distribution have caused ASR in concretes in Japan (Morino 1989). In the continental regions near the shield where geothermal gradient is low and the grade of diagenesis is accordingly low, even Palaeozoic cherts may have low crystallinity of quartz.

Carbonate rocks in such a geological setting contain cryptocrystalline quartz as an authigenic mineral in the argillaceous matrix and produce ASR. In the case of the most ACR-expansive aggregate, from Kingston, Ontario, cryptocrystalline quartz in the acid-insoluble residue presents a low CI of quartz (4.5), apparently indicative of the higher potential for ASR. This might explain why ASR takes place with this aggregate (Grattan-Bellew et al. 2008, 2010).

6.9.3.3 Uranyl fluorescence method

Fracture surfaces of concrete may be tested for the possible presence of gel with uranyl salts (for instance, $UO_2(C_2H_3O_2)_2 \cdot 2H_2O$), as proposed by Natesaiyer and Hover (1989), which is being adopted in ASTM C856 (2011). Using this test, Grattan-Bellew et al. (2008, 2010) confirmed a widespread occurrence of alkali-reactive gel on the fracture surfaces and within cracks of sidewalk (pavement) concrete in Ontario. Before applying this method, it should be kept in mind that this is an internationally regulated radioactive substance that is hazardous to health and the environment. In some countries, the waste solution cannot be transported or sent for disposal and must be permanently kept at the laboratory where it was used.

6.9.3.4 Chemical test

Testing dolomitic rocks with the ASTM C289 (2007) 'rapid chemical test' rarely gives a positive result, usually producing extremely high values of the 'reduction in alkalinity' (Rc), owing mainly to the reaction between the Mg of dolomite and OH ions in the NaOH solution. However, it is claimed that if the rock is first dissolved in acid and the insoluble residue then tested, this gives more reliable results (Katayama 1992). This assumes that expansion in concrete is essentially caused by ASR, because the acid-insoluble residue of the ACR-reactive dolomitic limestone in Kingston is tested as 'deleterious' (Swenson and Gillott 1960). Since

Rc values are affected by many factors, only 'dissolved silica' (Sc) in a modified chemical test is used to assess ASR reactivity of the carbonate aggregates in Québec, Canada, where ASR is generally believed to be caused by siliceous carbonate aggregates (Berard and Roux 1986, Fournier and Bérubé 1990), although dolomitic rocks are also invariably present.

6.9.3.5 Rock cylinder test

Until recently, the most accepted test for ACR potential was to immerse rock cylinders or prisms in a 1-molar solution of sodium hydroxide NaOH and measure the volume change, as specified in ASTM C586 (2011). This method well detects the expansive nature of the ACR rocks on a bed-by-bed basis within the same quarry (Rogers 1988; Williams and Rogers 1991). However, the method gives false results for some types of dolomitic rocks, the so-called 'late-expansive' type (Dolar-Mantuani 1971, 1983), which do not expand in concrete owing to the creep of concrete that accommodates deformation produced by slow expansion. The rock cylinder test may be used for the 'slow reactive/expansive' argillaceous aggregates, formerly called 'alkali–silicate reaction' (see Section 6.8.1), with *a modified criterion* for the deleterious expansion, whereas typical ASR-reactive rock types (chalcedony, agate, flint, etc.) contract through partial dissolution of silica into NaOH solution (Duncan et al. 1973). Petrographic examination of the tested rock cylinder, using polarising microscopy and SEM observations on polished thin sections, is useful for confirming the reaction site and mechanism of chemical reaction that takes place in the carbonate rock during exposure to alkali solution.

6.9.3.6 Concrete and mortar expansion tests

In ACR, the coarse aggregate is typically found to expand more than the fine aggregate. For this reason, concrete prism tests, such as CSA A.23.2-14A (1994, 2009), ASTM C1105 (2008) or RILEM AAR-3 (2000b), have sometimes been used to assess ACR-reactive carbonate aggregates, but they mainly take too long to complete (12 months or more). Accordingly, an accelerated concrete 'microbar' test was proposed by Xu et al. (2000), using only a single fraction of coarse aggregate (5–10 mm) for the concrete 'microbar', which is then immersed in 1-molar alkaline solution at a high temperature (80°C). This has been modified, using a 4–8 mm aggregate fraction, and developed into RILEM AAR-5 of wider use (Sommer et al. 2004, 2005). Another version has also been proposed to assess deleterious expansion of both ACR and ASR carbonate aggregates (Grattan-Bellew et al. 2003).

To shorten further the testing period, an autoclave test was proposed by Tang et al. (1994). However, the very high temperature (150°C) produces an irreversible effect of thermal expansion on well-crystallised carbonate aggregate and thus causes misleading results in the test (Mu et al. 1996). For a coarse-grained limestone, expansion was independent of both the cement alkali level and the curing conditions, but proportional to the size of the aggregate: the coarse aggregate (5–10 mm) gave the largest expansion (0.09%), essentially equal to the threshold (0.1%) given for reactive carbonate aggregates in the autoclave method (Tang et al. 1994). With the same total autoclaving time, the more frequent the heat treatment for interim measurements, the higher the expansion, which could be about three times. This is a heat-induced expansion along the cleavage planes of well-crystallised calcites, associated with the formation of portlandite. Hence, to validate the expansion data in the autoclave method, checking by petrography is also necessary (Mu et al. 1996).

With the same autoclaving condition (150°C), calcite and magnesite $MgCO_3$ decompose into portlandite and brucite, respectively (Tong 1994; Tang and Deng 2004). Although the total solid molar volumes decrease in these reactions ($CaCO_3 + 2OH^- \rightarrow Ca(OH)_2 + CO_2^{2-}$, $MgCO_3 + 2OH^- \rightarrow Mg(OH)_2 + CO_2^{2-}$), powders of pure limestone and magnesite produce a significant expansion when tested as a compacted mortar. Tong (1994) and Tang and Deng (2004) attribute this expansion to a *topochemical reaction* that takes place within a confined space and the crystallisation pressure of portlandite and brucite. They claim that the expansion of ACR is similarly due to the topochemical reaction of brucite formed during the de-dolomitisation. However, if this mechanism is to be accepted, then pure limestones should also be regarded as deleteriously expansive aggregates, which obviously contradicts the reality. The magnesite specimen tested was a well-crystallised one (Tong 1994), and the heat treatment (measurements) is so frequent that a thermal effect as observed by Mu et al. (1996) might also be intense. Therefore, in this autoclave method, care should be taken in interpreting the expansion data and in examining the test specimens. The original version of the autoclave test, with the same accelerating condition (150°C) but using a mortar 'microbar', has been used successfully in China (Tang et al. 1983) and France (Criaud et al. 1992) for typical ASR-reactive aggregates containing reactive silica, because quartz has a smaller thermal expansion than the carbonate minerals which develop cleavage.

6.10 DAMAGE FROM THERMAL CYCLING AND FIRE

Although concrete can occasionally deteriorate as a result of ambient thermal cycling, especially with aggregate types that can exhibit 'thermal hysteresis' (Senior and Franklin 1987) and/or in extreme climate conditions, the most common thermal threat to concrete arises from direct and exceptional heating and cooling cycles, which are usually associated with unintentional exposure to fire and its quenching. A recent overview of the design process for structural fire engineering will be found in Lennon (2011). Dedicated texts relating to concrete fire performance and fire damage may be found in the Concrete Society Technical Reports 33 (1990) and 68 (2008), Lin et al. (1996), Smart (1999), Cather (2003), Matthews et al. (2006), Khoury (2006, 2007) relating to tunnels and Ingham and Tarada (2007). Understanding about the performance of buildings in major fires has been greatly improved by a recent full-scale research programme conducted inside a former airship hangar at Cardington, near Bedford in the United Kingdom, including a seven-storey concrete building. The overall project is summarised by Lennon (2011), and the findings for the concrete building, which demonstrated the potentially excellent performance of concrete structures in fire, are described by Chana and Price (2003).

Normal concretes are incombustible and exhibit good resistance to fire, although they are not classed as refractory materials. Nevertheless, heating due to fire or other causes will induce alterations to the mineralogy and fabric of the material. These alterations are dependent on the maximum temperature reached, the length of time at that temperature and the rate of heating and cooling or quenching if water is used to extinguish the fire.

The severity of the deterioration will also be partly dependent on the cross sections of the concrete elements involved, because of the thermal gradients developed and the differential thermal expansions between the various components within the concrete, including steel reinforcement if present. The effects of fire damage can include discolouration, surface cracking, surface spalling and dehydration of calcium hydroxide and other cementitious phases. Internal cracking within the cement paste, cracking around and across aggregate particles and distortion and loss of bonding to reinforcing steel also occur.

Factors such as these, together with melting points of external objects, provide important evidence concerning the maximum temperature that the damaged portions of a structural element have attained in a fire. This allows the engineer to estimate the amount of concrete

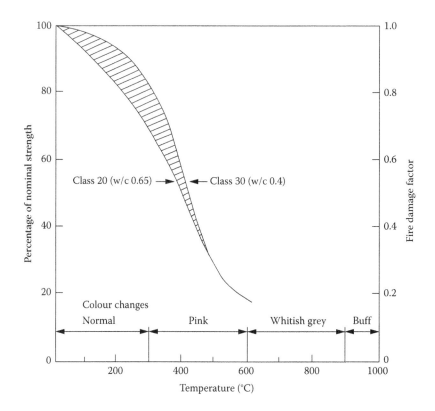

Figure 6.144 Reduction in strength with temperature of normal weight concrete and relation to colour changes. (From Concrete Society, Assessment and repair of fire-damaged structures, Concrete Society Technical Report No 33, The Concrete Society, Crowthorne, U.K., 1990.)

to be removed and replaced to allow reinstatement of the member, since strength loss can be correlated with temperature as illustrated by Figure 6.144. A review of methods of assessing temperatures reached during fire damage to concrete is given by Kramf and Haksever (1986) and the Concrete Society (1990, 2008).

Smart (1999) provided an excellent simplified illustration of how site and petrographic observations can be used to classify the condition of concrete in the vicinity of a fire and thus indicate the likely repair options (Figure 6.145).

6.10.1 Thermal expansion and cracking of concrete

A cement paste sample first expands on heating as a result of normal thermal expansion, but this effect is counteracted by shrinkage of the hydrates as water is driven off with this effect eventually becoming dominant. The temperature at which maximum expansion occurs depends on specimen size and heating conditions, but may be as high as 300°C for rapidly heated air-dried specimens. At a higher temperature, shrinkage continues and can produce reductions in dimension of 0.5% or more resulting in severe cracking. Calcium hydroxide loses water above 400°C–450°C to form calcium oxide. If this calcium oxide rehydrates owing to the presence of moisture, there is a considerable volume expansion which may disrupt the concrete.

The aggregates present in mortars and concretes typically undergo a progressive expansion as temperature rises, opposing the shrinkage induced in the cement paste. These opposing effects tend to weaken and crack the concrete. There is a considerable variation

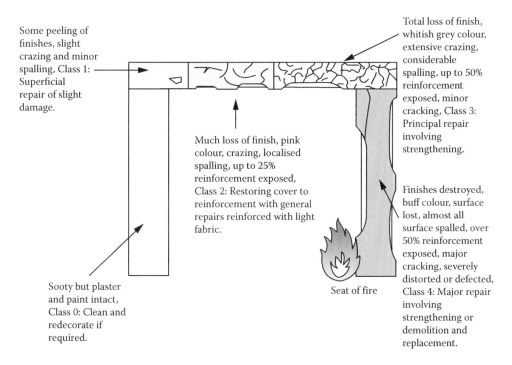

Some peeling of finishes, slight crazing and minor spalling, Class 1: Superficial repair of slight damage.

Total loss of finish, whitish grey colour, extensive crazing, considerable spalling, up to 50% reinforcement exposed, minor cracking, Class 3: Principal repair involving strengthening.

Much loss of finish, pink colour, crazing, localised spalling, up to 25% reinforcement exposed, Class 2: Restoring cover to reinforcement with general repairs reinforced with light fabric.

Finishes destroyed, buff colour, surface lost, almost all surface spalled, over 50% reinforcement exposed, major cracking, severely distorted or defected, Class 4: Major repair involving strengthening or demolition and replacement.

Sooty but plaster and paint intact, Class 0: Clean and redecorate if required.

Seat of fire

Figure 6.145 Simplified illustration of the damage classification and consequent repair options for concrete investigated for the effects of fire damage. (From Smart, S., Concrete, 33(7), 30, 1999.)

in the thermal expansive behaviour of the aggregate types (Sims and Brown 1998), with fine-grained basic rocks such as basalt or dolerite typically showing the best fire-resistant properties. Individual crystal grains in aggregate particles can exhibit anisotropic thermal behaviour. Quartz, for example, has a coefficient of linear expansion, α, of 7.5×10^6 parallel to and 13.7×10^6 perpendicular to the crystallographic axis. Such variations can lead to thermal stressing and cracking of aggregate particles, particularly if the crystal grains are relatively large. Calcite crystals also display such anisotropic behaviour, and Senior and Franklin (1987) have described how this can sometimes lead to 'thermal hysteresis' in the event of heating and cooling cycles.

These effects can be detected in concretes which have been heated to temperatures in excess of about 180°C. There is an additional problem with quartz, which is a common constituent of sand and gravels, in that there is a significant increase in volume associated with the phase transition from α to β quartz at 573°C. This expansion has a disruptive effect on a concrete containing quartz, chert or flint aggregates. Sandstone aggregates do not usually cause such severe cracking because the intergranular cement between the quartz grains tends to shrink on heating, thus in part compensating for the α to β expansion. Schematically, the various expansion and shrinkage effects in a concrete are illustrated in Figure 6.146.

The differences in the thermal properties of the cement paste and aggregate cause stresses to be developed in the concrete. Normally, for temperatures below about 250°C, these stresses are not large enough to cause significant cracking, though on cooling there is a residual contraction which is essentially a drying shrinkage of the cement paste. At higher temperatures, cracking becomes increasingly severe so that on cooling from high temperature, a residual expansion and irreversible damage remain.

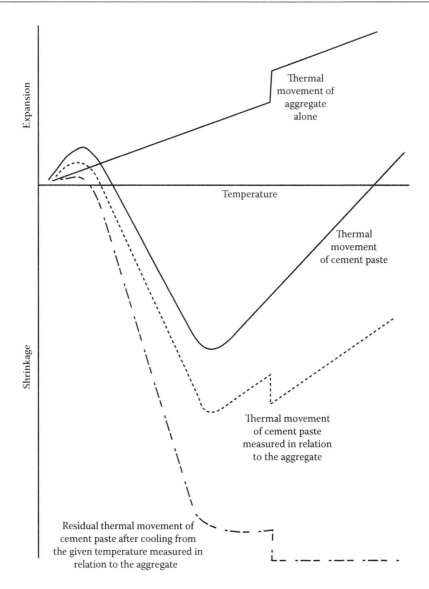

Figure 6.146 Expansion and shrinkage of cement paste and aggregate when heated. (From Zoldners, N.G., Thermal properties of concrete under sustained elevated temperatures, *Temperature and Concrete*, ACI SP-25, 1–32, Paper SP25-01, American Concrete Institute, Detroit, U.K., 1971.)

6.10.2 Effects of fire damage on concrete

The cement hydrates in the hardened matrix of concrete undergo a progressive loss of water on heating, with a resultant loss of strength. Initially, only the free water is lost with minimal loss in strength, but with increasing temperatures over 200°C, the cement hydrates start to lose their bound water and, by 400°C, this loss is serious and calcium hydroxide starts to dehydroxylate. By 500°C, the cement hydrates are largely dehydrated and, by 650°C, over 80% loss in strength may have occurred (Malhotra 1956). As well as this loss in strength, the concrete surfaces exposed to fire and heat will exhibit spalling or 'shelling', which will vary according to various factors including the concrete composition and quality (Cather 2003).

Depending on the severity of the heating, the nature of the aggregates and the concrete mix design, a number of fire damage features may be observed, principally various colour changes and surface spalling.

6.10.2.1 Colour changes

There is a change in concrete colour from grey to pink or red, which occurs at temperatures in the range 250°C–300°C and gives a visual indication of a significant reduction in the residual strength. The colour change is thought to be dependent on the presence of iron compounds, such as limonite, goethite and lepidocrocite in the aggregate, which all dehydrate and oxidise in this temperature range. If these minerals are not present, the colour change may not occur. The intensity of pink colouration is dependent on aggregate type, with flint aggregate concretes perhaps producing the most striking colour change. At temperatures in the region of 500°C–600°C, a further colour change from pink/red to purple/grey occurs. Bessey (1950) suggests that this change results from the reaction between ferric oxide and lime, forming lighter-coloured calcium ferrites. This suggested mechanism is supported by the observation that concretes that do not contain much lime remain pink up to 1000°C.

The colour changes that occur are illustrated by Figure 6.147, which is of test pieces of concrete heated to a series of different temperatures in a furnace. Smart (1999) well illustrated the type of temperature contouring that can be achieved by the petrographic examination of concrete samples (Figure 6.148). Sibbick (2011) has found that fire damage to HAC concrete creates broadly similar colour changes and matrix durability issues at similar temperatures.

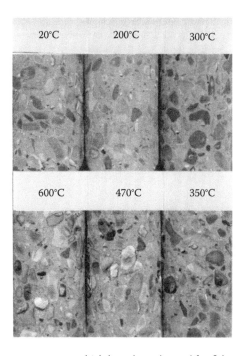

Figure 6.147 Flint aggregate concrete cores which have been heated for 2 h at the temperatures indicated.

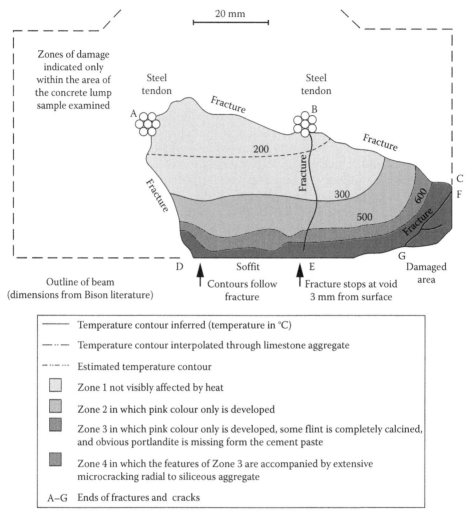

Figure 6.148 Sketch cross section of a concrete beam, showing temperature contouring based on petrography of a sample obtained from the soffit exposed to fire. (After Smart, S., Concrete, 33(7), 30, 1999.)

6.10.2.2 Surface spalling

Spalling of the surface layers of concrete is a common result of fire damage and often tends to give an exaggerated impression of its true extent. Surface spalling may be grouped into two types:

1. Explosive spalling is erratic and typically occurs during the first 30 min or so of exposure to fire (Figure 6.149);
2. Slower spalling occurs as cracks form parallel to the fire-affected surfaces, leading to a gradual separation of concrete layers and detachment of a section of concrete along some plane of weakness, such as a plane of steel reinforcement (Figure 6.150).

According to Castillo and Durrani (1990), spalling of high-strength concrete under load is most likely in the 320°C–360°C temperature range. The mechanisms of spalling are

Figure 6.149 Example of explosive spalling of concrete soffit caused by formwork catching fire within a residential apartment block being constructed. (Photograph courtesy of Paul Bennett-Hughes, RSK Environment Ltd.)

Figure 6.150 Example of slow spalling on a concrete portion of a wall exposed to fire. (Photograph courtesy of Paul Bennett-Hughes, RSK Environment Ltd.)

discussed by Harmathy (1968), who suggests that the thermal gradients will drive water vapour inwards from the surface, leading to a moisture-saturated layer parallel to the heated surface. If this saturated layer cannot migrate inwards quickly enough, it will be overtaken by the advancing heat front vaporising the moisture, with a consequent pressure build-up and spalling. The decomposition of limestone to calcium oxide, with the liberation of carbon dioxide, takes place at about 900°C, and this has been cited as the reason for the degradation of concretes containing limestone aggregates. However, the reaction is slow and fire tests indicate that unless the heating is prolonged, only the surface material is decomposed.

6.10.3 Investigation of fire-damaged concrete

6.10.3.1 Petrography

The differential expansions and contractions within the fabric of a concrete that has been exposed to fire introduce cracking on various scales, loss of bond to steel reinforcement and decomposition of the cement matrix. Depending on the scale of the cracking, the nature and pattern of the cracks may be studied using either cut and polished slab specimens or appropriate petrographic thin sections prepared from samples obtained from the affected structure. Suitable samples will typically include both intact pieces of spalled concrete, providing their source and context are clear, and core samples drilled from the exposed concrete left in situ behind the spalled material. Sampling locations must be selected to represent the various degrees of both apparent damage and of exposure to both fire and fire extinguishment measures, bearing in mind that the sudden quenching of heated concrete with cold water will cause damage over and above that of heating alone. Concrete Society Technical Report No 68 (2008) provides detailed guidance on the assessment of fire-damaged concrete structures.

The temperature gradient is steepest over the first 20–30 mm from the surface exposed to heating and consequently develops the most severe cracking. These near-surface cracks tend to reduce the thermal conductivity of this zone, which may then provide partial protection to the deeper concrete. Typically, in fire-damaged structures, surface spalling and the high thermal conductivity of reinforcing steel (48/W m^{-1} K^{-1} for carbon steel as against 1.5/W m^{-1} K^{-1} for concrete) reduce the effectiveness of a low-conductivity surface zone.

In addition to the development of cracking, changes to the mineralogical composition of the cement paste also occur as the concrete is exposed to heat, as described in subsection 6.10.2. The most important changes are the dehydration and later possible rehydration of the calcium hydroxide and cement hydrate phases. There is also the possible dehydration of iron compounds in aggregate particles and the calcining of limestone aggregate or carbonated concrete, if sufficiently high temperatures are maintained for a prolonged period. SEM and XRD techniques allow most of these changes to be identified. However, because of the fine-grained size of the minerals in a cement paste, only gross changes may be observed with the optical petrological microscope.

Exposure to temperatures below about 250°C causes little significant change in the general appearance of a concrete in petrographic thin sections, though SEM and observation of the effects of low-stress cycling of samples indicate that additional microcracks may develop in concrete heated at temperatures as low as 180°C. Above about 250°C, crack patterns develop and increase in severity as the temperature of exposure increases. Since the cracks develop principally as a result of differential thermal movements between the components of the concrete, the maximum temperature reached by a portion of concrete is more important than the length of exposure. However, prolonged exposure will allow the cracks to extend and widen and networks to become more extensive.

The severity of cracking will also depend on the thermal conductivities of the aggregate particles and the thermal movement mismatch between them and the paste. Flint aggregates have one of the highest thermal conductivities, and, as a consequence, concretes with this aggregate show severe cracking that cuts across both the paste and aggregate particles and also produces cracking at cement/aggregate interfaces, as illustrated in Figure 6.151. Similar effects are observed with granitic and sandstone aggregates, except there is less peripheral cracking at aggregate/cement interfaces, possibly because these aggregates can more readily accommodate differential expansion along intergrain boundaries of these polycrystalline materials. Limestone aggregates similarly appear

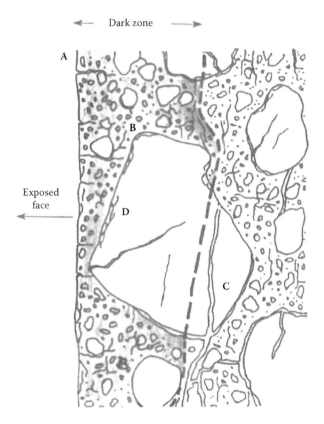

Figure 6.151 Cracking produced in a concrete with a flint coarse aggregate, which had been subjected to a fire with temperatures at the surface estimated to be between 300°C and 600°C. Note the 4 mm wide darkened zone where the cement paste has partly dehydrated. A, shrinkage cracks perpendicular to the outer surface. B, irregular cracking within the dark zone. C, cracking subparallel to the surface cutting both coarse aggregate and cement paste. D, cracking within the coarse aggregate due to thermal mismatch between the expansion of the aggregate and matrix. (Drawn from a thin section provided by courtesy of STATS Limited, now part of RSK Group plc.)

to accommodate thermal movement well, with the majority of cracking confined to the cement paste and the margins of fine aggregate particles, as shown in Figure 6.152. A general feature of this microcracking is the tendency to develop as a series of subparallel fractures parallel to the original heated surface. Locally, fracture orientations may be controlled by local modifications to the stress field in the concrete, for example, in areas close to air voids.

Although mineralogical changes are difficult to observe directly with the petrological microscope, the normally featureless cement paste begins to show a patchy anisotropy with pale yellow-beige birefringent colours at exposure temperatures over about 450°C (Riley 1991). The more or less complete development of this birefringence in the paste has been taken as an indication that the affected concrete has been exposed to temperatures in excess of 500°C. The birefringence indicates that the calcium silicate hydrate gels and the calcium hydroxide have developed significant amounts of new crystalline calcium silicate and calcium oxide phases as they have become dehydrated. A possible reason for this birefringence is the development of a 9 Å calcium silicate hydrate, known as riversideite (Heller and Taylor 1956), as the temperature is not sufficient to form β-wollastonite

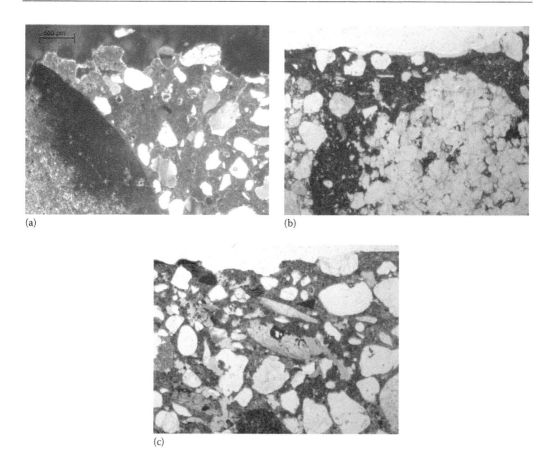

(a)

(b)

(c)

Figure 6.152 View of features observed in fire-damaged concrete, (a) a heated chert particle is shown red/ brown owing to mineralogical changes, (b) and (c) microcracking and voiding shown (yellow in [b] and blue in [c]) associated with thermal mismatch between the expansion of the aggregate matrix. (Courtesy of Paul Bennett-Hughes, RSK Environment Ltd.)

(Urabe 1990). However, the birefringence of riversideite is not large enough to give a yellow-beige colour under crossed polars, so that other minerals must be involved. XRD analyses confirm the disappearance of calcium hydroxide and the development of calcium silicates at around this temperature.

In 2011, a major pre-stressed concrete motorway bridge near London was subject to a fire caused by arson in a scrap yard beneath the bridge deck. The fire led to the busy motorway being closed for the following several days, before being partially reopened the following week, after the structure had been fully propped while an investigation was carried out (Figure 6.153).

The effects of the fire were obvious to see, with large sections of the pre-stressed beams that comprised the span of the bridge spalled and cracked, and, in some places, the pre-stressing reinforcement tendons were exposed. The M-beams were the main focus of the investigation, but the end beams and cantilever beams were also tested and examined. The M-beams comprised two layers of 7 mm diameter pre-stressed tendons in the lower portion of the concrete. Any significant damage to these tendons may have necessitated significant repairs and possibly demolition and reconstruction of parts of the structure.

(a)

(b)

Figure 6.153 Concrete motorway bridge near London affected by a major fire beneath. (a) General view. (Reproduced with the permission of PA Photos Limited, London.) (b) Portion of spalled soffit after propping. (Courtesy of Paul Bennett-Hughes, RSK Environment Ltd.)

A full site and follow-up laboratory programme was prepared to assess the condition of the bridge including:

- A geotechnical investigation of the ground beneath the props;
- A full visual, tapping and rebound (Schmidt) hammer survey of the extent of the damage to the concrete;
- Low-power and high-power microscopical examination of cores taken through the various concrete elements to determine the depth and condition of the effects of heating;
- Site and laboratory testing of the load capacity and characteristics of the steel reinforcement.

The Schmidt hammer survey was used to determine the extent of the damage beyond the spalled sections of the pre-stressed beams. This survey of the unspalled and non-hollow areas of the soffit of the M-beams suggested that the effects of the fire on the surface of the concrete extended beyond the visual and initial tapping survey evidence of damage. The relative rebound values were then mapped using Surfer software to show the degree of damage (Figure 6.154).

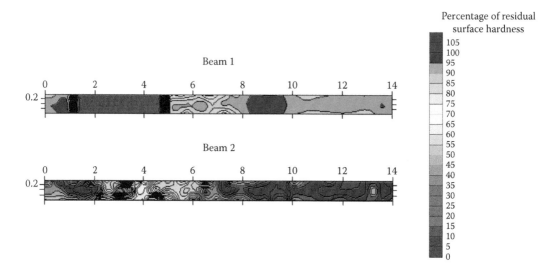

Figure 6.154 Fire damage investigation: example plots of relative damage to two beams of a motorway bridge near London, using a rebound hammer survey. (Courtesy of Paul Bennett-Hughes, RSK Environment Ltd.)

Cores and lump samples were taken from the M-beams in order to assess petrographically the depth of the damage caused by heating. The petrographic findings suggested that, in areas where the tendons had not been exposed by spalling, the effects of heating had not affected the concrete immediately adjacent to the lower layer of the tendons. The visual survey and the petrographic examinations suggested that in some places, the original surface of the beams had been subject to temperatures in excess of 600°C. Large proportions of the beams, the edge beam and the cantilever slab had been heated in excess of 300°C. An example summary of the petrographic examination of one of the cores from a beam is shown in Figure 6.155.

Site and laboratory analyses of the steel tendons also determined the residual stress within the beams exhibiting varying degrees of surface damage and the visual effects of heating. These investigations, including a petrographic assessment of the affected concrete units, helped the structural engineers to determine that the damaged beams could be safely repaired, so that the important structure would eventually be returned to an unrestricted service.

6.10.3.2 Other methods of investigation

A thermoluminescence test for fire-damaged concrete has been proposed by Placido (1980, also Placido and Smith 1981) and further developed by Chew (1988). This technique involves determining the residual thermoluminescence of sand extracted by acid from samples of fire-damaged concrete. It is claimed the method is simple and objective, but it seems to be rarely applied in practice, possibly because of its need for specialised equipment and operators (Concrete Society 2008); however, Smart (1999) considered that 'in special circumstances, thermoluminescence is invaluable, despite the expense'.

Harmathy (1968) investigated the use of thermogravimetric analysis and thermal dilatometry to determine the temperature history of concrete following fire exposures. Small powdered samples excluding as much aggregate as possible are used for the thermogravimetric measurements, and prisms of concrete cut parallel to the surface are measured for thermal dilation. All samples must be obtained within 1 or 2 days after the fire occurred, as rehydration of some of the components in the concrete is possible. It is claimed that fairly

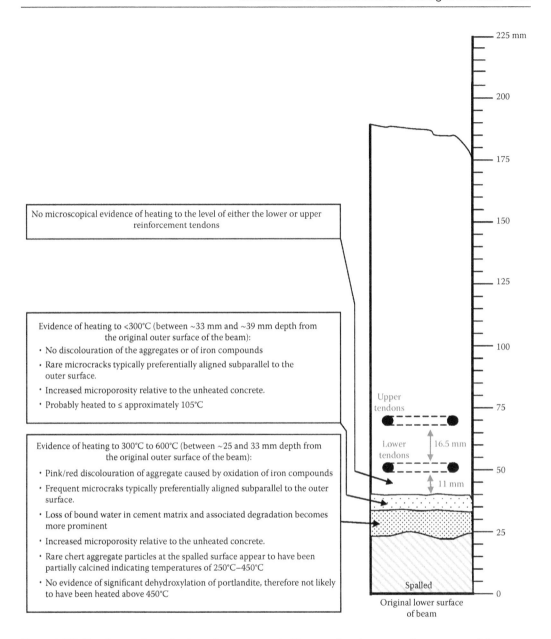

Figure 6.155 **Fire damage investigation of a motorway bridge near London: example summary of the petrographic examination of a core from a beam soffit, showing spalling to a depth of about 25 mm, then heating to up to 600°C for another 5–10 mm and then heating to less than 300°C for the next 5 mm; but both layers of pre-stressing tendons are contained within non-heat-affected concrete. (Courtesy of Paul Bennett-Hughes, RSK Environment Ltd.)**

accurate temperature estimates can be obtained in this way, even without carrying out tests on a reference specimen.

The earliest stages of thermal cracking in concrete may be observed directly using SEM of polished surfaces. Semi-quantitative comparisons with unaffected concretes from the same structural element may be accomplished using a traverse crack counting method (see Chapter 2) or by measuring the strain energy loss resulting from low-stress cycling tests.

REFERENCES

Addis, B. and Owens, G. (eds.). 2001: *Fulton's Concrete Technology*, 8th ed. Cement & Concrete Institute, Midrand, South Africa. CD-Rom version. See also Fulton.

Al-Adeeb, A.M. and Matti, M.A. 1984: Leaching and corrosion of asbestos cement pipes. *The International Journal of Cement Composites and Lightweight Concrete* 6(4), 233–240.

Al-Amoudi, O.S.B., Maslehuddin, M. and Saadi, M.M. 1995: Effect of magnesium sulfate and sodium sulfate on the durability performance of plain and blended cements. *ACI Materials Journal* 92(1), 15–24.

Alexander, M., Addis, B. and Basson, J. 1994: Case studies using a novel method to assess aggressiveness of waters to concrete. *ACI Materials Journal* 91(2), 188–96.

Alexander, S. 2002a: Understanding shrinkage and its effects: Part 1. *Concrete* 36(9), 61–63.

Alexander, S. 2002b: Understanding shrinkage and its effects: Part 2. *Concrete* 36(10), 38–41.

Al-Hashimi, W.S. and Hemingway, J.E. 1973: Recent dedolomitization and the origin of the rusty crusts of Northumberland. *Journal of Sedimentary Petrology* 43, 82–91.

Al-Kadhimi, T.K.H. and Banfill, P.G.F. 1996: The effect of electrochemical re-alkalisation on alkali-silica expansion in concrete. In *Proceedings of the 10th International Conference on Alkali-aggregate Reaction in Concrete*, Shayan, A. (ed.). Melbourne, Victoria, Australia, pp. 637–644.

Alvarez, M.G. and Galvele, J.R. 1984: The mechanism of pitting of high purity iron in NaCl solution. *Corrosion Science* 24(1), 27–48.

American Concrete Institute. 1980: *ACI SP-65 Symposium on Performance of Concrete in Marine Environment*, St. Andrews, New Brunswick, Canada, ACI.

American Concrete Institute. 1984: Causes, evaluation, and repair of cracks in concrete. ACI Committee 224. *ACI Materials Journal* 81(3), 211–230.

American Concrete Institute. 1994: Guide to the use of waterproofing, damp-proofing, protective and decorative barrier systems for concrete, ACI Committee 515, Report 515.1R-94, Detroit, MI.

American Public Health Association. 1992: *Standard Methods for the Examination of Water and Waste Water*, 18th ed. Washington, DC.

Andersen, K.T. and Thaulow, N. 1990: The study of alkali-silica reactions in concrete by the use of fluorescent thin sections. In *Petrography Applied to Concrete and Concrete Aggregates*, Erlin, B. and Stark, D. (eds.). ASTM STP 1061, American Society for Testing and Materials, Philadelphia, PA, pp. 71–89.

Anderson, W.A. 2008: *Foundation Problems and Pyrite Oxidation in the Chattanooga Shale, Estill County, Kentucky*. Report of Investigations 18, Series XII, Kentucky Geological Survey, Lexington, KY.

Asgeirsson, H. 1986: Silica fume in cement and silane for counteracting of alkali-silica reactions in Iceland. *Cement and Concrete Research* 16(3), 423–428.

Ash, J.E. 1972: Bleeding in concrete – A microscopy study. *ACI Journal* 69(4), 209–211.

Association of Consulting Engineers of Ireland. 2012: *Guidance Note for Members on Specification of Unbound Granular Fill Material Supporting Concrete Floors* (with particular reference to the pyrite issue). ACEI, Dublin, Ireland, 6pp.

ASTM C227-10, 2010: *Standard Test Method for Potential Alkali Reactivity of Cement-Aggregate Combinations (Mortar-Bar Method)*. American Society for Testing and Materials, West Conshohocken, PA.

ASTM C289-07, 2007: *Standard Test Method for Potential Alkali-Silica Reactivity of Aggregates (Chemical Method)*. American Society for Testing and Materials, West Conshohocken, PA.

ASTM C441-05, 2005: *Standard Test Method for Effectiveness of Mineral admixtures or Ground Blast-Furnace Slag in Preventing Excessive of Concrete Due to the Alkali-Silica Reaction*. American Society for Testing and Materials, West Conshohocken, PA.

ASTM C586-11, 2011: *Standard Test Method for Potential Alkali Reactivity of Carbonate Rocks as Concrete Aggregates (Rock-Cylinder Method)*. American Society for Testing and Materials, West Conshohocken, PA.

ASTM C823-95, 1995: *Standard Practice for Examination and Sampling of Hardened Concrete in Constructions*. American Society for Testing and Materials, Philadelphia, PA.

ASTM C856-11, 2011: *Standard Practice for Petrographic Examination of Hardened Concrete*. American Society for Testing and Materials, West Conshohocken, PA.

ASTM Cll05-08, 2008: *Test Method for Length Change of Concrete Due to Alkali-Carbonate Rock Reaction*. American Society for Testing and Materials, West Conshohocken, PA.

ASTM C1260-07, 2007: *Standard Test Method for Potential Alkali Reactivity of Aggregates (Mortar-Bar Method)*. American Society for Testing and Materials, West Conshohocken, PA.

Babushkin, V.I., Matveyev, G.M. and Mchedlov-Petrossyan, O.P. 1985: *Thermodynamics of Silicates*. Springler-Verlag, Berlin, Germany, 459pp.

Bager, D.H. and Sellevold, E.J. 1986: Ice formation in hardened cement paste. Part I Room temperature cured pastes with variable moisture contents, & Part 11 – Drying and resaturation of room temperature cured pastes. *Cement and Concrete Research* 16(5), 709–720; and 16(6), 835–844.

Bager, D.H. and Sellevold, E.J. 1987: Ice formation in hardened cement paste. Part III Slow resaturation of room temperature cured pastes. *Cement and Concrete Research* 17(1), 1–11.

Ballim, Y. and Alexander, M.G. 1991: Carbonic acid attack of Portland cement-based concretes. In *Proceedings of the 5th International Conference on the Durability of Materials and Components*, Brighton, U.K., E. & F. Spon, pp. 179–184.

Ballivy, G., Rivard, P., Pépin, C., Tanguay, M.G. and Dion, A. 2002: Damages to residential buildings related to pyritic rockfills: Field results of an investigation on the south shore of Montreal, Québec, Canada. *Canadian Journal of Civil Engineering* 29, 246–255.

Bamforth, P.B., Price, W.F. and Emerson, M. 1997: *International Review of Chloride Ingress into Structural Concrete*. TRL Contractor Report 359, Transport Research Laboratory, Edinburgh, U.K.

Barnard, J.L. 1967: *Corrosion of Sewers*. CSIR Research Report, Pretoria, South Africa, pp. 1–16.

Barnett, S.J. 1998: X-Ray powder diffraction studies of ettringite and related systems, PhD thesis, Staffordshire University, Staffordshire, U.K.

Barnett, S.J., Macphee, D.E., Lachowski, E.E. and Crammond, N.J. 2002: XRD, EDX and IR analysis of solid solutions between thaumasite and ettringite. *Cement and Concrete Research* 32(5), 719–730.

Bates, R.C. 1964: *Determination of pH – Theory and Practice*. Wiley, New York.

Bensted, J. 1999: Thaumasite – Background and nature in deterioration of cements, mortars and concretes. *Cement and Concrete Composites* 21(2), 117–121.

Bensted, J., Brough, A.R. and Page, M.M. 2007: Chemical degradation of concrete. In *Durability of Concrete and Cement Composites*, Page, C.L. and Page, M.M. (eds.). Woodhead and Maney (pp Institute of Materials, Minerals & Mining), Cambridge, U.K., Chapter 4, pp. 86–135.

Bensted, J. and Varma, S.P. 1971: Studies of ettringite and its derivatives. *Cement Technology* 73–76.

Bensted, J. and Varma, S.P. 1974: Studies of thaumasite. *Silicate Industries* 39, 11–19.

Bérard, J. and Roux, R. 1986: La viabilité des bétons du Québec: le role des granulats (The sustainability of concretes in Quebec: the role of aggregates). *Canadian Journal of Civil Engineering* 13, 12–24 (in French).

Berra, M. and Baronio, G. 1987: Thaumasite in deteriorated concretes in the presence of sulfates. In *Concrete Durability: Katharine and Bryant Mather International Conference*, Scanlon, J.M. (ed.). ACI SP-100, Paper SPI00-I06. American Concrete Institute, Detroit, MI, pp. 2073–2089.

Berra, M., Di Maggio, R., Mangialardi, T. and Paolini, A.E. 1995: Fused quartz as a reference reactive aggregate for alkali-silica reaction studies. *Advances in Cement Research* 7(25), 21–32.

Bérubé, M.A., Chouinard, D., Boisvert, L., Frenette, J. and Pigeon, M. 1996a: Influence of wetting-drying and freezing-thawing cycles, and effectiveness of sealers on ASR. In *Proceedings of the 19th International Conference on Alkali-Aggregate Reaction in Concrete*, Shayan, A. (ed.)., Melbourne, Victoria, Australia, pp. 1056–1063.

Bérubé, M.A., Duchesne, J. and Rivest, M. 1996b: Alkali contribution by aggregates to concrete. In *Proceedings of the 10th International Conference on Alkali-Aggregate Reaction in Concrete*, Shayan, A. (ed.), Melbourne, Victoria, Australia, pp. 899–906.

Bérubé, M.A. and Fournier, B. 1986: The products of alkali-silica reaction in concrete: Case study of Québec region. *Canadian Mineralogist* 24, 271–288 (in French).

Bérubé, M.A. and Fournier, B. 2004: Alkalis releasable by aggregates in concrete: Significance and test methods, In *Proceedings of the 12th International Conference on Alkali-Aggregate Reaction in Concrete*, Tang, M.S. and Deng, M. (eds.), Vol. 1, Beijing, China, International Academic Publishers/World Publishing Corporation, pp. 17–30.

Bérubé, M.A., Fournier, B. and Côté, T. 2012: Using the damage rating index for assessing the expansion of concrete affected by freeze-thaw, sulphate attack, or ASR. In *Proceedings of the 14th International Conference on Alkali-Aggregate Reaction*, Folliard, K.J. (ed.), Austin, TX. (proceedings available on a memory stick).

Bessey, G.E. 1950: *Investigations into Building Fires. Part 2: The Visible changes in Concrete or Mortar Exposed to High Temperatures*. National Building Studies Technical Paper No 18, HMSO, London, U.K., pp. 6–18.

Bickley, J.A., Hemmings, R.T., Hooton, R.D. and Balinksi, J. 1994: Thaumasite related deterioration of concrete structures. In *Proceedings of V.M Malhotra Symposium, Concrete Technology Past, Present and Future*, Detroit, MI, ACI SP144, American Concrete Institute, pp. 159–175.

Biczoc, I. 1967: *Concrete Corrosion and Concrete Protection*. Chemical Publishing Co., New York.

Blackwell, B.Q. and Pettifer, K. 1992: Alkali reactivity of greywacke aggregates in Maentwrog Dam (North Wales). *Magazine of Concrete Research* 44(161), 255–264.

Blaschke, R. 1987: Natural building stone damaged by slime and acid producing microbes. In *Proceedings of the Ninth International Conference on Cement Microscopy*, pp. 70–81.

Blight, G.E. and Alexander, M.G. 2011: *Alkali-Aggregate Reaction and Structural Damage to Concrete: Engineering Assessment, Repair and Management*. CRC Press, London, U.K., 250pp.

Bloom, M.C. and Goldberg, L. 1965: Fe_2O_3 and the passivity of iron. *Corrosion Science* 8(9), 623–630.

Bock, E. and Sand, W. 1986: Applied electron microscopy on the biogenic destruction of concrete blocks – Use of the transmission electron microscope for identification of mineral acid producing bacteria. In *Proceedings of the Eighth International Conference on Cement Microscopy*, pp. 285–302.

Bonen, D. and Cohen, M.D. 1992: Magnesium sulfate attack on Portland cement paste – II. Chemical and mineralogical analyses. *Cement and Concrete Research* 22(4), 707–718.

British Cement Association. 1992a: *Alkali-Aggregate Reaction 1977–1992,a Definitive Bibliography*. Ref: CIS/B8, BCA, Slough (later Crowthorne), U.K. See also Lee & Clowes.

British Cement Association. 1992b: *The Diagnosis of Alkali-Silica Reaction*. Report of a Working Party, 2nd ed. Ref: 45.042, BCA, Wexham Springs, U.K. See also Palmer, D.

British Cement Association. 1999: *Testing Protocol for Greywacke Aggregates (Protocol of the BSI B/517/1/20 ad hoc group on ASR)*. Ref 45.044, BCA, Crowthorne, U.K.

Broekmans, M.A.T.M. 2002: The alkali-silica reaction: Mineralogical and geochemical aspects of some Dutch concretes and Norwegian mylonites, PhD thesis, University of Utrecht, Geologica Ultraiectina, Utrecht, the Netherlands, Vol. 217, 144pp.

Broekmans, M.A.T.M., Fernandes, I. and Nixon, P.J. 15–19 September 2009: A global petrographic atlas of alkali-silica reactive rock types: A brief review. In *Proceedings of the 12th Euroseminar on Microscopy Applied to Building Materials*, Dortmund, Germany, pp. 39–50.

Bromley, A.V. 2002: *A Compendium of Concrete Aggregates Used in Southwest England*. Privately sponsored, Petrolab, Redruth (now Falmouth), Cornwall, U.K.

Bromley, A.V. and Pettifer, K. 1997: *Sulfide-Related Degradation of Concrete in Southwest England*. BRE Report, BR325, Building Research Establishment/CRC, Watford, U.K.

Bromley, A.V. and Sibbick, R.G. 1999: The accelerated degradation of concrete in Southwest England. In *Proceedings of the Euroseminar on Microscopy Applied to Building Materials*, Delft, Netherlands, pp. 81–90.

Brooks, J. 2003: Elasticity, shrinkage, creep and thermal movement. In *Advanced Concrete Technology – Concrete Properties*, Newman, J. and Seng Choo, B. (eds.). Elsevier, Oxford, U.K., Chapter 7, pp. 7/1–7/18.

Broomfield, J.P. and Atkins, C. 2007: Corrosion of steel in concrete (Repair Guidance Note 1). *Concrete* 41(4), 13–14.

Brown, P.W. and Bothe, J.V. 1993: The stability of ettringite. *Advances in Cement Research* 5(18), 47–63.

BS 812-104, 1994: *Testing Aggregates, Part 104, Method for Qualitative and Quantitative Petrographic Examination of Aggregates*. British Standards Institution (BSI), London, U.K.

BS 812-120, 1989: *Testing Aggregates, Part 120, Method for Testing and Classifying Drying Shrinkage of Aggregates in Concrete*. British Standards Institution (BSI), London, U.K.

BS 812-123, 1999: *Testing Aggregates, Part 123, Method for Determination of Alkali-Silica Reactivity, Concrete Prism Method*. British Standards Institution (BSI), London, U.K.

BS 5328, 1990: *Concrete, Part 1: Guide to Specifying Concrete*. British Standards Institution (BSI), London, U.K. See also BS EN 206-1 & BS 8500-1 & 2.

BS 7943, 1999: *Guide to the Interpretation of Petrographical Examinations for Alkali-Silica Reactivity*. British Standards Institution (BSI), London, U.K.

BS 8110, 1997: *Structural Use of Concrete, Part 1, Code of Practice for Design and Construction*. British Standards Institution (BSI), London, U.K. See also BS EN 1992-1-1.

BS 8500-1, 2006 (ind. A1: 2012): *Concrete, Complementary British Standard to BS EN 206-1, Part 1: Method of Specifying and Guidance for the Specifier*. British Standards Institution (BSI), London, U.K.

BS 8500-2, 2006 (ind. A1: 2012): *Concrete, Complementary British Standard to BS EN 206-1, Part 2: Specification for Constituent Materials and Concrete*. British Standards Institution (BSI), London, U.K.

BS EN 206-1, 2000: *Concrete, Part 1: Specification, Performance, Production and Conformity*. British Standards Institution (BSI), London, U.K. Incorporates A1:2004 & A2:2005.

BS EN 1367-4, 1998: *Tests for Thermal and Weathering Properties of Aggregates, Part 4, Determination of Drying Shrinkage*. British Standards Institution (BSI), London, U.K.

BS EN 1992-1-1, 2004: *Eurocode 2: Design of Concrete Structures, Part 1-1: General Rules and Rules for Buildings*. British Standards Institution (BSI), London, U.K. Incorporates corrigendum of January 2008.

BS EN 12620, 2002: *Aggregates for Concrete*. British Standards Institution (BSI), London, U.K. Incorporates corrigendum of May 2004 & A1:2008. See also national annex, BS PD 6682-1.

BS PD 6682-1, 2009: *Aggregates, Aggregates for Concrete, Guidance on Use of BS EN 12620*. British Standards Institution (BSI), London, U.K.

Buck, A.D. and Mather, K. 1978: Alkali-silica reaction products from several concretes: Optical, chemical and X-ray diffraction data. In *Proceedings of the Fourth International Conference on the Effects of Alkalies in Cement and Concrete*, West Lafayette, IN, Purdue University, pp. 73–85.

Buenfeld, N.R. and Newman, J.B. 1984: The permeability of concrete in a marine environment. *Magazine of Concrete Research* 36(127), 67–80.

Buenfeld, N.R. and Newman, J.B. 1986: The development and stability of surface layers of concrete exposed to sea-water. *Cement and Concrete Research* 16(5), 721–732.

Building Research Establishment. 1978: *Carbonation of Concrete Made with Dense Natural Aggregates*. Information Sheet, IS 14/78, Building Research Station, Watford, U.K., 3pp.

Building Research Establishment. 1991a: *Shrinkage of Natural Aggregates in Concrete*. Digest 357, BRE, Watford, U.K. Replaced BRS Digest 35: see Building Research Station.

Building Research Establishment. 1991b: *Sulfate and Acid Resistance of Concrete in Ground*. Digest 363, BRE, Watford, U.K. Replaced by BRE Special Digest 1.

Building Research Establishment. 1997 and 1999: *Alkali-Silica Reaction in Concrete*. Digest 330, Parts 1 to 4 (replacing 1988 edition), 1997 and 1999 editions, BRE, Watford, U.K.

Building Research Establishment. 2001: *Concrete in Aggressive Ground*. Special Digest 1, 1st ed., BRE, Watford, U.K. Replaced BRE Digest 263.

Building Research Establishment. 2002: *Minimising the Risk of Alkali-Silica Reaction: Alternative Methods*. BRE IP1/02, BRE, Watford, U.K.

Building Research Establishment. 2003: *Concrete in Aggressive Ground*. Special Digest 1, 2nd ed., BRE, Watford, U.K.

Building Research Establishment. 2004: *Alkali-Silica Reaction in Concrete*. Digest 330, Parts 1 to 4, current 2004 edition, BRE, Watford, U.K.

Building Research Establishment. 2005: *Concrete in Aggressive Ground*. Special Digest 1, 3rd & current ed., BRE, Watford, U.K.

Building Research Station. 1963: *Shrinkage of Natural Aggregates in Concrete*. BRS Digest 35, HMSO, London, U.K. See also later BRE Digest 35.

Building Research Station. 1968: *Shrinkage of Natural Aggregates in Concrete*. BRS Digest 35, 2nd ed., HMSO, London, U.K. See also later BRE Digest 35.

Burrows, R.W. January 2007: The life and death of Type II cement, *Concrete Construction*, 3pp. (download from http://www.concreteconstruction.net/concrete-construction).

Cahan, B.D. and Chen, C.-T. 1982: The nature of the passive film on iron, III. The chemi-conductor model and further supporting evidence. *Journal of the Electrochemical Society* 129(5), 921–925.

Castillo, C. and Durrani, A.J. 1990: Effect of transient high temperature on high-strength concrete. *ACI Materials Journal* 87(1), 47–53.

Cather, R. 2000: Concrete in chloride environments. *Concrete* 34(6), 42–43.

Cather, R. 2003: Concrete and fire exposure. In *Advanced Concrete Technology – Concrete Properties*, Newman, J. and Seng Choo, B. (eds.). Elsevier, Oxford, U.K., Chapter 10, pp. 10/1–10/13.

Cement Association of Japan. 1971: *Qualitative Analysis of Quartz (Phosphoric Acid Method)*. Standard Testing Method of the Association, CAJS 1-31, CAJ, Tokyo, Japan, pp. 103–119, (in Japanese).

Cement Association of Japan. 1996: *Quantitative Measurement of Quartz (Phosphoric Acid Method)*. Standard Testing Method of the Association, CAJS 1-31, CAJ, Tokyo, Japan, 11pp., (in Japanese).

Cement and Concrete Association of New Zealand. 1991: *Alkali-Aggregate Reaction. Minimising the Risk of Damage to Concrete: Guidance Notes and Model Specification Clauses*. Cement and Concrete Association Report No TR3. See also Munn 2003 for 2nd edition.

Chana, P. and Price, W. 2003: The Cardington test fire. *Concrete* 37(1), 28–33.

Chandra, S. 1988: Hydrochloric acid attack on cement mortar – An analytical study. *Cement and Concrete Research* 18(2), 193–203.

Charlwood, R. 2007: *Special Workshop on Chemical Expansion of Concrete in Dams and Hydro-Electric Projects*. ICOLD, Granada, Spain – http://www.dam-research.org/Granada-2007/index.html. Check that refs in text give date correctly as 2007 (& not 2008).

Charlwood, R. and Scrivener, K.L. 29 May to 3 June 2011: Expanding concrete in dams – Long term challenges, presented at the *79th ICOLD Annual Meeting*, Lucerne, Switzerland.

Charlwood, R., Scrivener, K.L. and Sims, I. October 2012: Recent developments in the management of chemical expansion of concrete in dams and hydro projects – Part 1: Existing structures, presented at *Hydro 2012: Innovative Approaches to Global Challenges*, Bilbao, Spain, 17pp.

Chatterji, S., Thaulow, N. and Jensen, A.D. 1987a: Studies of alkali-silica reaction. Part 4: Effect of different alkali salt solutions on expansion. *Cement and Concrete Research*. 17(5), 544–552.

Chatterji, S., Thaulow, N., Jensen, A.D. and Christensen, P. 1987b: Mechanism of accelerating effects of NaCl and Ca(OH)$_2$ on alkali-silica reaction. In *Proceedings of the Seventh International Conference on Alkali-Aggregate Reactions in Concrete*, Grattan-Bellew, P.E. (ed.), Ottawa, Ontario, Canada, pp. 115–119.

Chen, Z.Y. and Wang, J.G. 1987: Bond between marble and cement paste. *Cement and Concrete Research* 17, 544–552.

Chew, M.Y.L. 1988: Assessing heated concrete and masonry with thermoluminescence. *ACI Materials Journal* 85(6), 537–543.

Chrisp, T.M., Waldron, P. and Wood, J.G.M. 1994: Development of a non-destructive test to quantify damage in deteriorated concrete. *Magazine of Concrete Research* 45(165), 247–256.

CIRIA. 2007: *Early-Age Thermal Crack Control in Concrete*. Report C660, Construction Industry Research and Information Association, London, U.K.

Cocke, D.L. 1990: The binding chemistry and leaching mechanisms of hazardous substances in cementitious solidification/stabilisation systems. *Journal of Hazardous Materials* 24, 231–253. Check referred to in text.

Cohen, M.D. and Bentur, A. 1988: Durability of Portland cement-silica fume pastes in magnesium sulfate and sodium sulfate solutions. *ACI Materials Journal* 85(3), 148–157.

Cohen, M.D. and Mather, B. 1991: Sulfate attack on concrete – Research needs. *ACI Materials Journal* 88(1), 62–69.

Cole, W.F. and Beresford, F.D. 1980: Influence of basalt aggregate on concrete durability. In *Proceedings of the First International Conference on Durability of Building Materials and Components*, Sereda, P.J. and Litvan, G.G. (eds.), Ottawa, 1978, Philadelphia, PA, ASTM STP 691, American Society for Testing and Materials, pp. 617–628.

Collepardi, M. 2003: A state of the art review on delayed ettringite attack on concrete. *Cement and Concrete Composites* 25, 401–407.

Collepardi, M. and Monosi, S. 1992: Effect of the carbonation process on the concrete deterioration by CaCl$_2$ aggression. In *Proceedings of the Ninth International Congress on the Chemistry of Cement*, New Delhi, India, pp. 389–395.

Collett, G., Crammond, N.J., Swamy, R.N., and Sharp, J.H. 2004: The role of carbon dioxide in the formation of thaumasite. *Cement and Concrete Research*, 34(9), 1599–1612.

Collins, A.R. 1944: The destruction of concrete by frost. *Journal of the Institution of Civil Engineers* 23(1), 29–41.

Collins, R.J. and Bareham, P.D. 1987: Alkali-silica reaction: Suppression of expansion using porous aggregate. *Cement and Concrete Research* 17(1), 89–96.

Comité Euro-International du Beton. 1992: *Durable Concrete Structures, Design Guide*. Thomas Telford, London, U.K.

Commission of the European Communities Council. 1993: *Directive Draft of the Landfill of Wastes*. SYN 335, Brussels, WI.

Commission Internationale des Grands Barrages. 1989: *Exposure of Dam Concrete to Special Aggressive Waters: Guidelines*. Bulletin 71, CIGB/ICOLD, Paris, France. See also ICOLD.

Concrete Society. 1990: *Assessment and Repair of Fire-Damaged Structures*. Concrete Society Technical Report No 33, The Concrete Society, Crowthorne, U.K.

Concrete Society. 1992: *Non-Structural Cracks in Concrete*. Report of a Concrete Society Working Party. Concrete Society Technical Report No 22, 3rd ed., The Concrete Society, Crowthorne, U.K. Replaced by 2010 edition.

Concrete Society. 1999: *Alkali-Silica Reaction: Minimising the Risk of Damage to Concrete, Guidance Notes and Model Specification Clauses*. Report of a Concrete Society Working Party, Concrete Society Technical Report No 30, 3rd ed., The Concrete Society, Slough, U.K.

Concrete Society. 2008a: *Movement, Restraint and Cracking in Concrete Structures*. Concrete Society Technical Report No 67, The Concrete Society, Camberley, U.K.

Concrete Society. 2008b: *Assessment, Design and Repair of Fire-Damaged Concrete Structures*. Report of a Concrete Society Working Group, Concrete Society Technical Report No 68, The Concrete Society, Camberley, U.K.

Concrete Society. 2010: *Non-Structural Cracks in Concrete*. Report of a Concrete Society Working Party, Concrete Society Technical Report No 22, 4th & current edition, The Concrete Society, Camberley, U.K.

Conjeaud, M.L. 1980: Mechanism of sea water attack on cement mortar. In *Performance of Concrete in Marine Environment*, ACI SP-65, 39–61, American Concrete Institute, Detroit, MI.

Connor, J.R. 1990: *Chemical Fixation and Solidification of Hazardous Waste Forms*. Van Nostrand Reinhold, New York. Check referred to in text.

Costa, A. and Appleton, J. 1999: Chloride penetration into concrete in marine environment – Part 1: Main parameters affecting chloride penetration. *Materials and Structures* 32(218), 252–259.

Coutinho, A de S. 1979: Aspects of sulfate attack on concrete. *Cement, Concrete and Aggregates* 1(1), 10–12.

Coutts, R.S.P. 1983: Autoclaved beaten wood fibre-reinforced cement composites. *Composites* 18(2), 139–143.

Cowie, J. and Glasser, F.P. 1991/1992: The reaction between cement and natural waters containing dissolved carbon dioxide. *Advances in Cement Research* 4(15), 119–134.

Crammond, N.J. 1984: Examination of mortar bars containing varying percentages of coarsely crystalline gypsum as aggregate. *Cement and Concrete Research* 14(2), 225–230. (and personal communication 1990).

Crammond, N.J. 1985a: Quantitative x-ray diffraction analysis of ettringite, thaumasite and gypsum in concretes and mortars. *Cement and Concrete Research* 15(3), 431–441.

Crammond, N.J. 1985b: Thaumasite in failed cement mortars and renders from exposed brickwork. *Cement and Concrete Research* 15(6), 1039–1050.

Crammond, N.J. 1990: Personal communication. See Crammond (1984).

Crammond, N.J. June 2002: The thaumasite form of sulfate attack in the UK. In *Proceedings Issued to Delegates on CD-ROM The First International Conference on Thaumasite in Cementitious Materials*, Watford, U.K., Building Research Establishment.

Crammond, N.J. and Halliwell, M.A. 1995: The thaumasite form of sulfate attack in concretes containing a source of carbonate ions: A microstructural overview. In *Second Symposium on Advances in Concrete Technology*. Malhotra, V.M. (ed.). ACI SP 154, Paper SP 154–19, 357–80. American Concrete Institute, Detroit, MI.

Criaud, A., Vernet, C. and Defosse, C. 1992: A rapid test for detecting the reactivity of aggregates: the microbar method. In *Proceedings of the Ninth International Conference on Alkali-Aggregate Reaction in Concrete,* Poole, A.B. (ed.), London, U.K., pp. 201–209.

Crumpton, C.F. 1988: Lightweight aggregate concrete sometimes grows. Blame AAR if it's the most likely cause. *Concrete Construction* 33(6), 618–619 and 33(12), 1103–1105.

CSA A23.1-94. 1994: *Concrete Materials and Methods of Concrete Construction – Appendix B, Alkali-Aggregate Reaction.* Canadian Standards Association, Mississauga, Ontario, Canada, pp. 112–135.

CSA A.23.2-14A-09. 2009: *Methods of Test for Concrete, Potential Expansivity of Aggregates (Procedure for Length Change Due to Alkali-Aggregate Reaction in Concrete Prisms).* Canadian Standards Association, Mississauga, Ontario, Canada, pp. 207–216.

CSIR, South Africa. 1967: *Corrosion of Concrete Sewers.* CSIR Research Report 163, Pretoria, South Africa.

Czerewko, M.A., Cripps, J.C., Reid, J.M. and Duffell, C.G. 2003: The effects of storage conditions on the sulphur speciation in geological material. *Quarterly Journal of Engineering Geology and Hydrogeology* 36, 331–342.

Davidovits, J. 1994: Recent processes in concretes for nuclear waste and uranium waste containment. *Concrete International* 16(12), 53–58.

Davies, G. and Oberholster, R.E. 1986: The alkali-silica reaction product. A mineralogical and an electron microscope study. In *Proceedings Eighth International Conference on Cement Microscopy,* pp. 303–326.

Day, R. and Clarke, J. 2003: Plastic and thermal cracking. In *Advanced Concrete Technology – Concrete Properties,* Newman, J. and Seng Choo, B. (eds.). Elsevier, Oxford, U.K., Chapter 2, pp. 2/3–2/17.

Day, R.L. 1992: *The Effect of Secondary Ettringite Formation on the Durability of Concrete: A Literature Analysis.* Research and Development Bulletin, RD108T, Portland Cement Association, Skokie, IL.

De Belie, N. 2010: Microorganisms versus stony materials: A love-hate relationship. *Materials and Structures* 43(9), 1191–1202.

De Ceukelaire, L. 1991: Concrete surface deterioration due to expansion by the formation of jarosite. *Cement and Concrete Research* 21(4), 563–574.

De Ceukelaire, L. 1992a: Alkali-silica reaction in a lightweight concrete bridge. In *Proceedings of the Ninth International Conference on Alkali-Aggregate Reaction in Concrete,* London, U.K., pp. 231–239.

De Ceukelaire, L. 1992b: The effects of hydrochloric acid on mortar. *Cement and Concrete Research* 22(5), 903–914.

DETR. 1999: *The Thaumasite form of Sulfate Attack: Risks, Diagnosis, Remedial Works and Guidance on New Construction.* Report of the Thaumasite Expert Group (TEG), Department for the Environment, Transport and the Regions (DETR), London, U.K.

DETR. 8 May 2000: Thaumasite Expert Group one-year review, website download: http://www.construction.detr.gov.uk/thaumasite/oneyear/index.htm, Department for the Environment, Transport and the Regions (DETR), London, U.K. See also Thaumasite Expert Group.

DETR. March 2002: Thaumasite Expert Group Report: Review after three years experience, website download: www.safety.odpm.gov.uk/bregs/tegreports/index.htm.

Deng, M., Hong, D., Lan, X. and Tang, M. 1995: Mechanism of expansion in hardened cement pastes with hardburnt free lime. *Cement and Concrete Research* 25(2), 440–448.

Deng, M. and Tang, M. 1993: Mechanism of dedolomitization and expansion in dolomitic rocks. *Cement and Concrete Research* 23(6), 1397–1408.

Deng, M. and Tang, M. 1994: Formation and expansion of ettringite crystals. *Cement and Concrete Research* 24(1), 119–126.

Diamond, S. 1975: A review of alkali-silica reaction and expansion mechanisms. I: Alkalies in cements and in concrete pore solutions. *Cement and Concrete Research* 5, 329–345.

Diamond, S. 1989: Another look at mechanisms. In *Proceedings Eighth International Conference on Alkali-Aggregate Reaction,* Okada, K., Nishibayashi, S. and Kawamura, M. (eds.), Kyoto, Japan, pp. 83–94.

Diamond, S. 1992: *Alkali Aggregate Reactions in Concrete: An Annotated Bibliography 1939–1991.* Document SHRP-C/UPW-92-601, Strategic Highway Research Program, National Research Council, Washington, DC.

Diamond, S. 1996a: Delayed ettringite formation – Process and problems. *Cement and Concrete Composites* 18, 205–215.

Diamond, S. 1996b: Alkali silica reactions – Some paradoxes. In *Proceedings of the 10th International Conference on Alkali-Aggregate Reaction in Concrete*, Shayan, A. (ed.), Melbourne, Victoria, Australia, pp. 3–14.

Dolar-Mantuani, L. 1971: Late-expansion alkali-reactive carbonate rocks. Highway Research Record No.353, pp. 1–14.

Dolar-Mantuani, L. 1983: *Handbook of Concrete Aggregates – A Petrographic and Technological Evaluation*. Noyes Publications, Saddle River, NJ, 345pp.

Dron, R. and Brivot, F. 1996: Solid-liquid equilibria in K-C-S-H/H$_2$0 systems. In *Proceedings of the 10th International Conference on Alkali-Aggregate Reaction in Concrete*, Shayan, A. (ed.), Melbourne, Victoria, Australia, pp. 927–933.

Dubberke, W. and Marks, V.J. 1985: The effect of salt on aggregate durability, *Transportation Research Record*, 1031, 27–34. Transportation Research Board, National Research Council, Washington, DC.

Duchesne, J. and Berube, M.A. 1994: The effectiveness of supplementary cementing materials in suppressing expansion due to ASR: Another look at the reaction mechanisms, Part1: Concrete expansion and portlandite depletion, & Part 2: Pore solution chemistry. *Cement and Concrete Research* 24(1), 73–82, and 24(2), 221–230.

Dunbar, P.A., Mukherjee, P.K., Bleszynski, R. and Thomas, M.D.A. 18–23 August 1996: A comparison of Damage Rating Index with long term expansion of concrete prisms due to alkali-silica reaction. In *Proceedings of the 10th International Conference on Alkali-Aggregate Reaction in Concrete*, Shayan, A. (ed.), Melbourne, Victoria, Australia, pp. 324–331.

Duncan, M.A.G., Swenson, E.G. and Gillott, J.E. 1973: Alkali aggregate reaction in Nova Scotia, III Laboratory studies of volume change. *Cement and Concrete Research* 3, 233–245.

Dunstan, E.R. 1982: A spec odyssey – Sulfate resistant concrete for the 1980s. *George Verbeck Symposium on Sulfate Resistant Concrete*. ACI-SP77, American Concrete Institute, Detroit, MI, pp. 41–62.

Durand, B. and Bérard, J. 1987: Use of gel composition as a criterion for diagnosis of alkali-aggregate reactivity in concrete containing siliceous limestone aggregate. *Materials and Structures* 20, 39–43.

Eastwood, C. January 2007: Let's progress, not regress (a rebuttal to The life and death of Type II cement, by R W Burrows: see earlier) *Concrete Construction*, (download with Burrows' article at http://www.concreteconstuction.net/concrete-constuction).

Eden, M.A. 26–28 September 2011: Fire damaged concrete – The potential for on-going deterioration post-fire in concrete heated to temperatures of less than 300°C. In *Proceedings of the Fourth International Conference on Concrete Repair on Concrete Solutions 2011*, Grantham, M., Mechtcherine, V. and Schneck, U. (eds.), Technical University of Dresden, Dresden, Germany/ CRC Press, London, U.K.

Eden, M.A., White, P.S. and Wimpenny, D.E. 5–9 June 2007: A laboratory investigation of concrete with suspected delayed ettringite formation – A case study from a bridge in Malaysia. In *Proceedings of the 11th Euroseminar on Microscopy Applied to Building Materials*, Porto, Portugal.

Edwards, A.G. 1970: *Scottish Aggregates: Rock Constituents and Suitability for Concrete*. BRS Current Paper CP 28/70, Building Research Station, Ministry of Public Buildings and Works, HMSO, London, U.K.

Eglinton, M.S. 1987: *Concrete and its Chemical Behaviour*. Thomas Telford, London, U.K.

ENDS. February 1992: *Environmental News Data Services*. Report 205, pp. 11–33.

Erlin, B. and Stark, D.C. 1965/1966: Identification and occurrence of thaumasite in concrete (a discussion for the 1965 Highway Research Board Symposium on aggressive fluids). *Highway Research Record* 113, 108–113.

Evamy, B.D. 1967: Delolomitization and the development of rhombohedral pores in limestones. *Journal of Sedimentary Petrology* 37, 1204–1215.

Fairbanks, E.E. 1925: A modification of Lemberg's staining method. *American Mineralogist* 10, 126–127.

Fairhall, G.A. 1989: Effect of process operational variables on the product properties of encapsu-lated intermediate level wastes. In *Proceedings of the International Conference Organised by the British Nuclear Energy Society on Radioactive Waste Management 2*, London, U.K., Thomas Telford, pp. 79–84.

Fairhall, G.A. 1 April 1994: The treatment and encapsulation of intermediate level wastes at Sellafield. In *Proceedings of the Fourth International Conference on Nuclear Fuel Reprocessing and Waste Management Reconditioning*, British Nuclear Industrial Forum.

Fatttuhi, N.I. and Hughes, R.P. 1986: Resistance to acid attack of concrete with different admixtures or coatings. *The International Journal of Cement Composites and Lightweight Concrete* 8(4), 223–230.

Fernandes, I., Broekmans, M.A.T.M., Nixon, P.J., Sims, I., Ribeiro, M.D.A., Noronha, F. and Wigum, B.J. 20–25 May 2012: Alkali-silica reactivity of some common rock types – A global petrographic atlas. In *Proceedings of the 14th International Conference on Alkali-Aggregate Reaction*, Folliard, K.J. (ed.), Austin, TX.

Fernandes, I., Ribeiro, D.A., Broekmans, M.A.T.M. and Sims, I. 2015: Petrographic characterisation of aggregates regarding potential reactivity to alkalis – RILEM to AAR-1.2. Book published for RILEM by Springer.

Figg, J. 17 October 1983: Chloride and sulfate attack on concrete. *Chemistry and Industry* 770–775.

Figg, J.W. 1977: *Alkali-Aggregate (Alkali-Silica and Alkali-Silicate) Reactivity in Concrete, Bibliography*. Cembureau (The European Cement Association), Brussels, WI.

Figg, J.W. 1981: Reaction between cement and artificial glass in concrete. In *Proceedings of the Fifth International Conference on Alkali-Aggregate Reaction*, Cape Town, South Africa, paper S252/7.

Fiskaa, O., Hansen, H. and Mourn, J. 1971: *Betong i alunskifer*, Publication No. 86, pp. 1–32, Norwegian Geotechnical Institute, Oslo. (In Norwegian with summary).

Fookes, P.G. 1977: A plain man's guide to cracking in the Middle East. *Concrete in the Middle East*, Viewpoint Publication, Cement and Concrete Association, Slough, U.K., pp. 22–24. (originally published in *Concrete*, 10(9), 20–22, in 1976).

Fookes, P.G. 1993: Concrete in the Middle East – Past, present and future: A brief review. *Concrete* 27(4), 14–20.

Fookes, P.G. and Collis, L. 1975a: Problems in the Middle East. *Concrete* 9(7), 12–19.

Fookes, P.G. and Collis, L. 1975b: Aggregates and the Middle East. *Concrete* 9(11), 14–19.

Fookes, P.G. and Collis, L. 1977: Problems in the Middle East. Viewpoint Publication, Cement and Concrete Association, Slough, U.K., pp. 1–7. (originally published in *Concrete* 9(7), 12–19, in 1975).

Fookes, P.G., Simm, J.D. and Barr, J.M. September 1986: Marine concrete performance in different climatic environments. In *Marine Concrete '86 – International Conference on Concrete in the Marine Environment*, The Concrete Society, London, U.K.

Fookes, P.G. and Walker, M.J. 2011: Natural aggregates in the performance and durability of concrete: Physical characteristics. *Geology Today* 27(4), 141–148.

Fournier, B. and Bérubé, M.A. 1990: Evaluation of a modified chemical method to determine the alkali-reactivity potential of siliceous carbonate aggregates. In *Proceedings, Canadian Developments in Testing Concrete Aggregates for Alkali-Aggregate Reactivity*, Rogers, C.A. (ed.), Toronto, Ontario, Canada, Report EM-92, Ministry of Transportation, pp. 118–135.

Fournier, B. and Bérubé, M.A. 2000: Alkali-aggregate reaction in concrete: A review of basic concepts and engineering implications. *Canadian Journal of Civil Engineering* 27(2), 167–191.

Fournier, B., Bérubé M.-A., Folliard, K.J. and Thomas, M.D.A.T. 2010: *Report on the Diagnosis, Prognosis and Mitigation of Alkali-Silica Reaction (ASR) in Transportation Structures*. Report No: FHWA-HIF-09-004, Federal Highways Administration (FHWA), U.S. Department of Transportation, Washington, DC, 147pp.

Freitag, S.A. 1990: *Alkali-Aggregate Reactivity of Concrete Sands from the Rangitikei and Waikato Rivers*. Report, 90-B4207:27, New Zealand Works Consultancy Services (Opus International Consultants Ltd since 1996), Central Laboratories, Wellington, New Zealand.

French, W.J. 1980: Reactions between aggregates and cement paste, an interpretation of the pessimum. *Quarterly Journal of Engineering Geology* 13, 231–247.

French, W.J. 2002: The formation of thaumasite. In *First International Conference on Thaumasite in Cementitious Materials*, BRE, Watford, U.K. Not in proceedings: need to locate & complete citation – Cement & Concrete Composites?

French, W.J. 2005: Presidential address 2003: Why concrete cracks – Geological factors in concrete failure. *Proceedings of the Geologists' Association* 116(2), 89–105.

French, W.J. and Crammond, N.J. 2000: Sub-aqueous carbonation and the formation of thaumasite in concrete. In *20th Cement and Concrete Science Conference*, Institution of Materials, Sheffield University, Sheffield, U.K.

Friedman, G.M. 1959: Identification of carbonate minerals by staining methods. *Journal of Sedimentary Petrology* 29, 87–97.

Frondel, C. 1962: *Dana's System of Mineralogy, Volume III, Silica Minerals*. John Wiley and Sons, New York.

Fu, Y., Gu, P., Xie, P. and Beaudoin, J.J. 1995: A kinetic study of delayed ettringite formation in hydrated Portland cement paste. *Cement and Concrete Research* 25(1), 63–70.

Fulton, F.S. 1977: *Concrete Technology, South African Handbook*, 5th ed. Portland Cement Institute, Johannesburg, South Africa. See Addis & Owens for 8th edition, 2001.

Gabrisova, A., Havlica, J. and Sahu, S. 1991: Stability of calcium sulphoaluminate hydrates in water solutions with various pH values. *Cement and Concrete Research* 21(6), 1023–1027.

Gali, S., Ayora, C., Alfonso, P., Tauler, E. and Labrador, M. 2001: Kinetics of dolomite –portlandite reaction: Application to Portland cement concrete. *Cement and Concrete Research* 31, 933–939.

Garcia, E., Alfonso, P., Labrador, M. and Gali, S. 2003: Dedolomitization in different alkaline media: Application to Portland cement paste. *Cement and Concrete Research* 33, 1443–1448.

Gard, J.A. and Taylor, H.F.W. 1957: An investigation of two minerals: Rhodesite and mountainite. *Mineralogical Magazine* 31, 611–623.

Gerdes, A. and Wittmann, F.H. 1992: Electrochemical degradation of cementitious materials. In *Ninth International Congress on the Chemistry of Cement*, pp. 409–415.

Ghorab, H.Y., Heinz, D., Ludwig, U., Meskendahl, Y. and Wolter, A. 1980: On the stability of calcium aluminate sulphate hydrates in pure systems and in cements, In *Seventh International Congress on the Chemistry of Cement*, Paris, France, pp. 496–503.

Gillott, J.B. and Swenson, E.G. 1969: Mechanism of the alkali-carbonate rock reaction. *Quarterly Journal of Engineering Geology* 2, 7–24.

Gillott, J.E. and Rogers, C.A. 1994: Alkali-aggregate reaction and internal release of alkalis. *Cement and Concrete Research* 24(5), 99–112.

Glanville, J. and Neville, A.M. (eds.). 1997: *Prediction of Concrete Durability – Proceedings of STATS 21st Anniversary Conference*, The Geological Society, 16 November 1995, London, U.K., E & F N Spon, 185pp.

Glass, G.K. 2003: Reinforcement corrosion. In *Advanced Concrete Technology – Concrete Properties*, Newman, J. and Seng Choo, B. (eds.). Elsevier, Oxford, U.K., Chapter 9, pp. 9/1–9/27.

Glasser, F.P., Damidot, D. and Atkins, M. 1995: Phase development in cement in relation to the secondary ettringite problem. *Advances in Cement Research* 7(26), 57–68.

Godart, B., de Rooij, M. and Wood, J.G.M. (eds.). 2013: Guide to diagnosis and appraisal of AAR damage to concrete in structures – Part 1 Diagnosis (AAR-6.1), RILEM State-of-the-Art Report, Springer (Berlin and Heidelberg) pp RILEM, Paris, France, 91pp.

Goguel, R.L. 1996: Selective dissolution techniques in AAR investigation: Application to an example of failed concrete. In: *Proceedings of the 10th International Conference on Alkali-Aggregate Reaction in Concrete*, Shayan, A. (ed.), Melbourne, Victoria, Australia, pp. 783–790.

Goltermann, P. 1995: Mechanical predictions of concrete deterioration – Part 2: Classification of crack patterns. *ACI Materials Journal* 92(1), 58–63.

Gouda, G.R., Roy, D.M. and Sarker, A. 1975: Thaumasite in deteriorated soil-cements. *Cement and Concrete Research* 5(5), 519–522.

Gouda, V.K. and Monrad, H.M. 1973: Galvanic cells encountered in the corrosion of steel reinforcement: I, Differential pH cells; II, Differential salt concentration cells; III, Differential surface condition cells; IV, Differential aeration cells. *Corrosion Science*, 14, 687–690 and 15, 307–336.

Goudie, A. and Viles, H. 1997: *Salt Weathering Hazards*. John Wiley, Chichester, U.K., 241pp.

Grantham, M.G. 2005: External paving with polypropylene fibre concrete – A freeze/thaw problem? *Concrete* 39(9), 47–49.

Grantham, M.G., Gray, M.J. and Eden, M.A. 1999: Delayed ettringite formation in foundation bases – A case study. *Proceedings of Structural Faults and Repair*, Edinburgh, U.K., Edinburgh University Press.

Grattan-Bellew, P.E. 1989: Test methods and criteria for evaluating the potential reactivity of aggregates. In *Proceedings of the Eighth International Conference on Alkali-Aggregate Reaction*, Okada, K., Nishibayashi, S. and Kawamura, M. (eds.), Kyoto, Japan, pp. 279–294.

Grattan-Bellew, P.E. 1992: Microcrystalline quartz, undulatory extinction and the alkali-silica reaction. In *Proceedings of the Ninth International Conference on Alkali-Aggregate Reaction in Concrete*, London, U.K., pp. 383–394.

Grattan-Bellew, P.E. 1995: Laboratory evaluation of alkali-silica reaction in concrete from Saunders generating station. *ACI Materials Journal* 92 (2), 126–134.

Grattan-Bellew, P.E. 1996: A critical review of accelerated ASR tests. In *Proceedings of the 10th International Conference on Alkali-Aggregate Reaction in Concrete*, Shayan, A. (ed.), Melbourne, Victoria, Australia, pp. 27–38.

Grattan-Bellew, P.E. 2011: Personal communication + unpublished report: see text.

Grattan-Bellew, P.E., Cybaanski, G., Fournier, B. and Mitchell, L. 2003: Proposed universal accelerated test for alkali-aggregate reaction: the concrete microbar test. *Cement, Concrete and Aggregates* 25, 29–34.

Grattan-Bellew, P.E., Margeson, J., Mitchell, L.D. and Deng, M. 2008: Is ACR another variant of ASR? Comparison of acid insoluble residues of alkali-silica and alkali-carbonate reactive limestones and its significance for the ASR/ACR debate. In *Proceedings of the 13th International Conference on Alkali-Aggregate Reaction in Concrete (ICAAR)*, Broekmans, M.A.T.M. and Wigum, B.J. (eds.), Trondheim, Norway, pp. 706–716. Only available as a CD-ROM.

Grattan-Bellew, P.E. and Mitchell, L.D. May 2006: Quantitative petrographic analysis of concrete – the Damage Rating Index (DRI) method, a review. In *Marc-André Bérubé Symposium on Alkali-Aggregate Reactivity in Concrete*, Fournier, B. (ed.), Montreal, Quebec, Canada, pp. 321–334.

Grattan-Bellew, P.E., Mitchell, L.D., Margeson, J. and Deng, M. 2010: Is alkali-carbonate reaction just a variant of alkali-silica reaction – ACR = ASR? *Cement and Concrete Research* 40, 556–562.

Groves, G.W., Hodges, D.J. and Zhang, X. 1987: TEM studies of reactive aggregates and their reaction with cement. *Proceedings of the International Conference on Cement Microscopy*, pp. 458–65.

Grube, H. and Rechenberg, W. 1989: Durability of concrete structures in acidic water. *Cement and Concrete Research* 19(5), 783–792.

Guggenheim, S., Bain, D.C. et al. 2002: Report of the Association Internationale pour l'Etude des Argiles (AIPEA) Nomenclature Committee for 2001: Order, disorder and crystallinity in phyllosilicates and the use of the 'Crystallinity Index'. *Clay Minerals* 37, 389–393.

Haavik, D.J. and Mielenz, R.C. May 1991: Alkali-silica reaction causes concrete pipe to collapse. *Concrete International* 54–57.

Hadley, D.W. 1961: Alkali reactivity of carbonate rocks – Expansion and dedolomitization. *Proceedings Highway Research Board* 40, 462–474.

Hadley, D.W. January 1968: Field and laboratory studies in the reactivity of sand-gravel aggregates. *Journal of the PCA Research and Development Laboratories* 17–23.

Hagelia, P. and Sibbick, R.G. 2009: Thaumasite sulfate attack, popcorn calcite deposition and acid attack in concrete stored at the 'Blindtarmen' test site Oslo, from 1952 to 1982. *Materials Characterization* 60(7), 686–699.

Hagelia, P., Sibbick, R.G., Crammond, N.J., Gronhaug, A. and Larsen, C.K. 2001: Thaumasite and subsequent secondary calcite deposition in sprayed concrete in contact with sulfide bearing Alum shale, Oslo, Norway. In *Proceedings of the Eighth Euroseminar on Microscopy Applied to Building Materials*, Stamatakis, M., Georgali, B., Fragoulis, D. and Toumbakari, E.E. (eds.), Athens, Greece, pp. 131–138.

Hamilton, J.J. and Handegord, O.O. 1968: The performance of ordinary Portland cement concrete in Prairie soils of high sulfate content. *Performance of Concrete, Symposium* in *Honour of Thorbergur Thorvaldson* pp. 135–158.

Hansen, T.C. 1989: Marine concrete in hot climates – Designed to fail. *Materials and Structures* 22, 344–346.

Harboe, E.M. 1982: Longtime studies and field experience with sulfate attack. In *George Verbeck Symposium on Sulfate Resistant Concrete*, Detroit, MI, ACI-SP77, American Concrete Institute, pp. 1–20.

Harmathy, T.Z. 1968: Determining the temperature history of concrete constructions following fire exposure. *Proceedings of ACI Journal* 65(11), 959–964.

Harrison, T.A., Dewar, J.D. and Brown, B.V. 2001: *Freeze-Thaw Resisting Concrete – Its Achievement in the UK*, Report C559, CIRIA (Construction Industry Research and Information Association), London, U.K.

Harrison, W.H. 1992a: *Assessing the Risk of Sulphate Attack on Concrete in the Ground*. BRE Information Paper P 15/92, Building Research Establishment, Watford, U.K.

Harrison, W.H. 1992b: *Sulphate Resistance of Buried Concrete: The Third Report on Long-Term Investigations at Northwick Park and on Similar Concretes in Sulphate Solutions at BRE*. BRE Report, Building Research Establishment, Watford, U.K.

Harrison, W.H. and Teychenné, D.C. 1981: *Sulphate Resistance of Buried Concrete: Second Interim Report on Long-Term Investigation at Northwick Park*. BRE Report, Building Research Establishment, Watford, HMSO, London, U.K.

Hartshorn, S.A., Sharp, J.H. and Swamy, R.N. 1999: Thaumasite formation in Portland-limestone cement pastes. *Cement and Concrete Research* 29, 1331–1340.

Hartshorn, S.A. and Sims, I. 1998: Thaumasite – A brief guide for engineers. *Concrete* 32(8), 24–27.

Heinz, D. and Ludwig, U. 1986: Mechanisms of subsequent ettringite formation in mortars and concretes after heat treatment. *Eighth International Congress on the Chemistry of Cement*, Rio de Janeiro, Brazil, pp. 189–195.

Heinz, D. and Ludwig, U. 1987: Mechanisms of secondary ettringite formation in mortars and concretes subjected to heat treatment. In *Concrete Durability: Katharine and Bryant Mather International Conference*, Scanlon, J.M. (ed.), Detroit, MI, ACI SP-I00, Paper No. SPl00-105, American Concrete Institute, pp. 2059–2071.

Heller, L. and Taylor, H.F.W. 1956: *Crystallographic Data for the Calcium Silicates*. HMSO, London, U.K.

Helmuth, R., Stark, D., Diamond, S. and Moranville-Regourd, M. 1993: *Alkali-Silica Reactivity: An Overview of Research*. Document SHRP-C-342, Strategic Highway Research Program (SHRP), National Research Council, Washington, D.C., 105pp.

Henry, K. 2000: *A Review of the Thermodynamics of Frost Heave*, Technical Report ERDC/CRREL TR-00-16, U.S. Army Corps of Engineers, Cold Regions Research & Engineering Laboratory, NTIS, Springfield, VA, 25pp.

Higgins, D.D. 19–21 June 2002: Increased sulfate resistance of ggbs concrete in the presence of carbonate, In *Proceedings of the First International Conference on Thaumasite in Cementitious Materials*, Holton, I. (ed.), Watford, U.K., BRE, Building Research Establishment. Available as a CD.

Higgins, D.D. and Crammond, N.J. 19–21 June 2002: Resistance of concrete containing ggbs to the thaumasite form of sulfate attack. In *Proceedings of the First International Conference on Thaumasite in Cementitious Materials*, Holton, I. (ed.), Watford, U.K., BRE, Building Research Establishment. Available as a CD.

Highways Agency. 1987: *Early Thermal Cracking of Concrete*. BA24/87, Highways Agency, Department of the Environment, London, U.K.

Hills, C.D. 1993: The hydration of ordinary Portland cement during cement based solidification of hazardous wastes, PhD thesis, University of London (unpublished), London, U.K.

Hoar, T.P. and Jacob, W.R. 1967: The anodic behaviour of metals. *Corrosion Science* 7, 341–355.

Hobbs, D.W. 1978: Expansion of concrete due to the alkali-silica reaction: an explanation. *Magazine of Concrete Research* 30, 215–220.

Hobbs, D.W. 1981: The alkali-silica reaction – A model for predicting expansion in mortar. *Magazine of Concrete Research* 33, 208–220.

Hobbs, D.W. 1988: *Alkali-Silica Reaction in Concrete*. Thomas Telford, London, U.K., 183pp.

Hobbs, D.W. (ed.) 1998: *Minimum Requirements for durable Concrete – Carbonation – and Chloride-Induced Corrosion, Freeze-Thaw Attack and Chemical Attack*. BCA Ref 45.043, British Cement Association, Crowthorne, U.K., 172pp.

Hobbs, D.W. 2003: Thaumasite sulfate attack in field and laboratory concretes: Implications for specifications. *Cement and Concrete Composites* 25, 1195–1202.

Hobbs, D.W., Marsh, B.K. and Matthews, J.D. 1998b: Minimum requirements for concrete to resist freeze-thaw attack. In *Minimum Requirements for Durable Concrete – Carbonation – and Chloride-Induced Corrosion, Freeze-Thaw Attack and Chemical Attack*, Hobbs, D.W. (ed.). BCA Ref 45.043, British Cement Association, Crowthorne, U.K., Chapter 4, pp. 91–129.

Hobbs, D.W. and Matthews, J.D. 1998a: Minimum requirements for concrete to resist deterioration due to chloride-induced corrosion. In *Minimum Requirements for Durable concrete – Carbonation – and Chloride-Induced Corrosion, Freeze-Thaw Attack and Chemical Attack*, Hobbs, D.W. (ed.). BCA Ref 45.043, British Cement Association, Crowthorne, U.K., Chapter 3, pp. 43–89.

Hobbs, D.W. and Matthews, J.D. 1998c: Minimum requirements for concrete to resist chemical attack. In *Minimum Requirements for Durable Concrete – Carbonation – and Chloride-Induced Corrosion, Freeze-Thaw Attack and Chemical Attack*, Hobbs, D.W. (ed.). BCA Ref 45.043, British Cement Association, Crowthorne, U.K., Chapter 5, pp. 131–162.

Holland, T.C. 2005: *Silica Fume User's Manual*, Report No FHWA-IF-05-016, Silica Fume Association, Lovettsville, VA, for the Federal Highway Administration (FHWA), U.S. Department of Transportation, Washington, DC, 193pp.

Hudec, P.P. 1991: Common factors affecting alkali reactivity and frost durability of aggregates. In *Proceedings of the Fifth International Conference on the Durability of Materials and Components*, Baker, J.M., Nixon, P.J., Majumdar, A.J. and Davies, H. (eds.), 7–9 November 1990, Brighton, U.K., E. & F.N. Spon, London, U.K., pp. 77–86.

Hughes, P. 2012: A new mechanism for accelerated degradation of synthetic-fibre-reinforced marine concrete. *Concrete* 46(9), 18–20.

Idorn, G.M. 1961a: *Studies of Disintegrated Concrete – Part 1*, Progress Report N2, Committee on Alkali Reactions in Concrete, Danish National Institute of Building Research, Aalborg, Denmark.

Idorn, G.M. 1961b: *Studies of Disintegrated Concrete – Parts I–V*, Progress Report N2-N6, Committee on Alkali Aggregate Reactions in Concrete, Danish National Institute of Building Research, Aalborg and the Danish Academy of Technical Sciences, Lyngby, Denmark.

Idorn, G.M. 1967: *Durability of Concrete Structures in Denmark – A Study of Field Behavior and Microscopic Features*. Technical University of Denmark, Copenhagen, Denmark.

Ingham, J. and Tarada, F. 2007: Turning up the heat – Full service fire safety engineering for concrete structures. *Concrete* 41(9), 27–30.

Institution of Structural Engineers. 1992/2010: *Structural Effects of Alkali-Silica Reaction: Technical Guidance on the Appraisal of Existing Structures*. London, U.K. Addendum added 2010.

International Commission on Large Dams. 1989: *Exposure of Dam Concrete to Special Aggressive Waters: Guidelines*. Bulletin 71, ICOLD/CIGB, Paris, France. See also CIGB.

Ivanov, F.M. 1981: Attack of aggressive fluids. *Cement, Concrete, and Aggregates* 3(2), 105–107.

JCAS I-31. 1996: *Quantitative Measurement of Quartz (Phosphoric Acid Method)*. Japan Cement Association, Tokyo, Japan, 1996, 11pp (in Japanese). See also Cement Association of Japan.

Jensen, A.D., Chatterji, S., Christensen, P. and Thaulow, N. 1984: Studies in alkali-silica reaction – Part II. Effect of air entrainment on expansion. *Cement and Concrete Research* 14(3), 311–314.

Jensen, V. 1993: Alkali-aggregate reaction in Southern Norway, doctoral thesis, Technical University of Trondheim (NTH), Trondheim, Norway, 265pp +10 Appendices.

Jensen, V. 2004a: Alkali-silica reaction damage to Elgeseter Bridge, Trondheim, Norway: A review of construction, research and repair up to 2003. *Materials Characterization*, 53 (2–4: Special Issue – Selection of *Proceedings from the Ninth Euroseminar on Microscopy Applied to Building Materials*, Trondheim, Norway, 9–12 September 2003), pp. 155–170.

Jensen, V. 2004b: Measurement of cracks, relative humidity and effects of surface treatment on concrete structures damaged by Alkali-Silica Reaction. In *Proceedings of the 12th International Conference on Alkali-Aggregate Reaction in Concrete*, Tang M. and Deng, M. (eds.), Beijing, China, Vol. 2, pp. 1245–1253.

Jensen, V. 2005: Alkali-silica reaction (ASR) testing of deterioration in concrete. In *Inspection and Monitoring Techniques for Bridges and Civil Structures*, Gongkang, F. (ed.). Woodhead Publishing Limited, Cambridge, U.K., Chapter 4, pp. 22–64.

Jensen, V. 15–19 September 2009: The controversy of alkali carbonate reaction: State of art on the reaction mechanism. In *12th Euroseminar on Microscopy Applied to Building Materials*, Dortmund, Germany.

Jensen, V. 2011: Personal communication.

Jensen V. and Merz C. June 2008: Alkali-aggregate reaction in Norway and Switzerland. Survey investigations and structural damages. In *Proceedings of the 13th International Conference on Alkali-Aggregate Reaction in Concrete*, Broekmans, M.A.T.M. and Wigum, B.J. (eds.), Trondheim, Norway. Only available as a CD.

Johansen, V., Thaulow, N. and Idorn, G.M. 1994: Expansion reactions in mortar and concrete. *Zement-Kalk-Gips* 47(5), El50–E155.

John, P.H., Lock, M.A. and Gibbs, M.M. 1977: Two new methods for obtaining water samples from shallow aquifers and littoral sediments. *Journal of Environmental Quality* 6(3), 322–324.

Jones, A.E.K. and Clark, L.A. 1996: A review of the Institution of Structural Engineers report 'Structural effects of alkali-silica reaction (1992)'. In *Proceedings of the 10th International Conference on Alkali-Aggregate Reaction in Concrete*, Shayan, A. (ed.), Melbourne, Victoria, Australia, pp. 394–401.

Jones, F.E. and Tarleton, R.D. 1958: Reactions between aggregates and cement. Part VI. Alkali aggregate interaction. Experience with some forms of rapid and accelerated tests for alkali aggregate reactivity. Recommended test procedures, National Building Studies Research Paper 25, HMSO, London, U.K..

Jones, L.W. 1988: *Interference Mechanisms in Waste Stabilisation/Solidification Processes: Literature Review*. Hazardous Waste Engineering Research Laboratory, Office of Research and Development, Environmental Protection Agency (EPA), Cincinnati, OH.

Katayama, T. 1992: A critical review of carbonate rock reactions - is their reactivity useful or harmful? In: Poole, A.B. (ed), *Proceedings of the 9th International Conference on Alkali-Aggregate Reaction in Concrete*, London, U.K., pp. 508–518.

Katayama, T. 2004: How to identify carbonate rock reactions in concrete. *Materials Characterization* 53 (2–4), 85–104.

Katayama, T. 31 May–3 June 2006: Modern petrography of carbonate aggregates in concrete – Diagnosis of so-called alkali-carbonate reaction and alkali-silica reaction. In *Marc-Andre Berube Symposium on Alkali-Aggregate Reactivity in Concrete*, Fournier, B. (ed.), part of *Eighth CANMET/ACI International Conference on Recent Advances in Concrete Technology*, Montreal, Quebec, Canada, Supplementary paper (corresponding to 423–444).

Katayama, T. 2008: ASR gel in concrete subject to freeze/thaw cycles – Comparison between laboratory and field concretes from Newfoundland, Canada. In *Proceedings of the 13th International Conference on Alkali-Aggregate Reaction in Concrete (ICAAR)*, Broekmans, M.A.T.M. and Wigum, B.J. (eds.), Trondheim, Norway, pp. 174–183. Available only as a CD.

Katayama, T. 2010a: The so-called alkali-carbonate reaction (ACR) – Its mineralogical and geochemical details, with special reference to ASR. *Cement and Concrete Research* 40, 643–675.

Katayama, T. 2010b: Diagnosis of alkali-aggregate reaction – Polarizing microscopy and SEM-EDS analysis. In *Proceedings of the Sixth International Conference on Concrete under Severe Conditions (CONSEC'10)*, Castro-Borges, P., Moreno, E.I., Sakai, K., Gjørv, O.E. and Banthia, N. (eds.), Merida, Mexico, pp. 19–34.

Katayama, T. 2011a: So-called alkali-carbonate reaction – Petrographic details of field concretes in Ontario. In *Proceedings of the 13th Euroseminar on Microscopy Applied to Building Materials (EMABM)*, Ljubljana, Slovenia, 15pp.

Katayama, T. 2011b: Personal communication.

Katayama, T. 2012a: Rim-forming dolomitic aggregate in concrete structures in Saudi Arabia - Is dedolomitization equal to the so-called alkali-carbonate reaction? In *Proceedings of 14th International Conference in Alkali-Aggregate Reaction in Concrete*, Austin, TX, Paper 030411-KATA-01.

Katayama, T. 2012b: ASR gels and their crystalline phases in concrete – Universal products in alkali-silica, alkali-silicate and alkali-carbonate reactions. In *Proceedings of the 14th International Conference on Alkali-Aggregate Reaction*, Folliard, K.J. (ed.), Austin, TX, 10pp. Paper 030411-KATA-03.

Katayama, T. and Bragg, D.J. 1996: Alkali-aggregate reaction combined with freeze/thaw in New Foundland, Canada – Petrography using EPMA. In *Proceedings of the 10th International Conference on Alkali-aggregate Reaction in Concrete*, Shayan, A. (ed.), Melbourne, Victoria, Australia, pp. 243–250.

Katayama, T. and Futagawa, T. 1989: Diagenetic changes in potential alkali-aggregate reactivity of siliceous sedimentary rocks in Japan – A geological interpretation. In *Proceedings of the Eighth International Conference on Alkali-Aggregate Reaction*, Okada, K., Nishibayashi, S. and Kawaura, M. (eds.), Kyoto, Japan, pp. 525–530.

Katayama, T. and Grattan-Bellew, P.E. 2012: Petrography of the Kingston experimental sidewalk at age 22 years – ASR as the cause of deleteriously expansive, so-called alkali-carbonate reaction. In *Proceedings of the 14th International Conference on Alkali-Aggregate Reaction*, Folliard, K.J. (ed.), Austin, TX, 10pp. Paper 030411-KATA-06.

Katayama, T., Helgason, T.S. and Olafsson, H. 1996a: Petrography and alkali-reactivity of some volcanic aggregates from Iceland. In *Proceedings of the 10th International Conference on Alkali-Aggregate Reaction in Concrete*, Shayan, A. (ed.), Melbourne, Victoria, Australia, pp. 377–384.

Katayama, T., Ochiai, M. and Kondo, H. 1996b: Alkali-reactivity of some Japanese carbonate rocks based on standard tests. In *Proceedings of the 10th International Conference on Alkali-Aggregate Reaction in Concrete*, Shayan, A. (ed.), Melbourne, Victoria, Australia, pp. 294–301.

Katayama, T., Oshiro, T., Sarai, Y., Zaha, K. and Yamato, T. 2008: Late-expansive ASR due to imported sand and local aggregates in Okinawa Island, southwestern Japan. In *Proceedings of the 13th International Conference on Alkali-Aggregate Reaction in Concrete (ICAAR)*, Broekmans, M.A.T.M. and Wigum, B.J. (eds.), Trondheim, Norway, pp. 862–873. Available only as a CD.

Katayama, T., Sarai, Y., Higashi, Y. and Honma, A. 2004a: Late-expansive alkali-silica reaction in the Ohnyu and Furikusa headwork structures, central Japan. In *Proceedings of the 12th International Conference on Alkali-Aggregate Reaction in Concrete (ICAAR)*, Tang, M. and Deng, M. (eds.), Beijing, China, pp. 1086–1094.

Katayama, T. and Sommer, H. 2008: Further investigation of the mechanisms of so-called alkali-carbonate reaction based on modern petrographic techniques. In *Proceedings of the 13th International Conference on Alkali-Aggregate Reaction in Concrete (ICAAR)*, Broekmans, M.A.T.M. and Wigum, B.J. (eds.), Trondheim, Norway, pp. 850–860. Available only as a CD.

Katayama, T., St. John, D.A. and Futagawa, T. 1989: The petrographic comparison of some volcanic rocks from Japan and New Zealand potential reactivity related to interstitial glass and silica minerals. In *Proceedings of the Eighth International Conference on Alkali-Aggregate Reaction*, Okada, K., Nishibayashi, S. and Kawamura, M. (eds.), Kyoto, Japan, pp. 537–542.

Katayama, T., Tagami, M., Sarai, Y., Izumi, S. and Hira, T. 2004b: Alkali-aggregate reaction under the influence of deicing salts in the Hokuriku district, Japan. *Materials Characterization* 53(2–4), 105–122.

Kawamura, M., Takeuchi, K. and Sugiyama, A. 1996: Mechanisms of the influence of externally supplied NaCl on expansion of mortars containing reactive aggregates. *Magazine of Concrete Research* 48(176), 237–248.

Kay, K.A., Fookes, P.G. and Pollock, D.J. 1982: Deterioration related to chloride ingress. *Concrete in the Middle East.* Viewpoint Publication, 23–32, Eyre & Spottiswoode, London, U.K.

Kelham, S. 2003: Acid, soft water and sulfate attack. In *Advanced Concrete Technology – Concrete Properties*, Newman, J. and Ban Seng, C. (eds.). Elsevier, Oxford, U.K., Chapter 12, pp. 12/1–12/12.

Kennerley, R.A. and St John, D.A. 1988: Part I – Reactivity of New Zealand aggregates with cement alkalis. In *Alkali-Aggregate Studies in New Zealand*, St John, D.A. (ed.). DSIR (Department of Industrial & Scientific Research), New Zealand, Chemistry Division, Report No. CD 2390, pp. 60–61.

Keriene, J. and Kaminskas, A. 1993: Investigation of mineral wool fibre in alkaline solution and humidity using scanning electron microscopy. In *Proceedings of the Fourth Euroseminar on Microscopy Applied to Building Materials*, Lindqvist, J.E. and Nitz, B. (eds.), Sweden, Swedish National Testing and Research Institute, Building Technology SP Report, p. 15.

Kerrick, D.M. and Hooton, R.D. 1992: ASR of concrete aggregate quarried from a fault zone: results and petrographic interpretation of accelerated mortar bar tests. *Cement and Concrete Research* 22(5), 949–960.

Khoury, G.A. 2006: Tunnel concretes under fire: Part 1 – Explosive spalling. *Concrete* 40(10), 62–64.

Khoury, G.A. 2007: Tunnel concretes under fire: Part 2 – Toxicity. *Concrete* 41(1), 47–48.

Knights, J.C. and Baldwin, N.J.R. 2010: Diagnosis of delayed ettringite formation in precast concrete blocks. In *13th International Conference and Exhibition, Structural Faults and Repair*, 10–17 June, Edinburgh, U.K.

Knofel, D.F.E. and Rottger, K.G. 1985: Contribution on the influence of SO_2 on cement mortar and concrete. *Proceedings of the 7th International Conference on Cement Microscopy*, pp. 333–358.

Knudsen, T. 1992: The chemical shrinkage test: Some corrolaries. In *Proceedings of the Ninth International Conference on Alkali-Aggregate Reaction in Concrete*, London, U.K., pp. 543–549.

Koelliker, E. 1985: Method for microscopic observation of corrosion layers on concrete and mortar. *Cement and Concrete Research* 15(5), 909–913.

Kramf, L. and Haksever, A. 1986: Possibilities of assessing the temperatures reached in concrete building elements during fire. In *Evaluation and Repair of Fire Damage to Concrete*, Harmarthy, T.Z. (ed.), Detroit, MI, ACI SP-92, 115–42, American Concrete Institute.

Kropp, J. 1995: Chlorides in concrete. In *Performance Criteria for Concrete Durability – State of the Art Report Prepared by RILEM Technical Committee TC 116-PCD, Performance of Concrete as a Criterion of its Durability*, Kropp, J. and Hilsdorf, H.K. (eds.). RILEM Report 12, E & F N Spon, London, U.K., (pp RILEM, Paris), Chapter 6, pp. 138–164.

Kropp, J. and Hilsdorf, H.K. (eds.). 1995: *Performance Criteria for Concrete Durability – State of the Art Report Prepared by RILEM Technical Committee TC 116-PCD, Performance of Concrete as a Criterion of its Durability*. RILEM Report 12, E & F N Spon, London, U.K., (pp RILEM, Paris), 341pp.

Kuenning, W.H. 1966: Resistance of Portland cement mortar to chemical attack – A progress report. In *Symposium on Effects of Aggressive Fluids on Concrete*, Highway Research Record Number 113, National Research Council, Washington, DC, pp. 43–71.

Kuzel, H.J. 1990: Reactions of CO_2 with calcium aluminate hydrates in heat treated concrete. In *Proceedings of the 12th International Conference Microscopy*, Vancouver, British Columbia, Canada, pp. 219–227.

Laboratoire Central des Ponts et Chaussées (LCPC). 1991: *Recommandations provisoires pour la prevention des desordres dus à l'alcali-reaction*, LCPC, Ministère de l'Equipement, du Logement, des Transports et de la Mer, Paris, France. (in French)

Laboratoire Central des Ponts et Chaussées (LCPC). 1994: *Recommendations for the Prevention of damage by the Alkali-Aggregate Reaction*, Ministère de l'Equipement, du Logement, des Transports et de la Mer, Paris, France. (English translation of non-provisional version)

Lane, S.J., Bromley, A.V. and Sibbick, R.G. 1999: The application of microscopy and moisture sensitivity testing to the expansive 'Mundic' concrete block problem in Southwest England. In *Proceedings of the Euroseminar on Microscopy Applied to Building Materials*, Delft, Netherlands, pp. 557–566.

Lawrence, C.D. 1990: Sulfate attack on concrete. *Magazine of Concrete Research* 42(153), 249–264.

Lawrence, C.D. 1993: *Laboratory Studies of Concrete Expansion Arising from Delayed Ettringite Expansion*. BCA Publication C/16, British Cement Association, Crowthorne, U.K.

Lawrence, C.D. 1995: Mortar expansions due to delayed ettringite formation – Effects of curing period and temperature. *Cement and Concrete Research* 25(4), 903–914.

Lawrence, C.D., Dalziel, J.A. and Hobbs, D.W. 1990: *Sulphate Attack Arising from Delayed Ettringite Formation.* BCA Ref. ITN 12, British Cement Association, Slough, U.K.

Lea, F.M. 1968: Some studies on the performance of concrete structures in sulphate-bearing environments. In *Performance of Concrete, Resistance of Concrete to Sulphate and Other Environmental Conditions, a Symposium in Honour of Thorbergur Thorvaldson*, Swenson, E.G. (ed.), Canadian Building Series, University Press, Toronto, Ontario, Canada, pp. 56–65.

Lea, F.M. 1970: *The Chemistry of Cement and Concrete*, 3rd ed. Edward Arnold Limited, London, U.K. See also Hewlett for 4th edition.

Lee, E.J. and Clowes, H.M. 1992: *Alkali-Aggregate Reaction 1977–1992, a Definitive Bibliography.* BCA Ref: CIS/B8, British Cement Association, Slough (later Crowthorne), U.K. See also British Cement Association.

Leek, D.S. 1997: A study of the effects of chloride and sulphate on the hydration of Portland cement and the corrosion of carbon steel reinforcement using electron-optical techniques and energy dispersive X-ray analysis, PhD thesis, London University (unpublished), London, U.K.

Leek, D.S. 2011: Personal communication.

Leek, D.S. and Poole, A.B. 1990: The breakdown of the passive film on high yield mild steel by chloride ions. In *Corrosion of Reinforcement in Concrete*, Page, C.L., Treadaway, K.W.J. and Bamforth, P.B. (eds.), Elsevier Applied Science (pp Construction Materials Group, Society of Chemical Industry), pp. 65–73.

Leeming, M.B. 1990: *Concrete in the Oceans, Phase II, Coordinating Report on the Whole Programme.* CIRIA (Construction Industry Research and Information Association), London, U.K..

Leifeld, G., Munchberg, W. and Stegmaier, W. 1970: Ettringit und Thaumasit als Treibursache in Kalk-Gips-Puzen. *Zement-Kalks-Gips* (4), 174–177.

Lennon, T. 2011: *Structural Fire Engineering.* ICE Publishing, London, U.K., (with IHS BRE Press), 179pp.

Le Roux, A., Thiebaut, J., Guédon, J-S. and Wackenheim, C. 1999: *Pétrographie appliquée à l'alcali-réaction* (in French). Ouvrages d'art OA26, Laboratoire Central des Ponts et Chaussées (LCPC), Paris, France.

Lin, W-M., Lin, T.D. and Powers-Couche, L.J. 1996: Microstructures of fire-damaged concrete. *ACI Materials Journal* 93(3), 199–205, Title No 93-M22.

Lindgård, J., Andiç-Çakir, Ö., Fernandes, I., Thomas, M.D.A. and Rønning, T.F. 2012: Alkali-silica reactions (ASR): Literature review on parameters influencing laboratory performance testing. *Cement and Concrete Research* 42(2), 223–243.

Longworth, I. 2011: *Hardcore for Supporting Ground Floors of Buildings.* BRE DG 522, Parts 1 & 2, IHS BRE Press, Bracknell, U.K.

López-Buendía, A.M., Climent, V., Mar Urquiola, M. and Bastida, J. 2008: Influence of dolomite stability on alkali-carbonate reaction. In *Proceedings of the 13th International Conference on Alkali-Aggregate Reaction in Concrete (ICAAR)*, Broekmans, M.A.T.M. and Wigum, B.J. (eds.), Trondheim, Norway, pp. 233–242. Available only as a CD.

López-Buendía, A.M., Climent, V. and Verdu, P. 2006: Lithological influence of aggregate in the alkali-carbonate reaction. *Cement and Concrete Research* 36(8), 1490–1500.

Lorenzi, G., Jensen, J., Wigum, B.J., Sibbick, R., Haugen, M., Guédon, S. and Åkesson, U. 2006: *Petrographic atlas of the Potentially Alkali-Reactive Rocks in Europe (PARTNER programme).* Professional Paper 2006/1-N.302, Geological Survey of Belgium, Brussels.

Ludwig, U. 1980: Durability of cement mortars and concretes. In *Durability of Building Materials and Components.* ASTM STP 691, American Society for Materials and Testing, Philadelphia, PA, pp. 269–281.

Ludwig, U. and Mehr, S. 1986: Destruction of historical buildings by the formation of ettringite or thaumasite, *Eighth International Congress on the Chemistry of Cement*, Rio de Janeiro, Brazil, pp. 181–188.

Ludwig, U. and Rudiger, I. 1993: Quantitative determination of ettringite in cement pastes, mortars and concretes. *Zement-Kalk-Gips* 46(5), E153–E156.

Lukas, W. 1975: Betonzerstörung durch SO$_3$ – Angriff unter bildung von thaumasite und woodfordite. *Cement and Concrete Research* 5(5), 503–518.

Lumley, J.S. 1989: Synthetic cristobalite as a reference reactive aggregate. In *Proceedings of the Eighth International Conference on Alkali-Aggregate Reaction*, Okada, K., Nishibayashi, S. and Kawamura, M. (eds.), Kyoto, Japan, pp. 561–566.

McBee, C.L. and Kruger, J. 1974: Events leading to the initiation of the pitting of iron. In *Localized Corrosion*, Staehle, R.W. (ed.), NACE-3, 252–260, NACE (National Association of Corrosion Engineers), now NACE International, Houston, TX.

McKenzie, L. 1994: The microscope: A method and tool to study lime burning and lime quality. *World Cement* 25(11), 56–59.

McLaren, F.R. 1984: *Design Manual, Sulfide and corrosion Prediction and Control*, American Concrete Pipe Association, Vienna, VA (now Irving, Texas), 80pp.

Malhotra, H.L. 1956: The effect of temperature on the compressive strength of concrete. *Magazine of Concrete Research* 8(23), 85–94.

Marchand, J., Pigeon, M. and Setzer, M. (eds.). 1997: *Freeze/thaw Durability of Concrete*. E & F N Spon, London, U.K.

Marchand, J., Sellevoid, E.J. and Pigeon, M. 1994: The deicer salt scaling deterioration of concrete - An overview. In *Third International Conference on Durability of Concrete*, Malhotra, V.M. (ed.), Nice, France, ACI SP-145, American Concrete Institute, Detroit, MI, pp. 1–46.

Marchand, J. and Skalny, J.P. (eds.). 1999: Sulfate attack mechanisms (*Proceedings from Seminar on Sulfate Attack Mechanisms*, Montreal, Quebec, Canada, 5–6 October 1998), *Materials Science of Concrete*, Special Volume, American Ceramic Society, Westerville, OH.

Marfil, S.A. and Maiza, P.J. 1993: Zeolite crystallization in Portland cement concretes. *Cement and Concrete Research* 23(5), 1283–1288.

Marinoni, N. and Broekmans, M.A.T.M. 2013: Microstructure of selected aggregate quartz by XRD, and a critical review of the crystallinity index. *Cement and Concrete Research* 54 (December), 215–225.

Martin, R-P. 2010: Experimental analysis of the mechanical effects of delayed ettringite formation on concrete structures, PhD Thesis, Université Paris Est, Paris, France, 577pp (in French).

Martin, R.-P., Bazin, C. and Toutlemonde, F. 2012a: Alkali aggregate reaction and delayed ettringite formation: common features and differences. In *Proceedings of the 14th International Conference on Alkali-Aggregate Reaction*, Folliard, K.J. (ed.), Austin, TX. (proceedings available on a memory stick).

Martin, R-P., Renaud, J-C., Multon, S. and Toutlemonde, F. 2012b: Structural behavior of plain and reinforced concrete beams affected by combined AAR and DEF. In *Proceedings of the 14th International Conference on Alkali-Aggregate Reaction*, Folliard, K.J. (ed.), Austin, TX. (proceedings available on a memory stick).

Martineau, F., Guedon-Dubied, J.S. and Larive, C. 1996: Evaluation of the relationships between swelling, cracking, development of gels. In *Proceedings of the 10th International Conference on Alkali-Aggregate Reaction in Concrete*, Shayan, A. (ed.), Melbourne, Victoria, Australia, pp. 798–805.

Massazza, F. 1985: Concrete resistance to seawater and marine environment. *Il Cemento* 82, 3–26.

Mather, K. 1982: Current research in sulfate resistance at the Waterways Experiment Station. In *George Verbeck Symposium on Sulfate Resistant Concrete*, ACI-SP77, American Concrete Institute, Detroit, MI, pp. 63–74.

Mather, K., Buck, A.D. and Luke, W.I. 1964: Alkali-silica and alkali-carbonate reactivity of some aggregates from South Dakota, Kansas, and Missouri. In *Symposium on Alkali-carbonate Rock Reactions*Highway, Research Record, 45, Washington, DC, pp. 72–109.

Matsumoto, R., Iijima, A. and Katayama, T. 1988: Mixed-water and hydrothermal dedolomitization of the Pliocene Shirahama Limestone, Izu Peninsula, central Japan. *Sedimentology* 35, 979–998.

Matthews, J.D. 1992: The resistance of PFA concretes to acid groundwaters. In *Ninth International Conference on the Chemistry of Cement*, New Delhi, India, pp. 355–362.

Matthews, S., Rupasinghe, R. and Lennon, T. 2006: Concrete structures in fire – Performance in Fire, Design and Analysis Methods. *Concrete* 40(5), 42–43.

Mays, G. (ed.). 1992: *Durability of Concrete Structures*. E & F N Spon, London, U.K.

Mehta, P.K. 1993: Sulfate attack on concrete – A critical review. In *Materials Science of Concrete,* Skalny, J. (ed.), III, American Ceramic Society, Westerville, OH, pp. 105–130.

Mehta, P.K., Schiessi, P. and Raupach, M. 1992: Performance and durability of concrete systems. In *Ninth International Congress on the Chemistry of Cement,* New Delhi, India, pp. 571–659.

Merriaux, K., Lecomte, A., Degeimbre, R. and Darimont, A. 2003: Alkali-silica reactivity with pessimum content on Devonian aggregates from the Belgian Arden massive. *Magazine of Concrete Research* 55(5), 429–437.

Meyer, A. 1968: Investigations on the carbonation of concrete. *Proceedings 5th International Symposium Chemistry of Cement* 3, 394–401.

Midgley, H.G. 1958: The staining of concrete by pyrite. *Magazine of Concrete Research* 10(29), 75–8.

Mielenz, R.C., Greene, K.T. and Benton, E.J. 1947: Chemical test for reactivity of aggregates with cement alkalies: Chemical processes in cement-aggregate reaction. *Journal of the American Concrete Institute* (Proceedings, volume 44), 19(3), 193–221.

Milanesi, C.A., Marfil, S.A., Batic, O.R. and Maiza, P.J. 1996: The alkali-carbonate reaction and its reaction products an experience with Argentinean dolomite rocks. *Cement and Concrete Research* 26(10), 1579–1591.

Milestone, N.B., St John, D.A., Abbott, J.H. and Aldridge, L.P. 1986: CO_2 corrosion of geothermal cement grouts. In *Proceedings of the Eighth International Congress on the Chemistry of Cement,* Rio de Janeiro, Brazil, pp. 141–144.

Mindess, S. and Gilley, J.C. 1973: The staining of concrete by an alkali-aggregate reaction. *Cement and Concrete Research* 3(6), 821–828.

Mitchell, M. 2000: Freeze/thaw protection: a comparative study of polypropylene fibres and air entrainment. *Concrete* 34(6), 26–27.

Miyagawa, T., Seto, K., Sasaki, K., Mikata, Y., Kuzume, K. and Minami, T. 2006: Fracture of reinforcing steels in concrete structures damaged by alkali-silica reaction – Field survey, mechanism and maintenance. *Journal of Advanced Concrete Technology* 4(3), 339–355.

Moll, H.L. 1964: *Uber die Korrosion von stahl in beton.* DAfS Heft, Berlin, Germany.

Monteiro, P.J.M. and Mehta, P.K. 1986: Interaction between carbonate rock and cement paste. *Cement and Concrete Research* 16(2), 127–134.

Moriconi, G., Castellano, M.G. and Collepardi, M. 1994: Mortar deterioration of the masonry walls in historic buildings. A case history: Vanvitelli's Mole in Ancona. *Materials and Structures* 27(171), 408–414.

Morino, K. 1989: Alkali aggregate reactivity of cherty rock. In *Proceedings of the Eighth International Conference on Alkali-Aggregate Reaction in Concrete,* Okada, K., Nishibayashi, S. and Kawamura, M. (eds.), Kyoto, Japan, pp. 501–508.

Morrey, C. (ed.). 1991: *Digest of Environmental Protection and Waste Statistics,* No. 14. Department of the Environment, HMSO, London, U.K.

Mu, X., Xu, Z., Deng, M. and Tang, M. 1996: Abnormal expansion of coarse-grained calcite in the autoclave method. In *Proceedings of the 10th International Conference on Alkali-Aggregate Reaction in Concrete,* Shayan, A. (ed.), Melbourne, Victoria, Australia, pp. 310–315.

Mullick, A.K. and Samuel, G. 1987: Reaction products of alkali-silica reaction\a microstructural study. In *Concrete Alkali-Aggregate Reactions, Proceedings of the 7th International Conference on Alkali-Aggregate Reaction in Concrete, Ottawa, Canada,* 1986, Grattan-Bellew, P.E. (ed.), Park Ridge, NJ, Noyes Publications, pp. 381–385.

Munn, C.J. (Chairman) 2003: *Alkali Silica Reaction – Minimising the Risk of Damage to Concrete – Guidance Notes and Recommended Practice,* 2nd ed. C&CANZ Technical Report 3, Cement & Concrete Association of New Zealand, New Zealand.

Murata, K.J. and Norman, M.B. 1976: An index of crystallinity for quartz. *American Journal of Science* 276, 1120–1130.

Natesaiyer, K.C. and Hover, K.C. 1988: In situ identification of ASR products in concrete. *Cement and Concrete Research* 18(3), 455–463.

Natesaiyer, K.C. and Hover, K.C. 1989a: Further study of an in-situ identification method for alkali-silica reaction products in concrete. *Cement and Concrete Research* 19, 770–778.

Natesaiyer, K. and Hover, K.C. 1989b: Some field studies of the new insitu technique for the identification of ASR products. In *Proceedings of the Eighth International Conference on Alkali-Aggregate Reaction in Concrete,* Okada, K., Nishibayashi, S. and Kawamura, M. (eds.), Kyoto, Japan, pp. 555–560.

Neck, U. 1988: Auswirkungen de Warmebehandlung auf Festigkeit und Dauerhaftigkeit von Beton. *Beton* V(38), 488–493.

Neville, A.M. 2001: Consideration of durability of concrete structures: Past, present and future. *Materials and Structures* 34(236), 114–118.

Neville, A.M. 2006: *Concrete: Neville's Insights and Issues.* Thomas Telford, London, U.K., 314pp.

Neville, A.M. and Brooks, J.J. 1987: *Concrete Technology.* Longman, Harlow, U.K.

Neville, A.M. and Brooks, J.J. 2010: *Concrete Technology,* 2nd ed. Prentice Hall (Pearson), Harlow, U.K., 442pp.

Newman, J. and Ban Seng, C. 2003: *Advanced concrete Technology – concrete Properties.* Elsevier, Oxford, U.K., Part 4, Durability of concrete and concrete construction, 8/3–14/9.

Nielson, A. 1994: Development of alkali silica reactions on concrete structures with time. *Cement and Concrete Research* 24(1), 83–85.

Nielsen, J. 1966: Investigation of resistance of cement paste to sulphate attack. In *Symposium on Effects of Aggressive Fluids on Concrete,* Highway Research Record, No. 113, (Publication - National Research Council No. 1335), pp. 114–117.

Nijland, T.G. and Siemes, A.J.M. 2002: Alkali-silica reaction in the Netherlands: Experience and current research. *Heron* 47(2), Special issue on ASR, 81–85.

Nixon, P.J. 1978: Floor heave in buildings due to the use of pyritic shales as fill material. *Chemistry and Industry* 4, 160–164.

Nixon, P.J., Canham, I. and Page, C.L. 1987: Aspects of the pore solution chemistry of blended cements related to the control of alkali-silica reaction. *Cement and Concrete Research* 17(5), 839–844.

Nixon, P.J., Canham, I., Page, C.L. and Bollinghaus, R. 1987: Sodium chloride and alkali-aggregate reaction. In *Concrete Alkali-Aggregate Reactions, Proceedings of the 7th International Conference, 1986, Ottawa, Canada,* Grattan-Bellew, P.E. (ed.), Park Ridge, NJ, Noyes Publications, pp. 110–114.

Nixon, P.J., Collins, R.J. and Rayment, P.L. 1979: The concentration of alkalies by moisture migration in concrete – A factor influencing alkali aggregate reaction. *Cement and Concrete Research* 9(4), 417–423.

Nixon, P.J. and Longworth, I. 2003: Thaumasite Expert Group Report: Review after three years' experience. *Concrete* 37(4), 22–23. See also DETR.

Nixon, P.J. and Sims, I. 1992: RILEM TC106 alkali aggregate reaction – Accelerated tests, Interim report and summary of survey of national specifications. In *Proceedings of the Ninth International Conference on Alkali-Aggregate Reaction in Concrete,* London, U.K., pp. 731–738.

Nixon, P.J. and Sims, I. June 2000: Universally accepted testing procedures for AAR, the progress of RILEM Technical Committee 106. In *Alkali-Aggregate Reaction in Concrete, Proceedings, 11th International Conference,* Bérubé, M-A., Fournier, B. and Durand, B. (eds.), Montreal, Québec, Canada, Centre de Recherche Interuniversitaire sur le Béton (CRIB), Laval & Sherbrooke Universities, pp. 435–444.

Norwegian Concrete Association (Norsk Betongforening). 2008: *Durable Concrete with Alkali Reactive Aggregates,* NCA Publication No 21, Norwegian Concrete Association, Oslo, Norway.

Novak, G.A. and Colville, A.A. 1989: Efflorescent mineral assemblages associated with cracked and degraded residential concrete foundations in Southern California. *Cement and Concrete Research* 19(1), 1–6.

Novokshchenov, V. 1987: Investigation of concrete deterioration due to sulfate attack – A case history, *Durability of Concrete,* ACI SP-100, 1979–2006, American Concrete Institute, Detroit, MI.

Oberholster, R.E. 2001: Alkali-silica reaction. In *Fulton's Concrete Technology,* 8th ed., Addis, B. and Owens, G. (eds.). Cement & Concrete Institute, Midrand, South Africa, Chapter 10, pp. 163–189. CD-Rom version.

Oberholster, R.E. and Davies, G. 1986: An accelerated method for testing the potential alkali reactivity of siliceous aggregates. *Cement and Concrete Research* 16(2), 181–189.

Oberholster, R.E., Du Toit, P. and Pretorius, J.L. 1984: Deterioration of concrete containing a carbonaceous sulphide-bearing aggregate. In *Proceedings of the Sixth International Conference on Cement Microscopy*, pp. 360–73.

Oberholster, R.E., Maree, H. and Brand, J.H.B. 1992: Cracked prestressed concrete railway sleepers: alkali-silica reaction or delayed ettringite formation. In *Proceedings of the Ninth International Conference on Alkali-aggregate Reaction in Concrete*, London, U.K., pp. 739–749.

Oberste-Padtberg, R. 1985: Degradation of cements by magnesium brines. In *Proceedings of the Sixth International Conference on Cement Microscopy*, pp. 24–36.

Odler, I. and Chen, Y. 1995: Effect of cement composition on the expansion of heat-cured cement pastes. *Cement and Concrete Research* 25(4), 853–862.

Odler, I. and Gasser, M. 1988: Mechanism of sulfate expansion in hydrated Portland cement. *Journal of the American Ceramic Society* 71(11), 1015–1020.

Ohfuji, H., Boyle, A.P., Prior, D.J. and Rickard, D. 2005: Structure of framboidal pyrite: An electron backscatter diffraction study. *American Mineralogist* 90, 1693–1704.

O'Neill, R.C. 1992: Identification and effects of sulfate materials in hardened concrete. In *Proceedings of the 14th International Conference on Cement Microscopy*, pp. 198–208.

Osborne, G.J. 1991: The sulphate resistance of Portland blastfurnace slag cement concretes, *Durability of Concrete*, ACI SP-126, American Concrete Institute, Detroit, MI, pp. 1047–1071.

Page, C.L. and Page, M.M. (eds.). 2007: *Durability of Concrete and cement Composites*. Woodhead and Maney (pp Institute of Materials, Minerals & Mining), Cambridge, U.K., 404pp.

Palmer, D. 1992: *The Diagnosis of Alkali-Silica Reaction*, 2nd ed. Report of a Working Party, British Cement Association, Wexham Springs, Slough, U.K. See also British Cement Association.

Parrott, L.J. 1987: *A Review of Carbonation in Reinforced Concrete*. Ref C/1-0987, Cement and Concrete Association, Slough, U.K., 69pp.

Parrott, L.J. 1990: Damage caused by carbonation of reinforced concrete. *Materials and Structures* 23(135), 230–234.

Parrott, L.J. 1991: Assessing carbonation in concrete structures. In *Proceedings of the Fifth International Conference on the Durability of Building Materials and Components, Brighton, 1990*, E. & F.N. Spon, London, U.K., pp. 575–585.

Pavlik, V. 1994: Corrosion of hardened cement paste by acetic and nitric acids. Part 11: Formation of chemical composition of the corrosion products layer. *Cement and Concrete Research* 24(8), 1495–1508.

Pearson-Kirk, D., Callanan, T. and Croke, K. 2004: Field investigations to determine the occurrence of alkali-silica reaction. *Concrete* 38(6), 40–41.

Piasta, J., Piasta, W. and Sawicz, Z. 1987: Sulfate resistance of mortars and concretes with calciferous aggregates, *Durability of Concrete*, ACI SP-100, American Concrete Institute, Detroit, MI, pp. 2153–2170.

Pigeon, M. and Pleau, R. 1995: *Durability of Concrete in Cold Climates*. E. & F.N. Spon, London, U.K.

Pigeon, M., Zuber, B. and Marchand, J. 2003: Freeze/thaw resistance. In *Advanced Concrete Technology – Concrete Properties*, Newman, J. and Ban Seng, C. (eds.), Elsevier, Oxford, U.K., Chapter 11, pp. 11/1–11/17.

Placido, F. 1980: Thermoluminescence test for fire-damaged concrete. *Magazine of Concrete Research* 32(11), 112–116.

Placido, F. and Smith, L. February 1981: Thermoluminescence: A way of testing fire-damaged concrete. *Fire*, 464–465.

Pomeroy, R.D. 1977: *The Problem of Hydrogen Sulphide in Sewers*. Clay Pipe Development Association, Chesham, U.K.

Poole, A.B. 1981: Alkali carbonate reactions in concrete. In *Proceedings of the Sixth International Conference on Alkali-Aggregate Reaction*, Cape Town, South Africa, S252/13, pp. 1–13.

Poole, A.B. 1992: Introduction to alkali-aggregate reaction in concrete. In *The Alkali-Silica Reaction in Concrete*, Swamy, R.N. (ed.), Blackie, Glasgow, Chapter 1, pp. 1–29.

Poole, A.B., McLachlan, A. and Ellis, D.J. 1988: A simple staining technique for the identification of alkali-silica gel in concrete and aggregate. *Cement and Concrete Research* 18, 116–20.

Poole, A.B., Patel, H.H. and Shiekh, V. 1996: Alkali silica and ettringite expansions in 'steam-cured' concretes. In *Alkali-Aggregate Reaction in Concrete, Proceedings of the 10th International Conference*, Shayan, A. (ed.), Melbourne, Victoria, Australia, pp. 943–948.

Poole, A.B. and Sotiropoulos, P. 1980: Reaction between dolomitic aggregate and alkali pore fluids in concrete. *Quaternary Journal of Engineering Geology, London* 13, 281–287.

Poole, A.B. and Thomas, A. 1975: A staining technique for the identification of sulphates in aggregates and concrete. *Mineralogical Magazine* 40, 315–316.

Popovics, S. 1987: A classification of the deterioration of concrete based on mechanism, In *Concrete Durability – Katherine and Bryant Mather International Conference*, ACI SP-lOO, 1, American Concrete Institute, Detroit, MI, pp. 131–42.

Portland Cement Association (PCA). 2001: *Solidification/Stabilization with Cement: Turning Environmental Liabilities into Economic Opportunities*. PCA code SP384, PCA, Skokie, IL.

Portland Cement Association (PCA). 2007: *Effects of Substances on Concrete and Protective Treatments*. PCA Concrete Information, Skokie, IL.

Pourbaix, M. 1974: Applications of electrochemistry in corrosion science and practice. *Corrosion Science* 14(1), 25–82.

Powers, T.C. 1949: The non-evaporable water content of hardened Portland cement paste; its significance for concrete research and its method of determination. *ASTM Bulletin* 158, 68–76, American Society for Testing and Materials, Philadelphia, PA.

Powers, T.C. 1956: Resistance to weathering – Freezing and thawing. In *Significance of Tests and Properties of Concrete and Concrete Aggregates*, ASTM STP No. 169, American Society for Testing and Materials, Philadelphia, PA, pp. 182–187.

Prince, W., Perami, R. and Espagne, M. 1994: Une noevelle approche du mecanisme de la reaction alcali-carbonate (A new approach of the mechanism of the alkali-carbonate reaction). *Cement and Concrete Research* 24(1), 62–72 (in French).

Pullar-Strecker, P. 1987: *Corrosion Damaged Concrete: Assessment and Repair*. Butterworths/CIRIA (Construction Industry Research and Information Association), London, U.K.

Quillin, K. 2001: *Delayed Ettringite Formation: In-situ Concrete*. BRE Information Paper IP11/01, Building Research Establishment, Watford, U.K. (published by CRC Limited).

Ramlochan, T., Thomas, M.D.A. and Hooton, R.D. 2004: The effect of pozzolans and slag on the expansion of mortars cured at elevated temperature, Part II: Microstructural and microchemical investigations. *Cement and Concrete Research* 34, 1341–1356.

Ramlochan, T., Zacarias, P., Thomas, M.D.A. and Hooton, R.D. 2003: The effect of pozzolans and slag on the expansion of mortars cured at elevated temperature, Part I: Expansive behaviour. *Cement and Concrete Research* 33, 807–814.

Rayment, P.L. 1992: The relationship between flint microstructure and alkali-silica reaction. In *Proceedings of the Ninth International Conference on Alkali-Aggregate Reaction in Concrete*, London, U.K., pp. 843–850.

Rayment, P.L. and Haynes, C.A. 1996: The alkali-silica reactivity of flint aggregates. In *Proceedings of the 10th International Conference on Alkali-Aggregate Reaction in Concrete*, Shayan, A. (ed.), Melbourne, Victoria, Australia, pp. 750–757.

Rayment, P.L., Pettifer, K. and Hardcastle, J. 1990: *The Alkali-Silica Reactivity of British Concreting Sands, Gravels and Volcanic Rocks*. Contractor Report 218, Department of Transport, London, U.K.

Reardon, E.J. 1990: An ion interaction model for the determination of chemical equilibria in cement/water systems. *Cement and Concrete Research* 20(2), 175–192.

Regourd-Moranville, M. 1989: Products of reaction and petrographic examination. In *Proceedings of the Eighth International Conference on Alkali-Aggregate Reaction*, Okada, K., Nishibayashi, S. and Kawamura, M. (eds.), Kyoto, Japan, pp. 445–456.

Rendell, F., Jauberthie, R. and Grantham, M. 2002: *Deteriorated Concrete – Inspection and Physicochemical Analysis*. Thomas Telford, London, U.K., 194pp.

Richardson, M.G. 2003: *Alkali-Silica Reaction in Concrete – General Recommendations and Guidance in the Specification of Building and Civil Engineering Works – A Report Prepared by a Joint Working Party*, Institution of Engineers of Ireland, Dublin, and the Irish Concrete Society, Drogheda, Ireland.

Richardson, M.G. 2007: Degradation of concrete in cold weather conditions. In *Durability of Concrete and Cement Composites*, Page, C.L. and Page, M.M. (eds.). Woodhead and Maney (pp Institute of Materials, Minerals & Mining), Cambridge, U.K., Chapter 8, pp. 282–315.

RILEM. 1995: Draft recommendations for test methods for the freeze-thaw resistance of concrete – Slab test and cube test (RILEM 117-FDC, Freeze-thaw and deicing resistance of concrete). *Materials and Structures* 28(180), 366–371.

RILEM. 1997: Freeze-thaw and deicing resistance of concrete (Technical Committee TC 117-FDC). *Materials and Structures, Research and Testing* 30(196), S3–S6.

RILEM. 2000a: Recommended test method TC 106-2 (now AAR-2), Detection of potential alkali-reactivity of aggregates: A – The ultra-accelerated mortar-bar test. *Materials and Structures* 33(229), 283–289. New edition in preparation.

RILEM. 2000b: Recommended test method TC 106-3 (now AAR-3), Detection of potential alkali-reactivity of aggregates: B – Method for aggregate combinations using concrete prisms. *Materials and Structures* 33(229), 290–293. New edition in preparation.

RILEM. 2003a: Recommended test method AAR-0: Detection of alkali-reactivity potential in concrete – Outline guide to the use of RILEM methods in assessments of aggregates for potential alkali-reactivity (prepared by I. Sims and P. Nixon). *Materials and Structures* 36(261), 472–479. See also Sims & Nixon. New edition in preparation.

RILEM. 2003b: Recommended test method AAR-1: Detection of potential alkali-reactivity of aggregates – Petrographic method (prepared by I. Sims and P. Nixon). *Materials and Structures* 36(261), 480–496. See also Sims & Nixon. New edition in preparation.

RILEM. 2005: AAR-5: Rapid preliminary screening test for carbonate aggregates (RILEM TC 191-ARP: Alkali-reactivity and prevention – Assessment, specification and diagnosis of alkali-reactivity (prepared by H. Sommer, P.J. Nixon and I. Sims). *Materials and Structures* 38(282), 787–792. See also Sommer et al. Check that correct date (2005) is cited in text.

RILEM. 2012b: Guide to diagnosis and appraisal of AAR damage to concrete in structures – Part 1 Diagnosis (AAR-6.1), RILEM State-of-the-Art Report, Springer (Berlin & Heidelberg) RILEM, Paris, France, 91pp. See also Godart et al. (2013).

RILEM. 2015: RILEM recommendations for the prevention of damage by alkali-aggregate reactions and new concrete structures: AAR-0, new AAR-1, new AAR-2, new AAR-3, AAR-4.1, AAR-5, AAR-7.1, AAR-7.2, & AAR-7.3 (to be published by Springer as a separate RILEM book edited by PJ Nixon and I Sims, pp. RILEM Tc 219-Ac.).

Riley, M.A. 1991: Possible new method for the assessment of fire-damaged concrete. *Magazine of Concrete Research* 43(155), 87–92.

Rivard, P. and Ballivy, G. 2005: Assessment of the expansion related to alkali-silica reaction by the Damage Rating Index method. *Construction and Building Materials* 19(2), 83–90.

Rivard, P., Fournier, B. and Ballivy, G. June 2000: Quantitative assessment of concrete damage due to alkali-silica reaction (ASR) by petrographic analysis. In *Proceedings of the 11th International Conference on Alkali-Aggregate Reaction*, Bérubé, M.A., Fournier, B. and Durand, B. (eds.), Québec City, Quebec, Canada, pp. 889–898.

Roberts, M.H. 1986: *Determination of the Chloride and Cement Contents of Hardened Concrete*. BRE Information Paper IP 21/86, Building Research Establishment, Watford, U.K.

Robery, P. 2009a: Enhancing the natural durability of concrete for extreme exposure, Part I. *Concrete* 43(2), 38–39.

Robery, P. 2009b: Enhancing the natural durability of concrete for extreme exposure, Part II. *Concrete* 43(3), 41–42.

Rodrigues, A., Duchesne, J., Fournier, B., Durand, B., Rivard, P. and Shehata, M. 2012: Mineralogical and chemical assessment of concrete damaged by the oxidation of sulphide-bearing aggregates: Importance of thaumasite formation on reaction mechanisms. *Cement and Concrete Research* 42(10), 1336–1347.

Rogers, C.A. 1988: *General Information on Standard Alkali-Reactive Aggregates from Ontario, Canada*. Ministry of Transportation, Toronto, Ontario, Canada, 59pp.

Rogers, C.A. and Hooton, R.D. 1989: Leaching of alkalies in alkali-aggregate reaction testing. In *Proceedings of the Eighth International Conference on Alkali-Aggregate Reaction*, Okada, K., Nishibayashi, S. and Kawamura, M. (eds.), Kyoto, Japan, pp. 327–332.

Rogers, D.E. 1973: Determination of carbon dioxide and aggressive carbon dioxide in waters. *New Zealand Journal of Science* 16, 875–983.

Roper, H., Cox, J.E. and Erlin, B. 1964: Petrographic studies on concrete containing shrinking aggregates. *Journal of the PCA Research and Development Laboratories* 6(3), 2–18.

Roy, D.M. 1988: Cementitious materials in nuclear waste management. *Cements Research Progress* 261–91.

Royal Institution of Chartered Surveyors (RICS). 2002 and 2005: *The 'mundic' problem – Supplement to Second Edition* – Stage 3 expansion testing*, January 2002, revised February 2005, RICS Books, Coventry, U.K. (*see Stimson, 1997).

Royal Institution of Chartered Surveyors (RICS). 2015: The mundic problem, 3rd edition, RICS guidance note, RICS, London (technical editors: Brindle, I., Santo, P., Sims, I.), 73pp.

Samarai, M.A. 1976: The disintegration of concrete containing sulfate-contaminated aggregates. *Magazine of Concrete Research* 28(96), 130–142. (important to note the published corrigenda)

Schimmel, G. 1970a: Basische calciumcarbonate (Basic calcium carbonates). *Naturwissenschaften* 57(1), 38–39.

Schimmel, G. 1970b: Elektronenmikroskopische Beobachtungen an basischen Calcium Carbonaten. *Physikalische Blutter* 26, 213–216.

Schwartz, D.R. 1987: D-cracking of concrete pavements, Synthesis of Highway Practice, No 134, Transportation Research Board, Washington, DC.

Scrivener, K. and Skalny, J.P. 2005: Conclusions of the international RILEM TC 186-ISA workshop on internal sulfate attack and delayed ettringite formation (4–6 September 2002, Villars, Switzerland). *Materials and Structures* 38(280), 659–663.

Scrivener, K.L. and Taylor, H.F.W. 1990: Microstructural development in pastes of a calcium aluminate cement. In *Calcium Aluminate Cements, Proceedings of the International Symposium (dedicated to the late Dr H.G. Midgley)*, Mangabhai, R.J. (ed.), London, U.K., Queen Mary and Westfield College, University of London, E. & F.N. Spon (Chapman & Hall), pp. 41–51.

Scrivener, K.L. and Taylor, H.F.W. 1993: Delayed ettringite formation: A microstructural and microanalytical study. *Advances in Cement Research* 5(20), 139–146.

Senior, S.A. and Franklin, J.A. 1987: *Thermal Characteristics of Rock Aggregate Materials*. Report RR 241, Ontario Ministry of Transportation and Communications, Downsview Toronto, Ontario, Canada.

Shayan, A. 1988: Deterioration of a concrete surface due to the oxidation of pyrite contained in pyritic aggregates. *Cement and Concrete Research* 18(5), 723–730.

Shayan, A. and Quick, G.W. 1991/1992: Relative importance of deleterious reactions in concrete: Formation of AAR products and secondary ettringite. *Advances in Cement Research* 4(16), 149–157.

Shayan, A. and Quick, G.W. 1992: Microscopic features of cracked and uncracked railway sleepers. *ACI Materials Journal* 89(4), 348–361.

Shayan, A., Quick, G.W. and Lancucki, C.J. 1993: Morphological, mineralogical and chemical features of steam-cured concrete containing densified silica fume and various alkali levels. *Advances in Cement Research*, 5(20), 151–62.

Sheppard, W.L. 1982: *Chemically Resistant Masonry*, 2nd ed. Marcel Dekker, New York. (revised edition of 'A handbook of chemically resistant masonry', 1st edition, 1977).

Sherwood, W.C. and Newlon, Jr. H.H. 1964: Studies on the mechanisms of alkali-carbonate reaction, part 1. chemical reactions. In *Symposium on Alkali-carbonate Rock Reactions*, Highway Research Record, No.45, Washington, DC, pp. 41–56.

Shrier L.L., Jarman R.A. and Burstein G.T. (eds.). 1994: *Corrosion, Volume 1, Metal/Environment Reactions*, 3rd ed. Butterworth-Heinemann, Oxford, U.K.

Shrimer, F.H. May 2006: Development of the Damage Rating Index method as a tool in the assessment of alkali-aggregate reaction in concrete: A critical review. In *Marc-André Bérubé Symposium on Alkali-Aggregate Reactivity in Concrete*, Fournier, B. (ed.), Montreal, Quebec, Canada, pp. 391–401.

Sibbick, R.G. 2009: Microscopic examination of degradation processes in cement based materials associated with the marine environment, Presented at the *31st International Cement Microscopy Association (ICMA) Conference*, St Petersburg, FL. (available via http://www.cemmicro.org/icma/index.html).

Sibbick, R.G. 2011: Personal communication.

Sibbick, R.G. and Crammond, N.J. 2001: Microscopical investigations into recent field examples of the thaumasite form of sulfate attack (TSA). In *Proceedings of the Eighth Euroseminar on Microscopy Applied to Building Materials*, Athens, Greece, pp. 261–269.

Sibbick, R.G. and Crammond, N.J. 2003: The petrographical examination of popcorn calcite deposition (PCD) within concrete mortar, and its association with other forms of degradation. In *Ninth Euroseminar on Microscopy Applied to Building Materials*, Trondheim, Norway.

Sibbick, R.G., Crammond, N.J. and Metcalf, D. 19–21 June 2002: The microscopical characterisation of thaumasite. In *Proceedings of the First International Conference on Thaumasite in Cementitious Materials, BRE, Garston, UK,* Holton, I. (ed.), Watford, U.K., Building Research Establishment, & CRC Limited. Available on CD.

Sibbick, R.G. and Page, C.L. 27–31 July 1992: Susceptibility of various UK aggregates to alkali-aggregate reaction. In: *The Ninth International Conference on Alkali-Aggregate Reaction in Concrete, Westminster, London*, Conference Papers, Vol. 2, Ref CS 104 (two volumes), The Concrete Society, Slough, U.K., pp. 980–987.

Sibbick, R.G. and Page, C.L. 1998: Mechanisms affecting the development of alkali-silica reaction in hardened concretes exposed to saline environments. *Magazine of Concrete Research* 50(2), 147–159.

Sims, I. 1992: Alkali-silica reaction – UK experience. In *The Alkali-Silica Reaction in Concrete*, Swamy, R.N. (ed.), Blackie & Son Ltd., Glasgow, U.K., Chapter 5, pp. 122–187.

Sims, I. 1994: The assessment of concrete for carbonation. *Concrete* 28(6), 33–37.

Sims, I. 1996: Phantom, opportunistic, historical and real AAR. Getting diagnosis right. In *Proceedings of the 10th International Conference on Alkali-Reaction in Concrete*, Shayan, A. (ed.), Melbourne, Victoria, Australia, pp. 175–182.

Sims, I. 2003: Learning to avoid concrete degradation. *Concrete* 37(1), 34–37.

Sims, I. 2006: Selection of materials for concrete in a hot and aggressive climate. In *Proceedings of the 8th International Conference on Concrete in Hot and Aggressive Environments*, Manama, Bahrain, Bahrain Society of Engineers.

Sims, I. and Brown, B.V. 1998: Concrete aggregates. In *Lea's the Chemistry of Cement and Concrete*, 4th ed., Hewlett, P.C. (ed.). Edward Arnold, London, U.K., Chapter 16, pp. 903–1011.

Sims, I. and Hartshorn, S.A. 1998: Recognising thaumasite. *Concrete Engineering International* 2(8), 44–48.

Sims, I., Hunt, B.J. and Miglio, B.F. 1992: Quantifying microscopical examinations of concrete for alkali aggregate reactions (AAR) and other durability aspects. In *Durability of Concrete*, Holm, J. (ed.), ACI SP-131, American Concrete Institute, Detroit, MI, pp. 267–287.

Sims, I. and Huntley, S.A. 2004: The thaumasite form of sulfate attack – Breaking the rules. *Cement and Concrete Composites* 26(7), 837–844. Originally included in Holton, I. (ed.), *Proceedings of the First International Conference on Thaumasite in Cementitious Materials*, BRE, Garston, U.K., 19–21 June 2002: available on CD.

Sims, I. and Nixon, P.J. (prepared by) 2003a: RILEM recommended test method AAR-0: Detection of alkali-reactivity potential in concrete – Outline guide to the use of RILEM methods in assessments of aggregates for potential alkali-reactivity. *Materials and Structures* 36(261), 472–479. See also RILEM. New edition in preparation.

Sims, I. and Nixon, P.J. (prepared by) 2003b: RILEM recommended test method AAR-1: Detection of potential alkali-reactivity of aggregates – Petrographic method. *Materials and Structures* 36(261), 480–496. See also RILEM. New edition in preparation.

Sims, I. and Nixon, P.J. May 31–June 3, 2006: Assessment of aggregates for alkali-aggregate reactivity potential: RILEM international recommendations. In *Marc-André Bérubé Symposium on Alkali-Aggregate Reactivity in Concrete*, Fournier, B. (ed.), 71–91, part of the *Proceedings of the 8th CANMET/ACI International Conference on Recent Advances in Concrete Technology*, Montréal, Quebec, Canada, CANMET-MTL (Canada Centre for Mineral and Energy Technology – Materials Technology Laboratory), Department of Natural Resources, Hamilton, Ontario, Canada.

Sims, I., Nixon, P.J., Blanchard, I.G. and Bennett-Hughes, P. 16–20 June, 2008: Assessment of concrete in service in the UK – Remembering the value of practical petrography. In *Proceedings of the 13th International Conference on Alkali-Aggregate Reaction (ICAAR)*, Broekmans, M.A.T.M. and Wigum, B.J. (eds.), Trondheim, Norway. Available only on CD.

Sims, I., Nixon, P.J. and Godart, B. 15–20 October 2011: Eliminating alkali-aggregate reaction from new dams. In *CDA 2011 Annual Conference, Fredericton, New Brunswick, Canada*, Canadian Dam Association, Moose Jaw, Saskatchewan, Canada (available on CD).

Sims, I., Nixon, P.J. and Godart, B. 20–25 May 2012a: Eliminating alkali-aggregate reaction from long-service structures. In *Proceedings of the 14th International Conference on Alkali-Aggregate Reaction*, Folliard, K.J. (ed.), Austin, TX.

Sims, I., Nixon, P.J., Godart, B. and Charlwood, R. October 2012b: Recent developments in the management of chemical expansion of concrete in dams and hydro projects – Part 2: RILEM proposals for prevention of AAR in new dams, presented at *Hydro 2012: Innovative Approaches to Global Challenges*, Bilbao, Spain, 12pp.

Sims, I., Nixon, P.J. and Marion, A.-M. 15–19 October 2004: International collaboration to control alkali-aggregate reaction: The successful progress of RILEM TC 106 and TC 191-ARP. In *Proceedings of the 12th International Conference on Alkali-Aggregate Reaction in Concrete*, Tang, M.-S. and Deng, M. (eds.), Vol. 1, Beijing, China, International Academic Publishers/World Publishing Corporation, pp. 41–50.

Sims, I. and Poole, A.R. 1980: Potentially alkali-reactive aggregates from the Middle East. *Concrete* 14(5), 27–30.

Sims, I. and Poole, A.B. 2003: Alkali-aggregate reactivity. In *Advanced Concrete Technology – Concrete Properties*, Newman, J. and Ban Seng, C. (eds.), Elsevier, Oxford, U.K., Chapter 13, pp. 13/1–13/37.

Skalny, J. 1993: Simultaneous presence of alkali-silica gel and ettringite in concrete. *Advances in Cement Research* 5(17), 23–29.

Skalny, J., Clark, B.A. and Lee, R.J. 1992: Alkali-silica reaction revisited. In *Proceedings of the 14th International Conference on Cement Microscopy*, pp. 309–324.

Skalny, J., Marchand, J. and Odler, I. 2002: *Sulfate Attack on Concrete*, Modern Concrete Technology, 10, Spon Press, London, U.K., 217pp.

Smart, S. 1999: Concrete – The burning issue. *Concrete* 33(7), 30–34.

Smith, A.S., Dunham, A.C. and West, G. 1992: Undulatory extinction of quartz in British hard rocks. In *Proceedings of the 9th International Conference on Alkali-Aggregate Reaction in Concrete, Westminster, London*, Vol. 2, Ref. CS.104 (Two Volumes), Slough, U.K., The Concrete Society, pp. 1001–1008.

Sommer, H., Grattan-Bellew, P., Katayama, T. and Tang, M. 2004: Development and inter-laboratory trial of the RILEM AAR-5 rapid preliminary screening test for carbonate aggregates. In *Proceedings of the 12th International Conference on Alkali-Aggregate Reaction in Concrete*, Tang, M. and Deng, M. (eds.), Vol. 1, Beijing, China, International Academic Publishers/World Publishing Corporation, pp. 407–412.

Sommer, H. and Katayama, T. 20–23 September 2006: Screening carbonate aggregates for alkalireactivity. In *Ibausil 16 Internationale Baustofftagung*, Weimar, Germany, Tagungsbericht – Band 2, 2-0461-2-0468.

Sommer, H., Nixon, P.J. and Sims, I. (prepared by) 2005: AAR-5: Rapid preliminary screening test for carbonate aggregates (RILEM TC 191-ARP: Alkali-reactivity and prevention – Assessment, specification and diagnosis of alkali-reactivity). *Materials and Structures* 38(282), 787–792. See also RILEM.

Sorrentino, D.M., Clément, J.Y. and Goldberg, J.M. 1992: A new approach to characterize the chemical reactivity of the aggregates. In *Proceedings of the 9th International Conference on Alkali-Aggregate Reaction in Concrete, Westminster, London*, Vol. 2, Ref. CS.104, Slough, U.K., The Concrete Society, pp. 1009–1016.

Spence, R.D. (ed.). 1993: *Chemistry and Microstructure of Solidified Waste Forms*. Oak Ridge National Laboratory, Lewis Publishers, Boca Raton, FL.

Stanton, T.E., Porter, O.J., Meder, L.C. and Nicol, A. 1942: California experience with the expansion of concrete through reaction between cement and aggregate. *Journal American Concrete Institute* 13(3), 209–236 (and discussion in Supplement to November 1942).

Stark, D. 1982: Longtime study of concrete durability in sulfate soils. In *George Verbeck Symposium on Sulfate Resistant Concrete*, ACI-SP77, American Concrete Institute, Detroit, MI, pp. 21–40.

Stark, D. 1985: *Alkali-Silica Reactivity in Five Dams in the Southwestern United States.* REC-ERC-85-1O, U.S. Department of the Interior, Bureau of Reclamation, Washington DC.

Stark, D. 1987: Deterioration due to sulphate reactions in Portland cement-stabilized slag aggregate concrete. In *Durability of Concrete,* ACI SP-100, American Concrete Institute, Detroit, MI, pp. 2019–2102.

Stark, D. 2002: *Performance of Concrete in Sulfate Environments.* Portland Cement Association, Research and Development Bulletin, RD 129, Skokie, IL.

Stark, D., Morgan, B. and Okamoto, P. 1993: *Eliminating or Minimising Alkali-Silica Reactivity.* Report SHRP-C-343, Strategic Highway Research Program, National Research Council, Washington, DC.

Stark, J., Bollman, K. and Seyfarth, K. 1992: Investigation into delayed ettringite formation in concrete. In *Ninth International Conference on the Chemistry of Cement,* New Delhi, India, V, pp. 248–354.

Stark, J., Freyburg, E., Seyfarth, K., Giebson, C. and Erfurt, D. 2010: 70 years of ASR with no end in sight? (Parts 1 & 2). *ZKG International* (formerly Zement-Kalk-Gips) (4), 86–95, & (5), 55–70.

Stimson, C.C. (Chairman) 1997: *The' Mundic' Problem – A Guidance Note – Recommended Sampling, Examination and classification procedure for Suspect Concrete Building Materials in Cornwall and Parts of Devon,* 2nd ed. The Royal Institution of Chartered, Surveyors (RICS), London, U.K. See Royal Institution of Chartered Surveyors for 3rd edition (2015).

St John, D.A. 1982: An unusual case of ground water sulfate attack on concrete. *Cement and Concrete Research* 12(5), 633–639.

St John, D.A. 1985: The application of microscopy in the determination of alkali-aggregate reaction. In *Proceedings of the 7th International Conference on Cement Microscopy,* pp. 395–406.

St John, D.A. 1990: The use of large-area thin sectioning in the petrographic examination of concrete. In *Petrography Applied to Concrete and Concrete Aggregates,* Erlin, B. and Stark, D. (eds.), ASTM STP-1061, American Society for Testing and Materials, Philadelphia, PA, pp. 55–70.

St John, D.A. 1991: *Petrographic Examination of corroded Stormwater Pipe Samples from the Business District,* BM290, New Zealand DSIR, Chemistry Division, Internal Report, Rotorua, New Zealand.

St John, D.A. 1992: AAR investigations in New Zealand – Work carried out since 1989. In *Proceedings of the 9th International Conference on Alkali-Aggregate Reaction in Concrete,* Vol. 2, Slough, U.K., The Concrete Society, pp. 885–893, Check referred to in text.

St John, D.A. and Abbott, J.H. 1987: The microscopic textures of geothermal cement grouts exposed to steam temperatures of 150°C and 260°C. In *Proceedings of the 9th International Conference on Cement Microscopy,* pp. 280–300.

St John, D.A. and Goguel, R.L. 1992: Pore solution/aggregate enhancement of alkalies in hardened concrete. In *Proceedings of the 9th International Conference on Alkali-Aggregate Reaction,* Vol. 2, London, U.K., The Concrete Society, pp. 894–901.

St John, D.A., McLeod, L.C. and Milestone, N.B. 1993: *An Investigation of the Mixing and Properties of DSP Mortars made from New Zealand Cements and Aggregates.* Industrial Research Limited, Report No. 41, The New Zealand Institute for Industrial Research and Development, Lower Hutt, New Zealand.

St John, D.A. and Penhale, H.R. 1980: Estimation of ground water corrosion of concrete sewer pipes in the Hutt Valley, New Zealand. *Durability of Building Materials and Components,* ASTM STP-691, American Society for Testing and Materials, Philadelphia, PA, pp. 377–387.

St John, D.A., Poole, A.B. and Sims, I. 1998: *Concrete Petrography – A Handbook of Investigative Techniques,* 1st ed. Edward Arnold, London, U.K., 474pp.

St John, D.A. and Smith, L.M. September 1976: Expansion of concrete containing New Zealand argillite aggregate. In *The Effect of Alkalies (sic) on the Properties of Concrete, Proceedings of a Symposium held in London,* Poole, A.B. (ed.), (now recognised as the *3rd International Conference on Alkali-Aggregate Reaction*), Slough, U.K., Cement and Concrete Association, pp. 319–352.

Strunge, H. and Chatterji, S. June 1989: Microscopic observations on the mechanism of concrete neutralization. In *Durability of Concrete: Aspects of Admixtures and Industrial by-Products, 2nd International Seminar,* Berntsson, L., Chandra, S. and Nilsson, L.-O. (eds.), Swedish Council for Building Research, Stockholm, Sweden, pp. 229–235.

Stumm, W. and Morgan, J.J. 1970: *Aquatic Chemistry*. Wiley Interscience, New York.

Stutterheim, N. 1954: Excessive shrinkage of aggregates as a cause of deterioration of concrete structures in South Africa. *Transactions of the South African Institution of Civil Engineers* 4(12), 351–367.

Stutterheim, N., Laurie, J.A.P. and Shand, N. 1967: Deterioration of a multiple arch dam as a result of excessive shrinkage of aggregate. In *Proceedings of the 9th International Conference on Large Dams*, Istanbul, Turkey, pp. 227–4.

Swamy, R.N. (ed.). 1992a: *The Alkali-Silica Reaction in Concrete*. Blackie, Glasgow, U.K., 336pp.

Swamy, R.N. 1992b: Role and effectiveness of mineral admixtures in relation to alkali silica reaction. In *Durability of Concrete*, ACI SP-131, 219–54, American Concrete Institute, Detroit, MI.

Swenson, E.G. 1957: A reactive aggregate undetected by ASTM Tests. *American Society for Testing & Materials Bulletin* 226, 48–51.

Swenson, E.G. and Gillott, J.E. 1960: Characteristics of Kingston carbonate rock reaction. *Highway Research Board Bulletin* 275, 18–31.

Swenson, E.G. and Gillott, J.E. 1967: Alkali reactivity of dolomitic limestone aggregate. *Magazine of Concrete Research*, 19(59), 95–104.

Sylla, H.M. 1988: Reaktionen im Zementstein durch Warmebehandlung. *Beton* V(38), 449–454.

Talvite, N.A. March–April 1964: Determination of free silica: Gravimetric and spectrophotometric procedures applicable to air-borne and settled dust. *Industrial Hygiene Journal*, 169–178.

Tang, M. 27–31 July 1992: Classification of alkali-aggregate reaction. In *Proceedings of the 9th International Conference on Alkali-Aggregate Reaction in Concrete, Westminster, London*, Vol. 2, Ref. CS.104 (Two Volumes), Slough, U.K., The Concrete Society, pp. 648–653.

Tang, M. and Deng, M. 2004: Progress on the studies of alkali-carbonate reaction. In *Proceedings of the 12th International Conference on Alkali-Aggregate Reaction in Concrete*, Tang, M. and Deng, M. (eds.), Vol. 1, Beijing, China, International Academic Publishers/World Publishing Corporation, pp. 51–59.

Tang, M., Deng, M., Lan, X. and Han, S. 1994: Studies on alkali-carbonate reaction. *ACI Materials Journal* 91(1), 26–29.

Tang, M., Deng, M., Xu, Z., Lan, X. and Han, S. 1996: Alkali-aggregate reactions in China. In *Proceedings of the 10th International Conference on Alkali-Aggregate Reaction in Concrete*, Shayan, A. (ed.), Melbourne, Victoria, Australia, pp. 195–201.

Tang, M., Han, S. and Zhen, S. 1983: A rapid method for identification of alkali reactivity of aggregate. *Cement and Concrete Research* 13, 417–422.

Tang, M., Lan, X. and Han, S. 1994: Autoclave method for identification of alkali-reactivity of carbonate rocks. *Cement and Concrete Composites* 16, 163–167.

Taylor, H.W.F. 1990: *Cement Chemistry*. Academic Press, London, U.K.

Taylor, H.F.W. 1997: *Cement Chemistry*, 2nd ed. Thomas Telford, London, U.K.

Tepponen, D. and Ericksson, R.E. 1987: Damage in concrete railway sleepers in Finland. *Nordic Concrete Research* (6), 199–209.

Terzaghi, R.D. 1949: Concrete deterioration due to carbonic acid. *Journal of the Boston Society of Civil Engineers* 36, 136–60.

Thamasite Expert Group. 2000: Thaumasite Expert Group –one-year review. *Concrete* 34(6), 51–53. See also DETR.

Thaulow, N. and Geiker, M.R. 27–31 July 1992: Determination of the residual reactivity of alkali silica reaction in concrete. In *Proceedings of the 9th International Conference on Alkali-Aggregate Reaction in Concrete, Westminster, London*, Vol. 2, Ref. CS.104, Slough, U.K., The Concrete Society, pp. 1050–1057.

Thaulow, N. and Jacobsen, U.H. 1997: Deterioration of concrete diagnosed by optical microscopy. In *Proceedings of the 6th Euroseminar on Microscopy Applied to Building Materials*, Sveinsdóttir, E.L. (ed.), Reykjavik, Iceland, pp. 282–296.

Thaulow, N., Johansen, V. and Jakobsen, U.H. 1997: What causes delayed ettringite formation? In *Mechanisms of Chemical Degradation of Cement-Based Systems*, Scrivener, K.L. and Young, J.F. (eds.). E & F L Spon, London, U.K., Chapter/Paper 26, pp. 219–226.

Thaulow, N., Lee, R.J., Wagner, K. and Sahu, S. 2001: Effect of calcium hydroxide on the form, extent, and significance of carbonation. In *Materials Science of Concrete: Calcium hydroxide in Concrete*, Skalny, J., Gebauer, J. and Odler, I. (eds.), Special Volume, American Ceramic Society, Westerville, OH, pp. 191–202.

Thistlethwayte, D.K.B. 1972: *The Control of Sulphides in Sewerage Systems*. Butterworths, London, U.K.

Thomas, M.D.A. 2001: Delayed ettringite formation: Recent developments and future directions. In *Materials Science of Concrete*, Mindess, S. and Skalny, J. (eds.), VI, American Ceramics Society, Westerville, OH, pp. 435–482.

Thomas, M.D.A. and Folliard, K.J. 2007: Concrete aggregates and the durability of concrete. In *Durability of Concrete and Cement Composites*, Page, C.L. and Page, M.M. (eds.), Woodhead & Maney (pp Institute of Materials, Minerals & Mining), Cambridge, U.K., Chapter 7, pp. 247–281.

Thomas, M.D.A., Folliard, K.J., Drimalas, T. and Ramlochan, T. 5–9 June 2007: Diagnosing delayed ettringite formation in concrete structures. In *11th Euroseminar on Microscopy Applied to Building Materials*, Porto, Portugal, 11pp.

Thomas, M.D.A., Folliard, K., Drimalas, T. and Ramlochan, T. 2008: Diagnosing delayed ettringite formation in concrete structures. *Cement and Concrete Research* 38, 841–847.

Thomas, M.D.A., Fournier, B. and Folliard, K.J. 2008: *Report on Determining the Reactivity of Concrete Aggregates and Selecting Appropriate Measures for Preventing Deleterious Expansion in New Concrete Construction*. Report No. FHWA-HIF-09-001, Office of Pavement Technology, Federal Highway Administration, Washington, DC, 20pp.

Thomson, M.L., Grattan-Bellew, P.E. and White, J.C. 1994: Application of microscopic and XRD techniques to investigate alkali-silica reactivity potential of rocks and minerals. In *Proceedings of the 16th International Conference on Cement Microscopy*, pp. 174–192.

Thornton, H.T. 1978: Acid attack on concrete caused by sulfur bacteria action. *ACI Journal* 75, 577–584.

Tong, L. 1994: Alkali-carbonate rock reaction, PhD thesis, Nanjing Institute of Chemical Technology, Nanjing, China, 219pp.

Tong, L. and Tang, M. 1996: Concurrence of alkali-silica and alkali-dolomite reaction. In *Proceedings of the 10th International Conference on Alkali-Aggregate Reaction in Concrete*, Shayan, A. (ed.), Melbourne, Victoria, Australia, pp. 742–749.

Torii, K., Kawamura, M., Matsumoto, K. and Ishii, K. 1996: Influence of cathodic protection on cracking and expansion of the beams due to alkali-silica reaction. In *Proceedings of the 10th International Conference on Alkali-Aggregate Reaction in Concrete*, Shayan, A. (ed.), Melbourne, Victoria, Australia, pp. 653–660.

Torres, S.M., Sharp, J.H., Swamy, R.N., Lynsdale, C.J. and Huntley, S.A. 19–21 June 2002: Long-term durability of Portland limestone cement exposed to magnesium sulfate attack. In *Proceedings of the First International Conference on Thaumasite in Cementitious Materials*, Holton, I. (ed.), held at BRE, Garston, U.K., Building Research Establishment, Watford, U.K. Available as a CD.

Tovey, A.K. (Chairman) 1990: *Assessment and Repair of Fire-Damaged Concrete Structures*, Technical Report No 33, The Concrete Society, Slough, U.K. See also Concrete Society and updated version, TR No 68.

Tremper, B. 1932: The effect of acid waters on concrete. *Proceedings American Concrete Institute* 28(1), 1–32.

Tuohy, B., Carroll, N. and Edger, M. 2012: *Report of the Pyrite Panel*, prepared for the Department of Environment, Community and Local Government, Dublin, Ireland, 174pp. (Brendan Tuohy was Chairman of the Pyrite Panel).

Tuutti, K. 1982: *Corrosion of Steel in Concrete*, No 4–82, Swedish Cement and Concrete Institute (CBI), Stockholm, Sweden.

Urabe, K. 1990: Phase changes of calcium silicate hydrate on heating. *Proceedings of Cement and Concrete, Cement Association of Japan (CAJ)* (44), 36–41. (In Japanese with English abstract.)

Valenta, O. and Modry, S. 1969: A study of the determination of surface layers of concrete. *International Symposium on the Durability on Concrete*, Prague, Czech Republic, Final Report, 1, pp. A55–A64.

Van Aardt, J.H.P. and Visser, S. 1975: Thaumasite formation: A cause of deterioration of Portland cement and related substances in the presence of sulfates. *Cement and Concrete Research* 5(3), 225–232.

Van Dam, T.J., Sutter, L.L., Smith, K.D., Wade, M.J. and Peterson, K.R. 2002: *Guidelines for Detection, Analysis, and Treatment of Materials-Related Distress in Concrete Pavements*, Vol. 1: Final Report, FHWA-RD-01-163, Federal Highway Administration, Washington, DC.

Varma, S.P. and Bensted, J. 1973: Studies of thaumasite. *Silicates Industriels* 38, 29–32.

Vennesland, O. 1994: Electrochemical chloride removal and realkalisation – Maintenance methods for concrete structures. In *Improving Civil Engineering Structures Old and New*, French, W.J. (ed.), Geotechnical Publishing, Basildon, U.K., pp. 113–124.

Venu, K., Balakrishnan, K. and Rajagopalan, S. 1965: The potentiokinetic polarization study of the behaviour of steel in NaOH-NaCl system. *Corrosion Science* 5, 59–69.

Verbeck, G.J. 1958: Carbonation of hydrated Portland cement. *Cement and Concrete*, ASTM STP 205, American Society for Testing and Materials, Philadelphia, PA, pp. 17–36.

Vieira, A.P., Barbosa, N.P., Torres, S.M. and Leal, A.F. 2010: Thaumasite form of sulphate attack in a tropical climate. *Concrete* 44(6), 53–54.

Villeneuve, V., Fournier, B. and Duchesne, J. 2012: Determination of the damage in concrete affected by ASR – The Damage Rating Index (DRI). In *Proceedings of the 14th International Conference on Alkali-Aggregate Reaction*, Folliard, K.J. (ed.), Austin, TX. (proceedings available on a memory stick).

Vivian, H.E. 1947: The effect of void space on mortar expansion. *Commonwealth of Australia, CSIR Bulletin*, 229, Part III, 47–54.

Wakely, L.D., Poole, T.S., Weiss, C.A. and Burkes, J.P. 1992: Petrographic techniques applied to cement-solidified hazardous wastes. *Proceedings of the 14th International Conference on Cement Microscopy*, pp. 333–350.

Walker, M.J. (ed.). 2000: *Diagnosis of Deterioration in Concrete Structures – Identification of Defects, Evaluation and Development of Remedial Action*. Report of a Working Party, Technical Report No 54, The Concrete Society, Crowthorne, U.K.

Walker, M.J. (ed.). 2002: *Guide to the Construction of Reinforced Concrete in the Arabian Peninsula*, CIRIA Publication C577 & Concrete Society Special Publication CS136, Published jointly by CIRIA and the Concrete Society, Crowthorne & London, U.K.

Walker, M.J. (ed.). 2012: *Hot Deserts: Engineering, Geology and Geomorphology – Engineering Group Working Party Report*, Engineering Geology Special Publication No 25, Geological Society, London, U.K.

Walton, P.L. 1993: *Effects of Alkali-Silica Reaction on Concrete Foundations*, BRE Information Paper, IP 16/93, Building Research Establishment, Watford, U.K., 4pp.

Webb, J. 6 November 1993: The battle of Britain's nuclear dustbin. *New Scientist*, 14–15.

Werner, D. and Giertz-Hedstrom, S. 1934: Physical and chemical properties of cement and concrete. *Engineer* 135, 235–238.

West, G. 1996: *Alkali-Aggregate Reaction in Concrete Roads and Bridges*. Thomas Telford, London, U.K., 163pp.

Wigum, B.J. 1995: Examination of microstructural features of Norwegian cataclastic rocks and their use for predicting alkali-reactivity in concrete. *Engineering Geology* 40, 195–214.

Wigum, B.J. 1996: A classification of Norwegian cataclastic rocks for alkali-reactivity. In *Proceedings of the 10th International Conference on Alkali-Aggregate Reaction in Concrete*, Shayan, A. (ed.), Melbourne, Victoria, Australia, pp. 758–766.

Wilkins, N.J.M. and Lawrence P.F. 1980: Fundamental mechanisms of corrosion of steel reinforcement in concrete immersed in seawater. Technical Report No. 6, *Concrete in the Oceans*, Cement and Concrete Association, Crowthorne, U.K.

Willams, D.A. and Rogers, C.A. 1991: *Field Trip Guide to Alkali-Carbonate Reactions in Kingston, Ontario*. Report MI-145, Ontario Ministry of Transportation, Downsview, Kingston, Ontario, Canada, 26pp.

Winter, N.B. 2012a: *Scanning Electron Microscopy of Cement and Concrete*. WHD Microanalysis Consultants Ltd, Woodbridge, U.K., 192pp.

Winter, N.B. 2012b: Personal communication.

Winter, N.B. 2012c: *Understanding Cement – An Introduction to Cement Production, Cement Hydration and Deleterious Processes in Concrete.* WHD Microanalysis Consultants Ltd, Woodbridge, U.K., 206pp.

Wood, J.G.M., Nixon, P.J. and Livesey, P. 1996: Relating ASR structural damage to concrete composition and environment. In *Proceedings of the 10th International Conference on Alkali-Aggregate Reaction in Concrete*, Shayan, A. (ed.), Melbourne, Victoria, Australia, pp. 450–457.

Woolf, D.O. 1952: Reaction of aggregate with low-alkali cement. *Public Roads* 27(3), 49–56.

Xu, Z., Lan, X., Deng, M. and Tang, M. 2000: A new accelerated method for determining the potential alkali-carbonate reactivity. In *Proceedings of the 11th International Conference on Alkali-Aggregate Reaction in Concrete, Québec, Canada*, Bérubé, M.-A., Fournier, B. and Durand, B. (eds.), Sherbrook, Québec, Canada, Centre de Recherche Interuniversitaire sur le Béton (CRIB), Laval & Sherbrook Universities, pp. 129–138.

Yamanaka, T., Miyasaka, H., Aso, I., Tanigawa, M., Shoji, K. and Yohta, H. 2002: Involvement of sulfur- and iron-transforming bacteria in heaving of house foundations. *Geomicrobiology Journal* 19, 519–528.

Yang, R., Lawrence, C.D., Lynsdale, C.J., and Sharp, J.H. 1999: Delayed ettringite formation in heat-cured Portland cement mortars. *Cement and Concrete Research*, 29(1), 17–25.

Zappia, G., Sabbioni, C., Pauri, M.G. and Gobbi, G. 1994: Mortar damage due to airborne sulfur compounds in a simulation chamber. *Materials and Structures* 27, 469–473.

Zhang, X., Blackwell, B.Q. and Groves, G.W. 1990: The microstructure of reactive aggregates. *British Ceramic Transactions and Journal* 89, 89–92.

Zoldners, N.G. 1971: Thermal properties of concrete under sustained elevated temperatures. In *Temperature and Concrete*, ACI SP-25, Paper SP25-01, American Concrete Institute, Detroit, U.K., pp. 1–32.

Zubkova, N.V., Filinchuk, Y.E., Pekov, I.V., Pushcharovsky, Y. and Gobechiya, E.R. 2010: Crystal structures of shlykovite and cryptophyllite: Comparative crystal chemistry of phyllosilicate minerals of the mountainite family. *European Journal of Mineralogy* 22(4), 547–555.

Zubkova, N.V., Pekov, I.V., Pushcharovsky, Y. and Chukanov, N.V. 2009: The crystal structure and refined formula of mountainite, $KNa_2Ca_2[Si_8O_{19}(OH)] \cdot 6H_2O$. *Zeitschrift für Kristallographie*, 224, 389–396.

Chapter 7

Precast and special concretes

7.1 STANDARD PRECAST CONCRETE UNITS

A wide range of materials, products and techniques involve the use of Portland-type cements and a number of these are discussed in this chapter. The related materials, lime and gypsum, both of which pre-date Portland cement, are considered in Chapter 9. The choice and grouping of the topics chosen illustrate how petrography can be applied to such materials is not intended to be exhaustive, while some overlap with previous chapters has been necessary to ensure that each topic is adequately covered.

The term precast concrete applies to products that are made as modules or units that are later fitted together in construction. Beams, railway sleepers, panels, pavers, blocks, tiles and pipes which may, or may not contain steel reinforcement together make up the bulk of the precast industry output. To provide economies of scale, organisation and insure better product control they are usually made in factories rather than on site. The types and methods of manufacture of precast products are discussed by Glanville (1950), Childe (1961), Richardson (1991, 2003) and in the proceedings of triennial International Congresses of the Precast Concrete Industry organised by the Bureau International du Beton Manufacture (www.bibm.eu). A recent conference was held in Istanbul, Turkey in 2014. Descriptions of specifications, manufacturers and products intended as summary information for architects and engineers have been outlined in the annual publication 'Specification' (e.g., Williams 1994). Its final issue was produced in 1995 and is now only available on a CD-ROM basis. A general review of typical methods and materials used in the precast industry is given by Levitt (1982). General texts on finishes and materials intended mainly for architects have been written by Deane (1989) and Everett (1994).

The petrographer may not only be required to examine recent laboratory test specimens and construction units, but may also be required to investigate structures that date back to the introduction of Portland cement in the mid-nineteenth century. Consequently, an attempt has been made to detail some of the older techniques and products. However, pre-nineteenth-century materials have been largely excluded as these are close to archaeology and are outside the scope of this book.

The key to successful petrographic investigation of the manufacturing techniques and products produced relies on an understanding of the chemistry and physics of the materials and of the fabrication processes used. When these are understood petrographic examination of texture may confirm whether the claimed processes and materials specified have been used. It also may be possible to identify changes in the cementious fabric that may have occurred with time since manufacture and the causes of any deterioration.

In many respects precast concrete is identical in appearance to concrete cast on-site, nevertheless it is usually a simple matter to identify precast concrete units during an on-site inspection, or from building records. In a few cases additional information is cast or printed onto the unit itself and may allow source, batch and manufacture date to be determined. Since precast concrete is usually cast under standardised factory conditions with attention to quality control for both the raw materials and the finished product the individual units from a given batch are likely to be closely similar in mix design, appearance and in their physical and mechanical properties. In general the uniformity of precast units will allow comparative assessment of any localised external causes of damage such as frost or fire, but where the deterioration observed relates to the materials used, or to the manufacturing process, then similar damage may be expected to occur in other units from the same batch.

In general, as a result of the manufacturing process, the low water/cement ratios used and the quality control, precast concrete may be expected to be better compacted, with no honeycombing or bleeding, although the dry mixes used tend to increase the number entrapped air voids present and they tend to be evenly distributed. The aggregate also tends to be more evenly distributed, compared to cast in-situ concrete.

7.2 PRECAST, BLOCK, BRICK, TILE AND PAVERS

These products represent a major section of the precast concrete industry. This discussion has been limited to those containing normal weight aggregates and conventional reinforcement. Most of the differences between them are the result of variations in mix design, or are due to the method of fabrication used. In spite of automated methods being readily available, a surprising proportion of production is still manufactured by older methods, and typically, the more unusual or special items involve hand or machine tamping methods to consolidate the concrete into the mould.

Concrete masonry units are produced in a variety of sizes and shapes but may be divided into four broad groups, solid and hollow load bearing blocks, and solid or hollow non-load bearing blocks, as specified in ASTM C129 (2006). The load bearing blocks may be further subdivided according to their intended use and degree of exposure. In the United Kingdom precast concrete masonry units are specified in BS EN 771-3 (2003).

Concrete blocks, bricks, tiles and pavers are now commonly manufactured using a single casting machine for the whole range of products. These machines are proprietary equipment which use efficient, high energy vibration combined with pressure to compact or extrude semi-dry mixes of concrete either into an appropriate mould, or in the case of tiles onto a pallet mould. After casting/extrusion, the smaller concrete units though fragile, are stiff enough to be immediately demoulded. Cement contents typically range from 230 kg/m^3 for blocks and bricks and up to 350 kg/m^3 for tiles and pavers and water/cement ratios range from about 0.45 to 0.35. A typical maximum size of aggregate used is about 8 mm and the fineness moduli range from 3.5 to 5.0 indicating that very coarse sand gradings are used. This combination of a coarse sand and a low water/cement ratio results in a semi-dry mix. The unit may be cured at ambient temperature, in low-pressure steam at about 80°C, or in high-pressure steam at 180°C (see Section 7.5). In the last case silica flour may have been added to ensure the formation of the higher strength tobermorite type cement hydrates. The partial replacement of the cement by fly ash, slag or pozzolana is also a common practice.

The textures observed in the thin section of concrete block, concrete brick, tile and pavers reflect the semi-dry mixes used. At small scale the external surfaces are irregular with many re-entrant pores and aggregates which protrude to the outer surface of the paste. Internally, because of the aggregate gradings used there are numerous large, irregular pores, as well as smaller pores if air entrainment was used (see Figures 7.1 and 7.2).

Uncarbonated, crystals of calcium hydroxide may line and protrude into the large pores, line aggregate margins and often partially fill the smaller pores. In the hardened paste the

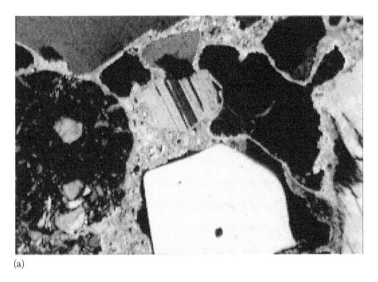

(a)

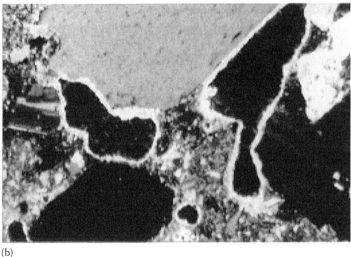

(b)

Figure 7.1 The internal texture of concrete brick. Width of fields 0.8 mm. (a) Crossed polars. One of the typical large pores that are scattered throughout the texture of a concrete brick are outlined by the intense carbonation near the surface. This brick was made by a Columbia machine and cured at 80°C in a carbon dioxide enriched atmosphere. There are remnants of a glass rim around the feldspar aggregate at the bottom of the micrograph. (b) Crossed polars. A few mm inside the brick the carbonation curing has outlined the margins of pores. The hardened paste fills the remaining space between the densely packed aggregate and is only partially carbonated. Carbonation is absent deeper than a few mm from the surface. The texture is of a well-compacted concrete with insufficient cement paste to fill all the space between the aggregates.

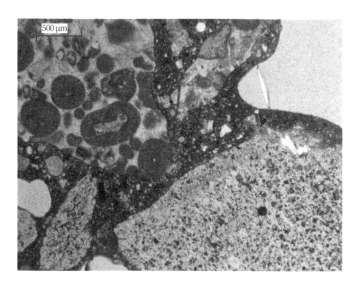

Figure 7.2 Concrete block containing coarse aggregate particles of limestone and chert bound in a Portland cement matrix. Note rounded pale straw air voids. Plane polarised illumination. (Courtesy of RSK Environment Ltd., Hemel Hempstead, U.K.)

crystals of calcium hydroxide tend to be minute and platy rather than elongate while the aggregate appears closely packed so that no large areas of cement paste are observable. This is the texture of harsh mortar with insufficient cement paste to fill all the space between the aggregates. The addition of silica flour, fly ash and slag is unlikely to materially change this texture, except for the presence of their residues and some reduction in the amount of calcium hydroxide observed. Drying shrinkage cracks are unlikely in these products as is to be expected given the cement contents and low water/cement ratios used.

Concrete block masonry made with plastic mixes present no specific features for petrographic study different from those already considered elsewhere in this text. Similarly, pavers and tiles compacted by hydraulic pressure present textures typical of dense concrete. These products are normally plain but special surface finishes are sometimes used, including 'glazes' formed by mixing selected aggregates materials and pigments with a thermosetting binder which may be hardened by a high-temperature high-pressure curing process. The principal properties of interest in these products are usually strength, dimensional stability, moisture movement and appearance. Petrographic examination of the concrete is not directly relevant to the investigation of these properties apart from appearance. A common problem with unsightly concrete block is caused by efflorescence (see Figure 7.3. This problem is also discussed in Section 5.5.6). A more unusual case of sulphate efflorescence is shown in Figure 7.4.

An investigation of unacceptable colour variation is described in a case study reported by D. St John. Intermittent subtle colour variations were occurring in a concrete brick made with ordinary Portland cement coloured with a titanium pigment. The bricks were initially low-pressure steam cured in an atmosphere enriched in carbon dioxide. It was claimed that the bricks were all fabricated under exactly the same conditions. Usually the subtle colour variation was not detected until the bricks were used in construction when the variation became apparent. The differences in shade could be clearly seen when bricks were compared side by side and could also be measured on a colour comparator. It was established that the change in colour affected the whole surface and was not an edge effect. Comparison of bricks of different colours in thin section failed to reveal any significant differences. Examination of fracture surfaces by scanning electron microscopy and EDS analysis showed that surfaces

Figure 7.3 Calcium carbonate formed from calcium hydroxide leached by rain onto the surfaces of a vertical concrete blockwork wall.

Figure 7.4 Calcium sulphate and calcium sodium sulphate efflorescence on new load-bearing precast concrete blocks.

of the bricks which were different in colour from 'normal bricks' contained less calcium. The inference drawn was that undetected variations in the production process were intermittently reducing the amount of efflorescence present on some of the bricks and allowing more of the titanium pigment to become visible.

'No fines concrete' blocks are made using a nominal single-sized coarse aggregate mixed with cement and water but contain no fine aggregate. Such blocks have a moderate compressive strength (1.5–14 MPa) but high porosity and good thermal and sound insulating properties. The water/cement ratio required is critical and minimal compaction is used only to avoid the formation of cavities. Reviews of no-fines concrete are given by Malhotra (1976), Spratt (1980), Neville (1995) and Neville and Brookes (1987). No-fines concrete, because of its high void content, is typically well carbonated but except for the absence of fine aggregate presents no petrographic differences from normal concrete which has carbonated. The texture of a no-fines concrete is shown in Figure 5.50a and also in Figure 8.7.

7.3 PRECAST CONCRETE PIPES

Concrete pipes were originally made by hand methods, or machine tamping to consolidate the concrete into the mould. The mechanical processes used for making pipe can be divided into horizontal and vertical forming methods (Dutton 1977). A large diameter horizontal machine is illustrated in Figure 7.5. Around the 1870s the first pipe-making machine was developed which employed a rotating dolly to simultaneously form the inner wall and compact the concrete in a vertical mould. In 1910 the Hume process of centrifugally spinning a

Figure 7.5 A 3600 mm diameter centrifugal spun concrete pipe being manufactured in a temporary fac-tory near the project. The inside is being finished and the mould is undergoing its final spinning sequence to ensure compaction. The concrete feeder system used to place the concrete is in the foreground. (Photograph courtesy of D. St John.)

horizontal mould in which concrete is placed to form the pipe wall was introduced. In spite of being labour intensive it remains one of the principal methods of fabrication and is still used in many parts of the world. This versatile process is used to make pipes and hollow poles and allows the use of steel reinforcing and prestressing. Around 1960 the 'Roller Suspension' process originally known as the Rocla process was introduced whereby concrete placed in a slowly rotating, horizontal mould is compacted by a high-pressure roller. In some respects it is a variant of the Hume process where the pressure roller replaces the centrifugal forces to compact the concrete.

There is a range of national specifications covering concrete pipes among which the fol-lowing are applicable. Pipes are described in general terms by BS 5911 (2002) and the com-plimentary European standards BS EN 1916 (2002) and BS EN 1917 (2002), and also in ASTM C76-11 (2011) and ASTM C497 (2005). Prestressed concrete pipes are described in BS EN 639 (1995) and BS EN 642 (1995) and ASTM C1089 (2006). The standards gener-ally require the mix design to be such that specified absorption, hydrostatic head and load test values can be attained. In ASTM C1089 (2006) a minimum concrete cover of 13 mm over reinforcement is required and provision is made for the use of galvanised steel. As with the majority of precast products the quality of the concrete is controlled by performance tests on the product. Levitt (1982) recommends a typical mix design suitable for spun con-crete as 3 parts by weight of clean 10 mm crushed rock, 1.5 parts of sharp concreting sand to 1 part of a coarsely ground cement and a water/cement ratio of 0.32. Where segregation is a problem he claims that replacement of 20%–40% of the cement by fly ash is beneficial. Glanville (1950) states that aggregate/cement ratios vary between 3.5:1 and 2.5:1 depending on the class of pipe and indicating that in practice a wide range of cement contents may be found in older pipes. BS 5991 (1984) allows the use of ordinary, rapid-hardening, Portland blastfurnace and sulphate-resisting cements and requires a minimum cement content of 360 kg/m^3. In practice the water/cement ratio appears to be closer to 0.4 than 0.3. It is not clear as to how much excess water is removed during the spinning process, but reasonably wet mixes are required to get good compaction. The use of low frequency, high amplitude vibration enables dryer mixes to be used. Detailed petrographic examination of the internal textures of pipe walls suggests that excess water is largely removed from the layers close to the internal surface of the pipe but no quantitative data is available.

Details of the internal texture of concrete pipe cast by the old method of compaction using a rotating dolly in a vertical mould are not available. The texture of pipes made by the

Figure 7.6 The concrete near the spigot area of this centrifugally spun pipe has not adequately compacted during spinning. This problem can occur unless care is taken during this type of manufacturing process. (Photograph by D. St John.)

'Roller Suspension' process is typically a well compacted low water/cement ratio concrete rich in cement, while the texture of pipes made in vibratory vertical casting machines is similar to that of modern concrete block. St John gives a case study example of a pressure pipe made by the 'Roller Suspension' method incorporated a pre-stressed steel strand wound round the fully cured pipe which was then protected by an external layer of pneumatically applied mortar. Experience indicated that the applied mortar was permeable to chlorides resulting in some failures due to corrosion of the steel reinforcement.

The internal texture of the wall of a centrifugally spun pipe is zoned because the fine material in the concrete mix partially segregates. Some of the excess water is forced to the inner surface as a thin slurry and is removed during the spinning process. Segregation at the inner surface leads to a mortar rich layer being formed which enables this surface to be smoothed with a forming bar. Where necessary, localised rough patches on the inner surface are sometimes repaired by sprinkling neat cement over the rough area during the final forming process.

Some other faults can occur in centrifugal concrete pipe production many of which are obvious on stripping the mould. Unless care is taken compaction around the spigot area can be a problem (Figure 7.6). Spun concrete pipe is commonly steam cured at 80°C–90°C to give adequate early strength for stripping but care in handling is still required to prevent cracking at this stage of production. External surfaces of spun pipe are usually excellent but can be marred by inadequate mould release agents (Figure 7.7).

The commonest fault in a centrifugally spun pipe is caused by inadequately graded and wrongly shaped aggregate which allows too much segregation to occur. This can result in a 10–15 mm deep zone on the inside of the pipe of almost neat cement paste grading into a rich mortar containing only the finest fraction of the sand. This texture is prone to shrinkage cracking. In a well-made pipe the inner zone is still present but the almost neat cement paste is present only as a thin skin and the dense mortar contains closely packed sand which forms an effective barrier to the ingress of fluids to the underlying concrete in the pipe wall. The overall depth of the segregated zone may be as little as 5 mm while the remainder of the wall consists of a dense concrete containing well-packed aggregate. Where the concrete mix is too harsh, excessive fissuring along margins of aggregate and reinforcing occur because of aggregate interlock and the inner surface may require excessive use of neat cement for smoothing.

Figure 7.7 The outer surface of this spun pipe has been damaged due to inappropriate use of release agent causing surface damage. This damage is superficial and after repair is unlikely to affect the durability of the pipe.

Many aspects of segregation and zonation can be seen by examining a finely ground section cut from across the wall of a pipe. If further detail is required it may be necessary to examine large-area thin sections. Both the depth and nature of the segregated zone are important. It is desirable but not critical for the segregated zone to be limited in depth as this indicates good mix design. It is more important that the segregated mortar is densely packed with sand to provide good durability. Fissuring along aggregate margins and reinforcing steel caused by lockup and localised inhomogeneity of the hardened cement paste due to the bleeding of water during the spinning of the pipe are features present in most pipes. This is especially the case where double reinforcing cages are used. The petrographer should be cautious in condemning a pipe on these grounds especially where the pipe complies with the requirements of the absorption, hydrostatic and load tests of the appropriate standard. A further indication that most of the channels and fissures in spun concrete pipes are not critical is that, although pipes that fail the hydrostatic test are left full of water for a number of days they then usually pass the hydrostatic test. Thus, channels and fissures present are easily blocked off by the formation of calcium hydroxide and autogenous healing. While concrete pipe rarely fails underground unless it is overloaded or attacked by aggressive fluids the same pipes used above ground may exhibit serious circumferential cracking. A number of cases of cracking in exposed pipe lines and pipes used as permanent formwork have occurred in New Zealand. The reason for this type of cracking has not been investigated but it is thought to be due to longitudinal rods in the reinforcing cage being inadequate to resist differential temperature stresses.

7.4 REINFORCED PRECAST CONCRETE UNITS

The majority of structural concrete worldwide contains steel reinforcement. A review of reinforcement types are given in Walker (2002). In the precast industry smaller products such as tiles and pavers are typically unreinforced, but larger precast units such as large diameter pipes, beams, lintels, railway sleepers and floor panels usually contain reinforcement which in some cases is also pre-stressed.

Although with normal steel-reinforced concrete the concrete is poured and compacted around the prepared reinforcing cage, with pre-stressed reinforced precast concrete the high tensile steel tendons are held in tension while the concrete sets around them. In post-tensioned

Figure 7.8 Spalling and cracking caused by corrosion of longitudinal steel reinforcement rods in a precast gate post.

units the tendons run through ducts in the concrete and are stressed once the concrete has achieved sufficient strength. The ducts are then typically filled with grout to prevent corrosion of the steel and to provide a bond between the tendon and the concrete.

Steel reinforcement in precast concrete presents no particular difficulties different from reinforced concrete generally. Corrosion of reinforcement is the most common problem and is discussed in Section 5.6.2 and more fully in Section 6.3. However, since many precast products have narrow cross sections, cover to the reinforcement is sometimes inadequate, and consequently corrosion is common in items such as thin panels and concrete fence posts (see Figures 5.56 and 7.8).

As has been noted previously, the mix design, the factory conditions of manufacture and quality control usually result in dense well compacted, relatively impermeable concrete units being produced. Consequently, the petrographic features of such concrete will reflect the details of the mix design, admixtures if used and the particular manufacturing process used. If any features of deterioration are present these will relate to the materials and/or manufacture, its usage, age and/or the local environmental conditions.

7.5 STEAM-CURED PRECAST CONCRETE UNITS

The objective of heat curing or steam curing of concrete is principally to increase the rate of strength gain, in order to speed up demoulding and to allow post-stressing to be undertaken earlier. In general terms there are three methods in common use. Systems that produce steam at atmospheric pressure from a steam boiler, hot air/misting systems and directly fired vapour generators. This last allows water vapour to mix directly with the combustion gases. The carbon dioxide by-product of combustion increases the concentration in the curing chamber atmosphere typically to about 3% and will enhance carbonation during curing. The temperatures used for the steam curing depends on the particular method used and can range from 60°C upwards to high-pressure steam curing (autoclaving) at 150°C or 200°C (Lea 1970). The temperature and the details of the curing cycle timings (particularly the initial curing at normal temperature prior to the steam cure) have significant effects on the hydration products formed in the cement paste and the eventual properties of the concrete product (Neville 1995).

Allowing temperature of setting and curing to rise above 55°C generally reduces the ultimate compressive strength of the concrete or mortar. This is most probably due to the development of microcracking resulting from the differential thermal expansion and contraction of the components and particularly the pore water during the heat cycle. There is a general decline in 28 day compressive strength as curing temperature is raised from 20°C up to 95°C (Lawrence 1998).

Microstructural changes due to heat curing lead to densification of CSH surrounding cement grains due to accelerated hydration and a coarsening of the hydration products. High-pressure high-temperature steam curing for normal periods of 6–12 h leads to formation of poorly crystallised CSH gels or tobermorite, $C_6S_2H_3$ and $C_2SH(A)$ (Lea 1970, Taylor 1990). Reaction with pfa, ggbs and other pozzalanic mineral admixtures accelerate with increasing temperature which may also result in reduced calcium hydroxide content. A laboratory study of mortars by Lewis and Scrivener (1997) showed that no ettringite was present after heat curing at 90°C and most of the aluminates and sulphate were absorbed into the CSH gels. However, ettringite then reformed over time within the material.

The modifications to the hydration of cement and the changes to mineralogy and crystallinity resulting from elevated temperature curing are best investigated using a combination of scanning electron microscopy, EDS microanalysis and x-ray diffraction techniques. However, the more general aspects of changes to the cement paste fabric, such as changes to the morphology and size of calcium hydroxide crystals and development of CSH around hydrating cement grains may be observed using the polarising microscope.

7.6 COMPOSITE PRECAST CONCRETE UNITS AND RECONSTITUTED (ARTIFICIAL) STONE

Both precast and placed in situ concrete may involve some form of composite construction so that surface layers differ in some way from the core material. This may range from thin coatings on decorative panels or floor screeds (see Chapter 8) to layers of up to 20 cm thick of artificial stone mortar laid against mould surfaces as a facing that are then backed by a concrete core material. Smaller decorative products such as garden ornaments and statuary may be composed entirely of the facing material. The majority are based on Portland type cements, though other materials including epoxy resins, ceramics and glazes may be used in more specialised applications.

Apart from the added complexity of forming composite materials an important concern is to insure that the interface does not introduce a plane of weakness between layers. Weakness and debonding may result from:

- Freeze–thaw damage in the surface layers
- Delays in placing the core material against the facing layer
- Inadequate compaction of core material against the facing layer
- Differences in thermal properties between layers
- Lack of adhesion between facing and core surfaces
- Moisture movement along the interface
- Mechanical stressing of the composite
- Development of secondary minerals at the interface

Modern caststone may best be referred to as architectural concrete which is manufactured to simulate natural cut stone and used in unit masonry applications, or as architectural features, trims, or ornament or facing for buildings and other structures. Cast stone can be made using carefully selected natural sands, crushed stone or well graded natural gravels with a white, or ordinary Portland cement and may contain a chemical admixture, such as a metal stearate waterproofing agent. Colouring agents are also sometimes used (usually metallic oxides) to achieve the desired colour and appearance while maintaining durable physical properties which may exceed most natural cut building stones. These additives may in favourable circumstances be identified using scanning electron microscopy microanalysis (EDS), or by infrared analysis (FTIR).

If carefully manufactured from well selected materials it may be difficult to differentiate this material from natural stone without recourse to petrographic microscopical examination. In thin section the differences in matrix mineralogy between the natural cements in rock and hydrated cementious materials are distinctive. Although carbonation of the matrix of a cast product using a limestone aggregate/filler can make identification difficult, the details of the mineral texture and the fabric of the matrix should remain distinctive. The typical appearance of the microstructure of cast stone is illustrated in Figures 7.9 and 7.10.

Reconstructed stone finishes may be made using mechanically compacted 'earth dry mix' for small elements and is sometimes referred to as semi-dry production. Compaction is typically achieved using pneumatic or electric sand rammers with two part composites the facing layer should have a 20 mm minimum thickness. An alternative is the wet cast production with higher water/cement ratio giving a plastic mix for veneered or cast panels, but wet-casting yields only one cast per mould per day. Wet cast products typically require more finishing work than the semi-dry units because of the laitance formed against the mould face. Finishing is usually achieved by grinding, chemical etching or hand rubbing. Specifications are given in

Figure 7.9 Cast stone microstructure with rounded fine aggregate particles of limestone and quartz in a white Portland-type cement matrix. Note air voids are shown by the pale yellow epoxy mounting medium. Plane polar illumination. (Courtesy of RSK Environment Ltd., Hemel Hempstead, U.K.)

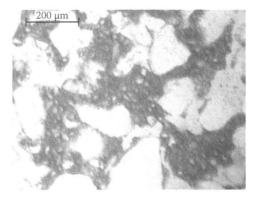

Figure 7.10 Cast stone microstructure at high magnification with quartz and limestone fine aggregate particles set in a white Portland-type cement matrix. Air voids appear pale yellow. Plane polar illumination. (Courtesy of RSK Environment Ltd., Hemel Hempstead, U.K.)

BS 1217 (2008) and recommend cement contents in excess of 400 kg/m^3 for facings with differences between facing and backing mixes being <80 kg/m^3 and a minimum of 40 mm cover to any steel reinforcement. A comprehensive review of cast stone, manufacture, design, installation, aftercare and the relevant British Standards is provided in the technical manual for cast stone (2011) by the United Kingdom Cast Stone Association.

7.7 FIBRE-REINFORCED PRODUCTS

The first widely manufactured fibre–cement mixture was developed around 1900 utilising asbestos as the fibre. Since the initial use of asbestos, other artificial fibres made from glass, steel, stainless steel, carbon, Kevlar, polypropylene, polyethylene, and nylon have been used (see also Section 5.6.3). The natural fibres sisal, jute, coir, bamboo, akwara, elephant grass (ACI Committee 544.2R-89 1989, ACI Committee 544.1R-96 2002) and kraft pulp (chemically pulped cellulose wood fibres) have also all found uses particularly in developing countries. Depending on the application and fibre type, lengths range from less than a millimetre to several centimetres and have diameters ranging from about 2 μm to 4 mm.

Reinforcing fibres are added to concrete in dosages that typically range from 0.6 to 60 kg/m^3 and will depend in part on the diameter of the fibre (Walker 2002). Dosages of 3–20 kg/m^3 for polypropylene and 20–40 kg/m^3 for steel fibres are common. The use of fibre reinforcement can enhance strain resistance, toughness, abrasion and impact resistance. Polypropylene fibres are used to suppress early age cracking while glass fibres are sometimes used in precast concrete and cast stone to provide high-quality finishes. Fibres are also commonly used in repair materials and mortars to suppress shrinkage cracking. A problem commonly encountered is the achievement of uniform dispersion of the fibres during mixing.

The petrographic examination of fibre-reinforced products in thin section requires some contrast between the fibre and the hardened cement paste if details of fibre distribution and fibre–matrix interface are to be observed. For instance, glass fibres may be difficult to differentiate in a cement matrix and thus are often better examined using the scanning electron microscope. In many cases the presence of fibres is most easily recognised in freshly broken surfaces using a binocular microscope. Many fibres such as asbestos and organic fibres in particular can be readily identified by their infrared 'fingerprint', since it is only necessary to extract a single fibre from the matrix to provide a specimen. Details of the volume proportion of fibres present, their distribution within the composite, the type of fibre, its average length and diameter are all of significance with respect to the physical and mechanical properties, particularly tensile strength, and the durability of the product.

The types and properties of fibre-reinforced cementitious products have been reviewed by Hannant (2003), Bentur and Mindess (1990) and Brandt (1995) and of natural fibre–reinforced cement and concrete by Swamy (1988). ACI Committee 544.1R-96 (2002) and BS 6432 (1990) specify the various tests applicable.

7.7.1 Asbestos–cement products

Products containing asbestos fibre are still manufactured today in the United States and elsewhere, but in very much smaller quantities than formerly because of the carcinogenic nature of asbestos. Crocidolite (blue asbestos) in particular, is banned from use in many countries. In the United Kingdom the use of all asbestos–cement products has been banned since 1999 (Health and Safety Executive 2005) and problems have sometimes been encountered with some non-asbestos fibre substitutes in products that formerly contained asbestos (see Chapter 4, Section 4.8.3 and Sims 2001). Materials containing asbestos, particularly in sheet or pipe form, are present in large numbers of older structures. Consequently, the petrographic examination of these materials is sometimes required. Preparation of asbestos products must be undertaken in such away as to prevent dust entering the atmosphere and all unwanted material and waste from the preparation has to be disposed of in a safe manner according to national guidelines. These safety criteria should also be applied to the preparation of material for use in laboratory investigations.

The term 'commercial asbestos' includes the following varieties:

- Chrysotile, $Mg_3[Si_2O_5](OH)_4$ (fibrous serpentine)
- Actinolite, $Ca_2Mg_5Si_8O_{22}(OH)_2$ (end member of the tremolite-actinolite series)
- Amosite, $(Fe^{2+},Mg)_7[Si_8O_{22}](OH)_2$ (fibrous grunerite)
- Crocidolite, $Na_2Fe^{3+}{}_2Fe^{2+}{}_3[Si_8O_{22}](OH)_2$ (a blue fibrous riebeckite)
- Anthophyllite, $(Mg,Fe^{+2})_7[Si_8O_{22}](OH,F)_2$ (end member of the anthophyllite-gedrite series)

This list does not include all fibrous amphiboles, but commercially, chrysotile and crocidolite were most widely used, while amosite and anthophyllite typically had more limited use.

Asbestos–cement products are composed of finely divided asbestos fibre dispersed in a matrix of hardened Portland cement. Where the product is high-pressure steam cured up to 40% of the cement may be replaced by silica flour. Typically flat and corrugated sheet contains 9%–12% asbestos fibre, pressure pipe 11%–14% and fire-resistant board 20%–30%.

Three principal methods of manufacture were used. The wet transfer roller or Hatschek process and its variant for making pipes known as the Mazza process, the semi-dry Magnani process principally used for making corrugated roofing sheet and the Manville dry extrusion process. In later manufacturing an injection moulding process using hydraulic pressure to de-water the product was used for making sheet from 10 to 100 mm thick. General details of these processes and the properties of the products are described by Concrete Society (1973), Ryder (1975), Coutts (1988) and Bentur and Mindess (2006).

The Hatschek and Mazza processes, were the two most commonly used methods of manufacture and lead to preferred orientation of the asbestos fibres. The predominant form of asbestos that was used in building products is chrysotile (Middleton 1979). Some amosite (grunerite) was added to improve the filtration properties of the wet slurry used in manufacture especially in the Hatschek process, while amosite predominates in fire-resistant boards because of its good high temperature resistance. Crocidolite was also added in the manufacture of pipe in the United States (Rajhans and Sullivan 1981). Since the early 1970s there has been a partial replacement of asbestos by either kraft pulp or glass fibre, and as is noted above, since the 1980s a ban on the use of asbestos, especially crocidolite has been in place in many countries. Thus, it is necessary when examining asbestos–cement products to be aware that other types of fibres may be present even for pipes dating back 50 years.

The appearance of the hardened matrix depends on the method of curing. Asbestos–cement products were usually steam cured to increase production rates. Products cured at ambient temperature, or with low pressure steam at about 80°C may only contain cement as the matrix. Where a more durable matrix is required, the products usually will have been cured in high-pressure steam at about 170°C. This can usually be detected as the matrix will contain the residues of the silica flour added to suppress the formation of the low strength αC_2SH in favour of the stronger tobermorite type phase.

In thin section the microscopic texture of a typical asbestos–cement consists of a hydrated matrix containing numerous partially hydrated and remnant grains of cement and well dispersed calcium hydroxide in which are embedded fibres of asbestos. Where silica flour has been added, remnant small patches of flour are not uncommon due to local imperfect mixing, coarser grains of silica, usually quartz, remain unreacted and the amount of calcium hydroxide present is considerably reduced, but is rarely completely absent (Figure 7.11) and

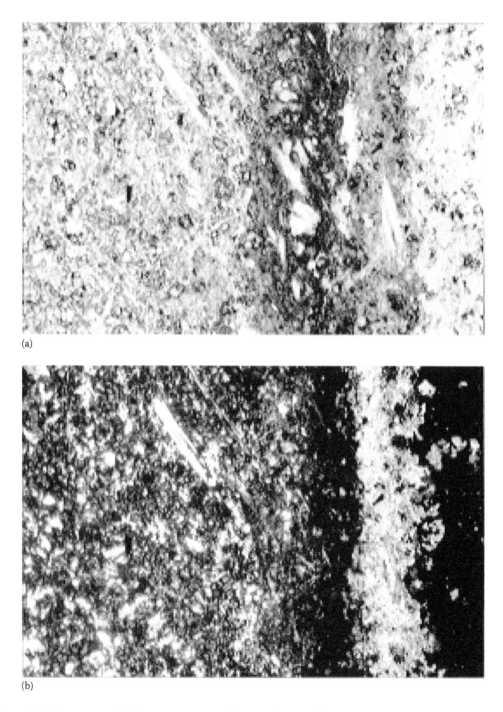

(a)

(b)

Figure 7.11 The texture of asbestos cement pipe. Width of fields 0.8 mm. (a) This is the region of an asbestos cement pipe which has been attacked by soft water. A few bundles of asbestos fibre are oriented sub parallel to the length of the pipe. The outer surface is to the top. (b) Under crossed polars the texture is seen to consist of quartz flour embedded in a dark matrix containing minimal calcium hydroxide, with the asbestos fibres being outlined by their birefringence. The texture of soft water attack consisting of corroded, carbonated and leached zones is clearly visible.

Section 5.6.3. The asbestos appears as bundles of fibres, usually about 3–10 µm in diameter and its optical characteristics clearly distinguish it from any other likely fibre to be used in these products (see Glossary of minerals) Where the product has been in use for a long time considerable depths of carbonation can occur. Thin sheets can be completely carbonated especially in the case of autoclaved products.

Reflected light microscopy of freshly broken surfaces of samples of asbestos products readily allows the fibres to be observed, but they are not easily identified by this method alone.

Identification of the asbestos types used in a particular product is sometimes required. Several methods are available including optical, chemical, infrared spectroscopy and x-ray diffraction (McCrone 1987). Fourier transform infrared analysis (see Section 2.5.4) provides a rapid means of identifying asbestos in a cement product, since identification can be undertaken on a single fibre taken from the sample. Figure 7.12 illustrates the infrared spectra in the 'fingerprint region' 500–1700 cm^{-1} and the OH region 2500–3900 cm^{-1} for a series of standard asbestos minerals compared to asbestos taken from a corrugated sheet and washed in weak acid to remove carbonates. The attenuated total internal reflection (ATR) procedure used for these spectra involved pressing the sample against a permanently aligned diamond window sampling attachment (Specac Golden Gate) which forms the flat face of a prism. The source radiation couples into the prism at the Brewster angle and is totally internally reflected. The contact face propagates an evanescent wave which interacts with the material to a depth of approximately 1 µm.

X-ray diffraction techniques allow definitive identification of asbestos minerals (see Glossary of minerals). Alternatively scanning electron microscopy with a microanalytical facility provides another method of identifying the fibres in a cement–asbestos composite. Key differences in chemical composition of the various asbestos minerals usually allows specific identification to be made, see Table 7.1.

Bentur and Mindess (2006) in reviewing the long-term performance state that in natural weathering performance the composite is excellent. However, there are two particular problems which the petrographer may have to investigate. These are carbonation leading to embrittlement and attack by aggressive water on pipes either due to soft water, or sulphates. It is necessary when considering these problems to distinguish between products cured below 100°C and autoclaved products as they are will respond differently to the environment. Opoczky and Pentek (1975) examined commercial sheet in thin section which had been weathered between 2 and 58 years. From about 15 years the fibres showed signs of corrosion from the alkaline pore solution with the possible formation of brucite. Magnesium carbonate and low lime calcium silicate hydrates were also identified. Testing showed that even where over 50% of chrysotile fibres were corroded the sheet still had excellent mechanical properties. However, Sarkar et al. (1987) failed to detect either corrosion of fibres or brucite and magnesite in sheet attacked by sulphate rich waters. There is little doubt that embrittlement of asbestos–cement due to carbonation frequently occurs with ageing and the extent of this will depend on the permeability of the product and how it was cured. In this respect while autoclaving gives enhanced mechanical properties its resistance to carbonation is likely to be less than a comparable product cured at 80°C or ambient temperature. Examination in thin section is an excellent method for determining the fine detail of carbonation texture but there appear to be no petrographic studies reported in the literature.

Asbestos–cement pressure pipe has been used extensively for water circulation in older structures. It is specified in BS EN 512 (1995) and BS 4624 (1981). Its ability to withstand soft water and soluble sulphates is of some importance. ASTM C500-07 (2007) gives guidelines for internal and external corrosion factors based on an aggressivity index of the water being transported and the acidity of the soil and soluble sulphate present in external water. There are additional recommendations that uncombined calcium hydroxide in the product should be <1%, and for aggressive conditions Type II cement should be used.

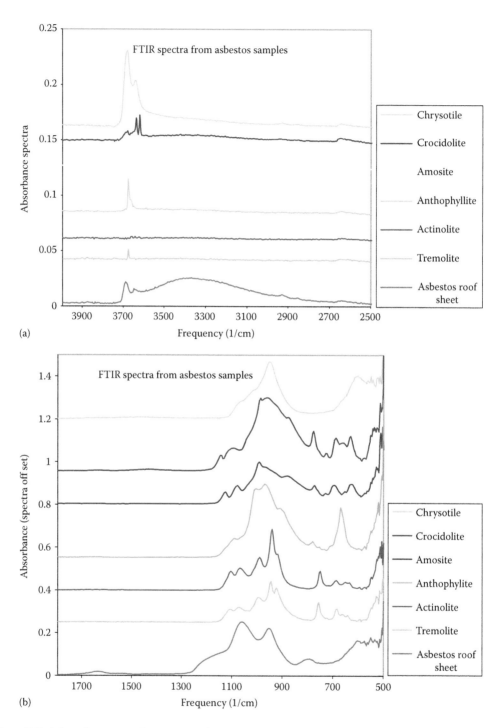

Figure 7.12 Infrared spectra of common asbestos minerals, chrysotile, antigorite, crocidolite, amosite, anthophylite, actinolite and tremolite compared to asbestos from a corrugated roof sheet. (a) Frequency range 2500–4000 cm⁻¹. (b) Frequency range 500–1800 cm⁻¹. (Courtesy of Dr A. Crossley, Department of Materials, University of Oxford, Oxford, U.K.)

Table 7.1 Typical chemical composition ranges of common asbestos minerals

	SiO_2	Fe_2O_3	FeO	MnO	MgO	CaO	Na_2O	K_2O
Chrysotile	40–43	0.2–0.3	0.05–1.6	0.3–0.4	35–43	0.0–0.1	0.0–0.3	0.0–0.1
Crocidolite	48–56	10–23	5–23	0.0–0.5	1–14	0.0–1	6–7	0.0–1
Amosite	47–50	3–4	36–37	0.3–0.6	6–7	0.5–1	0.0–0.1	0.0–0.2
Anthophyllite	50–60	0.2–3	5–20	0.0–0.4	20–30	0.0–1	0.0–1	0.0–0.4
Actinolite	50–59	2–4	2–11	0.2–10	8–20	9–11	0.0–1.5	0.0–0.3

Note: Data gathered from a range of published analyses.

It is reported by Al-Adeeb and Matti (1984) that asbestos–cement pipes in Kuwait have life expectancies of only 9–30 years. Tests showed the water to be moderately aggressive according to the ASTM C500 index and calcium hydroxide contents averaging 17% suggested that the pipes may not have been autoclaved. There are also reports from some areas of New Zealand that asbestos–cement water pipes are starting to fail after 25 years. Once again the water is moderately aggressive but in this case the pipes have been autoclaved which may be contributing to their longer working life.

There is an important difference between internal and external attack by aggressive water on asbestos–cement pipe. For significant external attack to occur the pH must be below 7 as discussed in Section 6.6. Internal attack sufficient to cause significant loss in strength takes place at pHs in the range of 7–9 and is dependent on the carbonate bicarbonate equilibria and its ability to dissolve calcium hydroxide. This is the basis of the ASTM aggressivity index and the limit of 1% uncombined calcium hydroxide. The limit on calcium hydroxide is probably more an indicator of pipe quality rather than having much effect on corrosion rate. One percent calcium hydroxide can only be attained in pipes where reactive silica flour and adequate autoclaving has been used so that a dense product results. There appears to be no reported petrographic studies on the failure of asbestos–cement pipe from internal aggressive water attack. Bursting of the pipes by splitting seems to predominate over beam type failure in these cases suggesting that the hardened cement matrix is weakened. The preferential alignment of fibres along the length of the pipe will also contribute to the splitting failure.

7.7.2 Synthetic fibre–reinforced concrete products

In this section, glass, steel, carbon and organic polymer fibres are grouped together for convenience. The properties and use of these fibres have been reviewed by Bentur and Mindess (2006) and discussion here will be mainly restricted to the use of glass fibre (see Figure 7.13). Initially the composition of glass used to make the fibres was an electrical grade of aluminosilicate glass known as E glass. A simple summary of reinforcement types and their typical uses is given in Walker (2002) with particular reference to their use in hot desert conditions. It was soon found that the alkali resistance of E glass fibre in a Portland cement matrix is limited in moist conditions and that corrosion and brittleness in the fibres over time can seriously weaken the product. The use of a chemical and alkali-resistant glass fibre containing zirconia known as CemFIL appeared to offer a solution to this problem, but it only slows the reaction down and does not prevent it. Similarly it was believed that the use of silica fume might have a beneficial effect but as reported by Yilmaz and Glasser (1991) silica fume inhibits notching damage from calcium hydroxide crystals but does not inhibit other attack mechanisms.

As a result of the potential long-term reduction in strength and increase in brittleness with ageing, fibre glass cement products have been limited to non-structural or semi-structural components such as thin sheet and claddings. According to Bentur and Mindess (1990) the principal cause of failure in panels is due to cracking. The shrinkage of the composite can

(a)

(b)

(c)

(d)

Figure 7.13 (a–d) Glass fibre reinforcement in mortar (a) a longitudinal section, (b) a transverse section. Inclined polar illumination, width of fields 1.3 mm. (Courtesy of M. Grove, private collection.) (c) and (d) Show the appearance of the E glass fibres in a cement–lime test prism in both longitudinal and cross section. Note the fibres are surrounded by calcium hydroxide and relicts of both cement and lime are visible. (c) plane polarised and (d) circular polarisation, width of fields 0.8 mm. (Photographs courtesy of D. St John.)

range from 0.05% to 0.25% depending on the sand content. Differential strains induced by thermal and moisture movements can lead to sufficient stress generation to cause cracking.

Most of the detailed examination of fibres in hardened cement has been carried out by scanning electron microscopy and EDS microanalysis and there appear to be few reported petrographic studies using thin-section examination (see also Section 5.6.3). Some observations with the optical microscope are possible in the cases of glass, asbestos and kraft pulp or other vegetable fibres (see Figure 7.14). As reported from scanning electron microscope studies (Barnes et al. 1978; Diamond 1986) unless the individual fibres of a glass strand debond, little if any paste or calcium hydroxide penetrates between the fibres.

Figure 7.13c and d illustrates the texture of a laboratory test prism made by mixing chopped strands of alkali-resistant glass with cement and lime. The individual glass fibres are about 15 μm in diameter and in longitudinal section have a thin dark margin

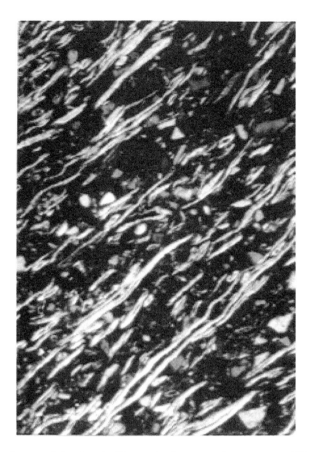

Figure 7.14 Paste interfaces with kraft pulp fibre and a low porosity aggregate. Width of fields 0.8 mm. The kraft pulp fibres have been aligned at 45 degrees to show maximum birefringence and the section cut in a plane to intersect the maximum number of fibres. Autoclaving at 180°C has reacted most of the calcium hydroxide with the silica flour of which the coarser relics are still visible. There appears to be no visible calcium hydroxide at the fibre interfaces. Polarised light.

which is the sizing. Being a glass they are isotropic apart from some minor strain birefringence. In cross section the strands of fibres that have not dispersed are partly penetrated by calcium hydroxide crystals but the space between the bunched fibres is otherwise empty. Where strands have opened up both cement paste and calcium hydroxide penetrate. In longitudinal section individual dispersed fibres are in intimate contact with the hardened cement paste and deposition of calcium hydroxide at the interface is not marked. In general the less the dispersion the more the deposition of calcium hydroxide at interfaces probably because of the space created by the strands starting to open up. Along the length of these strands, or debonded fibres copious amounts of calcium hydroxide can be crystallised.

In the case of asbestos fibre the high wetting capacity of this fibre results in an interface deficient in calcium hydroxide as shown in Figure 7.11. The small cross section of individual fibres appears to prevent crystallisation of any but isolate large crystals of calcium hydroxide at the fibre interface.

Many products containing fibrous reinforcement are steam cured at temperatures of around 180°C and also contain silica flour as a mineral addition. Effectively this removes much of the calcium hydroxide from the cement by reaction with the silica. Cement paste appears to be in direct contact with kraft pulp fibres with little calcium hydroxide present at the interface. At ambient temperatures, siliceous mineral additions can also have similar effects. The effect will be greatest for a highly active addition such as silica fume and will decrease with the coarser and less reactive additions.

The petrographic investigation of failed glass fibre cement composites requires a number of factors to be considered. It may be necessary to determine that an alkali-resistant glass and not E glass fibre has been used. The presence of zircon in the glass can be detected by EDS x-ray analysis on the scanning electron microscope but not by light microscopy as both E glass and alkali-resistant glass have similar refractive indices (see Glossary of minerals). It is advisable to use a low alkali cement to minimise alkaline attack on the fibre and a chemical analysis should indicate the alkali content to be <0.6% sodium oxide equivalent.

Thin-section examination will indicate the nature and characteristics of the fine aggregate and whether Portland type cement has been used, the presence of pozzolanas such as silica fume and the extent of carbonation. Some idea of fibre dispersion can be gained but generally this is not uniform on the local scale and requires a number of sequential thin sections, or large-area sections cut from orthogonal planes to have any meaning. Where high-alkali cement has been used a careful examination should be made for the presence of alkali-aggregate reaction. This may be difficult to observe as the amounts of alkali–silica gel involved may be quite small and an investigation using scanning electron microscopy may need to be undertaken. Examination of fibre matrix interfaces should also be carried out by scanning electron microscopy on fracture surfaces.

7.7.3 Natural fibre–reinforced concrete products

The range and properties of natural fibres used to reinforce cement composites has been reviewed by Bentur and Mindess (2006). Apart from the use of animal and vegetable fibres in historic mortars and similar materials (see Section 9.4) or the use of otherwise waste natural fibres to produce low cost cement products. The most important natural fibre is kraft pulp (chemically pulped cellulose wood fibres) currently used as a replacement for asbestos in sheet products manufactured in Australasia and some other countries. The discussion will be restricted to the use of kraft pulp fibre although the same investigative techniques used would be relevant to other types of natural fibres.

There are a number of problems in making a strong and durable sheet using kraft fibre. Because the effective density of natural fibres is about 1.2 compared with 2.6 for glass and chrysotile asbestos fibres, greater volumes of natural fibres have to be used to attain the same weight addition to the composite. This combined with other factors leads to flexural strengths of about 20 MPa as compared with 40 MPa for asbestos sheet. Unless the sugars are completely removed from the kraft pulp and it is adequately beaten, the setting of the cement and dispersion of the fibre can be affected (Coutts 1983). This may considerably reduce the flexural strength of the final product. Because of these effects kraft pulp fibre cement products are invariably autoclaved at about 170°C to attain maximum strength.

The durability of kraft pulp fibre cement sheet is affected by moisture conditions and carbonation. Bentur and Mindess (2006) report that shrinkage strains can exceed 0.15% which may be partly due to the presence of the kraft fibres. Sharman and Vautier (1986) found that alkaline corrosion is unlikely in autoclaved sheet because of the removal of lignin and hemicellulose during processing of the fibres. The data discussed by Bentur and Mindess (2006) on carbonation is confusing. However, for autoclaved product it can be assumed that thin sheet will normally be in an advanced state of carbonation and some increased brittleness will have occurred after exposure to the weather for any length of time. There appear to be no petrographic reports describing the texture of weathered sheet. The texture of autoclaved kraft pulp fibre cement sheet, fresh from the factory, is shown in Figure 7.14. Kraft pulp fibres varying from 10 μm to over 1 mm in length are preferentially aligned in a hardened matrix of hydrated cement containing fragments of quartz.

Longitudinal sections of kraft pulp fibres have parallel extinction, are length slow and show first-order interference colours which are typical of cellulose fibre. Fibres are of the order of 10–20 μm in width but this is variable as many fibres appear partially flattened with a tendency to undulate. They are rarely straight. In cross-section fibres are hollow with walls about 5 μm thick. Typically, considerable amounts of quartz flour remain in the matrix ranging from over 100 to 10 μm in size. Reaction rims may be present on some quartz grains but are not common. The hardened matrix consists of dark essentially isotropic material in which are embedded remnant grains of cement with reaction rims and some very fine grained low birefringent material which is probably calcium hydroxide. Otherwise calcium hydroxide appears absent. The outer margins of the sheet are rapidly carbonated to depths of 0.5–1.0 mm. This carbonation is reasonably intense at the surface but grades into scattered clumps of carbonation which give the texture a spotty appearance. Overall the texture is compact with few voids although in some samples there is a tendency for residual fissures to be present parallel to the length of the sheet derived from the original laying up of the product.

There is growing interest in the use of fibre-reinforced plastics (FRPs) in concrete as an alternative to steel reinforcement in that its use avoids corrosion possibilities. A review of the current use of FRPs has been presented by Clarke (2000).

7.8 POLYMER CEMENT PRODUCTS

In general polymer concretes once cured are chemically inert materials having higher tensile and compressive strengths than conventional concrete. However, they have a lower modulus of elasticity and higher creep. They may also be degraded by heat, oxidising agents, certain chemicals and solvents, UV light and micro-organisms. Careful selection, use of antioxidants and light stabilisers can overcome some of these disadvantages for particular applications (Neville and Brooks 1987).

Three broad groups of polymer concrete composites can be identified, polymer impregnated concrete, polymer-concrete and polymer Portland cement concrete. Polymer impregnated concrete (PIC) is a conventional Portland cement concrete which is saturated possibly under pressure and elevated temperature (150°C–300°C) with a liquid monomer such as methyl methacrylate (MMA), trimethylolpropane trimethacrylate (TMPTMA) or styrene (S). Polymerisation is achieved by gamma radiation or thermal-catalytic means which generate free radicals. These examples form polymethyl methacrylate (PMMA) or polystyrene (PS), respectively. The impregnated concrete, because of its lower permeability and porosity typically has higher strengths than untreated concrete (see Table 7.2) and higher resistance freeze–thaw.

The polarising microscope will allow identification of the normal components of the concrete and evidence of the impregnating polymer may be observed in voids and microcracks. Positive identification of the polymer type is best accomplished by preparing a microgram sample for Fourier transform infrared analysis (FTIR). A 'fingerprint' comparison with known polymer types is both rapid and definitive (see Section 2.5.4).

The second group, polymer-concrete (PC) is formed by polymerising a polyester, methyl methacrylate, or styrene monomer which is mixed with dry aggregate at ambient temperatures and cured with a promoter-catalyst (see Table 7.3). The addition of a silane acts as a coupling agent between aggregate and polymer improving the interfacial

Table 7.2 Typical mechanical properties of polymer impregnated concretes dried at 105°C

PIC	Polymer % mass	Strength (MPa) Compressive	Tensile	Flexural	Modulus of elasticity (GPa)
Unmodified	0	35	2	4	19
MMA	4.6–6.7	142	11	18	44
MMA + 10% TMPTMA	5.5–7.6	151	11	15	43
Styrene	4.2–6.0	99	8	16	44
60% S/40% TMPTMA[a]	5.9–7.3	120	6	—	44
Acronitrile	3.2–6.0	99	7	10	41
Chlorostyrene	4.9–6.9	113	8	17	39
Vinyl chloride	3.0–5.0	72	5	—	29
t-Butyl stryrene	3.5–6.0	127	10	—	45
Vinlidene chloride	1.5–2.8	47	3	—	21

Sources: After Neville, A.M. and Brooks, J.J., *Concrete Technology*, Longman Group Ltd., Harlow, U.K., 1987; Dikeau, J.T., Development in use of polymer impregnated concrete, in: Malhotra, V.M. (ed), *Progress in Concrete Technology*, Energy, Mines and Resources, CANMET, Ottawa, Ontario, Canada, 1980, pp. 539–582.

[a] Dried at 150°C overnight.

Table 7.3 Typical mechanical properties of polymer concretes after Neville and Brooks (1987) and Dikeau (1980)

Polymer/monomer	Ratio Polymer/aggregate	Strength (MPa) Compressive	Tensile	Flexural	Modulus of elasticity (GPa)
Unmodified	0	35	2	4	19
Polyester	1:10	117	13	37	32
Polyester	1:9	69	—	17	28
Epoxy/polyamide	1:9	95	—	33	—
Epoxy/polyaminoamide	1:9	65	—	23	32
NMA[a]-TMPTMA	1:15	137	10	22	35

[a] N-methylolacrylamide.

boding and hence the strength. Addition of Portland cement or silica flour improves workability and the polymer concrete can be placed and compacted in a similar way to conventional concrete.

Determination of aggregate type and modal proportions of components is readily undertaken using the polarising microscope, but identification of the polymer used is again best accomplished by infrared analysis.

The incorporation of polymer resins with aggregate, Portland cement and/or filler allows for rapid repairs to be undertaken or for the pressure moulding of relatively thin walled complexly shaped components such as pipes, cable ducts and roof gutters to be produced (Figure 7.15).

The third group, Polymer Portland Cement concrete (PPCC) is made by adding an aqueous solution of polymer or monomer to fresh concrete which is polymerised in situ. Rubber latexes, acrylics and vinyl acetates are typical materials used with antifoaming agents to minimise entrapped air. Polymer Portland cement concrete has better abrasion resistance, resistance to freeze–thaw and to impact loading compared with conventional concrete and is used typically as bridge deck overlays and for prefabricated curtain walls.

Figure 7.15 New polyester polymer-concrete precast thin walled drainage channels.

7.9 SPECIAL FLOOR COATINGS

Polymer floor coatings on concrete and mortar screed substrates are commonly used as floor surfaces (see also Sections 8.2 and 9.9). They usually consist of filled resins of either epoxy, polyester or polyurethane. Epoxy resins give the most resistant finish. Formulations are usually proprietary mixes and are applied by specialist contractors. More conventional surface finishes have used both natural and synthetic latex together with Portland or calcium aluminate cement to provide proprietary renders, screeds and surface coatings (see Figure 7.44).

St John illustrates a case study where the texture in thin section of a polymer floor finish is shown in Figure 7.16. The concrete floor has been roughly screeded and a mortar screed laid on top. Bond between the mortar screed and the concrete was variable with numerous microcracks and fissures present at the interface. In contrast the polymer bond to the mortar screed is excellent. The surface of the mortar screed is reasonably flat and is lightly carbonated to a depth of 1–2 mm. The polymer topping is in two layers. The bottom layer consists of a 1 mm thick layer of isotropic brown polymer which contains a limited amount of aggregate 100–150 μm in size and is full of voids 100–200 μm in diameter. The top layer is a colourless polymer about 0.5 mm thick and is heavily filled with quartz flour.

The polymers were not identified by infrared spectroscopy although they were reputed to be polyester. In thin sections, the resins in this example show similar moderate negative relief with respect to the epoxy mounting resin which had a refractive index of about 1.58. Polyester resins often have refractive indices of about 1.54 (see Table 3.4). The moderate negative relief observed suggests the polymer floor finish was based on polyester resin, but this could be readily confirmed by infrared analysis. This polymer floor was cored because the topping was lifting in patches. Coring revealed that no moisture barrier had been used in the floor and moisture movement up through the slab was the cause of the lifting. In spite of the excellent polymer bond to the screed it was insufficient to resist the movement of moisture through the slab.

A number of other proprietary floor coating treatments are available (see Section 8.2 for details). Others with self-levelling properties are based on Portland-type cements or calcium sulphates as binders to the fine aggregate materials (see also Sections 8.8 and 9.9).

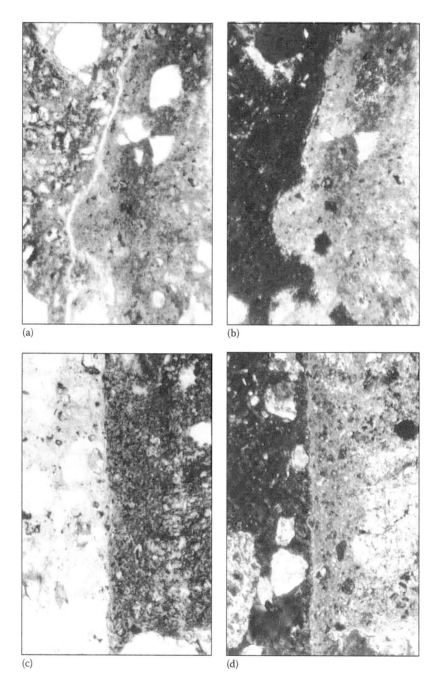

(a)   (b)

(c)   (d)

Figure 7.16 The interfaces between mortar and concrete and mortar and polymer topping. Width of fields 0.8 mm. (a) The mortar to the left is separated from the irregular surface of the concrete by a narrow fissure probably caused by differential movements. This type of fissuring at mortar/concrete interfaces is quite common and in this case extends along a considerable portion of the interface. It is possible that some interlock occurs across the fissure due to its irregular nature. (b) Under crossed polars the surface carbonation of the concrete clearly defines the interface. (c) The polymer topping is to the left and has clearly bonded to the mortar beneath without any fissuring. (d) Under crossed polars the lightly carbonated mortar defines the interface. The polymer topping is isotropic with numerous voids and quartz filler.

One surface coating treatment involves the use of lithium silicate which interacts with the calcium hydroxide resulting in the formation of insoluble tricalcium silicate hydrate which hardens and densifies the floor surface.

7.10 SELF-COMPACTING CONCRETE

Self-compacting concrete (SCC) was developed in the 1980s in order to achieve durable concrete structures without the need for vibration compaction. It has been used very successfully in Japan and is increasingly used elsewhere. A recent review of the materials and properties of self-compacting concrete is given by Gaimster and Dixon (2003).

Self compacting concrete has been defined as a highly flowable, yet stable concrete that can spread readily into place and fill formwork without any consolidation and without undergoing any significant segregation (Kayat et al. 2000). It requires a high-workability, high-strength mix which does not produce undue segregation or bleeding during placing.

These requirements are achieved by using a superplasticiser based on a polycarboxylic ether, naphthalene, or melamine formaldehyde (ASTM C494, Type F 2008), a filler, such as silica fume, fly ash or limestone powder and carefully selected and graded fine and coarse aggregates. Portland cements are used in most applications, the coarse aggregate sizes and gradings are similar to normal concrete, but the volume proportion is typically lower 30%–50%, while the proportion of fine aggregate forms 35%–45%. The filler may represent 20% or more of the mix. A range of mix designs have been used, but in principle the coarse and fine aggregate contents are fixed and the self-compactability of the mix is achieved by adjusting the water/binder ratio and the superplasticiser.

As with normal concretes a petrographic examination will allow the aggregate proportions to be estimated and the filler and cement type identified in the hardened concrete. The pozzalanic effects of the filler, particularly limestone dust will reduce or suppress the development of calcium hydroxide (see Section 5.2.5). Unless air entrainment has been employed air void content is typically 1.5% or lower. The presence of the superplasticiser may be inferred from an examination of the detailed texture of the paste matrix, but a chemical, or infrared analysis is required for its identification.

7.11 LIGHTWEIGHT AGGREGATES AND CONCRETES

In concrete construction the dead load of the structure forms a major part of the total loading so there are economic advantages in using lower-density concrete. The advantages are reduction in the size of foundations and structural members, of the formwork and greater ease in handling and placing the concrete. With blockwork, ease of handling and greater insulating properties are important advantages. The disadvantages are lower strength, higher moisture movement, increased carbonation and susceptibility to corrosion of reinforcing steel as the density of the concrete decreases.

Natural lightweight aggregates are pumice, scoria, volcanic tuffs and ashes. These names describe textures of volcanic materials and thus are not rocks of specific composition. Pumice is a white to grey coloured highly cellular glassy lava generally of rhyolitic composition while scoria is usually of basic composition and is characterised by marked vessicularity, dark colour and a texture that is partly glassy and partly crystalline. A tuff is a rock formed of compacted volcanic fragments usually varying from andesite to rhyolite in composition and ash is any loose unconsolidated volcanic material. The fragments making up these two latter materials are usually <4 mm in size.

Artificial materials are the most commonly used lightweight aggregates and are specified in BS EN 13055-1 (2002), ASTM C 330-09 (2009), 331-10 (2010) and 332-09 (2009). Their use in structural concrete is described by Clarke (1993). Furnace clinker and breeze (cinders) have been used in the United Kingdom for many years and make satisfactory aggregates provided they contain <10% of combustible material. Because their sulphur content causes corrosion problems with steel they should not be used in reinforced concrete. A more recent variant of this type of material is furnace bottom ash from power stations. It is somewhat variable in composition but the better grades may be crushed and screened to provide aggregate for block manufacture (Figure 7.17). Foamed and pelletised slags are manufactured from molten iron and steel slags. Molten slag is foamed with controlled amounts of steam and water and depending on the process aggregate can be produced as rounded pellets with sealed surfaces or material that requires crushing to size. Provided sulphate contents are kept below 1% these materials make reliable lightweight aggregates from 14 mm downwards.

As a result of the modern emphasis on re-cycling materials, in addition to the utilisation of slag, PFA and other furnace by-products a range of waste materials such as incinerator ash are now being processed to produce lightweight aggregates. Processes vary from pelletising and sintering the waste to using Portland cement as a binding agent, sometimes hydrating the cement at elevated temperature, or in an atmosphere enriched in carbon dioxide (Figures 7.18 and 7.19).

Many shales and clays can be expanded by heating to 1000°C–1200°C so that on fusion gases generated in the mass expand the material to give a texture that is similar to that formed in a loaf of bread. Dependent on the process, the product is either produced as a sintered cake which requires crushing to produce aggregate or as aggregate particles which mostly have smooth sealed surfaces (Figure 7.20). This process is often referred to as bloating. The clay mineral vermiculite has a composition and structure which exfoliates

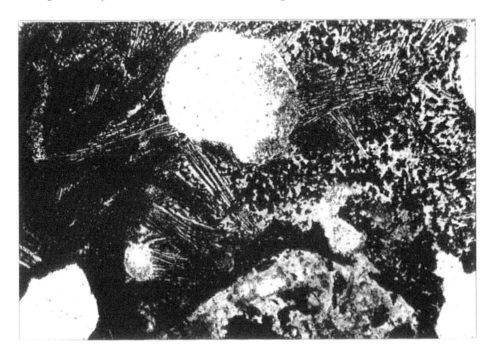

Figure 7.17 Porous boiler clinker with an opaque matrix showing development of acicular melilite crystals. Width of field 1.5 mm.

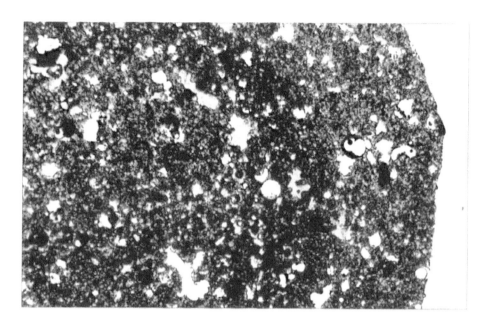

Figure 7.18 The texture of a 'Lytag' lightweight aggregate made from pulverised fly-ash. Some remnants of fly-ash spheres are still present. Note the zonal structure of the aggregate, the lack of interconnection between the pores and the outer sealed surface. This type of aggregate has only limited vesicularity making it suitable for structural lightweight concrete. Width of field 3.5 mm.

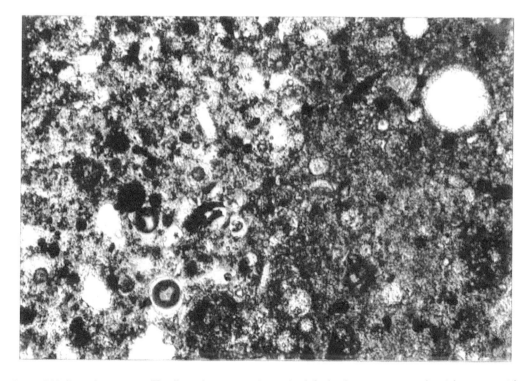

Figure 7.19 'Lytag' concrete. The 'Lytag' aggregate is to the left. In the concrete to the right some voids contain ettringite and possibly thaumasite crystals. Note the excellent contact between the 'Lytag' and the hardened cement matrix. Width of field 0.8 mm.

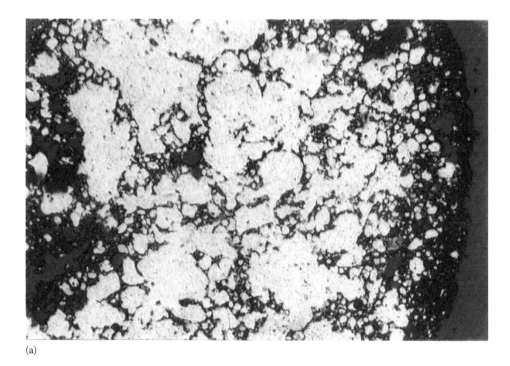

(a)

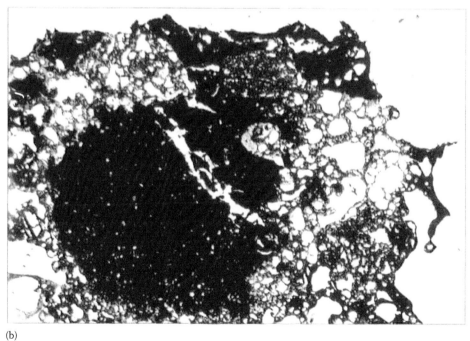

(b)

Figure 7.20 The textures of expanded clay aggregates. (a) The brown oxidised rim and the open vesicular structure are typical of expanded clay aggregates that are heavily bloated. Width of field 13 mm. (b) This expanded clay aggregate has been only partially bloated and contains much less vesicular structure and some areas of unexpanded clay. Width of field 5 mm.

when pellets are heated to 750°C–1100°C. Perlite is a glassy form of rhyolite containing 3%–5% of water. Graded particles heated to 900°C–1200°C decrepitate to form rounded aggregate which may be hollow.

In Table 7.4 the loose bulk volume densities of lightweight aggregates are listed with data derived mainly from Spratt (1980) and Short and Kinniburgh (1978). These should be considered as only indicative values. Other data is given by Rudnai (1963) but these have not been included. BS EN 13055-1 sets limits of 400–1200 kg/m³ for fine lightweight aggregate and 250–1000 kg/m³ for coarse aggregate. For insulating aggregates ASTM C332 specifies dry loose weights for perlite and vermiculite of 90–200 kg/m³ and a maximum density for other combined lightweight aggregates of 1040 kg/m³.

Fine aggregates have higher densities because they are often less vesicular than the comparable coarse aggregates. The wide range of bulk densities occurring in any one material are due to both this density difference between fine and coarse aggregates and the fact that most materials can be expanded over a considerable range of densities. However, all such materials have an optimum range of densities outside which they are not usable.

Lightweight aggregates typically are glassy and vesicular in texture and can easily be identified on cut surfaces with the hand lens. In thin section, the walls of the vesicles appear as thin rims of clear glass which often contain scattered crystallites which have failed to melt. In the case of scoria the glass will often be dark and much more crystallisation will be present. Aggregates that have been bloated and not crushed have a continuous dark-brown-coloured zone near the amorphous rims. For crushed and most natural lightweight aggregates there are no defined margins. As a general rule lower density aggregates have thinner glass walls and larger vesicles (Figure 7.20).

In concrete, lightweight aggregates with sealed rims behave in a similar manner to normal-weight aggregates. For those aggregates with porous rims the interfacial zone is more dense and homogeneous and there is better mechanical locking between cement paste and aggregate (Zhang and Gjorv 1990; Sarkar et al. 1992). The penetration of cement paste is limited to the outer porous margin of the aggregates in spite of lightweight aggregates usually having high water absorption (Zhang and Gjorv 1992).

Table 7.4 Approximate bulk volume densities of air dried aggregates

Aggregate type	kg/m³
Normal weight aggregates	1360–1600
Sintered PFA	770–1040
Brick rubble	760–1120
Scoria	720–1300
Volcanic slag	700–1200
Furnace clinker and breeze	720–1040
Expanded slag	700–970
Foamed slag	480–960
Pumice	480–880
Expanded slate	460–800
Sintered diatomite	450–800
Expanded shale and clay	320–960
Wood shavings and sawdust	320–480
Exfoliated vermiculite	60–160
Expanded perlite	50–240

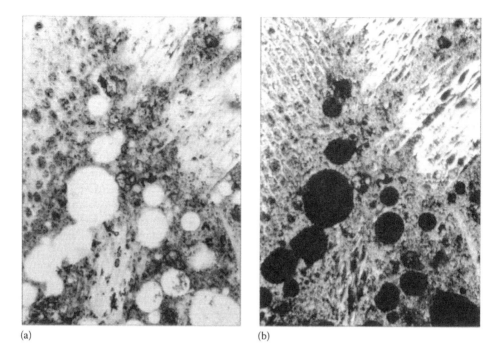

(a)   (b)

Figure 7.21 The texture of sawdust concrete. Width of fields 0.9 mm. (a) The chips of sawdust are in intimate contact with the hardened cement matrix which contains an extensive air void system.
(b) Under crossed polars the cellular structure of the wood and the typical birefringence of
the cellulose is shown. Because the product is porous the hardened matrix is well carbonated.

Wood particles and sawdust treated with lime are sometimes used as lightweight aggregates
for the manufacture of masonry blocks, wood wool slabs and for sawdust concrete (BS 1105
1981, BS EN 13168 2006). The texture of a sawdust concrete is illustrated in Figure 7.21.

The petrographer is more likely to examine structural lightweight concretes in densities
ranging from 1000 to 2000 kg/m³ rather than lightweight insulating concretes. The durability of lightweight structural concrete is affected by increasing levels of carbonation and
drying shrinkage as the density of the concrete decreases, as is indicated in Section 5.5.
To reduce these problems it is common practice to replace some of the fines with normal-
weight sand. Apart from these attributes the texture of structural lightweight concrete is
little different from normal-weight concretes and it appears that, providing cement contents in excess of 400 kg/m³ are used, it has similar durability (Mays and Barnes 1991).

Some petrographic textures are illustrated by St John in the following examples. Lytag,
which is manufactured by sintering pulverised fuel ash, forms readily identifiable spherical
particles 3–5 mm diameter with a rusty brown exterior and a dark almost black inner core.
In thin section, the zonal structure can be distinguished and the porous nature of the materials is clearly apparent (Figures 7.19 and 7.20) In one example of a Lytag concrete containing
fly ash as a cement replacement, some of the voids close to the carbonation front within the
concrete are partially rimmed by a straw coloured mineral with high birefringence which is
probably thaumasite (Figure 7.20). Some voids in the hydrated cement paste are partly filled
or wholly filled with acicular ettringite crystals which have grown in from the void margins.
An example of Leca expanded clay aggregate is illustrated by Figure 7.20a. The exterior
edge of the particle is denser than the vesicular interior. It is noticeable that relatively few of
the pores have been filled with the coloured impregnating resin clearly indicating that only

a small percentage of the pores in this expanded clay aggregate interconnect. Expanded clay aggregate varies considerably depending on the type and details of the manufacturing process. An example of an expanded clay where firing is incomplete is illustrated in Figure 7.20b where the opaque core of the particle has only partially bloated and contains areas of unexpanded clay. This is surrounded by a more porous outer zone.

REFERENCES

ACI Committee 212.3R-91, 1991: *Chemical Admixtures for Concrete.* American Concrete Institute, Detroit, MI, pp. 1–86.

ACI Committee 544.2R-89, 1989: *Measurement of Properties of Fibre Reinforced Concrete.* American Concrete Institute, Detroit, MI, pp. 1–12.

ACI Committee 544.1R-96, 2002: *State-of-the-Art Report on Fiber Reinforced Concrete* (Reapproved 2002). Concrete Institute, Detroit 5, Natural Fibre Reinforced Concrete.

Al-Adeeb, A.M. and Matti, M.A. 1984: Leaching and corrosion of asbestos cement pipes. *The International Journal of Cement Composites and Lightweight Concrete,* 6 (4), 233–240.

ASTM C76-11, 2011: *Standard Specification for Reinforced Concrete Culvert, Storm Drain, and Sewer Pipe.* ASTM, Philadelphia, PA.

ASTM C129, 2006: *Standard Specification for Non Loadbearing Concrete Masonry Units.* ASTM, Philadelphia, PA.

ASTM C330/C330M-09, 2009: *Standard Specification for Lightweight Aggregates for Structural Concrete.* ASTM, Philadelphia, PA.

ASTM C331/C331M-10, 2010: *Standard Specification for Lightweight Aggregates for Concrete Masonry Units.* ASTM, Philadelphia, PA.

ASTM C332-09, 2009: *Standard Specification for Lightweight Aggregates for Insulating Concrete.* ASTM, Philadelphia, PA.

ASTM C494/C 494-08a, 2008: *Standard Specification for Chemical Admixtures for Concrete.* ASTM, Philadelphia, PA.

ASTM C497-05, 2005: *Test Methods for Concrete Pipe, Manhole Sections, or Tile.* ASTM, Philadelphia, PA.

ASTM C500-07, 2007: *Standard Test Methods for Asbestos-Cement Pipe.* ASTM, Philadelphia, PA.

ASTM C1089-06, 2006: *Standard Specification for Spun Cast Concrete Poles.* ASTM, Philadelphia, PA.

Barnes, R.D., Diamond, S., and Dolch, W. 1978: The contact zone between Portland cement paste and glass aggregate surfaces. *Cement and Concrete Research* 8 (2), 233–243.

Bentur, A. and Mindess, S. 2006: *Fibre Reinforced Cementitious Composites,* 2nd edn. Taylor & Francis Group/E. & F.N. Spon, London, U.K., pp. 1–601.

BIBM 2014: Bureau International du Béton Manufacturé Congress, Istanbul, Turkey, European Federation of the Concrete Industry. www.bibmcongress.eu.

Brandt, A.M. 1995: *Cement-Based Composites-Materials, Mechanical Properties and Performance,* E. & F.N. Spon, London, U.K.

BS 1105, 1994: *Specification for Wood Wool Cement Slabs up to 125 mm Thick.* BSI, London, U.K.

BS 1217, 2008: *Cast Stone Specification.* BSI, London, U.K.

BS 4624, 1992: *Methods of Test for Asbestos-Cement Building Products.* BSI, London, U.K.

BS 5911-1, 2002: *Concrete Pipes and Ancillary Concrete Products. Specifications for Un-Reinforced and Reinforced Concrete Pipes (Including Jacking Pipes) and Fittings with Flexible Joints (Complimentary to BS EN 1916: 2002).* BSI, London, U.K.

BS 5991, 1984: *Precast Concrete Pipes and Fittings for Drainage and Sewerage,* Part 1. BSI, London, U.K.

BS 6432, 1990: *Methods for Determining Properties of Glass Fibre Reinforced Concrete Material.* (1984 Re-issued 1990). BSI, London, U.K.

BS EN 512, 1995: *Specification for Asbestos-Cement Pressure Pipes and Joints.* BSI, London, U.K.

BS EN 639, 1995: *Common Requirements for Concrete Pressure Pipes Including Joints and Fittings.* BSI, London, U.K.

BS EN 642, 1995: *Prestressed Concrete Pressure Pipes, Cylinder and Non-Cylinder Including Joints and Fittings and Specific Requirement for Prestressing Steel for Pipes.* BSI, London, U.K.

BS EN 771-3, 2003: *Precast Concrete Masonry Units. Specification for Precast Concrete Masonry Units.* BSI, London, U.K.

BS EN 1916, 2002: *Concrete Pipes and Fittings, Unreinforced, Steel Fibre and Reinforced.* BSI, London, U.K.

BS EN 1917, 2002: *Concrete Manholes and Inspection Chambers, Unreinforced, Steel Fibre and Reinforced.* BSI, London, U.K.

BS EN 13055-1, 2002: *Lightweight Aggregates. Lightweight Aggregates for Concrete, Mortar and Grout.* BSI, London, U.K.

BS EN 13168, 2006: *Thermal Insulation Products for Buildings—Factory Made Wood Wool (WW) Products—Specification.* BSI, London, U.K.

Childe, H.G. 1961: *Concrete Products and Cast Stone,* 9th edn. Concrete Publications Ltd., London, U.K., pp. 1–263.

Clarke, J.L. (ed.) 1993: *Structural Lightweight Aggregate Concrete.* Taylor & Francis Group, London, U.K., pp. 1–240.

Clarke, J.L. 2000: Fibre composite reinforcement and repair systems. In *Proceedings of the Sixth International Conference on Deterioration and Repair of Reinforced Concrete in the Arabian Gulf,* The Bahrain Society of Engineers, pp. 33–48.

Concrete Society. 1973: Fibre-reinforced cement composites. Technical Report of the Concrete Society. Concrete Society Technical Report No. 51.067.

Coutts, R.S.P. 1983: Autoclaved beaten wood fibre-reinforced cement composites. *Composites* 18 (2), 139–143.

Coutts, R.S.P. 1988: Wood fibre reinforced cement composites. In *Natural Fibre Reinforced Cement and Composites,* Swamy, R.N. (ed.). Blackie and Sons Ltd., Glasgow, U.K., pp. 1–62.

Deane, Y. 1989: *Finishes,* Mitchell's Building Series. Mitchell, London, U.K.

Diamond, S. 1986: The microstructure of cement paste in concrete. In *Eighth International Congress on the Chemistry of Cement,* Rio de Janeiro, Brazil, Vol. 1, pp. 123–147.

Dikeau, J.T. 1980: Development in use of polymer impregnated concrete. In *Progress in Concrete Technology,* Malhortra, V.M. (ed.). Energy, Mines and Resources, CANMET, Ottawa, Ontario, Canada, pp. 539–582.

Dutton, A.R. 1977: Manufacture of concrete pipe in South Africa. In *Concrete Technology. A South African Handbook,* Fulton, D.S. (ed.). The Portland Cement Institute, Johannesburg, South Africa, pp. 337–343.

Everett, A. 1994: *Materials,* Mitchell's Building Series. Mitchell, London, U.K.

Gaimster, R. and Dixon, N. 2003: Self-compacting concrete. In *Advanced Concrete Technology, 3. Processes,* Newman, J. and Choo, B.S. (eds.). Butterworth-Heinemann, Elsevier, pp. 9/1–9/22.

Glanville, W.H. (ed.) 1950: *Modern Concrete Construction.* The Caxton Publishing Co., London, U.K., Vol. 4, pp. 168–320.

Hannant, D.J. 2003: Fibre-reinforced concrete. In *Advanced Concrete Technology, 3. Processes,* Newman, J. and Choo, B.S. (eds.) Butterworth-Heinemann, Elsevier, pp. 6/1–6/17.

Health and Safety Executive. 2005: *Asbestos: The Analysts' Guide for Sampling, Analysis and Clearance Procedures* HGS 248, HSE Books (Available as a free download from HSE Books web-site).

Kayat, K.H., Ghezal, A., and Hardriche, M.S. 2000: Utility of statistical models in proportioning self-consolidating concrete. *Materials and Structures* 33, 391–397.

Lawrence, C.D. 1998: Physicochemical and mechanical properties of Portland cements. In *Lea's Chemistry of Cement and Concrete,* 4th edn., Hewlett, P.C. (ed.). Edward Arnold Ltd., London, U.K., Vol. 8, pp. 343–406.

Lea, F.M. 1970: *The Chemistry of Cement and Concrete,* 3rd edn. Edward Arnold Ltd., London, U.K.

Levitt, M. 1982: *Precast Concrete. Materials, Manufacture, Properties and Usage* 227. Applied Science Publishers, London, U.K.

Lewis, M.C. and Scrivener, K.L. 1997: *A Microstructural and Microanalytical Study of Heat Cured Mortars and Delayed Ettringite Formation*, 10th International Conference on Chemistry of Cement, Gothenburg,Vol. 4, pp. 409–416.

Malhotra, V.M. November 1976: No-fines concrete—Its properties and applications. *ACI Journal* (American Concrete Institute, Detroit, MI) 73 (11), 628–644.

Mays, G.C. and Barnes, R.A. 1991: The performance of lightweight aggregate concrete structures in service. *The Structural Engineer* 69 (20), 351–361.

McCrone, W.C. 1987: *Asbestos Identification*. Microscope Publications, Chicago, IL.

McCrone, W.C. and Delly, J.G. 1973: *The Particle Picture Atlas, 1 & 2*. Ann Arbour Science Publishers, Ann Arbour, MI.

Middleton, A.P. 1979: The use of asbestos and asbestos-free substitutes in buildings. In *Asbestos, Volume 1, Properties, Applications, and Hazards*, Michaels, L. and Chissick, S.S. (eds.). Wiley Interscience, Chichester, U.K., pp. 306–334.

Neville, A.M. 1995: *Properties of Concrete*, 4th edn. Longman Group Ltd., Harlow, U.K.

Neville, A.M. and Brooks, J.J. 1987: *Concrete Technology*. Longman Group Ltd., Harlow, U.K.

Opoczky, L. and Pentek, L. 1975: Investigation of the 'corrosion' of asbestos cement fibres in asbestos cement sheets weathered for long times. In *RILEM Symposium 1975, Fibre Reinforced Cement and Concrete*, Neville, A.M., Hannant, D.J., Majumdar, A.J., Pomeroy, R.D., and Swamy, R.N. (eds.). Paris, France. Construction Press, London, U.K., pp. 269–276.

Rajhans, G.S. and Sullivan, J.L. 1981: *Asbestos Sampling and Analysis*. Ann Arbor Science, Ann Arbor, MI.

Richardson, J.G. 1991: *Quality in Precast Concrete: Design—Production—Supervision*. Longman Scientific & Technical, Harlow, England.

Richardson, J.G. 2003: Precast concrete structural elements. In *Advanced Concrete Technology, 3. Processes*, Newman, J. and Choo, B.S. (eds.). Butterworth-Heinemann, Elsevier, pp. 21/3–21/46.

Rudnai, G. 1963: *Lightweight Concretes*. Publishing House of the Hungarian Academy of Sciences, Budapest, Hungary.

Ryder, J.F. 1975: Applications of fibre cement. In *RILEM Symposium 1975, Fibre Reinforced Cement and Concrete*, Paris, France. Neville, A.M., Hannant, D.J., Majumdar, A.J., Pomeroy, R.D., and Swamy, R.N. (eds.). Construction Press, London, U.K., pp. 23–35.

Sarkar, S.L., Chandra, S., and Berntsson, L. 1992: Interdependence of microstructure and strength of structural lightweight aggregate concrete. *Cement and Concrete Composites* 14, 239–248.

Sarkar, S.L., Jolicoeur, C., and Khorami, J. 1987: Microchemical and microstructural investigations of degradation in asbestos-cement sheet. *Cement and Concrete Research* 17 (6), 864–874.

Sharman, W.R. and Vautier, R.P. 1986: Accelerated durability testing of autoclaved wood fibre reinforced cement sheet composites. *Durability of Building Materials* 3, 255–275.

Short, A. and Kinniburgh, W. 1978: *Lightweight Concrete*, 3rd edn. Applied Science Publishers, London, U.K.

Sims, I. 2001: Material evidence: The forensic investigation of construction and other geomaterials for civil and criminal cases. In *Forensic Engineering—The Investigation of Failures, Proceedings of the Second International Conference on Forensic Engineering, Irrigation of Civil Engineers*, Neale, B.S. (ed.). 12–13 November 2001, Thomas Telford, London, U.K., pp. 39–50.

Spratt, R.H. 1980: *An Introduction to Lightweight Concrete*, 6th edn. Cement and Concrete Association, London, U.K.

Swamy, R.N. 1988: *Natural Fibre Reinforced Cement and Concrete*, Blackie and Sons Ltd., Glasgow, U.K., pp. 1–336.

Taylor, H.W.F. 1990: *Cement Chemistry*. Academic Press, London, U.K.

United Kingdom Cast Stone Association, 2011: *Technical Manual for Cast Stone* (full download). UKCSA, Northampton, U.K. (info@ukcsa.co.uk).

Walker, M. (ed.), 2002: *Guide to the Construction of Reinforced Concrete in the Arabian Peninsula*. Concrete Society Special Publication CS136, CIRIA C577.

Williams, A. (ed.), 1994: *Specification* 1994 (published in three volumes: *Technical*; *Products*; *Clauses*). Emap Business Publishing (Emap Architecture), London, U.K.

Yilmaz, V.T. and Glasser, F.P. 1991: Refraction of alkali-resistant glass fibres with cement. Part 1. Review, assessment, and microscopy. Part 2. Durability in cement matrices conditioned with silica fume. *Glass Technology* 32 (4), 91–98, 138–147.

Zhang, M. and Gjorv, O.E. 1990: Microstructure of the interfacial zone between lightweight aggregate and cement paste. *Cement and Concrete Research* 20 (4), 610–618.

Zhang, M. and Gjorv, O.E. 1992: The penetration of cement paste into lightweight aggregate. *Cement and Concrete Research* 22 (1), 47–55.

Chapter 8

Portland cement mortar, screeds, renders and special cements

8.1 MORTAR AND RELATED MATERIALS

Mortar is defined as a soft material or paste used as a jointing medium between bricks or masonry blocks which hardens in time. Mortars have been used since antiquity, with clays and bitumen being the earliest examples. The Egyptians used mixtures of sand with lime or gypsum in constructing the pyramids. Later, pozzolans were incorporated with lime mixes to speed up setting rates and improve durability.

Modern mortars are mixes of sand and cementitious binder, usually Portland cement and hydrated lime. They can be arranged into three broad groups: Portland cement–based mortar, lime-based mortars (see Section 9.1) and polymer cement mortars (see also Section 7.8). Pozzolans and chemical admixtures of various kinds are commonly used to assist setting, increase durability or impart specific properties such as water proofing to the mortar.

Although mortar by definition implies a jointing material between blocks, similar materials to mortar, containing sand and some types of cementitious binder, are frequently used for the top layers of floor screeds (see Section 7.9) or as renders for walls. Consequently, such materials are appropriately considered in this chapter.

8.2 FLOOR SCREEDS

The simplest method of finishing a concrete floor is to screed it level soon after placing the concrete and to wait until initial set has occurred and then trowel the surface to form a skin of compacted binder without the formation of laitance. This technique produces a thin skin of cement-enriched concrete with a low water/cement ratio on the surface. If required, the surface can be later treated with surface-hardening solutions which further increase hardness and impermeability. If the hand trowelling is replaced by power floating, even harder surfaces can be made. Provided that the surface is adequately cured, a hard-wearing, non-dusting surface is produced which is adequate for many purposes. A range of surface treatments, usually based on acrylics or polyurethane, may be used to improve durability and surface finish even further (see also Section 7.9).

The texture seen in a petrographic thin section of this type of surface finish will consist of a narrow band of cement-enriched concrete, but the use of surface hardeners, such as lithium silicate, may be difficult to observe and would require special techniques (see Chapter 2) for their identification. Carbonation should not penetrate much beyond the enriched surface zone, and drying shrinkage cracks should be only a few microns wide. A typical thin section of a concrete floor surface is shown in Figure 8.1 and an example of a failed four-layer surface finish in Figure 8.2.

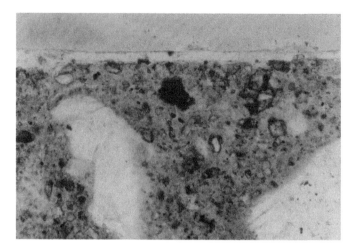

Figure 8.1 A photomicrograph of a thin section from a concrete floor with a well-bonded surface treatment, surface uppermost. Inclined polar illumination, width of field 2 mm. (Photo courtesy of M. Grove, private collection.)

Figure 8.2 A photomicrograph showing a four-layer surface coating. Shrinkage of the coating has caused a crack failure in the outer layer of the concrete. Inclined polar illumination, surface uppermost, width of field 2 mm. (Photo courtesy of M. Grove, private collection.)

Where the floor is intended for heavy-duty industrial use, increased abrasion and chemical resistance may be required. In this case, a special topping is used which may be laid over the still plastic floor as a thin monolithic screed or be placed as a thicker screed over an already cast floor. The topping is specifically designed for the conditions of use and may include selected aggregate and silica fume to increase abrasion resistance. Alternatively, a thin layer of an armour-plate finish containing iron particles may be laid on the surface. Because the topping is relatively thin, it is possible to use special materials, admixtures and low water/cement ratios. The surface can also be finished using a power compacter.

The texture of a proprietary industrial finish is shown in Figure 8.3. The black-coloured topping was variable in thickness and no more than 3–4 mm thick; it contained ragged iron particles 25–1000 μm in length and was well bonded to the concrete substrate. The carbonation of the surface varied from almost nil to a depth of 1–2 mm. The concrete underneath

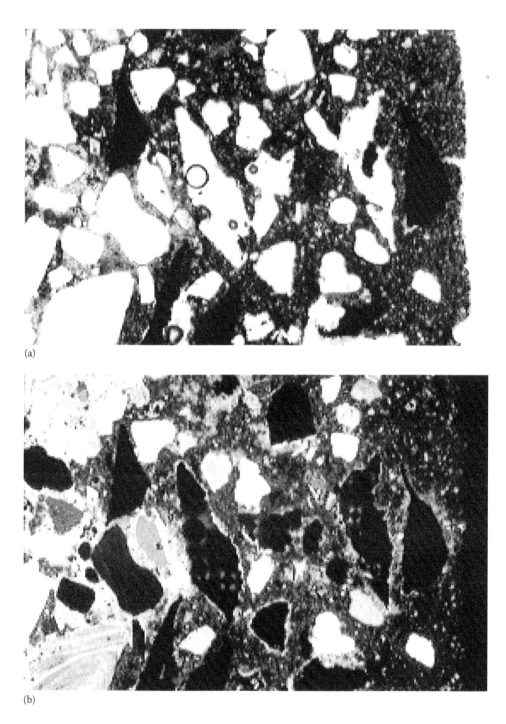

(a)

(b)

Figure 8.3 The texture of a proprietary floor finish. Width of fields 0.35 mm. (a) A thin layer of floor finish containing fragments of iron and aggregate has been worked into the surface of the concrete. (b) Under crossed polars it is evident that the surface of the original concrete is carbonated before the finish was laid and that the floor finish is partially porous and has allowed some carbonation to penetrate. Top surface on the right.

was also carbonated to a depth of about 2 mm indicating the topping was laid over a hardened surface. The black pigment was not apparent in thin section, but scattered throughout the topping were fine particles of moderate birefringent material. Overall, the texture is somewhat confused and has the appearance of a powdered material that has been smoothed over a moderately rough concrete surface.

Some floor finishes are built up in several layers as shown in Figure 8.4, and in some examples, the construction history becomes clear from the examination of the multiple layers (Figure 8.5). The wide range of finishes that are used on concrete floors are described by Deane (1989). Relevant standards on floor screeds, terrazzo, tiles and polymer coatings are BS EN 13318 (2000) that provides definitions and BS EN 13813 (2003) that covers properties and requirements. Information on the design of concrete floors for commercial and industrial use has been published by the New Zealand Portland Cement Association (1962).

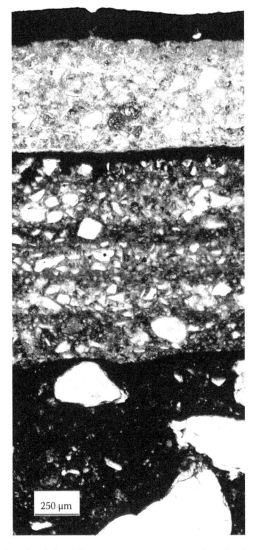

Figure 8.4 Multiple layers of each of the different compositions used to finish a floor screed surface. Top surface uppermost. NEC Pavillion. Plane polar illumination. (Photo courtesy of A. Bromley.)

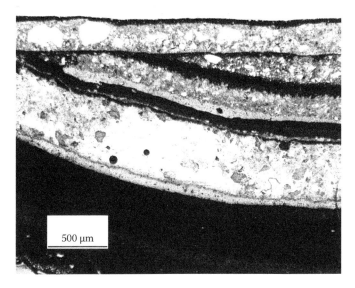

500 µm

Figure 8.5 Multiple layers of the different compositions used to finish a floor screed surface. Top surface uppermost with uppermost layer 'unconformable' on the lower layers, NEC Pavillion. Plane polar illumination. (Photo courtesy of A. Bromley.)

8.2.1 Terrazzo

Terrazzo, also known as Venetian mosaic, is a specialised concrete finish consisting of marble chips or other coloured particles set in a matrix of white or coloured cement, the surface of which has been ground to give a smooth finish. Modern variants known as 'thin-set terrazzo' originally used polyester and vinyl ester instead of cement mortar, but today epoxy resin is almost invariably used. Although the colour choice, impermeable characteristics, fewer shrinkage cracks and the lightweight and rapid construction are all advantages, thin-set terrazzo is only appropriate for internal use, as weathering may cause discolouration and degradation.

The process is described in Specification 94 (Williams 1994), and the properties of terrazzo tiles are specified in BS 4131 (1973), now replaced by BS EN 13748-1 and BS EN 13748-2 (2004). The marble chips are graded in size according to the surface appearance required and must be free of dust. A rich white or Portland cement mix varying between 1:2.5 and 1:3 cement–aggregate ratio is used, and up to 10% of pigment may be added to give the depth of colour required. Water/cement ratios are usually about 0.5 so that care is required in mixing and laying to ensure that segregation of the marble chips is avoided. Terrazzo floors are normally divided into panels no larger than one metre square to minimise shrinkage problems.

The thickness of the terrazzo varies according to the job. For floors, 20 mm is recommended but thinner dimensions for precast slabs and wall finishes are sometimes used. Terrazzo is laid as a finish either on a backing, such as concrete levelled with a mortar, or by using a cheaper backing mix, as is typical with precast tiles. In the case of floors, the terrazzo may be isolated either by a bitumen or by a sand layer. Alternatively, it may be firmly bonded by the use of grout or by directly laying on plastic concrete.

After placement, the terrazzo mix is screeded level without bringing laitance to the surface and wet cured for at least 7 days before coarse grinding. This removes the top few millimetres to reveal the aggregate and any unfilled voids. These are grouted with the same cement, and pigment used in the original mix and the surface is again wet cured to harden the grout. The surface is then ground with successively finer grades of carborundum wheels until the desired finish is achieved. In the case of precast tiles, the terrazzo mix is placed in the bottom of the mould and the backing mix laid over the top. Compaction is achieved by vibration and pressing.

The texture in thin section of a terrazzo topping is typical of an under-sanded concrete with a high cement content. If coloured cement has been used, much of the fine detail of the hardened cement paste may be obscured by the pigment. As a result of the rich wet cement mixes used, drying shrinkage cracks, about 10 µm in width, will always be present especially along the margins of aggregates. The extent of drying shrinkage present is largely a function of paste content. Surfaces are usually carbonated to a depth of 1–2 mm, and carbonation may extend up to 5 mm round aggregate margins and fill the entrances to shrinkage cracks. Isolated patches of carbonation may also be present depending on compaction.

One of the faults encountered with terrazzo is the presence of unsightly and excessive cracking. As with most types of concrete floor finishes, it is essential that sufficient joints in the backing are used to prevent movement in the structure from cracking the terrazzo. Additional joints restricting the area of the terrazzo topping are required to control cracking due to drying shrinkage. Some visible, isolated drying shrinkage cracks are inevitable, but if these become extensive, they soon fill with dirt and become unsightly. Larger cracks are almost invariable due to structural movement.

As the surface of terrazzo is smooth and appears polished, its appearance is sensitive to surface defects. Some problems are generated during construction. Reducing curing time before grinding can result in weakened surfaces which only become apparent with wear. New terrazzo is particularly prone to damage from other sub-trades and wherever possible should be laid last. There is also a problem of slippery surfaces. More coarsely ground surfaces are safer but lack a polished appearance, and consequently, they may be waxed to improve their appearance, which is counterproductive. Incorrect cleaning procedures can also damage terrazzo and are usually evident as pitting of the marble chips. Kessler (1948) investigated the effect of soda ash, tri-sodium phosphate and a synthetic sulphonate on terrazzo and found that tri-sodium phosphate was less injurious than soda ash and that the synthetic sulphonate was relatively free from injurious action.

Terrazzo is a decorative finish which the trade claims to be a hard and durable finish for walls and floors. These claims are true only if the terrazzo is correctly formulated and laid by skilled contractors. In addition, floors must be adequately maintained to prevent damage. With time, terrazzo floors become scratched and ingrained dirt in surface defects gives it an unacceptable appearance. Provided these defects are superficial, they can be removed by light regrinding. Where deeper pits and extensive cracking are present due to faulty construction and incorrect cleaning procedures, effective regrinding of surfaces may not be possible.

8.2.2 Tiled surface finishes

The tiles used on floors and walls are usually set in a cement mortar which now commonly contains bonding agents based on styrene–butadiene latex. The joints are filled with a grout which in many cases contains no Portland cement and may be based on polymers. Successful tiling depends on both the correct use of materials and the skill of the tiler. Current UK practice is described by Robinson et al. (1992) and Williams (1994). Further information on cementitious adhesives for tiling and other applications is given in Section 8.9.

The failure of tiled surfaces mainly involves loss of adhesion either between the tile and the bedding mortar or the bedding mortar and the substrate. Tiles and mosaics are dimensionally stable relative to the mortar and concrete substrate, and unless potential movements of the concrete and mortar are controlled by appropriate joints, lifting of tiles may occur. Water infiltration behind the tiles may also cause loss of adhesion and lifting. Coring of tiled surfaces and examination of thin sections provide information on composition of mortar and concrete but are justified only if compositions are in question. Chemical determination of cement content may be required where it is suspected that an over-rich mortar has been used. On walls where adhesion between the bedding mortar and the concrete has failed, the cause of the failure can be difficult to determine.

St John gives an illustrative case history with the texture in thin section of a floor tile setting as shown in Figure 8.6. Between the mortar bed and the tile, there is a layer about 0.5 mm thick of neat cement which was known to have been mixed with styrene–butadiene latex although there is no visible evidence for the presence of the latex. The hardened matrix of the bonding layer is very rich in cement grains, most of which do not show obvious signs of hydration rims. A limited amount of calcium hydroxide is present in the matrix with larger crystals present in the air voids. The bond to the tile and the underlying mortar bed is excellent where the mortar bed is adequately compacted and consists of closely packed sand particles with only a limited amount of cement paste. Portions of the bed were poorly compacted with large fissures. The hardened matrix contained minimal calcium hydroxide and remnant cement grains apart from large crystals of calcium hydroxide growing into the irregular voids.

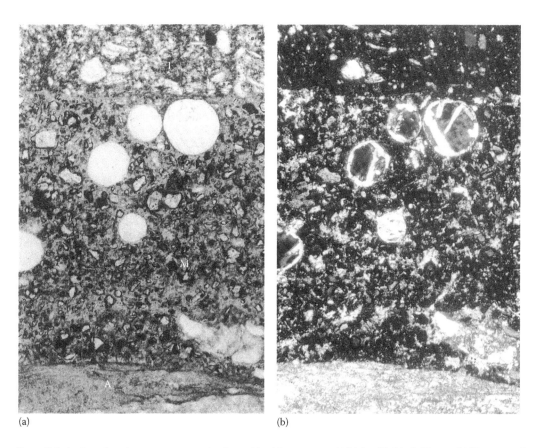

(a)   (b)

Figure 8.6 An interface between a ceramic tile and bedding mortar. Width of fields 0.65 mm. (a) The area of the rich bedding mortar (M) between the bottom surface of the tile (T) and a sand particle (A) contains some entrapped air and numerous partially hydrated and remnant cement grains. The paste is in close contact with the aggregate and the tile but in many other areas the interface contained voids and had parted from the tile. (b) At the tile interface a thin line of calcium hydroxide is just visible and to a lesser extent at the aggregate margin. The entrapped air voids contain laths of calcium hydroxide. It is possible the styrene-butadiene latex used in the mortar has had some effect on the way the calcium hydroxide has crystallised at the interface. Crossed polars.

The petrographic examination of this tile floor was carried out because of a 'drummy' sound caused by the detachment and cracking of the tiles. The chemical analysis showed that the cement content of the mortar varied between 1:5 and 1:10 and this, combined with poor workmanship in the laying of the tiles, resulted in them being damaged due to foot traffic and normal wear and tear.

8.3 RENDERS AND CEMENTITIOUS PLASTERS

Renders and plasters can be divided into those based on Portland cement, on lime or on gypsum. There are subdivisions within these groups and lime is also used in combination with both cement and gypsum plasters. This section is restricted to materials where Portland cement is the major cementing material in the mix. Lime and gypsum plasters are discussed in Chapter 9.

The term plaster is conventionally used to describe a material which coats and smooths internal surfaces, while the terms rendering or stucco is usually applied to external surface coatings, although in practice the term plaster is often used indiscriminately, as it was in historic times. The cause of failed interfaces between layers of plaster and plaster and substrate is a difficult area to investigate. The problem is not so much of the microscope examination which is quite capable of providing good data, but of the initial sampling method. It is difficult to obtain a satisfactory cross section of a plaster and substrate without coring, but there is a high probability of induced cracking which may make petrographic interpretation difficult. In many cases, the interface has already failed and coring is not practicable.

8.3.1 Cementitious plasters

Cement plaster is a mixture of cement and sand which is used to cover, seal and decorate surfaces (Figure 8.7). It is customary to use either lime or air-entraining agents to give water retention and workability. To improve adhesion, additions such as polyvinyl acetate, styrene–butadiene and acrylic copolymers may be added (Ramachandran 1984). The plaster may be applied traditionally either in two or three layers or as a single layer (Mehlmann 1989), depending on the job and local practice. The range of finished surfaces can vary from smooth to rough cast and from trowelled to thrown. There is a considerable variation from country to country in plastering methods. Current UK practice is described by Wickens (1992) and Deane (1989) and specified in BS EN 13914-1 (2005), BS 8481 (2006), BS EN 13914-2 (2005) and, for polymer plasters, PD CEN/TR 15123 (2005). American practice is covered in ASTM C 926 (2011) and ACI Committee 524R-93 (1993). Given the range of surfaces and materials that are plastered, it is not surprising that the skill and experience of the plasterer are the most important factors in producing durable plastered surfaces. Plaster mixes vary from proportions of 1:3 to 1:8 of cement and sand, and 25%–200% of lime by weight of cement may be added to the mix. The leaner the cement mix, the larger the amount of lime added. Air entrainment is often used in place of lime. One of the key ingredients is the sand. The types and gradings of sands previously used in the United Kingdom have been discussed by Cowper (1950), and some current gradings are specified in BS EN 13139 (2002) and ASTM C 897-05 (2009).

The durability of a plastered surface depends on the ability of the plaster to adhere to surfaces and to provide a barrier against moisture. Some cracking is inevitable due to moisture movement, but this can be controlled by construction joints. In general, more problems are encountered with external renderings than internal plaster. A plaster that has been applied with a trowel and finished with a smooth surface is usually less durable than thrown plasters with a rough cast or heavily textured finish.

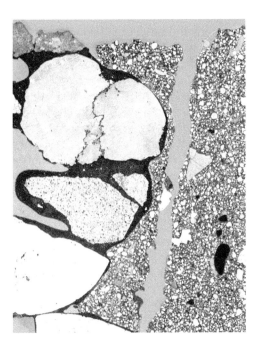

Figure 8.7 A photomicrograph of a crack in render against no-fines concrete in a high-rise apartment block, probably due to structural movement. High-resolution low-magnification digital image of a thin section. Width of field 40 mm. (Photo courtesy of A. Bromley, private collection.)

8.3.2 Petrographic methods of examination

Lime-based plasters tend to display fewer cracks due to differential movement as the binder is more ductile, while dissolution and re-precipitation of lime allow healing of fractures if they form. Petrographic analysis can, as in cementitious plasters, be carried out to identify failure of bond between successive layers and also between the basal layer and substrate. The petrography of lime mortars is discussed further in Chapter 9.

Petrographic examination may be useful where plaster has failed because of lack of adhesion and for cracking and surface problems such as efflorescence and discoloration. BS 4551 (2005) + A1 (2010) describes the sampling, the physical and chemical methods required to determine the composition of the hardened matrix and the grading of the sand, and these should form part of a full petrographic examination. It is important in any examination that the proportions of materials in the plaster be determined as these have a decisive effect on relative moisture movements. In some cases, it will be found that where tradesmen hand batch their materials by volume, it is possible for serious mistakes to be made in the batching, and this may be the cause of the problem.

BS 5262 (1991), now replaced by BS EN 13914-1 (2005), has a section on durability of plaster and discusses a number of factors which influence it. These are the nature and durability of the background, type of rendering and proportions of the mix, method of application, moisture movement in both background and rendering and differential shrinkage within the plaster itself. This is a formidable list of factors to be evaluated, to which the skill of the tradesmen must be added. It is not surprising that the diagnosis of a plastering problem depends more on experience and often it is not possible to establish the cause of the failure beyond doubt. While petrography can provide some data on the integrity of bond surfaces and low-power stereo, microscopic examination of the plaster can be helpful; thin-section

examination is unlikely to provide useful information that cannot better be gained by chemical analysis or electron microscopy unless identification of the components is also required.

The modern practice of casting walls in fair face shuttering is common. The dense smooth
surfaces which result are often sealed by curing agents designed to weather away before any
further treatment of the surfaces is applied. This weathering effect is dependent on the curing membrane being correctly applied and for sufficient time to elapse for weathering to take
place which may vary according to the season. The plasterer may try to combat the effect
of possible residues on such surfaces by the use of bonding agents rather than laboriously
preparing the surface by removing the surface skin from the concrete. Unless the bonding
agents are correctly used, they can also cause adhesion failure as was noted in BS 5262
(1991). It is possible to analyse for residues of curing agents and the presence of bonding
agents by solvent extraction of samples followed by appropriate methods of organic or FTIR
analysis. The results may indicate possible reasons for the failure but are rarely sufficient
proof of cause. At a fundamental level, plaster cracks fail, either because the adhesion is
broken by differential movements exceeding the strength of the bond or because adhesion
has not been adequately developed due to incompatibility of the substrate with the plaster.
The two factors are closely interrelated. Ignatiev and Chatterji (1992) identified the high
shrinkage strains and elastic moduli of modern mortars as the principle cause of cracking
and stress the need for using mortar with a low modulus of elasticity. As already discussed,
residues on concrete surfaces can prevent adequate adhesion ensuing failure irrespective of
the plaster formulation.

Problems can also arise from efflorescence due to the movement of soluble salts to external surfaces and unsoundness of lime causing cracking and popping of surfaces.

Chemical analysis of the efflorescence is usually sufficient to identify the source of the
salts. Detection of unsound lime (uncarbonated calcium hydroxide or unhydrated calcium
oxide) in a hardened plaster is difficult, and it is really necessary to obtain a sample of the
lime used and test it for unsoundness according to BS EN 459-1 (2010) or ASTM C 110-10
(2010) to clearly identify this problem.

8.4 JOINTING AND BEDDING MORTARS

Although bricklaying mortar is sometimes made as a simple mix of sand and cement
(see Figure 8.8), usually jointing and bedding mortars are traditionally mixes of sand,
cement and lime. The lime addition is used to slow the setting and improve workability.
In more recent times, a wide variety of mineral or chemical admixtures such as glass fibre,
air entrainers and silane waterproofers have been used to modify properties such as setting, workability and adhesion or to reduce cracking and other durability characteristics
of mortars. In addition to mortars based on Portland cements, epoxy-based varieties are in
common use and are specified in ASTM C395-01 (2001). Cement-based mortar mixes are
specified in BS EN 998-2 (2003) and ASTM C270-10 (2010), while aggregate specifications
are given in ASTM C144-04 (2004). More specialised mortars containing mineral or chemical admixtures have specifications in ASTM C1384-06a (2006).

Portland cement mortars for masonry work are today generally preferred over lime mortars for their more rapid strength gain characteristics. However, masonry mortars must
not be over-strong, and a reduction in the cement content to control the final strength can
lead to the production of harsh and unworkable mixes. Lime cement mortars overcome this
problem and are widely used; this is often achieved by adding Portland cement to a factory
pre-mixture of sand and lime. It is common for mortars to be gauged such that the lime and
Portland cement are present in equal volumes to give the traditional 1:1:6 mortar which

Figure 8.8 A photomicrograph of a typical sand/cement bricklaying mortar. Yellow areas are voids infilled with yellow-dyed resin. Inclined polar illumination, width of field 3 mm. (Photo courtesy of M. Grove, private collection.)

is most frequently encountered, but other mixes, for example, 1:2:9, are sometimes used. Mixes of sand and Portland cement only, ranging from a strong 3:1 mix to a weak 7:1 mix, are used for particular applications. The mortars that are based on lime, limestone dust and gypsum are discussed in Chapter 9.

In the chemical analysis of hardened mortars, the presence of building lime is indicated by an increase of the calcium/silica ratio typical of Portland cement provided that limestone aggregate and shell are not present. Where limestone aggregate is present, only microscopy can unequivocally identify the presence or absence of building lime in the hardened mortar, but quantification of the building lime is not possible by chemical analysis in such cases. Identifying the presence of building lime in a cement/lime/sand mortar by microscopy may not always be straightforward, especially if the mortar is heavily carbonated. Pockets of trapped calcium hydroxide, whether or not subsequently carbonated, can also indicate the presence of building lime, providing the observation established that the pocket is not a void infilled with portlandite as a result of leaching.

The most positive evidence particularly in older building limes is the identification of relict hard-burnt lime particles of calcium oxide (Figure 8.9) and the presence of unburnt fragments of fuel. Spheroidal pulverised-fuel ash (pfa) particles may also occur sporadically in the matrix and could be derived from either the Portland cement or the lime. (Modern kiln manufacture of both lime and Portland cement frequently uses pfa leaving residual traces of fuel in these materials.)

8.5 SPECIAL CEMENTS AND GROUTS

There are very large numbers of specialist and proprietary cements manufactured covering a very wide range of applications. Bensted (1998) provides a review of four of the major types: oil-well, decorative, chemical and non-standard Portland cements. This section is concerned with grouts, oil-well and decorative cements and a few specialised cements, while some of the other special cements and their applications are discussed elsewhere in this book. Information concerning particular specialist cementitious materials is best obtained from the manufacturer and from the technical literature.

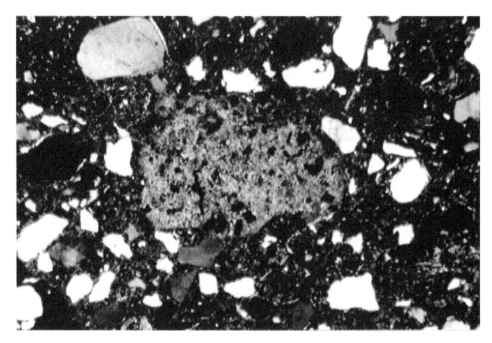

Figure 8.9 Photomicrograph of lime bound mortar in thin section, containing a relief particle of burnt limestone. Width of field 4 mm. Crossed polars. (Photo courtesy of RSK Environment Ltd., Hemel Hempstead, U.K.)

8.5.1 Cementitious grouts

The term grout describes cementing materials that are fluids rather than plastic mixes. They are intended to fill spaces, such as cracks, joints and pre-stressing ducts, to bed machinery and make pre-placed concrete. Chemical grouts are used extensively in geotechnology to consolidate ground for large works. Grouts are based on Portland cement with additions and sand-filled polymers such as epoxy resin or other proprietary mixtures of chemicals. Apart from simple cement grouts used for masonry and tiling, most grouts are proprietary formulations and usually put in place by specialist firms. Chemical and polymer grouts are outside the scope of this book, and texts such as those by Karol (1990) should be consulted for further information. A review of Portland-based special cements and grouts is given by Bensted (1998) in the fourth edition of *Lea's Chemistry of Cement and Concrete*.

Masonry grout is specified in ASTM C 476-09 (2009) and consists of Portland cement plus 0%–10% of hydrated lime mixed with either two to three parts of fine sand or one to two parts of coarse aggregate by volume. The minimum compressive strength required is 14 MPa but there is no limit on the drying shrinkage specified. ASTM C 938-02 (2002) outlines the practice for proportioning grout mixtures of pre-placed aggregate concrete which includes admixtures containing cement, pozzolana, fine aggregate, fluidifier and other chemical admixtures. Tests for fluidity, expansion, bleeding, water retentivity and unit weight are specified without performance requirements. Three types of packaged non-shrink hydraulic cement grouts are specified in ASTM C 1107/C1107M-08 (2008) intended for use under applied loads such as bedding machinery or structures. These are pre-hardening volume adjusting, post-hardening, volume adjusting and other combinations with volume adjusting types. Compressive strengths of 34.5 MPa are required, and limits of 0%–4% presetting and 0%–0.3% post-hardening expansions are specified.

Since the principal purpose of a grout is to fill a space so that loads can be transferred across the space, any shrinkage that reduces the load capacity is undesirable. Grouts based on Portland cement will shrink on drying unless additions are used to counteract this effect. Older anti-shrink additions are based on the expansion of rusting iron powder or the use of gas grouts based on aluminium powder, azides and other chemicals and fluid coke additions (Shaw 1985). Modern cement grouts often use shrinkage-compensating additions based on either calcium sulphoaluminates or lime or mixtures of Portland cement, high-alumina cement and gypsum (Mailvaganam 1984). In practice, proprietary formulations are used on which precise information is difficult to obtain.

There seems to be general agreement that cement-based grouts are satisfactory provided that they do not have to withstand vibrating or rolling loads. In these cases, epoxy resin–based grouts are usually recommended. Petrographic examination of grout is usually limited to examination of texture unless a sample can be cut across the interfaces with material being grouted.

8.5.2 Oil-well cements

A wide range of special cements and cement/additive systems are principally used in connection with the drilling of wells associated with the exploration and production of oil and gas but are also sometimes used for water, waste disposal and thermal wells. There will be considerable variations in individual well conditions, and cements may have to withstand extreme conditions such as elevated temperatures, sulphate attack and aggressive fluids of various kinds. They are generally required to form linings and seals to prevent the flow or leakage of fluids from the walls of a borehole or to seal or plug a redundant bore.

They are required to form low-viscosity slurries which can be pumped to considerable depths and may be pumped down a steel casing to displace water and fill and seal the annular space between the casing and borehole wall. Depending on circumstances, the slurries may range in relative density 1.0–2.1 kg/L or higher, require pumping times from 2 to 6 h and work in a range of temperatures from 0°C to 200°C. Details of these requirements and a general review of oil-well cement are given by Bensted (1998) and Bensted and Barnes (2001).

The specifications for these cements are provided by the American Petroleum Institute (API), are in common use worldwide and are referred to as API Specifications. Classes A through H relate to oil-well cements in common use and are based on Portland cement clinker interground with a calcium sulphate to various grades of fineness and may contain optional admixtures/additives of various kinds. The classes A, B and C are Portland-based cements interground with calcium sulphate and may contain set-modifying additives. They are similar to specifications in ASTM C465-10 (2010), B and C have sulfate-resisting or high early strength properties that are similar to ASTM C150-09/C150-09M (2009) types I and II. The American Petroleum Institute (API) specification classes D, E and F are similar to A, B and C but contain chemical retarders. They meet the requirements of ASTM C465 and are intended for use in conditions of moderately elevated temperatures and pressures though they are of little use today. Classes G and H are in common use and are intended as basic well cements. They are based on Portland cements without additions other than interground calcium sulphate. They are available in moderate and high sulphate-resistant grades.

In addition to these Portland cements, a range of non-standard cements are used in specialist applications. These non-standard cements contain admixtures of various kinds as summarised in Table 8.1.

A wide range of additives that are used with these cements include retarders to extend the pumping time and flow properties. Examples of these are calcium or sodium lignosulphonates, sugars, gluconates, citric or tartaric acids, gums and numerous others (see Bensted 1998). Accelerators are also required for some applications, for example, to increase early

Table 8.1 Typical compositions and applications of non-standard oil-well cements

Typical basic composition	Application	Comments
50/50 pozzolana, Portland cement + ~2% bentonite	Lightweight slurry	sg. = 1.6–1.8
Portland cement + ggbs + ~4%–7% gypsum	General cementing	Seldom used in the west
Portland cement + ggbs, or alkali active ggbs, ultra-fine particle size	High penetrability, e.g., repairs to casing leaks	Particle sizes < 15 μm
Sand + ggbs	High-temperature cementing above 110°C	Forms tobermorite at high temperature
Calcium aluminate cement	Refractory cement low-temperature work	HAC, see Section 5.4
Magnesium oxychloride	Temporary plugging refractory to 850°C	Sorel cement, see Section 9.6
Calcium phosphate cement	Protecting casing from aggressive fluids	Rapid setting, suitable for geothermal wells
Resin/plastic Portland based cement	Plugging holes, waste aggressive and acidic conditions	See polymer cements, Section 7.8
Portland cement + expansive admixture	Expansive sealing of cavities	Expansion begin within days
Portland cement + gypsum	Well plugging, fluid circulation loss, sealing weak formations	Rapid setting, expansive, thixotropic
Portland, class G or H + gypsum + ferrite phase + sodium chloride or HAC + pfa/pozzolana	Low temperature or arctic permafrost conditions	Low heat generation

Source: Modified after Bensted, J., Special cements, in: Hewlett, P.C., ed, *Lea's Chemistry of Cement and Concrete*, 1998, Chapter 14, pp. 779–835.

compressive strength, and may include calcium chloride, sodium chloride (sometimes from seawater), sodium silicate or sodium aluminate.

Other additives/admixtures include 'weighting agents' which typically have particle sizes similar to the cement and include haematite, baryte, galena and spent iron-based catalyst. Fluid loss of drilling fluids may be controlled by the inclusion of such materials as granulated walnut shell, nylon and other fibres and cellophane flakes. The swelling properties of bentonite (smectite) is used as a 2% to 12%+ addition to cement as a lightweight extender to absorb additional water giving slurry densities between 1.44 and 1.56 kg/L (Bensted 1998).

Unless foaming is specifically required to produce a lightweight slurry with relative densities down to 0.7 kg/L (this may be produced by the addition of a small amount of cationic or anionic surfactant), its use is avoided because it may be detrimental in causing pump cavitation and loss of suction. Salts of organic acids as additives may cause undue foaming which will then be controlled using a de-foaming agent such as lauryl alcohol or lower sulphonate oils which lower surface tension but remain insoluble in water.

8.5.3 Petrographic investigation of oil-well cements

The most frequent forms of failure in oil-well cementing arise from inadequate mixing of the components and additives used or by using excessive water in the mix. This leads to reduced compressive strength, settling and a lower than expected relative density. Both inhomogeneous mixes and water/cement ratios may be investigated using thin sections with the polarising microscope.

The majority of oil-well cements are based on Portland cement/gypsum mixtures and consequently hydrate with the formation of portlandite and calcium silicate hydrate (CSH) gels which may become crystalline in higher temperature applications. In higher-temperature applications above ~120°C, an addition of silica flour to minimise permeability and maximise compressive strength may be used; this allows the formation of tobermorite, though this will transform to xonotlite and gyrolite above about 150°C. Portlandite is uncommon at higher temperatures since it tends to form CSH with the silica. At normal temperatures, ettringite is initially formed during hydration from the interground gypsum, but after time and once the gypsum is fully consumed, it will change to monosulphoaluminate hydrate. Ettringite is not normally stable at elevated temperatures, and at least some of the SO_4^{2-}, Al^{3+} and Fe^{3+} are absorbed into solid solution with CSH.

Mineral additions such as weighting minerals, granulated blast-furnace slag (ggbs) or fly ash may be readily identified by optical microscopy, though particle size will be small. The various cement hydrates are better investigated using x-ray diffraction techniques or electron microscopy.

8.5.4 White and coloured Portland cements

Decorative concretes and mortars containing white and coloured Portland cements are widely used on building surfaces, to improve the aesthetic appearance of structures. The white colour results from the manufacture of the cement using chalk and china clay, both of which contain very low iron contents, intergrinding the clinker with a pure white gypsum set regulator and avoiding all iron contamination during processing. It should be noted that white grades of calcium aluminate cement (CAC) are also sometimes used in decorative applications. Coloured finishes are usually achieved by the inclusion of pigments (see Section 4.6.5) with either white or ordinary Portland cements depending on the colour required. 'Brightening agents' (such as titanium dioxide) are sometimes added to white cement clinkers prior to grinding, but because they have no cementitious properties themselves, overdosing can cause problems of surface durability. The raw materials used to manufacture white Portland cement are selected so that very little ferrite phase (<1%) will be formed in the clinker. Since this phase augments the formation of ettringite, little will be formed during hydration. Petrographic observations of hydrated white cement concretes and mortars will be similar to ordinary Portland cement concrete except for the absence of the ferrite phase (see Figure 8.10).

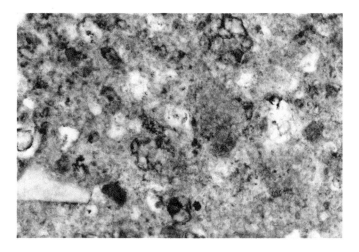

Figure 8.10 A photomicrograph showing residual unhydrated clinker grains in a white cement. These grains are composed of the three major phases, C_3S, C_2S and C_3A; there is no ferrite phase. Plane polar illumination, width of photomicrograph 1.3 mm. (Photo courtesy of M. Grove, private collection.)

8.5.5 Other Portland-based cements

A wide range of very specialised cements of 'Portland type' that cannot be classified into the standard categories are produced. Many are only for experimental studies because of cost. These include cements where the calcium carbonate is replaced by strontium or barium carbonate and the silica by germanium, tin or lead as raw materials. These cements have good cementitious and sulphate-resisting properties, but magnesium carbonates are not used because the magnesium silicates and aluminates have no hydraulic binding properties.

Gypsum as a set regulator in Portland cement is well established, but other regulators such as calcium lignosulphate/sodium hydrogen carbonate mixes have been proposed. This gypsum-free Portland cement has a low-water demand and high-early-strength-gain characteristics. The cement matrix morphology differs in that CSH gel phases tend to form equant rather than elongate particles which rapidly fill the limited spaces available, no ettringite forms and the matrix microstructure becomes massive after about 48 h. This type of cement can harden at low temperatures down to $-20°C$ without calcium chloride additions (Bensted 1998).

Modifying the mineral composition of Portland cements either by modifying the raw materials prior to firing in the kiln or by adding additional amounts of particular phases such as β-dicalcium silicate or tricalcium silicate will alter the setting and strength characteristics of the cement and find application in manufacturing high-early-strength cements. Enhanced early strength characteristics can also be achieved by the inclusion of calcium fluoride or other suitable mineraliser to the raw feed prior to firing (Bensted 1998). Sulphoaluminate cements (Klein's cement) based on belite and ferrite compositions, interground with 16%–25% gypsum, are manufactured in China but are gaining interest elsewhere because of the lower carbon dioxide emissions during manufacture. In China, they are used in mortars and concretes where applications requiring rapid strength gain, sulphate resistance and reduced shrinkage are important. Elsewhere this type of cement is used in proprietary mixtures with Portland cement and other admixtures for screeds and specialist repair materials. Discussion and evaluation of this cement type is given in Zhang and Glasser (1999) and Glasser and Zhang (2001).

Expansive cements are usually based on Portland cement with an expansive mineral component and have a number of specialised applications, for example, special oil-well cements (Section 8.5.2). Type K cement is manufactured by intergrinding an expansive clinker containing kleinite ($3CaO \cdot 3Al_2O_3 \cdot CaSO_4$) with Portland cement clinker and gypsum (or gypsum/anhydrite mixes). Other expansive cements may contain calcium oxide or possibly magnesium oxide. Type M expansive cement is based on the addition of aluminous cement or aluminous slag plus lime to Portland cement (Bensted 1998).

8.6 SPRAYED CONCRETE

Sprayed concrete is applied by feeding a specially designed concrete mixture into a stream of high-velocity air so that it is blasted onto the surface. If the maximum aggregate size is <10 mm, it is referred to as gunite and if over 10 mm as shotcrete. The cement and aggregate may be premixed dry and the water added at the spray nozzle, or a wet pre-mix is pumped to the spray nozzle and sprayed in a stream of high-velocity air. The dry process gives better compaction and higher strengths than the wet process and according to Tabor (1992) giving strengths of 40–50 and 20–30 MPa, respectively.

Schrader and Kaden (1987) state that the strength of concrete mixes used for shotcrete varies from 20 to 80 MPa and give examples of water/cement ratios varying from 0.37 to

0.42 and cement contents from 500 to 400 kg/m³. Set accelerators (Mailvaganam 1984), styrene–butadiene latex, polyvinylidene chloride and silica fume may be added, but these must be accurately gauged and steel, glass and other fibres may also be included in the mix. Schrader and Kaden (1987) stress that the most important factor in the application of sprayed concrete is the skill and experience of the nozzleman. They also provide data on durability. A review of the properties and applications is given by Austin and Robins (1995).

Sprayed concrete is used for building up and repairing surfaces where the concrete has had to be removed because of failure, fabricating underground works, for applying a protective layer to large surface areas and in the stabilisation of natural rock faces. It is usually applied by specialist operators. Specifications and codes of practice for sprayed concrete have been published by the Concrete Society (1979, 1980), ACI Committee 506 (1990), American Society of Civil Engineers (1995), ASTM C1116/C1116M-09 (2009) for fibre reinforcement and ASTM C1141/C1141M-08 (2008) for admixtures.

Localised lack of bond and the formation of large voids are common problems, and the in situ methods for investigating these are given in a report by ACI Committee 506 (1990). Adhesion problems can be caused by the relatively impermeable sprayed concrete locking in moisture in the substrate. The high cement content of sprayed concrete results in shrinkage values up to double that of typical concrete although this can be more than halved by the use of additions such as styrene–butadiene latex. Because of the strengths of both the substrate and the sprayed concrete, shrinkage rarely causes loss of adhesion but manifests itself as closely spaced fine cracking which can allow the penetration of moisture and salts. A mix with lower strength and higher permeability may be more durable and does not develop cracks so readily as stronger mixes which are more prone to cracking (Schrader and Kaden 1987).

Petrographic examination of thin sections cut from cores drilled through the sprayed concrete into the substrate can provide information on hardened paste volume, air content, type, distribution and volume of aggregate, number of applied layers, cracking, carbonation of the surfaces of both the sprayed concrete and the substrate and alkali–aggregate reaction identified if present. St John cites the examination of three cases of sprayed concrete to provide illustrations of some of the textures which result. The first case dates from the 1950s and is of a gunite used to repair an exfoliating tunnel lining attacked by sulphates. The second is of a new shotcrete used to protect and decorate a new concrete surface, and the third is of the gunite covering over the wound post-stressing wires of a Rocla pipe dating from the 1960s which had failed due to chloride attack.

In the first case, the gunite was built up in four layers totalling about 100 mm in depth. The interfaces of the layers were clearly demarcated by a 50–100 μm wide bands of diffuse carbonation and subtle changes in texture across the interfaces indicating some time lapse between applications (Figure 8.11). The outer surface of the gunite was carbonated to an average depth of 100 μm and up to 1.5 mm in the areas of porosity that was scattered along the outermost zone. The interface with the concrete was excellent in spite of the concrete being carbonated to a depth of 2 mm. The band of gunite adjacent to the concrete was cement rich compared to the remainder of the texture. The hardened cement paste was extremely rich in partially hydrated and remnant cement grains in which clusters of belite predominated. Abundant calcium hydroxide was present as large clusters of crystals and in the smaller voids indicating a reasonably wet mix was used. A limited number of small air voids were present together with larger entrapped air voids but overall the void content was low. The bond between interfaces was excellent apart from one interface which was narrowly fissured for much of its length. Isolated drying shrinkage cracks were present in the body of the gunite and one larger shrinkage crack extended from the surface through the two outer layers. The maximum aggregate size was about 3 mm, was well graded and gave a compact texture to the gunite.

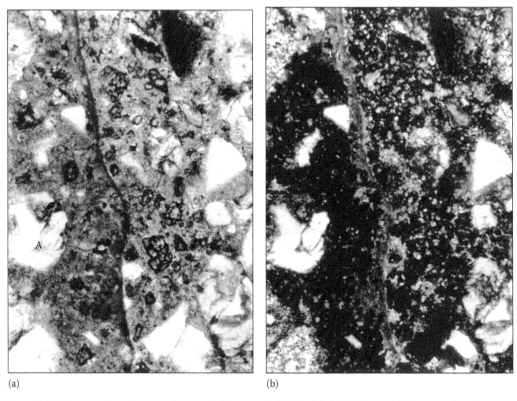

(a)   (b)

Figure 8.11 The interface between layers of shotcrete. Width of fields 0.7 mm. (a) The interface between the two layers of shotcrete is delineated by a diffuse dark band. In the area shown contact is excellent but voids and fissures can be present at these interfaces. Remnant cement grains are dotted through the paste (C) among the greywacke aggregate (A). (b) The dark band is now seen to be the lightly carbonated surface (CC) of the previous layer of shotcrete. Note the difference in the amount of calcium hydroxide present between layer 1 and layer 2. This is probably a segregation effect that occurs as the layers are built up during spraying. Crossed polars.

In the second case, the cores of the shotcrete did not include the concrete substrate. The shotcrete varied between 100 and 140 mm in depth and was built up in three layers which can be seen by examining the section with the naked eye. However, in thin section, these layers were difficult to detect under the microscope apart from bands of poor compaction and very light carbonation. The outer surface was carbonated to about 1 mm and was characterised by poor compaction and numerous large irregular voids. The hardened cement paste was rich in remnant cement grains in which the ferrite phase was predominant and the abundant calcium hydroxide was reasonably well dispersed indicating a lower water/cement ratio was used. Moderate amounts of calcium hydroxide had crystallised in pores and voids. The maximum aggregate size was about 10 mm, gap graded to a well-graded sand. There were numerous cracks and fissures associated with the interface zones, and the overall impression was of dense shotcrete which had been incorrectly applied in the interfacial zones.

In the third case, the gunite had been applied as one 25 mm thick layer over a pre-stressing strand. The texture was dense apart from the zones at the surface and adjacent to the pre-stressing strand. These were porous and full of large irregular voids. Carbonation had penetrated from 3 to 6 mm from the outside. The texture of the hardened matrix was similar to the first case in many respects.

The three textures described illustrate that in the second and third cases, the operators applying the sprayed concrete were having problems in obtaining adequate compaction at the beginning and end of each layer. While the textures are only described qualitatively, there would have been few problems in making quantitative measurements.

8.7 CEMENTITIOUS REPAIR MATERIALS

The important requirements for a repair material for concrete or other related materials are ease of mixing and placing on damaged areas including vertical surfaces and inverts. Also, good adhesion to damaged surfaces, no drying shrinkage, rapid setting and hardening properties, low permeability, good resistance to weathering, chemical attack and resistance to cracking are all required.

Numerous proprietary materials are available designed specifically to meet a range of particular repair requirements. Their formulations are commonly based on Portland cement or a calcium sulphoaluminate cement together with a quartz sand aggregate, one or more of a wide range of polymer admixtures and sometimes mineral fibres or mineral additions such as calcium carbonate. A standard specification for these materials is provided in ASTM C928/C928M-09 and relates to cement/aggregate mixes supplied as dry powder which only require to be mixed with the appropriate, but critical, volume of water.

The polymers used depend on the particular application but include vinyl esters, polyesters, epoxy, styrene–butadiene, acrylics, polyurethanes and methyl acrylates. Some may be included in the mix as emulsions. Further information on the use of polymers in repair materials and concrete is available from the American Concrete Institute, ACI 584 1R-92 (1993).

Shrinkage cracking may sometimes cause difficulties. Examination of thin sections will allow the identification of cement type, fine aggregate and mineral admixtures. The texture and fabric will also provide an indication of degree of compaction and the water/cement ratio, but identification of the polymers used requires investigation by other methods such as FTIR or chemical extraction and analysis (see also Sections 2.5.4 and 2.5.6).

8.8 CEMENTITIOUS LEVELLING COMPOUNDS

The self-levelling cementitious materials are similar in many respects to the specialised repair materials in that they are commonly a mix of Portland cement and fine quartz sand aggregate with additions such as latex or epoxy. Other types based on hemihydrate or other calcium sulphates are discussed in Section 9.3. They all require a rheology that when first mixed, they must allow flow under gravity to form a level surface. They need to remain fluid for sufficient time for a level to be achieved, typically 15–30 min, after which they need to harden rapidly to provide a durable, crack-free hard surface so that tiles or other decorative flooring finish can be applied. Some proprietary compounds claim that the surface becomes hard enough to be trafficked after 2 or 3 h and tiling or other finishing layers can be applied after 12–24 h.

In many cases, there is a recommendation that a priming sealing coat of a polymer such as acrylic is placed before the levelling layer is applied because any porosity in the base layer material will contain trapped air which will migrate through the levelling layer to cause porosity and pinholes with detrimental effects on the durability.

8.9 CEMENTITIOUS ADHESIVE COMPOUNDS

A major application of cement-based adhesives is in fixing wall and floor tiles (see Section 8.2.2). Less commonly cementitious adhesive mixtures are used to form a priming layer between an existing floor surface and a screed or decorative overlay. A simple commonly used primer slurry of this kind is composed of a 50/50 mix of Portland cement and fine plastering sand containing a polymer emulsion which is mixed with water to a paint consistency and laid with a roughened upper surface to provide a key for the top layer.

Cement-based polymer-modified adhesives for tiling are classified in BS EN 12004 (2007). The polymers used include PVA, latex and acrylics, and most of the proprietary adhesive mixes also contain fine quartz sand. Mixes intended for interior use may contain an antibacterial agent. Some proprietary mixes claim that no sealing coat primer is required, while others suggest sealing porous substrates and gypsum plaster surfaces is necessary.

Difficulties are usually associated with debonding from the tiling and sometimes debonding from the substrate. Sampling is best achieved by careful coring through the tile adhesive layer and into the substrate with a core also taken from a sound area for comparison to be made. Optical microscopy will provide a great deal of information concerning the cement, water/cement ratio, fine aggregate and in many cases the type of failure. In some cases, it is possible to identify the polymer used from the cement matrix morphology in thin section (see Figure 8.6), but in others, it may be necessary to use FTIR analysis or chemical extraction methods (see Section 2.5.6.3).

REFERENCES

ACI 548 1R-92, 1993: *Guide to the Use of Polymers in Concrete.* American Concrete Institute. ACI 548 1R-93, Detroit, MI.

ACI Committee 506, 1990: American Concrete Institute. Specification for materials, proportioning and application of shotcrete. *ACI Manual of Concrete Practice* 1995, Part 5, American Concrete Institute. ACI 506R-90, Detroit, MI. Reapproved 1995.

ACI Committee 524R-93, 1993: *Guide to Portland Cement Plastering.* American Concrete Institute. ACI 548 1R-93, Detroit, MI.

American Petroleum Institute, API, 2010: *Specifications for Cements and Materials for Well Cementing. Part 1: Specifications,* 24th edn. API, Washington, DC.

American Society of Civil Engineers, 1995: *Standard Practice for Shotcrete,* Vol. 11: *Technical Engineering & Design Guides as Adapted from the US Army Corps of Engineers.* ASCE.org.

ASTM C110-10, 2010: *Standard Test Methods for Physical Testing of Quicklime, Hydrated Lime, and Limestone.* ASTM, Philadelphia, PA.

ASTM C110-95, 1995: *Standard Test Methods for Physical Testing of Quicklime, Hydrated Lime, and Limestone.* ASTM, Philadelphia, PA.

ASTM C144-04, 2004: *Standard Specification for Aggregate for Masonry Mortar.* ASTM, Philadelphia, PA.

ASTM C150-09/C150-09M, 2009: *Standard Specification for Portland Cement.* ASTM, Philadelphia, PA.

ASTM C270-10, 2010: *Standard Specification for Mortar for Unit Masonry.* ASTM, Philadelphia, PA.

ASTM C395-01, 2001: *Standard Specification for Chemical Resistant Resin Mortars.* ASTM, Philadelphia, PA.

ASTM C465-10, 2010: *Standard Specification for Processing Additions for Use in the Manufacture of Hydraulic Cements.* ASTM, Philadelphia, PA.

ASTM C476-09, 2009: *Standard Specification for Grout for Masonry.* ASTM, Philadelphia, PA.

ASTM C897-05, 2009: *Standard Specification for Aggregate for Job-Mixed Portland Cement-Based Plasters.* ASTM, Philadelphia, PA.

ASTM C926, 2011: *Standard Specification for Application of Portland Cement Based Plasters.* Revision edition 11A. ASTM, Philadelphia, PA.

ASTM C928/C928M-09, 2009: *Standard Specification for Packaged, Dry Rapid Hardening Cementitious Materials for Concrete Repairs.* ASTM, Philadelphia, PA.

ASTM C938-02, 2002: *Standard Practice for Proportioning Grout Mixtures for Pre-Placed Aggregate Concrete.* ASTM, Philadelphia, PA.

ASTM C1107/C1107M-08, 2008: *Standard Specification for Packaged Dry, Hydraulic-Cement Grout (Nonshrink).* ASTM, Philadelphia, PA.

ASTM C1116/C1116M-09, 2009: *Standard Specification Fibre-Reinforced Concrete.* ASTM, Philadelphia, PA.

ASTM C1141/C1141M-08, 2008: *Standard Specification for Admixtures for Shotcrete.* ASTM, Philadelphia, PA.

ASTM C1384-06a, 2006: *Standard Specification for Admixtures for Masonry Mortars.* ASTM, Philadelphia, PA.

Austin, S. and Robins, P. 1995: *Sprayed Concrete Properties, Design and Applications.* Whittles Publishing, Caithness, Scotland.

Bensted, J. 1998: Special cements. In *Lea's Chemistry of Cement and Concrete,* Hewlett, P.C. (ed.). Chapter 14, pp. 779–835.

Bensted, J. and Barnes, P. (eds.). 2001: *Structure and Performance of Cements.* Spon Press, Taylor & Francis Group, London, U.K.

BS 4131, 1973: *Specification for Terrazzo Tiles* (now superseded by BS EN 13748). BSI, London, U.K.

BS 4551 2005 + A1, 2010: *Mortar, Methods of Test for Mortar Chemical and Physical Testing.* BSI, London, U.K.

BS 5262, 1991: *Code of Practice for External Renderings.* BSI, London, U.K.

BS 8481, 2006: *Design, Preparation and Application of Internal Gypsum, Cement, Cement and Lime Plastering Systems. Specification.* BSI, London, U.K.

BS EN 459-1, 2010: *Building Lime. Definitions, Specifications and Conformity Criteria.* BSI, London, U.K.

BS EN 998-2, 2003: *Specification for Mortar for Masonry Mortar.* BSI, London, U.K.

BS EN 12004, 2007: *Adhesives for Tiles. Requirements, Evaluation of Conformity, Classification and Designation.* BSI, London, U.K.

BS EN 13139, 2002: *Specifications for Building Sands from Natural Sources.* BSI, London, U.K.

BS EN 13318, 2000: *Screed Material and Floor Screeds. Definitions.* BSI, London, U.K.

BS EN 13748-1, 2004: *Terrazzo Tiles for Internal Use.* BSI, London, U.K.

BS EN 13748-2, 2004: *Terrazzo Tiles. Terrazzo Tiles for External Use.* BSI, London, U.K.

BS EN 13813, 2003: *Screeded Material and Floor Screeds – Properties and Requirements.* BSI, London, U.K.

BS EN 13914-1, 2005: *Code of Practice for External Renderings.* BSI, London, U.K.

BS EN 13914-2, 2005: *Design, Preparation and Application of External Rendering and Internal Plastering. Design Considerations and Essential Principles for Internal Plastering.* BSI, London, U.K.

Concrete Society. 1979: Specification for sprayed concrete. Publication 53.029. The Concrete Society, England, U.K.

Concrete Society. 1980: Code of practice for sprayed concrete. Publication 53.030. The Concrete Society, England, U.K.

Cowper, A.D. 1950: Sands for plasters, mortars and external renderings. *National Building Studies, Bulletin No. 7,* HMSO, London, U.K., pp. 20.

Deane, Y. 1989: *Finishes,* Mitchell's Building Series. Mitchell Publishing, London, U.K.

Glasser, F.P. and Zhang, L. 2001: High performance cement matrices based on calcium-sulfoaluminate-belite compositions. *Cement and Concrete Research* 31 (12), 1881–1886.

Hewlett, P.C. (ed.) 1998: *Lea's Chemistry of Cement and Concrete,* 4th edn., Arnold Hodder Headline Group, London, U.K.

Ignatiev, N. and Chatterji, S. 1992: On the mutual compatibility of mortar and concrete in composite members. *Cement & Concrete Composites* 14, 179–183.

Karol, R.H. 1990: *Chemical Grouting*, 2nd edn., Vol. 8. Civil Engineering Series, Marcel Dekker, New York.

Kessler, D.W. 1948: Terrazzo as affected by cleaning materials. *Journal of the American Concrete Institute* 20 (1), 33–40.

Mailvaganam, N.P. 1984: Miscellaneous admixtures. In *Concrete Admixtures Handbook*, V.S. Ramachandran (ed.), Noyes Publications, Park Ridge, NJ, pp. 480–557.

Mehlmann, M. 1989: Current state of research on renderings and mortars. *Zement Kalk Gips* (3), 70–72, translation (1/89).

New Zealand Portland Cement Association, 1962: Design of concrete floors for commercial and industrial use. *N.Z. Portland Cement Association Technical Bulletin ST 26*, Wellington, New Zealand.

PD CEN/TR 15123, 2005: Design, preparation and application of internal polymer plastering systems. BSI, London, U.K.

Ramachandran, V.S. (ed.) 1984: *Concrete Admixtures Handbook: Properties, Science and Technology*. Noyes Publications, Park Ridge, NJ.

Robinson, G., Vaughan, F., and Hordell, G. 1992: *Tiling. Specification 92*. MBC Architectural Press and Building Publications, London, U.K.

Schrader, E. and Kaden, R. 1987: Durability of shotcrete. In *Concrete Durability, Katherine and Bryant Mather International Conference*, Scanlon, J.M. (ed.), ACI SP-100 2, American Concrete Institute, Detroit, MI, pp. 1071–1101.

Shaw, J.D.N. 1985: Cementitious grouts – An update. *Civil Engineering* (London) (October) 50–59.

Tabor, L.J. 1992: Repair materials and techniques. In *Durability of Concrete Structures*, Mays, G. (ed.). E. & F.N. Spon, London, U.K.

Wickens, D. 1992: *Plasterwork and Rendering. Specification 92*. MBC Architectural Press and Building Publications, London, U.K.

Williams, A. (ed.) 1994: *Specification 1994* (published in three volumes: *Technical; Products; Clauses*). Emap Business Publishing (Emap Architecture), London, U.K.

Zhang, L. and Glasser, F.P. 1999: New concretes based on calcium sulfoaluminate cement. In *Creating with Concrete. Proceedings of International Conferences (and Seminars)*, Dyer, T.D., Henderson, N.A. and Jones, R.M. (eds.). Thomas Telford Ltd., London, U.K.

Chapter 9

Non-Portland cementitious materials, plasters and mortars

9.1 LIME-BASED MATERIALS AND PRODUCTS

Both lime and gypsum are among the oldest of construction materials. Various forms of lime were used in mortar and plaster and to make concrete by the early Egyptians and Greeks. It was also used extensively by the Romans, in China and in other societies (Lea 1970). The problem of adequately discussing lime arises from the large number of ways this material has been used for construction in the past, many of which have now lapsed in the West although some of these older techniques are still used in other parts of the world (Hill et al. 1992). The use of lime in repair to historic buildings, however, is increasing (Leslie and Gibbons 2000) and petrographic analysis of historic mortars has become a significant tool in conservation work. Lime is also discussed in Chapter 4, Section 4.2.1.

9.1.1 Limestone and lime

Both limestone and limes are used extensively in industry and agriculture as well as a component in construction materials. The manufacture and uses of lime are discussed by Searle (1935) and Boynton (1980). Construction uses include the intergrinding of limestone with Portland cement as a diluent (see Section 5.3.1). Hydrated lime is used in mortars and plasters and also used with a range of siliceous materials to manufacture sand-lime bricks, blocks and other products. Both quick and hydrated lime (CaO and $Ca(OH)_2$) are used extensively to stabilise soils, especially for road construction.

Hydrated lime is manufactured by crushing either limestone or magnesian/dolomitic limestones and heating them to between 1000°C and 1200°C to produce quick lime which is subsequently slaked with water to produce hydrated lime (see also Chapter 4.2.5). The slaking process disintegrates the lumps of quick lime to a powder, putty or slurry depending on the amount of water used and grinding is not normally required. In modern production, slaking is carried out under pressure which ensures full hydration and the product may also be ground to control particle size. In general terms the reactions taking place in the production hydrated lime are as follows:

$$CaCO_3 + heat \rightarrow CaO + CO_2$$

$$CaCO_3 \cdot MgCO_3 + heat \rightarrow CaO \cdot MgO + 2CO_2$$

$$CaO + H_2O \rightarrow Ca(OH)_2$$

$$CaO \cdot MgO + 2H_2O \rightarrow Ca(OH)_2 \cdot Mg(OH)_2$$

The mechanism of hardening of a hydrated lime is simply the formation of calcium and magnesium carbonates by absorption and reaction with carbon dioxide from the air as follows:

$$Ca(OH)_2 + CO_2 \rightarrow CaCO_3 + H_2O$$

$$Mg(OH)_2 + CO_2 \rightarrow MgCO_3 + H_2O$$

Atmospheric carbonation is slow and usually partial because the carbonated margins of plaster or mortar impede the ingress of carbon dioxide to the interior. A better idea of conditions controlling carbonation can be derived from artificial carbonation as discussed by Moorhead (1986). Maximum rates of carbonation were attained when the product was exposed to an atmosphere of 100% carbon dioxide and the capillary pores in the lime mortar were half filled with water. Carbonation virtually ceased when the capillaries in the lime mortar were either saturated or dry. Carbonation was also controlled by the thickness of mortar and it was found that rapid artificial carbonation was impractical over depths greater than 25 mm. It is of interest that Moorhead was able to produce within days fully carbonated lime mortars with compressive strengths of 50 MPa. In practice lime mortars are commonly kept damp for the first 2–3 days after application to aid the carbonation process.

The reactions given earlier are applicable to 'fat' limes. These are limes consisting of 90%–95% calcium and magnesium hydroxides as specified in BS EN 459-1 (2001), ASTM C 206-03 (2003) and ASTM C 207-06 (2006). The production of these limes requires a fairly pure limestone as the starting material. However, the majority of limestones are of lesser grade and those traditionally used for building lime production could contain from 5% to 40% of clay, silica and other minor components. When these impure limestones are calcined, both calcium and magnesium oxides and phases similar to those found in Portland cement are formed, such as the di- and tricalcium silicate, aluminate and ferrite phases, which give the lime some hydraulic properties and the ability to set under water (Roberts 1956). It is now known that tricalcium silicate, while uncommon, can be present in hydraulic lime mortars (Elsen 2006). Hydraulic limes are specified in BS EN 459-1 and ASTM C 141/141M-09 (2009). Lime for use with pozzolanas is specified in ASTM C 821-09 (2009).

Burning of an impure limestone creates a natural hydraulic lime (NHL), but combinations of the appropriate quantities of pure limestone and clay ± silica will create an artificial hydraulic lime (AHL). Both products are in current production.

Underburning and overburning occur not only where there is excessive temperature variation across the calcination zone but also if the raw feed is variable in composition. Each limestone has an optimum burning condition and unless variability in composition can be controlled the calcination cannot be optimised. According to Boynton (1980), variability of composition and properties has been a persistent problem in the production of hydraulic limes. Microscopic methods for determining lime quality and burning characteristics of high-grade pebble lime are given by McKenzie (1994, 1995). If the hydrated lime has not been adequately calcined or matured, it may contain particles of unslaked lime which will gradually absorb water from the atmosphere and will slake and 'blow' after the plaster has been incorporated in the building. Soundness tests for hydrated lime are provided by BS 6463-103 (1999) and ASTM C110-09a (2009).

The mechanism of hardening of hydraulic limes is a combination of carbonation and hydration of calcium silicates. Where the lime is weakly hydraulic, the lime may have good slaking characteristics and still retain some slow hydraulic activity even after slaking. Typically, hydraulic limes that may contain up to 26% silica (ASTM C 141/C141M 2009) may set within hours but will generally not slake easily and carbonation hardening is limited. There is another mechanism of setting whereby a 'fat' lime (high-calcium lime) is mixed with a pozzolanic material and the slow formation of a calcium silicate hydrate occurs to harden the mass into a dense and durable product, in effect a hydraulic lime that sets even under water. This is one of the types of cement used by the Romans and other historic civilisations and is so durable that it is still occasionally used for marine works. The modern autoclaved lime silica product is merely a modification of the old Roman cement.

Unlike the production of Portland cement, the techniques of lime burning varied widely. Modern calcining techniques are efficient and use high-grade limestone to produce quality, pure or 'fat' lime, often intended for industrial use as well as building. The least controlled process is the old and simple technique of using a local limestone of indeterminate composition and stacking the crushed limestone and wood in layers in a ventilated pit and burning the wood to provide the heat for calcination. This results in a range of slaking characteristics, of 'fatness' and of hardening properties of the lime putties used for construction. The wide diversity of burning techniques employed, especially in the past, is beyond the scope of this text and details can be found in Searle (1935), Boynton (1980) and Holmes and Wigate (2002).

The main issues of interest to the petrographer that arise from lime burning are caused by variability of raw material and insufficient control of calcination resulting in underburning and overburning. In many limes, these problems are interrelated and both underburnt and overburnt particles may be present together in one sample.

Underburning causes 'core' which is often removed in later processing. This is where the exterior of the limestone particle is calcined and the interior remains as calcium carbonate. Underburnt particles can be used by the petrographer to identify the source limestone burnt in the kiln (Elsen 2006; Carran et al. 2012). Overburning can make the calcium and magnesium oxides resistant to slaking and result in undue later expansion as the calcium and magnesium oxides slowly hydrate in the mortar. Traditionally, this problem was avoided by slaking the lime for weeks or even months before use. Lime putty was traditionally stored for a minimum of 3 months before use. Another problem arising from overburning is clinkering. This is where fusion occurs in particles containing fluxes so that they form a 'natural cement'. Traditionally, clinker is removed from quick lime as it does not slake well and will not form a powder. Ground-up clinker was used as a 'natural cement' and apparently was still being made for this purpose in some countries (Lea 1970). It may set within hours but will generally not slake properly and carbonation hardening is limited.

A common traditional technique to create good quality mortar is 'hot lime' mixing, where damp aggregate was mixed with quicklime, causing hydration of the calcium oxide during mixing of the mortar. There is anecdotal evidence that the temperature improves the bond between aggregate and binder.

9.2 LIME PLASTERS, MORTARS AND SCREEDS

According to Wickens (1992), lime plastering was the most commonly accepted system of plastering in the United Kingdom before 1939 but is now virtually obsolete. Some idea of the range of limes used for plaster, renders and mortars is given in Figure 9.1.

Currently, BS EN 459-1 (2001) specifies only high-calcium, semi-hydraulic and magnesian limes (see also Chapter 4, Section 4.2.5.1). Methods of test for chemical composition, fineness, soundness, standard consistency and hydraulic strength are now specified in BS 6463-102 (2001) and 103 (1999). Codes of practice for external renderings are given in BS EN 13914-1 (2005) and design and application principles of external and internal plastering in BS EN 13914-2 (2005). The details of recommended mixes for masonry mortars are specified in BS EN 413-1 (2004) and ASTM C 91-05 (2005).

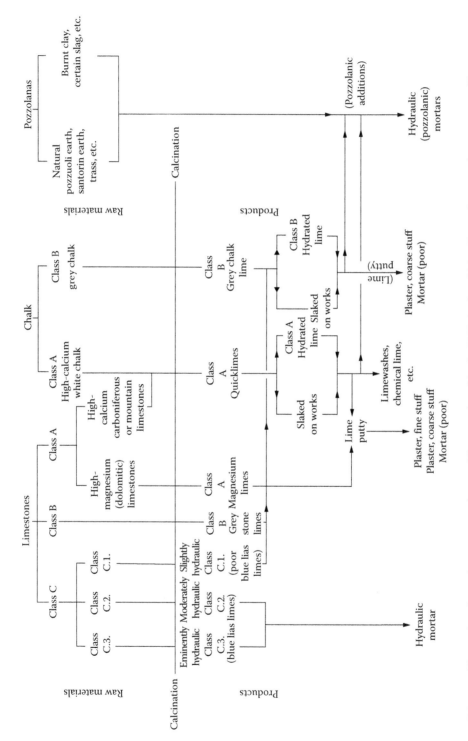

Figure 9.1 Diagram illustrating the various types and uses of lime produced in the U.K. in the 1920s. (From Cowper, A.D., Lime and lime mortars. DSIR Building Research Special Report No. 9. HMSO, London, U.K., 1927.)

The varieties of lime that have been used in the United Kingdom are shown in Figure 9.1. In Europe, the types of lime binder used are generally similar, but in the United States, dolomitic lime is commonly used. The 'fat' limes should be composed primarily of calcium hydroxide or carbonate with magnesium hydroxide and carbonate also being present in magnesian limes. However, the possibility of remnant particles of hard-burned calcium and magnesium oxides should always be considered. Other materials likely to be found in lime mortar and plaster are aggregate, Portland cement and fibre reinforcement (see Sections 5.6.3 and 7.7). The most common aggregate used is sand, but both perlite and vermiculite are used in sprayed insulating plaster.

In hydraulic limes, silica and alumina in the limestone form dicalcium silicate and calcium aluminates on calcination which will give rise to hydrated phases similar to those found in hydrated Portland cement. This poses a problem in differentiating between hydraulic lime and lime intermixed with Portland cement. However, careful microscopic examination of the material in thin section, or preferably in polished section, should allow a petrographer to distinguish if any remnants of alite grains are present. If remnants of alite are present, it indicates the use of Portland cement, since alite is not a common component of hydraulic lime. Since the amount of alite residue will be small, identification may be difficult so that using x-ray diffraction analysis and scanning electron microscopy/X-ray energy discussion analysis (SEM/EDS) methods may be required (see also Chapter 4, section 4.2.5.2).

As already discussed in Section 9.1.1, the rate of carbonation of lime is dependent on both the concentration of carbon dioxide and the amount of water present in the capillary space inside the plaster or mortar. Because most of the precipitation of the carbonate initially occurs in the capillary space, the permeability of the carbonated zone to carbon dioxide is reduced. Thus, in most mortars, even those that are hundreds of years old, there may be a central core of uncarbonated lime remaining.

The mean particle sizes of hydrated limes vary between about 3 and 7 μm with few particles exceeding 30 μm (Boynton 1980). This indicates that a proportion of remnant calcium hydroxide crystals will be large enough to be identifiable under the microscope unless the hydrated lime has been ground. Similarly, particles of hard burnt calcium oxide and magnesium oxide should be readily identifiable as both are isotropic and have distinctive shapes (see Glossary). Some particles become zoned as a result of partial hydration (Figure 9.2) and may become carbonated over time (Figures 9.3 and 9.4)

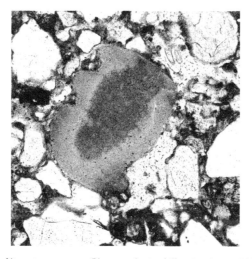

Figure 9.2 **A zoned nodule of lime in a mortar. Plane-polarised illumination; width of field, 2 mm. (Courtesy of A. Bromley, private collection.)**

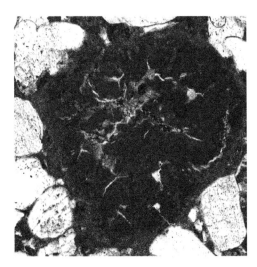

Figure 9.3 An uncarbonated nodule of lime with quartz fine aggregate grains. Plane-polar illumination; width of field, 2 mm. (Courtesy of A. Bromley, private collection.)

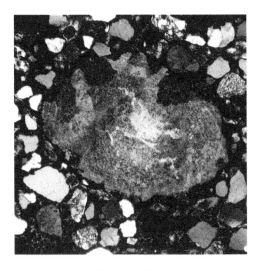

Figure 9.4 A large carbonated lime nodule with quartz fine aggregate grains. Crossed polar illumination; width of field, 2 mm. (Courtesy of A. Bromley, private collection.)

The texture of the carbonates formed depends on the conditions of carbonation. At one extreme, the artificially carbonated lime described by Moorhead (1986) consisted of crystals generally below 1 μm and there was evidence to suggest that about half the carbonate crystals were less than 0.1 μm in size and even possibly amorphous in character. This is the texture of very rapid carbonation where crystals have had insufficient time to develop. At the other extreme will be plaster and mortar that have been damp but not saturated for very long periods. It would be expected that in these cases, large crystals of carbonate would develop in void space. A general view of a plaster finishing coat is shown in Figure 9.5. Shrinkage cracking of this kind is often present in older lime mortars and an extreme example of it is shown in Figure 9.6.

Figure 9.5 A thin section showing the texture of a gypsum plaster finishing coat but is too fine-grained to show discrete crystals. Curving fissures developed at the margins of rounded dense hydrated zones; width of field, 2 mm. (Courtesy of A. Bromley, private collection.)

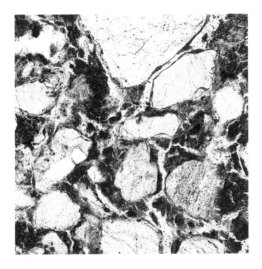

Figure 9.6 Hydrated lime mortar with extensive shrinkage microcracks outlined in yellow-dyed resin. Plane-polarised illumination; width of field, 1 mm. (Courtesy of A. Bromley, private collection.)

9.3 GYPSUM-BASED WALL PLASTERS AND PLASTERBOARD

Hemihydrate ($CaSO_4 \cdot 0.5H_2O$) is principally used to cast wall boards, building components and for plastering internal surfaces of buildings (Figure 9.5). It has some unique properties. On setting, it undergoes expansion which allows accurate replication of very fine details when used for moulding. Subsequent drying may reverse much of this setting expansion (Blakey 1959). Hemihydrate can be made to set in times that vary from a few minutes to several hours which enable it to be used for a wide range of casting and internal plastering purposes. Once it has set, it attains its full strength on drying so that products do not have to be cured like Portland cement. It is non-toxic and has excellent fire resistance. Its main drawback is that gypsum has a solubility of 0.27% in water at 20°C and thus it is not durable in moist conditions. It is one of the most traditional building materials and its use extends back at least 4000 years to the construction of the Egyptian pyramids.

The traditional production of hemihydrate, also known as plaster of Paris, or bassanite, is to heat ground gypsum in a large kettle at about 170°C to drive off 1.5 molecules of water to form coarse building plaster which is designated J-hemihydrate. Modern production methods now use counter-current rotary kilns and other energy-efficient devices (Daligand 1985). The anhydrous group is produced by calcining gypsum to temperatures of 400°C or greater to produce anhydrite as indicated by the following idealised equations. This particular group is restricted to plastering applications:

$$2(CaSO_4 \cdot 2H_2O) + heat \rightarrow 2(CaSO_4 \cdot 0.5H_2O) + 3H_2O$$

$$CaSO_4 \cdot 2H_2O + heat \rightarrow CaSO_4 + 2H_2O$$

The setting of hemihydrate and anhydrite with water is the reverse of these equations reforming the gypsum as a set crystalline mass. The natural mineral anhydrite can be used directly by grinding with the addition of a set accelerator to produce an anhydrous plaster without the necessity of heating in its manufacture.

There is considerable complexity in the types of gypsum plasters that can be made. Formation of any particular phase is mainly due to subtle changes in crystallinity. In spite of subdivisions such as α and β hemihydrate based on setting and physical properties, there are only four clearly defined crystalline phases. These are gypsum, hemihydrate, anhydrite and the highly metastable soluble anhydrite (Ridge and Berekta 1969). The range of gypsum plasters used in the early part of this century (Pippard 1938) is shown in Figure 9.7. With the exception of the hemihydrate, other types now find little use.

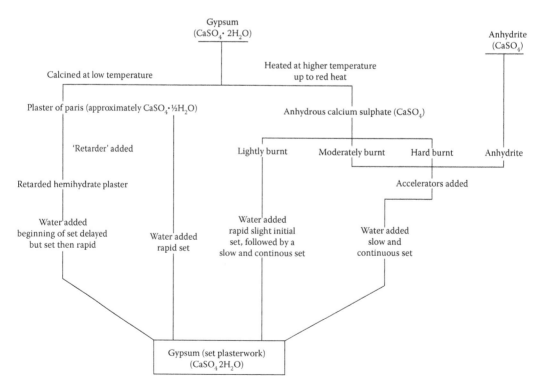

Figure 9.7 The types of hemihydrate and anhydrite plasters in use in 1938. The accelerated anhydrite plasters were introduced in the early nineteenth century. (From Pippard, W.R., Calcium sulphate plasters. DSIR Building Research Bulletin No. 13. HMSO, London, U.K., 1938.)

The duration and temperature of calcination affects the setting time and physical characteristics of gypsum plaster. For instance, β-hemihydrate can be produced to give setting times between about 10 and 30 min without the use of accelerators or retarders. A high grade of hemihydrate can also be made by autoclaving ground gypsum to produce α-hemihydrate commonly known as dental plaster. This plaster requires about half the amount of water as β-hemihydrate to give a fluid mixture and will produce much stronger casts.

The manufacture and technology of gypsum plaster is much more complex than is generally realised, and much of the industrial information is available only through the patent literature or published work from eastern Europe and Japan. There are few texts, and the translation of a Russian text by Volzhenskii and Ferronskaya (1974) is difficult to obtain. Kuntze (1976–1986) annually reviewed the literature on gypsum technology, and technical papers are regularly published in the journals *Zement-Kalk-Gips* and *Ciments, Bétons, Platres, Chaux*. The changes in the types and uses of calcium sulphate plasters in the United Kingdom can be seen from comparing the data given by Pippard (1938), Andrews (1948) and Wickens (1992). The properties of cast gypsum as a structural material have been reviewed by Blakey (1959) and the historical use of gypsum plaster is discussed by Davey (1965) and Ashurst and Ashurst (1988).

In the United Kingdom, gypsum plasters are divided as follows into four classes as specified in BS 1191, Part 1 (1973), with only the first two classes being currently manufactured in the United Kingdom:

Class A: Plaster of Paris ($CaSO_4 \cdot 0.5H_2O$) is used extensively for casting and stopping plasters. It sets too rapidly for most plastering purposes but is sometimes used for gauging lime.

Class B: Retarded hemihydrate gypsum plaster is available in several grades for undercoats and final coats. It is used neat or as a 50:50 mix gauged with lime putty.

Class C: Anhydrous gypsum plaster ($CaSO_4$) is used for final coats, sanded coats or undercoats where its gradual and progressive setting allows a high standard of finish.

Class D: Keene's plaster is an anhydrous final coat plaster that is used in specialist applications and can give a very hard smooth surface.

It should be noted that BS 1191(1973) has now been replaced by BS EN 13279-1 (2005) and BS EN 13279-2 (2004). The current standards include a threefold classification: Class A plasters, which are used in further construction processing; Class B, gypsum plasters and Class C, plasters for special purposes. All three classes contain several subdivisions.

A variety of premixed lightweight gypsum plasters conforming to BS EN 13279-1 (2005) includes mixtures of the retarded hemihydrate gypsum (Class B) and lightweight aggregates such as perlite and vermiculite. ASTM C28/C28M (2005) specifies a range of four plasters including one containing wood fibre. Gypsum projection plasters are a blend of hemihydrate and anhydrous gypsum plaster formulated to be used with mechanical plastering machines which mix and spray the plasters as a one-coat application on the substrate before being trowelled by conventional methods.

Paper-faced gypsum board is manufactured by automated continuous processes where an accelerated gypsum plaster is rapidly mixed and aerated in a pin mixer and fed onto a continuous belt covered by paper backing. Another layer of paper is rolled onto the top surface to form a sandwich. The plaster sets quickly and the boards are cut to length and kiln dried. Apart from the kiln drying, the whole process only takes a few minutes and large production rates are achieved so that this product constitutes a major use of building gypsum. A range of stopping plasters is associated with the installation of paper-faced gypsum board. The simplest is based on coarse building plaster with possibly some added methyl cellulose as a trowelling agent. The fine cornice plaster and similar materials in common use typically contain a fine filler together with a polymer addition in order to provide the necessary hardening.

The strength of paper-faced gypsum wall board lies in the paper facings and the gypsum is only required to be a low-strength inert filling preferably with a low density. Density is reduced usually by entraining air but other materials such as wood fibre and pumice have been used. Gypsum crystals typically grow perpendicular to the surface as shown in Figure 9.8. A typical modern paperbacked Gyprock plaster board is illustrated in Figure 9.9a and b.

There is also a range of handmade cast gypsum products such as fibrous plaster board, ceiling tiles and ornamental cornices, coves and ceilings. These products were originally reinforced with natural fibres, commonly sisal, but over the last few decades, this has been supplanted by 'E glass' fibre. The products containing fibre glass have superior flexural strength and unlike sisal, glass fibre does not interfere with the setting time of plaster. In this respect, the use of glass fibre mats in board faces is very effective (King et al. 1972).

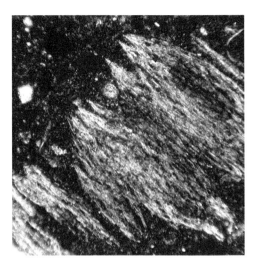

Figure 9.8 Gypsum crystals growing perpendicular to the plaster surface. Crossed polar illumination; width of field, 1 mm. (Courtesy of A. Bromley, private collection.)

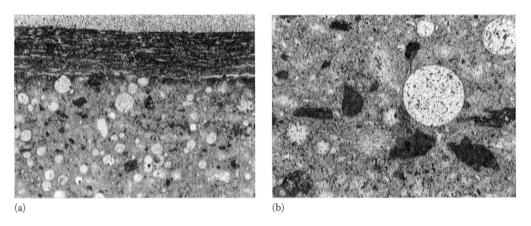

(a)   (b)

Figure 9.9 Texture of a paper-faced gypsum plaster board. (a) The paper facing stained red by dye and the air void system in this Gyproc plaster board are illustrated. Width of field 2 mm. (b) At higher magnification details of the air voids and the brown aggregate are shown. Crystals of gypsum are not large enough to be visible. Width of field 0.8 mm.

The development of gypsum structural members using low water/plaster ratios achieved by vacuum casting and chopped fibre glass as reinforcement was developed at the Building Research Station (Ali and Grimer 1969). Strengths of 30 MPa were attained enabling the casting of wall floor and ceiling units.

9.3.1 Petrographic investigation of hemihydrate and anhydrite products

The hydration of hemihydrate and anhydrite produces gypsum so that this is the major component of the hardened matrix of products. BS 1191 (1973) requires plaster of Paris to contain not less than 85% hemihydrate. In practice, many plasters contain over 95% of hemihydrate. The impurities usually consist of calcium carbonate, clay and more rarely organic matter. Because contaminants are finely divided, they may not be detectable by light microscopy and may require chemical and x-ray analysis or scanning electron microscopy for their detection and estimation. Chemical additions such as sugar, starch, alum and potassium and zinc sulphates used as accelerators and proteinaceous retarders are difficult to detect chemically as they are present only at the 0.5% to 1% level.

In thin section, gypsum plasters have a matrix dominated by an interlocking mass of acicular gypsum crystallites which may just be differentiated at the highest magnification. These acicular crystallites are set in anhedral masses of gypsum. Larger gypsum crystals are able to grow into voids (see Figure 9.10) and can be used for optical identification. With Class C and D gypsum plasters, some residual anhydrite remains in the set and hardened matrix and can be used to identify this type of plaster. Anhydrite is easily recognised by its high birefringence and tendency to form rectangular crystals (see Figure 9.11). In many examples, it is seen in petrographic thin section as discrete single grains (Figure 9.12), but euhedral grains also occur within larger, brown-stained impurity particles (see Figure 9.11a).

9.3.2 Gypsum and plaster finishes

Gypsum-plastered finishes, especially older finishes, may contain 50%–75% of lime in the hardened matrix and may be mixed with sand. Some neat finishes were reinforced with hair or natural fibre. The presence of lime, sand and fibre reinforcing are easily identifiable in thin section, and it should be possible to determine the mineralogy of the sand and type of fibre used.

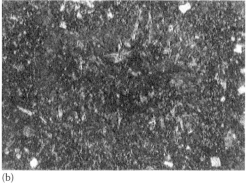

(a)   (b)

Figure 9.10 Visible gypsum crystals in a void. Width of fields 0.8 mm. (a) This void is large enough to allow the gypsum crystals to grow into visible fibrous laths which is their typical form. Around the void the gypsum crystals are too small to be identifiable. (b) Under crossed polars the low birefringence of the gypsum is shown.

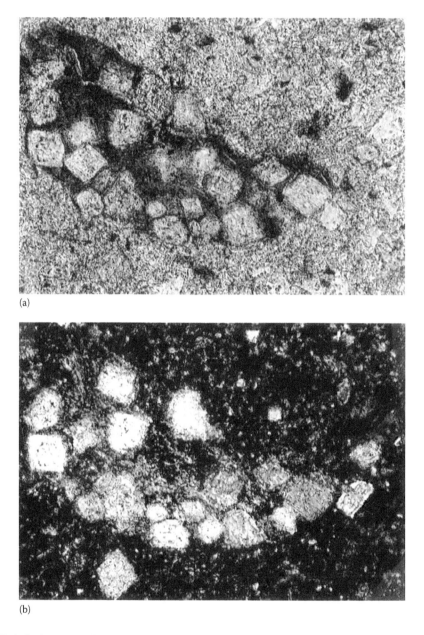

(a)

(b)

Figure 9.11 Anhydrite crystals in hardwall plaster. Width of fields 0.45 mm. (a) The iron-stained particle contains well-formed rectangular-shaped crystals of anhydrite. (b) Under crossed polars the anhydrite crystals show moderate birefringence.

There are a number of problems that can occur with gypsum plaster finishes and may involve both the materials used and the workmanship. Clay and loam in sand can affect both strength and setting, while magnesian limes may cause defects due to delayed hydration of magnesia. Under some conditions, magnesian limes may give rise to efflorescence due to the formation of magnesium sulphate. The set of Class A and B plasters can be 'killed' by working the surface after the initial set has occurred, but this gives low-strength surfaces prone to dusting. Anhydrous plasters which may take two to three days to set must not be

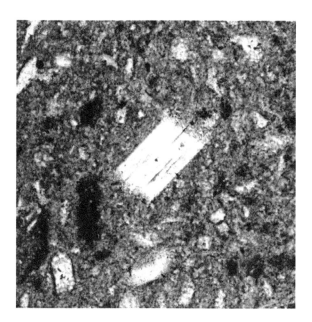

Figure 9.12 Large euhedral anhydrite crystal in a lime mortar. Plane-polarised illumination; width of field, 1 mm. (Courtesy of A. Bromley, private collection.)

allowed to dry out during this period. Otherwise low-strength and delayed setting expansion can result. Calcium sulphate plasters may with time corrode iron, steel, lead, zinc and aluminium, and these metals should be protected from direct contact. Even where lime is added, this problem can still occur especially in damp conditions, and in damp conditions, gypsum also loses strength. The absorption of as little as 1% of water can reduce strength by up to 50%. Long-term exposure allows a process known as 'rotting' to occur which involves a breakdown of the crystal interlock. These possible problems need to be taken into account by the petrographer when investigating gypsum plaster failures.

Cast gypsum products are invariably made from hemihydrate and do not suffer from many of the problems encountered with plaster finishes. In hand-cast products, manufacturers may use excess water for gauging the plaster to obtain increased fluidity for casting and larger output volumes. Coarse casting plaster ideally should be mixed at a powder/water ratio of 0.7 to 0.8, but ratios as high as 1.0 may be used resulting in low-strength porous casts. This is difficult to detect and measure (St John 1975). A less common problem can occur where products are overheated during kiln drying causing calcination of the plaster with reduction in strength and cracking. This can usually be detected by x-ray diffraction analysis. Another difficulty can arise from the high water/powder ratios used for hand-cast gypsum products in that they are porous and prone to absorption of smoke and stains which can migrate through the thickness of the product (St John and Markham 1976; Wilson and St John 1979).

9.4 HISTORIC MATERIALS

In this section, historic materials is taken to mean the binders, both lime based and cementitious, that were used before the final development of what is now known as Portland cement. These binders include non-hydraulic and NHL, artificial hydraulic lime AHL and the various early cementitious binders that were used predominantly in the nineteenth century.

It has been reported by Malinowski and Garfinkel (1991) that a type of dense lime concrete was used to make polished floors in Jericho dated around 7000 BC. Sims (1975) found that ancient lime mortars and concretes are quite distinctive in thin section under the petrological microscope, with the matrix being predominantly finely crystalline calcite (Figure 4.25) together with residual particles of partially burnt limestone (Figures 9.13 through 9.15) and charcoal contamination from the original lime kiln and patches of more coarsely crystallised calcite (Figure 4.26) deriving from the slow carbonation of pockets of trapped slaked lime (Sims 1975).

The Greeks and Romans both appear to have understood the differences between hydraulic and non-hydraulic limes and the use of pozzolanas to give added strength and resistance against water (Vitruvius 1962). The pozzolanas they used included not only materials such as the volcanic sands from Pozzuoli but also calcined clays usually in the form of ground-up bricks, tiles and potsherds. Lea (1970) points out that much of the excellence of Roman mortars depended not on any secret in the slaking or composition of the lime but in the thoroughness of mixing and ramming. The Romans also understood the need to use clean, graded sand.

Figure 9.13 Lime lump (brown) showing deformation top left to bottom right. Elongation due to deformation of the malleable (hydrated putty) lump during mixing and placing. Plane-polarised illumination; width of field, 2.5 mm. (Photomicrograph courtesy of the Scottish Lime Centre Trust.)

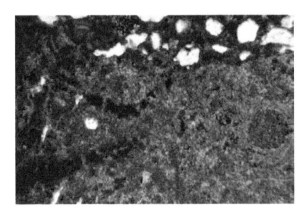

Figure 9.14 Big inclusion of unburned carboniferous limestone (round structure at the centre right is a bioclast). The limestone is burnt on the left where it breaks down in fractures and has porosity. It appears that some bioclasts in the limestone have preferentially burned in the kiln, leaving 'moulds' now filled by bonder. Mortar with quartz aggregate at the top of the image. Plane-polarised illumination; width of field, 2.5 mm. (Photomicrograph courtesy of the Scottish Lime Centre Trust.)

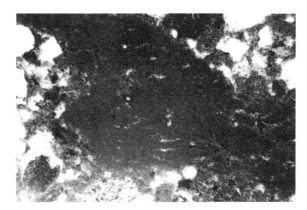

Figure 9.15 Lime inclusion from a Charlestown mortar made by the SLCT showing dark, lozenge-shaped relict crystals. These are probably the relicts of dolomite crystals in the original dolomitic limestone. Plane-polarised illumination; width of field, 2.5 mm. (Photomicrograph courtesy of the SLCT, Scottish Lime Centre Trust.)

According to Davey (1965), Roman mortars and concretes in England contained sand and coarse material which was selected for size and may have been specially screened. The proportions of lime to sand are quite similar to those used in modern lime mortars.

From ancient times until the mid-nineteenth century, lime was used in a similar fashion to the Romans in many parts of the East and in both the New and the Old World although in medieval times the use of hydraulic limes seems to have lapsed (Carran et al. 2012). While there is a wide variation in materials, petrographically, this variation can be assessed on the basis of texture and chemistry. Where a 'fat' lime has been used, the only two phases possible are calcium hydroxide and calcium carbonate, unless a sand with pozzolanic properties is present. Any fine crystals of calcium hydroxide dispersed among calcium carbonate will be impossible to distinguish unless x-ray diffraction analysis is used. Thus, only calcium hydroxide present in discrete areas or pockets will be visible in thin section, although staining with phenolphthalein can also be used to identify the presence of calcium hydroxide. In historic materials relict calcium hydroxide is very uncommon but can be preserved in overburnt particles where a crust of calcium carbonate has prevented further carbonation.

The presence of pozzolanic sand or calcined clay will alter the hardened matrix which may however still be predominantly calcium hydroxide and calcium carbonate though the quartz and other contaminants may be readily identified (Figure 9.16). In the matrix reasonably well-crystallised calcium silicate hydrate CSH(I), which can be identified only by careful x-ray diffraction analysis, will form as a result of the pozzolanic reaction. Reaction rims around some pozzolanic grains of sand should be visible as the reaction is slow unless the pozzolana is very finely ground.

Hydraulic mortars contain calcium silicates (dicalcium and uncommonly tricalcium silicate) with some aluminate and ferrite phases. The coarser portions of both calcium silicate and the ferrite phases can be surprisingly resistant to hydration, and remnants of these phases should be identifiable in thin section provided that sufficient cross section is examined. X-ray diffraction analysis may be insufficient to identify these phases as residual concentrations of the cementing phases may be below detection limit. CSH(I) is also formed from the hydration of dicalcium silicate and x-ray analysis will not distinguish this from CSH formed by pozzolanic reaction. If both a hydraulic lime and a pozzolan have been used, which apparently was not uncommon, the attributes of both textures described earlier should be present. Where chemical analysis is required, the techniques investigated by Stewart (1982) should be used.

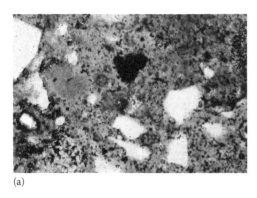

(a)   (b)

Figure 9.16 (a) Early eighteenth-century mortar with crushed chert (white, angular) and slaked lime. Lime particle, centre left; grass pollen grains (reddish/brown), top right. (b) Enlarged view of grass pollen grains. PPL illumination; width of field (a), 2 mm. (Courtesy of M. Grove, private collection.)

The crystallinity of the calcium carbonate will depend on both the age and the amount of moisture that has been present in the plaster or mortar during its lifetime. Initially, the calcium carbonate crystals will be small, but provided the moisture conditions are favourable, coarser crystals will grow at the expense of finer crystals. A young lime mortar might also contain dehydration shrinkage cracks, lenticular in shape (Figures 9.6 and 9.17). These are not observed in older mortars, and it appears that in a mortar exposed to wetting cycles, there can be significant changes to pore structure through time (Figure 9.18). Until the industrial era, wood was used to burn lime so that fragments of charcoal are almost invariably present in the ancient lime materials. In some cases, hair, reeds or other reinforcing material may have been used which can be identified in thin section (McCrone and Delly 1973). The heterogeneity of historic materials appears to have contributed to their durability, as the various impurities contributed to the setting and air-entraining properties of the mortar.

The mobilisation of calcium carbonate binder is a common occurrence in lime-based materials and can cause a significant change in the bulk chemical composition of the sample (Leslie and Hughes 2002). If this is not identified, then there is a risk of significant error in the reporting of mortar composition. Particular care should be taken to identify areas of binder where there has been either dissolution or reprecipitation of calcium carbonate (Figures 9.19 and 9.20)

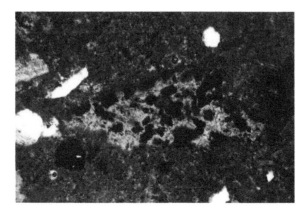

Figure 9.17 Relatively young mortar showing lenticular cavities in the binder, caused by early dehydration shrinkage. These cracks will heal over decades by reprecipitation of the binder. Plane-polarised illumination; width of field, 2.5 mm. (Photomicrograph courtesy of the Scottish Lime Centre Trust.)

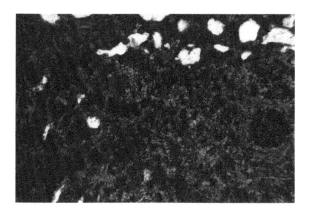

Figure 9.18 Lime inclusions within mortar showing the binder has evolved through dissolution and reprecipitation to form menisci at aggregate contacts. Plane-polarised illumination; width of field,
2.5 mm. (Photomicrograph courtesy of the Scottish Lime Centre Trust.)

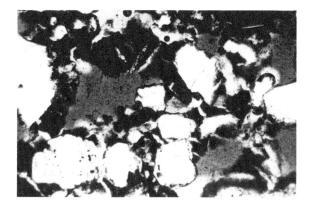

Figure 9.19 Porosity in mortar, probably caused by secondary dissolution of the calcium carbonate binder.
Secondary precipitation of calcium carbonate has caused a rim of calcium carbonate crystals
around the pores. Both processes have implications for binder content analysis. Plane-polarised
illumination; width of field, 2.5 mm. (Photomicrograph courtesy of the Scottish Lime Centre Trust.)

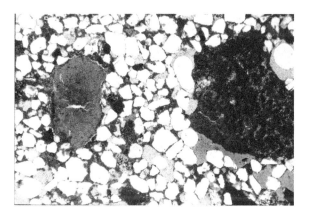

Figure 9.20 Partial dissolution of binder (left) with intact binder (right) showing the variability of binder
content within the mortar sample. Plane-polarised illumination; width of field, 2.5 mm.
(Photomicrograph courtesy of the Scottish Lime Centre Trust.)

as significant error in the reporting of mortar composition can result. This mobilisation cannot be identified through wet chemical analysis, and petrographic analysis is essential to ascertain and semi-quantify the dissolution and reprecipitation of binder. In a similar fashion, analyses of mortars made with calcium carbonate sand (Figure 9.21) can lead to misleading results.

9.5 CALCIUM SILICATE PRODUCTS

Quartz sand will react with lime in the presence of water if heated under pressure in an autoclave and is the basis of sand-lime brick manufacture. Calcium silicate products consist of a mixture of either siliceous sand or flour and about 10%–15% of a well-slaked high-calcium lime or pulverised quick lime. The mixture is pressure moulded while still 'green' and the product autoclaved for about 16 h at around 170°C. Sand-lime and flint-lime products use either natural siliceous sand, ranging from 2 mm down to 20 μm in size (Quincke 1967) or crushed flint as the siliceous component and often incorporate pigments. The reaction rate is largely controlled by the quartz sand surface area, so some European countries add up to 15% of silica flour to accelerate reaction and improve strength and also they may add small quantities of clay to aid moulding. These products are specified in BS EN 771-2 (2003). There is a range of materials and processes used in manufacturing calcium silicate products, but the commonest products are brick and block (Bessey 1967).

The unpigmented bricks produced are normally grey in colour, compact and are composed of calcium silicate hydrates which carbonate rapidly with time. They are relatively cheap to manufacture. However, shrinkage with time is a common feature and may lead to cracks developing, usually stepwise through the mortar bonding between the blocks or bricks.

Aerated calcium silicate concrete block and building components are used extensively in Scandinavia, where the process was developed, and also in eastern Europe. Many details of the patented manufacturing processes used are not available but it is known that the composition essentially consists of high-grade quick lime and silica flour with the possible inclusion of some fine sand and Portland cement. Aeration or foaming is usually achieved by the use of aluminium powder added to the mixture which generates hydrogen by reacting with calcium hydroxide to form a calcium aluminate. By varying the amount of foaming agent and mix proportions, lightweight products varying from a density of 0.2 to 0.9 can be made, with the range 0.2–0.5 being the most usual.

Silicate concrete was developed in the USSR and is based on milling high-grade lime and silica sand to increase reactivity and is used to make bricks, blocks and building units. It differs from sand-lime products in that compressive strength ranging from 50 to 80 MPa can be achieved compared with a strength for the sand-lime blocks of 10 to 50 MPa. Compressive strengths of aerated products range from 1.5 to 5 MPa, dependent on density (Short and Kinniburgh 1978).

Two important physical properties of calcium silicate products that affect durability are drying shrinkage and carbonation. BS EN 771-2 (2003) limits the drying shrinkage of sand-lime bricks to a maximum of 0.04%. According to Rudnai (1963), the shrinkage of autoclaved aerated products is usually less than 0.05%, but from saturated to fully dry condition, this may vary up to 0.09%. Precise details of carbonation are not available but it is generally agreed that all lime silica products carbonate with time given the appropriate conditions. With the exception of some shrinkage due to carbonation or leaching, overall the durability of sand-lime bricks appears adequate as indicated by the long-term tests carried out by Bessey and Harrison (1970). Short and Kinniburgh (1978) state that damp conditions lead

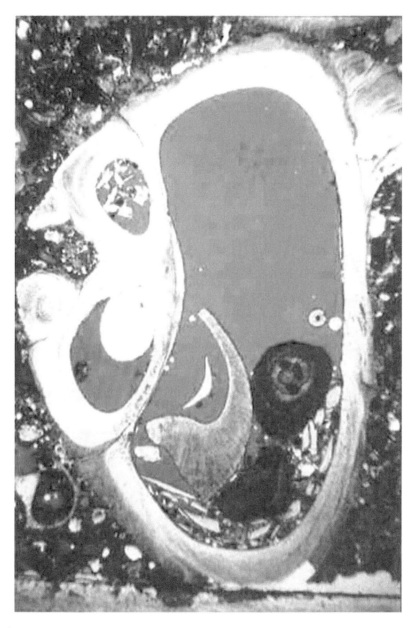

Figure 9.21 Mortar made from shell sand and lime binder, presumably created from burning of the same shell sand. Plane-polarised illumination; width of field, 2.5 mm. (Photomicrograph courtesy of the Scottish Lime Centre Trust.)

to an increase in strength of autoclaved aerated concrete, but in circulating water, strength may deteriorate because of the leaching of lime.

Petrographic examination will show that the siliceous aggregate is the major identifiable component in sand-lime bricks, and since the reaction with quartz or flint grains is only partial, they form the dominant constituent when viewed in thin section. In old bricks, the cementing material is partly or wholly carbonated so that at low magnifications, the aggregate grains are seen to be embedded in a highly birefringent fine-grained carbonate matrix.

Irregular patches of featureless cementing material between quartz grains vary in size with the extent of the carbonation. In some examples, small patches of the cementing medium exhibit low birefringence indicating it is more crystalline. Calcium hydroxide is present in these uncarbonated areas of matrix and forms large equant or elongate crystals up to 100 μm in size. The majority of these crystals form a discontinuous fringe around the quartz grains and remain even within areas of partial carbonation as shown in Figure 9.22a. Figure 9.22b shows an area of matrix with calcium hydroxide crystals bordering the quartz grains and

(a)

(b)

Figure 9.22 The textures of a 20-year-old sandlime brick. Width of fields 1.7 mm. (a) Under crossed polars bright calcium hydroxide crystals border the quartz grains and the low birefringent matrix consists of calcium hydroxide and CSH. Only minimal carbonation is present. (b) Under crossed polars the texture is almost wholly carbonated apart from areas marked X between some quartz grains which are assumed to be poorly crystalline CSH(I).

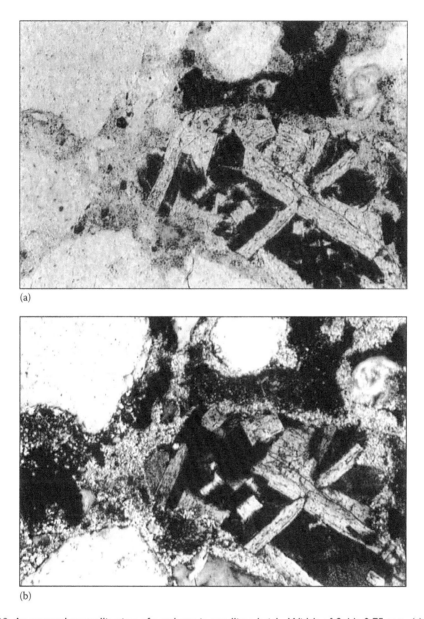

Figure 9.23 An unusual crystallisation of products in sandlime brick. Width of fields 0.75 mm. (a) A void is infilled with clear laths and plates and felted greenish brown crystals. (b) Under crossed polars the laths and plates have low birefringence and are possible dicalcium silicate hydrate and the acicular felted mass of green crystals are possibly actinolite. The matrix is partially carbonated.

the development of a small area of more crystalline matrix. Figure 9.23 shows an unusual crystallisation of dicalcium silicate hydrate and actinolite within a pore in a sand-lime brick.

The principal cementing phase formed in autoclaved calcium silicate products is a poorly crystallised CSH(I) with a Ca/Si ratio of 0.9–1.0 (Taylor 1990). In autoclaved aerated lime silica products, the pore system, silica flour and relics of the lime used together with associated carbonation and microcracking are visible as discussed and illustrated in Section 5.7.5. Textural details of calcium silicate concrete are not available but it may be assumed that the texture will be a denser form of that observed in aerated lime silica products.

9.6 SPECIAL FLOORING FINISHES

A range of decorative specialised and proprietary flooring and wall coating systems which are based on a variety of materials, for example, calcium sulphates rather than Portland cement, are in common use in construction throughout the world. A review of these and other chemical cements is given by Bensted (1998) (see also Glossary of minerals).

Perhaps the most commonly used of these is magnesium oxychloride or Sorel cement and related modifications of it. Sorel cement is made by mixing powered magnesia (magnesium oxide) with concentrated magnesium chloride solution to produce a range of hydrates with $3Mg(OH)_2 \cdot MgCl_2 \cdot nH_2O$ and $5Mg(OH)_2 \cdot MgCl_2 \cdot nH_2O$ typically formed at ambient temperatures. The five form is metastable and converts to the three form with time. The hydrates also slowly carbonate on exposure to air to form magnesium chlorocarbonate. The quality of the magnesia which should be adequately burnt is important to effective reaction; overburning produces insufficient reactive material, while underburning causes excessive reaction.

Resistance to water is dependent on the carbonation of the oxychloride hydrates to the less soluble carbonates. The water resistance and durability properties of Sorel cements can be further improved by the addition of inorganic substances such as borates, phosphates, or calcium sulphate–silicate mixtures or by addition of organic materials such as resins, melamine urea or formaldehyde. These admixtures typically slow the development of compressive strength.

A variant of Sorel cement, magnesium oxysulphate cement formed by treating magnesia with sulphuric acid or by mixing magnesium chloride solution with either calcium sulphate or calcium phosphate, leads to improved rheological properties and better water resistance. Another variant is zinc oxychloride cement which can be made by treating zinc oxide with zinc chloride to form zinc oxychloride hydrates. This material is resistant to both acids and boiling water, but is not used because of poor workability and chemical instabilities.

Sorel cements can be used with a variety of aggregates and fillers such as sand, crushed stone, glass fibres, expanded clays and wood chips. Sorel cements have good wear and fire resistance. The commonest use is for industrial flooring due to its high compressive strengths, resistance to accumulated loads and elasticity. Other applications are in refractory oil-well cements, temporary rapid curing highway patching and cast artificial stone, while oxysulphate cements have been used for lightweight insulating panels. Their main disadvantages are poor freeze–thaw resistance, dimensional instability, chemical instability in air and particularly their solubility in water with the release of corrosive solutions.

9.7 SURFACE COATINGS

Surface coatings of various types are widely used in the construction industry, ranging from decorative and protective paint coating through curing membranes for fresh concrete, sealants for brick and concrete surfaces to protective membranes for preventing water ingress or chemical attack.

The materials in common use include the following:

- Epoxy and polyurethane resin formulations
- Acrylic polymer resins
- Methylacrylate–silicone resins
- Bitumen and bitumen-latex formulations

The principal uses of coatings in construction are for water retention during curing in newly laid concrete, release agents for concrete formwork surface, crack impregnation and for surface protection against water or chemical ingress or attack of the hardened surfaces. Surface coatings for use on flooring are also discussed in Sections 7.9 and 8.2.

The petrographer can gain a little useful information from an examination of thin sections provided care has been taken to include the surface coating without introducing artefacts. The depth of penetration of an impregnating material (typically a few millimetres at most) can be assessed and the quality of adhesion between the coating and the underlying material examined. It is also possible to identify any mineral filler that has been included in the coating. Since the majority of coatings are based on organic materials and are non-crystalline, infrared spectroscopy or chemical microanalysis are the most appropriate methods for identifying the coating material.

REFERENCES

Ali, M.A. and Grimer, F.J. 1969: Mechanical properties of glass fibre-reinforced gypsum. *Journal of Materials Science* 4, 5, 389–395.

Andrews, H. 1948: Gypsum and anhydrite plasters. National Building Studies Bulletin No. 6, DISR. HMSO, London, U.K.

Ashurst, J. and Ashurst, N. 1988: *Practical Building Conservation. English Heritage Technical Handbook Volume 3. Mortars, Plasters and Renders*, U.K., Gower Technical Press.

ASTM C28/C28M, 2005: *Standard Specification for Gypsum Plasters*. ASTM, Philadelphia, PA.

ASTM C91-05, 2005: *Standard Specification for Masonry Cement*. ASTM, Philadelphia, PA.

ASTM C206-03, 2003: *Standard Specification for Finishing Hydrated Lime*. ASTM, Philadelphia, PA.

ASTM C207-06, 2006: *Standard Specification of Hydrated Lime for Masonry Purposes*. ASTM, Philadelphia, PA.

ASTM C110-09a, 2009: *Standard Test Methods for Physical Testing of Quicklime, Hydrated Lime, and Limestone*. ASTM, Philadelphia, PA.

ASTM C141/C141M-09, 2009: *Standard Specification for Hydraulic Lime for Structural Purposes*. ASTM, Philadelphia, PA.

ASTM C821-09, 2009: *Standard Specification for Lime for Use with Pozzolans*. ASTM, Philadelphia, PA.

Bensted, J. 1998: Special cements. In *Lea's Chemistry of Cement and Concrete*, Hewlett, P.C. (ed.). 14, Arnold, London, U.K., pp. 779–835.

Bessey, G.B. 1967: The history and present day development of the autoclaved calcium silicate building products industries. In *Symposium Autoclaved Calcium Silicate Building Products, London* 1965, Society of Chemical Industry, London, pp. 3–6.

Bessey, G.E. and Harrison, W.H. 1970: Some results of exposure tests on durability of calcium silicate bricks. *Building Research Current Paper 24/70*. Building Research Station, Watford, U.K.

Blakey, F.A. 1959: Cast gypsum as a structural material. *CSIR Division of Building Research Report* Z. 7, Melbourne, Australia.

Boynton, R.S. 1980: *Chemistry and Technology of Limestone*. John Wiley, New York.

BS 1191, 1973 1994: *Specification for Gypsum Building Plasters. Part 1: Excluding Premixed Lightweight Plasters. Part 2: Premixed Lightweight Plasters*. BSI, London, U.K.

BS 6463-102, 2001: *Quicklime, Hydrated Lime and Natural Calcium Carbonate. Methods for Chemical Analysis*. BSI, London, U.K.

BS 6463-103, 1999: *Quicklime, Hydrated Lime and Natural Calcium Carbonate. Methods for Physical Testing*. BSI, London, U.K.

BS EN 413-1, 2004: *Masonry Cement, Composition, Specifications and Conformity Criteria*. BSI, London, U.K.

BS EN 459-1, 2001: *Specification for Building Limes*. BSI, London, U.K.

BS EN 771-2, 2003: *Specification for Calcium Silicate (Sandlime and Flintlime) Bricks*. BSI, London, U.K.

BS EN 13279-1, 2005: *Specification for Gypsum Building Plasters. Premixed Lightweight Plasters*. BSI, London, U.K.

BS EN 13279-2, 2004: *Specification for Gypsum Building Plasters. Excluding Premixed Lightweight Plasters*. BSI, London, U.K.

BS EN 13914-1, 2005: *Design, Preparation and Application of External Rendering and Internal Plastering. External Rendering*. BSI, London, U.K.

BS EN 13914-2, 2005: *Design, Preparation and Application of External Rendering and Internal Plastering. Design Considerations and Essential Principles for Internal Plastering*. BSI, London, U.K.

Carran, D., Hughes, J., Leslie, A., and Kennedy, C. 2010: Lime lump development and textural alteration during the production of mortar. In *Proceedings of the Second Historic Mortars Conference*, Prague, September 2010, UNESCO-ICOM, Museum Information Centre, Edinburgh University Press, pp. 22–24.

Carran, D., Hughes, J., Leslie, A., and Kennedy, C, 2012: A short history of the use of lime as a building material – Beyond Europe and North America. *International Journal of Architectural Heritage* 6, 2, 117–146.

Cowper, A.D. 1927: Lime and lime mortars. DSIR Building Research Special Report No. 9. HMSO, London, U.K.

Daligand, D. 1985: Le platre et ses techniques de production. *Ciments, Betons, Platres, Chaux* 753, 83–88; 755, 222–230.

Davey, N. 1965: *A history of building materials*. Chapter 12, *Limes and Cements*. Phoenix House, London, U.K., 1961, reprinted 1965.

Elsen, J. 2006: Microscopy of historic mortars – A review. *Cement and Concrete Research* 36, 1416–1424.

Hill, N., Holmes, S., and Mather, D. (eds.). 1992: *Lime and Other Alternative Cements*. Intermediate Technology Publications, London, U.K.

Holmes, S. and Wigate, M. 2002: *Building with Lime: A Practical Introduction*, 2nd ed. ITDG Publishing, U.K. 309pp. www.fastonline.org.

King, G.A., Walker, G.S., and Ridge, M.J. 1972: Cast gypsum reinforced with glass fibres. *Building Materials and Equipment* 15, 41–43.

Kuntze, R.A. 1976–1986: *Gypsum and Plasters. Cements Research Progress*, Cements Division, American Ceramic Society.

Lea, F.M. 1970: *The Chemistry of Cement and Concrete*, 3rd ed. Edward Arnold Ltd., London, U.K.

Leslie, A.B. and Gibbons, P. 2000: Mortar analysis and repair specification in conservation of Scottish historic buildings. In *Historic Mortars: Characteristics and Tests*, Bartos, P. Groot, C., and Hughes, J. (eds.). RILEM International Workshop Proceedings PRO 12, pp. 273–280.

Leslie, A.B. and Hughes, J.J. 2002: Binder microstructure in lime mortars: Implications for the interpretation of analysis results. *Quarterly Journal of Engineering Geology and Hydrogeology* 35, 275–263.

Malinowski, R. and Garfinkel, Y. 1991: Prehistory of concrete. *Concrete International* 13, 3, 62–68.

McCrone, W.C. and Delly, J.G. 1973: *The Particle Atlas, 1 &2*. Ann Arbor Science Publishers, Ann Arbor, MI.

McKenzie, L. 1994: The microscope: A method and tool to study lime burning and lime quality. *World Cement* 25, 11, 56–59.

McKenzie, L. 1995: Lime particle size: Its effects on feed and products of a hydrator circuit. A plant case study. *Proceedings of the 17th Conference on Cement Microscopy*, Alberta, Canada, ICMA, pp. 402–423.

Moorhead, D.R. 1986: Cementation by the carbonation of hydrated lime. *Cement and Concrete Research* 16, 5, 700–708.

Pippard, W.R. 1938: Calcium sulphate plasters. *DSIR Building Research Bulletin No. 13*. HMSO, London, U.K.

Quincke, J.E. 1967: Broadening the particle size range of natural sands for calcium silicate manufacture. *Symposium Autoclaved Calcium Silicate Building Products, London 1965*. Society of Chemical Industry, pp. 11–17.

Ridge, M.J. and Berekta, J. 1969: Calcium sulphate hemihydrate and its hydration. *Reviews of Pure and Applied Chemistry* 19, 17, 17–44.

Roberts, M.H. 1956: The constitution of hydraulic lime. *Cement and Lime Manufacture* 29, 3, 27–36.

Rudnai, G. 1963: *Lightweight Concretes.* Publishing House of the Hungarian Academy of Sciences.

Searle, A.B. 1935: *Limestone and Its Products.* Ernest Benn, London, U.K.

Short, A. and Kinniburgh, W. 1978: *Lightweight Concrete,* 3rd ed. Applied Science Publishers, London, U.K.

Sims, I. 1975: Analysis of concrete and mortar samples from Sparsholt Roman Villa, near Winchester, Hampshire. Report prepared for the Department of the Environment. www.gov.uk.

Stewart, J. 1982: Chemical techniques of historic mortar analysis. *Bulletin-Association for Preservation Technology* 14, 1, 11–16.

St John, D.A. 1975: Some physical properties of fibrous plaster sheet manufactured in New Zealand. *New Zealand Journal of Science* 18, 417–431.

St John, D.A. and Markham, K.R. 1976: Paint-disfiguration by migration of antioxidant from wallboard adhesive. *Journal of the Oil Colour Chemist's Association* 59, 331–333.

Taylor, H.W.F. 1990: *Cement Chemistry.* Academic Press, London, U.K.

Vitruvius. 1962: *Vitruvius on Architecture,* 2 vols. Granger, F. (Trans.), William Heinemann, London, U.K.

Volzhenskii, A.V. and Ferronskaya, A.V. 1974: *Gypsum Binders and Articles.* Stroiizdat (Chem. Abst. 126219, vol. 81, 1974), Moscow.

Wickens, D. 1992: Plasterwork and rendering. *Specification* 92. MBC Architectural Press and Building Publications, London, U.K.

Wilson, R.D. and St John, D.A. 1979: The avoidance of staining in cast Australian gypsum. *Zement-Kalk-Gips* 68, 1, 41–43.

Glossary of minerals

G.1 INTRODUCTION

This glossary of minerals is intended to bring together optical data on some of the minerals mentioned in the text but excludes data on most rock-forming minerals in aggregates as these are already adequately covered in textbooks such as Kerr (1977), Deer et al. (1966/1992) and Battey (1981). Also many minerals are now described in the Imperial College (University of London) 'rock library'.[1] The mineral data given have been largely restricted to the type of information that can be gained from thin-section examination, though scanning electron microscopy (SEM), x-ray diffraction (XRD), Fourier transform infrared spectroscopy (FTIR) and thermal analysis data are included where appropriate. In the cases of efflorescences and precipitates, the appearance of the mineral in oil immersion has been included.

Many of the data on cement and concrete minerals are widely scattered or incomplete and can be difficult to access and, where this is the case, references are given. A synoptic listing is given in Highway Research Board (1972). In other cases, the general references used are Winchell (1951), Winchell and Winchell (1964), Kerr (1977), Kostov (1968), Troger (1971) and Rigby (1948). For illustrations of textures in colour, the reader is referred to McCrone and Delly (1973), Mackenzie et al. (1995), Mackenzie and Guilford (1980) and Campbell (1986, 1999). A brief, simplified description of the optical constants and other terminology used in this glossary is given here as an aid to the user, but more detailed descriptions are available in Winchell, Part 1 (1937), and Kerr (1977).

The symbols n, ε, ω, a, β, and γ used for the refractive indices have been chosen as they appear to be the ones most commonly used (Bloss 1966). For the reader who is unfamiliar with the practices of optical crystallography, Bloss (1966) is recommended as a simple and clear introduction to the subject.

G.2 SUMMARY OF OPTICAL PROPERTIES

Isotropic minerals: A limited group of minerals which only have one refractive index (*n*) and belong to the cubic crystal system. Glasses and amorphous substances are non-crystalline and are isotropic with a single refractive index.

Biaxial minerals: The bulk of naturally occurring minerals belong to this group which is distinguished by having two optical axes and three refractive indices *(α, β, γ)*. It consists of the triclinic, monoclinic and orthorhombic crystal systems.

---

[1] www2.imperial.ac.uk/earthscienceandengineering/rocklibrary.

α   The smallest refractive index
β   The intermediate refractive index
γ   The largest refractive index
γ-α   Birefringence
2 V   The optic axial angle
Sign   2 V is +ve when β – α is less than γ – β and –ve when β – α is greater.

Uniaxial minerals: A more restricted group of minerals in which the crystals only have one optical axis and two refractive indices. The tetragonal and hexagonal (which includes trigonal) crystal systems belong to this group.

ω Ordinary refractive index
ε Extraordinary refractive index which may be greater or less than ω and determines sign
Sign: ε > ω optically positive, ε < ω optically negative

Birefringence: The positive difference between the two refractive indices.

Sign of elongation: Applies to elongate crystals where one of the vibration directions is parallel to the length which is always the case for the hexagonal, tetragonal and orthorhombic systems. In some cases, this may apply to the monoclinic system. However, even in the monoclinic and triclinic systems, the extinction angle of the index in relation to the length of the crystal may often be small and allow a sign of elongation to be assigned.

Length slow: The vibration direction of the higher refractive index is parallel or subparallel to the length and the sign is positive.

Length fast: The vibration direction of the lower refractive index is parallel or subparallel to the length and the sign is negative.

Colour: The colour of the crystal as seen in thin section viewed in plane-polarised illumination. The colour in hand specimen may differ from that in thin section and they should not be confused.

Pleochroism: In plane-polarised light, some anisotropic crystals vary in colour as the stage is rotated due to differential absorption of light. This can be an important diagnostic feature.

Interference colour: The colour seen under crossed polar illumination in thin section. The maximum interference colour occurs when the crystal is observed perpendicular to the optic plane. If the birefringence is known, the interference colour can be estimated from the Michel-Lévy scale or vice versa provided the thickness of the crystal is known. Owing to differential absorption in some minerals, anomalous interference colours occur, which may be diagnostic.

Interference figure: Conoscopic examination of a biaxial crystal under crossed polars allows determination of optical sign and estimation of 2 V without the need to measure the refractive indices. Dispersion of the optical axes may also be observed under favourable circumstances.

Relief: When a crystal is immersed in a fluid of identical refractive index, its outline and details of relief disappear apart from colour fringes due to refraction at crystal edges. As the difference between the index of the immersion medium and that of the crystal is increased, the outline becomes darker and details of relief are highlighted. The amount of relief can be roughly estimated as low, medium or high and whether the immersion medium has a higher or lower index than the crystal can be determined from movement of the Becké line. If the mounting medium is less than the crystal, relief is positive, and if its index is higher, the relief of the crystal is negative. In the mineralogical literature, relief is usually quoted using an immersion medium index of 1.54 which has been adhered to in the data quoted in this glossary. Thin sections of concrete are often mounted in epoxy resin of which the index can vary between 1.54 and 1.61, so the index should be checked and allowance may need to be made for this change of index when estimating relief in thin sections of concrete.

Becké line: At vertical junctions between two transparent materials in a thin section, the total reflection produces a bright line at the boundary. This bright line moves into the material of higher refractive index as the microscope focus is raised.

G.3 EXAMINATION OF CRYSTALS IN THIN SECTION

It is possible to determine the optical properties of a crystal provided that it can be manipulated to measure its refractive indices. These properties are usually sufficient to identify the crystal unless it belongs to the cubic crystal system. However, in thin section, sufficiently precise measurement of refractive indices is rarely possible. In addition, many of the recrystallised products formed in hardened cement paste are too small to give interference figures so that the optic sign and 2 V cannot be determined. In many cases, observations are restricted to colour, estimates of refractive indices and to relief, sign of elongation, birefringence, pleochroism and anomalous interference colours, crystal habit, cleavage and extinction angle. Identification of a particulate mineral is often possible if at least some of the properties listed earlier can be observed.

XRD data: The first four strongest diffraction peaks are quoted from data published on the Powder Diffraction File (PDF) by the International Centre for Diffraction Data (ICDD), United States, unless otherwise stated.

Electron microscopy, thermal and infrared analysis: Where data are available, any diagnostic information is included with the individual mineral descriptions when this is appropriate.

G.4 MINERAL GROUPINGS

Although a number of minerals could appear under more than one major group heading. Their details will appear only once under the most appropriate heading for a petrographer concerned with concrete and related materials. The individual minerals are described here under the following major group headings:

1. Cement minerals
2. Calcium silicate hydrates
3. Aluminate hydrates
4. Other hydrates
5. Sulphate minerals
6. Carbonate minerals
7. Silica minerals and mineraloids
8. Magnesium minerals
9. Glassy materials
10. Mineral additions and their constituents
11. Pigments and colorant materials
12. Asbestos and other fibrous materials
13. Grinding and polishing compounds
14. Mounting media

Other reading
In addition to the publications to which reference has been made in this introductory section of the Glossary, useful general information will also be found in the following: Adams et al. (1984), Adams and MacKenzie (1998), Anonymous (1992), Chadwick (1976), Damgaard

Jensen et al. (~1984), Delvigne (1998), Fleischer et al. (1984), French (1991), Fundal (1980), Gille et al. (1965), Hills (1999), Hofmänner (1973), Ingham (2010), Kühnel et al. (1980), Lorenzi et al. (2006), MacKenzie and Adams (1994), Phillips (1971), Van Hees et al. (2008), Winchell (1937, 1951, 1954), Winchell and Emmons (1964), Winchell and Winchell (1961) and Yardley et al. (1990).

G.4.1 Cement minerals

Many of the anhydrous cement minerals associated with Portland, high-alumina and other types of hydraulic cements are principally of interest in concrete as unhydrated relics. There is a large literature pertaining to the microscopy of these minerals so only brief details are given. For more information on the mineralogy, see Campbell (1986, 1999), Fundal (1980), Hoffmanner (1973) and Gille et al. (1965).

Alite (tricalcium silicate, C_3S)
$3CaO \cdot SiO_2$
    Biaxial – ve
    Monoclinic
    $\alpha = 1.716–1.720$, $\gamma = 1.722–1.724$
The composition of alite in Portland cement is 96%–97% C_3S substituted with a mixture of 3%–4% of ferric, magnesium, aluminium, sodium, potassium and phosphorus ions. A number of triclinic, monoclinic and rhombohedral polymorphs exist. In commercial cements, the type I monoclinic polymorph predominates (Taylor 1990) (see also De Noirfontainea et al. 2006).
    Interference figure: –2 V varies from 20° to 60°.
    Relief: Relics stand out because of high positive relief.
    Birefringence: 0.005 ranging from 0.002 to 0.010 as a function of ionic substitution. Interference colours first-order grey.
    Form: Lath, tablet like or equant. Crystals usually six-sided in cross section. Normal crystal sizes range from 25 to 65 μm.
    Twinning: Rare polysynthetic twinning.
    Extinction: Wavy to straight extinction, length slow.
    Colour: Colourless to slightly coloured in transmitted light.
    XRD: 31–301: 2.789–100, 2.767–70, 2.754–65, 2.613–90 (also see Taylor 1990).
Occurrence: Alite is the principal constituent of Portland cement and ranges from 15% to 65% dependent on type. It also occurs in stabilised dolomite bricks and basic steel slags.
Comment: Alite is rare in adequately cured concretes except as crystals embedded in the matrix of large particles of clinker. At very low water/cement ratios, alite is quite common because of internal desiccation. Most alite crystals are broken during the milling of the cement so it is difficult to find complete crystals. Alite is readily identified in polished cement clinker samples using etching procedures (see Sections 3.5.3 and 4.2.2). In relict clinker grains within a hardened cement matrix, it tends to exhibit more euhedral forms than the rounded belite (C_2S) crystals.
    (Figures: *see* Figures 3.14, 3.19 and 4.2).

Belite (β-dicalcium silicate, larnite, C_2S)
$2CaO \cdot SiO_2$

    Uniaxial +ve

    Monoclinic

    $\alpha = 1.717$, $\beta = 1.722$, $\gamma = 1.736$

The composition of belite in Portland cement is 94%–96% C_2S substituted with a mixture of 4%–6% of aluminium, ferric, magnesium, potassium, sodium, phosphorus, titanium and sulphur ions. A number of trigonal, monoclinic and orthorhombic polymorphs exist. In commercial cements, β-dicalcium silicate is the second most common mineral (Taylor 1990).

Interference figure: +2 V varies from 64°–69°

Relief: In relics, C_2S crystals stand out because of high positive relief.

Birefringence: Usually second-order colours as relic grains in concrete are often larger than 40 μm in size.

Form: Rounded grains, often in clusters set in the clinker matrix.

Cleavage: Poor prismatic cleavage.

Twinning: Lamellar structure is distinctive and nearly always present. Further discussion of lamellar structures observed can be found in Campbell (1986, 1999).

Colour: Colourless, pale yellow, yellow, amber or shades of green dependent on substitution.

XRD: 33–302: 2.790–97, 2.783–100, 2.745–83, 2.610–42 (also see Taylor 1990).

Occurrence: Larnite occurs naturally in high-temperature contact zones but is rare. Belite is the second most common mineral in Portland cement clinker and ranges from 10% to 60% dependent on cement type (see also Bentor et al. 1963).

Comment: In concrete, relic grains often consist of almost entirely of rounded belites together with some adherent interstitial clinker matrix. The shape, birefringence and lamella twinning of belite are distinctive. Globules of belite are commonly found inside alite crystals. (Figures: *see* Figures 3.19, 4.4, 5.4 and 5.5).

Aluminate phases (tricalcium aluminate, C_3A, alkali aluminate)
$3CaO \cdot Al_2O_3$

Isometric

Cubic

$n = 1.710$

$Na_2O \cdot 8CaO \cdot 3Al_2O_3$ (alkali aluminate)

Biaxial –ve

Orthorhombic

$\alpha = 1.702, \gamma = 1.710$

Tricalcium aluminate typically contains a mixture of around 13% ferric, silicon, magnesium, sodium, potassium and titanium ions and in alkali aluminate up to 20%. The alkali content of the cubic form is around 1% and for the orthorhombic form 2%–4% which is less than required for NC_8A_3 (Taylor 1990) (see also Carlson and Berman 1960).

Interference figure: –2 V varies from 0° to 35°.

Relief: Stands out because of its high positive relief.

Birefringence: Nil to low.

Form: Cubic form typically fills interstices between crystals of belite and ferrite. Alkali aluminate occurs as tablets, laths and stave-like prismatic forms.

Twinning: May be present in alkali aluminate.

Extinction: Alkali aluminate has oblique extinction and is length slow.

Colour: Colourless in white cement. Tan to brown in Portland cements.

XRD: 38-1429: 2.699–100, 4.080–12, 1.5578–24 (also see Taylor 1990).

Occurrence: In ordinary Portland cement, C_3A averages about 12% but in special cements, it can range from 0% to 18%. Production clinkers may contain both the cubic lath-shaped forms especially in high-alkali cements.

Comment: Individual particles of the aluminate phases are not observed in concrete. The brown matrix in relic grains commonly seen in concrete probably contains some aluminate phase. In many clinkers, the aluminate and ferrite phases are often poorly differentiated and form the matrix in which the silicate phases are embedded. However, x-ray evidence clearly indicates that the matrix in commercial clinker is usually crystalline even when it appears undifferentiated by microscopical examination.
(Figures: *see* Figures 3.14 and 3.19)

Monocalcium aluminate, calcium aluminate 12:7, gehlenite and wüstite
$CaO \cdot Al_2O_3$

> Biaxial –ve
>
> Monoclinic
>
> $\alpha = 1.643$, $\beta = 1.655$, $y = 1.663$

$12CaO \cdot 7Al_2O_3 \cdot 0\text{-}1H_2O$ (mayenite)

> Isometric
>
> Cubic
>
> $n = 1.61 - 1.62$

$2CaO \cdot Al_2O_3 \cdot Si\,O_2$ (gehlenite)

> Uniaxial –ve
>
> Tetragonal
>
> $\Omega = 1.669$, $\varepsilon = 1.658$

$Fe(Ca)O$ (substituted wüstite)

> Isometric
>
> Cubic
>
> $n = 2.32$ lower with substitution

The typical composition of high-alumina cement is 40%–50% monocalcium aluminate, 20%–40% ferrite phase, together with smaller proportions of calcium aluminate 12:7 (mayenite, see Hentschel 1983), β-dicalcium silicate, gehlenite, pleochroite and wüstite (Taylor 1990). In transmitted light, most of the constituents are either glassy or dark coloured and almost opaque (Lea 1970, also see Section 4.2.4). Apart from pleochroite (see separate entry), which stands out clearly because of its pleochroism, the relics of high-alumina cement in concrete tend to be undifferentiated as shown in the figures indicated. Monocalcium aluminate is strongly hydraulic.
XRD: CA: 34–440: 2.97–100, 2.52–35, 2.41–18.
   Mayenite: 9–413: 4.89–95, 2.998–45, 2.680–100, 2.189–40.
   Gehlenite: 35–755: 3.063–21, 2.856–100, 1.523–60, 1.299–25.
   Wüstite: 6–615: 2.49–80, 2.153–100, 1.523–60, 1.299–25.
(Figures: *see* Figures 5.28, 5.29, 5.30 and 5.31).

Ferrite phases (brownmillerite, C_4AF, ferrite solid solution)
$4CaO \cdot Al_2O_3 \cdot Fe_2O_3$

Biaxial −ve

Orthorhombic

$\alpha = 1.98$, $\beta = 2.05$, $y = 2.08$

The composition of ferrite can vary from C_6A_2F to C_6AF_2. However, the typical composition of the ferrite phase in ordinary Portland cement is close to C_4AF containing up to 10% magnesium, silicon and titanium ions replacing the ferric ion (Taylor 1990) (see also Gollop and Taylor 1994).
Interference figure: −2 V moderate axial angle.
Relief: Very high positive relief.
Birefringence: White to yellow of the first order.
Form: Either bladed, prismatic, dendritic, fibrous, massive or infilling. Form varies from bladed to dendritic and infilling as the cooling rate of the clinker is increased. In modern cement production with fast efficient clinker coolers, the dendritic form is common.
Extinction: Crystals are length slow.
Colour: Brown to yellow. Pleochroic, varying from green to almost opaque.
XRD: 32–226: 7.25–45, 2.673–35, 2.644–100, 1.8149–45.
Occurrence: Brownmillerite is a rare natural mineral. It ranges from 5% to 15% in cements and is typically about 8% in ordinary Portland cement. A ferrite phase of more variable composition constitutes 20%–40% of high-alumina cement.
Comment: Ferrite and belite phases are commonly observed in concrete where fragments of clinker have failed to hydrate either because they are too large or the hardened paste has not fully hydrated.
(Figures: *see* Figures 3.14 and 3.19).

Pleochroite
$6CaO \cdot 4Al_2O_3 \cdot MgO \cdot SiO_2$

Biaxial −ve

Orthorhombic

$\alpha = 1.669$, $y = 1.673–1.680$

The simplified composition given earlier is variable and is modified by some substitution of calcium by ferrous ion and aluminium by the ferric ion (Taylor 1990).
Interference figure: +2 V = 45° approximately.
Relief: High positive relief.
Birefringence: First-order colours modified by pleochroism.
Form: Fibrous to acicular.
Colour: Strongly pleochroic. Distinctively violet.
XRD: 11–212: 3.70-vs, 2.87-vvs, 2.76-vvs, 1.76-vs.
Occurrence: Occurs in high-alumina cements made under reducing conditions.
Comment: The strong violet pleochroism identifies this mineral which can be observed in some concretes made with high-alumina cement (Ingham 2010).
(Figures: *see* Figures 5.28 and 5.29).

Calcium oxide (lime, burnt or quick lime)
CaO

> Isotropic

> Cubic

> $n = 1.838$

Form: Usually rounded grains in cement clinker and dolomite bricks. In calcined lime-stone, calcium oxide essentially retains the shape of the crushed limestone particles and will tend to be polycrystalline. Electron microscopy shows burnt lime to consist of agglomerates 0.5–5 μm in size consisting of minute equant to globular-shaped particles.

Cleavage: Perfect cubic cleavage but only apparent if material has been subjected to a high temperature.

Colour: Colourless but often stained yellow or brown by interaction with iron oxide.

XRD: 31–1497: 2.4059–100, 2.777–36, 1.7009–54.

Occurrence: Very rare natural mineral found in some high-temperature limestone contacts.

Comment: Present in cement clinker and dolomite brick and is the major constituent of burnt lime. Calcium oxide is not observable in thin sections of concrete. In cement clinker, it is best observed in polished section (Campbell 1986, 1999). The appearance of various lime powders in oil immersion is given by McKenzie (1994). It can be identified in powders and distinguished from periclase and spinel by White's test (White 1909), which is outlined by Rigby (1948).

Periclase (magnesium oxide)
MgO

> Isotropic

> Cubic

> $n = 1.736$ (1.732 – 1.738)

Relief: High positive relief.

Form: Periclase occurs as small crystals or anhedral crystal aggregates. Synthetic crystals are usually rounded and vary in size and may give a 'sugary' appearance but may also occur as closely packed hexagons or octagons. In cement clinker, it is often dendritic.

Cleavage: Perfect, parallel to cube faces. Not always present but will develop at high temperature.

Colour: Colourless when pure. In solid solution with FeO, colour changes from yellow to deep brown with increasing iron content. Also coloured red–brown due to minute crystals of magnesioferrite.

XRD: 4–289: 2.106–100, 1.489–52, 0.942–17, 0.860–15.

Occurrence: A fairly rare, relatively high-temperature natural mineral in dolomite contact zones which may be hydrated to brucite (see under Other hydrates). An important constituent of magnesite brick and present in dolomite and basic refractories. Periclase does not crystallise out in cement clinker until MgO exceeds 2%.

Comment: Magnesia slowly hydrates in concrete and magnesium lime plasters and increases in volume (Berry et al. 1998). Thus, magnesia is limited to 6% or less in Portland cement to prevent unsoundness due to this volume increase. Where problems of unsoundness occur in Portland cement and hydrated magnesium lime, they are monitored for unsoundness by the autoclave expansion test. Best examined using polished specimens in reflected light or by SEM.

(Figure: *see* Figure 4.4)

G.4.2 Calcium silicate hydrates

The calcium silicate hydrates may be conveniently divided into three groups, namely, (1) the hydrates that are found in hardened cement paste which has been cured at a temperature of less than about 60°C; (2) a range of hydrates that are either formed hydrothermally in autoclaved products and oil-well cements, geothermal grouts and other high-temperature applications; and (3) in non-Portland cement systems such as sand-lime bricks and refractories. In addition, there is a wide range of calcium silicate hydrates that may be synthesised or are found naturally but these are outside the scope of this book.

Calcium silicate hydrate formed at less than 60°C

Calcium silicate hydrate gel, C-S-H(II). In older texts is often referred to as 'tobermorite' gel. The term CSH is often used without discrimination.

$9CaO \cdot 4\text{-}6SiO_2 \cdot nH_2O$ is of variable composition. The CaO/SiO_2 ratio may vary between 1.5 and 2.0 and typically is about 1.8 in mature pastes in concrete. However, the ratio measured depends on the method of sample preparation and may also vary from point to point within the paste on the microscale. The water content in a fully saturated mature paste is 42%–44% and in a paste dried at –79°C is in the range 20.4%–22.0% (Taylor 1990).

Optical properties: Properties observable are those of the cement paste. Transparent, coloured, isotropic although vague indefinite birefringence may be observed. Mottled and/or wavy appearance to the texture. Where pozzolanic admixtures are present and have removed most of the calcium hydroxide (portlandite), the texture appears to become darker and more featureless. Refractive index of a mature paste is around 1.5.

Appearance: In mortars and concretes the gel appears as an indefinite mass infilling the space and surrounding components such as portlandite, remnant cement grains, pores, voids and aggregates. It may appear as 'hydration rims' around residual clinker particles, which stand out from the indefinite mass of hydrate material. C-S-H(II) is colourless but, because of its intimate mixture with the other hydrates, appears coloured. The colour of the paste containing C-S-H(II) is brownish to yellow often with some green tints and may have a mottled appearance.

Physical properties: Density is 1.85–2.45 dependent on state of drying. A fully hydrated cement paste has a density of about 2.13.

Structure: A nearly amorphous gel consisting of chains of polymerised silicate anions. Polymerisation increases with curing temperature. The structural appearance of the gel observed by electron microscopy varies according to the age of the sample and method of sample preparation. Soon after mixing a cement with water, the gel formed is fibrous or needle-like (type I) in appearance, which soon changes to a honeycomb structure (type II). As curing proceeds, the gel takes on a more massive appearance with equant grains (type III) and in the mature paste, that is at 28 days, becomes increasingly featureless (type IV) (Diamond 1986). The gel contains pores with equivalent radii between 0.6 and 1.6 nm and pores between gel particles that range from 1.6 to 100 nm. Although pore volumes are comparatively large, C-S-H(II) remains as a stable xerogel in concrete unless attacked by carbon dioxide.

XRD: 29-374: Diffuse peaks at 0.27–0.31 Å and a sharper peak at 0.182 Å. Peaks small and not well defined. Other peaks present are usually due to portlandite. Often in hardened concrete, only one broad shallow peak around the 0.3 Å area is observed and many of the peaks listed on Card no. 29–374 are absent. Peaks may become sharper as curing temperatures are increased, especially over 60°C.

Thermal analysis: TGA shows a continuous weight loss from 100°C to 500°C, with smaller losses to 800°C. Usually superimposed on these weight losses is the weight loss due to the dehydroxylation of portlandite between 400°C and 500°C. DTA does not give a useful diagnostic curve.

Infrared: Does not give useful diagnostic spectra.

Occurrence: Calcium silicate hydrate gel is formed from the hydration of the alite and belite phases present in Portland cement. Both alite and belite react to form similar hydrates, only differing in the amount of portlandite formed during hydrolysis. The gel is the main cementing phase present in all Portland cement concretes cured below 60°C and is intimately mixed with the aluminate and ferrite hydrates and interspersed with crystallites of portlandite. The reaction between portlandite and pozzolanas in concrete produces a similar gel, although the CaO/SiO_2 ratio may be reduced towards the lower end of the range.

Comment: The most important distinguishing feature is the association of the C-S-H gel with remnant cement grains, as even in a fully hydrated cement paste some relict grains remain. The appearance of C-S-H gel in mortars and concretes is best examined by electron microscopy, as optical examination rarely produces any useful data.

In addition to the publications already cited in this subsection, see Churakov (2009), El Hemaly et al. (1977), Henmi and Kusachi (1992), Henning (1974), Komarneni and Guggenheim (1988), Mamedov and Belov (1963), Mamedov et al. (1963), Merlino et al. (2000), Mitsuda and Taylor (1978), Richardson (2004), Suzuki and Sinn (1993), Taylor (1984 and 1992) and Wieker et al. (1982).

Calcium silicate hydrates formed at elevated temperatures in Portland cement concretes and from non-Portland cement systems.

Unlike the calcium silicate hydrates formed from the hydration of Portland cement cured in the normal range of temperatures, these hydrates are semi-crystalline to crystalline. However, with only one exception, they remain microcrystalline and require electron microscopy and XRD analysis for their observation and identification. They may be conveniently grouped into the high-silica and high-calcium types formed under hydrothermal conditions and another group formed by heating concrete in the absence of moisture (see Table G.1).

Many of these calcium silicate hydrates also occur naturally. In addition, there are other ranges of hydrates that can be synthesised or are found naturally (Taylor and Roy 1980, Coleman 1984). These are unlikely to occur in the types of building products encountered by the petrographer and are not further considered here.

Occurrence: *High-silica types*. These calcium silicate hydrates are formed in hydrothermally cured Portland cement pastes containing ground quartz, which is added to some concretes used in autoclaved products. In practical terms the minimum temperature of

Table G.1 The hydrothermal silicates in Portland cement

		CaO/SiO_2
High-silica types		
C-S-H(I)	$C_5S_6H_9$ approx.	0.83
11 Å tobermorite	$C_5S_5H_5$	1.0
Anomalous tobermorite	(incl. 0.5Al and 0.1Na)	0.76
Xonotlite	C_6S_6H	1.0
Gyrolite	$C_2S_3H_2$	0.66
Truscotite	$C_7S_{12}H_3$	0.58
High-calcium types		
α-dicalcium silicate hydrate	α-C_2SH	2.0
Tricalcium silicate hydrate (jaffeite)	$C_6S_2H_3$	3.0
Types formed on heating		
9 Å tobermorite, (riversideite)	$C_5S_6H_2$	0.83

Note: The 9 Å tobermorite is formed by heating.

formation for the 11 Å tobermorite starts between 130°C and 150°C and this phase is stable up to 180°C. From 180°C to 250°C, anomalous tobermorite may persist by substituting Al and Na into the lattice. Otherwise, the stable phase above 180°C is xonotlite, which converts to the calcium silicate, wollastonite, above 600°C. In oil-well cements, where the composition is rich in βC_2S and ground quartz is used, truscotite forms in the temperature range 180°C–350°C. Gyrolite has also been reported in some hydrothermal cementing systems between 140°C and 170°C and forms instead of tobermorite where a more reactive form of silica such as diatomite is used. Below 130°C, even when ground quartz is present, the α-dicalcium silicate hydrate may form together with C-S-H(I) and will only convert to 11 Å tobermorite after extended autoclaving or by increasing the temperature.

High-calcium types. These hydrates form above 110°C, where the cement does not contain the addition of ground quartz. Up to 180°C, α-dicalcium silicate hydrate forms. It has low strength and is the only hydrate that grows large enough to become optically visible. Between 180 and 400°C, tricalcium silicate hydrate is the stable phase. Like αC_2SH, it is also a low-strength material.

The degree of crystallinity achieved and the phases formed are dependent both on temperature and the time of heat treatment. For commercial autoclaving, typically carried out at 160°C–180°C for up to 18 h, the cement paste may only contain poorly crystalline C-S-H similar to 11 Å tobermorite. Similar phases will be formed for autoclaved products made from sand-lime formulations. Where the concrete is steam cured at 60°C–100°C, the phase formed will be C-S-H(II) but increased crystallinity may be detected by better definition of the XRD peaks.

The effect of heating concrete, for instance by exposure to fire, is firstly to increase the crystallinity of the phases and then with increasing temperature to contract the layer spacing until, above 300°C, the 9 Å tobermorite is the major phase which persists to 700°C (Taylor 1957). There is limited information on the composition of the phases in concretes exposed to fire. Xonotlite is also found in boiler scales and lightweight heat-insulating materials.

The following mineral descriptions are mainly derived from Heller and Taylor (1956):

Mineral descriptions

C-S-H(I)
Natural occurrence: Probably similar to poorly crystallised natural material from Ballycraigy, Larne, Northern Ireland.

Seemingly amorphous material with a flaky habit under SEM.

Orthorhombic, $n = 1.49$–1.530 ($n = 1.494$ when $d = 2.00$, $n = 1.525$ when $d = 2.20$)

XRD: 12.5-vs, 3.04-vs, 2.80-s, 1.82-s (Taylor 1990).

SEM: Usually appears as flocs, thin plates and twisted foils

11 Å tobermorite
Natural occurrence: Tobermory in Scotland, Northern Ireland and California.

Minute radiating aggregates of fibres with distinctive silky lustre.

Orthorhombic, biaxial +ve, low 2 V, both length slow and length fast

$\alpha = 1.570$, $\beta = 1.571$, $\gamma = 1.575$ (Ballycraigy)

$n = 1.558$ and 1.545–1.565 (for other samples)

$n = 1.55 - 1.565$ (synthetic material)

Density = 2.44.

XRD: 11 Å. 45–1480: 11.3–100, 3.08–92, 2.972–70, 2.806–68. Anomalous tobermorite exhibits similar XRD data.

SEM: Typically appears as a mass of long needle-like fibres.

Xonotlite
Natural occurrence: Tetela de Xonotla, Mexico.
Usually fibrous, colourless to pink.
Monoclinic, biaxial +ve, low 2 V, parallel extinction, length slow
$\alpha = 1.578–1.586$, $\beta = 1.583$, $\gamma = 1.590–1.595$
Density = 2.7.
XRD: 29–379: 3.085–100, 2.828–50, 2.697–40, 1.9468–40.
SEM: Typically appears as a mass of long needle-like fibres.

Truscotite
Natural occurrence: Benkulen, Sumatra.
White, spheroidal aggregates with fibrous, platy structure and pearly lustre on cleavage.
Hexagonal, uniaxial -ve, parallel extinction, length fast.
Mean $n = 1.560$ $\omega < 1.55$, weak birefringence
Density = 2.48.
XRD: 29–382: 4.21–70, 3.141–100, 2.840–80, 1.839–70.

Gyrolite
Natural occurrence: Skye, Scotland.
Colourless, lamellar crystals, often in radial aggregates.
Hexagonal, uniaxial –ve (may appear biaxial with small 2 V).
$\omega = 1.549$, $\varepsilon = 1.536$, $\omega = 1.540 – 1.549$
Density = 2.39.
XRD: 12–217: 22.0–90, 11.1–70, 4.20–70, 3.160–100, 3.097–90, 2.834–70.

α-dicalcium silicate hydrate
Colourless, transparent laths or prisms, with parallel growths and twinning.
Orthorhombic, biaxial +ve, 2 V = 68°, parallel extinction, length fast.
$\alpha = 1.614$, $\beta = 1.620$, $\gamma = 1.633$
Density = 2.8.
XRD: 29-373: 4.22–50, 3.302–45, 3.272–100, 2.816–45, 2.418–60, 2.606–45.
SEM: Appears as long needles or flat-ended rods (Grudemo 1965).

Tricalcium silicate hydrate (jaffeite)
Very small, colourless, anisotropic prisms or broad fibres.
Hexagonal, parallel extinction, length slow.
$\alpha = 1.589$, $\gamma = 1.633$
Density = 2.56.
XRD: 29–375: 8.6–95, 2.998–85, 2.898–95, 2.834–100.
SEM: Appears as long needles or flat-ended rods (Grudemo 1965).

9 Å tobermorite (riversideite)
Natural occurrence: Crestmore, California.
Appearance similar to 11 Å hydrate.
Orthorhombic, biaxial +ve, small 2 V, both length slow and length fast.
$\alpha = 1.600$, $\beta = 1.601$, $\gamma = 1.605$
Density = 2.6.
XRD: 29–329: 3.59–17, 3.15–19, 3.014–100, 2.784–18.
Thermal analysis: Refer to Ramachandran (1969).
Comment: Optically all the hydrates, with the exception of αC_2SH, do not form large enough crystals to be identifiable under the microscope. However, Hadley grains (Barnes et al. 1978b) are more common in the elevated temperature C-S-H matrix compared to concretes cured at normal temperatures. In geothermal samples exposed to downhole temperatures for up

to 9 months, the hardened paste may contain a background of minute, diffuse, birefringent crystals which it is believed are the crystallised hydrates. After three months' exposure down a geothermal well at 160°C, grouts without ground quartz contained laths of αC_2SH measuring up to 30 μm in length and 10 μm in width which tended to form radial groups with optical properties as listed earlier (St John and Abbott 1987).
(Figure: *see* Figure 5.15.)

In addition to the publications already cited in this subsection, see Bonaccorsi et al. (2004), Gard et al. (1977), Barnes et al. (1978a,b), Henry (1999), Kormarneni et al. (1985, 1987) and Merlino et al. (2000).

G.4.3 Aluminate hydrates

Amorphous hydrated alumina (pseudoboehmite, cliachite)
AlO(OH)n

$n = 1.56–1.61$

Form: Fine grained to gelatinous.
Colour: Grey, yellow, white or brown depending on the presence of iron.
XRD: Broad diffuse bands in the regions of the strongest lines of boehmite. Boehmite: 21–1307: 6.11–100,3.164/65,2.346–55, 1.860–30.
Thermal analysis: See Mackenzie (1970) and Brindley and Brown (1980) for discussion on XRD and thermal data.
Occurrence: Cliachite is the chief aluminium ore. Pseudoboehmite is the name given to the synthetic hydrous alumina used in medicine and industry.
Comment: Amorphous hydrated alumina together with gelatinous silica is the end products of the leaching and corrosion of hardened cement paste. While gelatinous silica is commonly observed, amorphous hydrated alumina is rarely identified. The hydrated aluminas are often poorly crystalline and may have some of the optical characteristics of boehmite.

Gibbsite (aluminium hydroxide)
$Al(OH)_3$

Biaxial +ve

Monoclinic

$\alpha \approx \beta = 1.565–1.577$, $y = 1.58–1.595$ (Winchell and Winchell 1951)

$\alpha = 1.55–157$, $\beta = 1.55–157$, $y = 1.57–1.59$ (Rigby 1948)

Interference figure: Crystals usually too small to obtain a figure. 2 V very small and may be zero.
Birefringence: Varies between 0.015 and 0.030, so interference colours range from high first order to second order.
Form: Variable. Hexagonal tablets, but not common. May be lamellar or fibrous or present mosaic structure similar to chalcedony.
Cleavage: Perfect basal cleavage.
Twinning: Polysynthetic twinning often observed.
Extinction: Fibrous crystals may be length slow or length fast and show inclined extinction.
Colour: Colourless to pearly white. Industrially may be pale brown.
SEM: Observed to form thin flaky plates (Taylor 1990)
XRD: 33–18: 4.849–100, 4.371–70, 4.382–50, 2.452–40 (see Brindley and Brown 1980 for further details)

Thermal analysis: See Mackenzie (1970).

Occurrence: A constituent of bauxite, laterites and aluminous clays. Occasionally found as a cementing mineral in sandstone (Wopfner and Höcker 1987).

Comment: Gibbsite is one of the hydration products of high-alumina cement, especially after conversion (see Section 4.2.4), and tends to be poorly crystalline, but, otherwise, is unlikely to be observed in concrete. However, Gillott and Rogers (1994, 2003) identified gibbsite as an alteration product of dawsonite in some concrete from the Montréal area of Canada (see also Ervin 1952, Ervin and Osborn 1951, Peryea and Kittrick 1988 and Trolard and Tardy 1987).

Calcium aluminium hydrates

Hydration of calcium aluminates in high-alumina cements at normal temperatures leads to the formation of a variety of metastable calcium aluminate hydrates, most notably CAH_{10} and C_2AH_8 together with alumina gel. CAH_{10} will convert to C_3AH_6 and alumina gel or gibbsite in warm moist conditions. C_3AH_6 is the only stable hydrate at ambient temperatures.

CAH_{10}
  Hexagonal
  XRD: 14.3, 7.16, 3.56, 2.55

C_2AH_8
  Hexagonal
  XRD: 10.70, 5.36, 2.86, 2.54

C_3AH_6
  Cubic
  XRD: 5.140, 2.300, 2.043, 2.816

C_4AH_3
  Hexagonal

Thermal methods: DTA or DSC thermograms in the 20°C–350°C temperature range give diagnostic absorption bands for calcium aluminate hydrates and alumina gel (Benge and Webster 1994, also see Section 4.2.4)

G.4.4 Other hydrates

Calcium hydroxide (Portlandite, slaked lime)
$Ca(OH)_2$

   Uniaxial –ve

   Hexagonal

   $\varepsilon = 1.545 - 47$, $\omega = 1.573 - 75$

Relief: Indices are close to the standard mounting medium index of 1.54 (note that epoxy may have higher indices of typically between 1.57–61). On rotation of stage relief is close to nil in one position.

Birefringence: In standard sections gives first-order red and second-order blue. In thinner sections of concrete more commonly gives first-order yellow, orange and red. Small crystals, such as occur in hydrated lime, appear white. In some hardened pastes, the yellow birefringence of small crystals of calcium hydroxide can be confused with the golden birefringence of small aggregates of carbonation.

Form: Crystals tend to occur as minute euhedral plates or as laths in hardened cement paste, but in thin sections of concrete, it may also be observed filling, or partially filling, void spaces and be irregular in form. Sizes range from microcrystalline to crystals as large as 50 μm in size (see Section 5.2.5).

Cleavage: Perfect basal cleavage which is rarely observed in thin sections of concrete.
Colour: Colourless
XRD: 4–733: 4.9–74, 2.628–100, 1.927–42, 1.796–36.
SEM: When observed by electron microscopy, occurs most commonly as hexagonal plates, tablets and rods as well as irregular shapes (Grudemo 1965).
Thermal analysis: Endotherm 500°C–600°C (Ramachandran 1969; Taylor 1997).
Occurrence: Very rare as a natural mineral found in some limestone contacts. Calcium hydroxide is ubiquitous in hardened cement paste (prior to carbonation) and is the principal constituent of slaked and hydraulic limes.
Comment: In hardened cement paste, calcium hydroxide is one of the few minerals that is not submicroscopic. The observation of calcium hydroxide (including form, crystal size and distribution) is an important diagnostic tool in concrete petrography. In slaked lime and its products, microscopical examination can give information on impurities, but the crystals of calcium hydroxide are too small to be readily observed. In hydraulic limes, the hydraulic components are best observed in polished section.
(Figures: *see* Figures 5.1, 5.2, 5.3, 5.13, 5.14)

Brucite
$Mg(OH)_2$

Biaxial +ve

Trigonal

$\varepsilon = 1.580–1.600, \omega = 1.560$

Magnesium hydroxide is a common alteration product resulting from sea water attack on Portland cement concrete. It is also believed to form during the de-dolomotisation of dolomitic aggregate in concrete in alkali–carbonate reaction (see Section 6.9).
Interference figure: Uniaxial, but sometimes anomalous biaxial with a small 2 V. Individual crystals are usually too small to give a figure.
Relief: Moderate, higher than standard mounting medium.
Birefringence: 0.012–0.020. First-order yellows and reds but may be anomalous red/ brown.
Form: Massive, foliaceous, or fibrous
Cleavage: Basal (0001) perfect
Extinction: Parallel on fibrous forms, length fast.
Colour: Usually colourless in thin section but can be pale yellow, greenish or pale brown.
XRD: 4.77, 2.35, 1.794, 1.537
Thermal and infrared methods: these methods can be diagnostic (see Farmer 1974).
Occurrence: Brucite is a major constituent of many bauxites and is often associated with gibbsite. It can form during the metamorphic alteration of dolomite and forms from the hydration of periclase (MgO) in cement clinkers and slags (especially steel slags).
Comment: It is a common alteration product resulting from the substitution of magnesium for calcium in portlandite and CSH in concretes exposed to sea water attack and other sulphate solutions. Brucite also forms as fine radiating crystals in the matrix of Sorel cement (see magnesium minerals, Section G.4.8 of this glossary) as it ages.
(Figure: *see* Figure 6.49a,b)

Hydrogarnets

Hydrogrossular is a solid solution series, which mineralogically is identified as katoite and hibschite as given earlier (see Table G.2). It is not an accepted mineral name. The refractive indices given will vary according to the composition.

Relief: Moderately high.

Birefringence: Hibschite may be weakly birefringent.

Twinning: Present in grossular.

Colour: Grossular ranges from colourless to coloured, similar to other garnets. Hibschite is colourless and it is probable that in cementing systems the other hydrogrossulars are also colourless to white.

XRD: Grossular 39–368: 2.962–35, 2.650–100, 2.418–18, 1.583–30.

Katoite 24–127: 5.13–90, 2.810–80, 2.295–100, 2.039–95.

Hibschite 45–1447: 3.072–50, 2.748–100, 2.244–60, 1.994–65.

Thermal analysis: Hydrogrossular decomposes at 200°C–250°C (Taylor 1990).

Occurrence: Grossular and hibschite occur naturally in contact altered limestones.

Comment: Minor quantities are formed from some composite cements. In Portland cement, it is a minor and poorly crystalline hydrated phase. Larger quantities were produced by some older Portland cements and it is a normal hydration product of autoclaved cement-based materials (Taylor 1990). In cementing systems, if sufficient is present, it is detected by XRD analysis and SEM. In addition to the publications already cited in this subsection, see Nijland et al. (1994), Schröpfer and Bartl (1993), Schwandt et al. (1996) and Van Aardt and Visser (1977).

Calcium chloroaluminate hydrate, Friedel's salt

$3CaO \cdot Al_2O_3 \cdot CaCl_2 \cdot 10\ H_2O$, $Ca_2Al(OH)_6$ $(ClOH) \cdot 2H_2O$

Layered double hydroxide, an anion exchanger. Main role in cement and concrete retention of chloride transport medium related to steel reinforcement corrosion. White solid.

Vivianite (Blue iron earth)

$Fe_3P_2O_8 \cdot 8H_2O$

Monoclinic

Form: Prismatic, radiating, reniform and encrusting

Cleavage: Perfect parallel to clinopinacoid

Colour: Usually deep blue/green but can be white.

Occurrence: Associated with iron, copper and tin ores. Sometimes in peat, clay and bog iron ore, also as encrustations on bones or shells.

Comment: Observed as a blue-green surface precipitate on concrete in waste-water treatment installations (Vogel 1995; Frossard et al. 1997).

Table G.2 The garnets grossular and the hydrogarnet solid solution series

Grossular (also grossularite)	
$3CaO \cdot Al_2O_3 \cdot 3SiO_2$	Cubic $n = 1.734$.
Hydrogrossular	
$3CaO \cdot Al_2O_3 \cdot 6H_2O - 3CaO \cdot Al_2O_3 \cdot 3SiO_2$	Cubic $n = 1.604–1.734$.
Katoite	
$3CaO \cdot Al_2O_3 \cdot 6H_2O - 3CaO \cdot Al_2O_3 \cdot 1.5SiO_2 \cdot 3H_2O$	Cubic.
Hibschite	
$3CaO \cdot Al_2O_3 \cdot 1.5SiO_2 \cdot 3H_2O - 3CaO \cdot Al_2O_3 \cdot 3SiO_2$	Cubic $n = 1.67–1.68$.

G.4.5 Sulphate minerals

In addition to the publications cited in this Section, see Alpers et al. (1999), Hina and Nancollas (1999), Myneni (2000) and Poole and Thomas (1975).

Anhydrite (insoluble anhydrite, hard-burnt gypsum)
$CaSO_4$

Biaxial +ve

Orthorhombic

$\alpha = 1.570$, $\beta = 1.576$, $\gamma = 1.614$

Interference figure: +2 V = 42°.
Birefringence: Strong polarisation colours up to the end of the third order.
Form: Fine- to medium-grained aggregates sometimes elongated. Well-formed crystals not common.
Cleavage: Three good cleavages intersecting at right-angles parallel to the sides of the crystals.
Twinning: Polysynthetic twinning common.
Extinction: Extinction is parallel to cleavages.
Colour: Colourless.
XRD: 37–1496: 3.499–100, 2.849–29, 2.3282–20, 2.2090–20.
Thermal analysis and infrared: Refer to gypsum.
Occurrence: Usually associated with gypsum and salt deposits and is a relatively widespread mineral. Principal constituent of hard-burnt gypsum plasters.
Comment: Green (1984) states that commercial anhydrites almost always consist of block crystals with square corners. The size of crystals and colourful appearance under crossed polars make anhydrite easy to identify
(Figure: *see* Figure 9.12)

Soluble anhydrite (γ-anhydrite, anhydrite III)
$CaSO_4$
Uniaxial +ve
Hexagonal
Indices are variable even from crystal to crystal and dependent on method and temperature of preparation.

$\omega = 1.50$, $\varepsilon = 1.56$ (Florke 1952)

Form: Hexagonal basal plates.
Colour: Colourless in thin section.
XRD: 45–157: 6.046–100, 3.485–22, 3.016–71, 2.794–22.
Thermal analysis and infrared: Refer to gypsum.
Occurrence: Formed by the low-temperature dehydration of hemihydrate. May be present in lightly burned gypsum plaster and Portland cement.
Comment: Appears to be a variable phase which is difficult to characterise under the microscope as in appearance it is similar to hemihydrate. Its presence in Portland cement can lead to premature stiffening and is best determined by a combination of thermal and chemical analysis (see also Chowns and Elkins 1979).

Hemihydrate (bassanite, plaster of Paris, gypsum plaster)
$CaSO_4 \cdot 0.5H_2O$

Biaxial +ve

Pseudo-hexagonal monoclinic

α form: α = 1.558–1.5599, β = 1.5595, γ = 1.5836–1.586 (Berg and Sveshnikova 1949).

β form: α = 1.550, γ = 1.556 (Berg and Sveshnikova 1949).

β: form: n = 1.540 variable (Green 1984).

For β-hemihydrate, the indices are variable and dependent on the method by which the hemihydrate is made.
Interference figure: Crystals usually too small to obtain an interference figure. +2 V = 14° but reported to vary between 10°–15°.
Birefringence: Varies from 0.006 to 0.025. Most commercial gypsum plaster will give grey polarisation colours.
Form: Acicular or fibrous. Particles retain the same shape as the gypsum particles from which they are formed. β-hemihydrate particles are porous. α-hemihydrate particles vary from blocky to acicular.
Twinning: Twinning may be present.
Extinction: Fibres have parallel extinction and are length slow.
Colour: Colourless. Some hemihydrates may appear brown.
XRD: 41–224: 6.013–80, 3.467–50, 3.006–100, 2.803–90.
Some variation in line intensity occurs between the α and β forms but no other significant differences are found.
Thermal analysis and infrared: Refer to gypsum.
Occurrence: Very rare mineral found in arid parts of Central Asia and in a few volcanic rocks. It is the chief component of gypsum plaster and forms in Portland cement due to the dehydration of gypsum during grinding. Can occur in thin sections of rocks containing gypsum due to heat generated by grinding the section.
Comment: The amount of molecular water present in hemihydrate can vary between 0.4 and 0.6 due to over or under dehydration of the gypsum and other factors which make exact characterisation difficult. The more crystalline α form appears to differ from the β form only in degree of crystallinity and crystal size (Taylor 1997). Commercial gypsum plaster consists mainly of the β form together with a little α.

Gypsum (selenite, alabaster, satin spar)
$CaSO_4 \cdot 2H_2O$

Biaxial +ve

Monoclinic

α = 1.520, β = 1.523, γ = 1.530

Interference figure: Biaxial positive with 2 V = 58°.
Birefringence: Weak first-order greys up to white.
Extinction: Parallel to the best cleavage for sections parallel to 010.
Form: Anhedral to subhedral aggregates often unevenly grained. May show fibrous structure.

Cleavage: Perfect in {010} direction.

Twinning: Simple twinning common caused by heating the section during preparation.

Orientation: Cleavage traces are parallel to both slow and fast rays.

Colour: Colourless in thin section.

XRD: 33–311: 7.63–100, 4.283–100, 3.065–100, 2.873–45.

Thermal analysis: On heating, gypsum gives two endothermic peaks, which begin at 125°C and 185°C owing to the loss of 1.5 and 0.5 molecules of water, respectively. The temperature of the peaks varies with heating rate and sample holder. The detailed thermal curves are described by Ramachandran (1969).

Infrared: See Farmer (1974).

Occurrence: The chief constituent of gypsum rock is widespread in many ore deposits and as crystals in sedimentary rocks.

Comment: The differences in the optical appearance of natural and synthetic gypsums have been described by Green (1984). Gypsum often forms in concrete undergoing sulphate attack. Much of the gypsum formed by sulphate attack is submicroscopic but veins of gypsum may be visible in the exfoliation texture of surfaces. These veins have a distinctive texture as shown in Figure 6.43 and Adams et al. (1984). Gypsum is interground with cement clinker to produce Portland cement. Due to the high operating temperatures of many grinding mills, most of the gypsum is converted either to hemihydrate or soluble anhydrite (see also De Meer and Spiers 1997 and De Meer et al. 1997).

(Figure: *see* Figure 6.43)

Thenardite (anhydrous sodium sulphate)

Na_2SO_4

Biaxial +ve

Orthorhombic

$\alpha = 1.464$–1.468, $\beta = 1.473$–1.474, $\gamma = 1.481$–1.485

Interference figure: Biaxial +2 V = 83°.

Relief: Moderate negative relief

Birefringence: Gives colours up to red of the second order.

Form: Thenardite is pyramidal, with short prismatic or basal plates.

Cleavage: Thenardite has distinct basal cleavage.

Colour: Colourless in thin section.

XRD: 37-1465: 4.657–71, 3.077–55, 2.784–100, 3.181–52, 2.648–52.

Thermal analysis: See Mackenzie (1970).

Occurrence: Both thenardite and mirabilite are found associated with hot springs and salt lake deposits.

Comment: Thenardite effloresces and converts to mirabilite below 32°C. Occurs as fine silvery needles deposited on surfaces as an evaporation product of groundwater sulphate (St John 1982). Sometimes found as efflorescence on protected surfaces of brick and mortar.

Mirabilite (glauber's salts, sodium sulphate decahydrate)

$Na_2SO_4 \cdot 10H_2O$

Biaxial –ve

Monoclinic

$\alpha = 1.394$, $\beta = 1.396$, $\gamma = 1.389$

Interference figure: Biaxial –2 V = 76°.
Relief: High negative relief compared to thenardite.
Birefringence: Gives first-order grey colours.
Form: Is commonly prismatic to acicular. Can form as massive efflorescent crusts of interlocking fibres.
Cleavage: Has a perfect 100 cleavage.
Colour: Colourless in thin section
XRD: 11-647: 5.49–100, 3.26–60, 3.21–75, 3.11–60.
Thermal analysis: See Mackenzie (1970).
Infrared: See Farmer (1974).
Occurrence: Both mirabilite and thenardite are found associated with hot springs and salt lake deposits.
Comment: Mirabilite effloresces and converts to thenardite above 32°C. Occurs as fine silvery needles deposited on surfaces as an evaporation product of groundwater sulphate (St John 1982). Sometimes found as efflorescence on protected surfaces of brick and mortar.

Syngenite
$K_2Ca(SO_4)_2 \cdot H_2O$

Biaxial –ve

Monoclinic

$\alpha = 1.500–1.501$, $\beta = 1.515–1.517$, $\gamma = 1.518–1.520$

Interference figure: –2 V = 28°.
Form: Tablets or prisms.
Twinning: Common.
Colour: Colourless.
XRD: 28–739: 9.49–40, 5.71–55, 4.624–40, 3.165–75, 2.835–100.
Thermal analysis: Endotherm at 260°C (Smillie et al. 1993).
Occurrence: Found in gypsum and anhydrite beds.
Comment: Forms in Portland cement and is one of the causes of premature stiffening (Smillie et al. 1993). Does not survive in concrete. Best detected by thermal analysis and XRD.
Arcanite K_2SO_4, calcium langbeinite $K_2SO_4 \cdot 2CaSO_4$, aphthitalite $(K_2Na)_2SO_4$
These sulphate minerals form in cement clinker but are not observable in concrete. Brief mineralogical details are given by Campbell (1986, 1999).

Magnesium sulphate (epsomite, hexahydrate, kieserite)
Epsomite $MgSO_4 \cdot 7H_2O$

Biaxial –ve

Orthorhombic

$\alpha = 1.433$, $\beta = 1.455$, $\gamma = 1.461$

Hexahydrate $MgSO_4 \cdot 6H_2O$

Biaxial –ve

Monoclinic

$\alpha = 1.426$, $\beta = 1.453$, $\gamma = 1.456$

Kieserite $MgSO_4 \cdot H_2O$

    Biaxial +ve

    Monoclinic

    $\alpha = 1.423$, $\beta = 1.535$, $\gamma = 1.586$

Interference figure: Epsomite $-2 V = 51°$, hexahydrate $-2 V = 38°$, kieserite $+2 V = 57°$.
Birefringence: Birefringence of kieserite is strong compared to others.
Form: Epsomite and hexahydrate commonly acicular to fibrous. Kieserite is usually prismatic.
Cleavage: Cleavages present.
Colour: Colourless to white or stained.
XRD: Epsomite 36–149: 5.98–30, 5.34–30, 4.216–100, 4.200–75.
Hexahydrate 33–882: 4.815–75, 3.405–100, 3.351–70, 3.313–70.
Kieserite 24–719: 5.45–50, 5.10–45, 4.39–100, 4.04–45.
Thermal analysis: Refer to Mackenzie (1970).
Occurrence: Epsomite is associated with gypsum and salt deposits and as an efflorescence in some ore deposits. Epsomite dehydrates to hexahydrate in air. Kieserite is associated with salt deposits.
Comment: These magnesium sulphate minerals may be associated with sulphate attack on concrete.

Ammonium sulphate (mascagnite)
$(NH_4)_2SO_4$

    Biaxial +ve

    Orthorhombic

    $\alpha = 1.521$, $\beta = 1.523$, $\gamma = 1.533$

Interference figure: $+2 V = 52°$.
Birefringence: 0.01. First-order greys and whites.
Form: Tablets or nearly equant.
Cleavage: Distinct.
Twinning: Twinning gives pseudo hexagonal habit.
Colour: Colourless or stained grey to yellow.
XRD: 40–660: 4.392–49, 4.335–100, 3.052–49, 2.096–22.
Thermal analysis: Decomposes at 350°C.
Occurrence: Found in guano and associated with volcanoes. Present in some fertilisers. Easily soluble in water.
Comment: One of the agents which causes sulphate attack on concrete. See Kloprogge et al. (2006) regarding the ammonium-sodium sulphate hydrate, lecontite.

Tricalcium aluminate monosulphate 12-hydrate (low sulfoaluminate, monosulphate)
$3CaO \cdot Al_2O_3 \cdot CaSO_4 \cdot 12H_2O$

    Uniaxial –ve

    Hexagonal

    $\omega = 1.504$, $\varepsilon = 1.488$

Interference figure: Crystals are too small to obtain a figure.

Relief: Moderate negative relief.

Birefringence: Maximum second-order red. In practice, crystals are too small for this colour to be obtained.

Form: Basal plates.

Extinction: Length slow.

Colour: Colourless in thin section.

XRD: 12-hydrate 45–158: 8.9–100, 4.45–70, 2.190–40, 2.070–40.

Thermal analysis: See Taylor (1990).

Occurrence: Does not occur naturally. The final product of the reaction between the aluminate and sulphate phases in the hydration of Portland cement.

Comment: Low sulfoaluminate is not visible optically and must be observed by SEM. It is metastable with respect to ettringite which will reform if additional sulphate enters the system. Its composition is variable and complex (see Taylor 1990 for further details).

Ettringite (calcium aluminium sulphate hydroxide, high sulfoaluminate)
$Ca_6Al_2(SO_4)_3(OH)_{12} \cdot 26H_2O$

Uniaxial –ve

Pseudo-hexagonal, trigonal

$\omega = 1.466, \varepsilon = 1.462$

Interference figure: Crystals usually too small to obtain an interference figure.

Relief: Moderate negative relief.

Birefringence: Maximum interference colour rarely above first-order grey–white owing to small dimension of needles.

Form: Always appears as needles, which under SEM may prove to be rods. Often present as randomly crossed needles in pores and fissures. In very old weathered concretes may form massive curtain-like material in large fissures in which needles are difficult to observe. Needles rarely more than a few microns in cross-sectional dimension. In hardened cement paste, as opposed to cracks, pores and fissures, ettringite is usually submicroscopic.

Extinction: Parallel extinction.

Orientation: Needles are length fast.

Colour: Colourless in thin section

XRD: 41–1451: 9.72–100, 5.61–76, 3.873–31, 2.569–29.

Thermal analysis: Strong endotherm at 125°C–130°C (Taylor 1990).

Infrared: See Bensted and Varma (1971) and Farmer (1974).

Occurrence: Found in metamorphic limestones. Forms as initial hydration product in Portland cement concretes and as a recrystallisation product in concrete associated with movement of water and alkali–aggregate reaction. Can be formed by delayed reaction in steam-cured products. One of the products formed by sulphate attack on concrete. Important phase in some types of shrinkage compensating cements. It is present as one of the hydration products of supersulphated cement. Optical data from Lerch et al. (1929).

Comment: Ettringite may convert to thaumasite due to prolonged sulphate attack on concrete. Loses crystallinity at 100°C (Buck 1986) and may become amorphous if heated above 60°C (Taylor 1990). Excessive grinding in preparation for XRD analysis can cause anomalous results. Ettringite is prone to atmospheric carbonation (Grounds et al. 1988). The morphology of ettringite is affected by organic additions (Baussant et al. 1989).

Ettringite can enter into solid solutions with hydroxyl and carbonate ions (Poellmann et al. 1990).

In addition to the publications already cited in this subsection, see Carpenter (1964), Cody et al. (2004), Deb et al. (2003), Fu and Beaudoin (1996), Kollman and Strübel (1978), Kühnel and Van der Gaast (1996), Lehmann (1874), McConnell and Murdoch (1962), Perkins and Palmer (1999) and Pöllmann et al. (1989).

(Figures: *see* Figures 5.8, 5.9, 5.10, 5.11)

Thaumasite (calcium carbonate silicate sulphate hydroxide hydrate)
$[Ca_3Si(OH)_6 \cdot 12H_2O](SO_4)(CO_3)$

Uniaxial −ve

Hexagonal

$\omega = 1.504$, $\varepsilon = 1.468$

Interference figure: Crystals usually too small to obtain an interference figure.
Relief: Relief varies with rotation of stage.
Birefringence: Maximum interference colour is first-order yellow but rarely above first-order white owing to small dimension of needles.
Form: Needles; can be threadlike. May be present as overgrowths on ettringite.
Extinction: Needles length fast with parallel extinction.
Colour: Colourless.
XRD: 25–128: 9.56–100, 5.51–40, 3.41–20, 3.18–16.
Thermal analysis: Strong endotherm at 150°C (Bensted and Varma 1974).
Infrared: Spectral data Bensted and Varma (1974).
Occurrence: Rare mineral found in altered limestones and basic rocks. Formed from solutions depositing calcite and zeolites in contact with anhydrite. Occurs in concrete and mortars due to sulphate attack (Crammond 1985a,b) where it is favoured by temperatures of 0°C–5°C and humidities greater than 90%. Often a close relationship observed between alkali–silica gel, ettringite and thaumasite. Good literature reviews by Brouxel and Valiere (1992) and Berra and Baronio (1987) but also see Section 6.5.6. Solid solution with ettringite and other data (Taylor 1990). Optical data from Erlin and Stark (1965/1966). Occurs as replacement of ettringite in cement mortars in contact with gypsum or lime gypsum mortars (Ludwig and Mehr 1986).
Comment: Appears similar to ettringite and is often misdiagnosed as such. May be present as overgrowths on ettringite. May require a higher optical magnification to distinguish from ettringite or the use of SEM. More thermally stable than ettringite. Usually requires XRD analysis for detection and easily missed in optical examination. See full discussion of the thaumasite form of sulphate attack in Section 6.5.6.

In addition to the publications already cited in this subsection, see Aguilera et al. (2001), Barnett et al. (1999), Carpenter (1964), De Ceukelaire (1990), Iden and Hagelia (2003), Kollman et al. (1976, 1977) and Sims and Hartshorn (1998).

(Figure: *see* Figure 6.57)

G.4.6 Carbonate minerals

See the following for information on staining methods for helping to identify carbonates: Dickson (1966), Friedman (1959, 1971), Hitzman (1999), Reid (1969) and Warne (1962).

Calcite (calcium carbonate)
$CaCO_3$

Uniaxial –ve

Trigonal (rhombohedral)

$\varepsilon = 1.486$, $\omega = 1.658$

Interference figure: A good uniaxial figure with several rings is obtained from suitable crystals. Occasionally, calcite gives a biaxial figure with a small angle.

Relief: Changes markedly on rotation of the stage as relief swings from positive to negative.

Birefringence: The maximum interference colours are pearl grey or white of the higher orders. Coloured fringes can often be seen around crystal edges. In carbonated hardened cement paste the calcium carbonate usually appears golden as it is present as microcrystalline aggregates which lower the birefringence considerably. In some hardened pastes, it can be difficult to differentiate carbonation from calcium hydroxide on birefringence alone.

Form: Fine to coarse aggregates, usually anhedral. Euhedral crystals are rare in rock and concrete sections. Fossiliferous limestones often show organic structures. Unless the calcium carbonate is redeposited on surfaces or in cracks, it is invariably present in hardened concrete as microcrystalline aggregates.

Twinning: Polysynthetic twinning is very common and may be generated by pressure on the crystal during grinding of the section. Twin lamellae may show first-order interference colours.

Extinction: Extinction is symmetrical with regard to cleavages.

Colour: White to grey. Colourless in thin section but is often cloudy.

XRD: 5–586: 3.035–100, 2.285–18, 2.095–18, 1.913–17, 1.875–18.

Thermal analysis: Endotherm at 800°C–925°C (Ramachandran 1969; Mackenzie 1970).

Occurrence: Calcite is the principal constituent of limestone and marble. It is a common vein mineral and is found in many rock types as a secondary mineral. Terminology for various forms of calcite (such as 'micrite' [grain size <0.01 mm] and 'sparite' [grain size >0.01 mm and forming a mineral rock cement]) has become established for the petrography of limestones and related carbonate sediments (Folk 1959, 1962; Adams and MacKenzie 1998).

Comment: Calcite is the principal raw component used to make Portland cement and is ubiquitous in the outer surfaces of hardened cement and concrete where it is produced by the carbonation of calcium hydroxide and the cement hydrates. Leaching of calcium hydroxide in solution and its subsequent carbonation by the atmosphere leads to the common development of carbonate exudations and deposits at concrete surfaces.

In addition to the publications already cited in this subsection, see Bertram et al. (1991), Broekmans (2009), Carlson (1983), Cooper (1962), Gaffey (1986), Gavish and Friedman (1973) and Kneller (1966).

(Figures: *see* Figures 5.32, 5.33, 5.34, 5.35, 5.38, 5.39 and 5.40)

Aragonite
$CaCO_3$

Biaxial –ve

Orthorhombic

$\alpha = 1.530$, $\beta = 1.682$, $\gamma = 1.686$

Interference figure: –2 V = 18°
Birefringence: Extreme. Interference colours pearl grey to high-order white.
Form: Columnar or fibrous.
Cleavage: Imperfect parallel to the length of the crystal.
Twinning: Fairly common both as contact and penetration twins.
Extinction: Parallel to fibres or columns.
Colour: White. Colourless in thin section.
XRD: 41–1475: 3.397–100, 3.274–50, 2.702–60, 1.9774–55.
Thermal analysis: See Mackenzie (1970).
Occurrence: Many shells are composed of aragonite. Occurs as a secondary mineral in cavities of basalts and andesites and as seams in limestones and sandstones.
Comment: Above 60°C aragonite forms in cementitious systems in preference to calcite. Common in lower-temperature geothermal cementing systems. May be found as silvery needles on concrete in exposure systems which cycle temperatures up to 60°C. Forms in preference to calcite in seawater attack because of the presence of Mg ions (Taylor 1990). Aragonite can alter to calcite (Perdikouri et al. 2008) (see also Zhou et al. 2009).

Vaterite
$CaCO_3$

Uniaxial +ve

Hexagonal

$\omega = 1.550, \varepsilon = 1.640–1.650$

Birefringence: Extreme colours, pearl grey to high-order white.
Colour: Colourless or white. Colourless in thin section.
XRD: 33–268: 3.294–100, 2.730–90, 2.063–60, 1.820–70.
Occurrence: Deposition of calcium carbonate in sea basins often follows the transition of calcium carbonate gel → vaterite → aragonite → calcite.
Comment: In the carbonation of hardened cement paste, vaterite is formed first and may go through the transition of vaterite → aragonite → calcite (Cole and Kroone 1959–1960). Is unlikely to be optically observable in thin section.

Magnesite
$MgCO_3$

Uniaxial –ve

Trigonal

$\omega = 1.700–1.726, \varepsilon = 1.509–1.527$

The crystal structure of magnesite is similar to calcite. There appears to a solid solution between magnesite and siderite ($FeCO_3$).
Interference figure: Uniaxial with many rings.
Relief: Very high, but changes on rotation like calcite.
Birefringence: Extreme, pearl grey and white of high order.
Form: Usually as aggregates of very small anhedral to subhedral crystallites (~1 μm).
Cleavage: Perfect rhombohedral similar to calcite.

Extinction: Symmetrical with respect to cleavages.

XRD: 8.479: 2.742, 2.102, 1.700, 2.503.

Occurrence: Most commonly as an alteration product of magnesium-rich igneous and metamorphic rocks such as serpentinite.

Comment: Occurs in carbonated magnesium limes and carbonation of brucite. It is difficult to distinguish from dolomite and calcite since it has similar optical properties, so best identified by XRD or SEM/EDX methods.

Dolomite

$Ca(Mg,Fe,Mn)(CO_3)_2$

Uniaxial –ve

Hexagonal

$\varepsilon = 1.500–1.526, \omega = 1.680–1.716$

Interference figure: Interference figure has many rings.

Relief: Changes markedly on rotation of stage. High positive relief parallel to the long diagonal of the rhomb.

Birefringence: Extreme and increases with increasing amounts of Fe and Mn.

Form: Fine to coarse grained. Well-formed rhombs are common and the crystals may be curved. Zonal structure is common.

Cleavage: Perfect rhombohedral cleavage which usually shows as two intersecting lines at oblique angles.

Twinning: Metamorphic dolomite usually shows polysynthetic twinning.

Extinction: Symmetrical to outlines of crystals and cleavages.

Colour: Cream to buff. Colourless to grey in thin section.

XRD: Dolomite: 34–426: 2.888–100, 2.193–19, 1.7870–13.

Magnesian calcite: 43–697: 3.004–100, 2.263–22, 1.8892–28, 1.8553–21.

Thermal analysis: Endotherms at 800°C and 925°C (Mackenzie 1970).

Occurrence: Chief constituent of dolomite rocks and a major constituent of dolomitic limestone; also occurs as replacement veins in limestones. Also a common mineral in some ore deposits.

Comment: Raw material used for dolomitic limes. Aggregates based on argillaceous dolomites interbedded with other types of dolomite and limestone may cause expansion in concrete due to the alkali–carbonate reaction (Dolar-Mantuani 1983, see also Section 6.9). The textures of some dolomites in thin section are given by Adams et al. (1998). It is sometimes difficult to differentiate from calcite in thin section but the staining procedures using Alizarin Red S can be used for this purpose (see Table 3.5 and Figure 3.18).

In addition to the publications already cited in this subsection, see Antao et al. (2004).

G.4.7 Silica minerals and mineraloids

Quartz

SiO_2

Uniaxial +ve

Trigonal

$\omega = 1.544, \varepsilon = 1.553$

Interference figure: Easily obtained if the crystal is large enough. May be pseudo-biaxial with a small 2 V angle if the crystal is strained.

Birefringence: In standard 30 μm thickness section, maximum birefringence gives an interference colour of white with a slight tint of yellow, but for concrete sections which are 25 μm or less in thickness, the colour is white to greyish white. Quartz is the main mineral used for estimating the thickness of a thin section under the microscope.

Form: Quartz occurs in a wide variety of forms which are well described in literature on optical mineralogy. There is an extensive literature relating to quartz, for example, Deer et al. (2001), Frondel (1962) and Winchell and Winchell (1961).

Twinning: Although twinning is common in quartz, it is rarely observed in thin section.

Orientation: Euhedral crystals are length slow.

Colour: Colourless

XRD: 33–1161: 4.257–22, 3.342–100, 1.8179–14, 1.5418–9.

Thermal analysis: Gives a sharp exothermic peak at 573°C due to the α to β inversion.

Occurrence: Quartz is a ubiquitous mineral found in many rocks as an essential, accessory, or secondary mineral. It is especially abundant in sandstones, arkoses, quartzites, granites, rhyolites and gneisses. It is one of the most common detrital minerals.

Comment: In loose detrital material, quartz can be confused with some of the feldspars. The forms of quartz of particular interest to the concrete petrographer are those exhibiting undulatory extinction and varieties of microcrystalline and disordered quartz, of which the latter are difficult to define either in thin section or by electron microscopy. The method for estimating the undulatory extinction quartz has been described by Smith et al. (1992). Measurement of undulatory extinction in quartz is no longer accepted as a method for predicting the reactivity of quartz-bearing rocks (Grattan-Bellew 1992). A method based on the estimation of total grain boundary areas of the quartz crystals appears to provide some correlation with reactivity in some metamorphic rocks (Wigum 1996).

In addition to the publications already cited in this subsection, see Broekmans (2004a,b), Flörke et al. (1991), Füchtbauer and Müller (1970) and Knauth (1994).

Chalcedony (very fine-grained, fibrous quartz)
SiO_2

Cryptocrystalline

Average n = 1.537 (1.530–1.539)

Birefringence: Weak, maximum colour is first-order grey.

Form: Fibrous, often spherulitic, may also show mosaic type structure.

Extinction: Usually parallel to length of fibres and rarely at 30°. Aggregate structures may fail to extinguish. Fibres usually length slow but may also be length fast.

Colour: Colourless in thin section.

XRD: Gives same pattern as quartz.

Occurrence: Present as sometimes the main constituent of flints, cherts, jaspers and as a cementing agent in some quartzites.

Comment: Formerly considered as a distinct mineral, but electron microscopy has shown chalcedony to consist of minute quartz crystals. Chalcedony is considered to be a potentially reactive material with cement alkalis, but work by Rayment (1992) has shown that other disordered forms of quartz are the reactive species in flints and cherts.

In addition to the publications already cited in this subsection, see Correns and Nagelschmidt (1933), Frondel (1978, 1982, 1985), Gíslason et al. (1993), Götze et al. (1998), Hattori et al. (1996), Heaney et al. (1994), Miehe et al. (1984) and White and Corwin (1961).

Moganite
SiO_2

Optical properties: Similar optical properties to chalcedony. Lamellar amethyst twinning at unit cell level.

XRD: Similar to quartz.

Occurrence: Originally identified as lutecite/lutecin from sandstones in the Paris Basin. Renamed after the Mogán Formation on Gran Canaria (International Mineralogical Association [IMA] 2007, Grice and Ferrari 2000). More widespread than originally believed owing to difficulties of identification. Iron plays an essential role in its formation.

Comment: Properties similar to opal. Higher solubility than quartz. It was only approved by the IMA as a separate mineral from quartz in 2007, but it's classification remains controversial.

In addition to the publications already cited in this subsection, see Bustillo (2001, 2002), Flörke et al. (1984), Gíslason et al. (1993, 1997), Godovikov and Ripinen (1994), Götze et al. (1998), Heaney (1995), Heaney and Post (1992), Miehe and Graetsch (1992), Michel-Lévy and Munier-Chalmas (1892), Moxon and Rios (2004), Murashov and Svishchev (1998) and Parthasarathy et al. (2001).

Chert/flint
$SiO_2 \cdot nH_2O$

Microcrystalline

$n = 1.53–1.54$

Chert is more accurately described as a chalcedonic rock type (Deer et al. 2004). Cretaceous chert (flint) typically contains about 1% water.

Birefringence: Weak, first-order whites and greys, sometimes masked by the body colour of the mineral.

Form: Typically shows a mosaic type structure of equant crystals but can be fibrous or and may contain fossil fragments.

Extinction: Usually parallel to length of fibres and rarely at 30°. Aggregate structures may fail to extinguish. Fibres usually length slow but may also be length fast.

Colour: Colourless to brown in thin section. Colour usually due to iron impurity. As aggregate particles colours may range from a dull translucent white, to greenish-grey, grey, brown or black.

XRD: Gives same pattern as quartz. Line broadening can provide an indication of crystallite size.

Occurrence: Commonly found as nodules or layers in limestones and chalks. Also is common as derived material in gravel aggregates and can form as a secondary vein infilling in other rock. As a chemically precipitated material, it may contain fossil fragments.

Comment: Considered to be a form of chalcedony and like chalcedony has been shown by electron microscopy to consist of small quartz crystals. Its mineral relative density of 2.57–2.64 is lower than quartz (2.65) indicative of microporosity. Like chalcedony, it is considered to be a potentially reactive mineral with cement alkalis, but work by Rayment (1992) has shown that other disordered forms of quartz are the reactive species within flints and cherts. It has also been suggested that cryptocrystalline quartz that may develop at the crystal grain boundaries within some flints and cherts provides the reactive component for its alkali reactivity in concrete.

Cristobalite
SiO_2

> Uniaxial –ve
>
> Pseudo-isometric (tetragonal)
>
> $\varepsilon = 1.484–1.487$

Interference figure: Crystals are usually too small to obtain a figure.

Birefringence: Variable. Cristobalite formed below 1470°C is practically isotropic. When formed at higher temperatures, both the synthetic and natural cristobalite have grey interference colours giving a characteristic mosaic effect.

Form: When formed below 1470°C, cristobalite is minutely crystalline and appears as an isotropic groundmass. Cristobalite exposed for long periods to higher temperatures exhibits a characteristic curved fracture, resembling overlapping roof tiles. Cristobalite crystallising from a glass may be dendritic, fernlike, needles or laths. In volcanic rocks, it is found in minute square crystals or aggregates in cavities. It also occurs intergrown with the feldspar fibres in spherulites.

(Figure: *see* Figure 6.122)

Twinning and extinction: Complex lamellar twinning is common with cristobalite formed at high temperature. Twinning is less common in natural cristobalite. Elongated crystals may show extinction at any angle.

Colour: Colourless in thin section.

XRD: 39–1425: 4.040–100, 3.136–8, 2.841–9, 2.487–13.

Thermal analysis: See Mackenzie (1970).

Occurrence: In silica and semi-silica bricks, cristobalite can occur around the periphery of quartz crystals and along cracks. Found in some high-temperature contact rocks and widely found as a late-forming metastable mineral in volcanic rocks. Often forms in cavities but is also present as small pools in the groundmass.

Comment: As little as 2% of cristobalite has been shown to be expansively reactive with cement alkalis (Kennerley and St John 1988) and its presence results in pessimum proportions in alkali–aggregate reaction (see Section 6.8). Thus, its detection is important when examining aggregates. It can be difficult to detect in basalts and requires the field diaphragm to be closed down to detect its negative relief and low refractive index. Chemical detection of the silica minerals such as cristobalite indicates that greater amounts are often present than detected in thin section (Katayama et al. 1989). Synthetic (calcined flint) cristobalite has been proposed by Lumley (1989) as a reference aggregate for use in the mortar-bar and other expansion tests for assessing ASR potential (see also Kühnel et al. 1987).

Tridymite
SiO_2

> Biaxial +ve
>
> Orthorhombic
>
> Synthetic $\alpha = 1.469$, $\beta = 1.470$, $\gamma = 1.473$.
>
> Natural $\alpha = 1.477$, $\beta = 1.478$, $\gamma = 1.481$. Indices may vary slightly.

Interference figure: Readily obtained unless crystals are very small which is usually the case in rocks.

Birefringence: Very weak. The maximum polarisation colour is grey.

Form: Wedged or arrowhead crystals are common but may be absent as tridymite may also crystallise as laths.

Twinning: Twinning is very characteristic in both wedge and lath forms.

Colour: Colourless in thin section.

Axial angle: Minimum 2 V = 35° with angles of up to 72° having been measured.

XRD: Natural orthorhombic 42–1401: 4.28–93, 4.08–100, 3.80–68, 3.242–48.

Thermal analysis: See Mackenzie (1970).

Occurrence: Tridymite is the principal constituent of silica brick with cristobalite as an associate. Natural tridymite occurs as a metastable mineral in a well-crystallised form in the cavities of acid and intermediate volcanic rocks and less frequently in basalt. In the groundmass, it is usually microscopic. Tridymite in general is much more abundant than cristobalite in volcanic rocks and in some cases can form up to 25% of the rock (Frondel 1962). May be associated with, or altered to, quartz or cristobalite.

Comment: Tridymite is usually much easier to detect in thin section than cristobalite because of its distinctive shape and twinning although greater amounts are usually present than are observed in thin section. Hornibrook et al. (1943) and Schuman and Hornibrook (1943) tested firebrick composed principally of tridymite and found it gave an expansion of 0.68% comparable to that found with opal. This means that like cristobalite and opaline materials, tridymite contributes to the effect of pessimum proportion and that only small percentages are required to cause expansion (see also Kühnel et al. 1987).

(Figure: *see* Figure 6.122c,f)

Opal

$SiO_2 \cdot nH_2O$

Isotropic

Water content commonly 4%–9% but can range up to more than 20%, influencing the refractive index (see Table G.3):

Birefringence: Usually nil but may show very weak birefringence due to strain.

Form: Often massive without any structure and can be present as a cementing material in sandstone. It is often found as rounded crusts, in veinlets and as a cavity filling or lining. It can replace wood and feldspars.

Colour: Colourless to pale grey or brown in thin section.

XRD: Gem quality opal 38–448: 4.08–100, 2.86–10, 2.51–60.

Elzea and Rice (1996) present XRD evidence for cristobalite and tridymite stacking sequences in opal.

Thermal analysis: Similar to cristobalite (Mackenzie 1970).

Occurrence: Occurs as the main mineral in opaline-bearing shales and rock. Opal is a secondary mineraloid in some volcanic rocks, where it is commonly associated with quartz, chalcedony and tridymite. It is the principal constituent of diatomite and siliceous sinter.

Comment: Opal occurs in several forms, truly amorphous opal shows an XRD 'typical glass bulge' around 4 Å (Sanders 1975). It can devitrify with age to a cristobalite–tridymite character (opal CT) (Graetsch 1994, Graetsch et al. 1994, Guthrie et al. 1995). Its detection

Table G.3 The relation between refractive index and water content in opals

Water content (%)	3.5	6.33	8.97	28.04
Refractive index (n)	1.459	1.453	1.447	1.409

in concrete aggregates is important as amounts as small as 1% can cause damaging expansion due to alkali–silica reaction (see Section 6.8, for detailed discussion on ASR) (see also Li et al. 1994 and Smith 1998).
(Figure: *see* Figure 6.122a,b)

Lechatelierite
SiO_2

Amorphous

$n = 1.458$

Rare fused silica glass formed at extreme high temperature for example in lightning strikes or extreme impact pressure such as meteor impact (see Kühnel et al. 1987).

Ignited silica hydrogel

Amorphous

Isotropic

$n = 1.48–1.485$

Formed from hydrated silica gel by ignition at high temperature.

Soluble silicates
Optical details of the anhydrous soluble silicates are given by Vail (1952). Details of some of the crystalline hydrated potassium and sodium silicates are given in Winchell and Winchell (1964). These soluble silicates are either orthorhombic or monoclinic with indices varying from 1.45 to 1.50 and birefringence from 0.007 to 0.009.

Alkali–silica gel
Sodium/potassium silicate gel: $n = 1.45–1.51$, mean ~1.48.
The composition of the gel is very variable and has been reported as
Na_2O = 1%–25%, K_2O = 0.5%–19%, CaO = 0.5%–23%, SiO_2 = 29%–85%, with ignition loss of 10%–25% (Idorn 1961a,b; Buck and Mather 1978). Older gels, gels remote from the site of reaction and crystallised gels may contain significant concentrations of Ca^{2+} ions within the gel/crystal structure.
Relief: Fairly low negative relief.
Birefringence: True gel is isotropic, but portions of gel often convert to birefringent material (crystalline gel) with a first-order white colour. May be carbonated near surfaces, when it will show higher-order interference colours more typical of carbonates.
Form: Gel often contains conchoidal drying cracks, especially where it lines cracks and pores. Has the appearance of a viscous material that has flowed and may be layered. Commonly present as tongues of gel extruding from the mouths of cracks in aggregates. Hardened cement paste saturated with gel at the margins of aggregates and cracks appear darkened under cross polars.
Colour: Can vary from colourless to yellow, brown and almost black. The colour does not appear to be related to crystallisation of the gel.
XRD: Indeterminate. Some data given by Buck and Mather (1978).
Occurrence: The product can be formed by the interaction of pore solution alkalis with siliceous or aluminosiliceous components in the aggregates in some concretes and mortars.

Sometimes observed where expansion and cracking of concrete has occurred. The various textures involving alkali–silica gel formed by ASR are described in Section 6.8.

Comment: Possible structural resemblance to kanemite ($NaHSi_2O_5.3H_2O$) (Benmore and Monteiro 2010). Alkali–silica gel is easy to recognise in cracks and pores, where it may sometimes be associated with ettringite. It is difficult to observe within the texture of aggregates. Where tongues of gel plug mouths of aggregate cracks, the tongue often sits on an area of finely crystalline birefringent material showing yellow and red colours of possibly the first order. This finely crystalline material, not to be confused with crystallised gel, may be an intermediate product and is claimed to be of the same composition as the adjacent alkali–silica gel (Andersen and Thaulow 1990).

In addition to the publications already cited in this subsection, see Almond et al. (1994), Annehed et al. (1982), Beneke and Lagaly (1977, 1983), Brindley (1969), Hou et al. (2004), Lagaly et al. (1975a,b), Marusin (1994), Moura and Prado (2009), Muraishi (1989), Peterson et al. (2006) and Sheppard et al. (1970).

(Figure: *see* Figures 6.117 and 6.118)

G.4.8 Magnesium minerals

Magnesium oxychloride hydrates (Sorel cement)

Formed by mixing calcined magnesite and calcium chloride solution. A range of hydrates are possible, with the most usual hydration products being $3Mg(OH)_2 \cdot MgCl_2 \cdot nH_2O$ and $5Mg(OH)_2 \cdot MgCl_2 \cdot nH_2O$ at ambient temperatures. The 5-form is metastable and converts to the 3-form. Also, these hydrates are unstable in air and slowly convert to the magnesium chlorocarbonate, $Mg(OH)_2 \cdot MgCl_2 \cdot 2MgCO_3 \cdot 6H_2O$. Formation of the calcium oxychloride, $3Ca(OH)_2 \cdot CaCl_2 \cdot 11H_2O$ also occurs in this system. At higher temperatures, more crystalline 9-form and 2-form hydrates occur. Data given are from Demediuk et al. (1955), Cole and Demediuk (1955) and Newman (1955) (see also Section 9.6).

Optical data: The 3-form occurs as a gel-like aggregate of possibly monoclinic, minute crystals which are irregular in shape.

The mean refractive index lies between 1.505 and 1.510.

The 5-form tends to be more crystalline forming well-shaped needles about 10 µm long. The mean refractive index is approximately 1.525.

Birefringence: Low for both forms.

Extinction: The needles of the 5-form have parallel extinction and are length fast. The calcium oxychloride is possibly the $Ca_4O_3Cl_2 \cdot 15H_2O$ listed by Winchell and Winchell (1964).

The $Ca_4O_3Cl_2 \cdot 15H_2O$ forms colourless, orthorhombic, acicular crystals with basal cleavage.

$$\alpha = 1.481,\ \beta = 1.536,\ \gamma = 1.543,\ +2\ V = 44°.$$

XRD: 3-form: 8.03-s, 3.83-s, 2.431-ms, 1.985-ms.

5-form: 7.57-s, 4.14-ms, 2.64-m, 2.420-s.

From Cole and Demediuk (1955). Data variable with hydration. JPCDS data difficult to correlate.

$Ca_4O_3Cl_2 \cdot 15H_2O$: 2–280: 4.08–100, 2.75–80, 2.61–80, 2.50–70.

Thermal analysis: See Cole and Demediuk (1955).

G.4.9 Glassy materials

In addition to the publications cited in this section, see Barboux et al. (2004), Bunker (1994), Fiore et al. (1999), Hunt et al. (1998), Jezek and Noble (1978) and Paterson (1982).

Table G.4 Composition, refractive index and relative density of some glasses

	SiO_2	Al_2O_3	B_2O_3	Li_2O	Na_2O	MgO	CaO	ZrO_2
Low-expansion borosilicate	81	2	13	—	4	—	—	—
Soda lime	72	1	—	—	15	3	9	—
E-glass	54	14	—	—	—	4.5	17.5	—
	Refractive index				Relative density			
Low-expansion borosilicate	1.47				2.23			
Soda lime	1.51–1.52				2.46–2.49			
E-glass	1.55				2.60			
Alkali resistant	1.54				2.62			

Source: After Winchell 1964.

Synthetic glasses

Of the synthetic glasses, only a few are of interest to the petrographer. The compositions given in Table G.4 are indicative of the compositional ranges used (Shand 1958; Boyd et al. 1994) of the glasses of interest. The relationship of optical properties to composition and density are summarised by Winchell and Winchell (1964).

Remarks: Crushed low-expansion borosilicate glass, commonly referred to as 'Pyrex' glass, can give a large expansion when tested as an aggregate using the ASTM C227-10 (2010) mortar-bar method and thus can be used therein as an upper limit. However, its reactivity in the mortar-bar test is variable and dependent on its thermal history. Like all glasses, crushed Pyrex exhibits conchoidal fracture. Coloured soda lime glass is occasionally used as an exposed aggregate in concrete or as glass tiles and can result in AAR (Figg 1981).

The common type of glass fibres used in building products is based on E-glass. Because of the poor alkali resistance of E-glass in concrete, an alkali-resistant glass containing zirconia was developed by Pilkingtons under the trade name of 'Cem–Fil'. To distinguish between this and E-glass, analysis is required as they cannot be differentiated in thin section. The glass fibres used for building products have filaments each of about 7 μm in diameter, which are combined together to form a strand. The strands are treated with sizes or coupling agents which are tailored to the end use of the fibreglass. The specific compositions of these agents are usually confidential to the manufacturer. Where a thick coating of sizing has been used, this may be visible in thin section especially where degradation of the fibre occurs. See also Sections 4.8.2 and 7.7.2, for discussions on glass fibres used in concrete and other cementitious products.
(Figure: *see* Figure 7.13)

Natural volcanic glasses

The common textural names used to describe volcanic glasses or rocks containing large amounts of glass are obsidian, perlite, pitchstone, pumice and scoria. Volcanic glasses can vary from acid to basic in composition and these glass textural terms need to be qualified; for example, rhyolitic obsidian and basaltic scoria.

Obsidian is usually a dark-coloured glassy rock, often banded, with conchoidal fracture. The term is now restricted to glasses of low water content and excludes those of higher water content such as perlite and pitchstone. Obsidians can range from acid to intermediate in composition, with rhyolitic obsidian being the most common. In thin section, the glass ranges from clear to brown in colour and usually contains crystallites. However, it is not uncommon to find obsidian fragments in concrete aggregates, which do not contain any crystallites and are wholly glassy.

Pumice is a light-coloured vesicular rock derived from frothed volcanic glass, typically of rhyolitic composition. Pumices of other compositions are not common. In thin section, its texture consists of a mass of vesicles separated by thin walls of clear or light-coloured glass. Scoria is a dark vesicular rock derived from basalt or andesite and is noticeably heavier than pumice. In thin section, the glass is always dark in colour and may contain considerable quantities of mineral dust and crystallites.

Perlite is typically the result of the hydration of obsidian. On heating, the water is driven off giving rise to highly vesicular expanded material sometimes used as lightweight aggregate.

Compositions: The chemical compositions of rocks comprising volcanic glass will be generally similar to the bulk composition of the equivalent crystalline rock. However, George (1924) states that natural glasses differ in composition from the related crystalline rocks in that they tend to be more acid and contain more alkali, with potash dominating over soda.

Refractive index and glass composition: A good correlation between the silica content of natural glass and refractive index has been claimed (George 1924) (see Table G.5). However, Tilley (1922) notes that this relationship is affected by the water content of the glass, as it increases the refractive index. Some comparative curves for silica content versus refractive index are given by Williams et al. (1954). An average extrapolation from the curves for glass made by whole-rock fusion gives the refractive indices indicated in Table G.5.

The information shown in Table G.6 is taken from Grout (1932) as it gives some idea of the range of indices to be expected from rocks. However, it is not clear how many of these data are applicable to glass groundmass, as opposed to essentially whole-rock glass.

Table G.5 Correlation of silica content and refractive index in some fused rock glasses

Silica content (%)	Refractive index
75	1.49
70	1.505
65	1.52
60	1.54
55	1.565
50	1.595
47	1.62

Source: After Williams, H. et al., *Petrography. An Introduction to the Study of Rocks in Thin Section.* W.H. Freeman and Co., San Francisco, CA, p. 406, 1954.

Table G.6 Refractive indices of some natural glasses

Name of rock	Average refractive index	Range recorded
Obsidian or rhyolitic glass	1.492	1.48–1.51
Pitchstone	1.500	1.492–1.506
Perlite	1.497	1.488–1.506
Pumice	1.497	1.488–1.506
Dacite	1.511	1.504–1.529
Trachyte	1.512	1.488–1.527
Andesite	1.512	1.489–1.529
Leucite tephrite	1.550	1.525–1.580
Tachylite, scoria, basalt glass including palagonite	1.575	1.506–1.562

Source: After Grout, F.F., *Petrography and Petrology.* McGraw-Hill Book Co., New York, 1932.

Glassy groundmass in volcanic rocks

The composition of the glassy groundmass of a volcanic rock is differentiated from the parent rock. It is the portion that remains after other components of the rock have crystallised. Thus, an important distinction must be made between natural whole-rock glasses, such as obsidian, and the glass which occurs in the groundmass of fresh volcanic rock. Glasses such as obsidian will approximate to the bulk composition of the rock type. However, the glass in the groundmass of a typical andesite, for example, may contain between 60% and 70% silica. This increase in silica content occurs because of differentiation, as the glass is the residual relatively acidic material remaining after the more basic minerals have crystallised.

Glass in granulated blastfurnace slag

$n = 1.57–1.66$ dependent on composition.

The refractive indices of a range of slag glasses have been investigated by Battigan (1986) who found these results. There appears to be a good relationship between refractive index and the calcium/silica ratio. Addition of alumina and magnesia into the ratio does not significantly affect the correlation, indicating that the combined effect of these ions appears to be neutral.

The range in composition of blast furnace slags is given by Lea (1970) and Taylor (1990). Glass contents vary dependent on method of manufacture. Granulated slags may contain in excess of 95% glass, while in pelletised slags it may be as low as 50%. Methods for estimating glass content of slag by transmitted light microscopy have been described by Drissen (1995) and Hooton and Emery (1983). Alternatively, glass content may be estimated by quantitative XRD techniques using internal standards and Rietveld analysis (Howard et al. 1988), but both microscopy and XRD methods will be unreliable if microcrystallinity is present (Taylor 1990). Ground-granulated blast furnace slag, which is frequently used in modern concrete as a cementitious addition and/or cement replacement, is addressed in Section 10 of this Glossary and also in Sections 4.2.4 and 4.6.2, and Section 5.3.3.

(Figures: *see* Figures 4.17, 4.18 and 4.19)

G.4.10 Mineral additions and their constituents

Mineral additions, or supplementary cementitious materials (SCMs) as they are termed in North America, are increasingly encountered in concrete petrography. The main types of these additions or SCMs are described in detail in Chapter 4 (Sections 4.2.4 and 4.6), with the essential features of selected examples also being included in this section of the Glossary, together with some of the mineral phases that are commonly found forming or associated with these materials.

Silica fume (microsilica, silicon dioxide)

SiO_2

Vitreous glass

Form: Spherical particles of reactive, amorphous silica which range from 20 nm to 1 μm in size.

Colour: Transparent to pale brown, individual particles are below the resolution of an optical microscope.

Occurrence: Silica fume is produced as a by-product of the silicon and ferrosilicon industries (Aitcin 1983) by reduction of high-purity quartz with coke in electric arc furnaces. It is collected as a loose powder with a bulk density of about 130 kg/m^3. It is 'densified' into a more manageable form by blowing air through the powder.

Comment: Silica fume is available for use as a pozzolana in concrete in both powdered form and as a 50% stabilised slurry. The densification of silica fume is caused by agglomeration of particles. Once agglomerated, it cannot be re-dispersed. In theory, the mean particle size of about 0.1 μm is about 100 times finer than any other pozzolana in use (American Concrete Institute 1987) and of cement grains. Although individual particles are below the resolution of the optical microscope, incomplete dispersion leads to agglomerations in concrete which are visible in thin section. Up to 10% of other material may be present mainly as coarser material which Bonen and Diamond (1992) report consists mainly of agglomerates of silica fume, quartz, cristobalite, silicon carbide, metallic silicon and carbon particles (see also Section 4.6.3). (Figure: *see* Figure 5.22)

Ground granulated blastfurnace slag

A complex material of calcium and aluminosilicates, partly crystalline and partly glassy (see Section G.9 in this Glossary for information on the glass in ground-granulated blast furnace slag [ggbs]), typically containing variable proportions of crystalline melilite (q.v.) and a calcium–aluminium–magnesium silicate vitreous glass which may represent 70%–90% of its volume. A typical chemical composition is approximately SiO_2, 35%; CaO, 40%; Al_2O_3, 13%; MgO, 8%.

Air-cooled slags from the iron making process are also discussed in Section 4.6.2, and are more crystalline than ggbs (Table G.7).

n = 1.57–1.66 dependent on composition.

Essentially isotropic

Relief: Relatively high

Form: The individual ground particles are typically angular shards ranging in size from 5 μm up to about 250 μm.

Colour: Off-white. Colourless to pale yellow–brown in thin section.

Occurrence: Blast furnace slag is developed as a by-product of industrial iron making, when it is granulated by quenching the molten slag in water and then ground to a powder for use in concrete.

Comment: Widely used as a cement replacement in the range 25%–70%, where it acts as a pozzolanic material. Concrete made using ggbs typically has a slower strength gain but has reduced heat of hydration and increased resistance to chloride and sulphate attack and its use reduces the risk of ASR. Some ggbs materials are 'activated' by treatment with alkali hydroxides. In thin section, ggbs is readily identified in a concrete or mortar because of its angular shard-like morphology, its transparency and mainly isotropic character. Minor crystalline components, especially melilite or merwinite depending upon the slag composition, usually appear as inclusions in the glassy phase. (Figures: *See* 5.26 and 5.27)

Table G.7 Mineralogical composition of an air-cooled blastfurnace slag

Main components	Melilite: Solid solution of gehlenite and åkermanite	$2CaO \cdot Al_2O_3 \cdot SiO_2$ and $2CaO \cdot MgO \cdot 2SiO_2$
	Merwinite (in basic slags)	$3CaO \cdot MgO \cdot 2SiO_2$
	Diopside and other pyroxenes (in acid slags)	$CaO \cdot MgO \cdot 2SiO_2$
Minor components	Dicalcium silicate, $\alpha, \alpha', \beta, \gamma$ (in basic slags)	$2CaO \cdot SiO_2$
	Monticellite	$CaO \cdot MgO \cdot SiO_2$
	Rankinite	$3CaO \cdot 2SiO_2$
	Pseudowollastonite	$CaO \cdot SiO_2$
	Oldhamite	CaS

Source: After Smolczyk, H.G., Slag structure and identification of slags, in *Seventh International Congress on the Chemistry of Cement*, Paris, France, vol. I, Principal reports, Sub-theme III, I/3 to I/17, 1980.

Melilite (åkermanite–gehlenite solid solution)
$2CaO \cdot Al_2O_3 \cdot SiO_2$ (gehlenite)

Uniaxial –ve

Tetragonal

$\omega = 1.669$, $\varepsilon = 1.658$ ($n = 1.638$ glass)

$2CaO \cdot MgO \cdot 2SiO_2$ (åkermanite)

Uniaxial +ve

Tetragonal

$\omega = 1.632$, $\varepsilon = 1.641$ ($n = 1.641$ glass)

Birefringence: Åkermanite is first-order grey, which for åkermanite/gehlenite 60/40 is almost isotropic and then rises to first-order pale yellow for gehlenite. Melilite in slags often has a distinctive anomalous blue interference colour as shown in Figure 5.26a.

Form: Short prisms or tablets often almost square in outline. In slags, distinct zonation is often present.

Extinction: Parallel to edges of crystals.

Colour: Colourless.

XRD: Åkermanite: 35–592: 3.087–25, 2.872–100, 2.478–15, 1.761–20

Gehlenite: 35–755: 3.0629–21, 2.845–100, 2.431–17, 1.754–25

Occurrence: Found in some contact zones, ores, and basic volcanic rocks. Formed by the action of lime on fire bricks and is a constituent of many blast furnace slags (see also Leslie and Hughes 2003).

Comment: The refractive indices given earlier are for the synthetic minerals and the natural minerals will vary because they contain other ions. The indices given for the synthetic glasses should not be used for the glasses found in blast furnace slags (see Section G.4.9 in this Glossary). (Figure: *see* Figure 5.26)

Merwinite
$3CaO \cdot MgO \cdot 2SiO_2$

Biaxial +ve

Monoclinic

$\alpha = 1.708$, $\beta = 1.711$, $\gamma = 1.724$

Interference figure: Biaxial positive with 2 V = approx. 70°.

Relief: Rather high.

Birefringence: Polarises in greys, whites and yellows. The maximum colour being first-order red.

Form: Crystals are granular or prismatic, tending to lozenge-shaped outline.

Cleavage: Poor.

Twinning: Polysynthetic twinning is common.

Extinction: Generally inclined, the maximum extinction angle is 36°.

Colour: Colourless in thin section.

XRD: 35-591: 2.756–26, 2.687–100, 2.671–65, 2.653–47.

Occurrence: Found in limestone contact zones. Occurs in some blastfurnace slags.

Comment: Can be distinguished by its weak birefringence, polysynthetic twinning and biaxial character. Merwinite is etched by nitric acid and stained by hydrofluoric acid.

Pulverised-fuel ash (pfa)
Silica (45%–50%) Alumina (25%–30%).

Alumosilicate glass + subordinate crystalline phases

'Class C' n = 1.56–1.70 dependent on composition.

'Class F' n = 1.53–1.64 dependent on composition.

Essentially isotropic

Relief: Relatively high.

The crystalline components of UK pfa materials contain variable amounts of mullite ($Al_6Si_2O_{13}$), haematite, quartz and magnetite, with about 5%–10% unburnt coal.

Form: Principally solid spherical glassy particles of which range from 1 to 150 μm in size (about 50% smaller than 20–30 μm). Also shards, irregular particles and hollow spheres (cenospheres).

Residual carbon particles and clusters, or cenospheres filled with, still non-reacted spheres and other less crystalline non-spherical 'contaminants' (quartz, calcite, lime, calcium sulphides, etc) may also be present in minor amounts.

Colour: Colourless, yellow, red, grey, brown and black depending on minor constituents present.

SEM: Broken surfaces allow confirmation of the presence of pfa, as betrayed by either unreacted spherical particles or relicts of mullite crystals and enable the original cement and pfa proportions to be estimated.

Occurrence: Pulverised bituminous coal is burnt at some electricity power stations and produces 75%–80% of a fine 'pulverised fuel ash' (pfa) dust by-product. Note that the term 'fly ash' has a wider generic application sometimes being used for the product of burning other materials.

Comment: Two main types are recognised (ASTM C618-12a, 2012):

'Class C' types are produced from sub-bituminous coal (or even lignite) and are relatively high in calcium content (>10%).

'Class F' types are produced from bituminous coals and are lower in calcium content (<10%). UK power stations produce only 'Class F' pfa. Substantial volumes of 'Class C' ash are produced in the United States and elsewhere and used as additions to concrete. 'Class C' ashes tend to be far more reactive and often leave little residual material when well activated. Further information on fly ashes is given in Section 4.2.4.

(Figures: *see* Figures 4.21, 4.22 and 5.25)

Natural pozzolanas

A very wide range of natural pozzolanic materials have been used with cements and concretes; they vary in composition, crystallinity, mineralogy and optical properties and are discussed in Section 4.6.4.

Mullite
$3Al_2O_3 \cdot 2SiO_2$

Biaxial +ve

Orthorhombic

α = 1.642, β = 1.644, γ = 1.654

Refractive indices increase with iron and titanium content.

Interference figure: +2 V = 45–50°.

Relief: Moderately high.

Birefringence: Maximum interference colour is yellow. Usually white to grey.

Form: Most characteristic form is needles or laths.

Extinction: Parallel extinction. Needles are length slow.

Colour: Colourless in thin section.

XRD: 15–776: 5.39–50, 3.428–95, 3.390–100, 2.206–60.

Occurrence: Rare natural mineral found in some high-temperature contact rocks. Mullite is a constituent of some porcelains. Crystallises out when slag reacts with fired clay materials. May be seen as fine needles crystallising out in the glass spheres in fly ash.

Comment: In some examples, mullite needles may be too small to be observed under the optical microscope and are best identified by XRD or SEM/EDS. Crystal structure is considered by Angel and Prewitt (1986).

Vermiculite
$(MgFe,Al)_3(Al,Si)_4O_{10}(OH)_2 \cdot 4H_2O$

Biaxial –ve

Monoclinic

$\alpha = 1.525–1.561$, $\beta = 1.545–1.581$, $\gamma = 1.545–1.581$

Interference figure: Small –ve 2 V, varies 0°–8°

Relief: Very low positive

Birefringence: High first order; variable depending on substitution.

Form: Small anhedral aggregates; scales rarely pseudohexagonal plates.

Cleavage: Perfect (001)

Colour: Colourless, white, pale yellow, green or brown.

XRD: 14.43, 2.392, 4.604, 2.866 (see Brindley and Brown 1980).

Occurrence: Weathering alteration product of biotite, phlogopite and mafic rocks.

Comment: When heated, exfoliates into a characteristic worm-like texture. It has excellent thermal, acoustic and insulation properties. The industrially produced expanded vermiculite is sometimes used as a lightweight concrete aggregate or in applications where its insulation or acoustic properties are required.

G.4.11 Pigments and colorant materials

Red (haematite, red ochres, burnt sienna, burnt umber)
$\alpha-Fe_2O_3$

Uniaxial –ve

Hexagonal

$\omega = 3.22–2.90$, $\varepsilon = 2.94–2.69$

Relief: Extremely high.

Form: Basal plates, rhombohedrons, platy, fibrous or massive.

Colour: Opaque unless crystals are very thin. Thin scales are coloured blood red, red–brown, orange or yellow. May be slightly pleochroic. Steel grey with metallic lustre in reflected light with a tendency to a marginal red.

XRD: 33-664: 2.74–100, 2.519–70, 1.841–40, 1.694–45.

Occurrence: A very important iron ore. Rather common as microscopic inclusions and alteration products in many minerals and rocks, colouring them red.

Comment: Haematite (also hematite) is the principal coloured mineral in Spanish and Persian and other red ochres and in burnt sienna and umber. Many of these are impure mixtures, especially some of the siennas and umbers, which may contain clays and manganese oxide. For industrial use and in concrete, red ochre is made synthetically where the quality is uniform and the colour shade is controlled by particle shape and size (Schiek 1982). Both synthetic and natural red iron oxides are classified in the International Colour Index as *CI pigment red 101*.

Haematite crystal structure is considered by Pauling and Hendricks (1925).

Black (magnetite)

Fe_3O_4, $(Fe,Mg,Zn,Ni,Mn)Fe_2O_4$

Isometric

Cubic

$n = 2.42$ varies with substitution

Many natural magnetites are substituted.

Relief: Very high.

Form: Octahedral or granular. Rarely cubes.

Colour: Opaque. Appears black in thin section. Grey metallic lustre in reflected light.

XRD: 19–629: 2.967–30, 2.532–100, 1.616–30, 1.485–40.

Occurrence: Widespread in igneous and metamorphic rocks and important in many black iron sands.

Comment: Is also synthesised for use as a pigment. Black colour may be tinged with red unless some manganese is present. Both natural and synthetic iron oxides are classified in the Colour Index as: *CI pigment black 11*. Present in slag and fly ash.

Carbon black

C

Amorphous

Opaque

Carbon black produced from oil feedstock varies in particle size from <0.5 to 4 μm. Particles less than 0.5 μm are rounded, but other particles are sharply angular or sliver-like without apparent agglomerates (McCrone and Delly 1973).

Comment: Carbon black is a superior pigment to iron black and is classified in the International Colour Index as *CI pigment black 6, 7 and 8*.

Yellow (goethite, lepidocrocite, yellow ochre, raw sienna)

FeO(OH)

Uniaxial –ve

Orthorhombic

α-FeO(OH) (goethite)

$\alpha = 2.275\text{--}2.15$, $\beta = 2.409\text{--}2.22$, $\gamma = 2.415\text{--}2.23$

γ-FeO(OH) (lepidocrocite)

$\alpha = 1.94$, $\beta = 2.20$, $\gamma = 2.51$

Interference figure: Goethite $-2\ V = 23°$ with strong dispersion.

Lepidocrocite $-2\ V = 83°$.

Relief: Very high.
Birefringence: Goethite crystals are length slow.
Form: Goethite, prismatic or striated or tabular or fibrous. Lepidocrocite scaly.
Extinction: Parallel.
Colour: Goethite has strong variable pleochroism from yellow to orange yellow. Lepidocrocite is also pleochroic ranging from colourless to yellow through to red orange.
XRD: Goethite: 29–713: 4.183–100, 2.693–35, 2.450–50, 1.792–20.
Lepidocrocite: 8–98: 6.26–100, 3.29–90, 2.47–80, 1.937–70.
For further x-ray data, see also Brindley and Brown (1980).
Thermal analysis: See Mackenzie (1970).
Occurrence: Goethite is a common mineral, present in limonite iron ore and is the commonest weathering product of siderite, pyrite, magnetite, glauconite and iron silicates. Lepidocrocite is often found associated with goethite. Goethite is the principal constituent of yellow ochre and raw sienna.
Comment: Limonite is the general term used for undifferentiated iron weathering products and usually consists of goethite and/or lepidocrocite. However, optically, it usually appears isotropic with a refractive index of 2.0–2.1. The yellow colour of ochres and synthetic iron oxide is caused by the presence of one of these hydrated iron oxides. The natural siennas are classified in the Colour Index as *CI pigment brown 7* and ferrite yellow as *CI pigment yellow 42*. Further information on goethite may be found in Balistrieri and Murray (1982), Das and Hendry (2011), Kühnel et al. (1975), Schwertmann and Fechter (1994) and Trolard and Tardy (1987).

Green (eskolaite, green cinnabar, phthalocyanine green)
Cr_2O_3 (eskolaite, green cinnabar)

Uniaxial –ve

Hexagonal

n = approximately 2.5

Relief: Very high.
Birefringence: Thin plates are anisotropic but polarisation colours are masked by natural green colour.
Form: Equant, tabular or scales.
Cleavage: Parallel to the faces of the rhomb.
Colour: Opaque in thick sections. Thin sections exhibit a bright emerald green colour.
XRD: Eskolaite: 38–1479: 2.665–100, 2.480–93, 1.6274–87, 1.4316–39.

Occurrence: Natural mineral known as green cinnabar. Occurs in refractory mixes containing chromic oxide.

Comment: Synthetic chromic oxide is the principal green colorant used with cement.

$C_{31}H_xCu.15\text{-}16Cl$ (phthalocyanine green)

n = about 1.40

Form: Laths (Gettens and Stout 1966).

XRD: 36–1870: 14.8–83, 13.2–57, 5.21–40, 3.34–100.

Comment: A chlorinated derivative of phthalocyanine blue classified in the Colour Index as *CI pigment green 7*. Variations also listed as *CI pigment green 42* and *43* are now listed as *CI pigment green 7* and *pigment green 36*. *CI pigment green 36* is a polybromochloro derivative of different shade. The systematic name is polychloro-tetrabenztetraazaporphyrin.

Blue (cobalt blue, phthalocyanine blue)

$Co \cdot Al_2O_3$ (cobalt blue)

Isometric

Cubic

$n > 1.78$ (red) $n = 1.74$ (blue)

XRD: 38–816: 2.439–100, 2.021–21, 1.557–36, 1.429–42.

Comment: This is Colour Index classification: *CI pigment blue 28*.

$C_{32}H_{16}Cu$ (copper phthalocyanine blue)

Biaxial

Monoclinic

n = approximately 1.38

Form: Laths (Gettens and Stout 1966).

XRD: 22–1686: 13.2–100, 21.1–100, 8.93–40, 5.71–40.

Comment: This is classified as *CI pigment blue 15* and forms α and β modifications which are slightly different in shade and stability. There are several variations of this pigment classified as 15.1, 15.2 and 15.3 which are phthalocyanine substituted with other ions such as nickel and cobalt. The systematic name is copper tetrabenztetraazaporphyrin.

White (rutile, anatase)

α-TiO_2 (rutile)

Uniaxial +ve

Tetragonal

$\omega = 2.609$–2.616, $\varepsilon = 2.895$–2.903

β-TiO_2 (anatase)

Uniaxial −ve

Tetragonal

$\omega = 2.561$–2.562, $\varepsilon = 2.488$–2.489

Interference figure: Both rutile and anatase may give anomalous biaxial figures with a small 2 V.

Relief: Highest relief of any rock-forming mineral.

Birefringence: Extreme birefringence, which does not show well on account of total reflection.

Form: Small prismatic to acicular crystals.

Cleavage: Parallel to length of crystals.

Twinning: Common.

Extinction: Parallel.

Colour: Yellowish to reddish brown in thin section. In reflected light shows adamantine lustre.

XRD: Rutile: 21–1276: 3.247–100, 2.487–50, 2.188–25, 1.687–60.

Anatase: 21–1272: 3.52–100, 2.378–20, 1.892–55, 1.700–20 (see Spurr and Myers 1957).

Occurrence: Widely distributed accessory minerals in metamorphic rocks and occur as detrital minerals. Rutile is much more widespread than anatase.

Comment: The synthetic pigment grades of rutile and anatase appear as dark agglomerates of very fine particles, which under crossed polars appear white with some reddish-brown coloration in the larger agglomerates. Individual fine particles exhibit Brownian motion and may show some visible birefringence (see Felius 1976). International Colour Index, *CI Pigment white 6*.

G.4.12 Asbestos and other fibrous minerals

Asbestiform minerals

A group of hydrous silicates primarily composed of fibrils ranging from 20 to 200 nm in thickness, which form durable, strong fibres that can be either woven or used as reinforcement in materials such as hardened cement and polymeric resins. The optical mineralogy of the asbestiform minerals is well covered in the literature (Ledoux 1979; Bergin 1989; Veblen and Wylie 1993; Hawthorne and Oberti 2006), so only a summary of optical data has been given. Owing to the serious health hazards posed by liberated asbestiform microfibres, asbestos is now banned from use in most countries. Initially, blue asbestos (crocidolite) was identified as being the most hazardous, but today most forms of asbestos are regarded as being too potentially hazardous to be used. However, petrographers will encounter asbestos in some historic concretes and especially some cementitious (asbestos-cement) products (see Section 7.7.1), when appropriate caution should be exercised if the materials are broken or degrading and also during their preparation. 'Chrysotile' (colourless or 'white' asbestos) is a fibrous form of the antigorite variety of serpentine and was probably the most common asbestos used in cementitious products. The other asbestiform minerals are fibrous varieties of various amphibole minerals. 'Crocidolite' (blue asbestos) is a fibrous form of the riebeckite variety of amphibole. 'Amosite' (brown asbestos) is a commercial name for a fibrous form of the grunerite variety of amphibole. The composition of 'tremolite' (another colourless or pale green asbestos) variety of amphibole is $Ca_2Mg_5Si_8O_{22}(OH)_2$, which forms a continuous solid solution with actinolite, which is the iron-rich member. Anthophyllite is another fibrous amphibole but was only rarely used as asbestos.

In summary, commercially, the important asbestos varieties are as follows:

Chrysotile: $Mg_3[Si_2O_5](OH)_4$ ('white', fibrous antigorite serpentine)

Crocidolite: $(Na \cdot Ca)_2(Fe \cdot Mn)_3Fe_2(Si \cdot Al)_8O_{22}(OH,F)_2$ ('blue' fibrous riebeckite amphibole)

Amosite: $(Fe^{2+},Mg)_7[Si_8O_{22}](OH)_2$ ('brown', fibrous grunerite amphibole)

Actinolite: $Ca_2(Mg,Fe)_5Si_8O_{22}(OH)_2$ (end member of the tremolite–actinolite amphibole series)

Anthophyllite: $(Mg,Fe^{+2})_7[Si_8O_{22}](OH,F)_2$ (end member of the anthophyllite–gedrite amphibole series)

Table G.8 Optical properties of asbestos minerals

	Chrysotile (antigorite)	Riebeckite (crocidolite)	Grunerite (amosite)	Anthophyllite (gedrite)	Tremolite/ actinolite
Composition	$Mg_3[Si_2O_5]$ $(OH)_4$	$(NaCa)_2(FeMn)_3$ $Fe_2(SiAl)_8O_{22}(OH_3F)_2$	$(Fe^{2+},Mg)_7$ $[Si_8O_{22}]$ $(OH)_2$	$(Mg,Fe^{+2})_5Al_2$ $[Si_6Al_2O_{22}]$ $(OH,F)_2$	$Ca_2(Mg,Fe)_8$ $Si_8O_{22}(OH)_2$
	Monoclinic	Monoclinic	Monoclinic	Orthorhombic	Monoclinic
α	1.493–1.546	1.693	1.657–1.667	1.598–1.652	1.600–1.628
β	1.504–1.550	1.695	1.684–1.697	1.615–1.662	1.613–1.644
γ	1.517–1.551	1.697	1.699–1.717	1.623–1.676	1.625–1.655
$\gamma - \alpha$	0.011–0.014	0.004	0.042–0.054	0.016–0.025	0.22–0.27
2V	0°–50° (+)	5° (+)	10°–14 ° (−)	70°–90° (+)	79°–85° (−)
Extinction	Parallel	5°	10°–15°	Parallel	10°–20°
Orientation	Length slow	Length fast	Length slow	Length slow	Length slow
Colour	Colourless	Blue	Colourless	Colourless or pale colour	Colourless to pale green
Pleochroism	None	Dark blue to grey blue, strong	Pale brown to yellow, slight	Slightly pleochroic	Trem., none Act., green
Mineral group	Serpentine	Clinoamphibole	Clinoamphibole	Orthoamphibole	Clinoamphibole

Table G.9 X-ray diffraction data for the asbestos minerals

Antigorite-1M	21–963	7.29–100, 3.61–80, 2.525–100, 2.458–60
Riebeckite	19–061	8.40–100, 3.12–55, 2.801–18, 2.726–40
Grunerite	31–631	8.33–100, 3.07–80, 2.766–90, 2.639–70
Anthophyllite	42–544	9.37–85, 4.59–80, 4.50–100, 3.23–70
Tremolite	13–437	8.38–100, 3.268–75, 3.121–100, 2.705–90
Actinolite	41–1366	8.42–75, 3.276–45, 3.117–100, 2.709–55

Provided that the fibres or fibre bundles are of the order of 1 μm or greater in thickness, small numbers of fibres can be detected by optical methods. However, the optical properties of chrysotile are difficult to determine because of the fineness of its fibres. When optical studies are made on fibre bundles, which is often the case for fibre-reinforced products, the optical properties are the integrated effects of the individual fibrils. This has the effect that the asbestiform minerals mostly show straight extinction in these cases (Zussman 1979). Special techniques are used for identifying asbestos types, for example, see McCrone (1987) and Millet (2006).

The optical properties of the common asbestos minerals are listed in Table G.8, and the X-ray diffraction data for these minerals are given in Table G.9.

XRD data also given by Zussman (1979), who notes that in powder diffraction of asbestos fibres, only the stronger reflections may be detected.

Thermal analysis: See Mackenzie (1970).

Infrared: The infrared spectra of the asbestiform minerals provide an unambiguous means of identification on the basis of characteristic absorption bands in the 200–1600 cm^{-1} frequency range (see also Farmer 1974 and Section 2.5.4).

(Figures: *see* 7.11, 7.12)

Synthetic and organic fibres

Many fibres are illustrated and discussed in volume 2 of McCrone and Delly (1973) and the Textile Institute (1985). Some optical data are provided in Table G.10. All the fibres have parallel extinction. The data for cellulose are relevant to flax, jute, kraft and sisal fibres.

Table G.10 Some optical properties of synthetic and natural fibres

	$n\parallel$	$n\perp$	Birefringence	Length
Acrylic	1.511	1.515	0.004	Fast
Aramid (Kevlar)	>2.00			
Carbon	Opaque black			
Nylon	1.58	1.52	0.06	Slow
Polypropylene	1.530	1.496	0.034	Slow
Cellulose	1.60	1.53	0.07	Slow

Source: After data from Textile Institute, *Identification of Textile Materials*, 7th ed., The Textile Institute, Manchester, U.K, 1985.

$\parallel$, parallel to fibre length; $\perp$, perpendicular to fibre length.

G.4.13 Grinding and polishing compounds

Diamond
C

Isometric

Cubic

$n = 2.414$

Relief: High positive relief.
Form: Cubes, octahedra, etc.
Colour: Colourless, white, yellow, orange, red, green, blue brown, black.
XRD: 6–675: 2.06–100,1.261–25,1.075–16,0.816–16.
Occurrence: Natural gemstone. Also manufactured.
Comment: Most diamond grinding and lapping compounds are made from synthetic diamond. Diamond particles are observed in polished and thin sections which have not been adequately cleaned during preparation. They stand out very clearly because of their high relief and colour.

Silicon carbide (moissanite)
SiC

Uniaxial +ve

Hexagonal

$\varepsilon = 2.689$–2.697, $\omega = 2.647$–2.654

Relief: Very high positive relief.
Birefringence: Strong with strong dispersion.
Form: Grinding powder often has splintered appearance, with conchoidal fracture showing radiating stress lines.
Colour: Colourless, green, blue or black. Industrial silicon carbide is usually colourless or bluish. Blue crystals are pleochroic.
XRD: Variable (see ICDD 1996 29–1126 through to 29–1131).

Occurrence: Rare natural mineral known as moissanite (Di Pierro et al. 2003). Manufactured by electric fusion of silicon and carbon for grinding media which can consist of mixtures of cubic to hexagonal silicon carbide in a number of polytypes.

Comment: Mohs hardness 9.5. Silicon carbide particles are observed in polished and thin sections which have not been adequately cleaned during preparation. They stand out very clearly because of their high relief and iridescent colours.

Aluminium oxide (corundum, α-alumina, γ-alumina)

Al_2O_3 (corundum, α-alumina)

> Uniaxial –ve
>
> Hexagonal
>
> $\varepsilon = 1.760$, $\alpha = 1.768$

γAl_2O_3 (γ-alumina)

> Isometric
>
> Cubic
>
> $n = 1.696$

Interference figure: Readily obtainable.
Relief: Very high positive relief.
Birefringence: First-order white for corundum.
Form: Natural corundum is often hexagonal in shape. Synthetic corundum consists of angular and irregular shaped particles, with some signs of conchoidal fracture. γ-alumina appears as irregular aggregations of fine particles.
Twinning: Simple or multiple twinning in corundum.
Colour: Colourless, white or grey.
XRD: Corundum: 13–373: 2.41–50, 2.12–50, 1.96–35, 1.395–100.
γ-alumina: 43–1484: 2.551–98, 2.086–100, 1.602–96, 1.374–57.

Occurrence: Found particularly in some contact rocks in low-silica environments, where it may be called sapphire (blue) or ruby (red). It is industrially synthesised for grinding media. Principal constituent of fused alumina brick and other products. γ-alumina is a synthetic precipitated product used as a polishing compound.

Comment: Synthetic corundum is widely used for grinding and polishing of cement clinker and concrete specimens. During grinding the corundum particles fissure and gradually break down to finer particles, which makes it an ideal grinding agent. Unless specimens are carefully cleaned afterwards, residual corundum particles may remain embedded in surfaces. It is not uncommon to find some corundum at the edges of the specimen embedded in the mounting media. γ-alumina is restricted to polishing compounds in the sub-micron range and is unlikely to interfere with optical examination.

Cerium dioxide (cerianite, ceria polishing compound)

> Isometric
>
> Cubic
>
> $n \approx 2.2$

Relief: Very high positive relief.
Form: As irregular shaped aggregates.
Colour: Yellow to brown.
XRD: 43–1002: 3.124–100, 2.706–27, 1.913–46, 1.632–44.
Occurrence: Extracted from bastnäsite (a cerium-carbonate-fluoride mineral).
Comment: The polishing compounds are a mixture of cerium dioxide and lanthanum and praseodymium oxides, together with other impurities. It is coloured brown by the other rare earth oxides (Kilbourn 1993). Used extensively for polishing glass and may be used to polish cement clinker. Is difficult to remove from surfaces of clinker and is observed as aggregations of orange–brown fine particles.

G.14.14 Mounting media

Canada balsam

Amorphous natural resin

$n = 1.54–1.55$

Relief: Close to the refractive index of optical glass (borosilicate crown glass: $n \sim 1.52$)
Form: On evaporation of solvent sets to an amorphous solid.
Colour: Colourless to pale yellowish. Tends to yellow with age.
Occurrence: Natural resin obtained from the Canadian balsam fir.
Comment: A thermoplastic natural resin soluble in xylene and similar solvents. It is stable and does not crystallise with age but has poor solvent resistance. Used extensively as a cement in optical applications and in the preparation of geological rock thin sections until the 1960s but now largely obsolete and replaced by various polymer resins.

Synthetic resins
Synthetic resins once hardened can exhibit a range of refractive index values depending on formulation, type and the type and proportion of hardener used. The petrographer requiring to know the refractive index of a particular resin formulation can determine this best in its fluid or solid state using a refractometer or refractive index liquids. It should be noted that the refractive index of a resin in its fluid state is likely to be lower than in its solid state. If the resin has been used in a thin section then the simplest method of estimating its refractive index is to compare it directly along the edges of the slide with minerals of known refractive index such as quartz or mica making use of the Becké line effect.

A summary of the properties of the common synthetic resins is given in Table 3.4, Section 3.4.4.

Epoxy copolymers

$n = 1.51–1.66$

Form: On polymerization sets to a hard amorphous transparent solid.
Colour: Transparent clear to transparent yellow
Comment: A modern two-part polymer resin which may be diluted in acetone, alcohol or amyl-acetate. The most commonly used material for mounting, impregnation and adhesive applications in thin-section preparation. The basis of various proprietary mixtures; some including acrylic are used in specimen preparation. Refractive index can be adjusted using mixtures of different epoxy resins (Augerson and Messinger 1993). The refractive index of

solidified epoxy resins may differ from their value when fluid, for example, Araldite 502 (Melmount) varies from 1.54 to 1.62 as it sets. In view of this possible variation in refractive index of synthetic resins, the petrographer would be wise to check the index of the mounting media used in a thin section before undertaking a petrographic examination.

Although the resins can be diluted, once set, they are unaffected by the usual solvents, although thin sections are usually cleaned using methylated spirits during manufacture. Extended exposure to light will cause some epoxy resins to yellow with age and this effect could have implications for methods involving fluorescent yellow dyes for determining porosity and water/cement ratio.

Polyester copolymers

$n \sim 1.5–1.6$

Form: On polymerization sets to a slightly flexible amorphous transparent solid.
Colour: Transparent clear to transparent yellow.
Comment: This family of polymers is most typically used for stabilising or encapsulating specimens prior to cutting and polishing, since it has the property of hardening in a period of minutes or hours depending on the proportion of hardener used. However, they remain sufficiently soft and plastic after curing to allow particles of grinding medium to become embedded in the ground surface. Once hardened, like the epoxy resins, it is little affected by the usual solvents, but yellowing on exposure to light is more marked in most formulations and in part is depends on the hardener used. Again, this effect could have implications for methods involving fluorescent yellow dyes for determining porosity and water/cement ratio.

Urethane/acrylic polymers

$n \sim 1.4–1.6$

Form: On polymerization sets to an amorphous transparent solid.
Colour: Transparent clear to transparent amber.
Comment: As with many other synthetic resins, refractive index changes during setting. An example is the alpha cyanoacrylate, Acrytol mounting media (Quickbond), which has a refractive index of 1.43 when fluid but alters to 1.49 when set.

A complex urethane polymethylmethacrylate (PMMA) formulation can be cured in a matter of a few seconds at room temperature on exposure to UV radiation with a spectral energy distribution around 365 nm. It is an alternative to epoxy adhesive that is now very widely used for mounting thin sections.

REFERENCES

Adams, A.E. and MacKenzie, W.S. 1998: *A Colour Atlas of Carbonate Sediments and Rocks under the Microscope*, Manson Publishing, London, U.K., p. 184.

Adams, A.E., MacKenzie, W.S. and Guildford, C. 1984: *Atlas of Sedimentary Rocks under the Microscope*, Longman, Harlow, U.K., p. 104.

Aguilera, J., Blanco Varela, M.T. and Vázquez, T. 2001: Procedure of synthesis of thaumasite. *Cement and Concrete Research* 31, 1163–1168.

Aïtcin, P-C. (ed.). 1983: *Condensed Silica Fume*, Les Editions de l'Universite de Sherbrooke, Sherbrooke, Quebec, Canada, p. 52.

Almond, G.G., Harris, R.K. and Graham, P. 1994: A study of the layered alkali metal silicate, magadiite, by one- and two-dimensional 1H and 29Si NMR spectroscopy. *Journal of the Chemical Society, Chemical Communications* 7, 851–852.

Alpers, C.N., Jambor, J.L. and Nordstrom, D.K. (eds.). 1999: Sulfate minerals – Crystallography, geochemistry, and environmental significance. *Reviews in Mineralogy & Geochemistry*, 40, Mineralogical Society of America, Washington, DC, pp. 608.

American Concrete Institute. 1987: Silica fume in concrete. Preliminary committee report, ACI Committee 226. *ACI Materials Journal*, 84(2), 158–66, Detroit, MI.

Andersen, K.T. and Thaulow, N. 1990: The study of alkali-silica reactions in concrete by the use of fluorescent thin-sections. In *Petrography Applied to Concrete and Concrete Aggregates*, Erlin, B. and Stark, D. (eds.). ASTM STP 1061, American Society for Testing and Materials, Philadelphia (now West Conshohocken), PA, pp. 71–89.

Angel, R.J. and Prewitt, C.T. 1986: Crystal structure of mullite: A re-examination of the average structure. *American Mineralogist* 71, 1476–1482.

Annehed, H., Fälth, L. and Lincoln, F.J. 1982: Crystal structure of synthetic makatite $Na_2Si_4O_8(OH)_2·4H_2O$. *Zeitschrift für Kristallographie* 159, 203–210.

Anonymous. 1992: *Le béton sous le microscope/Beton onder de microscoop*, Groupement Belge de Baton/Belgische Betongroep, Brussels, Belgium, pp. 61.

Antao, S.M., Mulder, W.H., Hassan, I., Crichton, W.A. and Parise, J.B. 2004: Cation disorder in dolomite, $CaMg(CO_3)_2$, and its influence on the aragonite + magnesite ↔ dolomite reaction boundary. *American Mineralogist* 89, 1142–1147.

ASTM C618-12a, 2012: *Standard Specification for Coal Fly Ash and Raw or Calcined Natural Pozzolan for Use in Concrete*. American Society for Testing and Materials, West Conshohocken, PA.

Augerson, C.C. and Messinger J.M. 1993: Controlling the refractive index of epoxy adhesives with acceptable yellowing after ageing. *Journal of the American Institute for Conservation* 32(3), 311–314.

Balistrieri, L.S. and Murray, J.W. 1982: The sorption of Cu, Pb, Zn, and Cd on goethite from major ion sea water. *Geochimica et Cosmochimica Acta* 46, 1253–1265.

Barboux, P., Laghzizil, A., Bessoles, Y., Deroulhac, H. and Trouvé, B. 2004: Paradoxical crystalline morphology of frosted glass. *Journal of Non-crystalline Solids* 345/346, 137–141.

Barnes, R.D., Diamond, S. and Dolch, W. 1978a: The contact zone between Portland cement paste and glass aggregate surfaces. *Cement and Concrete Research* 8, 233–243.

Barnes, R.D., Diamond, S., Dolch, W. 1978b: Hollow shell hydration particles in bulk cement paste. *Cement and Concrete Research* 8, 263–271.

Barnett, S.J., Adam, C.D., Jackson, A.R.W. and Hywel-Evans, P.D. 1999: Identification and characterisation of thaumasite by XPRD techniques, *Cement and Concrete Composites* 21, 123–128.

Battey, M.H. 1981: *Mineralogy for Students*. 2nd edition, Longman, London, U.K., p. 355 (1st edition, 1971).

Battigan, A.F. 1986: The use of the microscope for estimating the basicity of slags in slag-cements, *Eighth International Congress Chemistry of Cement*, Rio de Janeiro, Theme 3, 4, 17–21.

Baussant, J.B., Vernet, C., Defosse, C. 1989: Growth of ettringite in diffusion controlled conditions – Influence of additives on the crystal morphology. In *Proceedings of the 11th International Conference on Cement Microscopy*, New Orleans, LA, pp. 186–197.

Beneke, K. and Lagaly, G. 1977: Kanemite-inner crystalline reactivity and relations to other sodium silicates. *American Mineralogist* 62, 763–771.

Beneke, K. and Lagaly, G. 1983: Kenyaite – Synthesis and properties. *American Mineralogist* 68, 818–826.

Benge, C.G. and Webster, W.W. 1994: Evaluation of blast furnace slag slurries for oilfield application, Paper SPE/IADC 27449, *Society of Petroleum Engineers/International Association of Drilling Contractors Conference*, Dallas, TX.

Benmore, C.J. and Monteiro, P.J.M. 2010: The structure of alkali silicate gel by total scattering methods. *Cement and Concrete Research* 40, 892–897.

Bensted, J. and Varma, S.P. 1971: Studies of ettringite and its derivatives. *Cement Technology* 2, 73–76.

Bensted J. and Varma, S.P. 1974: Studies of thaumasite. *Silicate Industriels* 39, 11–19.

Berg, L.G. and Sveshnikova, V.N. 1949: On modifications of hemihydrate calcium sulfate. *Mineralogical Abstracts* 10, 464–465.

Bergin, T.F. 1989: The application of point counting procedures for the analysis of bulk asbestos samples. *The Microscope* 37, 377–387.

Berra, M. and Baronio, G. 1987: Thaumasite in deteriorated concretes in the presence of sulfates. In *Concrete durability: Katharine and Bryant Mather International Conference*, Scanlon, J.M. (ed.). ACI SP-100, 2073–2089. Paper SPI00-I06. American Concrete Institute, Detroit, MI.

Berry, A.J., Fraser, D.G., Grime, G.W., Craven, J. and Sleeman, J.T. 1998: The hydration and dissolution of periclase. In *Proceedings of the Eighth Annual V M Goldschmidt Conference*, Toulouse, France, *Mineralogical Magazine* 62a, 158–159.

Bertram, M.A., Mackenzie, F.T., Bishop, F.C. and Bischoff, W.D. 1991: Influence of temperature on the stability of magnesian calcite. *American Mineralogist* 76, 1889–1896.

Bhatty, J.I. 1991: A review of the application of thermal analysis to cement-admixture system. *Thermochimica Acta* 189, 313–350.

Bloss, F.D. 1966: *An Introduction to the Methods of Optical Crystallography*, Holt, Rinehart and Winston, New York.

Bonaccorsi, E., Merlino, S. and Taylor, H.F.W. 2004: The crystal structure of jennite, $Ca_9Si_6O_{18}(OH)_6 \cdot 8H_2O$. *Cement and Concrete Research* 34, 1481–1488.

Bonen, D. and Diamond, S. 1992: Investigation on the coarse fraction of a commercial silica fume. *Proceedings of the 14th International Conference on Cement Microscopy*, Costa Mesa, California, pp. 103–113.

Boyd, D.C., Danielson, P.S. and Thompson, D.A. 1994: Glass. *Kirk-Othmer Encyclopedia of Chemical Technology* 12, 551–627, Wiley Interscience, New York.

Brindley, G.W. 1969: Unit cell of magadiite in air, in vacuo, and under other conditions. *American Mineralogist*, 54, 1583–1591.

Brindley, G.W. and Brown, G. 1980: *Crystal Structures of Clay Minerals and their X-Ray Identification*, Monograph No. 5, Mineralogical Society, London, U.K.

Broekmans, M.A.T.M. 2004a: The crystallinity index of quartz by XRD, its susceptibility for ASR, and a brief methodological review. In *Proceedings of the 12th International Conference on Alkali-Aggregate Reactions in Concrete*, Tang, M. and Deng, M. (eds.), Beijing, China, pp. 60–68.

Broekmans, M.A.T.M. 2004b: Structural properties of quartz and their potential role for ASR. *Materials Characterization* 53(2–4), Special Issue 29, 129–140.

Broekmans, M.A.T.M. 2009: Petrography as an essential complementary method in forensic assessment of concrete deterioration: Two case studies. *Materials Characterization* 60(7), Special Issue 35, 644–655.

Brouxel, M. and Valiere, A. 1992: Thaumasite as the final product of alkali-aggregate reaction: A case study. In *The Ninth International Conference on Alkali-Aggregate Reaction in Concrete*, London, *Conference Papers*, Vol. 1, pp. 136–144, The Concrete Society (Ref CS 104), Slough (now Camberley), U.K.

Buck, A.D. 1986: Relationship between ettringite and chloroaluminate, strength and expansion in paste mixtures, In *Eighth International Congress Chemistry of Cement*, Rio de Janeiro, Brazil, Germany, Theme 2, Vol. III, pp. 417–424.

Buck, A.D. and Mather, K. 1978: Alkali-silica reaction products from several concretes: Optical, chemical and X-ray diffraction data, In *Proceedings of the Fourth International Conference on the Effects of Alkalies in Cement and Concrete*, Purdue University, West Lafayette, IN, pp. 73–85.

Bunker, B.C. 1994: Molecular mechanisms for corrosion of silica and silicate glasses. *Journal of Non-Crystalline Solids* 179, 300–308.

Bustillo, M.A. 2001: Cherts with moganite in continental Mg-clay deposits: An example of "false" Magadi-type cherts, Madrid Basin, Spain. *Journal of Sedimentary Research* 71, 436–443.

Bustillo, M.A. 2002: Aparición y significado de la moganita en las rocas de la sílice: una revision (Occurrence and significance of the moganite in silica rocks: a review). *Journal of Iberian Geology* 28, 157–166.

Campbell, D.H. 1986: *Microscopical Examination and Interpretation of Portland Cement and Clinker*. Construction Technology Laboratories, Portland Cement Association, Skokie, IL.

Campbell, D.H. 1999: *Microscopical Examination and Interpretation of Portland Cement and Clinker*, 2nd edn. Special Publication 30, 201pp, Portland Cement Association, Skokie, IL.

Carlson, E.T. and Berman, H.A. 1960: Some observations on the calcium aluminate carbonate hydrates. *Journal of Research of the National Bureau of Standards – A. Physics and Chemistry* 64A/4, 333–341.

Carlson, W.D. 1983: The polymorphs of $CaCO_3$ and the aragonite-calcite transformation. In *Carbonates: Mineralogy and Chemistry, Reviews in Mineralogy*, Reeder, R.J. (ed.). Vol. 11, pp. 191–225.

Carpenter, A.B. 1964: Oriented overgrowths of thaumasite on ettringite. *American Mineralogist* 48, 1396–1398.

Chadwick, P.K. 1976: Visual illusions in geology. *Nature* 260, 397–401.

Chowns, T.M. and Elkins, J.E. 1979: The origin of quartz geodes and cauliflower cherts through the silicification of anhydrite nodules. In *Society of Economic Paleontologists and Mineralogists, Reprint Series*, McBride, E.F. (ed.). Silica in sediments: Nodular and bedded chert, Vol. 8, pp. 120–128.

Churakov, S.V. 2009: Structural position of H_2O molecules and hydrogen bonding in anomalous 11Å tobermorite. *American Mineralogist* 94, 156–165.

Cody, A.M., Lee, H., Cody, R.D. and Spry, P.G. 2004: The effects of chemical environment on the nucleation, growth, and stability of ettringite $[Ca_3Al(OH)_6]_2(SO_4)_3 \cdot 26H_2O$. *Cement and Concrete Research* 34, 869–881.

Cole, W.F. and Demediuk, T. 1955: X-ray, thermal, and dehydration studies of magnesium oxychlorides. *Australian Journal of Chemistry* 8(2), 234–251.

Cole, W.F. and Kroone, B. 1959–1960: Carbon dioxide in hydrated Portland cement. *Proceedings of the American Concrete Institute* 56, 1275–1295.

Coleman, S.E. 1984: Calcium silicate hydrates formed under hydrothermal conditions. In *Proceedings of the 188th American Chemical Society National Meeting, Advances in Oil Field Chemicals and Chemistry*, St Louis, MO.

Cooper, R.E. 1962: Kinetics of twinning in calcite. *Proceedings of the Royal Society of London, Series A, Mathematical and Physical Sciences* 270/1343, 525–537.

Correns, C.W. and Nagelschmidt, G. 1933: Über faserbau und optische eigenschaften von chalcedon. *Zeitschrift für Kristallographie* 85, 199–213.

Crammond, N.J. 1985a: Quantitative X-ray diffraction analysis of ettringite, thaumasite and gypsum in concretes and mortars. *Cement and Concrete Research* 15(3), 431–441.

Crammond, N.J. 1985b: Thaumasite in failed cement mortars and renders from exposed brickwork. *Cement and Concrete Research* 15(6), 1039–1050.

Damgaard Jensen, A., Eriksen, K., Chatterji, S., Thaulow, N. and Brandt, I. 1984: *Petrographic Analysis of Concrete* (English version by C & N Meinertz-Nielsen), Danish Building Export Council, Copenhagen, Denmark, p. 12.

Das, S. and Hendry, M.J. 2011: Changes of crystal morphology of aged goethite over a range of pH (2–13) at 100°C. *Applied Clay Science* 51, 192–197.

De Ceukelaire, L. 1990: Ettringiet, thaumasiet en... kweethetniet? *Cement* 42(5), 30–33.

De Meer, S. and, Spiers, C.J. 1997: Uniaxial compaction creep of wet gypsum aggregates. *Journal of Geophysical Research* 102, 875–891.

De Meer, S., Spiers, C.J. and Peach, C.J. 1997: Pressure solution creep in gypsum: Evidence for precipitation reaction control. *Physics and Chemistry of the Earth* 22/1–2, 33–37.

De Noirfontainea, M.N., Dunstetter, F., Courtial, M., Gasecki, G., Signes-Frehel, M. 2006: Polymorphism of tricalcium silicate, the major compound of portland cement clinker, 2. Modelling alite for Rietveld analysis. An industrial challenge. *Cement and Concrete Research* 36, 54–64.

Deb, S.K., Manghnani, M.H., Ross, K., Livingston, R.A. and Monteiro, P.J.M. 2003: Raman scattering and X-ray diffraction study of the thermal decomposition of an ettringite group mineral. *Physics and Chemistry of Minerals* 30, 31–38.

Deer, W.A., Howie, R.A. and Zussman, J. 1978 (1997): *Rock-Forming Minerals, Volume 2A: Single-Chain Silicates*, 2nd ed., (reprinted with corrections), The Geological Society, London, U.K., p. 668.

Deer, W.A., Howie, R.A. and Zussman, J. 1982 (1997): *Rock-Forming Minerals, Volume 1A: Orthosilicates*, 2nd ed., (reprinted with corrections), The Geological Society, London, U.K., p. 919.

Deer, W.A., Howie, R.A. and Zussman, J. 1986 (1997): *Rock-Forming Minerals, Volume 1B: Disilicates and Ring silicates*, 2nd ed., (reprinted), The Geological Society, London, U.K., p. 629.

Deer, W.A., Howie, R.A. and Zussman, J. 1992: *An Introduction to the Rock-Forming Minerals*, 2nd ed., Pearson, London, U.K., p. 712. (1st edition, 1966, re-printed many times).

Deer, W.A., Howie, R.A. and Zussman, J. 1996: *Rock-Forming Minerals, Volume 5B: Non-Silicates: Sulphates, Carbonates, Phosphates, Halides*, 2nd ed., (by Chang, L.L.Y., Howie, R.A. and Zussman, J.), Longman, Harlow, U.K., p. 383.

Deer, W.A., Howie, R.A. and Zussman, J. 1997: *Rock-Forming Minerals, Volume 2B: Double-Chain Silicates*, 2nd ed., The Geological Society, London, U.K., p. 764.

Deer, W.A., Howie, R.A. and Zussman, J. 2001: *Rock-Forming Minerals, Volume 4A: Framework Silicates: Feldspars*, 2nd ed., The Geological Society, London, U.K., p. 972.

Deer, W.A., Howie, R.A. and Zussman, J. 2003: *Rock-Forming Minerals, Volume 3A: Sheet Silicates: Micas*, 2nd ed., (extensive revision by Fleet, M.E.), The Geological Society, London, U.K., p. 758.

Deer, W.A., Howie, R.A. and Zussman, J. 2004: *Rock-Forming Mnerals, Volume 4B: Framework Silicates: Silica Minerals, Feldspathoids and the Zeolites*, 2nd ed., (by Deer, W.A., Howie, R.A., Wise, W.S. and Zussman, J.), The Geological Society, London, U.K., p. 982.

Deer, W.A., Howie, R.A. and Zussman, J. 2009: *Rock-Forming Minerals, Volume 3B: Layered Silicates Excluding Micas and Clay Minerals*, 2nd ed., The Geological Society, London, U.K., p. 314.

Deer, W.A., Howie, R.A. and Zussman, J. 2011: *Rock-Forming Minerals, Volume 5A: Non-Silicates: Oxides, Hydroxides and Sulphides*, 2nd ed., (by Bowles, J.F.W., Howie, R.A., Vaughan, D.J. and Zussman, J.), The Geological Society, London, U.K., p. 920.

Delvigne, J.E. 1998: *Atlas of Micromorphology of Mineral Alteration and Weathering*, Special Publication 3, Mineralogical Association of Canada, Québec, Canada, p. 495.

Demediuk, T., Cole, W.F. and Hueber, H.V. 1955: Studies of magnesium and calcium oxychlorides. *Australian Journal of Chemistry*, 8(2), 215–233.

Diamond, S. 1986: The microstructure of cement paste in concrete, In *Proceedings of the 8th International Congress on the Chemistry of Cement*, Rio de Janeiro, Brazil, Germany, Vol. 1, pp. 123–147.

Dickson, J.A.D. 1966: Carbonate identification and genesis as revealed by staining. *Journal of Sedimentary Petrology* 361(2), 491–505.

Di Pierro, S., Gnos, E., Grobety, B.H., Armbruster, T., Bernasconi, S.M. and Ulmer, P. 2003: Rock-forming moissanite (natural α-silicon carbide). *American Mineralogist* 88, 1817–1821.

Dolar-Mantuani, L. 1983: *Handbook of Concrete Aggregates – A Petrographic and Technological Evaluation*. Noyes Publications, Park Ridge, NJ, p. 345.

Drissen, R. 1995: Determination of the glass content in granulated blastfurnace slag. *Zement-Kalk-Gips* 48(1), 59–62.

El Hemaly, S.A.S., Matsuda, T. and Taylor, H.F.W. 1977: Synthesis of normal and anomalous tobermorite. *Cement and Concrete Research* 7, 429–438.

Elzea, J.M. and Rice, S.B. 1996: TEM and X-Ray diffraction evidence for cristobalite and tridymite stacking sequences in opal. *Clays and Clay Minerals* 44(4), 492–500.

Erlin, B. and Stark, D.C. 1965/1966: Identification and occurrence of thaumasite in concrete (a discussion for the 1965 Highway Research Board Symposium on aggressive fluids). *Highway Research Record*, 113, 108–113.

Ervin Jr, G. 1952: Structural interpretation of the diaspore-corundum and boehmite-γ-Al$_2$O$_3$ transitions. *Acta Crystallographica* 5, 103–108.

Ervin Jr, G. and Osborn, F. 1951: The system Al$_2$O$_3$-H$_2$O. *Journal of Geology* 59, 381–394.

Everett, A. 1994: *Materials*, 5th ed. (revised by C.M.H.Barritt). Mitchell's Building Series, Longman (Pearson), Harlow, U.K., p. 272.

Farmer, V.C. (ed) 1974: *The Infrared Spectra of Minerals*, Monograph 4. Mineralogical Society, London, U.K., p. 539.

Felius, R.O. 1976: *Structural Morphology of Rutile and Tritutile Type Crystals*. PhD-thesis, University of Leiden, Leiden, the Netherlands, p. 101.

Figg, J.W. 1981: Reaction between cement and artificial glass in concrete. In *Proceedings of the Fifth International Conference on Alkali-Aggregate Reaction*, Cape Town, South Africa, Paper S252/7.

Fiore, S., Huertas, F.J., Tazaki, K., Huertas, F. and Linares, J. 1999: A low temperature alteration of a rhyolitic obsidian. *European Journal of Mineralogy* 11, 455–469.

Fleet, M.E. 2003: see *Deer, Howie & Zussman (2003)*, 2nd ed. of Volume 3A, Longman, London, U.K.

Fleischer, M., Wilcox, R.E. and Matzko, J.J. 1984: Microscopic determination of the nonopaque minerals. *United States Geological Survey Bulletin* 1627, 453.

Flörke, O.W. 1952: Kristallographische und rontgenometrische Untersuchungen in System $CaSO_4$-$CaSO_4.2H_2O$. *Neues Jahrbuch für Mineralogie, Abhandlungen* 84, 189–240; *Mineralogical Abstracts* 12, 94 (1953–1955).

Flörke, O.W., Flörke, U. and Giese, U. 1984: Moganite. A new microcrystalline silica mineral. *Neues Jahrbuch für Mineralogie, Abhandlungen* 149, 325–336.

Flörke, O.W., Graetsch, H., Martin, B., Röller, K. and Wirth, R. 1991: Nomenclature of micro- and noncrystalline silica minerals, based on structure and microstructure. *Neues Jahrbuch für Mineralogie, Abhandlungen* 163, 19–42.

Folk, R.L. 1959: Practical petrographic classification of limestones. *American Association of Petroleum Geologists, Bulletin* 43, 1–38.

Folk, R.L. 1962: Spectral subdivision of limestone types, In *Classification of Carbonate Rocks*, Ham, W.E. (ed.). Memoir 1, American Association of Petroleum Geologists, Tulsa, OK, pp. 62–84.

French, W.J. 1991: Concrete petrography: A review. *Quarterly Journal of Engineering Geology* 24(1), 17–48.

Friedman, G.M. 1959: Identification of carbonate minerals by staining methods. *Journal of Sedimentary Petrology* 29, 87–97.

Friedman, G.M. 1971: Staining, In *Procedures in Sedimentary Petrology*, Carver, R.E. (ed.). Wiley-Interscience, New York, pp. 511–530.

Frondel, C. 1962: *Dana's System of Mineralogy, Volume III, Silica Minerals*, 7th ed, John Wiley & Sons, New York.

Frondel, C. 1978: Characters of quartz fibers. *American Mineralogist* 63, 17–27.

Frondel, C. 1982: Structural hydroxyl in chalcedony (type B quartz). *American Mineralogist* 67, 1248–1257.

Frondel, C. 1985: Systematic compositional zoning in the quartz fibers of agates. *American Mineralogist*. 70, 975–979.

Frossard, E., Bauer, J.P. and Lothe, F. 1997: Evidence of vivianite in $FeSO_4$ flocculated sludges. *Water Research* 31(10), 2449–2454.

Fu, Y. and Beaudoin, J.J. 1996: On the distinction between delayed and secondary ettringite formation in concrete. *Cement and Concrete Research* 26, 979–980.

Füchtbauer, H. and Müller, G. (eds) 1970: Teil II. Sedimente und Sedimentgesteine. In *Sediment-Petrologie*, Engelhardt, W., Füchtbauer, H. and Müller, G. (eds.). E. Schweizerbart'sche Verlagsbuchhandlung, Stuttgart, Germany, p. 726.

Fundal, E. 1980: Microscopy of cement raw mix and clinker. *FLS-Review*, 25, F.L. Smidth Laboratories, Copenhagen, Denmark.

Gaffey, S.J. 1986: Spectral reflectance of carbonate minerals in the visible and near infrared (0.35–2.55 μm): Calcite, aragonite and dolomite. *American Mineralogist* 71, 151–162.

Gard, J.A., Taylor, H.F.W., Cliff, G. and Lorimer, G.W. 1977: A reexamination of jennite. *American Mineralogist* 62, 365–368.

Gavish, E. and Friedman, G.M. 1973: Quantitative analysis of calcite and Mg-calcite by X-ray diffraction: Effect of grinding on peak height and peak area. *Sedimentology* 20(3), 437–444.

George, W.O. 1924: The relation of the physical properties of natural glasses to their chemical composition. *The Journal of Geology* 32, 353–372.

Gettens, R.J. and Stout, G.L. 1966: *Painting Materials*. Dover Publications, New York.

Gille, F., Dreizler, I., Grade, K., Krämer, H. and Woermann, E. 1965: Mikroskopie des zementklinkers – Bilderatlas (Microscopy of cement clinker Picture atlas), Verein Deutscher Zementwerke, Beton-Verlag GmbH, Düsseldorf, Germany, 75pp (English translation by P Schmid, 1980).

Gillott, J.E. and Rogers, C.A. 1994: Alkali-aggregate reaction and internal release of alkalis. *Magazine of Concrete Research* 46, 99–112.

Gillott, J.E. and Rogers, C.A. 2003: The behavior of silicocarbonatite aggregates from the Montreal area. *Cement and Concrete Research* 33, 471–480.

Gíslason, S.R., Heaney, P.J., Oelkers, E.H. and Schott, J. 1997: Kinetic and thermodynamic properties of moganite, a novel silica polymorph. *Geochimica et Cosmochimica Acta* 61, 1193–1204.

Gíslason, S.R., Heaney, P.J., Veblen, D.R. and Livi, K.J.T. 1993: The difference between the solubility of quartz and chalcedony: the cause? *Chemical Geology* 107, 363–366.

Godovikov, A.A. and Ripinen, O.I. 1994: Lutecin – nur eine varietät des chalcedons? (Lutecin - only a variety of chalcedony?). *Lapis Magazine* 19(6), 44–47.

Gollop, R.S. and Taylor, H.F.W. 1994: Microstructural and microanalytical studies of sulfate attack. II. Sulfate-resisting Portland cement: ferrite composition and hydration chemistry. *Cement and Concrete Research* 24, 1347–1358.

Götze, J., Nasdala, L., Kleeberg, R. and Wentzel, M. 1998: Occurrence and distribution of "moganite" in agate/chalcedony: A combined micro-Raman, Rietveld and cathodoluminescence study. *Contributions to Mineralogy and Petrology* 133, 96–105.

Graetsch, H. 1994: Structural characteristics of opaline and microcrystalline silica minerals. In *Reviews in Mineralogy*, Heaney, P.J., Prewitt, C.T. and Gibbs, G.V. (eds.) Silica: physical behavior, geochemistry and materials applications Vol. 29, Mineralogical Society of America, pp. 209–232.

Graetsch, H., Gies, H. and Topalovic, I. 1994: NMR, XRD and IR study on microcrystalline opals. *Physics and Chemistry of Minerals* (Springer) 21(3), 166–175.

Grattan-Bellew, P.E. 1992: Microcrystalline quartz, undulatory extinction and the alkali-silica reaction. In *The Ninth International Conference on Alkali-Aggregate Reaction in Concrete*, London, Conference Papers, Vol. 1, 383–94, The Concrete Society (Ref CS.104), Slough (now Camberley), U.K.

Green, G.W. 1984: Gypsum analysis with the polarizing microscope. In *The Chemistry and Technology of Gypsum*, Kuntze, R.A. (ed.). ASTM STP 861, 22–47, American Society for Testing and Materials, Philadelphia (now West Conshohocken, Pennsylvania), PA.

Grice, J.D. and Ferrari, G. 2000: New minerals approved in 1999 by the Commission on New Minerals and Mineral Names, International Mineralogical Association. [99–035: moganite]. *Canadian Mineralogist*, 38, 245–250.

Grounds, T., Midgley, H.G. and Nowell, D.V. 1988: The carbonation of ettringite by atmospheric carbon dioxide. *Thermochimica Acta* 135, 347–352.

Grout, F.F. 1932: *Petrography and Petrology*. McGraw-Hill Book Co., New York.

Grudemo, A. 1965: *The Microstructures of Cement Gel Phases*. Transactions of the Royal Institute of Technology, Stockholm, Sweden, Civil Engineering, 13(242).

Guthrie Jr, G.D., Bish, D.L. and Reynolds jr, R.C. 1995: Modeling the X-ray diffraction pattern of opal-CT. *American Mineralogist* 80, 869–872.

Hattori, I., Umeda, M., Nakagawa, T. and Yamamoto, H. 1996: From chalcedonic chert to quartz chert: Diagenesis of chert hosted in a Miocene volcanic-sedimentary succession, central Japan. *Journal of Sedimentary Research* 66, 163–174.

Hawthorne, F.C. and Oberti, R. 2006: On the classification of amphiboles. *Canadian Mineralogist* 44, 1–21.

Heaney, P.J. 1995: Moganite as an indicator for vanished evaporites: A testament reborn? *Journal of Sedimentary Petrology* A65, 633–638.

Heaney, P.J. and Post, J.E. 1992: The widespread distribution of a novel silica polymorph in microcrystalline quartz varieties *Science*, 255, 441–443.

Heaney, P.J., Veblen, D.R. and Post, J.E. 1994: Structural disparities between chalcedony and macrocrystalline quartz. *American Mineralogist* 79, 452–460.

Heller, L. and Taylor, H.F.W. 1956: *Crystallographic Data for the Calcium Silicates*. HMSO, London, U.K.

Henmi, C. and Kusachi, I. 1992: Clinotobermorite, $Ca_5Si_6(O,OH)_{18}\cdot5H_2O$, a new mineral from Fuka, Okayama Prefecture, Japan. *Mineralogical Magazine* 56, 353–358.

Henning, O. 1974: Cements; the hydrated silicates and aluminates, In *The Infrared Spectra of Minerals*, Farmer, V.C. (ed.). Monograph, Mineralogical Society, London, U.K., Vol. 4, pp. 445–463.

Henry, D.A. 1999: Cuspidine-bearing skarn from Chesney Vale, Victoria. *Australian Journal of Earth Sciences* 46, 251–260.

Hentschel, G. 1983: *Die Mineralien der Eifelvulkane*. Christian Weise Verlag, München, Germany.

Hewlett, P.C. (ed) 1998: *Lea's Chemistry of Cement and Concrete*, 4th ed. Arnold (Hodder Headline), London, U.K., p. 1053.

Highway Research Board, 1972: *Guide to Compounds of Interest in Cement and Concrete Research*. Highway Research Board Special Report 127, Washington, DC.

Hills, L.M. 1999: An introduction to clinker microscopy. *Cement Americas* 11, 38–44.

Hina, A. and Nancollas, G.H. 1999: Precipitation and dissolution of alkaline earth sulfates: Kinetics and surface energy. In Sulfate Minerals – Crystallography, Geochemistry, and Environmental Significance, Alpers, C.N., Jambor, J.L. and Nordstrom, D.K. (eds.). *Reviews in Mineralogy & Geochemistry*, 40, 277–301 (2000).

Hitzman, M.W. 1999: Routine staining of drill core to determine carbonate mineralogy and distinguish carbonate alteration textures. *Mineralium Deposita* 34, 794–798.

Hoffmanner, F. 1973: *Microstructure of Portland Cement Clinker*. Holderbank Management and Consulting Ltd (Holcim Group Support Ltd since 2001), Zurich, Switzerland, p. 48.

Hooton, R.D. and Emery, J.J. 1983: Glass content determination and strength development predictions for vitrified blast furnace slag. In *Proceedings of the CANMET/ACDI First International Conference on the Use of Fly Ash, Silica Fume, Slag and Other Mineral by-Products in Concrete*, Montebello, Québec, Canada, ACI Special Publication SP-79, 943–962, American Concrete Institute, Detroit (Farmington Hills), MI.

Hornibrook, F.B., Insley, H. and Schuman, L. 1943: Effects of alkalis in Portland cement on the durability of concrete, An appendix to the Report of Committee Cl. *Proceedings American Society of Testing Materials*, 43, 218.

Hou, X., Struble, L.J. and Kirkpatrick, R.J. 2004: Formation of ASR gel and the roles of C-S-H and portlandite. *Cement and Concrete Research* 34, 1683–1696.

Howard, C.J., Hill, R.J. and Sufi, M.A.M. October 1988: Quantitative phase analysis by Rietveld analysis of X-ray and neutron diffraction patterns. *Chemistry in Australia* 55, 367–369.

Hunt, J.B., Clift, P.D., Lacasse, C, Vallier, T.L. and Werner, R. 1998: Interlaboratory comparison of electron probe microanalysis of glass geochemistry, In *Proceedings of the Ocean Drilling Program, Scientific Results*, Saunders, A.D., Larsen, H.C. and Wise jr, S.W. (eds.). Vol. 152, pp. 85–91.

International Centre for Diffraction Data. 2012: *Powder Diffraction File*. (ICDD), Newtown Square, PA.

Iden, I.K. and Hagelia, P. 2003: C, O and S isotopic signatures in concrete which have suffered thaumasite formation and limited thaumasite form of sulfate attack. *Cement and Concrete Composites* 25, 839–846.

Idorn, G.M. 1961a: *Studies of Disintegrated Concrete. Part 1*. Progress report, Series N - No. 2, Committee on Alkali Reactions in Concrete, Danish National Institute of Building Research and the Academy of Technical Sciences, Copenhagen, Denmark.

Idorn, G.M. 1961b: *Studies of Disintegrated Concrete – Parts I-V*. Progress Report N2-N6. Committee on Alkali-Aggregate Reactions in Concrete, Danish National Institute of Building Research and the Academy of Technical Sciences, Copenhagen, U.K.

Ingham, J.P. 2010: *Geomaterials under the Microscope, a Colour Guide*, Manson Publishing, London, U.K., p. 192.

International Colour Index 2009: *Colour Index, 4th Edition Part 1. Pigments and solvent dyes*. Society of Dyers and Colourists SDC, Bradford, U.K. (now available only as a Web download).

International Mineralogical Association (IMA). 2007: *Official IMA/CNMNC* (Commission on New Minerals, Nomenclature and Classification) *List of mineral names* (compiled by Nickel, E.H. and Nichols, M.C.), Materials Data Inc., Livermore, CA.

Jezek, P.A. and Noble, D.C. 1978: Natural hydration and ion exchange of obsidian: an electron microprobe study. *American Mineralogist* 63, 266–273.

Katayama, T., St John, D.A. and Futugawa, T. 1989: The petrographic comparison of some volcanic rocks from Japan and New Zealand – potential reactivity related to interstitial glass and silica minerals. In *Proceedings of the 8th International Conference on Alkali-Aggregate Reaction*, Okada, K., Nishibayashi, S. and Kawamura, M. (eds.). Kyoto, Japan, pp. 537–542.

Kennerley, R.A. and St John, D.A. 1988: Part I – Reactivity of New Zealand aggregates with cement alkalis. In *Aggregate Studies in New Zealand*, In St John, D.A. (ed.) *Alkali*-Report No. CD 2390, 60–1, DSIR (Department of Scientific and Industrial Research), Chemistry Division, New Zealand.

Kerr, P.F. 1977: *Optical Mineralogy*. 4th ed, McGraw-Hill, New York.

Kilbourn, R.T. 1993: Ceria and cerium products. In *Kirk-Othmer Encyclopedia of Chemical Technology*, 4th ed, Vol. 5, pp. 728–749, John Wiley & Sons, New York.

Kloprogge, J.T, Broekmans, M., Duong, L.V., Martens, W.N., Hickey, L. and Frost, R.L. 2006: Low temperature synthesis and characterisation of lecontite, $(NH_4)Na(SO_4) \cdot 2H_2O$. *Journal of Materials Science* 41, 3535–3539.

Knauth, L.P. 1994: Petrogenesis of chert. In: Heaney, P.J., Prewitt, C.T., Gibbs, G.V. (eds) Silica. Physical Behavior, Geochemistry and Materials Applications. *Reviews in Mineralogy* 29, 233–258.

Kneller, W.A. 1966: Significance of chert in limestone and dolomite used as concrete aggregate. *The Ohio Journal of Science* 66(2), 191.

Kollman, H. and Strübel, G. 1978: Experimental studies on the question of solid-solution formation between ettringite and thaumasite. *Fortschritte der Mineralogie, Beiheft* 56(1), 65–66.

Kollman, H., Strübel, G. and Trost, F. 1976: Synthesen von Ettringit und Thaumasit. *Fortschritte der Mineralogie, Beiheft* 54(1), 50–51.

Kollman, H., Strübel, G. and Trost, F. 1977: Mineralsynthetische Untersuchungen zu Treibursachen durch Ca-Al-Sulfat-Hydrat und Ca-Si-Carbonat-Sulfat-Hydrat. *Tonindustrie Zeitung* 101(3), 63–70.

Komarneni, S., Breval, E., Miyake, M. and Roy, R. 1987: Cation-exchange properties of (Al+Na)-substituted synthetic tobermorites. *Clays and Clay Minerals* 35, 385–390.

Komarneni, S. and Guggenheim, S. 1988: Comparison of cation exchange in ganophyllite and [Na+Al]-substituted tobermorite: Crystal-chemical implications. *Mineralogical Magazine* 52, 371–375.

Komarneni, S., Roy, R., Roy, D.M., Fyfe, C.A., Kennedy, G.J., Bothner-By, A.A., Dadok, J. and Chesnick, A.S. 1985: ^{27}Al and ^{29}Si magic angle spinning nuclear magnetic resonance spectroscopy of Al-substituted tobermorite. *Journal of Materials Science* 20, 4209–4214.

Kostov, I. 1968: *Mineralogy*. Oliver and Boyd, Edinburgh, Scotland.

Kühnel, R.A., Prins, J.J. and Roorda, H.J. 1980: *Opaque Minerals, The "Delft" System for Mineral Identification*. Delft University Press, Delft, the Netherlands, p. 204.

Kühnel, R.A., Roorda, H.J. and Steensma, J.J. 1975: The crystallinity of minerals – A new variable in pedogenetic processes: A study of goethite and associated silicates in laterites. *Clays & Clay Minerals* 23, 349–354.

Kühnel, R.A. and Van der Gaast, S.J. 1996: Humidity sensitive minerals, In *Proceedings of the 14th Conference on Clay Mineralogy and Petrology*, 2–6 September, Banska Stiavnica, Slovakia, *Acta Carpathica Series Clays* 4/2, 100–101.

Kühnel, R.A., Verkroost, T.W. and Chaudry, M.A. 1987: Thermal behavior of amorphous silica. In *Proceedings of the Conference on Geochemistry and Mineral Formation in the Earth Surface*, Rodríguez-Clemente, R. and Tardy, Y. (eds.). Madrid, Spain, pp. 863–875.

Lagaly, G., Beneke, K. and Weiss, A. 1975a: Magadiite and H-magadiite: I. Sodium magadiite and some of its derivatives. *American Mineralogist* 60, 642–649.

Lagaly, G., Beneke, K. and Weiss, A. 1975b: Magadiite and H-magadiite: II. H-magadiite and its intercalation compounds. *American Mineralogist* 60, 650–658.

Lea, F.M. 1970: *The Chemistry of Cement and Concrete*, 3rd ed, Edward Arnold Limited, London, U.K.. See Hewlett (1998) for 4th ed.

Ledoux, R.L. (ed) 1979: *Short Course in Mineralogical Techniques of Asbestos Determination*. Mineralogical Association of Canada (MAC), Québec, Canada, p. 278.

Lehmann, J. 1874: Über den Ettringit, ein neues Mineral in Kalkeinschlüssen der Lava von Ettringen. *Neues Jahrbuch für Mineralogie, Monatshefte, Abhandlungen*, Germany, 273–275.

Lerch, W.L., Ashton, F.W. and Bogue, R.H. 1929: The sulphoaluminates of calcium. *Journal of Research, National Bureau of Standards* 2, 715–731.

Leslie, A.B. and Hughes, J.J. 2003: High temperature slag formation in historic Scottish mortars: evidence for production dynamics in 18th-19th century lime production from Charlestown. In *Proceedings of the Ninth Euroseminar on Microscopy Applied to Building Materials*, Broekmans, M.A.T.M., Jensen, V. and Brattli, B. (eds) Trondheim, Norway, CD-ROM, p. 8.

Li, D., Bancroft, G.M., Kasrai, M., Fleet, M.E., Secco, R.A., Feng, X.H., Tan, K.H. and Yang, B.X. 1994: X-ray absorption spectroscopy of silicon dioxide (SiO_2) polymorphs: The structural characterization of opal. *American Mineralogist* 79, 622–632.

Lorenzi, G., Jensen, J., Wigum, B., Sibbick, R., Haugen, M., Guédon, S. and Åkesson, U. 2006: *Petrographic Atlas of the Potentially Alkali-Reactive Rocks in Europe*. Professional Paper 2006/1 (302), Geological Survey of Belgium, Brussels, Belgium, p. 63.

Ludwig, U. and Mehr, S. 1986: Destruction of historical buildings by the formation of ettringite or thaumasite. In *Eighth International Congress on the Chemistry of Cement*, Rio de Janeiro, Brazil, pp. 181–188.

Lumley, J.S. 1989: Synthetic cristobalite as a reference reactive aggregate. In *Proceedings of the 8th International Conference on Alkali-Aggregate Reaction, Kyoto, Japan*, Okada, K., Nishibayashi, S. and Kawamura, M. (eds.). The Society of Materials Science, Kyoto, Japan, pp. 561–566.

Mackenzie, R.C. (ed.) 1970: *Differential Thermal Analysis, Vol. 1, Fundamental Aspects*. Academic Press, London, U.K. (nb *Vol. 2, Applications*, published 1972).

MacKenzie, W.S. and Adams, A.E. 1994: *A Colour Atlas of Rocks and Minerals in Thin Section*. Manson Publishing, London, U.K., p. 192.

MacKenzie, W.S., Donaldson, C.H. and Guildford, C. 1995: *Atlas of Igneous Rocks and Their Textures*. Longman (Pearson), Harlow, U.K. (originally published 1982).

MacKenzie, W.S. and Guilford, C. 1980: *Atlas of Rock-Forming Minerals in Thin-Section*. Longman (Pearson), Harlow, U.K., p. 98.

Mamedov, Kh.S. and Belov, N.V. 1963: The crystalline structure of tobermorites. In *Crystal-Chemistry of Large Cation Silicates*, Belov, N.V. (ed.). Consultants Bureau, New York, pp. 60–62.

Mamedov, Kh.S., Klevtsova, R.F. and Belov, N.V. 1963: The crystalline structure of tricalcium silicate hydrate, In: Belov, N.V. (ed) *Crystal*-Consultants Bureau, New York, pp. 79–81.

Marusin, S.L. 1994: A simple treatment to distinguish alkali-silica gel from delayed ettringite formations in concrete. *Magazine of Concrete Research* 46, 163–166.

McConnell, D. and Murdoch, J. 1962: Crystal chemistry of ettringite. *Mineralogical Magazine* 33, 59–64.

McCrone, W.C. 1987: *Asbestos Identification*. Microscope Publications, Chicago, IL.

McCrone, W.C. and Delly, J.G. 1973: *The Particle Atlas, 1 & 2*. Ann Arbor Science Publishers, Ann Arbor, MI.

McKenzie, L. 1994: The Microscope: A method and tool to study lime burning and lime quality. *World Cement* 25(11), 56–59.

Merlino, S., Bonaccorsi, E. and Armbruster, T. 2000: The real structures of clinotobermorite and tobermorite 9Å: OD character, polytypes, and structural relationships. *European Journal of Mineralogy* 12, 411–429.

Michel-Lévy, M.M., Munier-Chalmas, 1892: Mémoire sur diverses formes affectées par le réseau élémentaire du quartz. *Bulletin de Société Française de Minéralogie* 15, 159–190.

Miehe, G. and Graetsch, H. 1992: Crystal structure of moganite: A new structure type for silica. *European Journal of Mineralogy* 4, 693–706.

Miehe, G., Graetsch, H. and Flörke, O.W. 1984: Crystal structure and growth fabric of length-fast chalcedony. *Physics and Chemistry of Minerals* 10, 197–199.

Millet, J.R. 2006: Asbestos analysis methods. Chapter 2 in *Asbestos. Risk Assessment, Epidemiology and Health Effects*, Dobson, R.F. and Hammar, S.P. (eds.). Taylor & Francis Group, pp. 9–37.

Mitsuda, T. and Taylor, H.F.W. 1978: Normal and anomalous tobermorite. *Mineralogical Magazine* 42, 229–235.

Moura, A.O. and Prado, A.G.S. 2009: Effect of thermal dehydration and rehydration on Na-magadiite structure. *Journal of Colloid and Interface Science* 330(2), 392–398.

Moxon, T. and Rios, S. 2004: Moganite and water content as a function of age in agate: An XRD and thermogravimetric study. *European Journal of Mineralogy* 16, 269–278.

Muraishi, H. 1989: Crystallization of silica gel in alkaline solutions at 100 to 180°C: Characterization of SiO_2-Y by comparison with magadiite. *American Mineralogist* 74, 1147–1151.

Murashov, V.V. and Svishchev, I.M. 1998: Quartz family of silica polymorphs: Comparative simulation study of quartz, moganite, orthorhombic silica, and their phase transformations. *Physical Review B* 57(10), 5639–5647.

Myneni, S.C.B. 2000: X-ray and vibrational spectroscopy of sulfate in earth materials. *Reviews in Mineralogy & Geochemistry* 40(1), 113–172.

Newman, E.S. 1955: A study of the system magnesium oxide-magnesium chloride-water and the heat of formation of magnesium oxychloride. *Journal of Research of the National Bureau of Standards* 54(6), 347–355.

Nijland, T.G., Verschure, R.H. and Maijer, C. 1994: Catalytic effect of biotite: formation of hydrogarnet lenses. *Comptes Rendues de l'Academie des Sciènces Paris, Sciences de la Terre et des Planètes* 318(II), 501–506.

Parthasarathy, G., Kunwar, A.C. and Srinivasan, R. 2001: Occurrence of moganite-rich chalcedony in Deccan flood basalts, Killara, Maharashtra, India. *European Journal of Mineralogy* 13, 127–134.

Paterson, M.S. 1982: The determination of hydroxyl by infrared absorption in quartz, silicate glasses and similar materials. *Bulletin de Minéralogie* 105, 20–29.

Pauling, L. and Hendricks, S.B. 1925: The crystal structures of hematite and corundum. *Journal of the American Chemical Society* 47, 781–784.

Perdikouri, C., Kasioptas, A., Putnis, C.V. and Putnis, A. 2008: The effect of fluid composition on the mechanism of the aragonite to calcite transition. *Mineralogical Magazine* 72, 111–114.

Perkins, R.B. and Palmer, C.D. 1999: Solubility of ettringite ($Ca_6[Al(OH)_6]_2(SO_4)_3 \cdot 26H_2O$ at 5–75°C. *Geochimica et Cosmochimica Acta* 63, 1969–1980.

Peryea, F.J. and Kittrick, J.A. 1988: Relative solubility of corundum, gibbsite, boehmite, and diaspore at standard state conditions. *Clays and Clay Minerals* 36, 391–396.

Peterson, K., Gress, D., Van Dam, T. and Sutter, L. 2006: Crystallized alkali-silica gel in concrete from the late 1890s. *Cement and Concrete Research* 36, 1523–1532.

Phillips, W.R. 1971: *Mineral Optics: Principles and Techniques*. W H Freeman & Company, San Francisco, CA, p. 249.

Poellmann, H., Kuzel, H.J. and Wenda, R. 1990: Solid solution of ettringites, Part I: Incorporation of OH^- and CO_3^{2-} in $3CaO \cdot Al_2O_3 \cdot 3CaSO_4 \cdot H_2O$. *Cement and Concrete Research* 20, 941–947.

Pöllmann, H., Kuzel, H.J. and Wenda, R. 1989: Compounds with ettringite structure. *Neues Jahrbuch für Mineralogie, Abhandlungen*, 160, 133–158.

Poole, A.B. and Thomas, A. 1975: A staining technique for the identification of sulphates in aggregates and concretes. *Mineralogical Magazine* 40, 315–316.

Ramachandran, V.S. 1969: *Applications of Differential Thermal Analysis in Cement Chemistry*. Chemical Publishing Co., New York.

Rayment, P.L. 1992: The relationship between flint microstructure and alkali-silica reaction. In *The Ninth International Conference on Alkali-Aggregate Reaction in Concrete, London, Conference Papers*, CS.104, Vol. 2, pp. 843–50, The Concrete Society, Slough (now Camberley), U.K.

Reid, W.P. 1969: Mineral staining tests. *Colorado School of Mines Mineral Industries Bulletin*, 12(3), 1–20.

Richardson, I.G. 2004: Tobermorite/jennite- and tobermorite/calcium hydroxide-based models for the structure of C-S-H: Applicability to hardened pastes of tricalcium silicate, β-dicalcium silicate, Portland cement, and blends of Portland cement with blast-furnace slag, metakaolin, or silica fume. *Cement and Concrete Research* 34(9), 1733–1777.

Rigby, G.R. 1948: *The Thin-Section Mineralogy of Ceramic Materials*. British Refractories Research Association (later British Ceramic Research Association and now British Ceramic Research Limited, or 'Ceram'), Stoke-on-Trent, U.K.

Sanders, J.V. 1975: Microstructure and crystallinity of gem opals. *American Mineralogist* 60, 749–757.

Schiek, R.C. 1982: Pigments (Inorganic), *Kirk-Othmer Encyclopedia of Chemical Technology*, 3rd ed., 17, 788–838, John Wiley & Sons, New York.

Schröpfer, L. and Bartl, H. 1993: Oriented decomposition and reconstruction of hydrogarnet, $Ca_3 \cdot Al_2 \cdot (OH)_{12}$. *European Journal of Mineralogy* 5(6), 1133–1144.

Schuman, L. and Hornibrook, F.B. 1943: Discussion to paper by T.E. Stanton: Studies to develop an accelerated test procedure for the detection of adversely reactive cement aggregate combinations. *Proceedings of the American Society of Testing Materials* 43, 875–904.

Schwandt, C.S., Cygan, R.T. and Westrich, H.R. 1996: Ca self-diffusion in grossular garnet. *American Mineralogist* 81, 448–451.

Schwertmann, U. and Fechter, H. 1994: The formation of green rust and its transformation to lepido-crocite. *Clay Minerals* 29, 87–92.

Shand, E.B. 1958: *Glass Engineering Handbook*. McGraw-Hill Book Co., New York.

Sheppard, R.A., Gude, A.J. and Hay, R.L. 1970: Makatite, a new hydrous sodium silicate from Lake Magadi, Kenya. *American Mineralogist* 55, 358–366.

Sims, I. and Hartshorn, S. 1998: Recognising thaumasite. *Concrete Engineering International* 2(8), 44–48.

Smillie, S., Moulin, E., Macphee, D.E. and Glasser, F.P. 1993: Freshness of cement: conditions for syn-genite $CaK_2(SO_4)_2 \cdot H_2O$ formation. *Advances in Cement Research* 5(19), 93–96.

Smith, A.S., Dunham, A.C. and West, G. 1992: Undulatory extinction of quartz in British hard rocks. In *The Ninth International Conference on Alkali-Aggregate Reaction in Concrete, London, Conference Papers*, CS.104, Vol. 2, 1001–8, The Concrete Society, Slough (now Camberley), U.K.

Smith, D.K. 1998: Opal, cristobalite, and tridymite: Non-crystallinity versus crystallinity, nomenclature of the silica minerals and bibliography. *Powder Diffraction* 13, 2–19.

Smolczyk, H.G. 1980: Slag structure and identification of slags. In *Seventh International Congress on the Chemistry of Cement, Paris,* Volume I, Principal Reports, Sub-Theme Ill, Editions Septima, Paris, France, 1/3 to 1/17.

Spurr, R.A. and Myers, H. 1957: Quantitative analysis of anatase-rutile mixtures with an X-ray diffractometer. *Analytical Chemistry* 29, 760–762.

St John, D.A. 1982: An unusual case of ground water sulfate attack on concrete. *Cement and Concrete Research* 12(5), 633–639.

St John, D.A. and Abbott, J.H. 1987: The microscopic textures of geothermal cement grouts exposed to steam temperatures of 150°C and 260°C. In: *Proceedings of the Ninth Annual International Conference on Cement Microscopy*, Reno, Nevada, pp. 280–300, International Cement Microscopy Association (ICMA), MI.

Suzuki, S. and Sinn, E. 1993: 1.4nm tobermorite-like calcium silicate hydrate prepared at room temperature from $Si(OH)_4$ and $CaCl_2$ solutions. *Journal of Materials Science Letters* 12(8), 542–544.

Taylor, H.F.W. 1957: The dehydration of tobermorite. *Clays and Clay Minerals* 6(1), 101–109.

Taylor, H.F.W. 1964: The calcium silicate hydrates. In Taylor, H.F.W. (ed.). *The Chemistry of Cements,* Vol. 1, Chapter 5, pp. 167–232, Academic Press, London, U.K.

Taylor, H.F.W. 1984: Studies on the chemistry and microstructure of cement pastes. *British Ceramic Proceedings* 35, 65–82.

Taylor, H.W.F. 1990: *Cement Chemistry*. Academic Press, London, U.K.

Taylor, H.F.W. 1992: Tobermorite, jennite, and cement gel. *Zeitschrift für Kristallografie* 202, 41–50.

Taylor, H.W.F. 1997: *Cement Chemistry*, 2nd ed., Thomas Telford, London, U.K., p. 459.

Taylor, H.W F. and Roy, D.M. 1980: Structure and composition of hydrates. In *Proceedings of the Seventh International Congress on the Chemistry of Cement, Paris*, 1, II-2, Editions Septima, Paris, France, p. 13.

Textile Institute. 1985: *Identification of Textile Materials*, 7th ed., The Textile Institute, Manchester, U.K.

Tilley, C.E. 1992: Density, refractivity, and composition relations of some natural glasses. *The Mineralogical Magazine* 19(96), 275–294.

Troger, W.E. 1971: *Optische bestimmung der gesteinsbilderenden minerale, Teil 1: Bestimmungstabellen* (Part 1: Identification tables), Newly revised 4th edition by Bambauer, H.F., Taborsky, F. and Trochim, H.D.E. Schweizerbart'sche Verlagsbuchhandlung, Stuttgart, Germany.

Trolard, F. and Tardy, Y. 1987: The stabilities of gibbsite, boehmite, aluminous goethites and aluminous hematites in bauxites, ferricretes and laterites as a function of water activity, temperature and particle size. *Geochimica et Cosmochimica Acta* 51, 945–957.

Vail, G.Y. 1952: *Soluble Silicates – Their Properties and Uses, Volume 1: Chemistry*. Reinhold Publishing Corporation, New York.

Van Aardt, J.H.P. and Visser, S. 1977: Formation of hydrogarnets: Calcium hydroxide attack on clays and feldspars. *Cement and Concrete Research* 7(1), 39–44.

Van Hees, R.P.J., Nijland, T.G., Quist, W.J., Naldini, S. and Lubelli, B. 2008: *Stone Atlas, Version 2.0*, Department ®MIT, Technical University Delft, the Netherlands, p. 63.

Veblen, D.R. and Wylie, A.G. 1993: Mineralogy of amphiboles and 1:1 layer silicates, In: Guthrie jr, G.D. and Mossman, B.T. (eds.), Health effects of mineral dusts, *Reviews in Mineralogy* 28, 61–137.

Vogel, A.I. 1995: *Vogel's Textbook of Quantitative Chemical Analysis*, 5th ed., (revised by G. H. Jeffery et al.), Longman, Harlow, U.K.

Volzhenskii, A.V. and Ferronskaya, A.V. 1974: *Gypsum Binders and Products*. Stroiizdat, Moscow (Chemical Abstracts, 81, 126219, American Chemical Society, Washington, DC).

Warne, S.StJ. 1962: A quick field or laboratory staining scheme for the differentiation of major carbonate minerals. *Journal of Sedimentary Petrology* 32(1), 29–38.

White, A.H. 1909: Free lime in Portland cement. *The Journal of Industrial and Engineering Chemistry* 1(1), 5–11.

White, J.F. and Corwin, J.F. 1961: Synthesis and origin of chalcedony. *American Mineralogist* 46, 112–119.

Wieker, W., Grimmer, A.R., Winkler, A., Mägi, M., Tarmak, M. and Lippmaa, E. 1982: Solid-state high-resolution 29Si NMR spectroscopy of synthetic 14Å, 11Å and 9Å tobermorites. *Cement and Concrete Research* 12(3), 333–339.

Wigum, B.J. 1996: A classification of Norwegian cataclastic rocks for alkali-reactivity, In *Proceedings of the 10th International Conference on Alkali-Aggregate Reaction in Concrete*, Shayan, A. (ed.). Melbourne, Victoria, Australia, pp. 758–766.

Williams, H., Turner, F.J. and Gilbert, C.M. 1954: *Petrography. An Introduction to the Study of Rocks in Thin Section*. W.H. Freeman and Co., San Francisco, CA, p. 406.

Winchell, A.N. 1937: *Elements of Optical Mineralogy: An Introduction to Microscopic Petrography, Part I. Principles and Methods* (originally with father, Winchell, N.H.), 5th ed., John Wiley & Sons, New York and Chapman & Hall, London, U.K., p. 262. (various later printings)

Winchell, A.N. 1939: *Elements of Optical Mineralogy: An Introduction to Microscopic Petrography, Part III. Determinative Tables*, 2nd edn., John Wiley & Sons, New York and Chapman & Hall, London, U.K., P. 228. (various later printings)

Winchell, A.N. 1951: *Elements of Optical Mineralogy: An Introduction to Microscopic Petrography, Part II. Descriptions of Minerals* (with son, Winchell, H.), 4th ed. John Wiley & Sons, New York and Chapman & Hall, London, U.K., P. 551. (various later printings).

Winchell, A.N. 1954: *Optical Properties of Organic Compounds*, 2nd ed. Academic Press, New York, p. 487.

Winchell, A.N. and Emmons, R.C. 1964: *Microscopic Characters of Artificial Inorganic Solid Substances, or Artificial Minerals*, 2nd ed. John Wiley & Sons, New York, p. 439.

Winchell, A.N. and Winchell, N. H. 1961: *Elements of Optical Mineralogy. Part II. Description of Minerals*, 4th ed. John Wiley & Sons, New York, P. 551.

Winchell, A.N. and Winchell, N. H. 1964: *The Microscopical Characters of Artificial Inorganic Solid Substances: Optical Properties of Artificial Minerals*. Academic Press, New York, p. 439.

Wopfner, H. and Höcker, C.F.W. 1987: The Permian Groeden sandstone between Bozen and Meran (northern Italy), a habitat of dawsonite and nordstrandite. *Neues Jahrbuch für Geologie und Paläontologie, Monatshefte* 3, 161–176.

Yardley, B.W.D., MacKenzie, W.S. and Guilford, C. 1990: *Atlas of Metamorphic Rocks and Their Textures*. Longman (Pearson), Harlow, U.K., P. 126.

Zhou, G.T., Yao, Q.Z., Ni, J. and Jin, G. 2009: Formation of aragonite mesocrystals and implication for biomineralization. *American Mineralogist* 94, 293–302.

Zussman, J. 1979: The mineralogy of asbestos. In Michaels, L. and Chissick, S.S. (eds.). *Asbestos: Properties, Applications and Hazards*. John Wiley & Sons, Chichester, U.K., pp. 45–56.

Index

The starting page of main entries is shown in **bold**. Referenced author entries are given in *italics*; references with more than one author are included under the first author. Glossary entries are shown in parentheses.

Printed and bound by CPI Group (UK) Ltd, Croydon, CR0 4YY

01/11/2024

01782606-0001